INTERNATIONAL SEMINAR ON
NUCLEAR WAR AND PLANETARY EMERGENCIES
48th Session

THE ROLE OF SCIENCE IN THE THIRD MILLENNIUM

Why Science is Needed for the Culture of the Third Millennium / The New Manhattan Project / The USA-Iran Nuclear Agreement / Toward Global Eradication of Infectious Disease / Renewable and Nuclear Energy / Energy End-Use Management / IPCC: Issues and Alternatives / Dying Oceans / Countering Terrorism with Technology and Data Cyber Security Issues / Systematic Errors in Climate Measurements / Energy for the Poor / Terrorist Motivations / Emerging Risk of Endocrine Disrupting Herbicide / Mechanisms of Susceptibility/Resistance to Human Pathologies / Cosmic Objects Update / Water Pollution and Development Issues — EU Energy / Transition Plans and Ship-Based Nuclear Energy

THE SCIENCE AND CULTURE SERIES

Nuclear Strategy and Peace Technology*

Series Editor: Antonino Zichichi

2012 — International Seminar on Nuclear War and Planetary Emergencies — 45th Session: Lectio Magistralis Global Warming Emergency / Why Science is Needed for the Culture of the Third Millennium / Mitigation of Terrorist Acts – Improvised Nclear Devices – Gap Analysis of EU – Precenting Nuclear Explosive Terrorism – Public Perception / Climate & Climate Economics / Global Nuclear Energy Issues – Outlook in Japan, China, Europe and USA – Sustainable Nuclear Energy Systems – Economics of Nuclear Power / Energy & Sustainability in Cities – Global Sustainability – Smart Grids – Complexity / Global Food Production & Forest Dynamics / Water, Pollution & Terrorism / Brain Aging & Behavior / Frontiers in Fast Computing Informatics/ Information Security

2013 — International Seminar on Nuclear War and Planetary Emergencies — 46th Session: The Role of Science in the Third Millennium / Why Science is Needed for the Culture of the Third Millennium / Lectio Magistralis: Alliance between Politics and Science / Nuclear Power Safety and Development / Water Pollution and Terrorism / Climate and the Limits of Geoengineering / Mitigation of Megaterrorism and Hybrid Threats / Pollution Innovative Cleanup Technologies/Information Security: Cybersecurit at Cross-Roads / Energy — Key to the Evolution of Cities / Energy as a Planetary Emergency/AIDS and Infectious Diseases

2015 — International Seminar on Planetary Emergencies — 48th Session: The Role of Science in the Third Millennium / Why Science is Needed for the Culture of the Third Millennium / The New Manhattan Project / The USA-Iran Nuclear Agreement / Toward Global Eradication of Infectious Disease / Renewable and Nuclear Energy / Energy End-Use Management / Ipcc: Issues and Alternatives Dying Oceans / Countering Terrorism with Technology and Data Cyber Security Issues / Systematic Errors in Climate Measurements / Energy for the Poor / Terrorist Motivations / Emerging Risk of Endocrine Disrupting Herbicide / Mechanisms of Susceptibility / Resistance to Human Pathologies / Cosmic Objects Update / Water Pollution and Development Issues — Eu Energy / Transition Plans and Ship-Based Nuclear Energy

*For the complete list of titles in this series, please go to:
http://worldscientific.com/series/scsnsp

THE SCIENCE AND CULTURE SERIES
Nuclear Strategy and Peace Technology

INTERNATIONAL SEMINAR ON NUCLEAR WAR AND PLANETARY EMERGENCIES

48th Session

THE ROLE OF SCIENCE IN THE THIRD MILLENNIUM

Why Science is Needed for the Culture of the Third Millennium / The New Manhattan Project / The USA-Iran Nuclear Agreement / Toward Global Eradication of Infectious Disease / Renewable and Nuclear Energy / Energy End-Use Management / IPCC: Issues and Alternatives / Dying Oceans / Countering Terrorism with Technology and Data Cyber Security Issues / Systematic Errors in Climate Measurements / Energy for the Poor / Terrorist Motivations / Emerging Risk of Endocrine Disrupting Herbicide / Mechanisms of Susceptibility/Resistance to Human Pathologies / Cosmic Objects Update / Water Pollution and Development Issues — EU Energy / Transition Plans and Ship-Based Nuclear Energy

"E. Majorana" Centre for Scientific Culture
Erice, Italy, 19 – 25 August 2015

Series Editor and Chairman: A. Zichichi

R. Ragaini (v. Ed.)

World Scientific

NEW JERSEY • LONDON • SINGAPORE • BEIJING • SHANGHAI • HONG KONG • TAIPEI • CHENNAI • TOKYO

Published by

World Scientific Publishing Co. Pte. Ltd.
5 Toh Tuck Link, Singapore 596224
USA office: 27 Warren Street, Suite 401-402, Hackensack, NJ 07601
UK office: 57 Shelton Street, Covent Garden, London WC2H 9HE

British Library Cataloguing-in-Publication Data
A catalogue record for this book is available from the British Library.

INTERNATIONAL SEMINAR ON PLANETARY EMERGENCIES — 48TH SESSION: THE ROLE OF SCIENCE IN THE THIRD MILLENNIUM / WHY SCIENCE IS NEEDED FOR THE CULTURE OF THE THIRD MILLENNIUM / THE NEW MANHATTAN PROJECT / THE USA-IRAN NUCLEAR AGREEMENT / TOWARD GLOBAL ERADICATION OF INFECTIOUS DISEASE / RENEWABLE AND NUCLEAR ENERGY / ENERGY END-USE MANAGEMENT / IPCC: ISSUES AND ALTERNATIVES DYING OCEANS / COUNTERING TERRORISM WITH TECHNOLOGY AND DATA CYBER SECURITY ISSUES / SYSTEMATIC ERRORS IN CLIMATE MEASUREMENTS / ENERGY FOR THE POOR / TERRORIST MOTIVATIONS / EMERGING RISK OF ENDOCRINE DISRUPTING HERBICIDE / MECHANISMS OF SUSCEPTIBILITY / RESISTANCE TO HUMAN PATHOLOGIES / COSMIC OBJECTS UPDATE / WATER POLLUTION AND DEVELOPMENT ISSUES — EU ENERGY / TRANSITION PLANS AND SHIP-BASED NUCLEAR ENERGY

ISBN 978-981-3148-64-2

Desk Editor: Rhaimie Wahap

Printed in Singapore

CONTENTS

CONCLUSIONS

PROJECT DIRECTORS AND SCIENTIFIC COMMITTEE MEETING

SESSION 1

OPENING SESSION

THE NEW MANHATTAN PROJECT TO FACE THE 72 PLANETARY EMERGENCIES

ANTONINO ZICHICHI
CERN, Geneva, Switzerland, INFN and University of Bologna, Italy,
Centro Enrico Fermi, Rome, Italy

Ladies and Gentlemen,

This year's Session of our Seminars is devoted to the role of Science in the Third Millennium.

I am pleased to start by reading the message of the President of the Italian Republic, H.E. Professor Sergio Mattarella, a strong supporter of the Ettore Majorana Centre for Scientific Culture. The other message I will read is from H.E. Professor Pietro Grasso, President of the Italian Senate.

Il Presidente della Repubblica

TELEGRAMMA

PROF. ANTONINO ZICHICHI
PRESIDENTE
"ETTORE MAJORANA" FOUNDATION AND CENTRE
FOR SCIENTIFIC CULTURE
VIA GUARNOTTA, 26
91016 ERICE (TP)

CARO PROFESSORE,

TI RINNOVO IL MIO RAMMARICO PER NON POTER ESSERE PRESENTE ALL'INAUGURAZIONE DEL SEMINARIO INTERNAZIONALE SULLE EMERGENZE PLANETARIE.

SONO CONSAPEVOLE DELL'IMPORTANZA DEGLI STUDI E DELLE RIFLESSIONI SCIENTIFICHE AL RIGUARDO E DI QUANTO SIA NECESSARIO CHE LE ISTITUZIONI NAZIONALI E INTERNAZIONALI VI PRESTINO ATTENTO ASCOLTO.

CON QUESTA CONVINZIONE ESPRIMO AI PARTECIPANTI AL SEMINARIO L'APPREZZAMENTO PER L'INCONTRO PROMOSSO DAL CENTRO ETTORE MAJORANA E L'AUGURIO DI BUON LAVORO

SERGIO MATTARELLA

Il Presidente
del Senato della Repubblica

Caro Professor Zichichi,

la "nostra" bellissima e difficile Sicilia si fa ancora una volta epicentro di una riflessione di altissimo valore scientifico e civile della quale, come cittadino e Presidente del Senato, le sono particolarmente grato. Desidero estendere questa gratitudine e un cordiale saluto a tutti i 120 scienziati giunti a Erice da 40 nazioni diverse per fornire il proprio contributo d'intelligenza e di impegno. La 48° sessione del Seminario Internazionale sulle Emergenze Planetarie si pone l'ambizioso obiettivo di affrontare temi di grande attualità e di cruciale rilevanza per il futuro del pianeta e dell'umanità come gli sviluppi dell'accordo nucleare tra gli Stati Uniti e l'Iran e le modalità attraverso le quali il mondo intero debba tentare di combattere e sconfiggere la diffusione di malattie infettive prodotte dalla globalizzazione. Nel 2013 ho avuto l'onore di partecipare alla 46°sessione del Seminario, nella quale sostenni che fosse di vitale importanza costruire un'alleanza duratura tra Scienza e Politica. Come ebbi modo di dire, "la politica, come la scienza, è una dimensione al servizio dei bisogni sociali, che trova il proprio obiettivo ultimo nella difesa della qualità della vita. Per risolvere le emergenze che minacciano il pianeta, la politica deve imparare a raccogliere le migliori risorse in campo e utilizzarle con il rigore del metodo scientifico, svincolata da precognizioni e conflitti d'interessi". Mi sembra che questa riflessione di allora possa essere pienamente estesa al lavoro che attende lei e i suoi colleghi nei prossimi giorni. Agli uomini e alle donne che in questi giorni animeranno il dibattito a Erice spetta il compito di fornire nuovi approcci per risolvere le emergenze che minacciano il futuro della Terra; alla politica e alle Istituzioni quello di tradurle in azioni concrete in grado di cambiare in meglio la vita di milioni di esseri umani. Concludo ringraziandola nuovamente per lo straordinario impegno e per aver messo la sua competenza al servizio del nostro Paese e della comunità scientifica internazionale. A lei e a ciascuno dei colleghi presenti i miei più calorosi auguri di buon lavoro, nella convinzione che, ancora una volta, esso saprà tenere alta l'attenzione, arricchire il dibattito pubblico e rappresentare un valido strumento di conoscenza e approfondimento per le Istituzioni nazionali, comunitarie e internazionali.

PIETRO GRASSO

PROFESSOR
ANTONINO ZICHICHI

I will try to synthesize the main points where our intellectual action of reflection is needed.

The first point is the recent past, including the risk of the nuclear war between the two superpowers. This is given in the following pages, starting with the Eugene Wigner statement in 1989, immediately after the collapse of the Berlin Wall (Figure 1). Then it follows the Enrico Fermi statement on the scientific value of the Ettore Majorana (Figure 2A).

> *«The Berlin Wall started to collaps many years* ***before 1989****,*
> *when in Erice the Seminars on Nuclear War*
> *started with the incredible presence of* ***Teller*** *and* ***Velikhov****».*
>
> Eugene Wigner, **1989**

Fig. 1: Eugene Wigner Statement.

When Ettore Majorana disappeared, Enrico Fermi said to his wife:

'Ettore was too intelligent. If he has decided to disappear, no one will be able to find him. Nevertheless, we have to consider all possibilities'.

In fact, he even tried to get Mussolini himself to support the search. On that occasion, Fermi said:

'There are several categories of scientists in the world; those of second or third rank do their best but never get very far. Then there is the first rank, those who make important discoveries, fundamental to scientific progress. But then there are the geniuses, like Galilei and Newton. Majorana was one of these.' (Rome 1938).

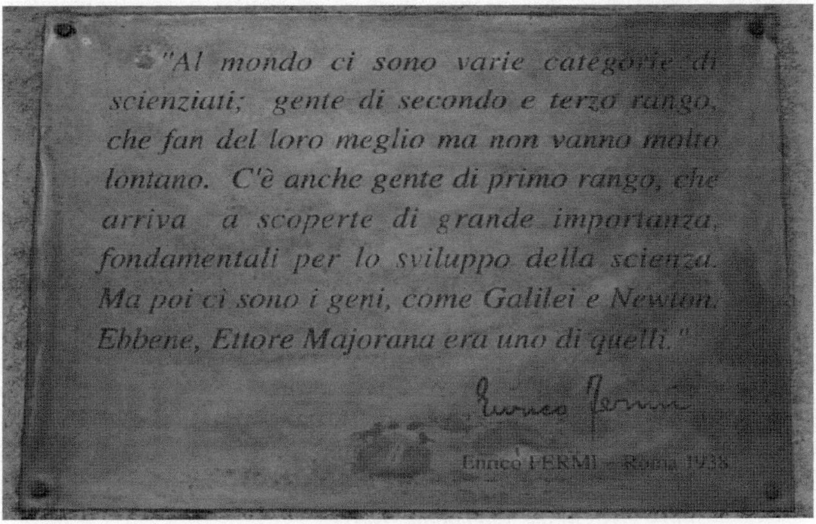

Fig. 2A: The Statement by Enrico Fermi at the Eugene Wigner Institute.

It follows the experimental result (Figures 3, 4, 5 and 6) which allowed me to personally interact with Dirac, thus starting the 1st Session of our Seminars in 1981. Let me recall what Robert Oppenheimer told me about Ettore Majorana (Figure 2B).

WHAT ROBERT OPPENHEIMER TOLD ME ABOUT FERMI QUOTING ETTORE DURING THREE CRISIS OF THE MANHATTAN PROJECT

After suffering heavy repercussions of his opposition to the development of weapons even stronger than those that destroyed Hiroshima and Nagasaki, Oppenheimer decided to get back to physics by visiting the biggest laboratories at the frontiers of scientific knowledge. This is how he came to CERN, the largest European Laboratory for Subnuclear Physics. At a ceremony organized by Weisskopf, for the presentation of the Erice School dedicated to Ettore Majorana, many illustrious physicists participated. I myself – at the time very young – was entrusted the task of speaking about the Majorana neutrinos. Oppenheimer wanted to voice his appreciation for how the Erice School and the Centre for Scientific Culture had been named. He knew the exceptional contributions Majorana made to physics from the papers he had read. This much, any physicist could do at any time. **What would have remained unknown is the episode he told me as a testimony of Fermi's exceptional esteem of '*Ettore*'**. He recounted the following episode from the time when the Manhattan Project was being carried out. The Project, in the course of just four years, transformed the scientific discovery of nuclear fission (heavy atomic nuclei can be broken to produce enormous quantities of energy) into a weapon of war. There were three critical turning points during this Project. During the executive meeting convened to address the first of these crises, Fermi turned to Wigner and said: '***If only Ettore were here***'. The Project seemed to have reached a dead end in the second crisis, during which Fermi exclaimed once more: '***This calls for Ettore!***'. Other than the Project Director himself (Oppenheimer), three people were in attendance at these meetings: two scientists (Fermi and Wigner) and a general of the US armed forces. Wigner worked with nuclear forces, like Ettore Majorana. **After the 'top-secret' meeting, the general asked the great Professor Wigner who this 'Ettore' was, and Wigner replied: '*Majorana*'. The general asked where Ettore was, so that he could try to bring him to USA. Wigner replied: '*Unfortunately, he disappeared many years ago*'**.

Fig. 2B: What Robert Oppenheimer told me about Fermi quoting Ettore during three crisis of the Manhattan Project.

Fig. 2C: Robert Oppenheimer (right) at CERN with Ed McMillan and Niels Bohr for the inauguration of the Proton Synchrotron (5 February 1960).

Fig. 2D: Robert Oppenheimer during an interview at the CERN Library (July 1962).

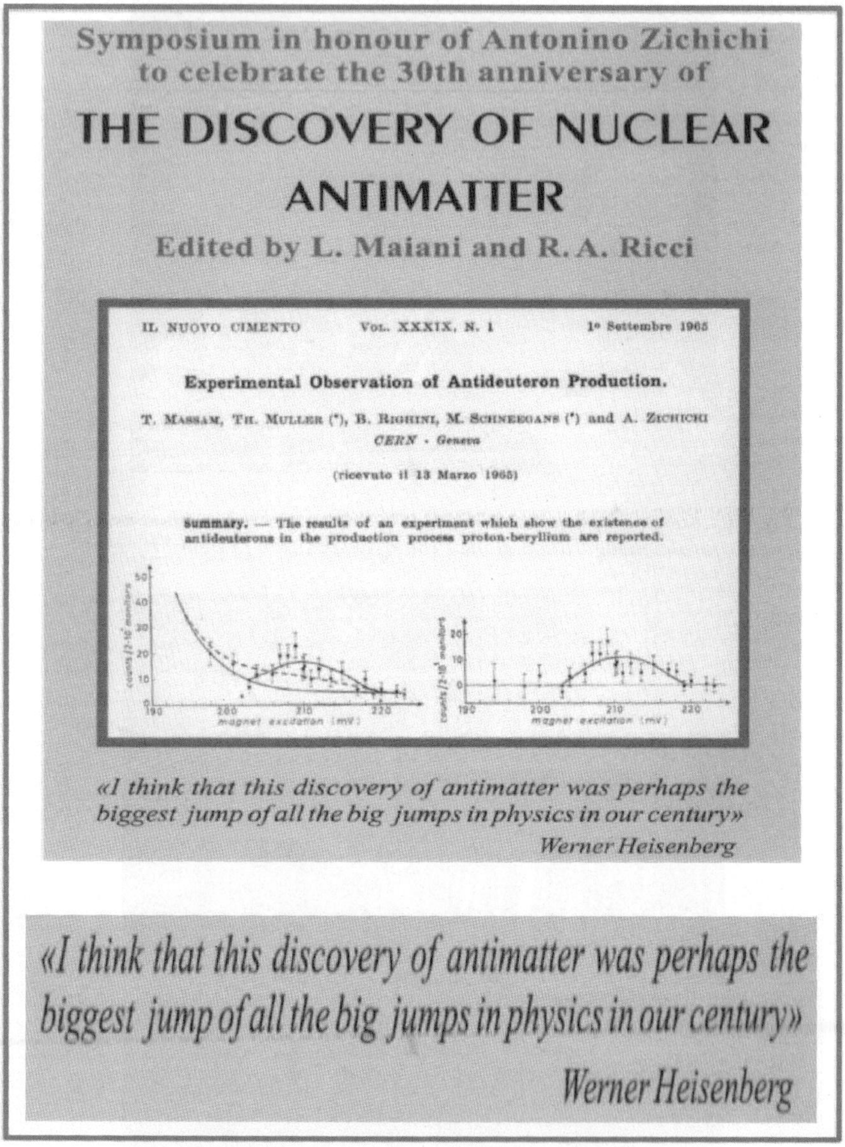
Symposium in honour of Antonino Zichichi
to celebrate the 30th anniversary of
THE DISCOVERY OF NUCLEAR
ANTIMATTER
Edited by L. Maiani and R. A. Ricci
IL NUOVO CIMENTO
VOL. XXXIX, N. 1
1° Settembre 1965
Experimental Observation of Antideuteron Production.
T. MASSAM, TH. MULLER (*), B. RIGHINI, M. SCHNEEGANS (*) and A. ZICHICHI
CERN - Geneva
(ricevuto il 13 Marzo 1965)
Summary. — The results of an experiment which show the existence of antideuterons in the production process proton-beryllium are reported.
magnet excitation (mV)
magnet excitation (mV)
«I think that this discovery of antimatter was perhaps the biggest jump of all the big jumps in physics in our century»
Werner Heisenberg
«I think that this discovery of antimatter was perhaps the biggest jump of all the big jumps in physics in our century»
Werner Heisenberg

Fig. 3.

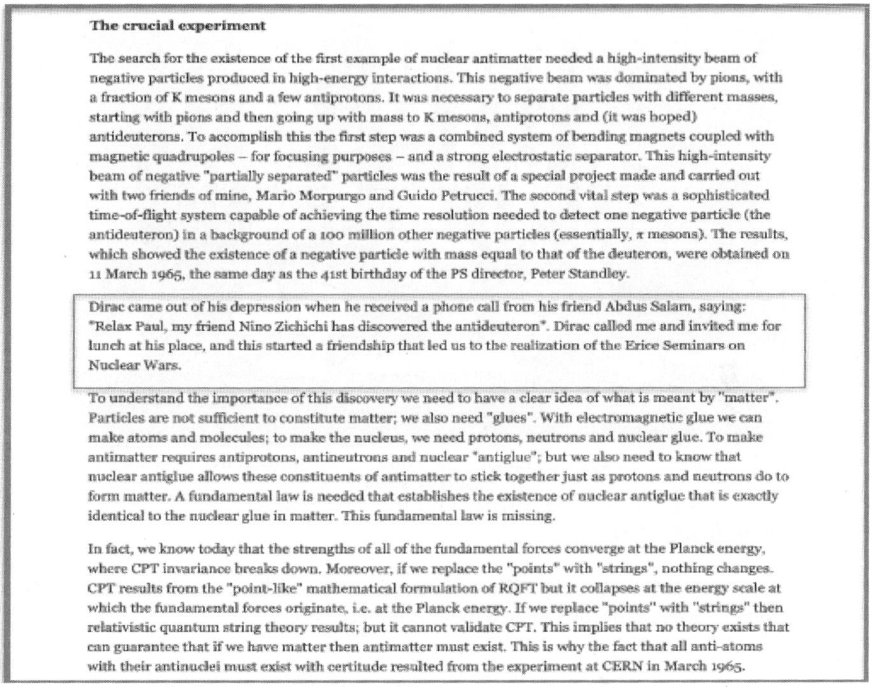

The crucial experiment

The search for the existence of the first example of nuclear antimatter needed a high-intensity beam of negative particles produced in high-energy interactions. This negative beam was dominated by pions, with a fraction of K mesons and a few antiprotons. It was necessary to separate particles with different masses, starting with pions and then going up with mass to K mesons, antiprotons and (it was hoped) antideuterons. To accomplish this the first step was a combined system of bending magnets coupled with magnetic quadrupoles – for focusing purposes – and a strong electrostatic separator. This high-intensity beam of negative "partially separated" particles was the result of a special project made and carried out with two friends of mine, Mario Morpurgo and Guido Petrucci. The second vital step was a sophisticated time-of-flight system capable of achieving the time resolution needed to detect one negative particle (the antideuteron) in a background of a 100 million other negative particles (essentially, π mesons). The results, which showed the existence of a negative particle with mass equal to that of the deuteron, were obtained on 11 March 1965, the same day as the 41st birthday of the PS director, Peter Standley.

Dirac came out of his depression when he received a phone call from his friend Abdus Salam, saying: "Relax Paul, my friend Nino Zichichi has discovered the antideuteron". Dirac called me and invited me for lunch at his place, and this started a friendship that led us to the realization of the Erice Seminars on Nuclear Wars.

To understand the importance of this discovery we need to have a clear idea of what is meant by "matter". Particles are not sufficient to constitute matter; we also need "glues". With electromagnetic glue we can make atoms and molecules; to make the nucleus, we need protons, neutrons and nuclear glue. To make antimatter requires antiprotons, antineutrons and nuclear "antiglue"; but we also need to know that nuclear antiglue allows these constituents of antimatter to stick together just as protons and neutrons do to form matter. A fundamental law is needed that establishes the existence of nuclear antiglue that is exactly identical to the nuclear glue in matter. This fundamental law is missing.

In fact, we know today that the strengths of all of the fundamental forces converge at the Planck energy, where CPT invariance breaks down. Moreover, if we replace the "points" with "strings", nothing changes. CPT results from the "point-like" mathematical formulation of RQFT but it collapses at the energy scale at which the fundamental forces originate, i.e. at the Planck energy. If we replace "points" with "strings" then relativistic quantum string theory results; but it cannot validate CPT. This implies that no theory exists that can guarantee that if we have matter then antimatter must exist. This is why the fact that all anti-atoms with their antinuclei must exist with certitude resulted from the experiment at CERN in March 1965.

Fig. 4: The crucial experiment.

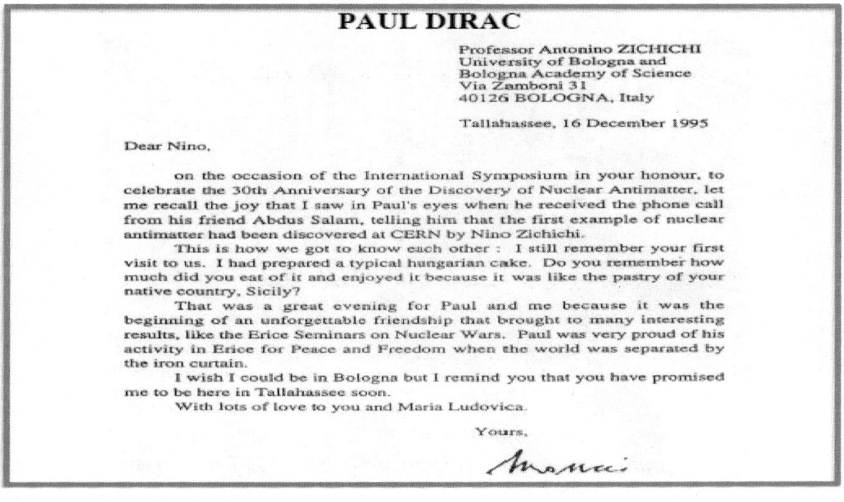

PAUL DIRAC

Professor Antonino ZICHICHI
University of Bologna and
Bologna Academy of Science
Via Zamboni 31
40126 BOLOGNA, Italy

Tallahassee, 16 December 1995

Dear Nino,

on the occasion of the International Symposium in your honour, to celebrate the 30th Anniversary of the Discovery of Nuclear Antimatter, let me recall the joy that I saw in Paul's eyes when he received the phone call from his friend Abdus Salam, telling him that the first example of nuclear antimatter had been discovered at CERN by Nino Zichichi.

This is how we got to know each other : I still remember your first visit to us. I had prepared a typical hungarian cake. Do you remember how much did you eat of it and enjoyed it because it was like the pastry of your native country, Sicily?

That was a great evening for Paul and me because it was the beginning of an unforgettable friendship that brought to many interesting results, like the Erice Seminars on Nuclear Wars. Paul was very proud of his activity in Erice for Peace and Freedom when the world was separated by the iron curtain.

I wish I could be in Bologna but I remind you that you have promised me to be here in Tallahassee soon.

With lots of love to you and Maria Ludovica.

Yours,

Manci

Fig. 5: Letter from Mrs. Manci Dirac.

Fig. 6: Erice, 1981.

These are our roots which allowed the presence of Teller and Velikhov to be in Erice for the 2nd Session of the Nuclear War Seminar (Figure 7).

Fig. 7: The Second Nuclear War Seminar—1982—Erice. Edward Teller (centre), scientific adviser to President Reagan and Yevgeniy Velikov (left), scientific adviser to President Gorbachev, together with Antonino Zichichi in Erice during the second International Seminar on Nuclear War in August 1982.

Please remember that the first name of our Seminar was not "Planetary Emergencies" but "Nuclear War".

In Figure 8 there is the statement by Dirac on the 2nd Session of our Seminars.

> *«Put together Teller and Velikhov*
> *was a fact that none of us had ever believed».*
>
> Paul Dirac, **1982**

Fig. 8: The statement by Dirac on the 2nd Session of our Seminars.

To our activity in Erice we had Dirac and Kapitza (Figures 9 and 10). This was the first time Kapitza could be out of the USSR borders to come to Italy.

Fig. 9: First row: Paul Dirac and Pyotr Kapitza. Second row: Eugene Wigner, John Eccles and R. Garwin. Erice, 1982.

Fig. 10: Pyotr Kapitza. If Kapitza had not refused to be the n. 1 of the USSR H-Bomb project the world would have had a millennium of Stalinistic Empire.

Kapitza, "pupillo" insieme a Dirac di Rutherford (il padre della Fisica Nucleare), era stato di fatto "sequestrato" alla fine degli anni trenta in USSR dove si era recato per una visita ai suoi.

Kapitza fu uno tra i più attivi estimatori di Giovanni Paolo II. Lo definì:

> "*La Luce del mondo accesasi per cacciare le tragiche tenebre del nazismo e dello stalinismo*".

Nella foto c'è Pëtr Kapitza a Erice nel 1982 per i Seminari sulle Guerre Nucleari. In seconda fila a destra Richard Garwin, il cervello numero uno con Teller, nel progetto per la Bomba H Americana.

During a Special Session of our International School in History the Figure 11 was elaborated in order to put on logical correspondence the two extreme values of Complexity:

1 – History, the example of highest level of Complexity;
2 – Science, the example of lowest level of Complexity.

The point I would like to call your attention is number ten: X. If Kapitza accepted to be the leader of the USSR H-Bomb Project, as wanted by Stalin, we would not be here today. Europe would have been for at least 50 years like East-Europe before collapse of the Berlin Wall.

	In History = EWRL		In Science = EBUS
I	What if Julius Caesar had been assassinated many years before?	***I***	What if Galileo Galilei had not discovered that $F = mg$?
II	What if Charles VII had not been able to win the 100 years war?	***II***	What if Newton had not discovered that $F = G\frac{m_1 \cdot m_2}{R_{12}^2}$?
III	What if America had been discovered a few centuries later?	***III***	What if Maxwell had not discovered the unification of electricity, magnetism and optical phenomena, which allowed him to conclude that light is a vibration of the EM field?
IV	What if Napoleon had not been born?	***IV***	What if Becquerel had not discovered radioactivity?
V	What if Louis XVI had been able to win against the 'Storming of the Bastille'?	***V***	What if Planck had not discovered that $h \neq 0$?
VI	What if the 1908 Tunguska Comet had fallen somewhere in Europe instead of Tunguska in Siberia?	***VI***	What if Lorentz had not discovered that space and time cannot both be real?
VII	What if the killer of the Austrian Archduke Franz Ferdinand had been arrested the day before the Sarajevo event?	***VII***	What if Einstein had not discovered the existence of time-like and space-like real worlds? Only in the time-like world, simultaneity does not change, with changing observer.
VIII	What if Lenin had been killed during his travelling through Germany?	***VIII***	What if Rutherford had not discovered the nucleus?
IX	What if Hitler had not been appointed Chancellor by the President of the Republic of Weimar Paul von Hindenburg?	***IX***	What if Hess had not discovered cosmic rays?
X	What if Pyotr Kapitza accepted to be the leader of the USSR H-bomb Project as wanted by Stalin?	***X***	What if Dirac had not discovered his equation, which opens new horizons, including the existence of the antiworld?
XI	What if Nazi Germany had defeated the Soviet Union?	***XI***	What if Fermi had not discovered weak forces?
XII	What if Karol Wojtyla had not been elected Pope, thus becoming John Paul II?	***XII***	What if Fermi and Dirac had not discovered the Fermi–Dirac statistics?
XIII	What if Gorbachev had not been defeated by Yeltsin?	***XIII***	What if Yukawa had not proposed the existence of a "meson" in order to have the nuclear glue?
XIV	What if the USSR had not collapsed?	***XIV***	What if the 'strange particles' had not been discovered in the Blackett Lab?

Fig. 11: UEEC Events in History and in Science.

Let us not forget that Edward Teller presented here at our Seminars in 1982 for the first time his project for the Strategic Defense System (SDI). In 1983, less than one year later, the President of USA, Ronald Reagan, presented to the world the SDI.

Another point I would like to recall: in 1985 Gorbachev in Geneva declared that they wanted to open the "Secret Labs". These were the biggest enemy for Peace in the world, as declared by the Erice Statement, signed in 1982 by 100,000 scientists from all over the word (Figure 12A).

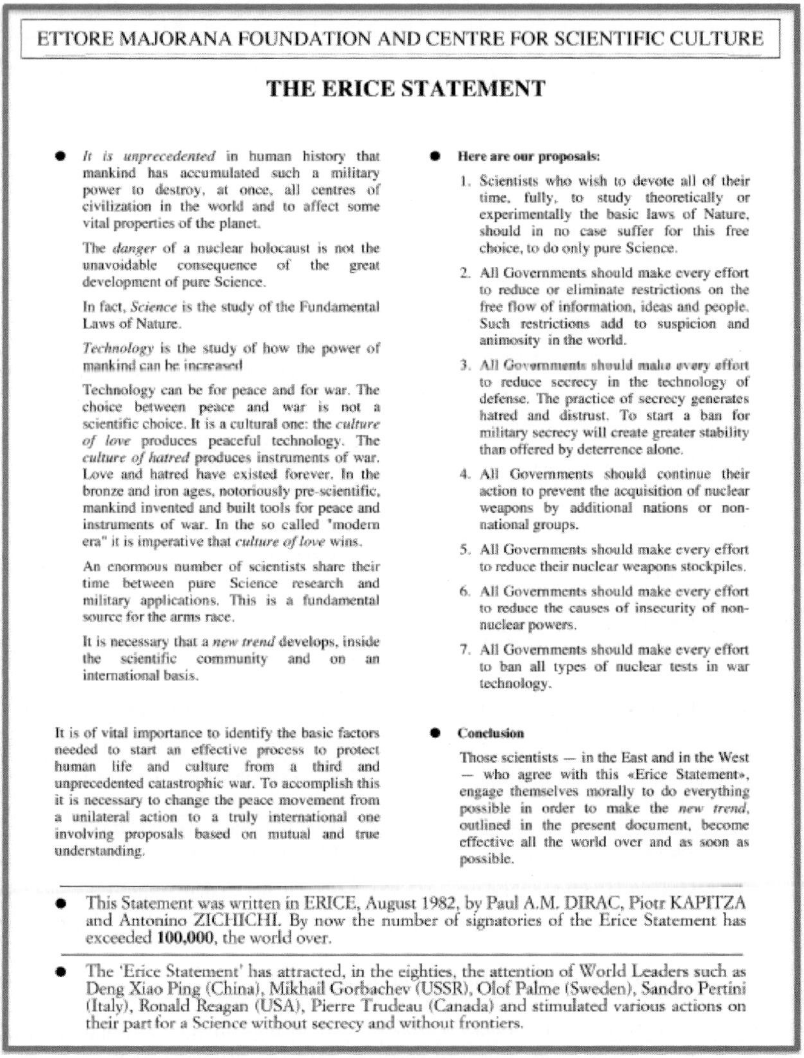

ETTORE MAJORANA FOUNDATION AND CENTRE FOR SCIENTIFIC CULTURE

THE ERICE STATEMENT

- *It is unprecedented* in human history that mankind has accumulated such a military power to destroy, at once, all centres of civilization in the world and to affect some vital properties of the planet.

  The *danger* of a nuclear holocaust is not the unavoidable consequence of the great development of pure Science.

  In fact, *Science* is the study of the Fundamental Laws of Nature.

  Technology is the study of how the power of mankind can be increased.

  Technology can be for peace and for war. The choice between peace and war is not a scientific choice. It is a cultural one: the *culture of love* produces peaceful technology. The *culture of hatred* produces instruments of war. Love and hatred have existed forever. In the bronze and iron ages, notoriously pre-scientific, mankind invented and built tools for peace and instruments of war. In the so called "modern era" it is imperative that *culture of love* wins.

  An enormous number of scientists share their time between pure Science research and military applications. This is a fundamental source for the arms race.

  It is necessary that a *new trend* develops, inside the scientific community and on an international basis.

It is of vital importance to identify the basic factors needed to start an effective process to protect human life and culture from a third and unprecedented catastrophic war. To accomplish this it is necessary to change the peace movement from a unilateral action to a truly international one involving proposals based on mutual and true understanding.

- **Here are our proposals:**

  1. Scientists who wish to devote all of their time, fully, to study theoretically or experimentally the basic laws of Nature, should in no case suffer for this free choice, to do only pure Science.
  2. All Governments should make every effort to reduce or eliminate restrictions on the free flow of information, ideas and people. Such restrictions add to suspicion and animosity in the world.
  3. All Governments should make every effort to reduce secrecy in the technology of defense. The practice of secrecy generates hatred and distrust. To start a ban for military secrecy will create greater stability than offered by deterrence alone.
  4. All Governments should continue their action to prevent the acquisition of nuclear weapons by additional nations or non-national groups.
  5. All Governments should make every effort to reduce their nuclear weapons stockpiles.
  6. All Governments should make every effort to reduce the causes of insecurity of non-nuclear powers.
  7. All Governments should make every effort to ban all types of nuclear tests in war technology.

- **Conclusion**

  Those scientists — in the East and in the West — who agree with this «Erice Statement», engage themselves morally to do everything possible in order to make the *new trend*, outlined in the present document, become effective all the world over and as soon as possible.

- This Statement was written in ERICE, August 1982, by Paul A.M. DIRAC, Piotr KAPITZA and Antonino ZICHICHI. By now the number of signatories of the Erice Statement has exceeded **100,000**, the world over.

- The 'Erice Statement' has attracted, in the eighties, the attention of World Leaders such as Deng Xiao Ping (China), Mikhail Gorbachev (USSR), Olof Palme (Sweden), Sandro Pertini (Italy), Ronald Reagan (USA), Pierre Trudeau (Canada) and stimulated various actions on their part for a Science without secrecy and without frontiers.

Fig. 12A: The Erice Statement signed in 1982.

Now we are working on the new project to defend our Planet against Cosmic Objects. An excellent example of the Erice-Spirit for Science without Secret and Frontiers. This series of events is in Figure 13. The Reagan and Gorbachev event in Geneva needs to be better illustrated. We will see that the most important component for "Science" in the New Manhattan Project is the ELN (the supercollider which will be mentioned in my Closing Lecture). The reason is straightforward.

In Geneva in 1985, Reagan and Gorbachev declared that secret labs are the number one enemy of peace and so we will open all secret labs. Since this dream turns out to be a very difficult one to come true, the only way out is to destroy the secret at its origin.

The motor of progress is scientific discovery, which generates technological inventions and here lies the problem. Inventions can be both for peaceful and for war technology. We must stop the use of technology for war and build the most powerful machine in the world, the ELN, Euroasiatic Long Intersecting Storage Accelerator. If the Berlin Wall had not fallen, ELN would have been built since Reagan, Gorbachev and Deng Xiao Ping had all agreed (see Figure 12B).

Fig. 12B: From left to right (first row): Professor Zhou Guang Zhao (Scientific Advisor to Premier Deng Xiao Ping), Professor Edward Teller (Scientific Advisor to President Reagan), Professor Antonino Zichichi (Chairman of the International Committee "Science for Peace") and Professor Yevgeniy Velikhov (Scientific Advisor to President Gorbachev), shaking hands after reaching the Agreement for International Scientific Collaboration East-West-North-South without Secrecy and without Frontiers (1986).

Fig. 13.

After Figure 13 which is the end of my first point, let me go to the second point: The New Manhattan Project, whose effective name should be "Science for Peace the World over".

My first point represents the "credibility" needed for the NMP.

THE NEW MANHATTAN PROJECT

Science for Peace the World Over

TABLE OF CONTENTS

INTRODUCTORY SUMMARY

This book is a conclusive contribution to the THREE DAYS that the University of Bologna has dedicated to Science, Technology and Culture, so that they enter into the Heart of Everybody.

We are in the Cultural Hiroshima, as Fermi feared in the middle of the last Century, when (1945) we were in full Political Hiroshima.

The *Three Days* of Bologna had two objectives: to defeat the Cultural Hiroshima and to solve the problems of Planetary Emergencies, proposing the implementation of a project so that the great achievements of Science come soon in everyday life.

The book has some parts in Italian and others in English, which is the universal language of Science.

The parts in Italian will be translated into all the other languages by our colleagues of WFS.

Figure 1 describes the Political Hiroshima (1945) and the Cultural Hiroshima (today).

Figure 2 illustrates the logic of the Project (The New Manhattan Project).

Figure 1

- **POLITICAL HIROSHIMA (1945)**
  The existence of Science is totally forgotten (for decisions at very high political level) as if it were not **thanks** to Science that the **new** war **Technologies**, million times more powerful, had been implemented.

- **CULTURAL HIROSHIMA (NOW)**
  The existence of Science is totally forgotten for decisions (at any political level) and in every day life as if it is not **thanks** to Science that **new Technologies** have been invented: technologies which allow to win life-threatening illness, technologies that liberate from rough work and that allow humanity to have a standard of living never before achieved.

Fig. 1: The Political Hiroshima (1945) and the Cultural Hiroshima (now).

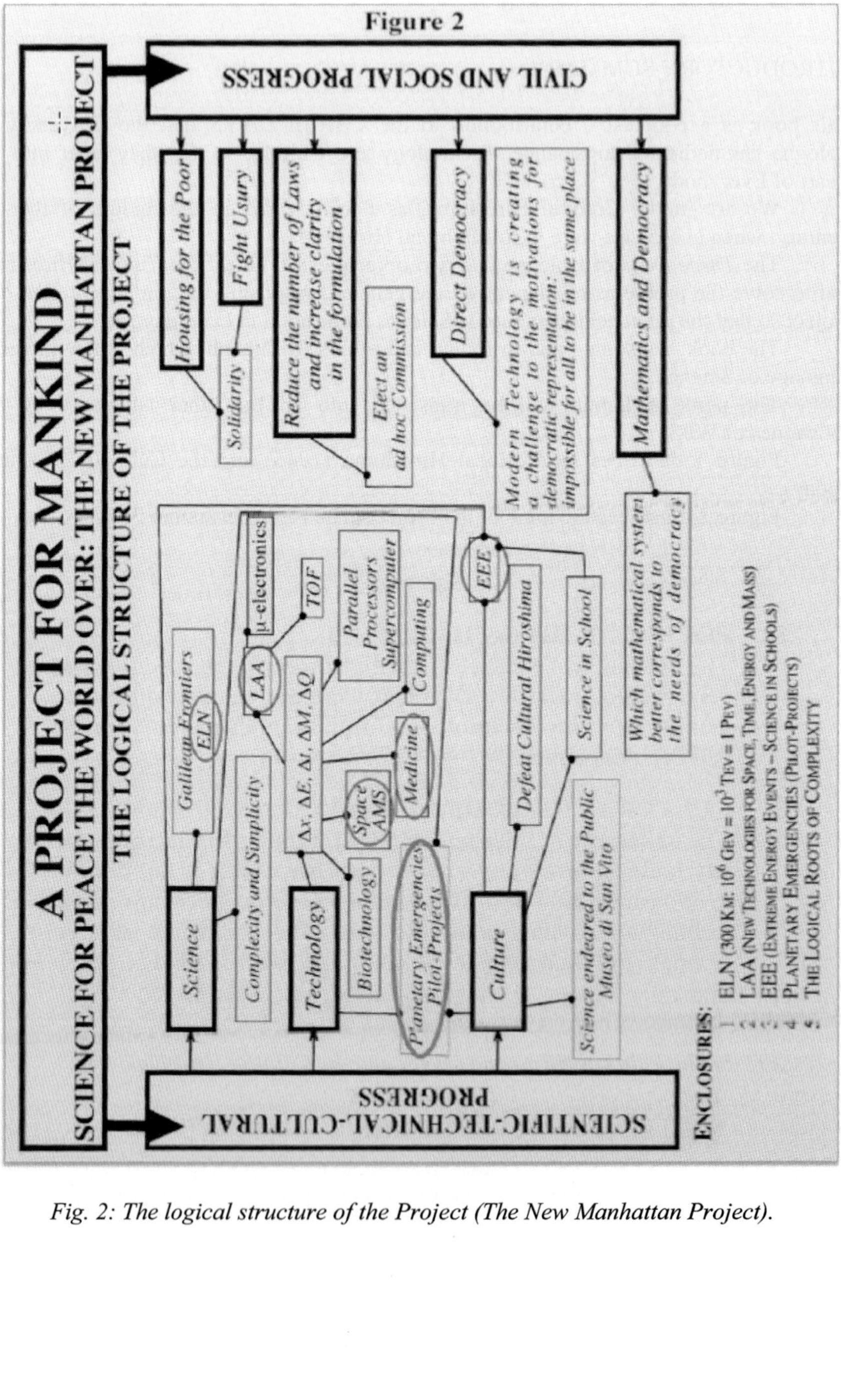

Fig. 2: The logical structure of the Project (The New Manhattan Project).

Suppose that the *3* Days of Bologna would take place on the 40 countries that you are coming from. If this hypothesis would become real the "New Manhattan Project" would become reality. From this year on, the contribution to our Seminar on Planetary Emergencies need to be the 1st step for a new Project or the report on the Status of an already existing Project. Let me give you three examples.

USA-Iran Nuclear Agreement
Professor Richard Garwin (USA)

As you know these Seminars began in 1981 as the International Seminar on Nuclear War, and the next talk addresses the Joint Comprehensive Plan of Action (JCPoA) achieved 14 July 2015 by two years of negotiations among Iran, France, Germany, the UK, the United States, China, and Russia (the E3/EU+3 - Iran Talks).

According to this "Iran Deal", Iran will limit its activities, accepting modern IAEA verification, to nuclear programs that are entirely civil, in return for the elimination of sanctions and for scientific and technical cooperation.

Professor Dick Garwin will speak about the details, in the context of an initiative of some in the American Physics community that characterizes this diplomatic achievement as effective, stringent, and innovative. It is extremely important that in the final negotiation of the Iran Agreement, nuclear physicists played a crucial role.

The Iran Deal verifiably blocks Iran's acquiring either plutonium or weapon-usable uranium and preparations for nuclear weapons, without impairing actions that might be taken in response to a future collapse of compliance with the Deal.

We look forward to welcoming scientists and engineers from Iran to our efforts against nuclear war and to improve the conditions of human life.

Global Eradication of Infectious Disease
Professor Lowell Wood and collaborators (USA).

Cosmic Objects Planetary Defence
Dr. Walter Huebner (*Los Alamos, USA*), Dr. Anatoly Zaitsev (*Moscow, Russia*) and Dr. Line Drube (*Berlin, Germany*).

The Cosmic Defense belongs to the Global Monitoring of the Planet

- In the Past
  Numerous Seminars were dedicated to this problem. Let me recall also the study of the search for solutions to the problems of Global Warming and the Ozone Hole.
  Another piece was the sophisticated seismological system put in place in the Mediterranean Basin, with stations on all sides of the Mediterranean and linked to worldwide networks. In addition, regional seismological networks were implemented in Countries prone to earthquakes.

The details are in the Proceedings of the International Seminars on Planetary Emergencies, published by World Scientific.

PROCEEDINGS
OF THE INTERNATIONAL SEMINARS
ON NUCLEAR WAR
AND PLANETARY EMERGENCIES

THE SCIENCE AND CULTURE SERIES
Nuclear Strategy and Peace Technology

Chairman and Series Editor: Antonino Zichichi

1981 International Seminar on Nuclear War — 1st Session: The World-wide Implications of Nuclear War. (World Scientific, Singapore - New Jersey - London - Hong Kong, 1992).
1982 International Seminar on Nuclear War — 2nd Session: How to Avoid a Nuclear War. (World Scientific, Singapore - New Jersey - London - Hong Kong, 1992).
1983 International Seminar on Nuclear War — 3rd Session: The Technical Basis for Peace. (World Scientific, Singapore - New Jersey - London - Hong Kong, 1992).
1984 International Seminar on Nuclear War — 4th Session: The Nuclear Winter and the New Defence Systems: Problems and Perspectives. (World Scientific, Singapore - New Jersey - London - Hong Kong, 1992).
1985 International Seminar on Nuclear War — 5th Session: SDI, Computer Simulation, New Proposals to Stop the Arms Race. (World Scientific, Singapore - New Jersey - London - Hong Kong, 1992).
1986 International Seminar on Nuclear War — 6th Session: International Cooperation: The Alternatives. (World Scientific, Singapore - New Jersey - London - Hong Kong, 1992).
1987 International Seminar on Nuclear War — 7th Session: The Great Projects for Scientific Collaboration East-West-North-South. (World Scientific, Singapore – New Jersey - London - Hong Kong, 1992).
1988 International Seminar on Nuclear War — 8th Session: The New Threats: Space and Chemical Weapons — What Can Be Done with the Retired I.N.F. Missiles-Laser Technology. (World Scientific, Singapore - New Jersey - London - Hong Kong, 1992).
1989 International Seminar on Nuclear War — 9th Session: The New Emergencies. (World Scientific, Singapore - New Jersey - London - Hong Kong, 1992).
1990 International Seminar on Nuclear War — 10th Session: The New Role of Science. (World Scientific, Singapore - New Jersey - London - Hong Kong, 1992).
1991 International Seminar on Nuclear War — 11th Session: Planetary Emergencies. (World Scientific, Singapore - New Jersey - London - Hong Kong, 1992).
1991 International Seminar on Nuclear War — 12th Session: Science Confronted with War (unpublished).
1991 International Seminar on Nuclear War and Planetary Emergencies — 13th Session: Satellite Monitoring of the Global Environment (unpublished).

1992 International Seminar on Nuclear War and Planetary Emergencies — 14th Session: Innovative Technologies for Cleaning the Environment. (World Scientific, Singapore - New Jersey - London - Hong Kong, 1993).

1992 International Seminar on Nuclear War and Planetary Emergencies — 15th Session (1st Seminar after Rio): Science and Technology to Save the Earth (unpublished).

1992 International Seminar on Nuclear War and Planetary Emergencies — 16th Session (2nd Seminar after Rio): Proliferation of Weapons for Mass Destruction and Cooperation on Defence Systems. (World Scientific, Singapore - New Jersey - London - Hong Kong, 1993).

1993 International Seminar on Planetary Emergencies — 17th Workshop: The Collision of an Asteroid or Comet with the Earth (unpublished).

1993 International Seminar on Nuclear War and Planetary Emergencies — 18th Session (4th Seminar after Rio): Global Stability Through Disarmament. (World Scientific, Singapore - New Jersey - London - Hong Kong, 1994).

1994 International Seminar on Nuclear War and Planetary Emergencies — 19th Session (5th Seminar after Rio): Science after the Cold War. (World Scientific, Singapore - New Jersey - London - Hong Kong, 1995).

1995 International Seminar on Nuclear War and Planetary Emergencies — 20th Session (6th Seminar after Rio): The Role of Science in the Third Millennium. (World Scientific, Singapore - New Jersey - London - Hong Kong, 1996).

1996 International Seminar on Nuclear War and Planetary Emergencies — 21st Session (7th Seminar after Rio): New Epidemics, Second Cold War, Decommissioning, Terrorism and Proliferation. (World Scientific, Singapore - New Jersey - London - Hong Kong, 1997).

1997 International Seminar on Nuclear War and Planetary Emergencies — 22nd Session (8th Seminar after Rio): Nuclear Submarine Decontamination, Chemical Stockpiled Weapons, New Epidemics, Cloning of Genes, New Military Threats, Global Planetary Changes, Cosmic Objects & Energy. (World Scientific, Singapore – New Jersey - London - Hong Kong, 1998).

1998 International Seminar on Nuclear War and Planetary Emergencies — 23rd Session (9th Seminar after Rio): Medicine & Biotechnologies, Proliferation & Weapons of Mass Destruction, Climatology & El Nino, Desertification, Defence Against Cosmic Objects, Water & Pollution, Food, Energy, Limits of Development, The Role of Permanent Monitoring Panels. (World Scientific, Singapore - New Jersey - London - Hong Kong, 1999).

1999 International Seminar on Nuclear War and Planetary Emergencies — 24th Session: HIV/AIDS Vaccine Needs, Biotechnologies, Neuropathologies, Development Sustainability – Focus Africa, Climate and Weather Predictions, Energy, Water, Weapons of Mass Destruction, The Role of Permanent Monitoring Panels, HIV Think Tank Workshop, Fertility Problems Workshop. (World Scientific, Singapore – New Jersey - London - Hong Kong, 2000).

2000 International Seminar on Nuclear War and Planetary Emergencies — 25th Session: Water – Pollution, Biotechnology – Transgenic Plant Vaccine, Energy, Black Sea Pollution, Aids – Mother-Infant HIV Transmission, Transmissible Spongiform Encephalopathy, Limits of Development – Megacities, Missile Proliferation and Defence, Information Security, Cosmic Objects, Desertification,

Carbon Sequestration and Sustainability, Climatic Changes, Global Monitoring of Planet, Mathematics and Democracy, Science and Journalism, Permanent Monitoring Panel Reports, Water for Megacities Workshop, Black Sea Workshop, Transgenic Plants Workshop, Research Resources Workshop, Mother-Infant HIV Transmission Workshop, Sequestration and Desertification Workshop, Focus Africa Workshop. (World Scientific, Singapore - New Jersey - London - Hong Kong, 2001).

2001 International Seminar on Nuclear War and Planetary Emergencies — 26th Session: AIDS and Infectious Diseases – Medication or Vaccination for Developing Countries; Missile Proliferation and Defense; Tchernobyl – Mathematics and Democracy; Transmissible Spongiform Encephalopathy; Floods and Extreme Weather Events – Coastal Zone Problems; Science and Technology for Developing Countries; Water – Transboundary Water Conflicts; Climatic Changes – Global Monitoring of the Planet; Information Security; Pollution in the Caspian Sea; Permanent Monitoring Panels Reports; Transmissible Spongiform Encephalopathy Workshop; AIDS and Infectious Diseases Workshop; Pollution Workshop. (World Scientific, Singapore - New Jersey - London - Hong Kong, 2002).

2002 International Seminar on Nuclear War and Planetary Emergencies — 27th Session: Society and Structures: Historical Perspectives – Culture and Ideology; National and Regional Geopolitical Issues; Globalization – Economy and Culture; Human Rights – Freedom and Democracy Debate; Confrontations and Countermeasures: Present and Future Confrontations; Psychology of Terrorism; Defensive Countermeasures; Preventive Countermeasures; General Debate; Science and Technology: Emergencies; Pollution, Climate – Greenhouse Effect; Desertification, Water Pollution, Algal Bloom; Brain and Behaviour Diseases; The Cultural Emergency: General Debate and Conclusions; Permanent Monitoring Panel Reports; Information Security Workshop; Kangaroo Mother's Care Workshop; Brain and Behaviour Diseases Workshop. (World Scientific, Singapore - New Jersey - London – Hong Kong, 2003).

2003 International Seminar on Nuclear War and Planetary Emergencies — 29th Session: Society and Structures: Culture and Ideology – Equity – Territorial and Economics – Psychology – Tools and Countermeasures – Worldwide Stability – Risk Analysis for Terrorism – The Asymmetric Threat – America's New "Exceptionalism" – Militant Islamist Groups Motives and Mind-sets – Analysing the New Approach – The Psychology of Crowds – Cultural Relativism – Economic and Socio-economic Causes and Consequences – The Problems of American Foreign Policy – Understanding Biological Risk Chemical Threats and Responses – Bioterrorism – Nuclear Survival Criticalities – Responding to the Threats – National Security and Scientific Openness – Working Groups Reports and Recommendations. (World Scientific, Singapore - New Jersey - London - Hong Kong, 2003).

2003 International Seminar on Nuclear War and Planetary Emergencies — 30th Session: Anniversary Celebrations: The Pontifical Academy of Sciences 400th - The "Ettore Majorana" Foundation and Centre for Scientific Culture 40th – H.H. John Paul II Apostolate 25th – Climate/Global Warming: The Cosmic Ray Effect;

Effects on Species and Biodiversity; Human Effects; Paleoclimate Implications; Evidence for Global Warming – Pollution: Endocrine Disrupting Chemicals; Hazardous Material; Legacy Wastes and Radioactive Waste Management in USA, Europe; Southeast Asia and Japan – The Cultural Planetary Emergency: Role of the Media; Intolerance; Terrorism; Iraqi Perspective; Open Forum Debate – AIDS and Infectious Diseases: Ethics in Medicine; AIDS Vaccine Strategies – Water: Water Conflicts in the Middle East – Energy: Developing Countries; Mitigation of Greenhouse Warming – Permanent Monitoring Panels Reports – Workshops: Long-Term Stewardship of Hazardous Material; AIDS Vaccine Strategies and Ethics. (World Scientific, New Jersey - London - Singapore - Beijing - Shanghai - Hong Kong - Taipei - Chennai, 2004).

2004 International Seminar on Nuclear War and Planetary Emergencies — 31st Session: Multidisciplinary Global Approach of Governments and International Structures: Societal Response – Scientific Contributions to Policy – Economics – Human Rights – Communication – Conflict Resolution – Cross-Disciplinary Responses to CBRN Threats: Chemical and Biological Terrorism – Co-operation Between Russia and the West – Asymmetrical Conflicts – CBW Impact – Cross-Disciplinary Challenges to Emergency Management, Media Information and Communication: Role of Media in Global Emergencies – Emergencies Responders – Working Groups' Reports and Recommendations. (World Scientific, New Jersey - London - Singapore - Beijing - Shanghai - Hong Kong - Taipei - Chennai, 2004).

2004 International Seminar on Nuclear War and Planetary Emergencies — 32nd Session: Limits of Development: Migration and Cyberspace; in Europe; Synoptic European Overview; From and Within Asia; Globalization - Climate: Global Warming; a Chronology; Simple Climate Models; Energy and Electricity Considerations – T.S.E.: CJD and Blood Transfusion; BSE in North America; Gerstmann-Straussler-Scheinker Disease – The Cultural Emergency: Innovations in Communications and IT – Cosmic Objects: Impact Hazard; Close Approaches; Asteroid Deflection; Risk Assessment and Hazard Reduction; Hayabusa and Follow Up – AIDS and Infectious Diseases: Ethics in Medicine; International Co-operation; Laboratory Biosecurity Guidelines; Georgian Legislation; Biosecurity Norms and International Organizations, Legal Measures Against Biocrimes – Water and Pollution: Cycle Overview; Beyond Cost and Price; Requirements in Rural Iran; Isotope Techniques; Clean and Reliable Water for the 21st Century – Permanent Monitoring Panels Reports – Workshops: Global Biosecurity; Cosmic Objects. (World Scientific, New Jersey - London - Singapore - Beijing - Shanghai - Hong Kong - Taipei - Chennai, 2005).

2005 International Seminar on Nuclear War and Planetary Emergencies — 34th Session: Energy: Nuclear and Renewable Energy; Energy Technologies for the 21st Century; Repositories Development; Nuclear Power in Europe and in Asia; The Future of Nuclear Fusion – Climate: Global Warming; Celestial Climate Driver; Natural and Anthropogenic Contributions; Climate Data and Comparison with Models; Understanding Common Climate Claims – AIDS and Infectious Diseases: New Threats from Infectious Agents – SARS Epidemic; Vaccines Development; Transmissible Spongiform Encephalopathies Update – Limits of

Development: International Points of View on Migration – Pollution: Science and Technology; Subsurface Laser Drilling – Desertification: A Global Perspective; Integrated Approach – Disarmament and Cultural Emergencies: A WFS Achievement in China; Non-Proliferation – Permanent Monitoring Panel Reports – Workshops: Energy; Information Security; Building Resilience Associated with the Third Meeting on Terrorism. (World Scientific, New Jersey - London - Singapore - Beijing - Shanghai - Hong Kong - Taipei - Chennai, 2006).

2006 International Seminar on Nuclear War and Planetary Emergencies — 36th Session: Energy: Global Nuclear Power Future; Global Monitoring of the Planet Proliferation: Nuclear Weapons; AIDS and Infectious Diseases: Avian Flu – Global Health; Climatology; Global Warming/Aerosols and Satellites; Pollution: Plastic Contaminants in Water; Information Security: Relevance of Cyber Security; Limits of Development: Development of Sustainability; Defence Against Cosmic Objects; WFS General Meeting; Cultural Energy-Focus: Terrorism; Permanent Monitoring Panel Reports; Limits of Development Permanent Monitoring Panel Meeting; World Energy Monitoring Workshop. (World Scientific, New Jersey - London - Singapore - Beijing – Shanghai - Hong Kong - Taipei - Chennai, 2007).

2007 International Seminar on Nuclear War and Planetary Emergencies — 38th Session: World Energy Crisis; Managing Climate Change; Mitigation of Greenhouse Gases; Geoengineering & Adaptation; Theoretical Alternatives to Climate Modelling; U.S. Missile Defence Shield; Global Monitoring of the Planet; Life Cycle Nuclear Energy Environmental Issues; The Epidemic of Alzheimer; Infectious Agents and Cancer. (World Scientific, New Jersey - London - Singapore - Beijing – Shanghai – Hong Kong - Taipei - Chennai, 2008).

2008 International Seminar on Nuclear War and Planetary Emergencies — 40th Session: Energy: Nuclear Power Present and Future; Sustainability of Biofuels; Resolving the Nuclear Waste – Climatology Model and Statistics; Ozone and Climate Change Interaction; Spatio-Temporal Field of Atmospheric CO_2; Forest Policies – Medicine: Vector-Borne Diseases; Screening Technology – Pollution: Air-Borne Particulates – Global Monitoring of the Planet: Disarmament and Non-Proliferation Regime; The Crisis in Internet Security; The Northern Sea Route – The Erice Science for Peace Award Scientific Session. (World Scientific, New Jersey - London - Singapore - Beijing – Shanghai - Hong Kong - Taipei - Chennai, 2010).

2009 International Seminar on Nuclear War and Planetary Emergencies — 42nd Session: World Energy Crisis – Energy & Pollution: Essential Technologies for Managing the Coupled Challenges of Climate Change and Energy Security, Energy, Water, Climate, Pollution & Limits of Development in Asian Countries; Global Monitoring of the Planet-Sensitivity of Climate to Additional CO_2 as Indicated by Water Cycle Feed-back Issued, Climate Uncertainties Addresses by Satellites, The Basic Mathematics Needed for all Models; Pollution and Medicine – The Revolution in the Environmental Health Sciences and the Emergence of Green Chemistry; Information Security – Cyber Conflict and Cyber Stability; Finding a Path to Cyber Peace; Cultural Pollution – The Erice Science for Peace

Award Scientific Session. (World Scientific, New Jersey - London - Singapore - Beijing - Shanghai - Hong Kong - Taipei - Chennai, 2010).

2010 International Seminar on Nuclear War and Planetary Emergencies — 43rd Session: The Role of Science in the Third Millennium. Energy: Nuclear Power Plants – Sustainable Systems – Pollution: Oil Spills – Air, Water and Waste Criteria – Climate: Paleoclimate Natural Variations and Implications – Sensitivity, Climategate, USA Legislation – Cosmic Objects: Countermeasures – Mitigation of Terrorist Acts: Scientific Contributions to Biosecurity – Lectio Magistralis: Science in Everyday Life – Information Security: Cyber Conflicts and Stability – The Erice Science for Peace Award Scientific Session. (World Scientific, New Jersey - London - Singapore - Beijing - Shanghai - Hong Kong - Taipei - Chennai, 2011).

2011 International Seminar on Nuclear War and Planetary Emergencies — 44th Session: The Role of Science in the Third Millennium Energy. World Energy Crisis – Global Nuclear Energy Issues after Fukushima – Energy Efficiency / Energy and Pollution – Unconventional Natural Gas: Benefits and Risks / Climate – Cosmic Rays and Climatic Processes / Water and Pollution – Water Scarcity and Pollution – Contaminants of Emerging Concern / Information Security – The Role of Science in Information Technology and Internet Tools in Developing Countries Food, Soil & Medicine – Greenhouse Gases Consequences – Evidence-Based Third Millennium Medicine / Lectio Magistralis – The Role of Science in the Third Millennium. (World Scientific, New Jersey - London - Singapore - Beijing - Shanghai - Hong Kong - Taipei - Chennai, 2012).

2012 International Seminar on Nuclear War and Planetary Emergencies — 45th Session: Lectio Magistralis Global Warming Emergency/Why Science is Needed for the Culture of the Third Millennium – Mitigation of Terrorist Acts – Improvised Nuclear Devices – Gap Analysis of EU – Presenting Nuclear Explosive Terrorism – Public Perception / Climate & Climate Economics / Global Nuclear Energy Issues – Outlook in Japan, China, Europe and USA – Sustainable Nuclear Energy Systems – Economics of Nuclear Power / Energy & Sustainability in Cities – Global Sustainability – Smart Grids – Complexity / Global Food Production & Forest Dynamics / Water, Pollution & Terrorism / Brain Aging & Behaviour / Frontiers in Fast Computing & Informatics / Information Security. (World Scientific, New Jersey - London - Singapore - Beijing - Shanghai - Hong Kong - Taipei - Chennai, 2013).

2013 International Seminar on Nuclear War and Planetary Emergencies — 46th Session: Why Science is Needed for the Culture of the Third Millennium – Lectio Magistralis: Alliance between Politics and Science / Nuclear Power Safety and Development / Water Pollution & Terrorism / Climate and the Limits of Geoengineering / Mitigation of Megaterrorism and Hybrid Threats / Pollution Innovative Cleanup Technologies / Information Security: Cybersecurity at Cross-Roads / Energy – Key to the Evolution of Cities / Energy as a Planetary Emergency / AIDS and Infectious Diseases. (World Scientific, New Jersey - London - Singapore - Beijing - Shanghai - Hong Kong - Taipei - Chennai, 2014).

THE THREE COMPONENTS OF "THE NEW MANHATTAN PROJECT": SCIENCE, TECHNOLOGY, CULTURE

The "New Manhattan Project" consists of two scientific-technological components and one component that is exclusively cultural. Each of these three components is characterized by great timeliness.

- Science. The first component concerns the most advanced frontier of Modern Physics—its goal is to realize the most powerful proton accelerator in the world, ELN: Eloisatron (Euroasiatic Long Intersecting Storage Accelerator). The machine consists of two rings, each of which is 300 km long and intersects with the other so as to enable interactions between protons at the maximum energies imaginable today with the most advanced sub-nuclear techniques.
- Technology. The second component involves a series of technological studies whose goal is to invent the most powerful instruments for measuring, with precisions never reached before, the fundamental quantities on which any set of physics variables is based, such as: space, time, quantity of motion, energy and charge (this is the LAA Project). In this Project there is a pilot-project for the Time of Flight (TOF) of Subnuclear particles which now holds the world record for accuracy: 15 picoseconds (one picoseconds corresponds to one thousandth of a billionth of a second). Our capacity to measure these fundamental variables of physics keeps improving through the invention of new technologies, at the heart of which are born new instruments capable of promoting progress in the most advanced sectors of application. This is why the technological component of the Project has diversified into many sectors of modern technology, sectors ranging from medical, biomedical, and genetic engineering research to the conquest of Space, to the tools necessary in everyday life, like computers, supercomputers and information networks like Internet and GRID.
- Culture. The third component. There is an additional component that includes the preceding ones but that is purely cultural in nature, in the sense that its role is to communicate the practical value of the results obtained in the projects on Science, Technology and Planetary Emergencies to everybody. Planetary Emergencies have a strong cultural component since the 72 Planetary Emergencies are relevant to our everyday life. To study these problems will enable us to give Governments the only way to resolve them despite they often appear deprived of short-term and even long-term solutions.

Some interesting details

To set out to confront and resolve the types of problems cited in the Three Components—Science, Technology, Culture—might appear utopian.

This is mistaken, however, because the Three Components already have roots in years of research carried out with the intent of creating the foundation on which to base an institution like "A Project for Mankind".

The Three Components of the Project were elaborated by an international scientific community that identifies itself with the ideals of the WFS (World Federation of Scientists, born during the times of the Cold War), such as a Science without secrets and without borders, as we have stated many times.

The roots of the PROJECT are on all those scientists who have contributed to the search for a Science without Secrets and Borders. Their names follow.

SCIENTISTS AND SPECIALISTS WHO HAVE CONTRIBUTED TO THE VARIOUS ACTIVITIES OF PILOT-PROJECTS

A. Abed Rabbo, S.V. Ablameyko, G. Adam, A. Adamishin, J. Adams, E. Adar, V. Adefela, K.S. Adhao, P. Adonnino, P.K. Aggarwal, K.B. Aglawe, C. Aglietta, P.B. Agrawal, A. Aguzzi, K. Ahmad, Z. Ahsan, R. Ajalov, A. Ajami, T. Akhmetov, J.P.D. Akue, H. Al Shahristani, M. Al-Ashkar Jabbour, H. Al-Shahristani, N. Al-Shalabi, K. Alberini, J. Alderdice, M. Alessi, R. Alexakhin, M. Ali, H. Alper, L. Alurralde, I. Amer, M.N. Amin, D.M. Anderson, R. Anderson, M.J. Antonovsky, G. Anzivino, A. Arefiev, I. Armillas, M. Arneodo, R. Arnowitt, C.J. Arntzen, G.T. Aronica, A. Aru, F. Arzarello, K. Asenso-Okyere, A.V. Asole, S. Atran, B.H. Aurade, A. Aureli, M. Awawdeh, R. Axelrod, V.B. Badhe, G. Badino, S.D. Bagchi, M.K. Bagde, M.A. Bagirov, K.R. Bagree, D.R. Bahekar, U.D.N. Bajpai, P. Bakaki, L. Bakanidze, J.S. Bakarc, G. Balamurugan, R. Balbinot, R. Baldini Ferroli, T.P. Ballestero, A.M. Band, G. Bandyopadhyay, S. Bandyopadhyay, H. Banerjee, S.K. Banerjee, A.N. Bannore, G. Barbagli, A. Barkat, W.A. Barletta, F. Barré-Sinoussi, L. Barrett, G. Barrocu, P. Bartel, M. Barthelemy, V. Baryakhtar, I. Bashmakov, M. Basile, G. Basile, A. Bassett, M. Bassetti, B. Basu, D. Basu, C.O. Bauer, A. Bazzani, E. Beach, A. Beauchamp, U. Becker, S.G. Bedi, S.Y. Bedore, R.B. Belorkar, N.G. Belsare, M.J.S. Belton, I. Ben Israel, M. Ben-Akiva, G. Benedek, G. Benfratello, M. Benot, R.W. Bentley, J. Berbiers, F. Bergsma, A. Bernotas, A. Bertin, R. Bertola, L.A.M. Bettencourt, V.S. Bhagde, D.R. Bhakre, P.D. Bhalcrao, K.S. Bhamra, N.C. Bhattacharya, T. Bhattacharya, S.N. Bhongle, M. Biagini, G. Biberfeld, S.S. Bijwe, A. Birkhofer, D. Birx, B. Blumberg, A. Boccafoschi, R.K. Bock, D. Bodansky, D.F. Boesch, R.D. Bojkov, L. Bolshov, S.M. Borikar, M.S. Borkar, A. Borthagaray, J.M. Borthagaray, O. Bosch, P. Bosch, E. Boschi, F. Bosio, J. Bouchard, R. Bouclier, A.F. Boukri, M.S. Boulahya, E. Braun, J.M. Braun, K. Brendlow, R.L. Breyer, V.B. Britkov, D. Brookshire, P. Brown, K.-M. Brust, R. Bruzzese, H. Budka, R. Budnitz, C. Bühlmann, F. Buonaguro, E. Buratti, J. Burger, D.K. Burghate, P.H. Burgi, D. Buriot, J. Bus, P. Cailleret, F. Calabrese, G. Caldarelli, K. Caldeira, G.M. Calderone, A. Caliandro, M. Campos, G. Canavan, A. Candela, S. Cannistraro, L. Caputi, G. Cara Romeo, T. Carettoni, G.R. Carmichael, L.V. Carmona, I. Carreca, S. Carruba, R.M. Carter, R. Casaccia, E. Cascetta, J.P. Casciano, A. Castellina, A. Castelvetri, A. Cavallini, A. Cellino, L. Cervenakova, J. Chahoud, D. Challoner, S.M. Chandekar, C. Chapman, G. Charpak, N. Charpak, Bofei Chen, Chaoqiang Chen, Huijun Chen, Jianpu Chen, Limin Chen, Pengri Chen, Senyu Chen, Siyu Chen, A.F. Cheng, D. Chereshkin, M. Cherian, M. Chiarini, A. Chiavassa, A.N. Chilingarov, S.P. Chiney, Yong-Sang Choi, N.H. Christ, S.A. Christopher, P. Chylek, O. Ciancio, L. Cifarelli, M. Cifarelli, F. Cindolo, M.V. Civita, A. Clare, K. Clark, R.A. Clark, J.H. Clarke, R. Cloninger, Zang Coa, E. Coccia, M. Cocco, T.B. Cochran, S. Cocina, D. Cohen, E. Colavita, W.E Colglazier, T.J. Collins,

P.M. Conneally, A. Contin, A.H. Cook, S. Cooper, S. Cornelius, W.J. Cosgrove, M. Creutz, D. Crosetto, F. Crosta, I. Crotty, Huachuam Cui, Ruyu Cui, Shiangzong Cui, K. Czerski, V. D'aragona, G. D'Alì, C. D'Ambrosio, M. D'Arcy, S. D'Auria, C. Daase, Guiliang Dai, C.K. Damle, D.N. Darane, M. Dardo, P. Darriulat, T.K. Das, M. Dasgupta, B.K. Datta, R. Datta, U.R. Datta, S. Datta Majumdar, D. Davasumony, R. Davis, J.W. Day, C. Dazzi, S. De Leon Mendoza, S. De Pasquale, V. De Sabbata, R. De Salvo, E. De Sanctis, G. De The, J.-F. Debroux, C. Decarli, S.T. Dekosky, C. Del Papa, A. Della Torre, Shusen Deng, Yanping Deng, S.L. Deoghare, V.S. Deogoankar, D.W. Deshkar, A.A. Deshmukh, B.T. Deshmukh, D.T. Deshmukh, K.C. Deshmukh, R.M. Deshmukh, V.R. Deshmukh, A.M. Deshpande, G.T. Deshpande, M.V. Deshpande, P.D. Deshpande, J.A. Dhanaraj, S.B. Dhannasare, G.B. Dhanokar, R.V. Dharaskar, M.V. Dhargave, M.T. Dharmadhikari, T.N. Dhobe, S. Di Liberto, M. Di Martino, R. Dierstein, J. Diez-Nicolas, C. Difiglio, G. Dinunzio, P. Diodati, M. Diop, A. Dirac, B. Dirac-Svejstrup, R. Dobinson, S. Dorey, D. Dorogan, P. Douglas, D.H. Douglass, J.E. Drewes, A. Driks, P. Droege, Wenfu Du, Xijiu Du, W. Duax, G. Duca, P. Duggal, A.G. Duncan, R. Duncan, J. Dupont, J. Dupraz, M. Dutta, T. Dyson, K. Eckhard, P.A. Eckhoff, V. Egikova, W. Eichhammer, T. Ekelöf, J. Ekmann, R.B. Elgira, M. Eliashvili, C. Ellis, C.D. Ellis, A.C. Elmore, Z. Eshhar, C. Essex, L. Everett, F. Fabbri, J.P. Fabre, N. Falsaperna, Cheng Fang, Shouxian Fang, Rong- Xiang Fang, M. Farah, M.R. Farlow, R.N. Faustov, Bai Fengning, C.D. Ferguson II, S. Ferrara, L. Ferrario, M. Ferreira Rea, A. Ferro-Luzzi, S. Fetter, G. Fierotti, P.J. Finck, J. Fischer, B. Fischhof, E. Fishbach, M. Fitzpatrick, L. Fog, M. Folstein, S.E. Folstein, D. Fong, P. Ford, D. Ford, R.E. Ford, S. Fortunato, R. Fox, I. Franca Junior, F. Frasconi, L. Friedl, R. Friedrich, E. Friis-Christensen, H. Fritzsch, M. Fukushima, F.P. Fulci, W. Fulgione, W. Fulkerson, M. Fusco Girard, P.M. Gadkari, A. Gagarinski, V.J. Gaikwad, C. Gajdusek, Y. Galaktionov, S. Galassini, P. Galeotti, M. Gallagher, B. Galland, R. Gallo, A. Galvez, M. Ganguli, R. Ganguli, Cuishan Gao, Wei Gao, Wenxiu Gao, P. Garderet, R.L. Garwin, M. Gasperini, J. Gaudacn, K.N. Gawande, S.G. Gawande, S.W. Gawande, G. Geiger, M. Gennari, M. Georgescu, P. Georgescu, H. Ghaffarzadeh, M.H. Ghate, S.M. Ghatole, H. Ghazi, S. Ghernaouti, B. Ghetti, G.A. Ghike, S.G. Ghosh, S.N. Ghosh, E. Giacobino, I. Gilada, G.T. Gillies, T.J. Gilmartin, F. Giorgi, M. Girard, P. Giusti, S. Gizowsky, K. Goebel, W.C. Goede, V.G. Gokhale, V.M. Gokhale, R. Goldoni, B.D. Goldstein, R. Goldstein, G. Gomes Serra, A. Gonzales-Pozo, N. Gopalkrishnan, Y. Gorodkov, M. Gourdin, T. Grahame, G. Gray, K. Gray, W.M. Grayman, A.R. Green, J.M. Greenberg, D. Greene, A. Grieder, D.W. Griffin, A. Grigoryan, J.G. Grimes, M. Grinius, C. Grinnel, A.T. Grove, A. Grübler, Mengping Gu, Shudi Gu, Songhua Gu, Fengqi Guang, I. Guastalla, B. Guerard, O. Guifo, L.J. Guillette, Weixin Guo, Yanan Guo, Zhiyuan Guo, V.D. Gupta, K. Gurney, B. Gurusamy, A. Gusev, T. Gys, M.J. Haddadin, S. Hadjitodorov, J.L. Haines, L. Hammarstrom, Qian Han, D. Hanson, W. Happer, J. Hardy, A.W. Harris, N. Harris, A. Hasanov, M.H.A. Hassan, M. Hassanizadeh, D. Hatzifotiadou, D. Hawkins, J.A. Haynes, R.B. Haynes, Mengjia He, B. Heap, E. Heijne, D. Heim, G. Heindel, S. Hellman, U. Helmbrecht, H.C. Hendrie, G. Herten, D. Hill, J. Hinkula, W.B. Hirurkar, E.K. Hocking, M. Holtzem, J.-M. Hombert, P. Hoodbhoy, H. Hoorani, R. Hoskins, M. Hourican, K. Howdeshell, Jialiu Hu, Jiawei Hu, Lidong Hu, Kaixi Huang, Li Huang, Nan Huang, Yinzhi Huang, R.K. Huber, G. Hubler, P.P. Huddar, W.F. Huebner, Yu-Ping Huo, C. Huraux, S.I. Hussain, C. Hutchinson, J.L. Hutton, N.T. Ibrahimov, G. Iklódy, M.F. Ingalgi, D. Ingersoll, A.G.

Ingle, B.G. Ingle, R.P. Ingole, A. Ioppolo, T. Isaacs, Y.Y. Israel, P.K. Iyengar, C. Izaurralde, Y.A. Izrael, P. Jackson, J.D. Jafar, J. Jain, S.K. Jain, R.K. Jambhorka, P.T. James, F. Janouch, J. Jarab, P. Jarron, M. Jefferson, P. Jenni, R.K. Jenny, M. Jensen, Yanwu Jia, Chunhua Jiang, Yanling Jiang, Qingshou Jin, Youkai Jin, Bai Jingzhi, A.S. Joalikar, S. Jobling, D.N. Joharapurkar, K. Johnsen, D. Johnson, V.B. Johri, L. Jones, S.V. Joshi, D. Jost, V. Journé, B.T. Judge, Yin Jun, D.K. Kadu, R.E. Kahn, B. Kai, L. Kairiukstis, J.M. Kale, M.G. Kale, A. Kamal, A. Kamalakar, M.A. Kamarulnizam, Y. Kamyshkov, T.M. Karadci, I. Karawan, M. Kareithi Fanin, U. Kasper, N. Kassis, W. Kastenberg, D.M. Katarla, T. Kawata, S.A.H. Kbatil, Songbai Ke, D.G. Kelkar, A.A. Keller, B. Kellman, D. Kennedy, G.M. Kesharwani, G.S. Khadekar, M.T. Khadse, G.A.N. Khan, M.Q. Khan, S. Khan, P. Khanal, A.T. Khandare, M.L. Khandelwal, R.B. Kharade, T.M. Kharade, R.B. Kharat, U.U. Kharat, R. Khatami, H. Khatib, S.M. Khodragade, N.W. Khodragade, P. Khosla, Do- Won Kim, D. Kimball, W. Kininmonth, T. Kinoshita, J. Kirkby, M.S. Kishirsagar, V. Klaus, A. Klimburg, E. Klushin, G. Knies, E. Knupp, E. Kobal, L.P. Koh, A. Konovalov, S. Kopits, M.S. Korde, I. Koren, A. Korshenko, C.P. Korthals-Altes, A.P. Kowale, B.P. Kowale, S.J. Kowall, J.-P. Kraehenbuhl, M.N. Krasnoperov, J. Krause, V. Kremenyuk, W. Krisher, J.E. Kristjansson, V. Krivokhizha, D.J. Kroll, N. Kroo, A.V. Krutskikh, Benlin Kuang, J.-F. Kuka, V.P. Kukhar, A.L. Kulikov, A.B. Kulkarni, D.K. Kulkarni, R.S. Kulkarni, A. Kumar, G.N. Kumbhare, R. Kuschel, A. Kutanov, I. Laakso, J.C. Labbè, G.R. Labde, P. Lachmann, K.S. Lackner, S.A. Ladhake, E.F. Lambin, L. Lane, Pengfei Lang, B. Langford, R.B. Lanjewar, R. Lanza, M.V. Lapsa, H. Larsen, S. Lau, G. Laurenti, N. Lawson, Tsung-Dao Lee, A. Lee, A. Lehmann, S. Leivesley, T. Lenton, A.R. Leo, M. Leo, G. Leoni, A. Lerner- Lam, G. Lester, H. Leutz, G. Levi, M. Levine, S. Levisley, Binfang Li, Bomin Li, Dashi Li, Dezhong Li, Guanglin Li, Jia Li, Jiacai Li, Jianguo Li, Jin Li, Youmeng Li, M. Li, V. Liakhovets, A.A. Lindberg, J. Lindsten, R.S. Lindzen, Dachun Ling, R. Linsky, Dekang Liu, Diankui Liu, Shiyao Liu, Tai Liu, Yucheng Liu, C. Liverani, M. Locati, P. Lock, V. Loginov, G.C. Lokre, L. Lone, Shida Long, A.D. Lopez, Changguo Lu, Jihua Lu, Weida Lu, G. Luches, D. Luckey, S. Lueders, H.S. Lunge, Yingxiong Luo, Zihua Luo, M.B. Lyles, P. Lyons, Jimao Ma, G. Maccarrone, M.C. Maccracken, J.S. Mackenzie, J.K. Madhugiri, H. Madsen, A. Maggiora, R.K.W. Mahalle, S.N. Mahapatra, M.A. Mahure, C.K. Majumdar, K. Makino, L. Malagnini, M. Malavasi, F. Mancini, B.N. Mandal, R. Manley, S.V. Manmode, M.E. Mann, J.L. Manputra, C. Mantegna, J.C. Manuguerra, Huishun Mao, B. Maraviglia, A. Margotti, A. Marino, G. Marland, S. Marmi, S. Martellucci, A. Martin, J.-F. Martin, V. Martyniuk, D. Masiga, R.A. Mason, T. Massam, T. Masuda, R. Materia, C.G. Mathe, K. Matsui, B.R. Maurya, C. Mayrhofer, S. Mazurek, P. Mazzanti, G. Mazzari, A.D. Mbewu, S. Mbulaiteye, C. McCombie, G. McConnell, C.A. Mcharc, S. Mcintyre, R. Mckitrick, M. Mckubre, F.J. Mcmahon, F. Mehr, K.H. Meier, H. Meinhard, V.N. Melnikov, S. Mendoza, M.G.K. Menon, A. Merari, M. Mesbahi, V.R. Metkar, R. Meunier, O. Meuvret, G. Meyer, Jianlin Mi, Rongsheng Mi, E.V. Miasnikov, A. Micallef, G. Miceli, L.M. Michaud, V. Mikhailov, A.I. Miller, B.L. Miller, G. Million, M. Milosevska, Dong-Pil Min, G.M. Mirdal, V. Mirianashvili, K.D. Mishra, R.B. Mishra, R.S. Mishra, S.B. Mishra, V.G. Miskin, I.V. Mitrofanov, J. Mittelstrass, A.Miyahara, S.G. Modak, A.K. Modi, G. Mohanty, S.C. Moholkar, C. Monckton of Brenchley, M. Moodie, A.L. Moore, C. Moore, L. Moore, S. Morain, A. Morelli, K. Morey, D. Morrison, F. Motta, A.E. Motter,

R. Mount, W. Mualla, W. Mueller Seedorf, A. Mukherjee, S. Mukherjee, D. Mulenex, K.V. Muley, V.D. Muley, K.A. Müller, W. Muller-Seedorf, P.M. Mullineaux, A.M. Mundhada, A. Muriel, J.R. Murrell, S. Muthiah, B. Myers, A.B. Nadange, A. Nagurney, P.T. Naiekar, S.Y. Nandanwar, R. Nania, E. Nappi, G. Nardoni, N. Narjoud, R.Y. Narkhede, S.G. Nasare, K. Nash, V. Nassisi, L. Natrajan, A. Naudi, G. Navarra, B. Ndiaye, R. Nelson, K.A.N. Nerkar, S. Netesov, N. Neureiter, H. Newman, R. Newton, T.N. Nguyen, A.M. Niang, S. Nickovic, J. Niemela, N.A. Nik Mahmud, B.R. Nikhade, F.M. Nirwan, W. Noh, A. Nolan, D. Norman, M. Ó Cinnéide, V. O'Shea, J. Oehlmann, Sun Kun Oh, E. Okandan, N. Olea, A. Oliva, L. Olsson, J. Ongena, C.R. Ordonez, J. Orear, I. Ortalli, A. Osterhaus, G. Ourisson, E. Özsoy, H.P. Paar, M. Pagliai, R.S. Pagrut, J.P. Pal, M. Pal, P. Palanza, F. Palmonari, G. Palshin, G.W. Paltridge, D. Palumbo, Huibao Pan, N.K. Pande, S.N. Pandey, T.N. Pandey, U.S. Pandey, Jiabiao Pang, K.G. Pangarkar, J.M. Pap, C. Papadia, G. Papini, S. Paradiso, S.C. Paranjape, Y.R. Paranjape, G. Parisi, D.E. Parker, F.L. Parker, R. Parker, T. Parkhalina, S. Parmigiani, D.V. Parwate, G. Passotti, C.B. Patil, M.B. Patil, P.D. Patil, S.P. Patil, B.A. Patki, A.A.N. Patrinos, A. Paul, K.V. Pawar, S.P. Pawar, K. Peabody O'brien, P. Pelfer, A.W. Pendharkar, C.R. Penn, K.D. Pennell, M. Pericak- Vance, G. Perkovich, C. Peroni, E. Perotto, J.S. Perry, V. Peskov, M.S. Petersen, A.T. Peterson, J. Peterson Myers, R. Petroski, A. Peyraube, M. Pezzotti, Duy Hien Pham, D. Piazza, D. Piedigrossi, L. Pietronero, A. Piontkovsky, G. Piragino, V. Pjidaev, C.S. Plesko, V. Plyaskin, G. Pocci, R.B. Pode, J. Podkanski, H.M. Poharkar, C.N. Potdar, N.B. Potdar, J. Pozela, V.K. Pratapwar, H. Pratomo, E. Predazzi, S. Prenafeta-Jenkin, E. Prescott, J. Price, S.D. Prince, G. Prisco, M. Puglisi, H.S. Pundkar, Zhuming Qian, Jimin Qiao, Jiu Qin, Hong Qiu, T. Quirk, R. Ragaini, V. Ragaini, V.D. Raghatate, S.B. Raichur, S. Raimondi, N. Rajabov, R. Rajaraman, V.B. Rajurkar, M. Rama Mohan Mao, A.D. Rangari, S.P. Rant, V.A. Ratate, G.D. Rathod, A.D. Raut, D.R. Raut, A. Razdan, V. Razumas, F. Re, A. Rebane, K. Rebane, B. Reitano, Wenbin Ren, A.C. Revkin, C. Reynolds, M. Ricci, P. Ricci, M. Ricketts, E. Ricotta, G. Righini, A. Rigoni, G. Rinaldi, J. Rispoli, C. Rizzuto, D.O. Rogers, H.-H. Rogner, F. Rohrbach, A. Ronne, N. Rosenberg, A.H. Rosenfeld, L. Rossi, G.T. Rossi, P. Rotelli, R. Roy, Tongze Ruan, E. Rubin, Z. Rudzikas, J.G. Ruiz, J.M. Ruiz, N. Rus, A. Rybalchenko, R. Rydell, H. Rykaszewski, O. Saavedra, N. Sacchetti, A. Safaenili, E. Sagarra, M. Sageman, J. Saginor, L. Said Issa, F. Sala, R. Salerno, I. Salihoglu, M. Salman, N. Samios, N. Samman, T. Sampson, D.L. Samudralwar, N.M. Samuel, I. Sandrea, M. Sankhare, M. Sano, B. Santer, J.C. Santiard, P. Santos- Ocampo, A. Sarkissov, G. Sartorelli, N.N. Saste, K.K. Satpathy, R.V. Satpute, F. Sauer, F. Sauli, M. Saviola, J.B. Savy, A. Saykin, W. Scandale, G. Scarlatti, D. Scavia, D. Schaffer, A. Schaper, E. Schenvit, J. Schipper, H. Schmidt, L. Schneider, R. Schock, E. Schöll, H. Schubert, U. Schuklenk, B. Schürch, R. Schweickart, G. Sciarabba, D. Scigocki, G. Scioli, D.R. Scott, L. Seeber, D. Segre, J. Séguinot, S.H. Selukar, D. Sen, M. Sen Gupta, N.D. Sen Gupta, J.U. Seo, S. Serci, G.G. Serra, G. Servizi, D. Sevier, Hanying Sha, L. Shabanova, E. Shaffer Vergino, D.I. Shahare, V. Sharan, P. Sharp, U. Shavit, N.J. Shaviv, E. Sheffner, V.P. Shelest, Boahua Shen, Junpeng Shen, Tianji Shen, P. Shetty, V. Shevchenko, Yinsheng Shi, Y. Shibata, A. Shihab-Eldin, V.S. Shiramwar, H. Shönbacher, S.P. Shrinkahnde, Weisan Shu, Yuede Shu, H.H. Shugart, J.B. Shukla, E. Shumilov, Shyampati, R. Sibilia, S. Siboni, K.M.B. Siegbahn, G. Simbolotti, G. Simonet, S.F. Singer, M.D. Singh, R.D. Singh, T. Singh, S.K. Singh, J. Singh Rana, K.P. Sinha,

K.C. Sivaramakrishnan, A.W. Smith, D.K. Smith, A. Smithson, P.V. Soadekar, A. Sobel, T. Sofu, V.S. Soitkar, G. Soliani, P. Sonderegger, D.N. Soni, S. Sorek, S. Sorooshian, M. Spadoni, M.G. Spillantini, L. Sportelli, G. Sprehn, W.A. Sprigg, K.S. Sreenivasan, R.S.L. Srivastav, A. Starobinsky, L. Stein-Spencer, F. Steinhaeusler, G. Stephens, M. Steuer, W.M. Stigliani, E. Storm, B. Stram, G. Stratan, N. Stratan, C. Strenger, M. Stürmer, M. Suffert, Songlan Sun, Yuelin Sun, Honglie Sun, Yuliang Sun, V.J. Sundaram, S.N. Supe, G. Susinno, H. Svensmark, D.L. Swackhamer, S.P. Swami, H. Swan, S.H. Swan, K. Swanson, K. Szigethy, J. Szyszko, V.A. Tabhane, S. Tailhardat, Y. Takano, K. Talattof, V.K. Tale, G. Tallia, C. Talsness, R.K. Talwekar, M. Tamada, T.M. Tamhane, H. Tanaka, Baoping Tang, Cheng Tang, Shuming Tang, M.J. Tannenbaum, P. Tans, M.S. Tapi, G.G. Tappan, E. Tarkowski, A. Tavkhelidze, A.N. Tawfik, T. Taylor, S. Tazzari, U. Teipel, M.T. Teli, T.J. Telranhe, B.R. Tembhurne, Kejian Teng, A.E. Terraneo, B. Tertrais, S.C. Thaker, V.A. Thakhare, A.W. Thakre, R.V. Thakre, V.M. Thatte, W. Thielemans, T.L. Thomas, K. Thompson, R.C. Thompson, R. Thorstensson, Fang Tian, L. Tieszen, S.D. Tikar, D. Tillitt, R. Timerbaev, S.C.C. Ting, B.P. Tiwari, S.A. Tiwari, S. Tlou, A.F.B. Tompson, G. Torelli, B. Torto, T. Toth, H.I. Touré, G. Triesman, G.C. Trinchero, P. Tschakert, A.A. Tsonis, V. Tsygichko, V.S. Tumram, G. Turchetti, G. Tveit, W.E. Tyner, P.F. Uhlir, H. Ujita, D.U. Umak, T. Valencic, S. Valenti, P. Vallania, P. Van De Terre, R. Van Der Zwaan, W. Van Dieren, E. Vardas, A. Vasiliev, J. Veizer, Y.P. Velikhov, V. Vengrinovich, G. Venturi, E.S. Vergino, S. Vernetto, I. Vetliski, A.D. Vickery, K. Vidya Gaikwad, Z. Vilakazi, F. Villa, J. Villinger, A. Vishnoi, M. Vivargent, O. Vizitiu, V.D. Volkan, F.S. Vom Saal, B. Von Alvensleben, J.G. Von Der Schmitt, F. Von Hippel, L. Votano, D.C. Wade, D.G. Wadke, S.D. Wadke, F. Waelbroeck, R.V. Waghamare, W.U. Waghmode, R. Walgate, J.D. Walker, S.H. Walkey, A.B. Walkhade, Bosi Wang, Dianchen Wang, Feng Wang, Hengfeng Wang, Hengjiu Wang, Hui-jun Wang, Jin Wang, Linlin Wang, Linzhou Wang, Man Wang, Shuhong Wang, Shuqin Wang, Taijie Wang, Tao Wang, Yunyong Wang, P.B. Wani, A.M. Wankhede, P.C. Wankhede, T.V. Warhekar, J. Warner, A. Warren, W.M. Washington, T.S. Wasnik, H. Wegener, Cao Wei, Kaiyu Wei, Zhuangzi Wei, T. Weidberg, T. Wenaus, H. Wenninger, R. Wesson, J.R. Westby, M. Wheelis, M. White, L. White, T. Wigley, R. Wigmans, C. Wilfert, R.G. Will, C. Williams, M. Willutzky, J. Wilson, R. Wilson, G. Witschel, D. Witthaut, A.T. Wolf, G. Wolf, H. Wolf, L. Wood, E. Worcester, L.J. Wright, Baizhi Wu, Chuanchou Wu, Maw-Kuen Wu, Mian Wu, Weimin Wu, Wentai Wu, Yingzhi Wu, Zhendong Wu, B. Wyslouch, Beixing Xi, Chen Xi, Jiwei Xi, Jun Xia, Liangrong Xiao, Jialin Xie, Peipei Xie, Xiaoxi Xie, Yenyi Xie, Jianming Xu, Shaowang Xu, Xiaokang Xu, Zhongxiong Xu, Zheng Kai Xu, Jingxuan Xue, Shengtian Xue, R. Yagannathan, A. Yair, N. Yamamoto, Binshan Yan, Jie Yan, Taixuan Yan, Wuguang Yan, Dajiang Yang, Yun Yang, H.Yano, Xiaoguang Yao, C.H. Ye, K.E. Yeager, R. Yeager, S.H. Yeh, A.K. Yelne, S.V. Yenkar, D. Yeomans, D.A. Yerxa, Zeng Yi, Jun Yin, Zhaosheng Yin, Cai Yixing, Zhang Yiyan, H. Youn, T. Ypsilantis, Fangqun Yu, Qinchang Yu, Qingfu Yu, Zhenze Yu, Zhongqian Yu, Daoxian Yuan, Wen Yue, T. Yuhara, T.M. Zachariah, M. Zahrani, A.V. Zaitsev, G. Zanetti, I.S. Zektser, H. Zeller, R. Zetterstrom, Boachang Zhang, Chuang Zhang, Huashun Zhang, Pinghua Zhang, Yan Zhang, Yingping Zhang, Zhenjiu Zhang, Xiliang Zhang, Yiyan Zhang, Bin Zhao, Duo Zhao, Fuguang Zhao, Jijiu Zhao, Jingbao Zhao, Weiren Zhao, Yongjie Zhao, Guoqing Zheng, Linsheng Zheng, Shuchen Zheng, Youchun Zheng, Zhipeng Zheng, Cai Zhiguo, Shicai Zhong, Dadi Zhou, Jikang

Zhou, Shu Zhou, Xiaoguang Zhou, Yuehua Zhou, Fuquan Zhu, Liangsheng Zhu, Renquan Zhu, Ren-Yaun Zhu, Shangen Zhu, Yucan Zhu, Jie Zhuang, A. Zichichi, J. Ziegler, J. Zinn, G.M. Zinovjev, K. Zographos, H. Zvonkova.

I would like to close this Opening Lecture asking all of you to read the special volume published (Figure 14) by the University of Bologna (the 1st in the world established more than 900 years ago) on the *3 Days* devoted to let people know in Bologna with 60 events and 30,000 participants.

- Science
- Technology
- Culture

At the centre of all these events we had, have and will have the "New Manhattan Project".

Fig. 14. The New Manhattan Project.

GIORNALE DI SICILIA
VENERDÌ 21 AGOSTO 2015

Fatti&Notizie | 9

Erice

Mattarella e Grasso scrivono agli scienziati

●●● «Sono consapevole dell'importanza degli studi e delle riflessioni scientifiche e di quanto sia necessario che le istituzioni nazionali e internazionali vi prestino ascolto». Lo scrive il presidente della Repubblica, Sergio Mattarella, in un messaggio inviato al professor Antonino Zichichi, che ieri a Erice ha aperto i lavori della quarantottesima sessione dei Seminari internazionali sulle emergenze planetarie. Agli scienziati riuniti ad Erice ha fatto pervenire un messaggio anche il presidente del Senato, Piero Grasso: «Caro professore Zichichi, la "nostra" bellissima e difficile Sicilia si fa ancora una volta epicentro di una riflessione di altissimo valore scientifico e civile».

GIORNALE DI SICILIA TP
Cultura

EMERGENZE PLANETARIE

Scienziati a Erice, Mattarella: "Le istituzioni devono ascoltarvi"

21 Agosto 2015

ERICE. **«Sono consapevole dell'importanza degli studi e delle riflessioni scientifiche** e di quanto sia necessario che le istituzioni nazionali e internazionali vi prestino ascolto». Lo scrive il presidente della Repubblica, Sergio Mattarella, in un messaggio inviato al professor Antonino Zichichi, che ieri a Erice ha aperto i lavori della quarantottesima sessione dei Seminari internazionali sulle emergenze planetarie.

Agli scienziati riuniti ad Erice ha fatto pervenire un messaggio anche il presidente del Senato, Piero Grasso: «Caro professore Zichichi, la 'nostra' bellissima e difficile Sicilia si fa ancora una volta epicentro di una riflessione di altissimo valore scientifico e civile della quale, come cittadino e presidente del Senato, le sono particolarmente grato».

Sono 120, provenienti da 40 nazioni, i partecipanti ai Seminari sulle emergenze planetarie, i cui lavori si concluderanno domenica prossima. Tra i temi in scaletta, il recente accordo nucleare tra Stati Uniti ed Iran e la diffusione di malattie infettive prodotte dalla globalizzazione.

THE 14 JULY 2015 IRAN AGREEMENT: JOINT COMPREHENSIVE PLAN OF ACTION—JCPOA

RICHARD L. GARWIN
Thomas J. Watson Research Center, IBM Research Division
Yorktown Heights, NY, USA

THE IRAN AGREEMENT—THE JCPOA

On July 14, 2015, Iran and the EU+3 (France, Germany, UK, China, USA, Russia) reached a historic agreement on limiting Iran's activities in the field of nuclear energy to those that are entirely and transparently civil. This is the JCPoA (for Joint Comprehensive Plan of Action) colloquially known as the Iran Deal.[1] It is *not* primarily a USA-Iran Agreement.

Many, including myself, have been concerned that Iran would acquire nuclear weapons, and Iran, on the other hand, has been suffering under sanctions imposed by the United Nations, the United States, and other countries, primarily as a response to its activities in the nuclear area.

Iran has always maintained that its activities have been peaceful and oriented toward nuclear power for its grid, as exemplified by the reactor at Bushehr, built initially by Germany and completed and put into operation by Russia. Indeed, during the 1970s Iran under the Shah was seeking a full-scope nuclear power program, including 20 light-water power reactors, enrichment, and reprocessing.[2] The argument was that oil would not last forever and that Iran should use nuclear power—electrical energy from uranium—in order to spare its oil reserves for higher-value uses. Such a nuclear program was encouraged under the 1970 NPT, subject to safeguards implemented by the IAEA, but the U.S. administrations of Presidents Gerald Ford and Jimmy Carter attempted to impose further restrictions on Iran beyond the NPT, perhaps because in 1974 "...the Shah told *Le Monde* that one day 'sooner than is believed,' Iran would be 'in possession of a nuclear bomb.'"[3]

Iran's nuclear program was put on hold for at least two reasons. First came the revolution in 1979, after which the United States was not interested in having Iran acquire nuclear capabilities. Furthermore, the economics of nuclear power were never such that it could not compete against oil or gas production of electrical energy, especially in a resource-rich country such as Iran. Of course, if the cost of fossil fuel includes preventing industrial CO_2 from entering the atmosphere, nuclear power is advantaged economically.

---

[1] Announced at a joint press conference by Federica Mogherini and Iranian Foreign Minister Javad Zarif, http://eeas.europa.eu/statements-eeas/2015/150714_01_en.htm (Also posts the JCPoA and six annexes. The total is 159 pages, of which 62 are a list of people, entities, aircraft and ships related to sanctions.

[2] http://media.snn.ir/Original/archive/21-2-1393%5CFILE635354886585732024.pdf *TACKLING THE IRAN-U.S. CRISIS: THE NEED EOR A PARADIGM SHIFT*, by Mohammad Javad Zarif, Columbia University *Journal of International Affairs*, Spring/Summer 2007, vol. 60, no. 2.

[3] http://foreignpolicy.com/2010/12/29/the-shahs-atomic-dreams/

Iran joined the NPT in 1970, and was thus required to inform the IAEA of its activities in the nuclear power sector. It did not so fully inform, and although Iran has always maintained that its work with uranium, including enrichment of uranium was acceptable under the NPT for a non-nuclear weapon state, sanctions were imposed, as least in part for this failure to comply with the NPT.

In addition, Iran until 2003 explored the development and fabrication of nuclear weapons—the "possible military dimensions" (PMD) that figured prominently in the negotiations. Much of that is now ancient history, still heatedly disputed by many on either side.

Under the political leadership of Hassan Rouhani, who took office in August 2013, and his Foreign Minister Javad Zarif, diplomats and scientists and engineers from Iran negotiated over the last several years for an agreement that would remove the sanctions that have been burdening Iran's commercial and industrial sector, and that would allow movement toward the indigenous supply of nuclear power within Iran.

Although electrical energy from fission need not convey a capability to build nuclear weapons, it has in many cases been used to provide such a latent capability. A non-proliferant nuclear-power sector would also very likely be the most economical way to obtain nuclear energy—the use of light-water reactors either designed and built indigenously or bought under contract from abroad, with fuel enriched and fabricated abroad, which fuel would be shipped out of the using country after its typical four-year life in the reactor, plus a few years for "cooling" to reduce the fierce radioactivity and "decay heat" of the spent fuel.

The Iran Deal verifiably closes the two routes to obtaining weapon-usable fissile material—that of enriching uranium from the natural 0.71% U-235 concentration to the 80+% that would be useful for a nuclear explosive, and the acquisition of Pu-239 as a result of neutron capture on the abundant U-238 in a nuclear reactor.

Specifically, the plutonium route, which implies nuclear weapons only of the implosion type, and in which high explosive is used to assemble a sub-critical mass of plutonium metal into a more compact configuration so that it can explode, is blocked by Iran's commitment not to reprocess any fuel, to ship out of the country the spent power-reactor fuel from Bushehr that is supplied by Russia, and to completely reconfigure the planned heavy-water reactor at Arak that had been designed by Iran to use natural-uranium fuel, which is a copious producer of Pu-239. As initially designed, the 40-MWth Arak reactor was to be very similar to the Israeli reactor at Dimona that over the years secretly produced fuel for Israel's nuclear weapons. At its nominal 40 MW of fission power, Arak would have produced 40 grams of Pu-239 per day—a "significant quantity" (SQ) of 8kg of Pu in 200 days; enough for more than two Nagasaki type bombs per year.

Instead, Iran's nuclear specialists worked closely with the technologists of the negotiating partners, agreeing to remove the interior structure of the initial Arak design, fill that structure with concrete, replace it with a core of low-enriched uranium (<3.67% U-235), and limit the reactor power to 20MWth. Iran agreed not to reprocess the fuel or to have a capability to do so, and to send the spent fuel out of the country. The only chemical separation activity to be undertaken at Arak will be that to obtain isotopes of medical and industrial importance from the low-enriched uranium "targets" that will be irradiated eventually in the newly designed Arak reactor. In addition, the JCPoA bars Iran from working on metallic uranium or plutonium—an important limitation.

On this particular question, an interesting sidelight is provided in the *Science Magazine* interview[4] with Ali Akbar Salehi, president of the Atomic Energy Organization of Iran (AEOI) and the chief technical negotiator in the last months before agreement on 14 July 2015.

> Salehi: *"We said that if we need some depleted uranium, we may ask for it from [the P5+1] during this period. This is an implicit understanding. But we will not produce the metal. We did reprocessing, and produced plutonium. And we told the IAEA, that's it. That was in 2003."*

Uranium enrichment for nuclear energy has a long history, beginning with that performed on a massive scale at Oak Ridge, Tennessee, to produce the highly enriched uranium (HEU) for the gun-assembled weapon that destroyed Hiroshima. Although the gaseous diffusion process was used for most of the "separative work" required, the "electromagnetic process" (Calutrons) was used for the final step of enrichment to HEU of >80% U-235.

As an economic matter, both approaches have been largely replaced by the gas centrifuge, which in its modern, efficient design goes by the name of [Gernot] Zippe in the West and [Evgeni] Kamenev-centrifuge in Russia. Whatever the name, the gas centrifuge since the early 1960s is a marvel of invention and industrial design with its self-pumping vacuum, large open hole in the upper end cap of the rotor, and internal gas circulation that increases the separation factor in a centrifuge operating at a peripheral speed of 400 m/s from 1.02 to 1.12 or more. Note that the pressure of the UF6 gas in the rapidly rotating rotor is maximum at the wall and drops off rapidly within millimeters of the wall (the "law of atmospheres" but with Earth's gravity *g* replaced by *v2/R* which is typically 200,000 *g*). The UF6 pressure at the wall is typically estimated as 100 torr and in the interconnections between centrifuges about 4 torr,[5] allowing centrifuges to be connected to one another without pumping the UF6 gas.

Centrifuges are connected in a *cascade*, and Iran, following designs promulgated by A.Q. Khan of Pakistan who stole them from the European enrichment center, Urenco,[6] has used a cascade of modules of 164 identical machines—the IR-1. In the early 2000s, concern was raised about Iran's pursuit of enrichment, and difficult discussions begun. Part of the difficulty was that the United States had no diplomatic relations with Iran.

The concern that Iran might develop and stockpile nuclear weapons led to several informal discussions at the time. Iran wished to develop the capability to enrich uranium, nominally to supply fuel for the incomplete power reactor that had been built at Bushehr by Germany, and also for the Tehran Research Reactor that used 19.75% U-235—still, technically, low-enriched uranium (LEU);. Highly enriched uranium (HEU) is defined as U-235 content of 20% or above.

---

[4] http://news.sciencemag.org/people-events/2015/08/exclusive-iran-s-atomic-czar-explains-how-he-helped-seal-iran-nuclear?rss=1

[5] inmmvienna.org/Presentations/INMM12_Smith_GCEP_concepts.pptx (slide 5 shows "*P*~ 4 torr" for the gas centrifuge headers)

[6] http://www.nytimes.com/2006/04/16/world/asia/16chron-khan.html?_r=3& for a timeline of A.Q. Khan's activities.

Then-President Ahmadinejad inspecting Iran's IR-1 centrifuges.

It is often thought that producing enriched uranium for nuclear weapons is, somehow, a bigger task than producing fuel for a power reactor. In fact, quite the opposite is true. An enrichment facility to supply fuel for the single power reactor at Bushehr, reconfigured, could produce HEU for a nuclear weapon every two weeks.

Ignoring the enrichment required for the initial fuel load of a reactor producing almost 1000 megawatt electric (MWe), just supplying the annual reload of about 25 tons per year of ~4% U-235 fuel is a large job. Such a reactor fissions about a tonne of U-235 per year, and this requires about 1000 kg of U-235 in the annual supply of fuel to the reactor. You will see in a moment (Row 15 in the spreadsheet) that it takes about 151,000 SWU/yr to enrich that fuel from the NU feed of 0.711% to a product of 4.4%, with 0.25% "tails." On the other hand, to enrich to 90% U-235 one can see from Row 8 that only 231 SWU/kg of U-235 is required, so for an SQ of 25 kg of U-235 only about 5775 SWU would be needed.

The 150,000 SWU/yr plant needed to resupply fuel for the single power reactor at Bushehr could produce one bomb's worth of HEU every 5775/150,000th of a year or every two weeks. This was the problem in the very beginning, and it is the conceptual problem now for an entity that insists that it needs at some time in the future to be able to provide enriched uranium for its nuclear power plants. For Iran, that need is a long time in the future, and within the confines of the Agreement, enrichment can take place only at Natanz for more than 8 years, with only 5060 IR-1 centrifuges, and with none other. And Iran agrees for 15 years not to explore means of enrichment other than the gas centrifuge.

In connection with other work, I had long distributed a SWU calculator spreadsheet, for instance at http://fas.org/rlg/20.htm that enables anyone to provide the concentration of feed, “tails,” and product to determine how much separative work would be required in an ideal separation cascade. For proliferation-related questions I find it useful to discuss this in terms of SWU required per kg of U-235 in the product, which I think is much more revelatory than SWU per kg of product.

SWU per kg for various enrichment parameters. SWU_Calculations (version 3k).xls R.L. Garwin, 12/19/2007

Row												
3	Xp	Xw	Xf	P	W/P	F/P	Vp	Vw	Vf	ΔSWU/	ΔSWU/	ΔSWU/
4	(product)	(waste)	(feed)	kg	kg W/kg P	kg F/kg P	-------------value function------------------ (2x-1) ln [x/(1-x)] ---------------			kg product	kg U-235	P kg of
5	% U-235	% U-235	% U-235	kg of product			---				in product	product.
6												
7	95.000	0.250	0.711	1.00	204.53	205.53	2.65	5.96	4.87	220.75	232.37	220.75
8	90.000	0.250	0.711	1.00	193.69	194.69	1.76	5.96	4.87	208.03	231.15	208.03
9	80.000	0.250	0.711	1.00	171.99	172.99	0.83	5.96	4.87	183.46	229.32	183.46
10	19.900	0.250	0.711	1.00	41.62	42.62	0.84	5.96	4.87	41.35	207.77	41.35
11	90.000	0.400	0.711	1.00	287.10	288.10	1.76	5.47	4.87	170.42	189.36	170.42
12	3.500	0.400	0.711	1.00	8.97	9.97	3.08	5.47	4.87	3.64	103.89	3.64
13	90.000	0.400	3.500	1.00	27.90	28.90	1.76	5.47	3.08	65.33	72.58	65.33
14	3.500	0.360	0.710	1.00	7.97	8.97	3.08	5.58	4.87	3.89	111.22	3.89
15	4.400	0.250	0.711	1.00	8.00	9.00	2.81	5.96	4.87	6.66	151.41	6.66
16	3.500	0.400	0.710	1.00	9.00	10.00	3.08	5.47	4.87	3.64	104.02	3.64
17	95.000	0.500	0.711	1.00	446.87	447.87	2.65	5.24	4.87	163.79	172.41	163.79
18	95.000	0.500	19.900	1.00	3.87	4.87	2.65	5.24	0.84	18.85	19.84	18.85
19	19.900	0.711	4.400	1.00	4.20	5.20	0.84	4.87	2.81	6.69	33.62	6.69
20	90.000	0.400	3.500	1.00	27.90	28.90	1.76	5.47	3.08	65.33	72.58	65.33

Enter your desired set of enrichment parameters in Columns B-D for Xp, Xw, and Xf-- the U-235 concentrations in %, and in Col. E the kg of product.
Rows 23-30 are formatted for this purpose, as are Rows 7-20. If you corrupt the worksheet, download it afresh.

24	90.000	2.000	3.670	25.00	51.69	52.69	1.76	3.74	3.03	35.35	39.28	883.76
25	90.000	10.000	19.750	25.00	7.21	8.21	1.76	1.76	0.85	7.46	8.29	186.58
26	90.000	3.670	19.750	25.00	4.37	5.37	1.76	3.03	0.85	10.43	11.59	260.79
27	4.400	0.250	0.711	1.00	8.00	9.00	2.81	5.96	4.87	6.66	151.41	6.66
28	4.400	0.250	0.711	1.00	8.00	9.00	2.81	5.96	4.87	6.66	151.41	6.66
29	4.400	0.250	0.711	1.00	8.00	9.00	2.81	5.96	4.87	6.66	151.41	6.66
30	4.400	0.250	0.711	1.00	8.00	9.00	2.81	5.96	4.87	6.66	151.41	6.66

SWU_Calculations (version 3k).xls. This live spreadsheet, when downloaded to any computer with Excel or a compatible program, allows the user to specify the U-235 concentration (in per cent) for the “product”, “waste” or “tails”, and the “feed” to an ideal enrichment cascade.
Optionally, the mass of product (kg) in column E can be specified as well. User inputs in Columns B-E result in calculated outputs in Columns F-M as indicated in Rows 3-4.
Thus, Column F is the kg of waste per kg of product; Col. G is the kg of feed material per kg of product. Cols. H-J provide the value function (2x-1) in [x/(1-x)] for the product, waste, and feed concentrations, respectively.
The output columns K-L show the Separative Work Units that must be provided by the ideal cascade to produce 1 kg of product, or alternatively, per kg of U-235 contained in the product. The rate of SWU production (SWU/yr) of an ideal cascade of identical machines is the product of the unit SWU rating of a centrifuge and the number of centrifuges. Thus if a single centrifuge can be operated at 2 SWU/yr, an assembly of 3000 centrifuges could produce 2x3000 = 6000 SWU/yr, and if assembled into a suitable set of series and parallel configurations would produce enriched uranium at a rate illustrated in one of the rows of the spreadsheet. Row 12 shows that 3.64 SWU must be invested per kg of 3.5% U-235, for a cascade fed natural uranium with Xf = 0.711% U-235, and with a waste stream containing Xw = 0.40% U-235.
Thus a cascade rated at 6000 SWU/yr could produce 6000/3.64 = 1648 kg of 3.5% product per year, containing 6000/103.89 = 57.75 kg of U-235.

Iran had some centrifuges running at the time, and it was the declared purpose of the U.S. administration to achieve an agreement with Iran under which “not a centrifuge would turn.” For Iran, its rights or privileges under the NPT were sacred, and among those rights is to benefit from the non-military uses of nuclear energy, including civil nuclear power.

Under the IAEA, a Significant Quantity (SQ) of HEU is 25 kg of contained U-235, and from Row 11 of the SWU calculator, HEU having 90% U-235 contains 189 SWU/kg of U-235, if produced from 0.71% (natural uranium—NU) feed, with a tails concentration of 0.4% U-235. Thus, to produce an SQ in that way would require 25*189 = 4725 SWU, and at an expected rate of about 1 SWU/yr with the IR-1 centrifuge, that would correspond to 4725/500 = 9.5 years of operation of a bank of 500 centrifuges, which limit of 500 seemed acceptable to Iran. Of course, to conduct such enrichment as a member of the NPT, a state would need to justify it as producing fuel for a specialized reactor.

Note, however, that if a state had access to 3.67% U-235, and was willing to enrich to 90% U-235 with a tails concentration of 2%, then the additional SWU required would be 39.28 SWU/kg of U-235 (Row 24). Enrichment that would require 9.5 years from NU feed would thus require only (39.28*25/500) = 1.96 years with LEU feed at 3.67% to produce 25 kg of U-235 in 90% HEU.

Even more concerning is the minimal SWU requirement to obtain 90% HEU from 19.75% LEU feed. With a tails concentration of 10% U-235 this would be 8.29 SWU/kg of U-235 in 90% product, and with a tails concentration of 3.67% (Row 25), thus using (20-3.67)/20 = 83% of the feed) weapon HEU for a total of 207 SWU for 1 SQ of 90% HEU—just 5 months of operation of a 500-centrifuge establishment. For the 5060 centrifuges Iran would be allowed under the JCPoA, the enrichment would take 15 days—highlighting the necessity to restrict the amount of LEU available to Iran, which could serve as feed for enrichment to weapon-usable HEU.

> *2. Iran will begin phasing out its IR-1 centrifuges in 10 years. During this period, Iran will keep its enrichment capacity at Natanz at up to a total installed uranium enrichment capacity of 5060 IR-1 centrifuges. Excess centrifuges and enrichment-related infrastructure at Natanz will be stored under IAEA continuous monitoring, as specified in Annex I. [JCPoA]*

This exercise illuminates the complexity of the agreement, and its both innovative and stringent character. It involves limiting the amount of LEU on hand, abjuring any amount of HEU, and achieving these goals by shipping excess material out of the country, selling it on the market, and the like.

In return, of course, Iran gains relief from sanctions, which will let it sell its oil freely on the market, and trade more or less normally. It will still abide by many restrictions for 8, 10, or 15 years or more, including continued membership in the NPT, adoption of the Auxiliary Protocol (AP) and restriction to a tiny number of the advanced centrifuges that Iran has been working on—specifically the IR-6 and the IR-8. Only after 8.5 years will Iran be able to have more than 30 (each) of these more advanced centrifuges.

The centrifuges at Fordow will be reduced to 1000 IR-1 machines and will never see uranium—instead, they will be used to provide stable isotopes for medical and industrial purposes.

It is not very polite in diplomatic circles to speak of "breakout" or response to breakout, but surely such questions are taken into account on both sides. An array of 5060 IR-1 centrifuges, even if operating at 1 SWU/yr, never achieved thus far by Iran under strict reporting by IAEA, would take more than a year to produce 1 SQ from natural uranium. But if fresh fuel for Bushehr, delivered on site by Russia, were seized by Iran, converted to UF6 and fed even to the 5060 centrifuges, Row 20 in the Table shows that only 73 SWU need be added per kg of U-235 in the product, so on the order of 1825 SWU (4.3 months) to produce 25 kg of U-235 in weapon uranium starting from Bushehr fuel. And less than one ton of the 25-ton annual refueling would be consumed in this process.

Evidently, the continuous monitoring of the sanctity of Bushehr fresh fuel is important and interference with this reporting link would be taken as violation of the Agreement.

Iran would do well to provide instant reassurance that all was well rather than to delay an urgent inspection.

One should not imagine that the intelligence capabilities of interested parties to this agreement will deteriorate over the years, and means for intervention will also have a high priority, although one hopes they would never be invoked.

Other aspects of the Agreement are also important. Acquiring nuclear weapons would not improve the security of Iran, in my opinion, especially if it resulted in the acquisition of nuclear weapons by nearby states. One can be sure that the EU and the United States will attempt to use their good offices to try to limit proliferation of nuclear weapons to others in the area.

In addition to enrichment of non-uranium stable isotopes at Fordow, Iran will undertake other nuclear-energy research there, and will, one hopes, work with foreign scientists to contribute to the world's advancement in civil fusion energy, medical and industrial imaging, and the like.

It will be a boon to be able to work with our Iranian scientific colleagues on common problems.

APPENDIX

WHY I SUPPORT THE JCPoA

The Iran Deal is criticized primarily for what it doesn't do—restrain Iran's support of Hezbollah and the Syrian government, change an authoritarian sectarian supreme leadership, eliminate not only the government-sponsored utterance of "Death to Israel" and "Death to the United States" but eliminate the feeling and (some fear) the programs behind those statements. But that was never the aim of the negotiation. It is enough, in my opinion, that it eliminates for 10 years the "existential threat" of an Iranian nuclear weapon to Israel, leaving time and newly achieved level of interaction to continue to constrain that threat.

A bit of arithmetic counters the criticism that the removal of some of the sanctions (other terrorist-related sanctions remain) will allow Iran to have access to $100B in revenue that could be used to heighten the damage to Israel by rockets launched from Gaza and Lebanon. Is Iran now so poor that it can't afford $1B in rocketry? At $1000 per rocket that would be one million artillery rockets; at $10,000 each, $1B would be 100,000 long-range modern rockets. That Iran hasn't supplied them is not due to shortage of funds, and the availability of many tens of billions of dollars will not, in itself, add to the threat.

As Harold Brown, U.S. Secretary of Defense under Jimmy Carter, writes:[7]

> *"Failure of U.S. adherence to the deal would not slow but would hasten an Iranian nuclear weapon. The Russians and Chinese would certainly not sustain economic sanctions. Sanctions by the other countries would erode as well. With all of this in mind, approving the agreement is a no-brainer."*

---

[7] https://www.washingtonpost.com/opinions/why-accepting-the-iran-nuclear-deal-is-a-no-brainer/2015/08/13/06adba26-3c5c-11e5-b3ac-8a79bc44e5e2_story.html

TOWARD GLOBAL ERADICATION OF INFECTIOUS DISEASE

LOWELL WOOD
Hoover Institution, Stanford University, University of California and Lawrence Livermore National Laboratory, Stanford, CA, USA

PHILIP ECKHOFF
Institute for Disease Modeling, Intellectual Ventures LLC, Bellevue, WA, USA

ROBERT LANGER
Department of Chemical Engineering and Koch Institute, Massachusetts Institute of Technology, Cambridge, MA, USA

SUMMARY

Infectious diseases whose pathogens are specialized to a single host species—as most human pathogens are—are susceptible to global eradication by clearing all infected species-members of a targeted pathogenic agent for a brief interval of time: inter-host transmission chains are thereby broken and immune surveillance combined with medical interventions then annihilate the pathogens isolated in each infected individual, resulting in extinction of the pathogenic species by synchronized evaporation of the entirety of its ecological niches.

Two pathogens have been thereby annihilated over the past third-century: smallpox, a major episodic killer of mankind for more than the past millennium,[1,2] in the late '70s and rinderpest (a fearsome disease of several types of ruminants favored by pastoralists) during the past two decades.[3] Both of these historic achievements were accomplished by use of vaccines to immunize sufficiently large fractions of vulnerable populations at a given moment that all pathogen transmission chains were broken effectively simultaneously.

We discuss ongoing efforts aimed at global eradication of wild poliovirus, at worldwide extinction of the mosquito-vectored eukaryotic parasites responsible for malaria, and at global eradication of other human-infectious diseases, noting the challenges they face and some of the means of addressing them. "Combined arms" approaches—involving a wide variety of distinct, some quite novel means aimed in concert at breaking transmission chains for each of several major classes of pathogens—are central to these efforts, and extensive physical modeling is crucial to efficient planning of the associated campaigns synergistically applying these approaches in adequately simultaneous and worldwide fashions. We survey some of the potent tools anticipated to underlie these approaches, ranging from enduring polyvalent vaccination-at-birth through mass administration of 'specifics' and on to engineered parasitization of pathogens, and we provide estimates of their efficacies in pertinent global eradication scenarios.

The prospects for global eradication of many major infectious diseases of humankind over the next few decades, critically enabled by nascent and emerging understanding of key mechanisms and recently developed and still developing technical means, appear to be remarkably good. Being forever-free thereafter of the pathogens which have circumscribed and oppressed the human condition from the mists of prehistory may be expected to be of epochal significance.

INTRODUCTION

Of the two basic classes of acquired disease—infectious and degenerate—the former is arguably the more vexing, as it strikes with little-or-no warning, often 'early' in life, indiscriminately, too-often fatally and frequently imposes lasting damage. Infectious disease acts to make life forever-uncertain, too often painful, too frequently terrifying—and always too costly, usually in several distinct coins. It proffers no countervailing or offsetting benefits. The Human Condition would be greatly improved by its elimination, which condition however has been attainable only to limited degrees for selected peoples in restricted circumstances, and often with high associated costs.

The nature of infectious diseases has become truly apparent only since microorganisms were first unequivocally established as the source of many of them late in the 19th century, and submicroscopic agents—viruses—as the basis of many others, in the early 20th century. With growing and ever more quantitative understanding of the associated epidemiology, the issue has arisen in the past century as to whether it is feasible to globally eradicate at least some infectious diseases,[1] with the epochal global eradication of smallpox a third-century ago providing a proof-of-principle, the global eradication of rinderpest more recently establishing the non-uniqueness of global eradication to the smallpox case, and the still-on-the-cusp global eradication of poliovirus as a test-case for the elimination of a notorious pathogenic target, one that's the more challenging to engage due to its highly occult nature.

The question thus placed before Humanity by these recent and pending accomplishments is whether the systematic global eradication of at least the major infectious diseases should become a—perhaps even *the*—central public health endeavor of our times.

Infectious Disease as a Modeling-Susceptible Spatiotemporal Phenomenon

One of the key motivators of consideration of this matter is the very recent susceptibility of infectious diseases to mathematical modeling as space and time-dependent phenomenon in a great ecosystem of hosts and pathogens living in a time-dependent, three-dimensional world. Epidemiology as an intellectual discipline dates back at least to Hippocrates and as a quantitative one to the seminal studies of Bernoulli,[4] with its history since then leading through the monumental works of such luminaries as Semmelweis, Snow, Leveran, Ross,[5] McKendrick, Macdonald,[6] and Soper. Due to the enormous amounts of computing capacity, data storage and long-range/high-bandwidth communication which has become available over the past two decades, it has become feasible to model a rich variety of infectious diseases world-wide on single-person levels and on an hour-by-hour basis—which is quite adequate for the rational planning and optimal execution of global eradication campaigns against disease-engendering pathogens. Complementing such capabilities "in the small" are the present-day capabilities to conduct intra-host infection-modeling in exquisite detail, e.g., following the second-by-second interaction of all components of an human immune system with the millions of members of a viral quasi-species such as that of HIV-1 during the mature phase of an infection—and to do so over time-scales of decades.

In the context of eradication planning, mathematical modeling can examine each aspect of a disease, quantify the impact of uncertainties, and design a “combined arms” approach, one that aims to synergistically merge and employ all pertinent means for the destruction of a pathogen species in a human population, including its suppression in natural world or (usually, animal-) vector reservoirs which serve to conserve and convey it between individual human hosts. For instance, in the case of malaria, detailed mechanistic modeling of vectors,[7] within-host parasite dynamics,[8] and the population diversity of parasite antigens[9] each inform campaign planning, as well as impelling and guiding new research, toolset-development and data-collection efforts. Models can then be used to plan combined arms campaigns[10] which synthesize *ad hoc* approaches from available toolsets which are well-optimized to each particular disease-suppression challenge and which also evaluate the potential impact of new tools under development, such as malaria vaccines.[11]

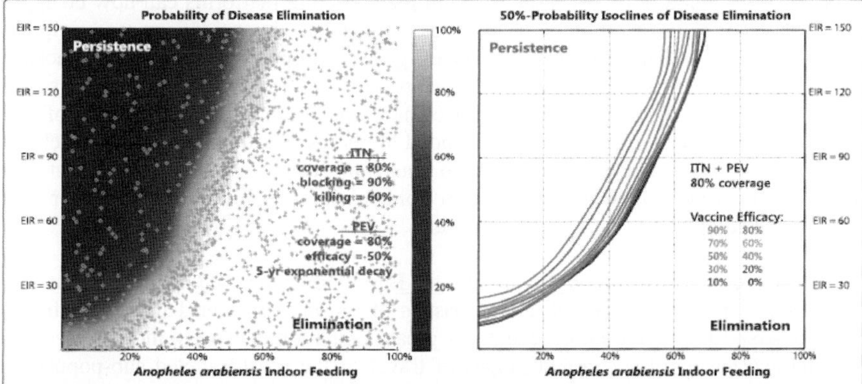

Figure 4 Regions of disease elimination and persistence after PEV and ITN distribution. Left panel: The separatrix between the regions of parameter space with disease persistence (dark blue) and local disease elimination (white). Individual simulations are marked either with a cross (for successful elimination) or a circle (disease persistence). ITNs and 50%-efficacy pre-erythrocytic vaccines (PEV) are distributed independently at 80% coverage. Right panel: The location of the separatrix (50% probability isocline) between regions of parameter space with disease persistence for PEV efficacy values ranging from 0% to 90%.

Fig. 1: Modeling the projected impact of different vaccines in a variety of settings can help inform product development and research.[11]

E.g., for polio, mathematical models can quantify and spatially-classify gaps in population immunity for various sub-populations and viral serotypes, allowing campaign operations to proactively respond to actual future risk—within uncertainty bounds which the model may be able to rather accurately estimate—rather than just reactively respond to cases-appearing.[12] Due to polio’s extraordinarily-high rate of asymptomatic infections, reactive case response may be rather severely effectiveness-limited compared to the well-known success of ring vaccination in suppressing smallpox, for which most cases were symptomatic and clearly identifiable.

The proactive strategy has made major impacts in reducing Nigeria's polio-case count, and Nigeria has now gone over a year since its last case. Elimination of wild polio virus in Nigeria may have been accomplished, and it's been enabled by mathematical modeling. Mathematical models can also help plan the eradication endgame—which is observed in smallpox and subsequently to have much different qualities from the early- and middle-games—and in optimizing surveillance strategies post-elimination.[13] While modeling was not a major component of the smallpox or rinderpest global eradication campaigns, a previous lack of sufficiently-powerful modeling has limited the extending of global eradication success beyond these two. The new generation of mathematical models exercised on contemporary petaFLOPS-scale computing systems readily capable of a quadrillion sophisticated mathematical computations each second—of the order of a million complicated arithmetic operations every second for every person now alive!—is already demonstrating real promise to help guide all human disease-suppressing endeavors—and especially the global eradication-directed ones—to lasting successes.

Such modeling has been critically enabled by the enormous strides made over the past two decades in digital information transmittal, processing and storage technologies, in both capacity and unit cost respects. Multiple modeling platforms can now be applied in parallel to similar problems, enabling independent investigations and improved confidence.[11,14-18] Sharing of very powerful, general purpose modeling code-packages, e.g., those of the Institute for Disease Modeling—have helped to accelerate and solidify these recent gains. Together, these advances have enabled modeling of degrees of detail which enable rational credulity to be extended to results of modeling studies which have single person/single dwelling resolution on hour time-scales—i.e., about the finest scales in space and time which are relevant to epidemiology—while extending over entire continents and indeed over the entire world.

The Present Global Burden of Disease and the Basic Parameters of Global Eradication

One of the most salient characteristics of the modern world from an infectious disease perspective is the unprecedentedly great population density in most every locale, and the extraordinary speed and ranges of travel of the most-traveled sub-populations. From an epidemiological perspective, these considerations concatenate to offer prospects of "unprecedentedly perfect storms" for commencement and propagation of epidemics and pandemics, as well as for the appearance of hitherto-obscure infectious disease agents such as *Ebolavirus* or the emergence of quite new pathogenic diseases of zoonotic origin, e.g., HIV-1, SARS or MERS, in recency of time-order. Indeed, this is routinely seen in the global propagation of the annual influenza-A pandemic of continuingly-dithered antigenic signature, which is a contributing cause-of-death for a few hundred thousand people each year.

Human movement is the primary spatial vector for disease, and the globe is now more densely connected than at any time in history, with denser populations than at any previous time. In fact, this density and interconnectedness was a principal difference between the recent Ebolavirus outbreak in West Africa and earlier outbreaks. On the other side of the ledger, cell phones provide a powerful new, highly-automated and real-time data source on human mobility, analyses of which can inform malaria transmission,[19] while other pertinent data sources such as border crossings are now highly digitized and can be used to plan the longer-range aspects of global eradication efforts, such as the nascent malaria one.[20] Population distributions are being mapped with high levels of accuracy and spatial resolution, with both traditional data sources and new tools such as remote sensing, from both aerial and Earth-orbiting platforms.[21] These new data sources enable new mathematical modeling tools ever more accurately to inform strategy and tactics of infectious disease-suppression efforts on all scales from local to global. Another set of data needs in support of optimized campaign operations are current vaccination levels for various sub-populations-at-risk, which may readily exhibit high levels of geographic and demographic heterogeneity, even within a single province. New tools to integrate and harmonize various sources of vaccination data to develop improved coverage estimates and drive new targets[12] are proving to be highly useful.

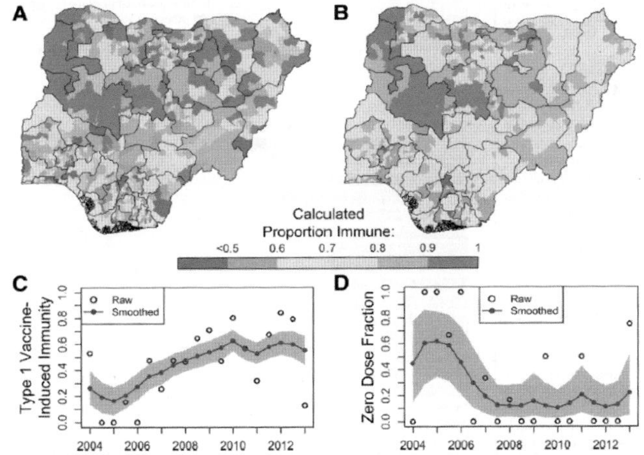

Figure 3 Impact of statistical smoothing model. (A), Map of district NP-AFP WPV1 vaccine-derived immunity, December 2012 through May 2013. Districts colored gray if no data NP-AFP cases reported in this interval. **(B)**, Map of smoothed district WPV1 immunity estimates for the same period. See Discussion for explanation of the relatively low calculated immunity in southern districts. **(C, D)**, Raw versus smoothed indicators over time in Maiduguri, Borno district for WPV1 immunity estimates **(C)** and zero-dose fraction **(D)**. The shaded region represents the point-wise 95% posterior credible interval (CI) of parameter mean estimates.

Fig. 2: Spatial risk modeling of polio risk allows targeting of vaccination efforts in Nigeria.[12]

In addition to these datasets that define the distribution of humans, their mobility and their induced resistance to pathogenic infections, it is important to understand the current burden and distribution of disease. Both of these areas have been advanced by improved digital data and advanced mathematical techniques. The global burden of disease has been quantified to unprecedented levels of detail and accuracy by the work of the Institute for Health Metrics and Evaluation,[22,23] with detailed analysis of the corresponding risk factors.[24] Knowing the past and present burdens of disease helps to quantify and to prioritize health goals and pathogen eradication targets, and the risk factors identify cross-cutting and disease-specific factors.

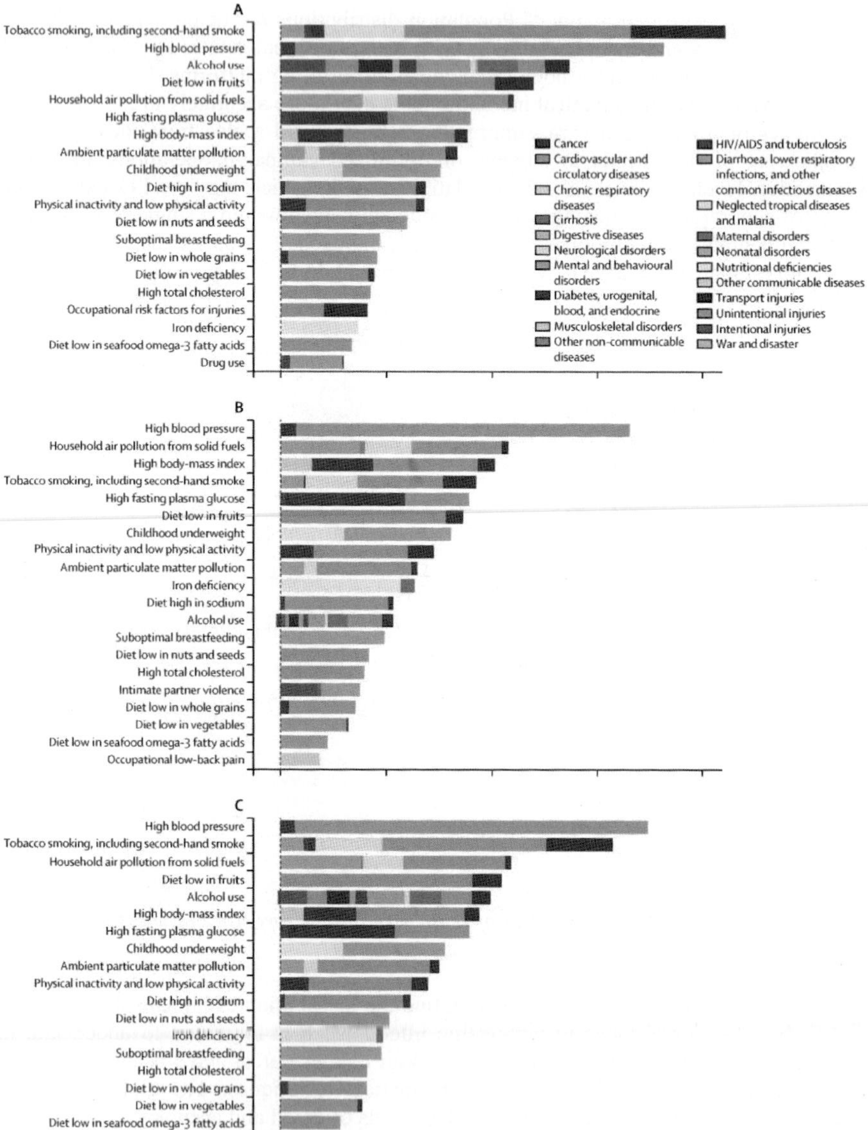

Fig. 3: Risk factors for different categories of disease burden in 2010 for men (A), women (B), and both sexes (C).[24]

Diseases of interest can then be mapped spatially and temporally, as has been done for malaria by the Malaria Atlas Project.[25,26] Such techniques can also map other eradication-relevant quantities such as the habitats of malaria vectors[27] and polymorphisms of the red blood cell relevant to malaria transmission.[28]

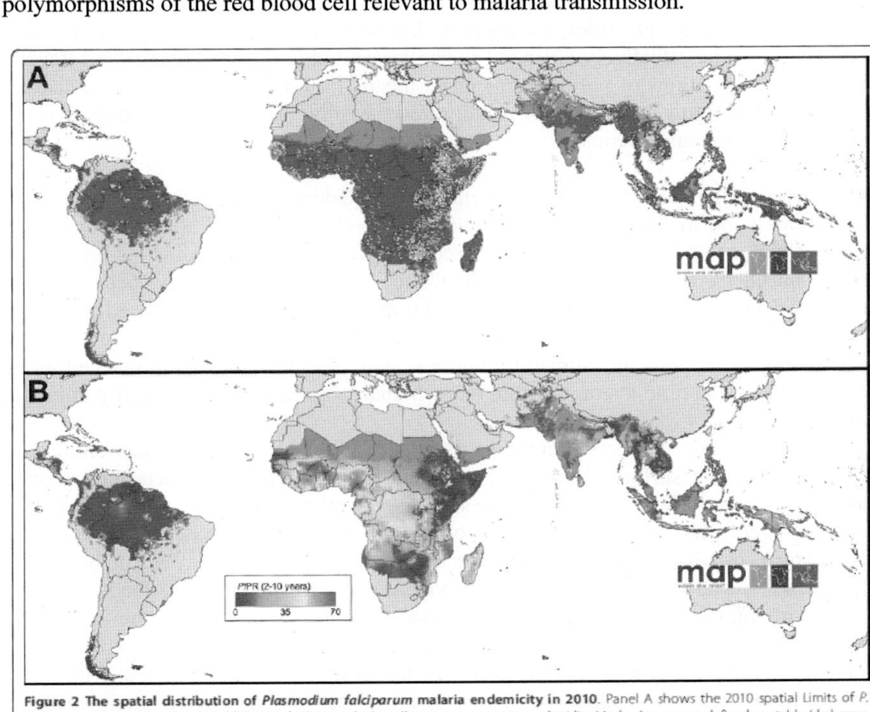

Figure 2 The spatial distribution of *Plasmodium falciparum* malaria endemicity in 2010. Panel A shows the 2010 spatial Limits of *P. falciparum* malaria risk defined by *Pf*API with further medical intelligence, temperature and aridity Masks. Areas were defined as stable (dark grey areas, where *Pf*API ≥0.1 per 1,000 pa), unstable (medium grey areas, where *Pf*API < 0.1 per 1,000 pa) or no risk (light grey, where *Pf*API = 0 per 1,000 pa). The community surveys of *P. falciparum* prevalence conducted between January 1985 and June 2010 are plotted. Of the 23,612 surveys collected, 22,212 satisfied the inclusion criteria for modelling (see Methods and Additional File 1, 2) and are shown here. The survey data are age-standardized [40] ($PfPR_{2-10}$) and presented as a continuum of blue to red from 0%-100% (see map legend), with zero-valued surveys shown in white. Panel B shows the MBG point estimates of the annual mean $PfPR_{2-10}$ for 2010 within the spatial limits of stable *P. falciparum* malaria transmission, displayed on the same colour scale. Areas of no risk or unstable risk are as in (A).

Fig. 4: Mapping of the distribution of P. falciparum by Malaria Atlas Project enables improved planning.[26]

Indeed, it's becoming increasingly feasible to significantly modify these disease maps, by shaping and sculpting of susceptible pathogens—especially sexually-reproducing ones such as the malaria parasites and the parasites responsible for most of the neglected infectious tropical diseases such as the several varieties of worms and flukes—and of their (typically, insect) vectors, via modern 'gene-drive' techniques. Particular note must be taken of ongoing efforts to shape the population structures of mosquitoes toward lower population sizes or less-efficient disease transmission via vertically-transmitting bacterial infections with selected, species-specific strains of

Wolbachia,[29] homing endonuclease gene-based approaches,[30] and of the far more powerful methods currently centered on vigorous exercise of CRISPR/Cas9 technologies to shape and tailor vector population numbers and properties in desired directions, regardless of 'raw' fitness considerations of the gene(s) being 'driven' through the targeted population, in order to support Eradication campaigns. If and as these approaches can be effectively extended to similar sculpting of the genomes and thus the phenotypic populations of pertinent *Bacteria*, means for coping with infectious diseases of bacterial origins may be revolutionized swiftly and cheaply, moreover in exceptionally-favorable manners.

Immune Surveillance, Instructed by Vaccination and Undergirded by Nutrition

Operation of the human immune system, a ~1 kg organ in adults comprised of largely-fluidized and widely-distributed tissues, to defend its host against pathogens and related intruders, is termed immune surveillance. This surveillance has been crucially augmented in recent decades, especially in the more economically-advanced portion of the world, by special 'instruction' in salient pathogen characteristics via the processes of vaccination with antigenic signature mimics, and has been greatly enhanced in its performance in the time-average within many human populations by drastically improved nutrition.

Vaccination 'instructs' immune surveillance in the characteristic macromolecular 'signatures' of the surface features of a foreign object, such as a bacterium or virus particle, by abruptly and briefly 'presenting' a microgram-scale sample of such particles—generally, a fraction of a picoMole of a pathogen's exterior structure—to surveillance, usually by sub-cutaneous or intra-muscular injection in a suitable carrier fluid (often sterile normal saline) with hypodermic needle and vaccine-loaded syringe. Such instruction is most effective when conducted at least twice, with time-intervals of weeks-to-months between each such 'instruction event'—in order to 'persuade' the immune surveillance system that the antigenic threat is "a persistent one that's not going to go away" and thus must be the subject of serious and indefinitely-sustained vigilance—and results in antibody concentrations in circulating blood typically rising to much higher levels and then persisting for much longer time-intervals after each successive 'instruction event.'

Due to the remarkable success of vaccines and vaccination in preventing serious and often-fatal infections, several dozen types of vaccines were created and brought into widespread use during the 20th and 21st centuries,[31,32] and attaining and maintaining world-wide vaccination coverage as ~130 million new humans appear each year is presently a great international public health effort.

Indeed, sufficiently-comprehensive vaccination is the *only* route to global eradication of a pathogen yet to be performance-proven in the physical—vs. the modeled—world. Human smallpox was globally eradicated in 1979 at the culmination of a world-wide vaccination campaign extending over a decade-and-a-half, and rinderpest, a disease of ruminants such as cattle and goats, was globally eradicated over the past two decades by mass vaccination of susceptible animals by their human keepers of sustained intensity sufficient to break all pathogen transmission chains at one time.[3] [It's worth noting-in-passing that both of these successful global eradication campaigns commenced with usage of a thermofragile vaccine that had to be maintained in cold-chain until time-of-injection, but that a thermodurable analog of each one had to be created in mid-campaign for practicality reasons—and observing that no thermostable vaccines against the major human pathogens currently exist.]

At the same time, immuno-chemoadjuvants developed in the later 20th century, such as antibiotics and antivirals, have not only greatly reduced losses to pathogens—by filling-in the chinks-in-armor of the human immune response, primarily in robustness and 'focus' of prompt-acting innate responses and timeliness-of-slowly-rising adaptive responses—but their sustained-and-widespread sub-optimal use (e.g., as monotherapies) has also acted to create entire new quasi-species of pathogens which manifest high degrees of robustness against the host-protective actions of these agents. The resulting *man-made* pathogenic diseases, ranging from highly drug-resistant tuberculosis mediated by mycobacteria effectively impervious to the once-potent antibiotics which served to wipe out TB in the Developed World to viral infections robust against initially highly-potent antivirals and ancient bacterial ones (e.g., *S. aureus*) currently essentially totally lacking susceptibility to once-powerful antibiotics, are presently threatening to inexorably reverse many of the medical-pharmacological gains of the last few decades—accentuating the key role to be played by global eradication of the underlying pathogens, regardless of their drug-resistance characteristics. Understanding the origin and spread of these resistant strains is another endeavor in which modern mathematical modeling can have a dramatic impact.[33]

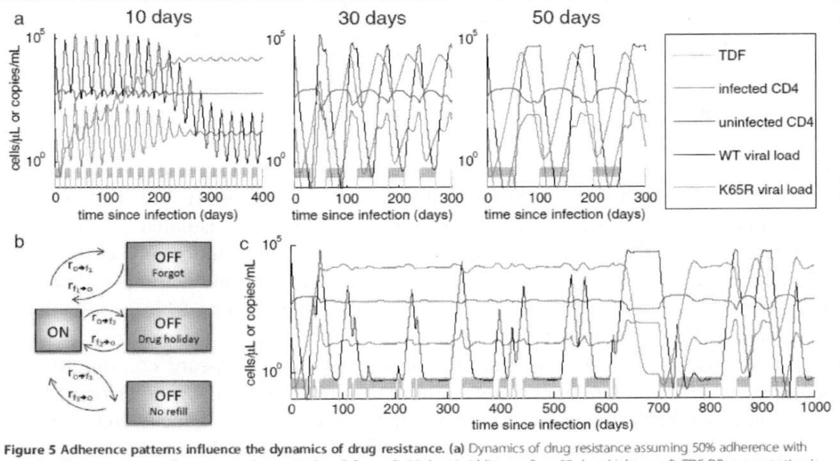

Figure 5 Adherence patterns influence the dynamics of drug resistance. (a) Dynamics of drug resistance assuming 50% adherence with periodic adherent and non-adherent spans lasting 10 days (left panel), 30 days (middle panel), or 50 days (right panel). TDF-DP concentration is overlaid in gray for reference. **(b)** Example of how a Markov model for drug adherence could represent underlying drivers of missed doses. **(c)** Dynamics of drug resistance using a two-state Markov model with exponentially distributed duration spent in the adherent and non-adherent states, each with an average duration of 30 days. This is the stochastic version of the periodic model in the middle panel of **(a)**, providing an example of how such a model could be driven stochastically to reflect this aspect of real-life adherence patterns.

Fig. 5: Simulation of the impact of adherence patterns on the development of drug resistance.[33]

Also during the 20th century, the life-essential nature of a few dozen substances present in foodstuffs in tiny quantities—typically ingested in fractional-gram quantities daily—was recognized, and the effective functioning of immune surveillance on the presence-in-the-body of adequate quantities of these materials—vitamins and trace minerals—was established beyond any doubt. The routine provision of a sufficiency of these vitamins and trace minerals, in conjunction with adequate quantities of 'complete' protein, essential fatty acids and dietary calories are now recognized as prerequisites of effective immune surveillance—and major deficiencies of one-or-more of these are the source of several million lives-lost each year—mostly those of children and mostly with one-or-more infectious diseases as the focal cause-of-death.[24] Indeed, grave dietary deficiencies of vitamin A, zinc and iron alone are estimated to result in ~300,000 such avoidable deaths annually, nearly all of them in the Developing World.[24]

The past century also saw the emergence of the Developed World from the pervasive nutritional depression of the previous era. As a Nobelist economist recently pointed out, only early in the 20th century did adult male stature in the USA return to the levels of the late 18th century, after the widespread under-nutrition of the 19th century, which was more representative of the Malthusian Middle Ages due to population surges unmatched by food-production ones—which 'trough' was inescapably associated with immunological under-performance[34] and corresponding major and continuing loss of life and life-expectancy to epidemic and endemic pathogenic disease. Indeed, a substantial fraction of the historic gains in life expectancy in late 19th and early 20th century in the now-Developed World are attributed to the immunological gains associated with steadily-improving nutrition leveraging concurrent sanitation and hygiene advances[35] a record which is being recapitulated in recent decades and through the present time in the now-Developing World.

Comprehensive and Enduring Immunization at Birth, and the Aging-Out of Infectious Disease

Most births in the current era are attended by a healthcare-giver, often with some training and carrying a "safe birth" packet, even in the most isolated portions of the Developing World. However, this may be the last time that too many children 'see' medical personnel thereafter, so that far too many pediatric immunizations are precluded and these myriad individuals remain as susceptible to succumbing to infectious diseases as were their ancestors in prehistory.

This poses a challenge to the Developing World health-care community which conceals an under-exploited opportunity: immunization at time-of-birth performed by the nearly-ubiquitous birth attendant, moreover in a manner which induces enduring, high-antibody-titer immunity, moreover against all pathogens for which vaccines exist.

The technological requirements of such comprehensive and enduring immunization-at-birth are basically of two distinct natures: the ability to present pulses of vaccine to immune surveillance at widely-separated points in time—weeks to months apart—and the ability to preserve usually-thermofragile vaccines against deterioration at body temperatures for time-intervals at least comparable to the total intervals-of-presentation. If both of these abilities could be attained simultaneously, depots of thermostabilized vaccine could be created subdermally which present their varied antigenic payloads at appointed times, potentially in manners which adequately reproduce the series of needle-mediated pediatric vaccinations conducted over the first year-or-so of postnatal life in Developed World circumstances. Potentially, all pediatric vaccines could be so emplaced in the literal hour of birth, so that an infant would then be 'automatically vaccinated' at appropriate times in the first year-or-so of life, against all pathogenic diseases for which vaccines exist—and thus be effectively immune to most major infectious diseases well into adulthood. [Note that those for whom such immunization was imperfect for some reason-or-another will still be highly-protected from infection by "herd immunity" effects, then having essentially no disease-susceptible peers from whom to contact infections...]

Interestingly, such technologies seem feasible to develop. Indeed such R&D endeavors are reasonably well underway already, and the first of them—deliberately aimed at realizing general-purpose 'platform technologies' of wide applicability—has demonstrated excellent results in animal trials and may be ready to commence initial human trials in a year-or-two.

Thus, it may become feasible during the next decade to vaccinate nearly every newborn against most every major infectious disease. At that point, the Final Clock starts to run for the large majority of human-specialized pathogens: on the time-scale of a single human lifespan, all of the ecological niches for all such pathogens will "age out" and the entire set of these niches—and presumably their pathogen-inhabitants—will cease to exist.

For pathogens which prey preferentially on infants or children—of which at least some serotypes of bacterial pneumonia seem exemplary in the present time—global eradication might thereby be attained in a single decade after universal immunization-at-birth gets well underway. Alternatively, universal application of such comprehensive and enduring vaccination to entire populations of all ages may attain global eradication of all vaccine-susceptible pathogens on time-scales of a handful of years.

Disease-Suppression Implications of Universal Omni-Nutrition

Synergistic with a full spectrum of aptly-'trained' memory B and cytotoxic-T cells which can be realized by omni-vaccination—either performed at birth or later-in-life—is a *soma* which supports operation of these immune surveillance cells to maintain robust immune surveillance against forever-opportunistic pathogens. Infectious disease transmission chains of several types are extensively documented to become more substantially more tenuous when facing immune surveillance in bodies which are replete in key micronutrients, e.g., vitamins A and D_3—each of which are required by human adults only in quantities less than a milligram per day and moreover are subject to storage in multi-month quantities in somatic depots, likely indicative of their essential contributions to sustaining of life against unceasing pathogenic threat.

It's therefore of very considerable interest in the global public health context to ensure that all people—most especially infants and children—are continually and fully supplied with all of the nutrients essential to robust resistance to infection—and to enabling prompt recovery from infections—until omni-vaccination may become fully implemented.

Thus it's heartening that R&D work aimed at sourcing all essential micronutrients in fully-adequate quantities by packaging them into quite ubiquitous foodstreams—such as ordinary table-salt—is already reasonably well-advanced. [Recall that, after breathing-air, water and solid food, salt is the most nearly-universal requirement for any person's physical existence, moreover in quite closely-fixed, time-independent amounts: ~4-12 grams of NaCl daily, on the time-average.] On time-scales of a relatively few years, it may become feasible to dispatch full quantities of all essential micronutrients in very common foodstuffs, moreover at remarkably low costs in terms of both economics and side-effects. Already, cooking conditions-stable formulations containing nominally-challenging key micronutrients such as zinc and iron have been demonstrated to be practically feasible. Moreover, the total costs-of-goods of sourcing the full spectrum of essential micronutrients in full recommended quantities to an adult is estimated to be at most $1 per year, tiny compared to virtually any other cost of staying alive and a fraction of 1% of the disposable income of the most wretchedly economically-positioned individual in the contemporary world.

It's thus feasible to look toward a none-too-distant future in which most-if-not-all humans are adequately micronourished in all essential respects, leaving the innately-more-challenging issue to be addressed of assuring adequate macronutrition, not merely in total calories but in key constituents thereof, specifically including sufficient quantities of the nine essential amino acids, total protein, essential fatty acids, total calories—and key mini-nutrients such as sodium, potassium, calcium, phosphorus and magnesium. The total cost-of-goods of providing the minimum essential nutrients to fully support a reasonably active adult's daily life is less than 0.2 USD per day—and working together with like-minded colleagues, we have demonstrated practical and palatable means of doing so. We estimate that similar means can see a young child perfectly nourished from birth through puberty—when sound nutrition is most developmentally critical—for a total cost-of-goods of about 500 USD.

Happily enough, reprogramming of as little as 3% of the present-day human-grade foodstreams will suffice to eliminate all pediatric malnutrition, if optimally done, and modest extension of such processes can suffice to eliminate all malnutrition of *all* human beings of *all* ages in *all* walks of life, *everywhere*. However, such optimal reprogramming includes extensive use of human-grade plant foodstuffs—e.g., compositionally-enhanced (e.g., high-lysine) cereal grains—for feeding people, rather than feeding agricultural animals which in turn are used to feed people—with remarkably-small throughput efficiencies in both key nutrients and total calories (as we've noted previously).

The remarkable fact is that young humans can be fed to adult size at unit costs of the same order-of-magnitude as calves can be grown to cattle, i.e., at food costs of the order of several dollars per pound. No, such food-cost-minimized human teenagers can't be fed extensive amounts of hamburger or beefsteak, but they can be literally-perfected nourished with a number of distinct mixes of plant products, and possibly more healthily so. Such alternative nutrition for children in the least economically-advanced portions of the Developing World may be an all-around-feasible path-forward toward full and timely attainment of the pertinent Millennium Development Goals.

It's crucial to note that such optimal nutrition of young humans throws sharply-etched shadows many decades into the future. Malnutrition of pregnant and lactating women and their infant and toddler children is extensively documented to impose permanent, non-recoverable physical and cognitive damage on the young, which is not only life-long and occasionally of utterly-ruinous effect but which is also epigenetically transferred to their children—and perhaps on to their grandchildren, as continuing studies of the *in utero* victims of the Dutch Famine at end-World War II—and of their descendants—have so graphically demonstrated.

Bluntly characterized, such nutritionally profoundly-abused infants and children become short, diseased, impoverished and cognitively impaired, i.e., shorter-lived and lower-performance, adults. These permanently individuals have sharply-curtailed human potentials—i.e., akin to the lower classes of peasantry of medieval times—and they tend to vertically transmit such profound, entirely-avoidable impairments to their children, generation-after-generation.

When the World's material means were far more meager and knowledge-of-consequences much more limited, such pervasive suppression of human potential might be implicitly excused along lines of *force majeure*. However, neither poverty of material means nor of pertinent knowledge of avoidable consequences any longer exist: persistence of large sub-populations of severely malnourished infants and children is a charge which History currently levies and records against the personal moral account of each and every educated adult—and most assuredly against their thought-leaders.

CONCLUSIONS

The simultaneous-on-historical timescales advents of high-efficacy anti-pathogenic vaccines, adequate nutrition of major sub-populations and high-performance planetary-scale information systems have jointly laid the foundation for global eradication of many major infectious diseases, by making sufficiently large population fractions sufficiently and simultaneously inhospitable to the causative pathogens that all infective-transmission chains may be broken at the same moment in time—and thus remain so forever-after.

Less than a century elapsed between the advent of a formal theory of infectious diseases due to Pasteur, Koch, Ehrlich and others and the first global eradication of smallpox, one of the hitherto most fearsome killers of mankind, by Henderson and his many colleagues. During the past half-century, mankind has unsuccessfully attempted the global eradication of the causative pathogen of malaria, which has arguably scarred the human genome more deeply than any other infectious disease, and presently appears about to succeed during the next several years in completely destroying wild polio virus, which was still paralyzing hundreds of thousands of children each year a mere quarter-century ago and whose annual incidence-rate is currently more than a 1000-fold lower; a second malaria global eradication campaign is now less than a decade in age and is building strength.

Other disability and death-dealing pathogens are even now under increasingly-intensive modeling studies and analyses aimed at their world-wide annihilation, and the advents of enabling toolsets continue at quickening paces. A number of serotypes of bacterial pneumonia and meningitis[32] appear to have surprisingly fragile transmission chains in today's world, and measles likewise appears susceptible to global eradication in the distinctly foreseeable future—even without omni-vaccination or thermostable vaccine technologies being brought-to-bear.

Both the failed and the successful eradication experiences and lessons-learned therefrom, the presently burgeoning anti-pathogen toolset springing forth from the modern biology revolution, the exceptional ways and means enabled by modern information technologies and the indefinitely-large future savings in human misery and death and in social-economic losses and disease-treatment costs combine to presently motivate an all-out campaign-of-annihilation against the major pathogens, exploiting the full set of synergisms implicit in such a historic undertaking.

Humanity may thereby look forward to completing the second century of the Pasteur-Koch-Ehrlich era with a string of global eradications of the major infectious diseases, and the commencement of a new epoch forever-free of the ancient terrors of suddenly-appearing "death, in the very midst of life."

REFERENCES

1. Henderson, D.A. (1980) *Smallpox eradication.* Public Health Reports. **95**(5): p. 422-426.
2. Henderson, D.A. and P. Klepac (2013) *Lessons from the eradication of smallpox: an interview with D.A. Henderson.* Philosophical Transactions of the Royal Society B: Biological Sciences. **368**(1623).
3. Klepac, P. et al. (2013) *Towards the endgame and beyond: complexities and challenges for the elimination of infectious diseases.* Philosophical Transactions of the Royal Society B: Biological Sciences. **368**(1623).
4. Dietz, K. and J.A.P. Heesterbeek (2002) "Daniel Bernoulli's epidemiological model revisited." *Mathematical Biosciences.* **180**(1-2): p. 1-21.
5. Ross, R. (1910) *The prevention of malaria.* London,: J. Murray. 2 p. l., vii-xx, 669 p., 11.
6. Macdonald, G. (1956) "Theory of the eradication of malaria." *Bull World Health Organ.* **15**(3-5): p. 369-87.
7. Eckhoff, P. (2011) "A malaria transmission-directed model of mosquito life cycle and ecology." *Malaria Journal.* **10**(1): p. 303.
8. Eckhoff, P. (2012) "P. falciparum Infection Durations and Infectiousness are Shaped by Antigenic Variation and Innate and Adaptive Host Immunity in a Mathematical Model." *PLoS ONE.* **7**(9): p. e44950.
9. Eckhoff, P. (2012 "Malaria parasite diversity and transmission intensity affect development of parasitological immunity in a mathematical model." *Malaria Journal.* **11**(1): p. 419.
10. Eckhoff, P. (2013) "Mathematical Models of Within-Host and Transmission Dynamics to Determine Effects of Malaria Interventions in a Variety of Transmission Settings." *The American Journal of Tropical Medicine and Hygiene.* **88**(5): p. 817-827.
11. Wenger, E. and P. Eckhoff (2013) "A mathematical model of the impact of present and future malaria vaccines." *Malaria Journal.* **12**(1): p. 126.
12. Upfill-Brown, A. et al. (2014) "Predictive spatial risk model of poliovirus to aid prioritization and hasten eradication in Nigeria." *BMC Medicine.* **12**(1): p. 92.
13. Eichner, M. and K. Dietz (1996) "Eradication of Poliomyelitis: When Can One Be Sure That Polio Virus Transmission Has Been Terminated?" *Am. J. Epidemiol.* **143**(8): p. 816-822.

14. Griffin, J.T. et al. (2010) "Reducing Plasmodium falciparum Malaria Transmission in Africa: A Model-Based Evaluation of Intervention Strategies." *PLoS Med.* **7(8)**: p. e1000324.
15. Smith, T. et al. (2006) "Mathematical modeling of the impact of malaria vaccines on the clinical epidemiology and natural history of Plasmodium falciparum malaria: Overview." *Am J Trop Med Hyg.* **75**(2 Suppl): p. 1-10.
16. Johnston, G.L. et al. (2014) "Modeling Within-Host Effects of Drugs on *Plasmodium falciparum* Transmission and Prospects for Malaria Elimination." *PLoS Comput Biol.* **10**(1): p. e1003434.
17. Maude, R. et al. (2009) "The last man standing is the most resistant: eliminating artemisinin-resistant malaria in Cambodia." *Malaria Journal.* **8**(1): p. 31.
18. Eaton, J.W. et al. (2012) "HIV Treatment as Prevention: Systematic Comparison of Mathematical Models of the Potential Impact of Antiretroviral Therapy on HIV Incidence in South Africa." *PLoS Med.* **9**(7): p. e1001245.
19. Wesolowski, A. et al. (2012) "Quantifying the Impact of Human Mobility on Malaria." *Science.* **338**(6104): p. 267-270.
20. Pindolia, D. et al. (2014) "Quantifying cross-border movements and migrations for guiding the strategic planning of malaria control and elimination." *Malaria Journal.* **13**(1): p. 169.
21. Linard, C. et al. (2012) "Population Distribution, Settlement Patterns and Accessibility across Africa in 2010." *PLoS ONE.* **7**(2): p. e31743.
22. Murray, C.J.L. et al. (2012) "Disability-adjusted life years (DALYs) for 291 diseases and injuries in 21 regions, 1990-2010: a systematic analysis for the Global Burden of Disease Study 2010." *The Lancet.* **380**(9859): p. 2197-2223.
23. Murray, C.J.L. et al. (2014) "Global, regional, and national incidence and mortality for HIV, tuberculosis, and malaria during 1990-2013: a systematic analysis for the Global Burden of Disease Study 2013." The Lancet. **384**(9947): p.1005-1070.
24. Lim, S.S. et al. (2012) "A comparative risk assessment of burden of disease and injury attributable to 67 risk factors and risk factor clusters in 21 regions, 1990-2010: a systematic analysis for the Global Burden of Disease Study 2010." *The Lancet.* **380**(9859): p. 2224-2260.
25. Hay, S.I. et al. (2009) "A world malaria map: Plasmodium falciparum endemicity in 2007." *PLoS Med.* **6**(3): p. e1000048.
26. Gething, P. et al. (2011) "A new world malaria map: Plasmodium falciparum endemicity in 2010." *Malaria Journal.* **10**(1): p. 378.
27. Hay, S.I. et al. (2010) "Developing Global Maps of the Dominant Anopheles Vectors of Human Malaria." *PLoS Med.* **7**(2): p. e1000209.
28. Howes, R.E. et al. (2011) "The global distribution of the Duffy blood group." *Nat Commun.* **2**: p. 266.
29. Dobson, S.L., C.W. Fox, and F.M. Jiggins (2002) "The effect of Wolbachia-induced cytoplasmic incompatibility on host population size in natural and manipulated systems." *Proceedings of the Royal Society of London. Series B: Biological Sciences.* **269**(1490): p. 437-445.

30. Windbichler, N. et al. (2007) "Homing endonuclease mediated gene targeting in Anopheles gambiae cells and embryos." *Nucleic Acids Research.* **35**(17): p. 5922-5933.
31. Sutter, R., O. Kew, and S. Cochi (2008) *Poliovirus vaccine-live*, in *Vaccines*, D. Plotkin, W. Orenstein, and P. Offit, Editors. Saunders Elsevier: Philadelphia. p. 631-686.
32. LaForce, F.M. and J.M. Okwo-Bele (2011) "Eliminating Epidemic Group A Meningococcal Meningitis In Africa Through A New Vaccine." *Health Affairs.* **30**(6): p. 1049-1057.
33. Bershteyn, A. and P. Eckhoff (2013) "A model of HIV drug resistance driven by heterogeneities in host immunity and adherence patterns." *BMC Systems Biology.* **7**(1): p. 11.
34. Floud, R. et al. (2011) *The Changing Body: Health, Nutrition, and Human Development in the Western World since 1700.* New York: Cambridge University Press.
35. Fogel, R.W. (2004) *The Escape from Hunger and Premature Death, 1700-2100.* New York, NY: Cambridge University Press.

SESSION 2

RENEWABLE AND NUCLEAR ENERGY

RENEWABLE AND NUCLEAR ELECTRICITY: OPPORTUNITIES, CHALLENGES AND POLICY RECOMMENDATIONS

CARMINE DIFIGLIO
Office of Energy Policy and Systems Analysis, Department of Energy
Washington, DC, USA

ROBERT BUDNITZ
Earth Sciences Division, Lawrence Berkeley National Laboratory
Berkeley, CA, USA

MICHAEL JEFFERSON
ESCP Europe, Business School
London, UK

HISHAM KHATIB
World Energy Council
Amman, Jordan

TATSUO MASUDA
The Graduate School, Nagoya University Commerce and Business
Tokyo, Japan

CHARLES MCCOMBIE
McCombie Consulting, Arius Association
Gipf-Oberfrick, Switzerland

JEF ONGENA
Plasmaphysics Laboratory, Ecole Royale Militaire
Brussels, Belgium

LARA PIERPOINT
Office of Energy Policy and Systems Analysis, U.S. Department of Energy
Washington, DC, USA

ADNAN SHIHAB-ELDIN
Kuwait Foundation for the Advancement of Sciences
Safat, Kuwait

BRUCE STRAM
Element Markets
Houston, TX, USA

1. INTRODUCTION

Economic growth and electricity supplies are strongly linked when economies are developing. While the developed world may be able to meet rising electricity service demand with improved end-use efficiency, worldwide power demand is, nonetheless, expected to significantly increase due to economic growth outside the OECD.[1] A failure to satisfy power demand in the developing world would not only constrain economic growth, but would also sustain "energy poverty" experienced today by 1.4 billion people with no access to electricity. The UN Millennium Development Goals, adopted in 2000, placed a high priority on eradicating energy poverty.

Also, in order to achieve the greenhouse gas emission goals endorsed at the 2009 G8 Summit,[2] the electric power sector must produce substantially less greenhouse gas emissions per kWh. Since nuclear and renewable power plants produce essentially no greenhouse gas emissions, the future world-wide fleet of power plants must include a substantial share of renewable and nuclear plants or other zero-emission technologies.[3]

The power sector investments needed to support both economic growth and emission reduction are formidable. The IEA estimates that $17 trillion must be spent for new power plant and transmission infrastructure between 2013 and 2035. While it is evident that zero-emission power plants must constitute a major part of this investment—$6.2 trillion for renewable power plants (IEA, 2013b)—there is a significant controversy whether it is practical for renewable energy to expand sufficiently to meet this need, or whether governments should give nuclear power support comparable to that provided to renewable energy to allow for more electricity and less greenhouse gas emissions. This paper is intended to inform this debate by examining the technical and economic challenges faced by nuclear and renewable power: cost, safety, environmental impacts, public acceptance and integration into the electrical grid.

2. KEY MESSAGES AND POLICY RECOMMENDATIONS

- World-wide economic growth will require continued power sector investments to meet growing electricity demand. While electricity demand growth in the OECD will be relatively modest, most serious studies, including

[1] The IEA estimates that electricity demand will increase by 2.3%/year through 2035 driven by demand growth of 3.3%/year outside of the OECD (IEA, 2013a).

[2] At the 2009 G8 Summit in L'Aquila, Italy, a goal to reduce world-wide greenhouse gas emissions in 2050, relative to 2005 emissions was endorsed by the G8 members. In addition, governments of the countries that constitute 80% of total GHG emissions have agreed to take measures to significantly cut GHG emissions. This is the best reason to explain why this paper does not address coal with CCS.

[3] Multiple studies by the IEA, IIASA and others conclude that 1) very significant cuts are necessary in the power sector to achieve deep greenhouse gas emission cuts, 2) the emission reductions in the power sector can be achieved at relatively low costs compared to other sectors, and 3) while there is some room for "transitional" low GHG technologies such as natural gas or coal with carbon capture and storage, it will be necessary to have a large share of renewable and nuclear power to reduce world-wide greenhouse gas emissions by 50% by 2050. Consequently, natural gas plants with waste heat recovery, coal plants with CO_2 capture and storage, or natural gas plants with CO_2 capture and storage can all contribute to significant power-sector emission reductions in many countries: the ultimate goal must be a high percentage of zero-emission renewable and nuclear plants.

the International Energy Agency's *World Energy Outlook 2013*, project that the developing economies will increase electricity demand by over 3 percent per year over the next several decades (IEA, 2013a).

- What should be pursued along with the above is the eradication of energy poverty, particularly that of electricity, which has a critical importance for human dignity and capacity building. As increased global electricity production does not necessarily resolve this problem, special efforts should be made for the redistribution of electricity in impoverished or detached areas in an environmentally compatible manner as much as possible. Recent advances in PV technology can greatly assist in this regard.
- Governments of the largest world economies are committed to substantial reductions of greenhouse gas emissions. For example, on September 6, 2012, nations of the world's leading economies ranked climate change, along with growth, jobs, investment and trade, as a key issue of the future; one that will extract ever higher costs if "we delay additional actions" (G20, 2013).
- Multiple studies have shown, world-wide, that the power-sector would have to achieve deep cuts in greenhouse gas emissions in order to make significant cuts in overall emissions. The studies show that a large share of energy-sector emissions come from the power sector. The studies' cost analyses also show that reducing greenhouse gas emissions in the power sector is relatively cost effective compared to reducing emissions in other sectors, especially the transportation sector.
- While replacement of coal-fired plants with natural gas or fossil-fuel plants with carbon capture and storage can facilitate progress in cutting power sector emissions, the share of carbon-free power sources, such as renewable energy and nuclear power, must constitute a much larger share of power sector generation if overall emission goals are to be achieved.
- New renewables (e.g., wind, solar, modern biomass and biofuels, tidal, wave and ocean energy) are widely regarded as clean, indigenous and sustainable sources of energy. Therefore, they are favoured by many governments and the public as a whole. However, their present contribution to global energy consumption is still limited (about 3% in 2013 and only about 10% even for heating) since their economics are often not yet favourable, especially at the margin. In most instances, they need to be supported by state subsidies and regulations. They suffer from high investment costs and, as a result of the intermittent and diffused nature of wind, solar, and tidal, relatively low utilisation factors.
- The costs of renewable technologies have come down as a result of research, development and demonstration and technology learning that is gained with commercial experience. Cost breakthroughs in several renewable technologies can be expected especially given the considerable investments government and industry are making to achieve lower costs.
- Government policies to promote the growth of renewable energy or nuclear power that result in high electricity costs are not sustainable as high electricity costs and limited electricity supplies are harmful to economic growth.

- If renewable energy programmes or initiatives for new nuclear power are based on overly optimistic technical assessments, costs are likely to exceed sustainable levels. The strongest proponents of both nuclear power and renewables tend to have an overly optimistic view of the contribution their favoured technology is likely to make to future energy demand. Nuclear growth is limited primarily by the long lead time and the public opposition to new nuclear plants. New renewables are limited primarily by the inherently lower energy densities and by the massive demands their large scale deployment puts on other parts of the electricity supply system.
- Policies should ensure the uptake of the most cost-effective technologies. Policies should not prematurely force the uptake of technologies that require technological advances or technologies that cannot be expected to experience significant cost reductions when applied on a commercial scale.[4]
- Policies should consider the full costs and benefits of power sector technologies. Important benefits may include the elimination of pollutants such as toxic particulates, sulphur, mercury, volatile organic compounds and nitrogen oxides from fossil-fuel power plants. Cost may include pollution, resources required to construct renewable or nuclear power plants and their necessary transmission grids, adverse impacts on wildlife or property values and, for nuclear plants, the possible release of radiation during plant operation or the disposition of spent reactor fuel.
- Cost-effective electricity storage would significantly improve renewable and nuclear power economics, reducing the costs for meeting peak power demands and reducing the need to dispatch power at low prices when supply outstrips demand. Storage benefits intermittent renewable technologies such as wind and solar plants.
- Storage also benefits base load power plants like nuclear which cannot be easily cycled to follow load.
- Research, development and demonstration to develop cost-effective storage should be a priority of governments.
- Likewise, research, development and demonstration to reduce the cost or address shortcomings of renewable and nuclear technologies should also be a governmental priority.
- Significant reductions in power sector greenhouse gas emissions will require reductions in renewable and nuclear power costs. Cost reductions can be achieved through commercial application but also require research, development and demonstration.
- Nuclear power plants are currently not likely to be built in most OECD economies without government backing as the financial risks to utilities, other energy companies and banks are too large. Building nuclear power plants in

---

[4] Wind energy is an example of an initially costly technology that achieved large cost reductions as a result of commercialization. Cost reductions from commercialization are often referred to as "technology learning." It can be used to justify subsidies to technologies that are not currently competitive if it is expected that, with expanding commercial production, costs will come down. A "learning ratio" is often used to measure the cost reduction achieved by a doubling of commercial capacity.

the United States is particularly challenging as long as low U.S. natural gas prices persist ($4.70 per million Btu as of 19 June, 2014).

- The prospects for nuclear power are much better outside the OECD in more centrally planned economies. The prevalence of state-owned power companies, generally lower construction costs, and the prevailing regulatory environments help explain why the bulk of planned nuclear power plants are in Russia, Asia and the Middle East.
- Modular nuclear power reactors may offer significantly lower costs, better scalability, improved safety, and reduced economic risk, and their development should be a governmental priority.
- Left to the private market, nuclear power investments are unlikely in the OECD. OECD governments should explicitly consider how and if they want nuclear power to help reduce national carbon emissions. The alternatives include:

  – Currently forgo significant reliance on light-water reactors as a means to achieve GHG goals and wait to see whether commercially available modular nuclear reactors would be competitive;[5] perhaps by waiting to see whether nuclear reactor projects outside the OECD (and those few within the OECD) achieve the necessary cost reductions that might make nuclear reactors more attractive to investors in competitive power markets;
  – Develop modular reactor designs that could greatly reduce on-site costs, provide a manufactured product with a predictable customer cost and improve the economic competitiveness of nuclear power;
  – Provide financial backing to enable the construction of light-water reactors, perhaps with the expectation that successful experience with the technology may eventually lead to commercial acceptance of the technology in competitive power markets;
  – Assume that development of nuclear power is not likely to be a means to achieve deep cuts in GHG emissions.

- If OECD governments choose to forgo support for nuclear power by either not assuming part of the financial risk involved in building current-technology reactors or by not participating in developing modular reactors for widespread deployment, then there is less likelihood that deep cuts in power sector carbon emissions will be achieved. If governments hope to achieve the desired increased electricity production and CO_2 reductions primarily by the (subsidised) expansion of renewable power, then there is a likelihood that this will not succeed and that increased use of fossil fuels and/or failure to meet the goals will result.
- If nuclear power is to be deployed successfully in countries without a current commercial nuclear program, several cultural attributes must be present. A political culture that can make a long-term commitment, provide for an

[5] Assuming that there would be an overall scheme such as cap-and-trade or GHG taxes to financially reward technologies that produce lower GHG emissions.

independent regulatory agency with both authority and resources, and assure independence of courts to enforce contracts is crucial. Equally crucial are a set of social-culture issues including freedom from corruption, holding safety as paramount, a commitment to transparency in management practices and communication, and a strong continuity of institutions. Without these, a nuclear-power program is less likely to achieve an adequate safety record.

- Nuclear power has always been linked to concerns with proliferation of nuclear weapons, specifically the potential misuse of nuclear power reactor or fuel-cycle technology to make weapon-usable material directly or to extract it from a power-reactor or a fuel cycle facility by either a government or a non-governmental group for malevolent purposes. If these concerns are to be managed in the future, it is necessary that the international protocols based on the Non-Proliferation Treaty be maintained and expanded. This includes mandatory agreement to the NPT's Additional Protocol, along with a world-wide commitment to ostracize states that openly flaunt the NPT.
- The development of fusion energy could offer future supplies of energy on a scale order of magnitude larger than current known energy resources. Considerable scientific and engineering progress has been made to develop fusion energy. While we are at least a few decades away from the first commercial scale fusion power plant, it is recommended that governments continue to provide the necessary research funds to advance fusion power.
- The support for fusion energy research, recommended above, should be part of a much larger R&D agenda to enable commercially significant supplies of GHG- free electricity, including advanced renewable power technologies, new energy storage technologies and modular nuclear reactors. While not discussed in this paper, research and development of carbon capture and storage technologies may also lead to significantly reduced power sector emissions while still utilizing the large world-wide reserves of coal and natural gas.

3. KEY CHALLENGES TO EXPANDING RENEWABLE ENERGY

The purpose of this section is to outline and evaluate the challenges presented by increasing penetrations of renewable electricity generation. These generation sources primarily include solar and wind, which are growing rapidly and are new enough to the grid that the impact of high penetrations is not fully understood. The challenges associated with solar and wind will be examined qualitatively, and to the extent possible, in terms of the costs they might imply to the system over and above the costs associated with deploying and operating the electrical generation equipment. Hydropower is generally well integrated into existing operations, but has much less potential for growth. Tidal, wave, run-of-river hydro and geothermal sources are currently much more limited both in quantity and growth. Energy efficiency is sometimes considered an energy "source," but is not usually combined with renewables; nonetheless, it is discussed below because the potential for more efficiency would obviate the need for renewables (or for more nuclear) to some extent. Special challenges posed in developing countries are presented as well.

The intrinsic nature of solar and wind power is very likely to present greater system challenges than "conventional" sources. Within limits, those challenges can be overcome, but at a cost. Later sections of the paper will draw on a variety of sources to identify a range of such costs, at least as they are foreseen by researchers helping prepare ambitious plans for grids to obtain high shares (30-50%) of their megawatt hours from primarily solar and wind generation.

Since a primary policy rationale to including renewables is reduction of carbon emissions, prospects for high (30-50%) penetrations of renewable power will be a key focus, even though they may require significant technology improvements. These issues also raise the question of what role is appropriate for nuclear power (a non CO_2 power source whose capacity can be expanded) in helping integrate renewables into grid operation.

3.1 Some Power System Fundamentals

Modern electric power grids are really a technological marvel, even if routine. Electric power, delivered as needed on demand, is a very valuable resource for industry, commerce and everyday household life. In addition to the simple availability of easy-to-use energy, the value is much increased by virtue of the reliability and the "quality" of the supply. Power must be supplied to user within voltage tolerance limits, and rigidly maintained frequency (60Hz in the U.S.).

Power systems are typically very large networks drawing on many generating sources of power and dispatching that power to many loads. This involves large scale "transmission" of power to nodes and "distribution" of power to users who range from very large industrial complexes to ordinary homeowners and everyone in-between. These systems are extensively thought out and planned.

The degree of difficulty for maintaining such a system is substantially exacerbated by the fact that power flows are virtually instantaneous and that power storage opportunities are very limited. Thus, any upsets in the system (large power plants or industrial users tripping offline, transmission line breakdowns, etc.) must be offset on the order of seconds. Further, like any mechanical equipment, power plants must be periodically shut down for planned maintenance. Additional reserve power must be available to supply power during these periods.

The two most basic measures of electricity are megawatts (MW) which connotes the flow rate of power or capacity of generation, and megawatt hours (MWh) which indicates energy usage.

Costs associated with power sources occur at three levels. First there are the costs associated with the generation facility itself. These are capital costs for land, equipment and construction, and ongoing operating and maintenance costs. The given facilities must be connected to the grid, and the grid must be capable of carrying sufficient power to desired nodes which gives rise to additional costs—both capital and operating. Finally, all power systems have substantial infrastructure whose primary purpose is to maintain grid stability and reliability. This need is acute because, at least for now, the possibility of storing electricity is very limited and ranges from very expensive to impossible.

Solar and wind power are fundamentally different in character from traditional fossil fuel sources (or nuclear and hydro.) Such "conventional" facilities are normally described by their capacity for power production in terms of MW. That capacity is generally available on a dispatchable basis for the system operator. Fuel (coal, oil, reservoir water, uranium) is stored on site or piped in (natural gas). While breakdowns do occur, they are rare.

Solar and wind are also rated according to capacity. However, these ratings are based on specified conditions: 1 KW/square meter insolation and 12 m/s wind, for example. But these conditions are not achieved at the discretion of the operator, but rather are subject to nature.

Consequently, both solar and wind energy present intermittency issues to system operators. These are time of day, seasonal and even idiosyncratic. Solar and wind produce energy according to sunshine and wind available. Either of those can change very suddenly. But power demand doesn't fall, so the shortage situation must be dealt with. Wind availability may be ill timed according to time of day or seasonal power needs. In addition, both solar and wind at utility scale require substantial expanses of land for deployment and good insolation or wind availability. Thus, these sources may require substantial transport of power from locations not convenient to the grid. Conventional sources have similar problems, but face fewer location constraints since their fuel is more transportable or ubiquitously available.

Up to a certain level of penetration of intermittent renewables, these issues don't present unique challenges for grid operators. While rare, breakdowns from conventional power generation equipment certainly occur and variations in load occur frequently. Operators have developed a number of tactics for dealing with these. Spinning reserves, for example, can be brought online very quickly to offset power sags or meet load coming onto the system. Power is already transported long distances in particular circumstances.

However, these features come at a cost. Currently such costs are often subsumed into aggregate system operation costs. Since renewable sources tend to draw more heavily on these resources, they are in effect being subsidized by grid operation. (They can be charged directly for such services, but currently, given their small portion of total supply, this is not often done.) This implicit subsidy is in addition to direct subsidies that renewables are often currently provided.

As the proportion of solar and wind power to conventional power grows, the operational realities of these issues come to light. To the extent possible, some exemplary costs imposed by renewables will be evaluated. A number of European nations are planning for substantial increases in solar and/or wind shares for their power sources. The California grid currently provides excellent examples which will be explored. Further, as current energy policy, particularly in California, dictates unprecedented shares of renewable power for future years within utility planning horizons, ultimate limitations of such power are being investigated and will be examined here.

3.2 System Costs versus Facility Costs

All large power plants require substantial capital investment for site locations, structures and equipment. In the case of "conventional" power plants (coal, oil, natural gas, hydro and nuclear), these are all directed toward turning some sort of turbine which is attached to generation machinery. These capital investments are usually expressed in terms of dollars per MW. These generators produce alternating current power. That power must be conditioned and transferred to the grid. Critically, all of these spinning generators must be synchronized, i.e., spin at the same speed.

Except for hydro, these conventional plants have fuel costs. All have operating costs similar to what would occur in any factory and require significant periodic maintenance.

3.2.1 Connection

Solar and wind power installations require high capital expenditure per MW of rated power (which does not have the same connotation as conventional plants). Solar PV power needs specific equipment to "condition" produced power to convert direct current to AC, increase voltage and maintain grid synchronization. Additional steps must be taken to ensure that solar PV installations meet grid requirements for upset conditions. Grid operators prefer that wind farms install additional equipment to support grid interaction, providing, for example, the ability to curtail wind output.

3.2.2 Transmission

Large scale power facilities generally need their produced power transmitted to the grid for ultimate delivery to system nodes and then distribution among users. Such transmission costs tend to be higher for solar and wind relative to their rated capacity because of the location limitations cited above, and because, relative to their rated power, they have relatively low load factors. Transmission is generally sized to accept the highest output likely from such facilities. This is especially true when plans for renewable energy set ambitious goals for production. Higher per MW costs spread over relatively fewer MWh means higher cost.

Nuclear facilities also often bear relatively high transmission costs. For safety considerations, they have a redundancy in connection to provide greater assurance for cooling water pumps. This is in addition to onsite backup diesel or other power.

3.2.3 Balancing and Adequacy of Supply

Over and above these facilities costs, power grids require system infrastructure to maintain the continued operation and stability of the grid. While solar and wind inherently place disproportionate demands on this infrastructure, these are not different in kind from conventional sources.

Solar and wind power availability may tend to vary on a predictable daily and/or seasonal cycle. Solar especially can be expected to have a daily cycle. During any off cycle, if power demand doesn't fall in concert (the solar cycle tends to almost coincide with peak daily power demand), standby power must be available (at least according to most current standard practice). This same requirement applies to conventional power which needs backup power (i.e., "reserve") for maintenance cycles. Conventional sources must be ramped down as solar and wind are ramped up to keep the system in balance. Some base load facilities, coal and nuclear primarily, have historically run more or less continuously at full power in most systems. System goals set for renewable power generation (i.e., California's 33% of electricity retail sales from renewables by 2020, but to a greater extent 50% in the future which is under consideration) require displacing power from these facilities.

Further, solar and wind have inherently more volatile output than conventional sources whose variability is tied basically to relatively rare breakdowns. For solar and wind respectively, sunlight may be obstructed, and wind may stop blowing. These events are likely to have some warning, but they are akin to breakdowns in mechanical equipment. The problem is exacerbated in that these effects may be correlated among a range of solar installations or wind farms.

The principal means of providing these needs is first spinning reserve, usually a coal or oil plant, gas turbine, or hydro, intentionally running at less than full capacity. So situated, these may be ramped up (or down) very quickly to respond to very short term situations.

With greater warning, provided by weather forecasts in the case of renewables, and maintenance planning as well as equipment monitoring, reserve facilities can be fired up and brought online to fill in gaps in demand. Again, these are typically gas-fired fossil fuel plants or hydro. Such units typically provide needed power at lower incremental cost through greater efficiency. Simple cycle gas turbines are used for spinning reserve because they have lower capital costs than most other capacity options.

3.3 Flexibility

The flexibility of the contribution of new renewables to the electricity supply varies. Although hydroelectricity is generally reliable and provides flexible supply (except in periods of prolonged drought), it can also store huge amounts of water/power to be used when necessary. This advantage is also assisted by pumped hydro storage, which involves pumping water uphill into the reservoir at off-peak and then releasing it when needed. Biomass (subject to land and water availability, and food requirements) and geothermal can provide reliable renewable sources. However, other new renewables (particularly solar and wind) present formidable challenges to the transmission supply operators due to their intermittency, unpredictability, and changing meteorological conditions. This intermittency is delaying or preventing electricity markets from accommodating and meeting the ambitious renewable goals set by energy planners in some countries.

The flexibility of national electricity grids to accommodate variable power sources is enhanced by: achieving geographic and technological diversification of variable energy sources, improving local energy management and increasing prospects of energy-storage schemes, and also by trading with other electricity grids.

Nuclear plants pose other types of system challenges in that they most often dispatch power below cost when power demand is low. They need some time to ramp up and down their load and follow system. Some existing LWR type plants have limited ability to significantly vary their output to match changing demand. PWRs, as well as CANDU and BWR, have load-following capability, which will allow them to fill more than baseline generation needs. Some newer reactors also offer some form of enhanced load-following capability. For example, the Areva EPR can slow its electrical output power between 990 and 1,650 MW at 82.5 MW per minute.

In the generation dispatching schedule, renewables production is "free," and it is a "must dispatch" source of generation. This is followed by nuclear. Correspondingly, if the system has minimum load, such as at weekends, and a surge of wind occurs, renewables and nuclear output may exceed demand. It is not easy to ramp down load on nuclear quickly or shut down other dispatchable facilities in the generating system. These events lead generation to significantly exceed load. If there are enough regional interconnections, neighbouring systems can accommodate some of this excess generation (as regularly happens with Denmark's wind energy generation); but what happens if those systems have the same problem? For example, it has been found that offshore North-West Europe periods of calm quite regularly coincide across a wide area of ocean. The least-cost solution is to offer incentives to consumers to take more loads by various means including negative pricing. This means that nuclear output is not only offered free, but consumers are actually paid incentives to take it. We are not talking here of a hypothetical case, but actual incidents experienced by the European system in recent years. In other cases, the power output of these plants is simply curtailed. If new renewables, particularly the wind component, is increased, then this dispatching problem is going to become more acute. Whether this will lead to an increase in the disadvantages and costs of nuclear generation or to the fuller exposure of the frailties of new renewables to the advantage of nuclear, remains to be seen.

3.4 System Cost and Renewable Share

A number of countries and states in the U.S. have put forward very ambitious goals for the share of power provided by renewable sources. California calls for 33% electricity retail sales from renewables by 2020 (originally 20% by 2017), not including hydro. Finland, France, Germany, Republic of Korea, and the UK have ambitious goals as well. Denmark is already deriving 30% of its power production from wind, but exports power to nearby countries in times of surplus. Maine, Massachusetts, New Hampshire, New York and Maryland have already cut their power sector emissions by 40% with renewables and gas. Consequently, much study and planning has been undertaken to understand the consequences of these goals. These are potentially two fold. First total cost of power per MWh increases (on a fully allocated cost basis which includes the cost of capital) because renewable generation is at least now relatively expensive, and because system costs per MWh increase as the share of renewables increases.

System costs increase because proportionately more balancing and back-up resources are required to compensate for the intermittency of those increased renewables.

The second concern is the extent to which there is a limit to the share of renewable power, at least as provided by wind and solar that can't be dealt with without some dispatchable fossil fuel power in the system.

A recent study illustrates these factors specifically with respect to Germany and presents data across nations. This is summarized in Figure 3.1.

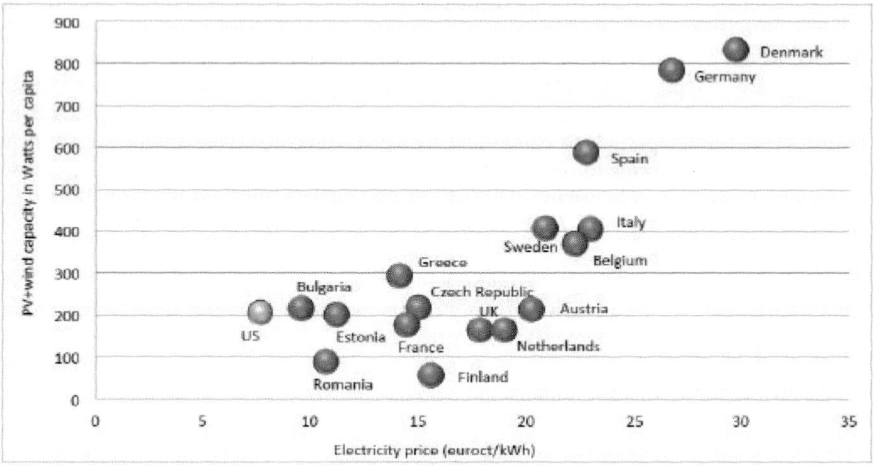

Fig. 3.1: Comparison of the amount of wind and solar capacity and electricity prices, selected countries, 2012.

As can be seen, there appears to be a strong positive correlation between consumer power costs and the percentage of solar and wind penetration. A more subtle insight is that the cost impact from penetration will vary by system.

Specifically in Germany this situation has also reduced revenues for "traditional" fossil fuel operators. On a real time basis, grid power is priced on the basis of what is in effect an essentially continuous auction based upon marginal generation costs. Solar and wind have zero marginal cost or are designated as "must run" facilities (perhaps for this reason). The availability of substantial "free" resources lowers prices overall. And in the case of solar, these reduced prices often coincide with peak power demand.

Given the nature of the auction, conventional suppliers bid their short run costs and don't generate unless these are covered and they are dispatched. However, they typically depend on "high" peak prices to make significant margins above cost in order to cover capital costs. These peak prices have been substantially reduced (Figure 3.2).

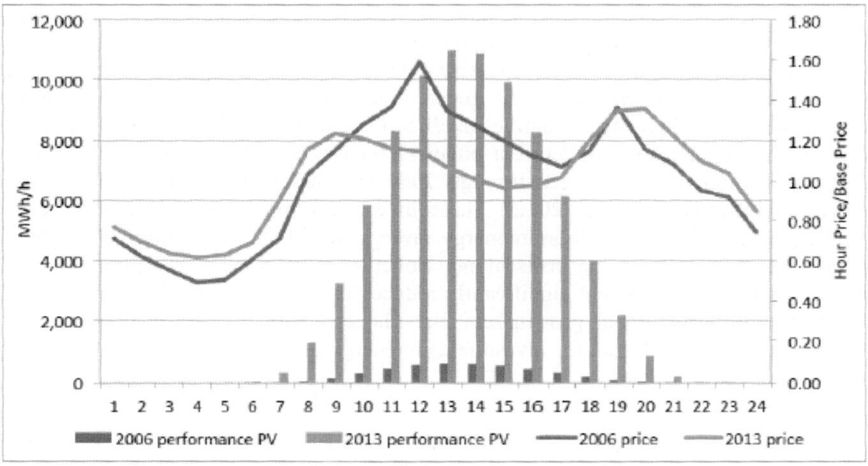

Fig. 3.2: Yearly performance of PV and electricity tariff (standardized).

Poser et al. note that:

> "This situation is making cost recovery even more difficult. This means that the owners of the plants have to write-off the money they originally invested...RWE [a major German electric utility] recently announced a write down of €4.8 billion on assets, mainly power stations, and a net loss of €2.76 billion ($3.8 billion) on total sales of €54.1 billion, its first full-year loss since 1949.59." This situation is not unique to Germany. One of Australia's largest electric utilities, Energy Australia, also announced a significant write down due, in part, to gas-fired generators becoming uneconomic because of the rapid growth of solar power and the rapid increase of electricity prices for customers.
>
> Generators' financial difficulties have already translated into lower stock prices and credit ratings. In terms of the magnitude of share price reductions, power utilities have fared badly. According to the MSCI (Morgan Stanley) European utilities share price index, the top 20 energy utilities have lost more than half a trillion dollars since their peak in September 2008."

Adding significant amounts of renewable power on grid have reduced wholesale power prices overall, and harmed the financial viability of utilities serving the grid.

3.5 Energy Efficiency

Multiple studies show that energy efficiency could be the most important method to reduce greenhouse gas emissions while maintaining energy service consumption. Under the IEA New Policies Scenario, total energy demand increases by one-third from 2013 to 2035, rather than 45% under current policies (see Figure 3.3). Efficiency, especially in end-use sectors, accounts for three-quarters of that reduction. The IEA

"Efficient World" scenario is even more aggressive, reducing global primary energy demand by 50% compared to the "new policies" case (IEA, 2013a).

Both the IEA and IIASA emphasize the importance of energy efficiency to reduce the required demand for new power generation. In IEA's "new policies" scenario an additional $3.4 trillion of cumulative investment through 2035 will be made in energy efficiency compared to their "current policies" scenario (IEA, 2013a). IIASA estimates that $15 trillion will be needed through 2050 (Johansson, Patwardhan, Nakicenovic, Gomez-Echeverri, eds. et al. 2012). These investments produce an impressive payoff with IEA estimating $6.1 trillion in energy savings, and IIASA estimating $57 trillion in avoided heating and cooling costs alone. Nonetheless, as shown in Figure 3.1, while energy efficiency investment significantly reduces the global growth in primary energy demand, significant global growth remains. The estimates of electric power demand growth cited above, account for investments in energy efficiency required by new energy policies.

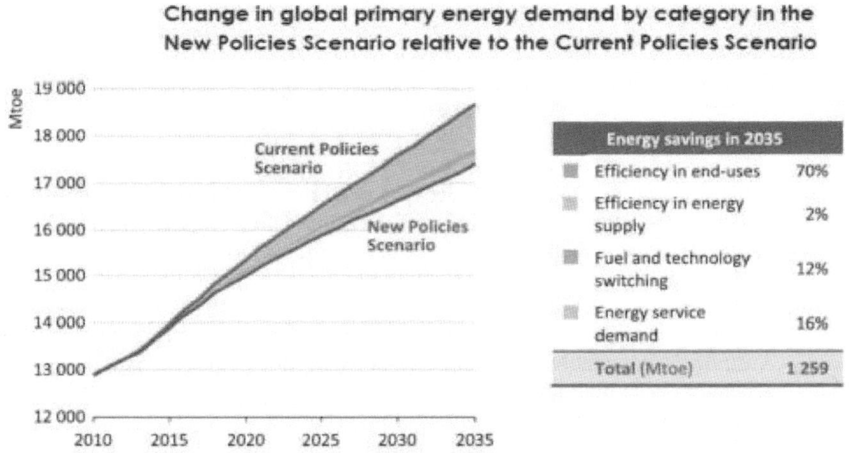

Fig. 3.3: Comparison of energy demand in IEA new policies scenario vs. current policies scenario.

Realizing this efficiency potential will require strong, sustained policies. The IIASA Global Energy Assessment suggests governments need to address split incentives, such as when energy users and building owners have different time horizons for recouping investments, and recommends strong building and appliance standards coupled with information and awareness programs. Policies like these would help nations reduce future energy demand, with benefits including reductions in GHG emissions and other pollutants and significant job creation.

3.6 Challenges Outside of the OECD

Outside the OECD, electrification started rather late, but is gaining momentum. The share of electricity in the Total Final Energy Consumption is 16% in 2011, which is expected to grow to 22% in 2035 according to the IEA New Policies Scenario. The annual growth rate of electricity demand between 2011 and 2035 is estimated at 3.3%, which is much higher than the growth rate of non-OECD Total Final Energy Consumption (2.0%).

Although energy consumption needs to grow at rates high enough to support economic development, rapid expansion of fossil fuel use will have deleterious effects on the environment. Therefore, in emerging countries, a major energy policy objective should be the "decoupling" of economic growth from energy consumption growth. Along with efforts to promote energy conservation and efficiency, clean energy supplies such as nuclear and renewables will be needed; gas is a cleaner-burning fossil fuel then coal, and will help reach GHG targets, but is not sufficient to decarbonize the electricity sector. However, there is uncertainty whether the electricity sector will attract enough investment to meet growing demand, especially in unstable countries. Business models as well as regulatory regimes tested and practiced in OECD countries will not necessarily be applicable to emerging economies due to the difference in not only developmental stage but also social, political and cultural background. Energy subsidies present a particular problem, as growing demand places burdens on national budgets and discourages energy efficiency. As several OECD energy companies have discovered, money will have to be found to fund new plants, and non-OECD countries in particular will be hard-pressed. Theft of power, corruption and other problems has caused several foreign power investments by independent power companies to fail.

3.7 Energy Poverty

However, the largest challenge, which may even dwarf the challenges discussed above, will be the "eradication of energy poverty," especially electricity poverty, in the least developed countries and regions. This is an indispensable part of the efforts under the first goal ("Eradicate extreme poverty and hunger") of the UN Millennium Development Goals adopted in 2000.

There are several causes of energy poverty, the most obvious of which is living in a community that has no, or very limited, access to electricity. Energy poverty can also exist in geographical regions that have adequate electricity supplies, especially urban areas in less-developed countries. In these circumstances, the inability to afford electricity and associated end-use equipment are among the many consequences of poverty that cause suffering to hundreds of millions of people including malnutrition and preventable disease. These are problems that have existed throughout human history but are incongruous with the opportunities provided by today's technologically advanced economies. As Pope Francis said recently in Rio de Janeiro, "No amount of peace-building will be able to last, nor will harmony and happiness be attained in a society that ignores, pushes to the margins or excludes a part of itself."

There are 1.4 billion people around the world that lack access to electricity, some 85% of them in rural areas (Figure 3.4). In some countries, this problem is widespread. For example, 74% of the population in Myanmar lacks access to electricity (OECD NEA,

2012). Because it is critical to the quality of human life, access to electricity should be facilitated above all else.

Without electricity, the opportunity for education of younger generations will be reduced severely, with serious negative implications on human capacity-building.

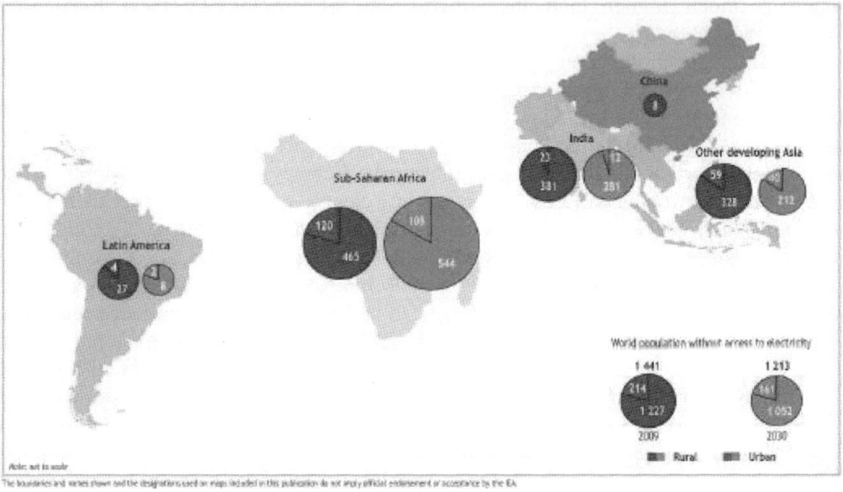

Fig. 3.4: Number of people without access to electricity in rural and urban areas in the IEA new policies scenario (unit: millions of people). (IEA, 2010).

In this regard, a leapfrogging as seen in telecommunication (bursting into mobile phone bypassing the stage of landline phone) should be widely materialized by making the best use of renewables. Micro-grids using locally available energy sources, such as solar and wind power, will be an important option rather than waiting for the investment in large-scale generation and centralized transmission systems. Low-cost, off-grid solar power, as seen in remote villages in India, is a good start. In a nutshell, every effort should be mobilized to overcome such chronic energy poverty as soon as possible, though the particular combination of solutions relevant to each country will vary.

While renewable energy technologies, micro-grids and interconnections to national power systems are able to provide electrical services to many communities that now do not have them, energy poverty will not be eliminated without outside intervention. Many governments are unable to finance these initiatives without outside aid. In addition, outside aid cannot solve problems in countries with corrupt governments or countries torn with ethnic or sectarian warfare, a condition too often present in communities that do not have access to electricity services.

It is beyond the scope of this paper to discuss the existing sources of funding for energy poverty projects, such as the World Bank, its Global Environmental Facility and other institutions except to note that much work needs to be done to finance energy poverty projects. We only note that renewable energy technologies have features that make them especially valuable for application in remote communities and to power micro-grids.

4. NUCLEAR POWER: STATUS REPORT AND FUTURE PROSPECTS

The scope of this section includes the current status and future prospects of commercial nuclear electric power, with emphasis on issues of safety, physical security, proliferation, and economics. (Issues related to radioactive waste are discussed elsewhere in this paper.)

Discussions of these issues are presented separately for the current fleet, for new reactor designs similar in size to the current fleet and for prospective new reactors of substantially smaller size. Also, this section discusses the issue of expansion of commercial nuclear power into new countries.

4.1 Reactor Safety

The fundamental issue that makes nuclear power reactor installations different from other industrial undertakings is, of course, the possibility that a major accident could release large amounts of radioactivity into the environment, endangering offsite populations and contaminating offsite land and property, possibly for a long time. As the world saw with the accident at Fukushima in Japan in 2011, this is a real possibility, even in the most advanced countries, and not merely a threat confined to countries with lesser technological prowess.

What do we know now about the likelihood of such accidents? The answer is that we know considerably more than was known two or three decades ago. We understand the origins of these accidents, we understand the variety of ways in which they might progress, and we can estimate the likelihood and the consequences of the many different accident scenarios. The advances in our analysis methods, and in the collection and use of the operating data needed to support them, have been remarkable world-wide in recent years.

We analyse "safety" by working out, in as realistic a way as we can, the likelihood per year that a large accident might occur. We now have methods that enable us to perform such an analysis, which is intrinsically probabilistic in character, and these analyses are now done routinely around the world. The community has settled on two different figures-of-merit, one being the annual frequency that a reactor will suffer a core-damage accident, and the other the annual frequency of a so-called "large" release of radioactivity, defined as sufficient to cause prompt radiation-induced fatalities offsite.

Based on what we know today, the annual frequency of core damage of one of the reactors in the worldwide operating fleet is estimated to be, on average, in the range of around a few times 10-5 per year. This is a typical value. It varies considerably from one to the next individual reactor even in the same country or even operated by the same company, and it is known only within a factor of three or so, or even less well, depending on the analyst and the design. However, based on what we know, the worldwide average is believed to be in the range just cited. The likelihood that any such accident will progress to a large release is small, perhaps in the range of a few percent to ten percent.

This means that the frequency of a large release is likely in the range at or above ~10-6 per year or so, as a worldwide average.

There is ample evidence, based on objective assessments of many different indicators that the general safety performance of today's operating fleet has substantially improved over the last 15-20 years. This is a worldwide trend. The indicators include rates of initiating events that might cause an accident if not mitigated, failure rates of vital safety equipment, rates and severity of operator and maintenance errors, major improvements in fire protection and changes in the designs to provide better safety system backup and reliability. Another significant improvement has been the worldwide sharing of operating experience and a diligent effort worldwide to learn from this experience. Although the improvements vary from plant to plant, they have occurred worldwide.

Whether today's performance is adequately safe will not be addressed here. This is both an individual judgment that people make differently and a societal judgment made differently by different countries. Thorough risk-based assessments, including assessing risks associated with alternative sources of energy, are to be encouraged.

There is a vital caveat that must be discussed, of course. Like any technology, nuclear-power technology can be mismanaged, and a major accident could occur at any time. (As an example, consider the fact that for a passenger flying in a commercial aircraft, the safety performance is 30 to 100 times better on average than it was 40 years ago, and incontrovertibly so. But a plane crash could occur on any day. Yet if it did, even though it would a tragedy, it would not contravene the strong evidence that commercial flight is indeed much safer on average than it was decades ago.)

So too with nuclear power reactors. If the worldwide performance is in the range of a few times 10-5 per year, and if there are close to 500 reactors today, one such core-damaging event might occur on average every several decades. Yet one could occur tomorrow. Crucially, what worries the reactor safety community is that if a major error of some kind were made, as happened in Japan with the decision to place the Fukushima reactors at a tsunami-vulnerable site without adequate protection, the likelihood of an accident would be higher. What to do?

It is important to note that we do know how to achieve a safety level like the above. A few vital attributes of the enterprise must all be present: (a) We must incorporate the appropriate safety features in all existing reactors, including improvements in equipment reliability and in the training of operators; (b) we must apply all lessons learned from operating experience worldwide; (c) we must emphasize safety culture everywhere; (d) we must remain attentive to aging issues, including equipment aging, staff aging, and institutional fossilization; (e) we must maintain the morale of staff, taking care to infuse the enterprise with a generational "mix" that includes both experience and youth; and (f) crucially, we must maintain and reinforce a vigorous and independent regulatory agency in every country, enhanced to the maximum extent possible by a strong international framework for establishing high safety standards, provision for technical assistance and peer reviews of design and operational safety and sharing and disseminating lessons learned and best practices experience.

If the above attributes are all present and attention is paid to maintaining each of them, the existing worldwide operating reactor fleet can continue to achieve the safety performance noted above. This is true not only for the light-water power reactors (LWRs) that comprise the bulk of today's operating fleet, but also for the others operating today: the heavy-water reactors, the gas reactors, and the Russian RBMK water-graphite reactors—but only if "operated well".

Why can't we do better with the existing large LWR reactors? Based on our understanding of the design of the existing LWR reactors and of how they are operated, certain limitations seem to make it difficult to achieve, say, an order of magnitude better fleet average than the above. These limitations involve the way the designs call on safety equipment when in trouble, the reliability of the equipment, the way operators are relied on, and the interplay among these factors.

However, for new LWR designs, especially those with more passive safety features and those whose design has taken maintenance and reliability into account from the start, the community of reactor safety experts has reason to believe that much better safety performance will ensue—perhaps an order of magnitude better. A few reactors with these advanced designs are now under construction around the world.

Why are these advanced LWRs better? There are many reasons besides the two crucial ones mentioned above (passive features, and design for reliability and maintenance). The new designs have greater embedded "engineering margin" in many different places; they have fewer and better components whose failure matters; they are easier to operate and more tolerant of operator error; and they are simpler to understand, and hence to analyse if an off-normal event were to occur. Crucially, they have included a number of advanced technologies that simply did not exist when the earlier nuclear plants were designed, such as advanced computer-based information systems, digital instrumentation and control systems, advanced operator-training simulators, and advanced materials in many different aspects of the design.

What about the safety of other advanced designs? There are two quite different types: (a) First are the proposed new smaller LWRs, the so-called "small modular" LWRs. There is reason to believe that their safety performance should be every bit as good as that of the new large LWRs, and in some ways perhaps better, because they have fewer components and simpler operations in many areas. Because most of the technologies used in these smaller LWRs are very similar to those in today's large LWRs, these smaller LWRs probably have a "head start" in reaching commercial viability, compared to other advanced reactor designs. (b) Second are various advanced reactors other than the smaller LWRs. Some are large and some are small. They include advanced gas reactors, homogeneous reactors, liquid-metal reactors, and a few others. Among these are a few advanced designs using fast-neutron cores; these offer the capability to use the energy in the fertile heavy metal in the fuel as well as the fissile content, along with the capability to destroy the actinides to reduce the residual high-level waste burden substantially. Effectively, such reactors would use a greater percentage of the original energy contained in the fuel, and would alleviate some of the longest-lived contributors to the nuclear waste burden.

All of the smaller new design concepts, whether they are thermal reactors or fast reactors, possess a set of safety-improvement features that are intrinsic to their smaller size. Specifically, the smaller size makes thermal transients and many other upset conditions easier to design against and also to control if an upset condition were to occur. Some smaller designs also require less standby safety equipment, which is an advantage for simplicity, for lower capital cost, and for easier and thus less costly maintenance. But even the largest of these advanced designs has several very attractive safety features, often involving passive safety design concepts and easier operation. The issue, of course, is that until any of these is actually designed in full and then built and operated, we cannot really know. (Admiral Hyman Rickover famously once remarked along the lines that a "paper reactor" is always safer and less expensive than a real one.)

What about the safety of power reactors potentially to be deployed in "newcomer" countries, meaning countries without a current commercial reactor program? The key issue is likely one of culture. First is the political culture, which involves a long-term national commitment and public acceptance if the launching of a nuclear–power program is to be successful. And crucially, a country needs a culture in which an independent regulatory agency with authority and independence from politics can operate appropriately. Second is the social culture. While many of these "newcomer" countries have strong social cultures, some do not. A culture of corruption, or a culture that does not hold safety as paramount, or a culture without commitment to transparency in management practices and communication, a strong continuity of institutions and a tradition of follow-through, is less likely to foster the environment needed to achieve an adequate safety record. Finally, in any "newcomer" country, there is the need for technical and financial infrastructure. This is true across the board, including routine technical infrastructure such as maintenance and electronics-instrumentation-electricity-grid support, the availability of competent craft labour, and spare-parts availability. A country also needs a financial infrastructure to support a large industrial installation like a nuclear power plant, including an insurance infrastructure and a trustworthy court system to enforce contracts. Without all of these various "cultural" attributes, power reactors in a "newcomer" country are unlikely to achieve an adequate safety record.

Thorium-fuelled reactors: All of today's operating reactors, and almost all of the advanced designs that are under development around the world, use uranium as the fuel. However, reactor designs that use thorium as the fuel are fully feasible, and several different such designs have been advanced over the years. Much of the excitement over using thorium relates to thorium's relatively high earth abundance, including in countries like India which possess little uranium. Using thorium for a power reactor is fully feasible, although there is almost no operating experience with them beyond a very few experimental facilities. There is no reason to believe that a thorium reactor cannot achieve the safety, security and non-proliferation record of the best of the uranium-fuelled designs, if "done right," while recognizing that many of the technical issues with both safety and non-proliferation differ in important ways. In India, a program to develop a thorium-fuelled power reactor has been in place for some years. However, only the future will tell us whether this development program will succeed, both technically and in terms of cost.

International institutions to promote reactor safety and security: A fundamental principle worldwide is that the regulation of the safety of power reactors is a national responsibility. However, because there is a worldwide recognition of a fact that is often phrased as "a nuclear reactor accident anywhere affects nuclear power programs everywhere," and because many power reactors are located near the border with another country, a number of international and multi-national institutions have arisen that work to promote nuclear reactor safety and security worldwide. Foremost among them, of course, is the International Atomic Energy Agency. One major IAEA initiative is the Nuclear

Safety Convention, to which all countries with power reactors are signatories. The IAEA has for decades also developed principles for safety and security, international safety and security standards, guidance documents, and programs of assistance that have had important impacts. The IAEA also collects, analyses, and disseminates information about reactor safety and security issues, organizes conferences and other meetings to promote international cooperation among experts, maintains an incident response and information centre that can respond if needed when an accident or incident occurs, and provides several services to countries that request them to provide on-the-site advice and review of safety programs, safety practices, and regulatory-agency performance. All of the above provide vital assistance worldwide, although taking advantage of these services is entirely voluntary on the part of any individual country.

Another organization that carries out a set of similar programs in many of the same areas is the OECD's Nuclear Energy Agency, although its programs are mostly restricted to the OECD's own member countries. The NEA's efforts have had great value. They have positively improved safety and security worldwide even beyond the borders of the NEA's own member countries. A strengthened international safety regime would bring about tangible enhancement to safety of nuclear power worldwide and increase acceptance for an expanded role as part of future energy mix. One initiative that might provide greater worldwide benefit would be if participation could be made mandatory for at least some of the IAEA's programs, such as those that do site peer-review visits to reactors worldwide to give advice and review of a site-specific program. Given the IAEA's current structure, it seems difficult to see how to bring about such a change to mandatory participation, but there is little doubt as to the benefits, if something along these lines could be put in place.

4.2 Reactor Physical Security

The issue here is the potential vulnerability of a reactor installation to an outside attack, or to a malevolent act by an "insider" team, or some combination. The current approach to security emphasizes "guns, guards, and gates" along with intelligence and counter-intelligence systems. The obvious concerns are the adequacy of the external defences, the adequacy of the on-site security force and system, and the effectiveness of systems that screen employees to detect those with relevant skills and malevolent motivations. Recently, concerns with cyber-security threats have also come into prominence.

On balance, the general feeling on the part of the community of nuclear power experts and many others worldwide is that the current schemes used to achieve the desired security regime seem adequate, everywhere around the world. (Even in countries with other problems, the military tends to be strong and well organized.) There are always ongoing concerns, which revolve around the screening of employees and guards, especially in countries where corruption is an issue. Also of concern is the potential decline in vigilance due to complacency and the potential for political interference.

One issue that continually arises is how to know whether today's physical security scheme is "adequate." No objective definition is accepted by all. Several methods have been developed and used to assess the extent to which a given security scheme can protect against a given postulated threat. International technical forums and projects have worked on this issue for many years, and some convergence has occurred on the main issues and analysis approaches. However, disagreements persist on how to do this evaluation, and part of the problem is that the values brought to the evaluation by different countries inevitably differ, and probably always will.

4.3 Proliferation Issues

The issue here is the potential misuse of nuclear power reactor or fuel-cycle technology to make weapons-usable material directly, or to extract it from a power-reactor or a fuel cycle facility by either a government or a non-governmental group for malevolent purposes.

Today, there is an international protocol to achieve the aims of deterring proliferation of nuclear weapons and weapons technology, under the Non-Proliferation Treaty, while fostering peaceful uses of nuclear energy including nuclear power. The treaty is enforced through a safeguards system administered by the UN's International Atomic Energy Agency. The central aspects of this NPT protocol include inspections both planned (through safeguards) and unplanned (through the Additional Protocols that are not agreed to yet by all NPT members), monitoring both within and from outside a given installation or country, the availability of extensive training, the use of exercises to test the functioning of the system, and periodic reviews. Other institutions, such as the Nuclear Suppliers Group, also contribute significantly. A vital aspect of the overall international regime is a set of restrictions on the export of so-called "dual use" technologies which have both civilian uses and potential uses that could enhance the proliferation of nuclear-weapons capability.

For many years, major emphasis was placed on the issues of proliferation arising if power-reactor fuel is reprocessed so that plutonium is extracted for re-use, making it "available" in "weapons-usable" form. The concern was mostly directed at government-sponsored reprocessing programs that are openly acknowledged. That emphasis has in the last decade or so been supplemented by emphasis on deterring the use by a government of uranium-enrichment technology, which can produce weapons-usable highly-enriched U-235 (HEU) under the guise of enriching to lower enrichment levels to make LWR fuel. Both technologies have dual use are permitted in principle under NPT if genuinely and demonstratively justified for a large nuclear power program and deployed transparently.

NPT has been relatively successful so far, but not fully. First, there are several examples in which a few members of NPT have developed dual use technologies under the pretext of a civilian national nuclear power program only to prove, or be suspected, to be otherwise (e.g., North Korea walking away on its NPT obligations and testing several nuclear weapons, and Iran widely suspected of aiming to develop at least a capability for nuclear weapons, especially because it initially planned and developed its enrichment program in secret). The second failure is evident in that several countries chose not to be party to NPT and went on to develop nuclear weapons, declared (India and Pakistan as well as South Africa), and undeclared (in the case of Israel). Another is the failure of the recognized nuclear weapons states under NPT to begin a credible program to systematically reduce, eliminate and renounce their arsenal of nuclear weapons.

Today, this concern about proliferation arising from openly acknowledged government-sponsored programs (whether a reprocessing program or an enrichment program) has been supplemented with a major parallel concern about possible clandestine government-sponsored programs, or about a program embarked upon by a sub-national group.

While theft of separated plutonium has always been an issue, theft of HEU has typically been of lesser concern because of its lesser availability -- although research and test reactors around the world that use HEU have always been a concern and remain so. Today, however, the theft of both separated plutonium and HEU are emphasized together as threats to peace.

The physical security aspects of this issue seem to be adequate—that is, adequate against theft by an unauthorized individual or non-governmental group. Less clear is the adequacy of the world's ability to intervene if a sitting government proceeds, as government policy, down a path toward proliferation (meaning a path leading to the development of an actual nuclear weapon). This includes using a commercial nuclear power program as a "cover" for an illicit weapons program, or using an ostensible power-reactor facility (such as a low-enriched-uranium enrichment plant) for those ends. This set of issues has political ramifications that are beyond the scope of this essay.

4.4 Recommendations

From the above discussion, a few recommendations seem to emerge naturally. Some are for technical work, and some concentrate on institutions and policies.

Recommendations (technical)

A. Small (modular) reactors show great promise to enable the expansion of nuclear power in many electricity markets that are otherwise inaccessible to the larger LWRs. These smaller reactors are likely to be especially attractive in many developing countries, although not exclusively in them. To make these reactors a genuine option, an emphasis on technical research and engineering development, and also on developing efficient means for regulatory agencies to review and approve the new technologies, is imperative. It should receive priority worldwide, as should identifying the most promising SMR designs and moving them toward prototype demonstrations.

B. Enhancing the training of technical personnel and developing technical infrastructure, especially in countries that today have little or no nuclear-power deployment and perhaps even little strong technical base at all, should be another worldwide priority.

Recommendations (policies and institutions)

C. A major barrier to widespread deployment of nuclear power is the set of "culture issues" that include corruption culture, safety culture, and the culture of institutional continuity and integrity. Policies that emphasize the positive attributes of these cultural issues, or discourage their opposite, are an urgent matter.

D. The need for an independent regulatory agency with authority and independence from politics cannot be overemphasized. No country without this should deploy nuclear-power technology. Policies, including policies in the developed world to provide assistance and mentoring, are a vital component of a successful worldwide nuclear power endeavour.

E. There is a need to strengthen the international safety regime, including, where feasible, making some current advisory services of IAEA mandatory.

F. In cases where policies are established to reduce the carbon footprint of the electricity grid, nuclear power should not receive discrimination in favour of renewables.

5. ECONOMICS OF NUCLEAR AND RENEWABLES

New renewables (NR: wind; solar; modern biomass and biofuels; tidal, wave and ocean energy) are widely claimed to be clean, indigenous and sustainable sources of energy. Therefore, they are favoured by many governments and the public as a whole. However, their present contribution to global energy consumption is still limited (about 3% in 2013 and only about 10% even for heating) since their economics are not yet favourable. In most instances, they need to be supported by state subsidies and regulations. They suffer from high investment costs and, as a result of the intermittent and diffused nature of wind, solar, and tidal, relatively low-utilisation factors.

This section focuses on the economic aspects of nuclear power and new renewables for the generation of electricity. Incorporating new renewables into power grids poses challenges due to dispatching problems and potential needs for expensive transmission extension and/or grid reinforcement; issues with intermittency and prospects for energy storage as a solution are discussed in other sections.

Wider introduction of smart grids and the likely demise of nuclear in some OECD countries (in the short and medium term, at least) will enhance the future prospects for new renewables. However, their immediate future expansion will depend on continued subsidies, which are becoming difficult to sustain in present economic circumstances. Development of large energy storage facilities and carbon pricing could significantly enhance future NRs prospects as indicated earlier. Correspondingly, NRs, in spite of their popularity with some governments and sections of the public, are likely to face challenges which will slow their present rapid progress.

Nuclear power plants face different challenges and prospects, being now shied away from in many industrialised countries and having mixed prospects in developing economies. In many instances, they suffer from high initial costs, long lead times and often excessive construction delays. Nuclear power also faces challenging risks—investment as well as regulatory—although the wider diffusion of small modular reactors (SMRs), when developed and deployed beyond 2025, may reduce these risks. The challenges are compounded by safety and proliferation considerations and a lack of technical skills in many countries as discussed in an earlier section. In contrast to renewables, their share of global energy consumption is declining, though nuclear electricity generation will continue to grow in absolute terms. Nuclear power's share of global electricity generation has declined from as high as 18% in 1990 to only 12% in 2013,—a percentage that is not likely to improve significantly in the near future.

5.1 The Economic and Financial Evaluation of Renewables and Nuclear

The levelised cost of electricity (LCOE) (DECC, 2013), is the traditional method for assessing and listing dispatchable generating facilities according to their annual costs, and it applies to assessment of the cost of nuclear. However, it does not directly apply to non-dispatchable technologies, like renewable, due to their intermittency. Developing the necessary algorithms for such a purpose, particularly in the case of wind energy, is not easy because of the difficulty in forecasting wind's intensity and duration. For example, account needs to be taken of capacity factors achieved by wind energy developments. In the UK, it has been confidently claimed that onshore wind energy developments achieve a capacity (or load) factor between 20% and 50%, with an average attained of 30%. The official statistics demonstrate that for onshore wind energy developments in England, the rolling average capacity factor achieved to early 2014 has been barely 24%, and in 2010, nearly 60% of these developments failed even to achieve 20%—yet capital costs will be similar wherever wind energy developments occur onshore regardless of mean wind speeds impacting performance. Solar and tidal forms of energy suffer from the same problem to some extent.

5.1.1 Assessing the returns on investment in renewables

To assess the viability and economics of NRs (particularly solar and wind), it is necessary to compute the future stream of the electrical system cost with NRs and compare it with the system cost without the incorporation of NRs.

This also has to be weighted with the carbon saving and other intangibles of the NRs. The electric system cost does not only apply to generation facilities, but also to the transmission system extension and grid reinforcement as well, which can be substantial in the case of wind technologies. In most cases, the electric system cost per kWh delivered with the presence of NRs is going to be higher than in the absence of NRs. The extra cost indicates the extent of the subsidy (feed-in tariff) which needs to be provided, or increase in tariff to consumers to compensate, for these extra costs (NRs' penalty). Calculating the present worth of system costs without NRs is straightforward. With the incorporation of NRs, there is a need for a more elaborate approach. Below we develop a simple, but effective, means to compute the financial and economic effect of incorporating NRs. We have, however, to distinguish between solar energy, which is relatively predictable in timing, duration and extent, and wind with its short term (if any) predictability in timing, duration and extent. In addition, we need to take account, in the case of solar, for the levels of direct and indirect solar variation which will vary between locations—generally higher in lower latitudes and lower in higher latitudes. Also, we need to differentiate if the investment is done by a regulated utility which can pass the extra cost of renewable electricity to consumers or it is executed by independent investors selling in the spot market. These need to comprehend fully the financial implications before committing themselves to a risky renewables investment.

The economic evaluation of power generating technologies should aim at evaluating their market value, that is, the revenue they generate to the provider. This particularly applies to NRs. The market value of NRs is lower than their LCOE due to integration costs. A new concept of 'System LCOE' has been developed. It is composed of generation cost plus integration costs. This new term system LCOE is the standard LCOE plus the indirect costs that occur at system level [Hirth (2013), Vithayasrichareon and Mac Gill (2013)]. The integration cost of NRs is additional system costs that are not direct generation cost of NRs. Integration costs of NRs have three cost drivers: variability, uncertainty, and location. All generation technologies have integration costs. This is true, but it is more specific and pronounced in NRs due to the variability and unpredictability in dispatching.

The three components of the integration cost (variability, uncertainty and location) mentioned above are shown in Figure 5.1 and are defined as follows:

a. Profile (variability) costs occur because wind and solar PV are variable. In particular, at higher shares this leads to increasingly inappropriate load-matching properties. Backup capacities are needed due to variability of NRs. The full-load hours of capital-intensive dispatchable power plants decrease while these plants need to ramp up and down more often.
b. Balancing (uncertainty) costs occur because renewable supply is uncertain. Day-ahead forecast errors of wind or solar PV generation cause unplanned intraday adjustments of dispatchable power plants and require operating reserves that respond within minutes to seconds.
c. Grid-related (location) costs occur not only because NRs are mostly located far from load centres and large investments in transmission might be necessary, but also because of grid constraints and congestion management.

System LCOE is defined by adding the three components of integration costs to standard LCOE that reflect generation costs. Such integration costs vary from one system to another depending on the extent of penetration of NRs, location and the composition of the dispatchable plant in the generation system. They need to be computed separately for any national grid in order to compute the true market value of NRs. Therefore, the system costs of NRs can be significantly higher than their LCOE as demonstrated in Figure 5.1, though they do come down in the long-term, as the system adapts and integration options increase.

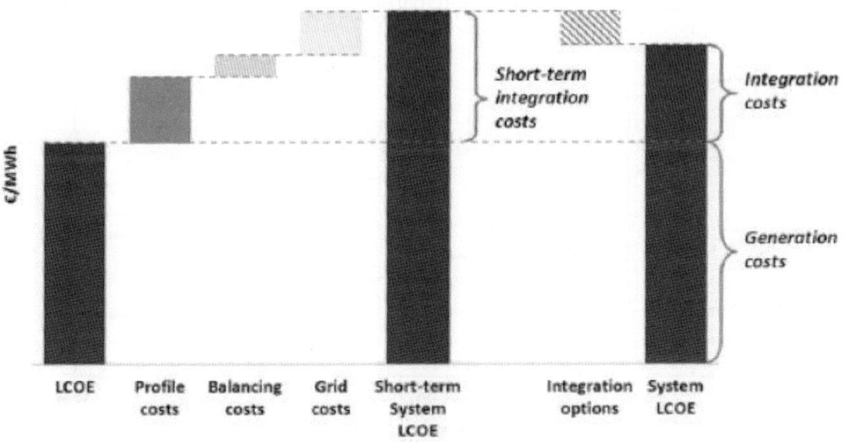

Fig. 5.1: System costs of new renewable. Source: Ueckerdt, Hirth, Luderer, Edenhofer (2013).

There are now many investors in NRs, ranging from regulated utilities to private investors in the spot market, and from large facilities and investments to individual households investing in small roof top PV installations. New and more advanced ways of investing in NRs are being developed: individual investors, leasing, net metering, etc. Therefore, it is not possible to deal with each case—but general and simple guidelines for assessment should be developed as indicated below.

The recent pace of change in the electric power sector has been little short of incredible, including significant increases in renewable generation in numerous countries prompted by rapidly falling prices just as electricity demand growth seems to have come to a virtual stand-still in many places, resulting in depressed wholesale prices and challenges of integrating growing amounts of intermittent renewable generation while maintaining grid's reliability.

Recently (2014-2015) there have been significant developments in the economics of renewable energy technologies (see Figure below). Both the cost of installation and production from both PV and wind installations has been significant. Recent contracts for PV, in favorable sites, were less than 6 cents/kWh and still falling. The prices quoted by independent power producers (IPPs) in windy sites both in north and South America is as low as 4.25 cents/kWh. In both cases system costs can be significantly higher as already expressed above. Still these are significant developments which are likely to impact the economics of renewables and their contribution to power production. Their impact on system operation is also to be felt. Renewables must dispatch electricity; this can only be at the expense of other base load generation, mainly nuclear and large base load coal firing plants. Correspondingly the need for solid interconnections and meshing of the neighbouring networks are becoming increasingly crucial to absorb and dispense with sudden variation in renewables generation, particularly the difficult to predict wind.[6]

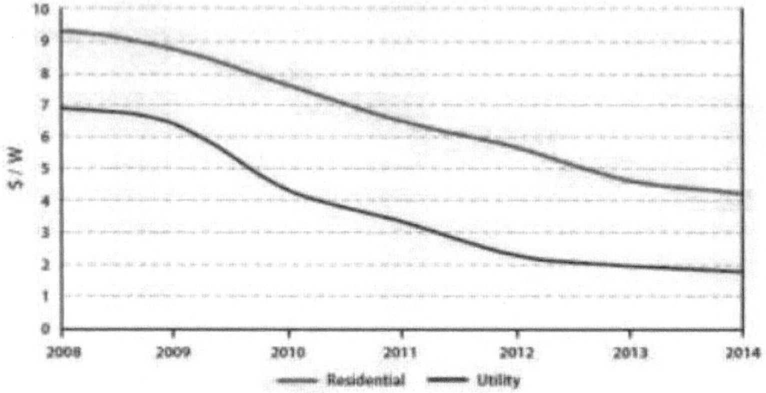

Fig. 5.2: Falling costs: MIT study estimates of the average U.S. prices for residential and utility-scale PV systems. Source: The future of solar energy: An interdisciplinary MIT Study, May 2015.

5.1.2 The Economics of Nuclear Power

Nuclear's global share as primary energy has been on a slow decline and is not expected to recover in the short to medium term despite a massive construction boom in China and a handful of other centrally-planned economies. This is in contrast to NRs. Hydro's global share will remain flat while NRs show greater promise, albeit starting from a small base. (See Figure 5.3)

With increasing privatisation and liberalisation of the power generation sector, private investors are generally not interested in investing in nuclear without one or another type of government assistance to reduce the financial risk, due to its capital intensive nature, escalating construction costs, long lead times and possible high future safety risks.

It is mostly in centrally-planned power systems that nuclear is still finding some favour. Among the many reasons that are keeping private investors from investing in nuclear power is that it is difficult to know how much a given project is going to cost by the time the plant goes operational, or how long it will take to complete the project. This alone explains why so few are being built in the West. In centrally-planned economies, the cost of capital can be disguised, thus making it difficult to estimate costs across different economic systems.

---

[6] The discussion in this and the preceding paragraph applies to countries that have an electricity industry and distribution system serving the bulk of its population.

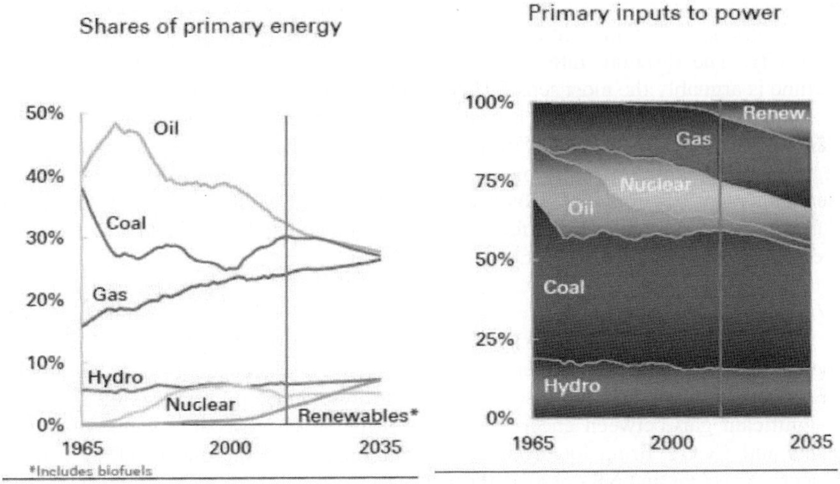

Fig. 5.3: One outlook for renewable and nuclear energy (BP). Source: (BP, 2014).

There are currently only a handful of new reactors under construction in the U.S. The UK is pushing hard to get two new ones built—with some difficulty—and only two others are currently under construction in France and Finland, both behind schedule and over budget. Over the next decade, more reactors may be retired in the OECD than new ones built.

The nuclear industry has not succeeded in achieving the cost reduction through commercialization that has often characterized other new technologies. Several factors help explain this including more stringent regulations and standards, insufficient standardization and modularisation, uncertain regulatory environments and difficulties faced during construction. This is coupled with safety concerns, particularly those resulting from the three major accidents which nuclear plants have suffered since their inception: Three Mile Island (United States), Chernobyl (the USSR, now the Ukraine) and the more recent example of Fukushima (Japan).

Analysis of the economics of nuclear power must take into account costs and accidents that foreshadow the risks of future uncertainties. To date all operating nuclear power plants were developed by state-owned or regulated utility monopolies where many of the risks associated with construction costs, operating performance, fuel price, and other factors were borne by consumers rather than suppliers. Many countries have now liberalized the electricity market where these risks, and the risk of cheaper competitors emerging before capital costs are recovered, are borne by plant suppliers and operators rather than consumers, which leads to a significantly different evaluation of the economics of new nuclear power plants (MIT, 2003).

Because of the large capital costs for nuclear power and the relatively long construction period before revenue is returned, servicing the capital costs of a nuclear power plant is the most important factor in determining the economic competitiveness of nuclear energy. The investment can contribute about 70% to 80% of the costs of electricity. The discount rate chosen to cost a nuclear power plant's capital over its lifetime is arguably the most sensitive parameter to overall costs, as detailed below.

The industry consensus is that a 5% discount rate is appropriate for fossil fuel plants operating in a regulated utility environment where revenues are guaranteed by captive markets, and 10% discount rate is appropriate for a competitive deregulated or merchant plant environment. However, an independent MIT (2003) study used a more elaborate finance model distinguishing equity and debt capital, and had a higher (11.5%) average discount rate for nuclear.

5.1.3 LCOE Comparisons

Both the U.S. DOE (EIA) and the UK publish estimates for cost of generation of nuclear and NRs. The competiveness with fossil fuel plants depends on carbon pricing. If this is not included, as is the case in most centrally-planned power systems, there will be a significant gap between cheap generation utilizing cheap coal in modern coal firing plants and CCGT firing locally available natural gas on the one hand, and the more expensive nuclear and NRs on the other.

The following figures exhibit projected U.S. LCOE costs for new generation in both year 2020 and 2035 (EIA, 2013a). Note that countries like China and the UAE experience lower capital costs for new nuclear, and have a lower figure.

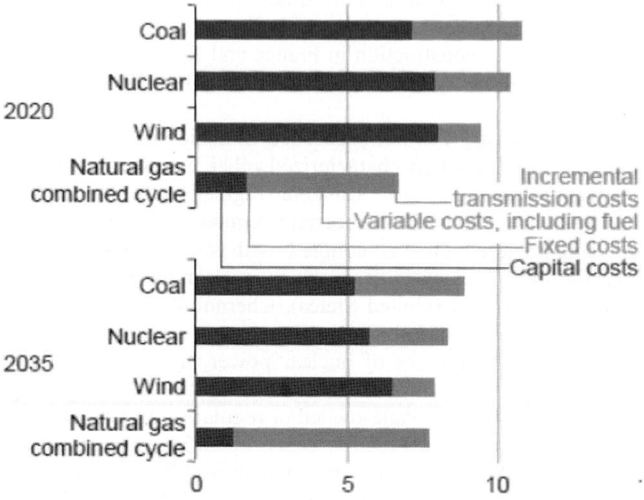

Fig. 5.4: Levelised costs of electricity in the USA for the years 2020 and 2035 (U.S. cents per kWh).[7]

[7] Coal plant costs include regulatory and environmental uncertainties (i.e., carbon pricing).

Table 5.5: Levelised cost of electricity generation estimates, UK Department of Energy and Climate Change (2012 £ per MWh).

	CCGT	OCGT	Nuclear-FOAK[8]	Onshore >5 MW (UK)	Biomass Conversion	Offshore R2	Offshore R3	Large scale solar PV
Pre-development Costs	0	5	6	7	1	4	6	0
Capital Costs	9	54	64	70	9	77	78	134
Fixed O&M	4	23	11	18	10	31	36	24
Fuel Costs	49	73	5	0	86	0	0	0
CO_2 Capture and Storage Costs	0	0	0	0	0	0	0	0
Decommissioning and Waste Fund	0	0	2	0	0	0	0	0
Total Levelised Costs (without carbon)	62	151	90	101	108	113	120	158
Carbon Cost	18	26	0	0	0	0	0	0
Total Levelised Costs (with carbon)	80	181	90	101	108	113	120	158

Source: DECC (2013)

5.1.4 Cost of Nuclear Generation

The life-cycle costs of a project and its feasibility, for a given output, depend on three factors: (i) the investment cost, (ii) the operational costs and (iii) the discount rate utilised. Many planners think that the discount rate is the most important of these three factors. It greatly affects the whole economics of the project and the decision making, particularly in capital-intensive projects like those of the nuclear industry. In spite of its crucial importance in project evaluation, it is surprising how little effort project evaluators exert to research the proper discount rate needed for their project evaluation. Simultaneously, all the efforts in estimating investment and operational costs are rendered untrustworthy by an incorrect choice of discount rates.

The aim of a nuclear power program is to generate energy, and more explicitly electricity, at a cost that fits the economic needs of the country at issue. In practice, because the cost of reactor fuel is typically a modest component of electricity cost (seldom as large as 20%, typically 10% or so), and operating costs are typically in a similar range, the capital cost of building a power reactor is the major economic issue standing in the way of a major worldwide expansion of nuclear power.

This has led to an emphasis on capital cost as the major design figure-of-merit of the new large LWR reactors. A few years ago, it appeared that in the developed world the capital cost of these new LWRs was in a competitive range with other new base load capacity. However, recent sharp declines in natural-gas prices (especially in North America but in some other locations as well) have made large-LWR capital costs less competitive. This is exacerbated to the extent that the subsidies or regulations aimed at

[8] First of a Kind.

reducing greenhouse gas emissions in the power sector are often not applied to nuclear energy (e.g., feed-in tariffs, clean energy standards, etc.).

The cost of constructing a new nuclear plant varies from one country to another, but appears to have increased over the last several years. Cost factors include location, availability of cooling water, seismic activity, labour and material costs and the cost of capital. Estimates of the cost of future nuclear power plants vary. According to EI Energy report, the average overnight capital cost (excluding interest during construction) of new nuclear plants in U.S. and Europe is around 5300$/KW in 2014 (in 2014 $) and expected to stay level and then decline slightly to around 5000 $/Kw by 2030 and 4500 by 2050 (EINE 2014). On the other hand, there are a variety of cost studies, some of which project higher costs due to the inclusion of interest costs during construction or other factors, especially those associated with unanticipated delays. Moody's Global (2009), in 2008, estimated the cost of nuclear plant in U.S. to be around $7,000 per KW. Some existing projects tend to support the higher cost estimates. The cost of the planned Hinkley Point 1,600 MW reactor, albeit a First of a Kind, is estimated to be $13.7billion (i.e., $8,300/KW), according to calculation by UK DECC. The strike price for this plant is expected to be $158.7/MWh, i.e., 15.4 cents/kWh. While this is double the UK wholesale electricity cost in 2013, it is in line with the subsidised cost of onshore wind energy developments in the UK, and two-thirds of the cost of OCGT. The UK government, in taking such a decision, is prompted by the need to comply with EU emissions' regulations and the phasing out of existing nuclear plants. It is likely that future cost of similar design reactors will be significantly lower. The cost of nuclear power plants in some developing countries, particularly with centrally-planned power systems, government financing and less stringent regulations, is likely to be much less: below $5,000 per KW (Schlissel and Biewald, 2008).

One reason there is a controversy about estimated nuclear plant costs is that the initial estimates for Generation III+ power plants were often based on vendor estimates rather than actual experience (Kessides, 2010). These estimates might underestimate future nuclear power plant costs compared to estimates for other power technologies that are based on numerous actual projects and contracts. Generation III plant designs emerged in the 1990s. The cost estimates for these designs were typically in the neighbourhood of $1,000/kWe. As these designs have evolved, current U.S. and European cost estimates have grown to be several times higher. Plant cost estimates in the United States currently range from $2,500/kWe to $10,000/kWe. If the outlier cost estimates are deleted, current U.S. cost estimates are reported from some sources to be from $4,200/kWe to $6,600/kWe (Thomas, 2010). Other sources, such as the U.S. Energy Information Administration report overnight costs within this range ($5,500/kWe). It should be noted that overnight costs are not a good guide to the true cost of nuclear power plants as they take a long time to build, usually much longer than initially assumed, and financing costs can significantly add to the actual capital cost of a nuclear power plant.

Unfortunately, there is no assurance that cost escalation will not continue. An important factor behind these high estimated (and, in many cases, actual) costs to build nuclear reactors is the delay that these projects often face during licensing and construction that increases the capital burden, often at high interest rates. This is also a reason why the economics of nuclear power may be more favourable in countries such as Russia, China, UAE and South Korea where projects tend to stay on schedule.

5.1.5 Technology Learning

Predictions of future energy technology costs usually employ the concept of technology learning. The general principal is: for each doubling of capacity of a new energy technology, cost goes down by a fixed percentage (IEA, 2000). Since the nuclear industry is well established, additions to capacity would fall well short of any opportunity for further technology learning unless the Generation III and III+ designs were judged to be sufficiently different than Generation II reactors to be new technologies. Most estimates of future nuclear technology costs do not view them as such and do not anticipate significant cost reductions in Generation III reactors.

An analysis of the cost trajectory of the French nuclear reactor fleet indicates that construction costs, in constant dollars, have increased over time (Grubler, 2010). While this is consistent with U.S. experience, it may be surprising since the French reactor fleet is generally regarded as the most successful and well-managed nuclear power deployment, in part, by avoiding the U.S. experience of building many different designs each requiring individual regulatory assessments and separate engineering design work.

Going forward, there may be reasons for nuclear reactor costs to systematically increase. About 60% of the cost of a nuclear reactor stem from on-site engineering as opposed to the major equipment items (reactor vessel, steam generators and turbines). Reactor heat-sink cooling is also an important cost component that has increased. In particular, environmental regulations to protect fish and other species have made it more likely that cooling towers would be required when reactors are located on rivers.

With high Asian economic growth that is not expected to lessen, the demand-supply balance for construction materials is likely to produce higher prices. Likewise engineering and labour costs are likely to increase with high economic growth. Compared to the opportunities to reduce the costs of manufactured products, large nuclear power plants appear to face the likelihood of higher costs, not lower costs.

5.1.6 Capital Cost Escalation and Other Risks in Liberalized Power Markets

Risks of cost over-runs are less likely to deter nuclear power plant investments in uncompetitive power markets or state-owned enterprises. A costly power plant in these markets will not necessarily have adverse consequences on investors as the higher cost can be averaged into the existing fleet of power plants. A private investor in an uncompetitive State-regulated power market would still receive the normal rate of return. A state-owned company likewise passes on the incremental costs to ratepayers or the government incurs higher electric subsidy costs.

A liberalized power market presents an entirely different picture to an investor. Delays and cost over-runs not only threaten to diminish their rate of return but put the investment itself at risk. This risk is compounded by five factors: 1) nuclear power has high capital costs per kilowatt of electrical generating capacity (kWe) compared to natural gas or coal power plants; 2) nuclear power has large economies of scale and require large investments for a unit of investment (1 gigawatt or more); 3) having high capital costs, nuclear power is more vulnerable to increases in material costs and construction delay; 4) construction delays are more likely due to uncertainties in the regulatory process, especially for new plant designs; and 5) because of potential public opposition to nuclear power, construction delays can also be caused by delays in issuing local permits, lawsuits and political pressure. Because of these risk factors, financing is often not feasible without government intervention to mitigate the financial risks. Equity financing is typically not an alternative to debt financing as it is difficult to ask shareholders to forgo dividends and accept risk that banks were unwilling to take.

Recognizing these barriers to nuclear power plant investments, the U.S. government established government subsidies and risk-sharing for nuclear power plants in the Energy Policy Act of 2005. These included production-tax subsidies for the earliest new nuclear plants, guarantees against regulatory delays and a loan-guarantee program to attract private financing. While no new nuclear plants yet built have taken advantage of the production tax subsidies, the U.S. Department of Energy (DOE) has issued two loan guarantees for two 1.1 gigawatt Westinghouse AP1000 nuclear reactors at the Vogtle generating plant.

Like all of the several loan-guarantee applications DOE received, these new plants (Vogtle units 3 and 4) are being built at an existing nuclear plant site. The project sponsors include the Georgia Power Company. The Vogtle plants conditionally received \$8.33 billion of DOE loan guarantees, or approximately \$3,800 /kWe capacity. The plants were initially estimated to have and overnight cost of \$11.9 billion, or \$5,400 per KWe capacity (Thomas, 2010). Prior to construction, the estimated cost, with financing, was \$14.1 billion and has since grown to \$16.5 billion, not including over \$1.1 billion in possible litigation costs. The project is now 42 months behind schedule. In sum, a project that was originally expected to have an overnight capital cost of \$5,400/KWe is now expected to have an actual cost of \$8,000/KWe with the possibility of further delays and cost overruns as construction continues. Had the project not received Federal loan guarantees, the accumulation of additional financing cost would have been considerably larger.

The Vogtle experience is instructive since, as discussed in Difiglio and Wanner (2012), Vogtle enjoys significant economic advantages relative to power plants built in liberalized power markets.[9] Typically, in liberalized markets, investors, not utility customers, accept the financial risk of cost over-runs and delays. Therefore, in addition to the Federal loan guarantees, Vogtle has lower-than-normal interest rates and a utility market regime that reduces financial risks to investors. Nonetheless, Standard and Poor's downgraded the Outlook on Southern Power Company and Georgia Power's credit ratings from "stable" to "negative (May, 2013) and other credit rating agencies have downgraded the credit rating of partners involved in the project including the Municipal Authority of Georgia bonds (Taxpayers for Common Sense, 2014). For U.S. nuclear power project that would not enjoy these advantages, the economic outlook is much more negative. For example, Moody's Investors Service has characterized a decision to build a nuclear power plant as a "bet-the-farm" risk. "From a credit perspective, companies that pursue new nuclear generation will take on higher business and operating risk profile, pressuring credit ratings over the intermediate- and long-term…Of the 48 issuers that we have evaluated during the last nuclear building cycle (1965-1995), two received ratings

---

[9] The Georgia Power Company is fully regulated by the Georgia Public Service Commission. Competition is limited to customers with manufacturing or commercial loads of at least 900 kW.

upgrades, six went unchanged and 40 had downgrades…the average downgrade fell four notches" (Moody's Global, 2009). It is notable that these credit downgrades affected utilities that were mostly operating in a state-regulated environment where regulators have effectively transferred the utilities' risks to ratepayers. Moody's concludes that utilities that pursue new nuclear power plants will experience credit downgrades and should: 1) rely on strategic partnerships; 2) increase reliance on equity financing; 3) moderate dividend policies to increase cash flow; and 4) adopt a "back-to-basics" focus on core electric utility operations. These recommendations are aimed at utilities that "can expect regulators to support the financial health of the utilities they regulate and will authorize recovery of investments and costs over a long time frame" (Moody's Global, 2009). In a liberalized power market where power companies and investors would not enjoy such protections, the risk of credit downgrade would be much higher.

5.1.7 Load-Following: An Operational Cost Issue

Because a power reactor's capital cost is such a large component of the total cost of electricity, it has seldom made economic sense to use such a reactor in a "load-following" mode—that is, in a mode in which its output would vary from hour to hour as the demand on the grid changes. However, deployment of large amounts of renewable electricity sources on a grid may make the desire for reactor load-following more important in the future. Besides the financial issues, there are also important technical considerations: frequent cycling of a reactor up and down in power can induce thermal and mechanical stresses on equipment for which some of them may not have been designed. It is definitely feasible to design a large power reactor to account for these technical issues, as the French have done, especially if the power excursions are limited to the 10% to 20% range for a modest fraction of the time (which would not affect overall economic viability very much.) However, for many of today's existing large LWRs the technical feasibility of significant load-following, especially if required on nearly a daily basis, has not been convincingly demonstrated yet. This option requires technical evaluation on a case-by-case basis. But going forward, new reactors can certainly be explicitly designed for this capability.

5.1.8 Countervailing Factors Supporting Nuclear Economics

The analysis suggests that nuclear power plants will be difficult to finance in liberalized power markets without government intervention to help mitigate the financial risks. If nuclear power plants built with government support or those built by state-owned power industries prove that nuclear power plants can be reliably built at a competitive cost, power plant economics will be a significant positive factor for the nuclear industry, especially taking into consideration national commitments to reduce greenhouse gas (GHG) emissions. If costs continue to increase over the estimates being used by the IEA and EIA, power-plant economics become a significant negative factor for the nuclear industry that may not be offset by pricing CO_2 emissions.

These findings do not apply to reactor projects in many countries outside of the OECD, especially in Asia and the Middle East. Nuclear power is estimated to enjoy a more favourable cost advantage than in the United States, Europe or Japan. In addition, private financing is less important in many Asian countries and delays due to licensing problems are less likely. Given the high electricity demand growth expected in Asia, more competitive costs, and government support, it is not surprising that over 60% of currently-planned nuclear reactors are in Asia.

5.1.9 Small Modular Reactors

The myriad financial challenges associated with nuclear projects that can exceed $10 billion have increased interest in the idea of deploying the proposed new smaller (modular) designs. If these were available, capital investments could be made in smaller-sized investment packages—capacity would be added as needed by building additional small reactors one-by-one on a given site. These features would be beneficial for both the developing and OECD economies.

The smaller (modular) reactor designs hold out a promise of lower costs due to factory fabrication, simpler operating features, and simpler, less capital-intensive design features. They also have the promise of being safer, and also requiring less safety equipment. The issue with these new smaller (modular) reactor designs is that, until they go into production, their potential cost advantages will not have been demonstrated. This is a classic "chicken-and-egg" dilemma, of course. How it will play out is still for the future.

5.2 Subsidies

Subsidies are an integral feature of the economics and costing of both NRs and nuclear. These subsidies are usually driven by two considerations: the first, "green" features with no carbon emissions during production; the other, that they provide a measure of local energy security being not dependent on imported fossil fuels or depletable resources.

5.2.1 Subsidies of renewables

Subsidies to NRs, mostly in OECD countries, take many forms both direct and regulatory. They are mainly in:

- Capital subsidies and loan guarantees to manufacturers.
- Feed-in tariffs (FiT) to consumers. These provide financial encouragement to convert small electricity consumers (mainly households) into producers.
- Production tax credits (PTC) for investors.
- Renewable portfolio standards (RPS) that ensures that a segment of electricity production goes to NRs (mandatory targets).
- Tradable energy certificates.
- Priority in dispatching, guarantee to access transmission lines and long term contracts.
- Carbon pricing and taxation.

Feed-in tariffs (FiT) are generally considered to provide the best inducement towards encouraging investment in NRs particularly to individual domestic suppliers. Feed-in tariffs are a generic class of polices that guarantee a set payment for electricity production from certain sources that feed the grid. This provides home owners as well as many establishments with an incentive to invest in solar panels, with a secured return on their investment. In 2004, the German government, which was in the forefront of such subsidy providers, guaranteed investors almost 60–70 cents per kilowatt-hour (kWh) they generate for the next two decades from photovoltaic generation (PV). This compares most favourably with German retail prices of 21–24 cents per kWh. Germany was not the only country to adopt FiT; similar practices were adopted by many European countries leading to expansion of uneconomical individual and community PV generation. However, such a bonanza cannot go on as market realities and economic hardships have begun to dawn. Recently, Germany decided that solar PV incentives will fall to 19.5 cents per kWh for smallish installations, 16.5 cents for installations up to 1 MW, and 13.5 cents for larger installations with gradual cuts of 0.15 cents per kWh per month.

The cost of solar panels, particularly rooftop systems has been dropping recently, particularly due to the reduction of the cost of Chinese panels. In Germany, the cost was around 2,900 euro per kW in 2010, this dropped to around to 1,900 euro in 2012. This was mainly due to the drop in the cost of manufactured panels which recently went down from 1,000 euro per kW to less than half of that. Yet it should be realised that the cost of labour, connections and other infrastructure did not significantly drop. These now constitute more than half cost of the system and are not likely to improve significantly in the future.

Subsidies for PV facilities have, from their inception until 2012, generated over 100 billion euros to consumers in Germany, despite the fact that PV plants only account for 3 per cent of electricity generation in Germany. Correspondingly, the cost of this huge subsidy increased the bill of German households by around 25 per cent between 2007 and 2011, putting a large burden on low income and seniors in the country. The domestic tariff in Germany is now 25.3 euro cents (32 U.S. cents) per kWh, one of the world's highest. Taxes and FiT account for almost half of this. Many wind energy investors now claim that they can produce electricity at a cost comparable to that of the national grid. This may be true in abstract terms, but it ignores system effects, the multiple restrictions and other expenses, particularly new transmission investments, grid reinforcement and dispatching restrictions, associated with such generation, some of these are already mentioned and further detailed below. What is important is system cost, how much (positively or negatively) is the system affected by the introduction of the NRs. These costs do not yet allow wind energy to provide competitive electricity without subsidy in one form or another of the many incentives mentioned above. Solarbuzz, a solar market research and analysis firm, provides data for solar electricity costs in the U.S. range from 28.91 cents per kWh in sunny climates to a very high 63.60 cents per kWh in cloudy ones. This is many times the 11-12 cents per kWh cost of grids utilising fossil fuels or nuclear generation.

There is also the issue of what we mean by costs. Where modern biomass or biofuels undermine ecosystems, biodiversity, agricultural land and water availability (with knock-on effects on food supplies and prices), as explained in a later section, it surely behooves policy makers to heed a broad definition of costs.

The fact remains that most NRs without governmental subsidies and intervention cannot compete, in most cases, with traditional thermal power generation practices and facilities. No new private NRs investment would venture into the market without support or encouragement of some form or another. With the present global economic challenges, future NR projects may not continue to enjoy the type of governmental favours of the past. The market future of NRs is challenging to new investments, and needs careful assessment as detailed below. With the increasing amount of variable-output NRs in the network, there is a need also to increase flexible generation. This can be not only in the form of hydroelectric plant or pump storage schemes if they exist, but also in the form of rapidly dispatchable single cycle gas turbines and diesel generation which can rapidly respond to load variation.

The WEO 2013 (IEA, 2013a) reckons that in 2011, renewables excluding large hydro received an estimated $88 billion in subsidies in various forms, up 24% from 2010, of which $64 billion went to electricity and the remainder to biofuels (Figure 5.6). Solar PV received more than any other renewable energy technology for electricity generation ($25 billion), followed by wind ($21 billion) and bioenergy ($15 billion). In the new Policies Scenario, total future subsidies to renewables will grow to about $185 billion in 2020 and reach almost $240 billion per year by 2035. Support provided to bioenergy for power generation continues to grow over time, reaching $69 billion in 2035, exceeding that received by any other technology. The amount received by solar PV grows rapidly in the medium term, reaching $77 billion in 2027, before falling to $58 billion in 2035, as retired installations are replaced by new less expensive capacity. Onshore wind power receives more support each year until around 2020, before falling to $14 billion by 2035, as this technology becomes increasingly competitive. Biofuels receive $24 billion in 2011, increasing to $46 billion in 2020 and $59 billion in 2035, with the vast majority going to conventional biofuels in 2035.

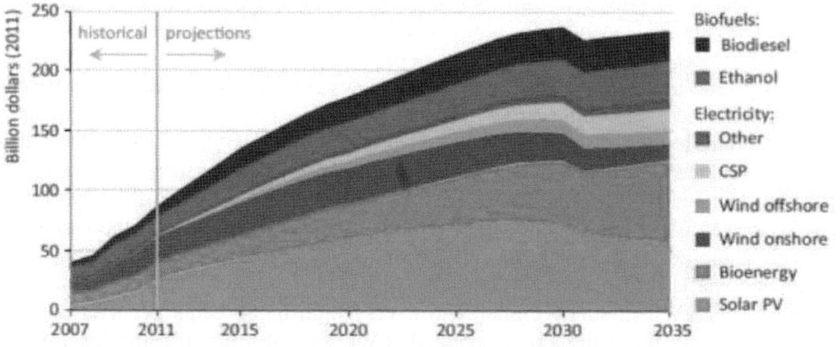

Fig. 5.6: Subsidies paid out, and anticipated, to new renewables (excluding hydro) in U.S.$ millions (2013). Source: (IEA, 2013a).

While subsides to NRs in the power sector increase in total, they decline on a per-unit basis as the cost of NR technologies fall and electricity prices increase, mainly due to higher fossil fuel prices and the introduction—in some regions—of a carbon price.

5.2.2 Subsidies to Nuclear Power

Nuclear power benefits from various forms of government support and subsidies involving liability limitations, preferential financing rates and export credit agency subsidies. However, the most important subsidies to the nuclear industry do not involve cash payments. Rather, they shift construction costs and operating risks from investors to taxpayers and ratepayers, burdening them with an array of risks including cost overruns. This approach has remained remarkably consistent throughout the nuclear industry's history and distorts market choices that would otherwise favour less risky energy investments. Nonetheless, there are many examples of governments providing or subsidizing electric power.

5.3 Advanced/Next Generation Technologies

We have already explained that the nuclear industry has not been able, for many decades, to reduce the cost of nuclear facilities. On the contrary, these have been increasing over time due to the rising cost of materials and difficulty of financing coupled with increasingly stringent regulations prompted by recent nuclear accidents and public awareness. Learning over the last few years did not assist in reducing the cost of nuclear due to lack of modularisation, standardisation, streamlining of regulations, and procurement/ construction process. Therefore the answer to nuclear's future lies in dealing with the above, and answering the following, cost considerations:

- Scale effects: Can the increase in the size of the plant lead to a reduction in the construction costs per MW installed? Alternatively, can the widespread diffusion of small modular reactors (SMRs) reduce construction costs per MW installed?
- Modularization: Can the building of more components in factories and less on site (for large plants or SMRs) reduce construction time and cost?
- Standardization and cumulative experience: Can capital cost reductions be achieved in standardizing plant designs and constructing similar plants in large series?
- Regulation: Can the regulatory framework reduce the risk of cost overruns while providing adequate safety levels?
- Procurement and competition: Can improved competition and procurement contracts result in significant cost reductions (D'haeseleer, 2013)?

5.4 Challenges to System Integration of Renewables and Nuclear

As discussed above, one of the main challenges to NRs and nuclear is their integration in the power generation systems. If the contribution from NRs is small, say less than 10%, these problems are manageable. However, when their penetration assumes significant proportions then these pose serious challenges to system operation and cause an increase in system costs.

5.4.1 Examples of Estimated System Costs

Many countries have enacted official targets for renewable power share for their grids, and others are considering them. U.S. states are doing so as well. This is leading to increasing shares in a number of jurisdictions. Consequently the challenges for grid operation caused by the intrinsic character of the intermittent nature of wind and solar are being faced by real world grid operators. Further, since these targets are typically still higher for future years, various planners have been engaged to forecast future operations and their costs.

The following are some costs calculated by the OECD NEA. They are presented with strong caveats in the original source. They are included here to be illustrative of the fact that these costs are believed to be real and significant. Further, they vary by power system.

Table 5.7 Grid-level system costs in selected OECD countries ($/MWH).

Finland

Technology	**Nuclear**		**Coal**		**Gas**		**Onshore wind**		**Offshore wind**		**Solar**	
Penetration level	**10%**	**30%**	**10%**	**30%**	**10%**	**30%**	**10%**	**30%**	**10%**	**30%**	**10%**	**30%**
Back-up costs (adequacy)	0	0	0.06	0.06	0	0	8.05	9.7	9.68	10.67	21.4	22.04
Balancing costs	0.47	0.3	0	0	0	0	2.7	5.3	2.7	5.3	2.7	5.3
Grid connection	1.9	1.9	1.04	1.04	0.56	0.56	6.84	6.84	18.86	18.86	22.02	22.02
Grid reinforcement and extension	0	0	0	0	0	0	0.2	1.72	0.12	1.04	0.56	4.87
Total grid-level system costs	**2.37**	**2.2**	**1.1**	**1.1**	**0.56**	**0.56**	**17.79**	**23.56**	**31.36**	**35.87**	**46.67**	**54.22**

France

Technology	**Nuclear**		**Coal**		**Gas**		**Onshore wind**		**Offshore wind**		**Solar**	
Penetration level	**10%**	**30%**	**10%**	**30%**	**10%**	**30%**	**10%**	**30%**	**10%**	**30%**	**10%**	**30%**
Back-up costs (adequacy)	0	0	0.08	0.08	0	0	8.14	8.67	8.14	8.67	19.4	19.81
Balancing costs	0.28	0.27	0	0	0	0	1.9	5.01	1.9	5.01	1.9	5.01
Grid connection	1.78	1.78	0.93	0.93	0.54	0.54	6.93	6.93	18.64	18.64	15.97	15.97
Grid reinforcement and extension	0	0	0	0	0	0	3.5	3.5	2.15	2.15	5.77	5.77
Total grid-level system costs	**2.1**	**2.1**	**1**	**1**	**0.5**	**0.5**	**20.47**	**24.1**	**30.83**	**34.5**	**43**	**46.6**

Germany

Technology	**Nuclear**		**Coal**		**Gas**		**Onshore wind**		**Offshore wind**		**Solar**	
Penetration level	**10%**	**30%**	**10%**	**30%**	**10%**	**30%**	**10%**	**30%**	**10%**	**30%**	**10%**	**30%**
Back-up costs (adequacy)	0	0	0.04	0.04	0	0	7.96	8.84	7.96	8.84	19.22	19.71
Balancing costs	0.52	0.35	0	0	0	0	3.3	6.41	3.3	6.41	3.3	6.41
Grid connection	1.9	1.9	0.93	0.93	0.54	0.54	6.37	6.37	15.71	15.71	9.44	9.44
Grid reinforcement and extension	0	0	0	0	0	0	1.73	22.23	0.92	11.89	3.69	47.4
Total grid-level system costs	**2.4**	**2.3**	**1**	**1**	**0.5**	**0.5**	**19.36**	**43.9**	**27.9**	**42.9**	**35.6**	**83**

Republic of Korea

Technology	Nuclear		Coal		Gas		Onshore wind		Offshore wind		Solar	
Penetration level	**10%**	**30%**	**10%**	**30%**	**10%**	**30%**	**10%**	**30%**	**10%**	**30%**	**10%**	**30%**
Back-up costs (adequacy)	0	0	0.03	0.03	0	0	2.36	4.04	2.36	4.04	9.21	9.4
Balancing costs	0.88	0.53	0	0	0	0	7.63	14.15	7.63	14.15	7.63	14.15
Grid connection	0.87	0.87	0.44	0.44	0.34	0.34	6.84	6.84	23.85	23.85	9.24	9.24
Grid reinforcement and extension	0	0	0	0	0	0	2.81	2.81	2.15	2.15	5.33	5.33
Total grid-level system costs	**1.7**	**1.4**	**0.5**	**0.5**	**0.3**	**0.3**	**19.64**	**27.8**	**35.99**	**44.2**	**31.4**	**38.1**

United Kingdom

Technology	Nuclear		Coal		Gas		Onshore wind		Offshore wind		Solar	
Penetration level	**10%**	**30%**	**10%**	**30%**	**10%**	**30%**	**10%**	**30%**	**10%**	**30%**	**10%**	**30%**
Back-up costs (adequacy)	0	0	0.06	0.06	0	0	4.05	6.92	4.05	6.92	26.08	26.82
Balancing costs	0.88	0.53	0	0	0	0	7.63	14.15	7.63	14.15	7.63	14.15
Grid connection	2.23	2.23	1.27	1.27	0.56	0.56	3.96	3.96	19.81	19.81	15.55	15.55
Grid reinforcement and extension	0	0	0	0	0	0	2.95	5.2	2.57	4.52	8.62	15.18
Total grid-level system costs	**3.1**	**2.8**	**1.3**	**1.3**	**0.6**	**0.6**	**18.6**	**30.2**	**34.05**	**45.4**	**57.9**	**71.7**

United States

Technology	Nuclear		Coal		Gas		Onshore wind		Offshore wind		Solar	
Penetration level	**10%**	**30%**	**10%**	**30%**	**10%**	**30%**	**10%**	**30%**	**10%**	**30%**	**10%**	**30%**
Back-up costs (adequacy)	0	0	0.04	0.04	0	0	5.61	6.14	2.1	6.85	0	10.45
Balancing costs	0.16	0.1	0	0	0	0	2	5	2	5	2	5
Grid connection	1.56	1.56	1.03	1.03	0.51	0.51	6.5	6.5	15.24	15.24	10.05	10.05
Grid reinforcement and extension	0	0	0	0	0	0	2.2	2.2	1.18	1.18	2.77	2.77
Total grid-level system costs	**1.7**	**1.7**	**1.1**	**1.1**	**0.5**	**0.5**	**16.3**	**19.8**	**20.51**	**28.3**	**14.8**	**28.3**

(Source: OECD NEA (2012))

These are rather heroic calculations given the paucity of sources (consequently the uncertainties are large), but they do indicate plausible effects. They illustrate the influence of back up and grid costs. They show the relatively benign impact of nuclear power on system costs compared to renewable power. They illustrate large differences among different countries. Back-up costs for renewables tend to increase in direct proportion to the renewable capacity (MW) that needs back up and the increased capacity adds proportional MWh. However, balancing costs increase because more spinning reserve capacity is required at lower load factors. As penetration increases, grid reinforcement and extension costs increase because the grid must be extended farther afield to find suitable renewable sources, and outer reaches of the grid must be reinforced to carry the increased load. As a side note, backup and balancing is less costly in the United States because the usual back-up power, natural gas, benefits from lower natural gas prices than most other countries.

A further inference can be suggested from these results: it is that the marginal system costs will generally increase with increased penetration of renewables essentially because of their intermittency and tendency for remote locations.

In addition to marginal system cost per MWh, there is another critical metric: marginal cost per ton of CO_2 emissions reduced by increased deployment of renewables. After all, that is a primary policy driver for renewable targets.

The state of California has set ambitious goals for renewable penetration: 33% by 2020, not including hydro. Even more has been considered (up to 51% in a legislative proposal) for the future. The 33% level is thought to engender an implicit cost of carbon reduction of $50/ton. A number of power generators commissioned a recent study to investigate the problems and costs associated with increasing that target up to 50%. As the following table shows, the resulting estimates are $340/ton for 40%, and a lowest cost $403/ton for 50%. However, the Diverse Scenario is based on assumed flexibility benefits occasioned by 5000 MW of storage or advanced demand response. Recent trends are that increased renewable capacity has come in the form of large solar. Extending this yields $636/ton.

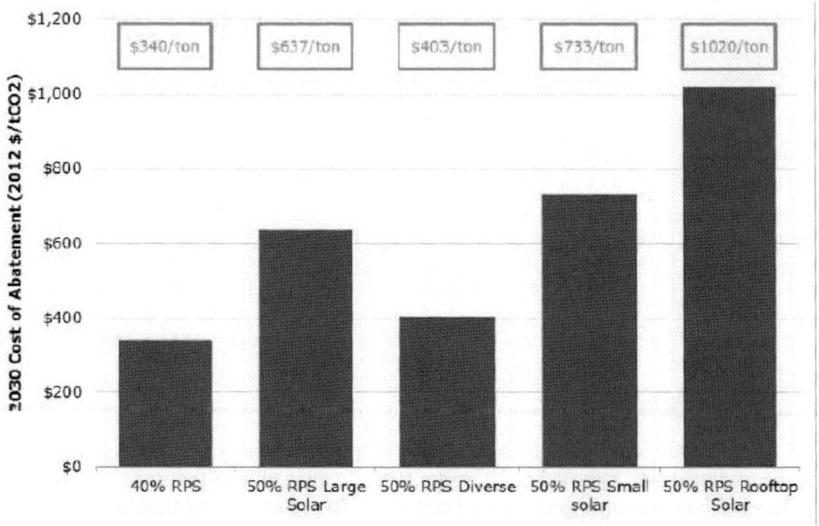

Fig. 5.8: Implied cost of carbon abatement in 2030 for each scenario relative to the 33% RPS Scenario. (Source: Energy and Environmental Economics, p. 144.)

5.5 Conclusions

Both renewables and nuclear face significant economic challenges. The costs associated with renewable technologies have consistently been higher than those for building fossil-fuelled power plants, although renewables costs (especially solar PV) are decreasing. Furthermore, renewables cost estimates may not fully include systems integration costs. For nuclear, the capital costs remain a major challenge; the cost of financing new nuclear plants is very high in many countries, owing to investors' fears about time, technology, regulatory, and safety risks. These issues may not be as large for state-owned companies or regulated markets for which utilities have ready access to inexpensive capital, and this partly explains why enthusiasm for nuclear reactors is much stronger in many developing countries than in the United States or Europe. Learning could help decrease costs for both types of technologies, but the track record for learning-by-doing in the nuclear sector is not good (though this may change with small modular reactors if they can be fabricated by manufacturing processes).

Both nuclear and renewables receive subsidies that vary by country and region. Some subsidies are direct, like feed-in-tariffs for renewables, while others shift risks from utilities to customers or taxpayers.

The following recommendations would help establish a better accounting for the costs of nuclear and renewables, and could help support cost reduction of both:

1. Comparisons of nuclear and renewables costs should account for systems integration and differences in capacity factors.
2. In order to estimate nuclear costs, more attention should be given to the sensitivity of the estimated cost to the discount rate, as the discount rate greatly affects the cost of production. Studies of the technical issues that may impede using a nuclear reactor in "load-following" mode are important and should be given high priority.
3. Priority should be given to new reactor technologies like SMRs and regulatory reform in order to reduce nuclear capital costs.

6. ELECTRICITY STORAGE

A central challenge for grid operators, described in detail in previous sections, is matching electricity supply to demand. The challenge increases when more of the electricity supply is composed of intermittent renewable resources, and the trend is up for renewables on many nations' electricity grids. Besides this operational challenge, intermittent renewables pose market challenges for base load resources. For example, at night when the wind blows strongly and electricity demand is low, prices are suppressed, but nuclear units generally cannot power down. If more renewables lead to more sustained price suppression, base load units have trouble recovering their costs (Renewable Analytics, 2013).

Several system options could help balance electricity supply and demand given different mixes of intermittent, base load and load-following generation capacity. One alternative is to match each renewable installation with full backup capacity (typically natural gas) such that all the combination of renewables and back-up is reliable and dispatchable. This, however, would increase the cost of renewable generation and reduce the environmental benefits of renewables. A second option is to build substantial transmission capacity, so a drop in renewable electricity generation in one area (e.g., due to clouds over solar panels) could be compensated for by electricity production in another region. In the U.S., studies have shown that 45% penetration of renewables is achievable with the current transmission system (IEA, 2011), and as much as 80% renewable penetration can be achieved by expanding transmission capacity, including increasing the number of ties between the Eastern and Western grids and building transmission backbone lines to bring Midwest wind to the coasts (Mai et al. 2012). Smaller grids, however, may not have sufficient capacity or geographic diversity to accommodate sudden, large shifts in supply. Furthermore, U.S., cost and permit issues are major barriers to achieving significant transmission capacity expansion. A third option for managing intermittency is to increase cycling of base load plants, but this can be challenging and costly, especially for nuclear power. Demand response (DR), or contracting with consumers to reduce demand when the grid is short on supply, is yet another option to ensure a supply and demand match; large DR potential remains in many countries and has relatively few drawbacks, other than challenges with consumer acceptance and eventual limitations for how much DR the system can sustain.

An important additional option to integrate varied generation sources, match electricity supply and demand, and allow generation sources to dispatch power at a profit is electricity storage.[10] In theory, if electricity storage could be deployed widely, grids of any size could sustain a wide range of profiles of intermittent and base load power. Storage systems enable arbitrage of renewable generation, in which power is bought at times when demand is low, power is cheap, and renewables are generating, and then sold back when power prices are high. They can also provide ancillary services to the grid. At the consumer level, they can provide backup power and help maximize use of distributed generation systems. Storage, however, is currently costly, and most market structures worldwide do not favour its adoption because they do not allow remuneration for the flexibility benefits storage provides. Because neither storage nor the other options described above are without drawbacks, integrating high levels of renewables with base load power may require a combination of strategies to meet multiple objectives. Among these strategies, energy storage is of particular interest because storage systems are flexible, may be carbon-free, and are scalable. However, high costs are currently a barrier.

The next section gives an overview of grid-scale storage technologies in various stages of development and deployment. Following sections discuss existing energy storage installations, valuing energy storage on the grid, and R&D efforts on storage. The emphasis throughout this report is on large storage systems that have the ability to provide grid-level renewables integration services, including electricity arbitrage and ancillary services like frequency regulation.

---

[10] Note that storage can provide a wide range of grid services on many timescales, and some systems can provide more than one type of service. For purposes of this study, we focus on grid-scale storage with the ability to provide electricity arbitrage and regulation services, as these are of especially high importance for grid integration of renewable and base load sources.

6.1 Overview of Selected Storage Technologies

Cost estimates can be challenging for storage technologies, as costs vary widely by geography (especially for pumped hydro and compressed air systems) and are hard to estimate for emerging technologies. Table 6.1 provides an overview of technology costs, and each is discussed further (with references) below.

Table 6.1 Summary of Storage Technology Costs*

Technology	Cost/kW (power)	Cost/kWh (storage)
Pumped Hydro	$1500-$4300	$250-430
Compressed Air	$800-$1500	$125/kWh
Flywheels	up to $4000	$200-$500
Lead-Acid	$1700 - $4900	$450-$950
Li-ion		$500-$850
NaS	$3100-$3300	$520-$550
Flow Batteries	$3100-$3700	$520-$550
Hydrogen	$1150	

*Note that these are published cost estimates for utility-scale installations in the U.S. from EPRI, BNEF, and peer-reviewed papers as of May 2014. Between May 2014 and May 2015, several companies have announced sales of battery systems available at lower cost; these examples are discussed below.

6.1.1 Pumped Hydroelectric

Pumped hydro is the most mature, widely-deployed energy storage technology. Pumped hydro facilities include high and low water reservoirs, typically with elevation differences in the hundreds of meters range. During periods of electricity storage or "charging," water is pumped from the lower to the higher reservoir, and then when electricity is needed, the water flows from the high reservoir to the lower through a hydroelectric turbine. According to the Electricity Storage Association, these processes lose very little electricity, reaching round-trip efficiencies as high as 80%. The major challenges for pumped hydro include dependence on geography and public concerns about environmental degradation where new reservoirs are created.

Current cost estimates for storage technologies vary by country and can vary by location within countries—especially for pumped hydro storage. EPRI estimates U.S. pumped hydro costs at $2500-$4300/kW (power) and $420-430/kWh (storage) for small facilities with capacities less than roughly 5500 MWh, and $1500-$2700/kW and $250-270/kWh for facilities up to 14000 MWh (EPRI, 2010a). The national hydropower association reports slightly lower capital costs for pumped hydro (NHA, 2010).

The potential for building new pumped hydro varies by country as well. In the U.S., DOE and Oak Ridge National Laboratory are surveying potential for new facilities. Elsewhere, over 60 pumped hydro projects are under construction, most in Europe, China, India, and Japan (NHA, 2010). The potential for pumped hydro could be large, even in Europe: Gimeno-Gutierrez and Lacal-Arategui (2013) indicated the possibility for a ten-fold increase in storage capacity, and MacKay (2009) estimates up to 400GWh of potential in the UK. However, realizing this potential will likely require overcoming environmental challenges for specific geographies. In addition, in some regions, the economics for new pumped hydro facilities may be at issue, as more demand-side management and solar deployment shaves peak electricity prices and reduces arbitrage opportunities for storage plants. Switzerland has good examples of both. Following the decision to discontinue nuclear, the Swiss government published ambitious plans for increased hydro—which were immediately opposed by the cantons. Plans for increasing the capacity of pumped storage at the Grimsel have been put on ice because of the cheap midday wind energy flooding in from Germany.

6.1.2 Compressed Air Energy Storage

Compressed air energy storage systems (CAES) pump air underground into geologic formations or (more rarely) into above-ground vessels and pipes. When electricity is needed, the air is released, heated, and expanded by natural gas to drive a turbine. Advanced CAES technologies include those that take advantage of electricity production during the compression phase, and those that recycle flue gas in order to minimize natural gas burns. Although CAES faces twin challenges of limited geography and greenhouse gas emissions, it is the third-most widely deployed storage resource. The IEA in conjunction with the International Renewable Energy Agency estimates CAES capex costs at $800 - $1000/kW with cheap, natural underground storage (IEA-ETSAP and IRENA, 2012), EPRI estimates costs up to $1250/kW and $125/kWh (EPRI, 2010b), and Wesoff (2011) estimates $1500/kW.

6.1.3 Flywheels

Flywheels are low-energy, high-power storage devices typically used for maintaining power quality and to provide black start capacity. Flywheels consist of rotors on electromagnetic bearings that increase spin speed while charging, then maintain spin with minimal frictional losses as an energy storage mechanism. When electricity is needed, the rotors generate electricity as they slow. Challenges include materials fatigue and limits on the system energy density; until flywheels can store more energy, they may not be a primary technology choice for peak-shifting applications needed to integrate intermittent renewables. IEA/IRENA estimate flywheels to cost up to $4000/kW for MW-sized systems, and a Lawrence Livermore National Laboratory study pegs storage costs at $200-$500/kWh (IEA-ETSAP and IRENA, 2012).

6.1.4 Batteries

Batteries consist of electrochemical cells, historically with an aqueous electrolyte and solid negative and positive electrodes. When charging, positive ions flow from the positive to the negative electrode while electrons flow in the opposite direction; these flows are reversed during discharge. In order to provide grid support services, batteries should be able to store significant amounts of energy at low cost (though unlike for consumer electronics and vehicles, grid-scale batteries may be large). For peak-shifting applications, grid-scale energy storage systems comprising batteries should be high-powered (e.g., MW-scale), able to respond to grid needs appropriately, and possess long cycle lifetimes. In general, batteries may be separated into sealed or flow designs. Lead-acid, lithium-ion, and sodium-sulphur technologies have, to date, been the most prominent battery chemistries for grid-scale storage, yet cost, durability and/or safety challenges remain.

Lead-acid batteries are generally considered the most mature of rechargeable battery technologies. Though they are deployed in smaller numbers than lithium-ion for grid support applications, they appear extensively in automobiles, telecommunications, and other systems. Challenges to expanded production and use of lead-acid for grid applications include battery toxicity, high costs, and low cycle life. EPRI estimates costs for commercial, advanced lead-acid batteries at $1700-$1900/kW for 200 MWh, 50 MW systems, and up to $4900/kW for 250 MWh systems with twice the cycling life. Costs per kWh nearly doubled for the larger, longer-cycle life system, from about $450/kWh to about $950/kWh (EPRI, 2010b).

Over the last few years, grid penetration of lithium-ion batteries has increased significantly. Lithium-ion batteries include a range of different chemistries, but all continue to face residual challenges with safety (e.g., reactivity and flammability) and cost. Both of these factors become more of an issue when the batteries are scaled from everyday applications in personal electronics devices to grid-scale energy storage systems. Bloomberg estimates lithium-ion battery capex costs at $850/kWh for hour-scale price arbitrage applications (Bloomberg, 2013a).

This particular discussion does not address residential-scale applications of storage, but in principle, if residential storage were widely deployed and grid-connected, the storage network could assist in balancing utility-scale renewables and base load power. Therefore, the recent announcement by U.S. company Tesla in that it would sell residential lithium-ion battery systems for $350/kWh could be significant. The cost is significantly lower than for other residential systems, due to Tesla's intention to build a large manufacturing facility taking advantage of economies of scale. But whether Tesla or its competitors achieve significant grid penetration depends on the demand for residential battery systems, which may not be high even at these lower price points (Warshay, 2015). Tesla has also suggested it will offer utility-scale applications of its technology at $250/kWh, offering some of the lowest costs Li-ion batteries available.

Sodium-sulphur batteries have been developed and deployed primarily in Japan. They generally consist of a high-temperature molten sulphur positive electrode and liquid sodium negative electrode separated by a solid, sodium-conducting ceramic. Because of the abundance and elemental nature of the battery chemistry, these systems have the opportunity for low cost; however, safety and cycle life challenges remain. EPRI estimates Sodium-sulphur battery costs at $3100-$3300/kW and $520-$550/kWh (EPRI, 2010b). To address safety and improve on cost, an analogous high-temperature molten $Na/NiCl_2$ has been developed and is entering early stages of commercial deployment.

Flow batteries separate the power and energy function of a battery by storing charge in solution (typically aqueous) electrolytes that are flowed over a stack of electrochemical cells for power generation and storage. Higher power requires a larger stack of cells, while higher energy requires a larger tank of electrolyte. The benefit of this power/energy separation is typically achieved at the expense of energy density; flow battery systems are typically 2-5X larger than sealed battery systems. Vanadium redox batteries are, so far, the most widely- deployed flow batteries and utilize a single electrolyte for both the positive and negative battery active material. Advantages of vanadium redox flow batteries include long cycle life and relative safety. The high cost of vanadium limits the potential for this flow battery chemistry; EPRI estimates the cost of a 250 MWh vanadium redox battery at $3100-$3700/kW and $620-$740/kWh (EPRI, 2010b).

Recent announcements by vanadium flow battery companies indicate further cost reductions may be imminent. Eos Energy Storage introduced a grid-scale zinc hybrid cathode battery with an aqueous electrolyte in January of 2015. The stated cost for the Eos system is \$1000/kW or \$160/kWh, making it the least expensive on the market. Through partnerships with utilities, Eos intends to deploy up to 1MW of storage capacity in the U.S. in 2015. Another company, Imergy, has announced vanadium flow batteries available for \$500/kWh. Whether these costs are achieved in near-term projects remains to be seen.

6.1.5 Hydrogen

Hydrogen energy storage systems are virtually negligible in terms of grid deployment; SBC estimates less than 2MW capacity is installed worldwide (SBC, 2013). Nonetheless, hydrogen fuel cells have the potential to provide significant amounts of electricity balancing for intermittent renewables. Most hydrogen in the U.S. is produced by steam reformation of hydrocarbons, but clean grid-scale battery technologies typically focus on hydrogen production from water. They operate by converting water into hydrogen and oxygen during charging (either through electrolysis, biological or catalytic processes), and they recombine hydrogen and oxygen in order to produce electricity. These systems have been developed for over 50 years, primarily for space applications but with an eye towards automotive. One major advantage is that hydrogen systems can store energy for long periods of time, but a major drawback is the low efficiency with which hydrogen conversion occurs. Sandia National Labs estimates current hydrogen systems to cost about \$1150/kW for 6 hours of storage (Schoenung, 2011).

6.2 Existing Storage Facilities

By far, the most widely-deployed storage technology is pumped hydro, owing to its historical cost effectiveness compared to other storage technologies. Pumped hydro, however, is limited by geography; it requires two large reservoirs with a significant altitude difference between them. Research is underway on using seawater and/or underground reservoirs to minimize the environmental footprint, but these options are likely to be significantly more costly (IEA, 2013d).

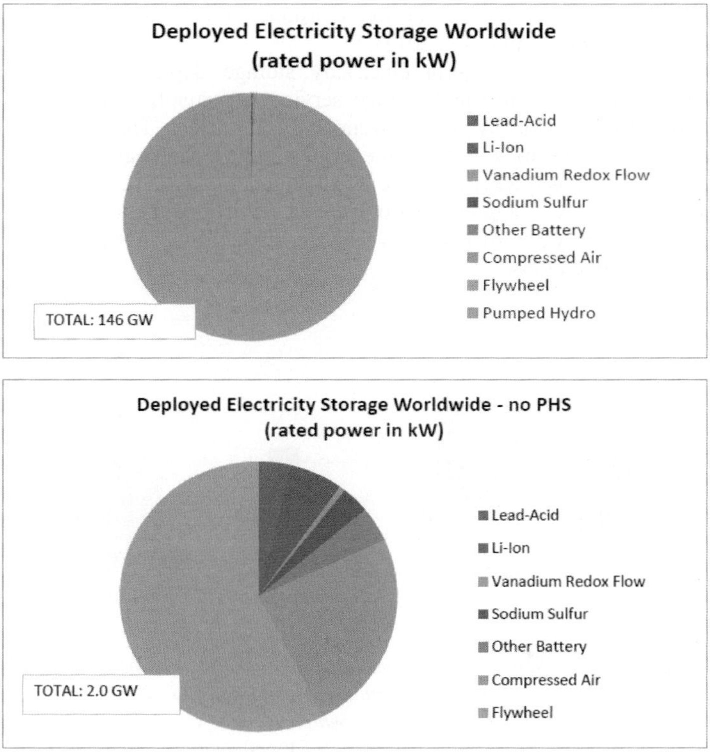

Fig. 6.2: Deployed storage capacity. SOURCE: U.S. DOE (2015).

Flywheels represent the next largest segment of the storage market; the installed capacity is mostly composed of two half-gigawatt scale projects in Europe. Compressed air energy storage has significant presence in the U.S. Japan has the largest sodium sulphur battery installation, at 34 MW, and the U.S. leads on installed capacity of lithium-ion batteries. Thermal energy storage, often associated with buildings, represents a growing market segment to shift electricity demand (IEA, 2013d). Because thermal storage is often used in this demand-response-like manner, it is not depicted on the charts above or discussed further here.[11]

Several nations and regions worldwide have adopted policies to increase storage deployment. California passed a bill requiring utilities to procure 1.3 GW of electrical and thermal storage by 2020 in order to mitigate grid disruptions that may result from its 33% renewable portfolio standard (GreenBizz, 2013). Japan and Germany have both

---

[11] Note, however, the significant exception of thermal storage coupled with concentrated solar power systems, rendering CSP a dispatchable renewable technology.

begun subsidize storage systems at the consumer level, to enhance operation of distributed PV systems (Wilkinson, 2013).

Forecasting even near-term electricity storage deployment is very difficult because the market structures and storage service opportunities vary widely, and the potential for cost reduction by storage technology is unclear. The IEA examines three storage deployment scenarios in its 2014 energy storage technology roadmap, showing a large increase in deployed energy storage by 2050 under a 2C GHG emissions target, and an even larger increase if aggressive storage cost reductions are achievable. This work clearly illustrates the value of activities that are directed towards reducing the cost of energy storage technologies.

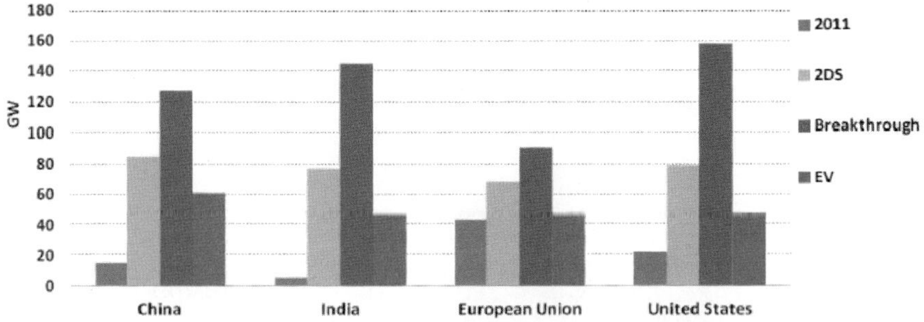

Fig. 6.3: Three 2050 Electricity storage deployment scenarios (IEA); including a scenario in which warming is limited to 2 degrees C (2DS), a "breakthrough" scenario in which storage costs come down substantially, and an "EV" scenario with high EV penetration. SOURCE: (IEA, 2014).

6.3 Valuing Storage in Electricity Markets

Ultimately, storage will only achieve high levels of penetration if it can compete for service provision in electricity markets. Markets are set up to value electric capacity and the generation that traditional power plants provide, but they do not generally value the high level of flexibility storage can provide. In some cases, regulatory decisions prevent storage systems from participating in ancillary service markets.

One way for storage systems to make money in electricity markets is with arbitrage, storing electricity from the grid when it is cheap and selling it back when it is expensive. Predicting precisely when to buy and sell, however, is difficult. A Bloomberg analysis determined that buying and selling at set times, i.e., buying at the 4th hour of the day and selling at the 12th, would yield nearly 30% less revenue than the perfect strategy in the German day-ahead market between 2008 and 2012 (Bloomberg, 2013b). In general, Bloomberg found that participation in the German ancillary services market was more economical than arbitrage. Besides providing arbitrage or ancillary services, a third way for storage systems to obtain market revenues is to receive capacity market payments, as PJM is currently considering in the U.S.

Rather than allow storage to compete in electricity, ancillary service, or capacity markets (or some combination), a different option involves valuing energy storage based on a regulated, cost-of-service or willingness-to-pay model. Regulators could ensure a rate of return for storage developers that could include valuation of avoided costs, like the ability for storage to obviate the need for additional transmission. Under a rate-recovery system, storage could be treated alongside transmission, and this may become possible in the U.S. FERC is currently creating accounting standards to facilitate rate-basing of storage assets in 2015 (Bloomberg, 2013a).

The particular combination of market and non-market alternatives most conducive for storage will vary depending on the particular needs of each grid and the existing market dynamics. Bloomberg's analysis of the market competitiveness of storage in Germany, North America, and the UK shows very few, but varying applications for which lithium-ion or sodium-sulphur batteries are economic. In Germany, only the provision of primary control reserve generated a positive IRR. In North America, arbitrage is profitable in some regions of Ontario, and economic opportunities exist for storage applications elsewhere (DOE, 2013a), especially if battery costs come down and storage becomes eligible for capacity payments. In the UK, the only currently profitable application for lithium-ion batteries is deferral of transmission upgrades (Bloomberg, 2013c). In general, storage systems are otherwise too expensive.

As alluded to above, the storage technologies most likely to achieve commercial success are those which can provide more than one service, "stacking" the payments they can receive (Kaun and Chen, 2013). The U.S. Department of Energy classifies technologies according to whether they can definitely, potentially, or never supply various electricity services. Table 6.4 describes commercial and in-development technologies and lists the renewable integration-related services they can provide.

Table 6.4: Technologies and energy services

Service	Peak-shifting (arbitrage)	Energy smoothing and shaping	Ancillary services (e.g. frequency regulation)
Timescale	hr.+	min. to hr.	sub-minute
Technology			
Pumped Hydro	[solid] *	[hashed]	[hashed]
Compressed Air	[solid] *	[empty]	[hashed]
Flywheels	[empty]	[hashed]	[solid] *
Lead-Acid	[solid]	[solid]	[hashed]
Li-ion	[solid]	[solid]	[hashed]
NaS	[solid]	[solid]	[hashed]
Flow Batteries	[solid]	[solid]	[hashed]
Hydrogen	[hashed]	[solid]	[hashed]

Solid box = can provide, Hashed box = potentially provide, Empty box = cannot provide SOURCE: [DOE (2013b), SBC (2013)].[12]

From the perspective of maximum opportunity for service provision, battery technologies are attractive because they can provide all three types of services. They are also more attractive than other listed technologies because they tend to be more scalable (for example, they are not bound by reservoir sizes and can be made quite small), and

[12] DOE lists these as "can potentially provide".

thus have applications at the distribution and consumer level. This larger market depth for batteries may mean they have more opportunities to come down the cost curve.

Lazard (2014) calculates the LCOE of batteries, and determines the range to be \$265-\$324/MWh for lead-acid at the low end and NaS at the high end. But Lazard also notes the costs may come down as low as \$168/MWh by 2017; at that point, they may be competitive with other generation technologies (Lazard, 2014).

The next section examines research, development and demonstration priorities that could enhance the ability of storage systems to provide a range of services at low cost.

6.4 Development of Advanced Electricity Storage Technology

Much of the R&D focus on storage is and should be on attaining cost-competitive systems. This means developing storage solutions with low per-kW power and per-kWh storage capital costs. DOE has a near-term target of achieving costs below \$250/kWh, and a long-term target of achieving capex under \$150/kWh. DOE estimates that achieving a levelised cost of 10 cents/kWh in the long term will make storage solutions economically scalable without subsidies; for reference, the levelised cost of storage for lithium-ion is currently on the order of 10 times that (EPRI, 2010b). Other research targets include reaching low O&M costs, related to achieving high efficiency, high energy density, high cycle life, and low capacity fade, among other parameters. Each of the technologies discussed so far faces different challenges achieving low cost targets.

Safety and environmental issues are also of importance to potential storage investors. Battery technologies, for example, contend with some safety and reliability issues, and some battery research focuses on reducing safety risks and/or demonstrating the reliability of battery systems. Compressed air energy storage presents environmental challenges because it typically requires a natural gas system to heat expanding air. New technologies, however, may obviate the need for a gas heat source. Table 6.5 describes the primary focus of R&D for each of the storage technologies.

Table 6.5: R&D thrusts by storage technology

Technology	**Primary R&D Challenges**
Pumped Hydro	Cost of underground/seawater reservoirs to remove geographic dependence & mitigate environmental harms*
Compressed Air	Adiabatic and isothermal processes to avoid use of gas to heat air upon decompression; artificial reservoirs to remove geo. dependence
Flywheels	Materials cost; flywheel stability
Lead-Acid	Improve cycle life and depth of discharge
Li-ion	Find lower-cost materials for the negative electrode, increase energy density & cycle life; confirm safety; scale-up from small, cylindrical to large-format, prismatic cells for energy storage applications
Molten Salt	Improve durability/ manufacturability of ion-conducting separator; improve corrosion resistance; alternative chemistries to reduce operating temperature, improve safety
Flow Batteries	Improve specific energy and cycle life; alternative chemistries to lower system cost
Hydrogen	Improve efficiency of oxygen reactions while reducing cost of oxygen electrodes

SOURCES: [DOE (2013b), SBC (2013)].

Two promising, very early-stage battery technologies aim to address some the challenges listed above. Lithium-air batteries could be a substitute for Li-ion, as they replace the cathode with oxygen and obviate the need for a cathode structure. Challenges with the designs remain, however, because the cathode degenerates due to encroaching humidity (SBC, 2013), and Li-air's reliance on oxygen electrochemistry means it suffers from efficiency challenges similar to those encountered by hydrogen batteries. Other flow battery chemistries may also be of interest; EPRI estimates, for example, that zinc-bromine batteries could cost as little as \$1450-\$1750/kW, or about half the cost of vanadium redox systems (EPRI, 2010b). Liquid metal batteries provide a second alternative to conventional molten salt batteries: the cathode, anode, and electrolyte are all liquid metal layers, separated due to their different densities. Research on liquid systems focuses on developing materials that can be kept at lower temperatures and sustain higher voltages (Wesoff, 2014a). The manufacturing, scale-up, and putative cost reduction of conventional, small cylindrical Li-ion batteries and systems comprising them is another interesting development (Wesoff, 2014b).

The U.S. and other countries are devoting substantial resources to storage technology development. The U.S. government spent about \$1.3 billion on storage initiatives between 2009 and 2012, of which the majority went through DOE programs and roughly 25% through the Department of Defence. DOE programs have included basic research, storage demonstration programs under the American Recovery and Reinvestment Act, energy storage "hubs," and ARPA-e storage projects aiming to develop new solutions and make them market-ready.

The European Electricity Grid Initiative mapped storage RD&D projects in Europe and found nearly \$1.72 billion spread across nearly 400 projects. About 1/5 of the funding for these projects came from the EU and the remainder came from national governments. The survey showed that Europe is betting heavily on batteries and thermal storage, especially non-centralized storage systems. Geth, Kathan, Sigrist and Verboven (2013) noted this funding largely applied at an early R&D stage, and that in the next few years, the EU may implement increasing numbers of follow-on demonstration projects. Japan continues to fund efforts toward demonstration and deployment of battery systems as well, leading the world on NaS.

Another frontier for storage technology research includes the use of electric vehicles (EVs) to provide grid services. Plug-in vehicles could provide frequency regulation and fast-response reserve power, competitive with spinning reserves. Some vehicle technologies could provide peak-shifting services useful for integrating intermittent generation sources. One study estimated that if wind supplied half of 2005 U.S. electricity generation (700 MW), 23% of the vehicle fleet could provide the needed firm backup capacity in the form of fuel cell cars with collective storage capacity of 1190 GWh (Kempton and Tomi´c, 2005). Similar studies have demonstrated promise for EV support of wind integration in the UK and in Germany (Loisel, Pasaoglu and Thiel, 2014).

Despite this promise, deployment of grid-connected vehicle systems remains minimal worldwide. A major issue is cost; the German study noted that due to battery degradation, battery-powered electric vehicle owners could have difficulty recouping costs in the primary ancillary services market (Loisel, Pasaoglu and Thiel, 2014). Another challenge includes developing a business model that is workable for grid operators and attractive for consumers. Grid operators will want maximum flexibility to use vehicle storage resources, while drivers need to be assured their cars will charged and available when needed.

The U.S. government is engaging in research to reduce these barriers to EV/grid systems. The Department of Energy has funded vehicle-to-grid demonstration projects, including one in the PJM territory with early results establishing revenue for grid-enabled vehicle owners of about $5/day (Shepard and Garther, 2013). The California ISO has developed a vehicle-grid integration roadmap as part of its Zero Emission Vehicles (ZEV) initiative, and the Department of Defence pledged $20 million in 2013 toward establishing grid-enabled plug-in electric vehicles on military bases. Navigant estimates that more than 250,000 grid-enabled of vehicles will be sold worldwide by 2022 (Shepard and Garther, 2013).

Deciding where countries should focus storage R&D efforts requires understanding which technologies are most likely to compete based on the services they can provide, and which technologies are closest to commercial maturity. Environmental and social impacts are important considerations as well, especially as nations move to internalize emissions costs.

At this stage, no single technology stands out as being clearly most worthy of RD&D support. Pumped hydro and compressed air storage are the most mature and cost-effective, but both come with geographic limitations and environmental drawbacks. They should be deployed as cleanly and widely as possible, and research should continue on options to mitigate the challenges associated with both technologies. Batteries remain expensive, so research should focus on ways to bring costs down (including by improving performance parameters like efficiency and shelf-life). Battery configurations relying on earth-abundant materials should be top priority. Otherwise, the suite of battery systems receiving support should remain broad and include advanced, less-mature designs, until more clarity emerges about which technologies will compete best on cost. Electrochemical storage, including flow batteries and hydrogen systems, should remain a focus as well. Scientists have examined electrochemical storage solutions for decades, and while smaller-scale storage and mobility applications have taken centre stage and a large amount of the hydrogen funding at least in the U.S., renewed focus should rest on inexpensive conversion processes for grid-scale electricity storage.

6.5 Conclusions

Electricity storage can provide vital grid services needed for integrating renewable and nuclear generation. Worldwide deployment is still relatively small and dominated by pumped hydro, but that picture may change with sustained R&D investment and electricity market reforms to ensure low-cost storage solutions can provide value. To that end, we recommend entities with regulatory jurisdiction or the ability to provide research support do the following:

1. Create a level playing field, allowing storage to participate in provision of all market services for which it is qualified,
2. Support deployment of mature storage technologies as widely as possible (especially pumped hydro), by characterizing resources to identify those with

the lowest environmental impact and highest potential to help integrate base load and intermittent generation, and

3. Continue to fund research, development, and demonstration projects across a wide range of advanced energy storage technologies, including grid-enabled electric vehicle systems.

7. ENVIRONMENTAL IMPACTS OF NUCLEAR AND RENEWABLES

There have been many studies on the environmental aspects of nuclear power generation and of renewable energy sources, for two particular reasons:

1. Nuclear energy production involves a series of processes from uranium mining through to final waste disposal, all of which are major engineering activities. These commonly require the production and assessment of an official Environmental Impact Assessment (EIA) before they can be licensed. Large scale use of renewables such as in wind farms or solar power plants also requires an EIA.
2. Nuclear energy production is a controversial subject in most countries, resulting in an active debate on the associated environmental issues, and on their impacts relative to alternative means of producing electricity. Increasing controversy also surrounds the environmental impacts of renewables as aggressive programmes to increase their market penetration have led to growing local opposition in some countries.

Most often in the current energy debate, the comparisons that have been made are between nuclear and fossil fuels or between renewables and fossil fuels. Given the widely acknowledged high impact of fossil fuels, both on human health and on potential climate change (Lynas, 2014), the more interesting comparison is between nuclear energy production and the use of other low carbon alternative energy sources such as hydropower, solar, wind and biomass.

In this section we consider the environmental impacts of nuclear and renewables in terms of a wide range of actual and potential impacts. We begin by using the literature data to summarise, as quantitatively as is feasible, the impacts of the different technologies in terms of their materials and energy requirements, their emissions during operation, their health effects during operation, the accident risks, and the associated waste streams. Recent studies referenced below provide important information for such comparisons, although challenges remain in gathering reliable data and in appropriate normalization of the impacts. We follow up these sections with some more anecdotal evidence on selected impacts that are either particularly topical or are important but less commonly addressed. These include impacts of wind turbines on persons and on bird life, the underestimated problems with biomass, and concerns about biodiversity reduction. Finally we address the public attitudes towards both renewable energy technologies and to nuclear power. The energy policies of many countries are perhaps more strongly influenced by public and political perceptions of available technologies than they are by rational assessment of the actual benefits and drawbacks.

The most transparent approach to comparing all such environmental impacts from various electricity production methods is within a Life Cycle Analysis (LCA) in which all impacts, including costs, are assessed and summed throughout all stages. These types of analyses were extensively studied within the ExternE Project of the EU which ran through to 2005. The website established for this project remains updated to include references and results for subsequent work up to the present (ExternE, 2014). This includes links to the NEEDS project (Lecointe et al. 2007) which extends the nuclear power plant studies to include estimates for newer reactor technologies that may be implemented up to 2050.

Many of the figures given in the texts below for nuclear power and for renewable technologies are based on LCA analyses, often making use of the Ecoinvent database produced in Switzerland (Ecoinvent, 2014). A large database was also assembled for the recent IPCC report on climate change (IPCC, 2014) and results from this comprehensive document are also included below.

7.1 Resource Requirements

7.1.1 Materials

The principal materials concerns with wind and solar energy technologies relate to the use of "rare earth" materials. The U.S. Department of Energy (U.S. DOE, 2011), and others, have identified neodymium, dysprosium, terbium, europium, and yttrium as substances which wind turbines and solar panels are currently dependent. Shortages in the short to medium term have been identified; these are likely to impact on the availability and cost of these materials. Particular concerns have arisen about the role of China which controls a large proportion of rare earth supplies.

The proportion has been estimated at over 90% of the world total production of rare earth metals (REM). Recently, China has cut its exports for the second time due, it is claimed, to environmental concerns (Rudarakanchana, 2015). This has stimulated those economies with the greatest demand for critical materials—the EU, the USA and Japan—to seek the implementation of a strategy which would encourage new mining activity, focused R&D, and other actions (including efforts to offer EU experience in the field of environmental protection as part of trade agreement arrangements with China) to help offset reliance on China (Baldi, Peri and Vandone, 2014).

A top capacity wind turbine may use 2000kg of neodymium-based permanent magnet material. Not all wind turbines use permanent induction generators (that currently require neodymium); some use Tesla induction generators that do not require rare earth magnets, but the latter require variable pitch control and, if HVDC transmission becomes a widespread reality, permanent induction generators are likely to be given preference.

Thin film solar cells, while still a small segment of the solar market, draw on three other "critical" (but not rare earth) materials—gallium, indium, and tellurium. Indium and tellurium are already exceptionally scarce. Issues related to the materials employed in the various power generation technologies relate, not just to the quantities of different elements but also to their toxicities, especially if toxic elements eventually land up in waste streams. This aspect is dealt with in Section 7.6.

Table 9.6 | Critical raw materials content of renewable resources technologies.

Application	Component	Critical raw materials content
Wind and hydropower plants	Permanent magnets of synchronous generator	Neodymium, dysprosium, praseodymium, terbium
	Corrosion-resistant components	Chromium, nickel, molybdenum, manganese
Photovoltaics	Transparent electrode	Indium
	Thin film semiconductor	Indium, gallium, selenium, germanium, tellurium
	Dye-sensitized solar cell	Ruthenium, platinum, silver
	Electric contacts	Silver
Concentrating solar power (CSP)	Mirror	Silver
Fuel cell-driven electric vehicles	Hydrogen fuel cell	Platinum
	Electric motor	Neodymium, dysprosium, praseodymium, terbium, copper
Biomass to liquid (BtL)	Fischer-Tropsch synthesis	Cobalt, rhenium, platinum
Electricity storage	Redox flow rechargeable battery	Vanadium
	Lithium-ion rechargeable battery	Lithium, cobalt
Electricity grid	Low-loss high-temperature super-conductor cable	Bismuth, thallium, yttrium, barium, copper

Fig. 7.1: Critical raw materials content of renewable resources.

There have been considerable efforts to portray the scope for deploying advanced materials for "sustainable energy" (Dusastre, 2011), and there has been great interest in the use of graphene for solar cells.

Renewable power sources also require significant quantities of conventional materials such as concrete. For the case of wind turbines, a 300MW installation using 152 wind turbines of 2MW each required 930,000m^3 of concrete (Lafarge, 2012). For a 1000MW plant, this implies that 310,000 m^3 of concrete would be required.

In nuclear power plants, the only material for which the availability of resources has been discussed is the uranium fuel itself. The world's present measured resources of uranium at prices around 1.5 times present prices, if used only in conventional reactors, are estimated to last for about 90 years (WNA, 2015), but uranium is ubiquitous and recovery even from sea water may become economically feasible. More practically at present, if uranium were used in breeder reactors in a closed fuel cycle the available resources could last 50 times longer. Nuclear power plants also require large quantities of steel and concrete, which together comprise over 95% of the material inputs (Peterson, Zhao and Petroski, 2005). The data of these authors imply that construction of older nuclear power plants with a 1000 MWe capacity required around 45,000t of steel and 100,000 m^3 of concrete. However, they show that the evolutionary Generation III plants—EPR and ABWR—use approximately 25% more steel and 70% more concrete than older LWRs. Other sources quote differing figures for a 1000 MWe plant, with estimates for steel ranging from 35,000 to 70,000t and for concrete up to 350,000m^3 (White and Kulcinski, 1999). Specific figures for a Swiss PWR given by Dones et al. (2007) are 40,000t for steel and 170,000 m^3 for concrete. It is apparent that the concrete needed per MW capacity is higher for wind power than for nuclear. If the normalization is to GWe years rather than maximum power, then the comparison looks even less flattering for wind power, given its lower average availability.

7.1.2 Water Use

Water usage by power plants is second only to that required by agriculture, but the major part of the water used for cooling a plant is returned to the river or sea from which it is taken. The figures given by the U.S. Nuclear Energy Institute (NEI, 2014) for water use by a nuclear plant normalise to 1500 l/MWh(e) for once through cooling and 2700 l/MWh(e) for a plant with wet cooling towers. For a specific example (the 1000MWe Leibstadt plant in Switzerland), the required cooling water throughput is 32 m^3/s and the losses from evaporation amount to $1m^3/s$. The nuclear plant requirements are stated to be 2 to 4 times lower than for geothermal or solar-thermal power plants. The highest water usage is by hydropower plants which can lose 170,00l/MWh(e) due to evaporation from reservoirs. The IPCC WG III report on renewables (IPCC, 2011) gives a comprehensive overview of water usage, as illustrated below.

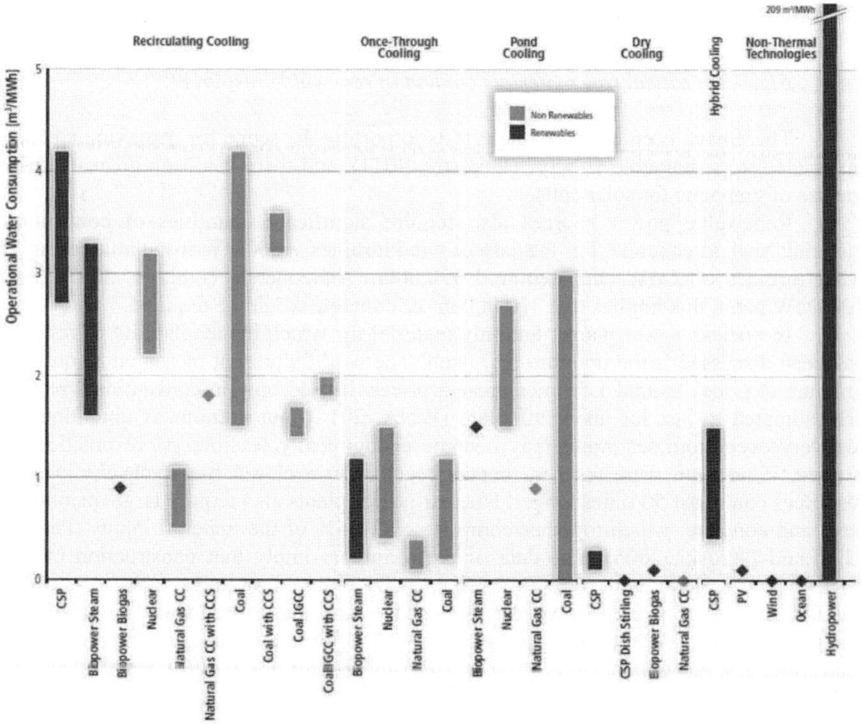

Fig. 7.2: Power plant water consumption.

7.1.3 Land Use

Hirschberg and Dones (2014) estimate that, for operation of nuclear power plants, the necessary land requirements correspond to only 0.6 m^2/GWh(e). The Swiss figures, given for hydro power and large solar, are respectively 49 and 1275 m^2/GWh(e). They point out, however, that solar panels on existing buildings need no additional land areas. Mackay (2009) states that wind farms need around 500 times as much land as a nuclear plant. This figure is in broad agreement with that of the NEI (2014) which estimates that wind farms need 300 times the area to produce the same output as a nuclear plant. The SBC Energy Institute (SBC, 2014) gives land use figures for the plant itself and also for all indirect land use. In normalised units m^2/GWe, these are for nuclear power 50/120; for photovoltaic 329/463 and for wind 1500 to 3200.

A further contentious point, however, is the potential for long-term loss of land use. This can be large for a major nuclear accident, as evidenced by Chernobyl and Fukushima. It can also be large for a dam; the extreme case is the Three Gorges dam in China which will eventually have a 660km long reservoir with an area of 630 km^2 and a capacity of 18GW.

7.3 Emissions

7.3.1 CO_2 Emissions

Mackay (2009) estimates that building a 1GWe nuclear power plant results in CO_2 emissions of 300,000 t CO_2. For a 40 year plant lifetime, this corresponds to around 1g CO_2/kWh(e). This is much lower than the figures that he gives for fossil fuels (400g CO_2/kWh(e)). For the total CO_2 emissions for all stages of the fuel cycle MacKay quotes an IPCC estimate of 40g CO_2/kWh(e). The Austrian study by Wallner et al. (2011) points out that below a certain ore concentration the life cycle emissions become dominated by the ore preparation; for a low concentration of 0.013%, the value given is 288g CO_2/kWh(e).

In general, however, nuclear power and all of the renewables considered result in much lower greenhouse gas emission than do fossil fuels.[13] This is illustrated clearly by the figure below which is based on IAEA data.

[13] The figures for wind and solar are somewhat dependent upon whether aluminium is used in the construction of wind towers or solar plants. The association of sulphur hexafluoride with aluminium, a so-called greenhouse gas having a Global Warming Potential in excess of 3,000 times an equivalent volume of carbon dioxide, makes this a potentially important consideration.

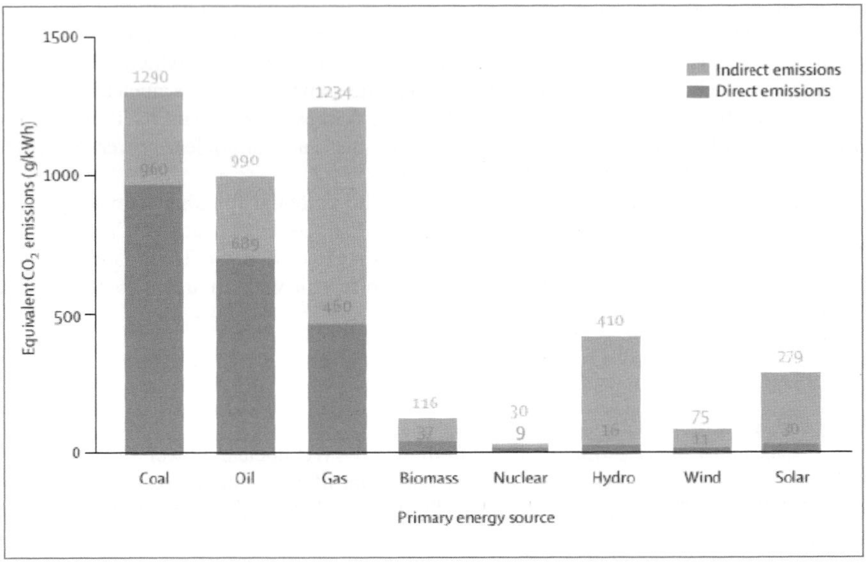

Fig. 7.3: Greenhouse gas emissions of primary energy sources. Source: Markandya and Wilkinson, 2007.

The IPCC 5th Assessment Report (IPCC, 2014) also quotes CO_2 emission data confirming that nuclear energy is among the lowest carbon forms of generation similar to wind turbines at 12g CO_2/kWh(e). Hydro and solar have emissions of 24g CO_2/kWh(e) and 28g CO_2/kWh(e) respectively. Biomass has much higher total emissions of 220gCO_2/kWh(e). For comparison, coal is at 920gCO_2/kWh(e), gas at 490gCO_2/kWh(e) and fossil fuels with CCS at 160-220gCO_2/kWh(e).

7.3.2 Toxic Emissions

All energy technologies also produce air pollutants at some points in their life cycle. These include particulates, nitrous oxides, sulphur dioxide and volatile organic compounds. Nuclear and fossil fuels also result in some radioactive releases. At the local and regional levels, air quality can be severely affected by carbon monoxide, and sulphur and nitrogen oxides. Per GWh(e) generated, solar thermal emits larger volumes of carbon monoxide (285 kg) than natural gas (190 kg) or oil (110 kg). Solar PV has been estimated to release 263 kg of nitrogen oxides and 731 kg of sulphur oxides per GWh(e) electricity generated, and wind energy 71 kg and 137 kg respectively (Hirschberg, 2007). The data gathered from numerous sources by the IPCC, show clearly that the SO_2 and NO_2 emissions per GWh(e) generated by fossil fuels and biomass far outweigh those from nuclear power and all other renewables. Nuclear power plants release radioactive gases such as Kr-85 and Xe-133 and I-131. These can result in very small doses to persons.

7.4 Effects on Health

Systematic studies of the effect of energy production on health were begun at Brookhaven National Laboratory in the early 1970s and at Lawrence Berkeley Laboratory soon thereafter. More recently, Canada's Atomic Energy Control Board studied the overall risks of 11 energy sources and found that total risk per unit of energy output of wind came third (after coal and oil), and was then followed by solar PV in fourth place, solar thermal in fifth, solar space heating in sixth, and hydro seventh. Nuclear and finally natural gas offered the lowest risks in this survey. Man-days lost per unit of energy output, though lower than for coal and oil, were not inconsiderable—with wind and solar PV taking the next highest slots. A 1991 IAEA study found somewhat similar rankings, with fossil fuels (coal, oil, and gas) posing the highest mortality risks, followed by renewables and finally nuclear (Haddad and Dones, 1991). Based on ExterE data, Starfeld and Wikdahl (2011) produced the chart below comparing deaths per KWh from various sources.

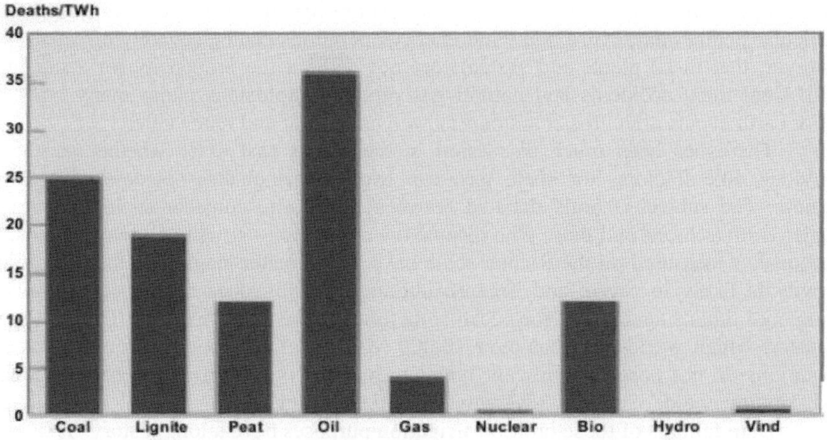

Fig. 7.4: Deaths per KWh by primary energy. Source: Starfeld and Wikdahl (2011).

The main cause of human fatalities in the solar PV sector arises from falls incurred by solar rooftop installers, and there have been estimates of these up to 150 per year worldwide. In the wind energy sector, human fatalities have been rising as the number of turbines installed has risen. In the USA 99 fatalities were reported as having occurred in the recent past (East Country Magazine, 2012). A UK source reported that there were between 157 and 166 wind energy-related accidents annually between 2011 and 2014 at least 1,660 since the 1980s, and 151 fatalities, of which over 60 were not engaged in the industry (CWIF, 2014). These figures are considered to be only "the tip of the iceberg". For example, in December, 2011, the UK's main wind energy industry association, RenewableUK, reported that there had been 1,500 accidents over the previous five years, although claiming these had resulted in only four fatalities and about

300 injuries. The industry association's then Director of Health and Safety further claimed: "No members of the public have ever been injured or harmed in the reports we have received." Nevertheless, there have been numerous near misses due to blades falling off, small turbines collapsing (one in a school playground), and ice shear (TDT, 2011). In Germany there were 880 ice shearing events alone between 1990 and 2003; 94 in 2005; and 27 in 2006. Many of the operational accidents have been transport-related.

Of course, public interest focuses primarily not on impacts of normal operation, but rather on potential consequences and probabilities of accidents. These are treated in the following section. But in the long run, given pressures of world population increase and land and water stresses, it is arguably modern biomass and biofuel developments which pose the greatest threat to sustainable development among the new renewables.

7.5 Accidents

In this section, we have in mind the potential impacts of earthquakes, tsunamis, volcanic eruptions, land slips, flooding and hurricanes. These potential impacts can have a bearing on some forms of new renewable energy operations as well as on nuclear. In particular geothermal, hydro and wind energy operations can be severely affected. Note, however, that fossil plants and systems are not without the propensity for accidents as well. Coal mine accidents and natural gas pipeline explosions claim many lives, and major earth events affect fossil systems as well as nuclear and renewables.

There has been much discussion in the recent past as to whether geothermal activities, like fracking for shale gas, can trigger earthquakes—or less severe earth tremors. The subject is hotly debated between those who consider such activities can trigger earth tremors, and those who dispute the contention—usually on two grounds, that earthquakes measured on the Richter scale have a far greater magnitude than any human activity is likely to cause, and because although drilling does not cause earthquakes, using cool liquids against hot rocks can sometimes cause minor tremors. The geothermal industry, which world-wide has over 10,000 MW installed capacity, would claim their operations do not pose a significant threat to humans or the environment. Earthquakes and flooding could, of course, adversely affect their operations.

Dam failures of those built for irrigation purposes have a long history. Those built for hydropower have a somewhat shorter one. About half the failures that have occurred in recent decades were the result of heavy rainfall and resultant flooding - such as Banqiao- Shimantan, China, in 1975; Machuchu- 2 in India, in 1979; Shakidor, Pakistan, in 2005; and Kopru, in Turkey (where the dam's diversion tunnel seal also failed), in 2012. Some 30 percent of failures are due to structural issues—the hydropower dam at Gleno, Italy, in 1923; Teton, Idaho, USA, in 1976; and Campos Novos, in Brazil, in 2006. No estimates of human fatalities match those at Banqiao-Shimantan, in Henan province, in which 171,000 people died. In excess of 1,800 people died as a result of the failure of Machchu-2 in 1979, since then, there have been only three cases where fatalities exceeded 100.

Hurricanes and strong winds more generally can cause wind turbine towers to collapse, blades to be spun off, turbines to catch fire, and ice shear to spread wider—potentially to areas where human life and property may be threatened. However, engineering solutions including providing the ability of wind towers to yaw can significantly reduce the risk of wind tower damage. The U.S. National Academy of

Sciences has looked into the potential for tower buckling, and found it high relative to other locations in Galveston County, Texas, and in Dare County, North Carolina.

Most energy sources and facilities are, of course, liable to be affected by earthquakes, strong winds, tidal waves, and other causes of flooding. From offshore oil and gas rigs to inadequately protected nuclear plants, and even coal mines let alone solar or wind power or biomass/biofuels, these facilities are not risk-free or able to withstand all possible natural disasters. Stefan Hirschberg has reported that in the period 1969-2000, hydropower was responsible for the highest number of fatalities among all forms of energy in non-OECD countries if the very high number of deaths following the Banqiao/Shirmantan dam accident is included. Even when this accident is excluded fatalities rank equal second, with non-OECD including China for coal, to those in the LPG sector ranked highest.

The IPCC (2011) report again presents relevant data. Results are presented for severe accidents with 5 or more fatalities since these tend to attract most public attention and to influence political actions most strongly. The authors give fatality rates normalized to the electricity generation in GW-years, and they estimate maximum consequences for a single accident of a specific energy technology. Coal has by far the highest fatality rate (especially in China) and hydropower and nuclear have the lowest of the large technologies. Hydropower can, however, have large numbers of fatalities due to dam failures. All other renewable energy technologies have low fatality rates and low maximum consequences, the latter because of their distributed nature. The authors point out, however, that other types of risks should also be considered and give as examples induced seismicity from enhanced geothermal systems and the risk of ship collisions with offshore wind turbines. The extract below from the IPCC report illustrates the fatality rates for nuclear and renewables:

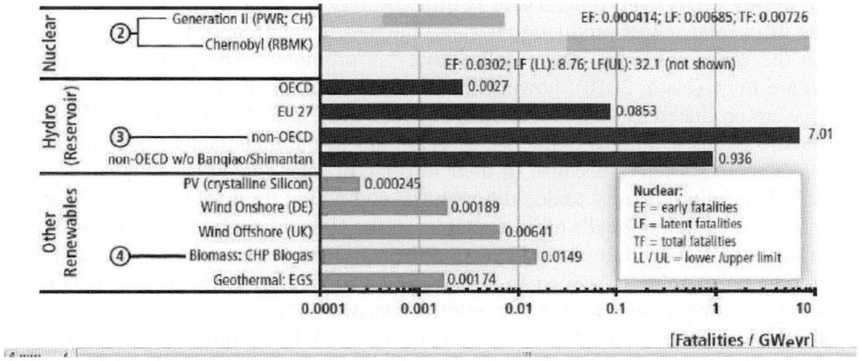

Fig. 7.5: Fatalities for nuclear and renewables. Source: IPCC, 2011.

<u>7.6 Wastes</u>

Because of the enormously higher energy density in nuclear fuels, nuclear power plants produce much smaller quantities of wastes than do fossil plants. But the more relevant comparison here is with renewable energy systems which are often thought to produce little or no wastes. Again the most objective comparison normalizes the wastes produced to the electricity generated.

A 1000 MWe light water reactor generates around 200-350 m^3 low- and intermediate- level waste and about 25 tonnes) of used fuel per year, or about 1'500 tonnes of used fuel over an assumed 60-year plant lifetime. (A coal fired plant of the same size produces about 400,000 tonnes of ash). While the total volume and mass of nuclear wastes produced per year is relatively small, it is the radiotoxicity and heat of nuclear waste that determines the relative burden in managing it. Nuclear waste requires isolation from the biosphere and sufficient spacing and cooling, so is per-tonne more expensive to manage than wastes from renewables. Rhodes and Beller (2000) state that a 1000MWe solar electric plant generates 6'850t of hazardous waste from metals processing over a 30y lifetime, and a 1000MWe solar thermal plant would generate 435,000t of manufacturing waste of which 16'300t would be contaminated by heavy metals.

Disposal of the wastes produced from electricity production is a high profile issue in the case of nuclear power. Despite the fact that geological disposal is widely recognised by the technical community as a suitable and safe approach, there is often strong public and political opposition to specific disposal projects, based primarily on the high toxicity and long lifetimes of the most active wastes. Around 5% of the 1500 tonnes of used fuel from the lifetime of a large reactor contain consists of highly radioactive fission products and plutonium; the rest is uranium which can be recycled into fresh fuel. The energy produced over the lifetime is around 50 GWe years, giving a specific spent fuel inventory of 30 tonnes per GWe year or a highly toxic waste inventory of around 1.5 tonnes per GWe year.

On the other hand, solar modules contain some potentially dangerous materials which do not decay with time. There is little awareness of the problems of dealing with wastes from solar power production because the panels are produced largely in China and not in the other countries where solar power is promoted. The potential problems in China are huge (Nath, 2010); however,; even in California around 20,000t of hazardous waste were produced by solar companies between 2007 and 2011 (Dearen, 2013). One of the most obvious examples is the use of cadmium in the manufacture of thin film solar panels. This is attractive because of their lower cost than silicon-based solar. However, elemental cadmium is very toxic, although the compound used CdTe may be less so. In the USA, Cd containing cells are acceptable; in the EU, although Cd is acknowledged as carcinogenic, it is widely forbidden—but exceptionally allowed in solar cells since promoting renewables is judged to be more important (EU Directive 2011/65/EU). A quantitative comparison with toxic nuclear wastes is of interest. The large Desert Sunlight Solar Farm in the Mojave Desert in the USA has 8 million solar panels, each with around 6g of Cd. This implies that the Cd inventory for the 550 MWe peak capacity plant is around 40 tonnes. Unless all this Cd is recycled after the 25y plant lifetime, the toxic waste inventory per GWe year is around 6 tonnes, assuming 50% availability.

7.7 Public Attitudes

Few robust general statements can be made about public attitudes to either new renewables or nuclear power. In both cases, there are strong proponents and the bitterly opposed. However, it is very clear from public polling that there is a fundamental difference in public attitudes to renewable energy sources and to nuclear power. The data below from 2011 (Wallard et al. 2012) illustrate this clearly:

Energy source	**Public Support (%)**
Solar power	97
Wind power	93
Hydroelectric power	91
Natural gas	80
Coal	48
Nuclear energy	38

Source: Wallard et al. 2012

Fig. 7.6: Public attitudes towards nuclear and renewable power.

7.7.1 Attitudes to Renewables

Public attitudes towards modern biomass and biofuels have been influenced by loss of traditional forests, reduced diversity of agricultural crops grown, concerns about impacts on wildlife, and the activities of environmental pressure groups. In general environmental pressure groups have been opposed to modern biomass and biofuel development on rational grounds. The same is not true for wind energy development and operation, large hydropower schemes, or estuarine barrages. In each of these cases, there are strong emotional issues involved in the argumentation. Support is normally high for solar PV at a small scale, but the acceptance of large area solar parks has not been widely tested.

Performance claims for renewables have been shown to be exaggerated (Jefferson, 2012). Negative impacts of wind energy developments on residential property prices, sleep patterns for some people, and visual outlooks are not regularly discussed. Atif Ansar and colleagues (Ansar et al. 2014) have noted that for large hydropower schemes: "Much of the data in existing literature are drawn from surveys and interviews of dubious validity. At times, interest groups, seeking to promote a particular kind of scale or technology, also report distorted data." Their first recommendation was: "Create transparency on risk profiles of serious energy alternatives, from not only the perspective of financial cost and benefit, but also the environmental and social impacts—hard evidence is a counter-point to experts' and promoters' oft-biased inside view."

In the case of estuarine barrages for tidal power, there have been statements such as: "The most detailed work…carried out on the Severn has suggested that birds may actually benefit in most cases to a (sic) barrage being present." Speakers at a tidal power conference covering estuarine barrages in the Bay of Fundy managed to avoid any reference to birds (until challenged from the audience), and time has proved them incorrect in their assessment of fish mortality at Annapolis (Clare, 1992).

Concern at challenges which may be posed by enhanced near-surface global warming also affects attitudes and support for new renewable energy developments in many quarters.

Public attitudes at the local level may also differ sharply. Many may find wind turbines aesthetically pleasing, while others do not. Some may be deeply disturbed by their visual intrusion, on historic assets or fine landscapes, while others may be indifferent. Some may have residential properties at risk of price falls, or are among those who suffer loss of sleep pattern due to aerodynamic modulation from wind turbine blades. Some may see job opportunities arising, or other financial benefits (farmers where wind turbines are located, for example), while others are concerned their vacation homes will be blighted (such a polarity has emerged in various European countries, including Finland and the UK). There are also many examples of public acceptance growing over time among those living in close proximity to wind turbines, solar 'parks', and nuclear plants. There has even been reported an "inverse NIMBY" (Not in My Back Yard) syndrome where those with wind energy developments in their "backyard" strongly support the technology (Warren et al. 2005). As Charles Warren and his colleagues point out, the media may often give disproportionate attention to vocal minorities—in this case those opposed to wind energy developments. But vocal minorities often demonstrate that they are more closely in touch with the local realities of costs and subsidies, impacts on residential property values and potential health impacts, and more sensitive to landscape impacts. Nevertheless, as Matthew Cotton and Patrick Devine-Wright have written:

> "The NIMBY label is problematic, as it is often used by proponents of development projects as a means to discredit all forms of project opposition, regardless of its motivation, and wrongly characterizes local people as worried, irrational, ignorant of scientific and technical facts, and selfishly unwilling to support projects that benefit the broader society.
>
> Even if one were to accept this characterization of citizens opposed to the siting, NIMBYism is neither irrational nor necessarily unethical" (Devine-Wright, 2011a).

There is a growing body of evidence that local people are frequently well-informed about a technology, its likely performance in the intended setting, and the wider implications of its usage. There is also growing recognition in the literature of the importance of a sense of place attachment, something which due to familiarity over time is more deeply entrenched in rural dwellers. Thus landscape values and historic assets may be more highly valued by a higher proportion of rural dwellers than urban dwellers. There may be significant differences in emotional attachment between individuals, wherever their location (Lynne and Devine-Wright, 2014) There also appear to be significant differences between countries when it comes to rural countryside protection concerns—stronger in England, for example, than in The Netherlands or in Germany (Devine-Wright, 2011b).

Beginning to emerge among the general public in some relatively prosperous countries, however, is an even more profound concern. This is that a "dominant faith that renewable energy alternatives exist in abundance, and all we need to do is to shift them, and then our energy and climate change problems can be solved, and we can go on enjoying affluence and growth forever" (Trainer, 2013). In light of the above, and given the significant economic challenges faced by countries that have adopted high penetrations of intermittent renewables (including Germany and Spain), this view deserves a long and hard look.

Thus even the aspiration of those who believe that the last remaining hope, on technical potential grounds, of meeting these challenges is solar energy has been challenged. That hope was, and for some still is, the use of concentrating solar power, with ultra-high voltage direct current transmission (CSP+UHVDC). This has the technical potential to meet the demand for energy services not just in Europe (the Desertec concept) but throughout Africa, as well as in the Americas and Asia. Proponents have had their optimism undermined by political and social turmoil in North Africa, the differing perspectives of European countries (especially between France, Germany and Spain), and failure to press forward determinedly with this technology (which has been around since 1912) in the USA.

7.7.2 Attitudes to Nuclear Power

It is interesting to study the fluctuations over time in public support for or rejection of nuclear power. There have been many studies over the decades since nuclear power was introduced which have discussed the public attitude to this energy source and, in particular, have analysed the reasons for the acknowledged anxieties of the public concerning radiation. Some have asserted that the nuclear bombing of Hiroshima and Nagasaki, Japan, near the end of WWII, has coloured public perceptions since then.

Others (Weart, 1988) claim that, even before then, there was an inbuilt fear of a hazardous agent that one "could not see, touch, taste or smell". It has also been claimed by von Hippel for example in GEA (2012), that "one abiding concern has been the connection with nuclear weapons". However, the weapons connection today seems more of an argument for specialists than for the general public. The potential for high consequence accidents influences people more than do the arguments that probabilities of these are very low; the risks imposed on individuals by others are less acceptable than voluntary risks that they take upon themselves and the potential for long-term and wide-spread contamination of land causes concern.

Nevertheless, support for nuclear energy increased slowly in many countries, with the strongest support often being in countries with significant numbers of nuclear plants. In numerous countries, however, there has long been extremely active opposition to nuclear power, the most pronounced examples being Germany, Austria, Italy and Australia. The safety-related arguments have alternated between concerns about power plant operation and radioactive waste disposal. Currently, in western countries moving to expand or introduce nuclear power, an additional major controversy is related to the cost of nuclear energy relative to the (subsidized) costs of wind or solar power.

In the first decade of this century, additional powerful arguments leading to increased support for nuclear were the need for expanded base load electricity in developing countries and also the growing fear of catastrophic climate change due to CO_2 emissions from fossil fuel plants. Countries like China, India and South Korea developed massive expansion plans. Countries like the UK and the USA have been, mainly for economic reasons, ambivalent about nuclear new build. Countries with ageing plants such as Switzerland and the Netherlands started planning for newer generation plants. Even in Germany, the Government reversed plans for shutting down nuclear and agreed to extend plant lifetimes.

Then in March 2011, an unprecedented large earthquake and resulting tsunami led to the Fukushima Daiichi plant in Japan losing on-site electrical power for so long that three reactor cores melted, hydrogen explosions wrecked the upper part of the reactor buildings where the spent fuel pools were, and widespread radiation releases resulted.

Unlike in the Chernobyl disaster decades before, this plant was of the same design as western reactors, the media coverage was intense and the nuclear consequences (which did not include any radiation induced deaths to members of the public) were accompanied by massive loss of human lives and of buildings due to the huge wave that swept inland. Since the Fukushima accident, there have been a number of studies and public polls aimed at assessing the public and political impacts. Questions have also been raised about the level of defences (e.g., height of containing walls to prevent incursion of tsunami driven water in an area known to be close to a tectonic plate) and quality of maintenance somewhat removed from reactor design issues.

The immediate impacts were naturally very negative. Rapid political decisions in Germany and Switzerland (taken before any proper assessment of public attitudes—and shortly before upcoming elections) led the governments of these countries to announce that they would renounce nuclear power. A WIN-Gallup International Global Poll within a month of the accident recorded the views of over 24000 persons in 42 countries. 52.7 percent favoured the use of nuclear energy before the accident, and only 45.4 percent favoured it after the accident (Younghwan et al. 2013), with the sharpest decrease in support (28%) being, as expected, in Japan.

In some countries, however, the dip in acceptance was short-lived or absent. In the USA, the 57% who favour nuclear power just one year after the accident was identical to the percentage measured just before the Fukushima incident (Newport, 2012). Equally striking are the results from the UK (Poortinga et al. 2013). The figure below shows that support for nuclear has remained rather constant throughout the Fukushima phase. In fact, the proportion wanting to phase out nuclear power (immediately or gradually) decreased from 50% in 2005 to 40% in 2013. It also shows, interestingly, that support for new renewables, solar and wind, although still much higher, has been reducing as their implementation has been spreading. Other countries where national nuclear support continues to be strong include China, Finland, India and South Korea.

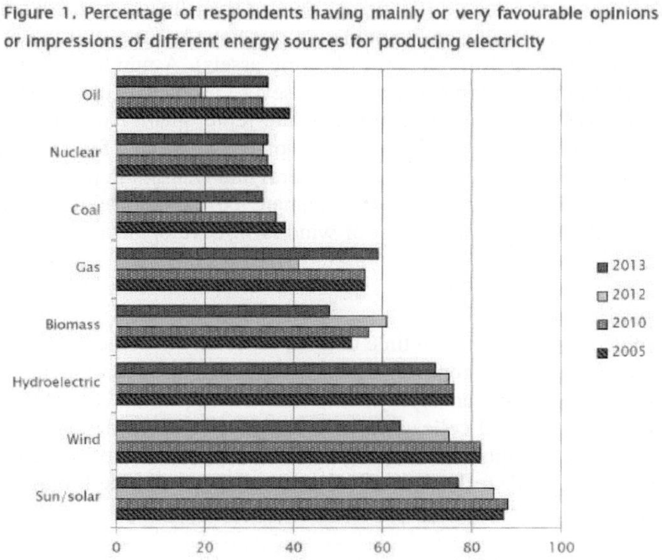

Fig. 7.7: Public attitudes about different energy sources.

It is interesting to speculate whether government support in these countries is a reason for the higher levels of public support or whether the greater public support makes it easier for governments to propose expansion.

7.8 Additional Issues

7.8.1 Wind Power

For some people the incidence of aerodynamic (or amplitude) modulation of wind turbines can cause sleep disturbance, and hence health problems, at distances up to 1.5 kilometres from their nearest turbine (Bowdler, 2008; Stigwood, 2009; and Pedersen and Persson, 2007). This is not turbine noise as such, but air disturbance (the swish-swish noise of blades turning) (Hanning, 2009). Hanning's paper reports on acoustics experts and doctors' case histories. At the March, 2014, Institute of Acoustics meeting held in Newport, U.K., examples of amplitude modulation and its impacts were provided from the USA, Sweden and Australia, as well as the UK. Concerns and complaints have continued to be reported, and it was therefore somewhat surprising that an MIT study funded by the European Wind Energy Association and its Canadian counterpart found there were "no clear or consistent associations between noise from wind turbines and any reported harm to human health". (December 4, 2014)

One consequence of the above phenomenon may be loss of residential property values. Wind energy developments can reduce residential property values by over 10% where located within 2 kilometres of the nearest turbine—in some cases resulting in properties becoming unsaleable (Gibbons, 2013). Often linked to these effects are the occasional impacts of aerodynamic modulation mentioned above, and visual impacts (on picturesque landscapes and in relation to historic assets). A number of cases have been reported where wind development has caused severe problems such as in Lincolnshire, UK (Gibbons, 2013). Whether such problems are representative of typical wind energy developments is not well known. For example, some studies have found a relatively weak statistical relationship between wind developments and changes in property values (Hoen et al. 2013; Laposa and Mueller, 2010; Hinman, 2010; and Heintzelman and Tuttle, 2012). More research about the impact of wind farm development on property values is needed.

7.8.3 Hydropower and Solar

On a much larger scale from time to time, especially in some developing countries, are the impacts of large hydropower schemes resulting in displacement of population, loss of agricultural land, unemployment, and consequential economic and social upheaval, as Atif Ansar and colleagues have suggested (2014). To date of only limited significance are the complaints from some aircraft pilots and passengers of the near blinding glare of the Sun's reflection off solar mirrors placed in the Mojave Desert on the California/Nevada border.

7.8.4 Impacts on Bird Life

The concern that modern biomass and biofuel developments will compromise biodiversity has led to increased interest in the study of bird life. This is because birds are an interesting indicator of biodiversity, being high in the food chain, sensitive to monoculture expansion, more mobile than most other groups, and therefore likely to respond to and reflect environmental quality on a broader scale than mammals or terrestrial insects. Hence they are often considered to be good indicators of wider ecosystem health (Gregory, 2005; Bateman, 2013).

There is a vast literature on the impacts of loss of biodiversity and of human-erected structures on bird species and mortality. Although bird mortality is strongly linked to windows and other parts of buildings, passing traffic, and domestic cats, there is also strong evidence that wind turbines have been the cause of large numbers of bird fatalities. Fatalities appear to be increasing with the growing height of towers and to blade tip.

In the USA and elsewhere bird fatalities are higher around migration routes, especially down the western and eastern sides of the continent. In Canada also, the placing of wind energy developments along bird migration routes has been strongly condemned—such as on the peninsula in Prince Edward County, on the eastern shore of Lake Ontario. In Australia a wind energy development in North-West Tasmania has been responsible for the deaths of endangered wedge-tailed eagles.

There has been strong criticism of poorly conducted site surveys—even when, for example, the UK's Royal Society for the Protection of Birds (RSPB) has had an involvement (for example, the Moorsyde Environmental Statement near Chillingham, Northumberland). Fewer site visits have been made in such instances than a proper survey would have required, there has been notable under-recording of bird fatalities, and faulty claims that the site was not on a bird "flyway" (Windbyte, 2014). This is despite the RSPB claiming: "We insist that wind farm proposals are subject to rigorous environmental assessment before development is permitted", and acknowledging that: "Some poorly sited wind farms have caused major bird casualties, particularly in Tarifa and Navarra in Spain, and the Altamont Pass in California" (RSPB, "Wind Farms", December 3, 2014). In Denmark, the authorities are accused of having bulldozed plans through to place seven 250-metre high wind turbines between the Thy National Park and the Veljerne RAMSAR site, which some estimate is Europe's largest breeding ground for migratory birds. There are also instances, however, where concerns about bird species have caused the planned expansion of wind energy developments to be abandoned—as in the case of the offshore London Array, where the proximity of a significant red-throated diver colony was a major factor in 2014.

Efforts to reduce casualties have arrived at the following conclusions. A prior environmental assessment should be carried out to ensure wind energy developments are not placed on or close to bird migration routes. Start-up wind speeds should be increased from 4 metres per second to 5.5 metres per second (this has been found to result in little loss of overall wind turbine performance—in load/capacity factor terms). A radar look-out for the approach of large numbers of birds has also been recommended, although research findings suggest a poor link between number of birds in the vicinity of turbines and fatalities (Manuela, 2008). Stopping wind turbines within three minutes of the sighting of the approach of birds has been estimated to reduce fatality rates of griffon vultures by up to 50 percent.

It is not just bird fatalities that have given cause for concern. In 2010 420 wind turbines in Pennsylvania killed more than 10,000 bats according to the State Game Commission. At Tug Hill Plateau in New York State there are 195 wind turbines, with bat fatalities estimated at close to 60 per turbine per year. At Canada's Ontario Wolf Island Eco- Power Centre there were nearly 2,000 bird and bat deaths in the first eight months of the wind development's operation, with 33 bird species involved and five bat species—bat fatalities accounted for 1,270 of that total number. Then in December, 2013, a study from the University of Colorado at Denver appeared which concluded that: "well over 600,000 bats may have been killed at wind energy facilities in 2012" (Hayes, 2013). Bat fatalities due to wind turbines have been reported in many European countries.

To date little action has been taken against wind energy developers and operators to discourage placing turbines where bird and bat fatalities arise. In the USA, however, there are environmental laws which can be applied to discourage this under the Migratory Bird Treaty Act. Duke Energy was fined $1 million in November, 2013, when eagles died at Top of the World and Campbell Hill wind energy developments outside Casper, Wyoming. Since then Exxon Mobil have been fined under this Act for bird fatalities at U.S. oil production fields; and PacificCorp for electrocuting 232 Golden Eagles and other migratory birds on power lines in Wyoming, so the problem is not confined to wind turbines.

Two other forms of new renewable energy have attracted adverse criticism on grounds of their impact on birds. On February 12, 2014, The Wall Street Journal published an article headed: "The $2.2 Billion Bird-Scorching Solar Project". This related to the large solar thermal (CSP or concentrating solar power) project in the Mojave Desert on the Nevada/California border. Dozens of birds have died there in recent months apparently because of the heat emitted from the tower-based structures (as opposed to the more traditional, though still grossly under-exploited, parabolic mirrors). U.S. Federal wildlife investigators visiting the BrightSource plant at Ivanpah Dry Lake, California, reported an average of one bird igniting in midair due to the solar plant every two minutes. These dead, or dying, birds, are called "streamers" and whereas BrightSource estimate the fatalities at 1,000 per year, the Centre for Biological Diversity estimates the figure at 28,000 (Syracuse.com, August 18, 2014).

The other form is estuarine barrages. These have been under discussion since Thomas Fulljames proposed a barrage across the Severn Estuary, between England and Wales, in 1849 (it is now the site of the Severn Bridge). The most widely known example in existence is near St. Malo, in North West France, where an estuarine barrage runs across the La Rance River. M. Rodier, of the developer Electricite de France, has stated that "The total closure of the estuary between 1963 and 1966 during the construction of the plant caused the almost complete disappearance of the original species." (Rodier and Clare, 1992) Subsequently, Canada has closed its Annapolis barrage in the Bay of Fundy due to concerns about bird and fish mortality, and not proceeded with earlier plans for the Minas and Cumberland Basins, and Shepody Bay, for the same reason (Jefferson, 2008). In 1999 the Cardiff Bay Barrage, in the Severn Estuary between Wales and England, was completed. The loss of inter-tidal mudflats (access to which is essential if wading and some other bird species are to survive, as they need to access the invertebrates there) led to almost all the Common Shelduck and shorebirds that used the Bay to be greatly reduced. Some initially used nearby sites, but this proved unsustainable. Only two major studies have been published, both with the same lead author (Niall Burton, of the British Trust for Ornithology, 2000 and 2006), and focused mainly on the Redshank (Tringa tetanus). Although this species has relatively high site-fidelity, there are numerous other species which exhibit site-fidelity but which have not been studied. It is known that weight loss and lower survival rates have resulted at nearby settlement areas. This all raises the issue of whether objective assessments have been carried out and, if not, why not. There have also been ongoing concerns of anglers about the impacts of the barrage on fish. On March 18, 2015, in his Budget Speech, the UK's Chancellor of the Exchequer announced that negotiations were opening for a tidal lagoon at Swansea Bay. The would-be developers are seeking £168 per MWh (nearly U.S.$ 248) for 35 years), supported by a (electricity consumer) subsidy of around £92 MWh ($136) per year. The proposed site includes the Blackpill Site of Special Scientific Interest; the Burry Inlet Special Protection Area; and would threaten one bird species (the sanderling) where it occurs in nationally important numbers; and several other species currently present in regionally important numbers (e.g., the ringed plover, the dunlin, the great crested grebe, and the oystercatcher).

7.9 Conclusions and Recommendations

In setting preferences for the choice of electricity generation systems, many aspects have to be considered and, as Hirschberger and Dones (2014) point out in their comprehensive evaluation of systems, no single system performs best under all criteria. For the health and safety and environmental aspects discussed in this Chapter, renewables and nuclear energy both outperform by far fossil fuel sources. The relative advantages and drawbacks of renewables and nuclear are perceived or weighted differently by different technical, public or political stakeholders. Large hydropower is in many aspects superior, but opportunities for its use are often limited by topography and land use conflicts. The new renewables, solar and wind, can supply plentiful but dilute energy; they use large amounts of special materials and can require substantial tracts of land. Biomass can contribute, but the impacts on food production and on biodiversity need to be taken more into account. For nuclear power, the major hurdles are the fear of low-probability, high-consequence accidents and concerns about radioactive waste disposal. All these feed into the further hurdle that costs are high and uncertain. In practice, the dogmatic positions taken by extreme supporters on renewables or of nuclear power may serve only to prolong and increase the use of cheap fossil fuels, despite the acknowledged higher risks associated with these. An "all of the above" approach encouraging the use of all energy sources that have low output of carbon and other emissions appears to be the most rational approach.

Key conclusions and recommendations include:

- In the absence of carbon-capture-and storage, the negative environmental impacts of fossil fuels are so pronounced compared to those of renewables and of nuclear power, that electricity generation strategies should be based on reducing fossil fuel usage by promoting all low carbon technologies over the highest- carbon technologies—where they can be soundly located and ideally with the consent of any local populations.
- Sound analysis of actual and potential technical options should inform energy policymaking, yet in some significant cases is overlooked. Energy policies of many governments in those cases appear to be based on subjective judgements of risks, influenced by powerful energy lobbies of all kinds, and/or designed primarily to appeal to the public.
- Life cycle analyses of energy systems are the most promising approach to developing objective criteria that can aid future decision making. Unfortunately, too little attention is paid to this approach by decision makers.
- The strongest proponents of both nuclear and renewables tend to have an overly optimistic view on the ability of their favoured technology to provide all of the low carbon electricity that is required.
- Nuclear power faces special challenges because of the widely recognised "radiation phobia" in the public. The scientific community in general, and the nuclear industry in particular, have not been very successful at responding to emotionally based public fears with rationally based science data. The greatest public concern about nuclear power is the potential for high consequence accidents, even if their estimated frequency is very low. The ultimate disposal of long-lived wastes also continues to be a concern for the public in most countries.
- The environmental impacts of some renewable electricity sources may be underestimated by the public. Examples include:

    - The hazards of large dams
    - The requirements for rare and toxic materials
    - Failure to consider their efficacy in terms of power density, EROI, sound location (in relation to wind, solar, land and water availability).

- Limitations on sustainability posed by the environmental and human impacts of energy technologies should be factored in to policies that support certain technologies over others.

8. NUCLEAR FUSION: STATUS REPORT AND FUTURE PROSPECTS

The number of conceivable non-fossil candidates that could replace the current massive use of fossil fuels is very limited: essentially renewables and nuclear fission at the moment. A third option for use in the future is nuclear fusion. It is still in the development phase, but has particularly valuable environmental and safety advantages and virtually inexhaustible resources. This chapter discusses the current status of worldwide fusion research, resources, safety, environmental and economic aspects of fusion energy.

8.1 Principles

Replicating the fusion reaction in the sun would be a first possible approach to realize fusion on Earth. However, the p-p reaction in the sun essentially converts 4 protons into a Helium-4 (4He) nucleus containing 2 neutrons. This reaction requires thus the conversion of protons into neutrons, implying a weak interaction with a very low probability and therefore not suited for an economical process on earth. A much more 'simple' solution is offered using hydrogen isotopes, already containing the necessary numbers of neutrons and protons from the start, thus resulting in a reaction where essentially a rearrangement of the nuclides takes place, with a 10^{24} times higher reaction rate than the p-p process in the sun. From all possible reactions involving H isotopes, the least difficult fusion reaction is the one between the hydrogen isotopes deuterium (D) and tritium (T):

$$D + T \rightarrow {}^{4}He\ (3.5MeV) + n\ (14.1MeV)$$

To produce sufficient fusion reactions, the core temperature of a D-T plasma has to be about 150-200 million C.

The reaction products are very energetic (one 3.5 MeV Helium particle and one 14.1 MeV neutron). This energy will be converted into heat in a blanket that surrounds the reactor, extracted by a cooling system and converted into steam to produce electricity using conventional technology. As about one million times more energy is released from a fusion reaction compared to an average chemical reaction, this then explains why so little fuel can produce so much energy: when burnt in a fusion reactor, the deuterium contained in 1 liter of sea water (about 30 mg) will produce as much energy as burning 250 liters of gasoline.

8.2 Status of Magnetic Fusion Research

There exist presently two approaches to realise nuclear fusion on earth: inertial and magnetic fusion. Magnetic fusion uses magnetic fields to confine the fuel. Three classes of devices in use are tokamaks, stellarators and reversed field pinches and many devices exist around the globe. Strong magnetic fields are used to keep the hot particles away from the walls of the confinement device. This method uses the property of charged particles to follow a helical path around magnetic field lines (caused by the Lorentz force) and possible movements perpendicular to the field are thereby highly restricted. The interested reader can find a wealth of additional information in Cook (2001).

Inertial fusion consists of micro-explosions of small fuel pellets by means of powerful lasers or particle beams. Confinement of the fuel is based on the inertia of the pellet fuel mass, which resists the natural expansion when it is heated to thermonuclear fusion temperatures. To achieve sufficient fusion reactions in this short time requires a compression of the fuel to a density of about 1 kg/cm^3, a 10,000-fold increase over the original density and about 100 times the density of lead or 1000 times that of water (Tabak et al. 1994). This approach is being investigated in far fewer countries largely because of the relation of target physics and design to military programs. This is also the reason why some aspects of inertial fusion research remain classified in several countries.

Substantial progress has been made in magnetic fusion in the last decades. Three generations of magnetic devices (mainly tokamaks) led to a 10,000 times better performance (characterized by the value of the fusion triple product—density times temperature times confinement time) in the last 40 years, approaching now reactor conditions (Wesson, 2004).

First large scale fusion experiments took place in the early 1990s. Several MW of fusion power have been released in a controlled way in deuterium-tritium experiments in JET (Joint European Torus, Culham, UK) and TFTR (Tokamak Fusion Test Reactor, Princeton, USA). Peak values of about 16 MW have been obtained on JET in 1997 corresponding to values for the power amplification QDT (i.e., the ratio of the power released from deuterium-tritium fusion reactions to the power applied to heat the fuel) close to 0.7. A summary of what has been achieved in high power D-T experiments is shown in Figure I. In addition, ion temperatures up to 45 keV, which is about 30 times higher as that in the centre of the sun, have been achieved in Japanese tokomak JT-60U (Japan Atomic Energy Agency, Naka Fusion Institute, Naka, Japan) (Ishida et al. 1997).

Magnetic fusion research has thus now arrived at the point where large amounts of fusion energy can be produced in a controlled way. The next step is to maintain a steady power output from fusion reactions in long pulses. To make this possible, superconducting coils for the magnetic fields will have to be used allowing in principle unlimited pulse lengths (evidently, as long as the coils are cooled below 4K). The use of superconducting coils has been very successfully demonstrated e.g., in Tore Supra (CEA-Cadarache, France) and the Large Helical Device LHD (National Institute for Fusion Science, Toki, Japan). Both machines have record length pulses at relevant plasma parameters in hydrogen and deuterium plasmas: 6 minutes and 30 seconds on Tore Supra in 2004 (Bucalossi et al. 2005) and 54 minutes on LHD in 2007 (Mutoh et al. 2007).

Magnetic fusion research entered a new era in 2005 with the international agreement (28 June 2005) on the construction site of ITER at Cadarache (close to Aix-en-Provence in France). Seven large nations are participating in the project: Europe, Russia, India, China, Japan, South Korea and the USA. Construction started in August 2010. First low fusion power experiments are projected for 2023, with high performance fusion plasmas around 2030 and pulse durations of 400s and longer, producing up to 500 MW of fusion power. In addition, Europe and Japan have launched a joint project called "Broader Approach" for exploring the next step device after ITER, which should be a first prototype of a fusion power reactor, producing power for the grid, and is called DEMO. The "Broader Approach" includes three research projects (JAEA, 2014):

A. The Engineering Validation and Engineering Design Activities (EVEDA) for the International Fusion Materials Irradiation Facility (IFMIF), a dedicated laboratory to be built in the near future to test candidate first wall materials/alloys for a future fusion power reactor (Knaster et al. 2013)
B. The International Fusion Energy Research Centre (IFERC) in Rokkasho, close to Aomori, in Japan, for DEMO design and R&D (Nakajima et al. 2012)
C. An upgrade of the tokamak JT-60U, termed JT-60 Super Advanced or JT-60SA (Kamada et al. 2013)

Progress in inertial fusion progress has also been impressive. The National Ignition Facility in the U.S. (NIF, Livermore, California), hosts one of the most powerful laser systems in the world. The NIF laser system consists of 192 individual beamlets that deposit the huge energy needed for the compression of the pellet in one thousandth of a billionth of a second. NIF was completed in 2009 and a recent important milestone has been reached on this facility (Hurricane et al. 2014). The European counterpart to NIF is the Laser Megajoule (LMJ), an experimental ICF device being built near Bordeaux, France by CEA, the French nuclear science directorate. Laser Mégajoule is the largest ICF experiment being built outside the U.S. and uses a series of 240 laser beam lines, organized into eight groups of 30 beams. GEKKO XII is a high-power 12-beam laser system at the Institute for Laser Engineering in Osaka, Japan in use since 1983. The Osaka group has proposed a new concept, called fast ignition, using a Petawatt laser, and is now promoting this as the FIREX project (Azechi et al. 2013).

Future plans in high power short pulse laser systems are ELI (Extreme Light Infrastructure) in Europe and GEKKO EXA in Japan. Both systems are intended to deliver the extreme energies in pulses that are thousand to one million times shorter than in NIF.

8.3 Fusion Fuel Resources

Deuterium, a non-radioactive isotope of hydrogen is extremely plentiful as it can be obtained from ordinary water (about 30 g from 1 ton (Friedman, 1953)) with cheap extraction techniques using conventional technology. Complete burning of deuterons and the first generation fusion products (T and ^{3}He) results in the overall equation:

$$6D \rightarrow {}^4He + 2H + 2n + 43.3\ MeV$$

The amount of deuterium in seawater on earth is estimated at 4.6 10^{13} tons, equivalent to a potential energy production of about 5 1011 TWyr, i.e., 10 billion times the current world consumption per year.

Tritium is the radioactive isotope of hydrogen. It decays to ^{3}He by emission of an electron:

$T \rightarrow {}^{3}He + e^- + 18.7$ keV

with the rather short half-life of 12.3 years. The quantities available in nature are not sufficient for technical applications and tritium will be produced in-situ using from a blanket containing Li compounds, according to the reactions:

$${}^{6}Li + n \rightarrow {}^{4}He\ (2.05MeV) + T\ (2.73MeV)$$

$${}^{7}Li + n \rightarrow {}^{4}He + T + n - 2.47\ MeV$$

Thus the real consumables in the D-T fusion process are D and Li, while T is an intermediate product burned in the fusion reaction.

Lithium, like deuterium, is a widely available element. Estimated reserves of natural Li are about 29 million tons in known ore deposits and brines (Evans, 2008) and about 200 billion tons dissolved in sea water (0.1-0.2ppm) (Evans, 1978), equivalent to about thousand to a million times the current yearly energy consumption.

8.4 Safety Aspects

8.4.1 Inherent and Passive Safety

First, the amount of fuel available at each instant is sufficient for only a few tens of seconds, in sharp contrast with a fission reactor where fuel for several years of operation is stored in the reactor core. Second, fusion reactions take place at extremely high temperatures and the fusion process is not based on a neutron multiplication reaction. With any malfunction or incorrect handling the reactions will stop. An uncontrolled burn (nuclear runaway) of the fusion fuel is therefore excluded on physical grounds. Even in case of a total loss of active cooling, the low residual heating excludes melting of the reactor structure (Cook, 2001).

8.4.2 Minimal Radioactivity

The basic fuels (D and Li) as well as the direct end product (4He) of the fusion reaction are not radioactive. However, a fusion reactor will require confinement and control of radioisotopes since it has a radioactive inventory consisting of (i) tritium and waste contaminated by tritium and (ii) reactor materials activated by the neutrons of the fusion reaction. Studies (Cook, 2001) indicate, however, that an adequate choice of the latter can minimise the induced radioactivity such that recycling should become possible after some decades to a century. Thus, radioactivity does not have to be inherent to nuclear fusion, in contrast to nuclear fission where the fission reaction itself leads to dangerous long-lived radioactive products.

The tritium cycle is internally closed, and the total tritium inventory in the fusion power plant will be on the order of a few kg. An evacuation of the public might not even be needed in case of an accident if a proper detritiation system is implemented. Special permeation barriers will have to be used to inhibit discharge into the environment of tritium diffusing through materials at high temperature (Cook, 2001). As tritium is chemically equivalent to hydrogen, it can replace normal hydrogen in water and all kinds of hydrocarbons. It could thus contaminate the food chain when released in the atmosphere. The absorption of tritium contaminated food and water by living organisms is a potential hazard. However, possible damage is reduced owing to the short biological half-life of tritium in the body of about 10 days.

8.4.3 Reduced Proliferation Risk

The operation of fusion reactors is not accompanied by the production of fissile materials required for nuclear weapons. Only a significant modification of the fusion reactor—the introduction of a special breeding section containing fertile material—would make the production of weapons grade fissile materials possible. However, the presence of such a section (in an environment where none at all should be present) could be easily discovered by qualified inspectors (IFRC, 1990).

8.5 Environmental Aspects

8.5.1 Environmental Pollution?

The primary fuels (D and Li) and the direct end product (4He) are not radioactive, do not pollute the atmosphere, and do not contribute to the greenhouse effect or the destruction of the ozone layer. Helium is in addition chemically inert and indispensable for superconducting applications. There are no problems with mining (Li) and fuel transportation. No ecological, geophysical and land-use problems exist such as those associated with biomass energy, hydropower and solar energy.

Measures for tritium containment and detritiation of substances contaminated with tritium will have to be taken. During normal operation the dose for the public in the neighbourhood of the plant will only be a fraction of the dose due to natural radioactivity.

8.5.2 Dangerous Waste?

An important advantage of fusion is the absence of direct radioactive reaction products, in contrast to fission, where radioactive waste is unavoidable since the products of the energy releasing nuclear reaction are radioactive.

Adequate disposal of radioactive waste is especially difficult if the products are volatile, corrosive or long-lived. The neutron-activated structural materials of a fusion reactor would not pose such problems and because of their high melting point and their low decay heat, will not necessitate active cooling during decommissioning, transport or disposal. Studies (Cook, 2001) show that over their life time, fusion reactors would generate, by component replacement and decommissioning, activated material similar in volume to that of fission reactors but qualitatively different in that the long-term radio toxicity is considerably lower (no radioactive spent fuel).

Fusion could be made even more attractive by the use of advanced structural materials with low activation as e.g., vanadium alloys or silicon carbides. These materials offer in principle the prospect of recycling in about 100 years after the shutdown of the reactor as the radioactivity would fall to levels comparable to those of the ashes from coal-fired plants (Cook, 2001) (which contain always small amounts of thorium and other actinides). As already discussed above, there is a large research programme set up jointly between the EU and Japan to prepare the experimental investigation of properties of candidate alloys, with the IFMIF/EVEDA program.

8.6 Economic Aspects

It is obviously difficult to estimate with any useful precision the cost of a system that will only be put into service several decades from now. In comparison with other energy sources, environmental and safety-related advantages and the virtual inexhaustibility of the fuel sources should be taken into account, as well as the evolution of the cost of electricity based on (exhaustible) resources. It is difficult at this point of time to estimate the likely cost of a commercial fusion power plant. Investment costs (reactor chamber, blanket, magnets, and percentage of recirculating power...) will probably be higher than current power plants, but the fuel is inexpensive and abundant. On the basis of present knowledge, technologically sophisticated power plants will probably have an electrical output larger than 1 GW to be economic. The fast neutrons produced in the D-T reaction could be used to produce fissile material in fusion-hybrid breeder reactors. This complementary role for fusion might improve system economics compared with pure fusion systems; however, it would increase societal concerns related to safety, environment and weaponry.

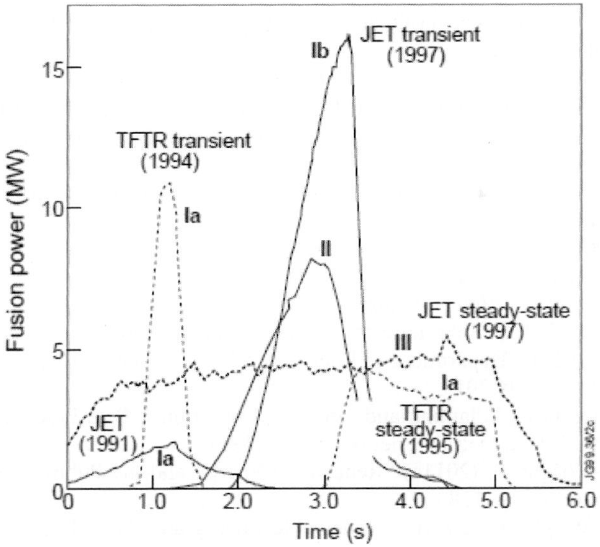

Fig. I: Fusion power development in the D-T campaigns of JET (full and dotted lines) and TFTR (dashed lines) (Jacquinot and JET Team, 1999).

REFERENCES

1. Ansar, A., Flyvbjerg, B., Budzier, B., and Lunn, D. (2014) "Should we build more large dams? The actual costs of hydropower megaproject development." *Energy Policy*. pp.1-14, March 2014.

2. Azechi, H. et al. (2013) “Present status of fast ignition realization experiment and inertial fusion energy development,” *Nuclear Fusion*, 104021 pp. 53.
3. Baldi, L., Peri, M., and Vandone, D. (2014) “Clean energy industries and rare earth materials: Economic and financial issues.” *Energy Policy*. March, 2014, pp. 53-61.
4. Barta, P. (2007) “Jatropha Plant Gains Steam in Global Race for Biofuels.” *Wall Street Journal online*.
5. Bateman, I. (2013) “The UK National Ecosystem Assessment: Valuing Changes in Ecosystems Services.” *Helm and Hepburn*. pp. 86.
6. Bertrand, K., G. Lindsay and Y. Bar-Yam (2012) *NECSI Food Briefing*. September 12, 2012.
7. Bloomberg (2013a) “Energy Storage Valuation Study: North America.” *Bloomberg New Energy Finance*. March 20, 2013.
8. Bloomberg (2013b) “Energy Storage Valuation Study: Germany.” *Bloomberg New Energy Finance*. July 9, 2013.
9. Bloomberg (2013c) “Energy Storage Valuation Study: UK.” *Bloomberg New Energy Finance*. January 9, 2012.
10. Bowdler, D. (2008) “Amplitude Modulation of Wind Turbine Noise: A Review of the Evidence.”
11. BP (2014) *2014 BP Energy Outlook 2035*. January 2014.
12. Bucalossi, J. et al. (2005) 32nd EPS Conference on Plasma Phys. Tarragona, ECA Vol. 29C, O-4.005 (2005); G.Giruzzi et al. *Plasma Phys. Contr. Fusion*, 47, 12B, B93 (2005), June 27 - July 1 2005.
13. Clare, R. (1992) “Tidal power: trends and developments.” contributions by R.G. Rice and G.C. Bakker pps. 3-21; and by S. Muirhead at p. 258.
14. Cook, I. et al. (2001) “Safety and Environmental Impact of Fusion.” Report EFDA-S-RE-1, EUR(01) CCE-FU/FTC 8/5, April 2001.
15. CWIF (2013) *Caithness Windfarm Information Forum.*
Dearen, J. (2013) “Solar industry grapples with hazardous wastes.”
16. DECC (2013), “Department of Energy & Climate Change: Electricity Generation Costs 2012.” July 2013.
17. DEFRA (2007), “Planting and Growing Miscanthus: Best Practice Guidelines—for Applicants to Defra’s Energy Crops Scheme.”
18. Devine-Wright, P. (2011a) “Renewable Energy and the Public: from NIMBY to Participation.” pp. 118.
19. Devine-Wright, P. (2011b) “Renewable Energy and the Public: from NIMBY to Participation.” pp. 83.
20. Devine-Wright, P. (2011a) “Renewable Energy and the Public: from NIMBY to Participation.” pp. 118.
21. D’haeseleer, W. (2013) “Synthesis of the economics of nuclear energy, for the EC.” Nov 2013.
22. Difiglio, C. and B. Wanner (2013) “Economics of Nuclear Power in Liberalized Power Markets.” *World Scientific Publishing Co.*
23. DOE (2013a) “Assessment of Energy Storage Alternatives in the Puget Sound Energy System” DOE/ PNNL Report DE-AC05-76RL01830, December 2013.
24. DOE (2013b) “Grid Energy Storage,” December 2013.

25. Dones, R. et al. (2007) "Life Cycle Inventories of Energy Systems."
26. E3 (2014a) "Investigating a Higher Renewables Portfolio Standard in California." *Energy and Environmental Economics.*
27. *East County Magazine* (2012) San Diego, April 4, 2012.
28. *Ecoinvent* (2014) online-data.
29. Energy Informer (2014) *The International Energy Newsletter*. March 2014.
30. EIA (2013a) *2013 Annual Energy Outlook*, United States Energy Information Administration.
31. EINE (2014) "EI New Energy: The Energy Cost Report." Volume 3 Issue 25, June 19.
32. EPRI (2010a) "Electricity Energy Storage Technology Options: A white paper primer on applications, costs, and benefits." *Technical Update*. December 2010.
33. EPRI (2010b) "Electricity Energy Storage Technology Options: A white paper primer on applications, costs, and benefits." *Technical Update*. December 2010.
34. Evans, R. (1978) "Lithium Reserves and Resources." *Energy*. 3, pp. 379-385.
35. Evans, R. (2008) "An abundance of lithium." July 2008.
36. ExternE (2014) http://www.externe.info/externe_d7/.
37. Fouquet, R. (2013) "A Handbook on Energy and Climate Change, 2013." pp. 258-259.
38. Friedman, I. (1953) *Geochimica and Cosmochimica Acta*. 4, pp. 89-103.
39. G20 (2013) *G20 leaders declaration*, Saint Petersburg Summit, St.Petersburg, Russia (Sep. 5-6).
40. GEA (2012) "2012: Global Energy Assessment—Toward a Sustainable Future." Cambridge University Press, Cambridge UK and New York, NY, USA and the International Institute for Applied Systems Analysis, Laxenburg, Austria.
41. Geth, F., J. Kathan, L. Sigrist, and P. Verboven (2013) "Energy Storage Innovation in Europe: A Mapping Exercise." *GRID+ and European Electricity Grid Initiative*, October 2013.
42. Gibbons (2013) "Gone with the wind: valuing the local impacts of wind turbines through house prices." London School of Economics, preliminary draft.
43. Gimeno-Gutierrez, M. and R. Lacal-Arategui (2013) "Assessment of the European potential for pumped hydropower energy storage." *European Commission Joint Research Centre.*
44. GreenBizz (2013) http://www.greenbiz.com/blog/2013/12/11/inside-california-energy-storage-mandate.
45. Gregory, R.D et al. (2005) "Developing Indicators for European Birds." *Philosophical Transactions of the Royal Society,* 269-288. pp. 360.
46. Grubler, A. (2010) "The costs of the French nuclear scale-p: A case of negative learning by doing." *Energy Policy.*
47. Haddad, S. and R. Dones (1991) "Comparative health and environmental risks for various energy sources." IAEA, Bulletin-board.
48. Hall, C. et al. (2014) "EROI of Different Fuels and the Implications for Society." *Energy Policy*. 64, January 2014, pp. 141-152.
49. Hanning, C. (2009) "Sleep disturbance and wind turbine noise." http://docs.wind-watch.org/Hanning-sleep-disturbance-wind-turbine-noise.pdf. July, 2009.

50. Harding, G. (1993) "Living Within Limits: Ecology, Economics, and Population Taboos."
51. Hayes, M. (2013) "Bats Killed in Large Numbers at United States Wind Energy Facilities." *Bioscience*, (63) 12, December 2013.
52. Heintzelman, M.S., and C. Tuttle (2012) "Values in the Wind: A Hednoic Analysis of Wind Power Facilities." *Land Economics*. August (88): 571-588.
53. Hind, A. (2007) "Could jatropha be a biofuel panacea?" http://news.bbc.co.uk/2/hi/business/6278140.stm.
54. Hinman, J.L. (2010) "Wind Farm Proximity and Property Values: A Polled Hedonic Regression Analysis of Property Values in Central Illinois." Illinois State University Ph.D. Thesis, Normal, Illinois.
55. Hirschberg, S. (2007) "Comparative Analysis of Energy Systems."
56. Hirschberg, S. and R. Dones (2014) "Evaluation of Sustainability of Electricity Supply Systems." *Paul Scherrer Institut*. Switzerland, April 2014.
57. Hirth, L. (2013) "The market value of variable renewables, the effect of solar wind power variability on their relative price." *Energy Economics*. 38, pp. 218-236.
58. Hoen, B. et al. (2013) "A spatial hedonic analysis of the Effects of Wind Energy Facilities on Surrounding Property Values in the United States." Lawrence Berkeley National Laboratory BLNL-6362E.
59. Hurricane, O.A. et al. (2014) "Fuel gain exceeding unity in an inertially confined fusion implosion." *Nature*. vol. 506, pp. 343.
60. IEA (2000) *Experience Cures for Energy Technology Policy*. Paris.
61. IEA (2010) *Energy poverty*. 2010.
62. IEA (2011) "Harnessing Variable Renewables: A Guide to the Balancing Challenge."
63. IEA (2013a) *World Energy Outlook*. 2013, Paris.
64. IEA (2013b) *2013 World Energy Outlook*. pp.226-227.
65. IEA (2013c) "Renewable Energy Mid-term Market Report." OECD/IEA.
66. IEA (2013d) "Thermal Energy Storage: Technology Brief." *International Energy Agency-Energy Technology Systems Analysis Programme & International Renewable Energy Agency*. January 2013.
67. IEA (2014) "Energy Storage Technology Roadmap Launch slide presentation." Paris, France, March 19, 2014.
68. IEA-ETSAP and IRENA (2012) "Electricity Storage: Technology Brief." April 2012.
69. IEA-NEA (2010) "Projected Costs of Generating Electricity, 2010 Edition."
70. *International Energy Agency and Nuclear Energy Agency*. Paris.
71. IFRC (1990) "Status Report on Controlled Thermonuclear Fusion, Executive Summary and General Overview." *International Fusion Research Council (IFRC)* International Atomic Energy Agency, Vienna.
72. IPCC (2011) "Strategic Report on Renewable Energy Sources." Chapter 9.
73. IPCC (2014) *Fifth Assessment Report.*
74. Ishida, S. et al. (1997) "High performance experiments in JT-60U high current diverter discharges." Proc. 16th Int. Conf. Plasma Physics and Controlled Nuclear Fusion Research, Vol. 1 (1997) 315, Montreal, Canada, 1996, IAEA, Vienna.

75. Jacquinot, J. and JET Team (1999) "Deuterium-tritium operation in magnetic confinement experiments: results and underlying physics." *Plasma Phys. Control. Fusion.* 41, A13-A46.
76. JAEA (2014) "Japan Atomic Energy Agency." http://naka-www.jaea.go.jp/english/ba/index.html.
77. Jefferson, M. (2008) "Accelerating the transition to sustainable energy systems." *Energy Policy.* 36 (2008) pp. 4116-4125.
78. Jefferson, M. (2012) "Capacity concepts and perceptions—Evidence from the UK wind energy sector." *IAEE Energy Forum.* pp. 5-8. Spring 2012.
79. Jefferson, M. and R. Fouquet (ed.) (2013) "Handbook on Energy and Climate Change." *Edward Elgar Pub.* ISBN-13: 978-0857933683, June 30, 2013.
80. Johansson, T., A. Patwardhan, N. Nakicenovic, and L. Gomez-Echeverri, eds. et al. (2012) "Global Energy Assessment: Toward a Sustainable Future." *International Institute for Applied Systems Analysis (IIASA)* Laxenburg, Austria.
81. Joskow, P.L. and J.E Parsons (2009) "The economic future of nuclear power." *Daedalus.* Fall 2009.
82. Kamada, Y. et al. (2013) "Progress of the JT-60SA project." *Nucl. Fusion.* 53 104010.
83. Kaun, B. and S. Chen (2013) "Cost-Effectiveness of Energy Storage in California: Application of the EPRI Energy Storage Valuation Tool to Inform the California Public Utility Commission Proceeding R. 10-12-007." *EPRI,* Technical Update, June 2013.
84. Kempton, W. and J. Tomi´c (2005) "Vehicle-to-grid power implementation: From stabilizing the grid to supporting large-scale renewable energy." *Journal of Power Sources 144.* pp. 280–294.
85. Kessides, I. (2010) "Nuclear Power: Understanding the economic risks and uncertainties." *Energy Policy.* April 2010.
86. Knaster, J. et al. (2013) "IFMIF: overview of the validation activities." *Nucl. Fusion.* 53 116001.
87. Laposa, S.P. and Mueller, A. (2010) "Wind Farm Announcements and Rural Home Prices: Maxwell Ranch and Rural Northern Colorado." *Journal of Sustainable Real Estate.* 2(1): 383-402.
88. Lazard, "Lazard's Levelised Cost of Energy Analysis—Version 8.0." September, 2014.
89. Lecointe, C., D. Lecarpentier, V. Maupu, D. Le Boulch, and R. Richard (2007) "Final report on technical data, costs and life cycles inventories of nuclear power plants." EDF.
90. Loisel, R., G. Pasaoglu, and C. Thiel (2014) "Large-scale deployment of electric vehicles in Germany by 2030: An analysis of grid-to-vehicle and vehicle-to-grid concepts." *Energy Policy 65.* pp. 432-443.
91. Lynas, M. (2014) "Nuclear 2.0: Why a Green Future Needs Nuclear Power." *UIT Cambridge.* April 2014.
92. Lynne, W. and P. Devine-Wright (2014) "Place Attachment: Advances in Theory, Methods and Applications."
93. MacKay, D. (2009) "Sustainable Energy—without the hot air." *Uit Cambridge Limited.*

94. Mai, T., R. Wiser, D. Sandor, G. Brinkman, P. Denholm, D.J. Hostick, N. Darghouth, A. Scholsser, and K. Strzepek (2012) "Renewable Electricity Futures Study; Vol. 1: Exploration of High-Penetration Renewable Electricity Futures." National Energy Laboratory, Volume 1.
95. Manuela, L. et al. (2008) "Collision fatality of raptors in wind farms does not depend on raptor abundance." *Journal of Applied Ecology*. 45, pp. 1695-1703.
96. MIT (2003) "The Future of Nuclear Power." Massachusetts Institute of Technology.
97. Moody's Global (2009) "New Nuclear Generation: Ratings Pressure increasing." *Moody's Investors Service*, New York, June 2009.
98. Morris, C. (2012) "German Solar Bubble." *European Energy Review*. July 5.
99. Mutoh, T. et al. (2007) "Steady-state operation and high energy particle production of MeV energy in the Large Helical Device." *Nuclear Fusion*. 47, pp. 1250-1257.
100. Nakajima, N. et al. (2012) "Overview of IFERC project in Broader Approach activities." Paper OV/P-05 Proc., 24th IAEA Fusion Energy Conference, San Diego, USA, October 8-13, 2012.
101. Nath, I. (2010) "Cleaning Up After Clean Energy: Hazardous Waste in the Solar Industry." *Stanford Journal of International Relations*. Spring 2010.
102. NEI (2014) "Water Use and Nuclear Power Plants."
103. Newport, F. (2012) "Americans Still Favor Nuclear Power a Year After Fukushima."
104. NHA (2010) "Challenges and Opportunities for New Pumped Storage Development: A White Paper Developed by NHA's Pumped Storage Development Council."
105. OECD NEA (2012) "Nuclear Energy and Renewables: System Effects in Low-carbon Electricity Systems." *Nuclear Energy Agency*. pp. 127-128
106. Pedersen, E. and W.K. Persson (2007) "Response to wind turbine noise in different living environments." *Occup. Env. Medicine*. (64).
107. Peterson, P., F. Zhao and R. Petroski (2005) "Metal And Concrete Inputs For Several Nuclear Power Plants." Report UCBTH-05-001
108. Pimentel, D. and M.H Pimentel (2008) "Food, Energy, and Society." 3rd edition.
109. Poortinga, W., N. Pidgeon, S. Capstick and M. Aoyagi (2013) "Public Attitudes to Nuclear Power and Climate Change in Britain Two Years after the Fukushima Accident."
110. Renewable Analytics (2013) "Effects of PV Generation on Wholesale Power Prices Summary 2012 and January 2013." *Renewable Analytics LLC*.
111. Rhodes, R. and D. Beller (2000) "The Need for Nuclear Power." *Foreign Affairs*. January 1, 2000.
112. Rodier, M. and R. Clare (ed.) (1992) "Tidal power: trends and developments." *Thomas Telford*. pp. 307-308.
113. SBC (2013) "Leading the Energy Transition: Electricity Storage Factbook." *SBC Energy Institute*. September 2013.
114. SBC (2014) *Fact Sheets on Renewables*.

115. Schenler, W. et al. (2009) "Project no: 502687, NEEDS, New Energy Externalities Developments for Sustainability, Deliverable D10.2 - RS2b, Final report on sustainability assessment of advanced electricity supply options."
116. Schlissel, D. and B. Biewald (2008) "Nuclear Power Plant Construction Costs." *Synapse Energy Economics Inc.* Cambridge, July 2008.
117. Schoenung, S. (2011) "Economic Analysis of Large-Scale Hydrogen Storage for Renewable Utility Applications." *Sandia National Laboratories*. August 2011.
118. Searchinger, T. et al (2008) "Use of U.S. croplands for biofuels increased greenhouse gases through emissions from land-use change." *Science*. 319 (5867) pp.1238-1240.
119. Shepard, S. and J. Garther (2013) "Vehicle to Grid Technologies: V2G Applications for Demand Response, Frequency Regulation, Microgrids, Virtual Power Plants, and Renewable Energy Integration." *Navigant Research Report.*
120. Smith, J. W. and S. Positano (2010) "Self-Destructive Affluence of the First World: The Coming Crisis of Global Poverty and Ecological Collapse."
121. Sovacool, B. (2011) "Contesting the Future of Nuclear Power: A Critical Global Assessment of Atomic Energy." *World Scientific*. pp. 126.
122. Stephenson, A. (2013) "DECC's BEaC Model (Bioenergy Emissions and Counterfactual Model)."
123. Stigwood, M. (2009) "Large wind turbines—are they too big for ETSU-R-97?" *Institute of Acoustics*. January 16.
124. Tabak, M. et al. (1994) "Phys. Plasmas 1, 1626".
125. Taxpayers for Common Sense, "DOE Loan Guarantee Program: Vogtle Reactors 3&4." February 2014, http://www.taxpayer.net/library/article/doe-loan-guarantee-program-vogtle-reactors-34.
126. TDT (2011) The Daily Telegraph, December 11, 2011.
127. Thomas, S. (2010) "The Economics of Nuclear Power: An Update." *Heinrich Boll Stiftung Foundation*. Berlin, March 2010.
128. Trainer, E. (2007) "Renewable Energy Cannot Sustain a Consumer Society." pp. 8.
129. Trainer, E. (2013) "Some critical notes on the Global Energy Assessment's Renewable Energy chapter." January 29, 2013.
130. Ueckerdt, F., L. Hirth, G. Luderer, O. Edenhofer (2013) "System LCOE: What are the costs of variables renewables?" *Potsdam-Institute for Climate Impact Research*, Figure 1.
131. U.S. DOE (2011) "Department of Energy Critical Materials Strategy, 2011."
132. U.S. DOE (2014) *Global Energy Storage Database*, on-line data.
133. Vithayasrichareon, P. and I. Mac Gill (2013) "Assessing the value of wind generation in future carbon constrained electricity industries." *Energy Policy.*
134. Warshay, B., Goldie-Scot, L., and Morsy, S., "Tesla's 'home battery': ripples not waves." *Bloomberg New Energy Finance*. April 2, 2015.
135. Wallard, H., B. Duffy, P. Cornick (2012) "After FukushimaI, Global Opinion on Energy Policy." *IPSOS Social Research Institute.*
136. Wallner, A. et al. (2011) "Analyse von Energiebilanz und CO_2 Emissionen der Nuklearindustrie über den Lebenszyklus."

137. Warren, C. et al. (2005) "'Green on Green': Public Perceptions of Wind Power in Scotland and Ireland." *Journal of Environmental Planning and Management.* (48) 6, 853-875, November 2005.
138. Weart S. (1988) "Nuclear Fear: A History of Images." *Harvard University Press.* Cambridge, MA.
139. Wesoff, E. (2011) "Compressed Air Storage Beats Batteries at Grid Scale." *Greentech Media.* March 3, 2011.
140. Wesoff, E. (2014a) "Slideshow: Update on Ambri's Liquid Metal Grid-Scale Battery." *GreenTech Media.* February 19, 2014.
141. Wesoff, E. (2014b) "Tesla Giga Factory: $4B to $5B Price Tag, With Battery Production Slated for 2017." *GreenTech Media.* February 27, 2014.
142. Wesson, J. (2004) Third Edition, Oxford Science Series nr. 118, Oxford (2004) *Clarendon Press.* ISBN-0-19-850922-7.
143. White, S. and G. Kulcinski (1999) "Birth to Death Analysis of Energy Payback Ratio and CO_2 Gas Emission Rates for Coal, Fission, Wind and DT Fusion Electric Power Plants."
144. Wilkinson, S. (2013) "PV Storage Market Set to Explode to $19 Billion in 2017; Germany Leads Again." *IHS,* April 25, 2013.
145. Windbyte (2014) http://windbyte.co.uk/birds.html. July 31, p. 63-67.
146. WNA (2012) *World Nuclear Association.* WNA Reactor Data Base, June 25.
147. Younghwan, K., K. Minki and K. Wonjoon (2013) "Effect of the Fukushima nuclear disaster on global public acceptance of nuclear energy."

ECONOMICS OF RENEWABLE AND NUCLEAR POWER

HISHAM KHATIB
World Energy Council
Amman, Jordan

Both nuclear and renewable energies are considered to be carbon free and encouraged on environmental and energy security considerations. However they markedly differ in their economics, system integration, capacity factor and dispatching (intermittency); also in public acceptance and foreseeable future growth prospects.

Nuclear represents 4.4% of global primary energy (2014) and 11% of electricity (18% in 1996).

New Renewables (mainly PV and wind) recently grew significantly, reaching 3% of global energy (0.9% a decade ago). In power generation, they recently grew by 12% annually, accounting for a record of 6% of global electricity.

Evaluation of the economics of nuclear and renewables are significantly different from other power generation technologies.

In nuclear, the actual investment cost can be significantly higher than the overnight cost, due to long execution times, delays and financing cost over runs.

In renewables, the actual system costs can also be significantly higher due to integration cost, land cost, and (often well-informed) NIMBY considerations.

The economics of both will significantly benefit if there is a price for carbon.

The economics of nuclear suffer from the high investment cost. The industry has not succeeded in achieving cost reduction through commercialization. Therefore its global energy share is slowly declining despite new construction, mainly in centrally planned economies.

In OECD private investors—without firm governmental guarantees and subsidies which are not easily come by—will not invest in nuclear due to high costs and risks.

Worldwide, the nuclear fleet—particularly outside OECD—is primarily owned by governments.

This is in contrast to new renewables, which are increasingly favoured by governments and the public all over the world, due to its clean nature, low (though often heavily subsidised with hidden costs) and decreasing investment requirements and risk.

Newly quoted LCOE costs (mainly subsidised) are now less than conventional generation: 6 c/kWh for PV solar and 4.2 c for wind. Actual system costs are higher.

For renewables, beside the LCOE, there are significant integration costs:

- Profile Costs,
- Planning Costs and
- Grid Costs.

These vary from one system to another, but they significantly increase the real cost of renewables.

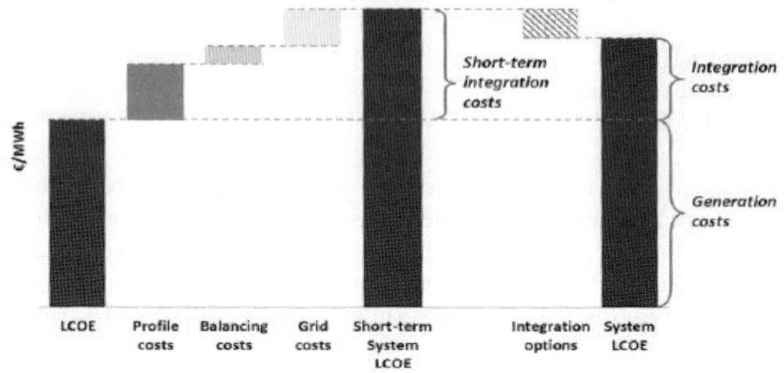

The most significant factor in economics of power generation is the "discount rate" utilized in LCOE evaluation. This is greatly influenced by risk.

Nuclear suffers from high discount rate (over 10%) due to its many investment and political risks. This has a significant and negative influence on its economics.

Renewables are mainly risk free, correspondingly they enjoy the least discount rates (5- 8%) which contributes in their favour.

Both nuclear and NRs will continue to be significant in the global energy/power generation scene (say to 2040).

Nuclear contribution is expected to remain stagnant, below 4.5% of global energy.

Renewables will be expected to continue to grow significantly year after another. But, in spite of their intermittent nature, low capacity factor, limited storage technologies and the huge inertia of existing systems, their contribution to future power generation will increasingly become significant (but limited).

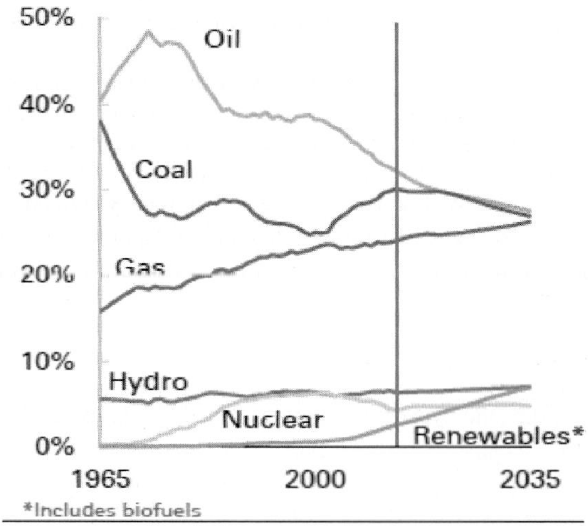

ENERGY STORAGE FOR ELECTRIC POWER GENERATION

LARA PIERPOINT
Office of Energy Policy and Systems Analysis, U.S. Department of Energy
Washington, DC, USA

Options for ensuring electricity demand meets supply:

- Generation reserves (especially natural gas)
- Build additional transmission capacity
- Base load cycling
- Demand response
- Electricity Storage

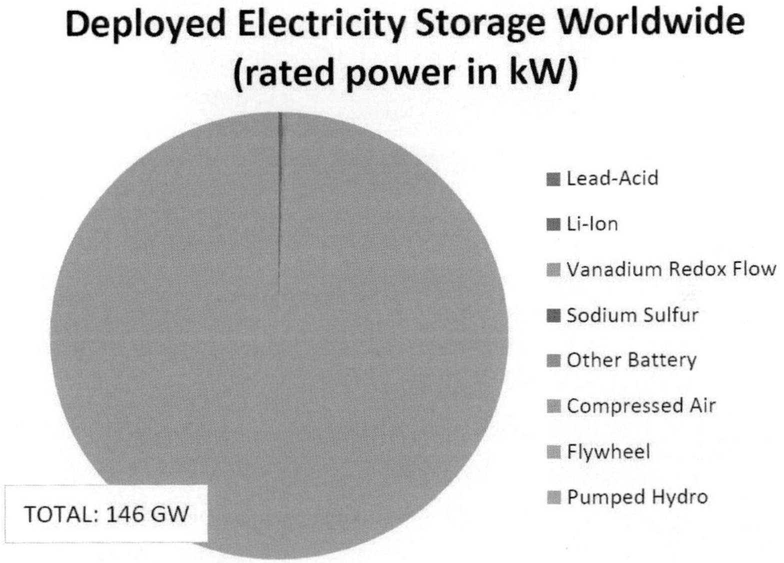

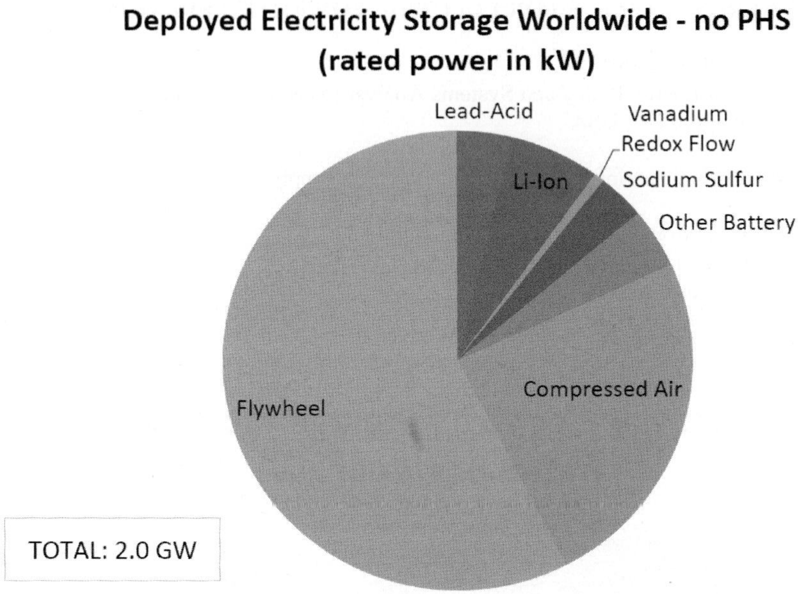

Storage Costs

Technology	Cost/kW (power)	Cost/kWh (storage)
Pumped Hydro	$1500-$4300	$250-430
Compressed Air	$800-$1500	$125/kWh
Flywheels	up to $4000	$200-$500
Lead-Acid	$1700 - $4900	$450-$950
Li-ion*		$500-$850
NaS	$3100-$3300	$520-$550
Flow Batteries**	$3100-$3700	$520-$550
Hydrogen	$1150	

*Tesla plans to sell Li-ion batteries at $250/kWh (utility-scale).

**Eos energy storage claims it has a vanadium redox flow battery at $1000/kW or $160/kWh; another company, Imergy, has announced vanadium flow batteries available for $500/kWh.

Lazard (2014) estimates the range to be $265 -$324/MWh for lead-acid at the low end and NaSat the high end.

Storage Deployment Scenarios

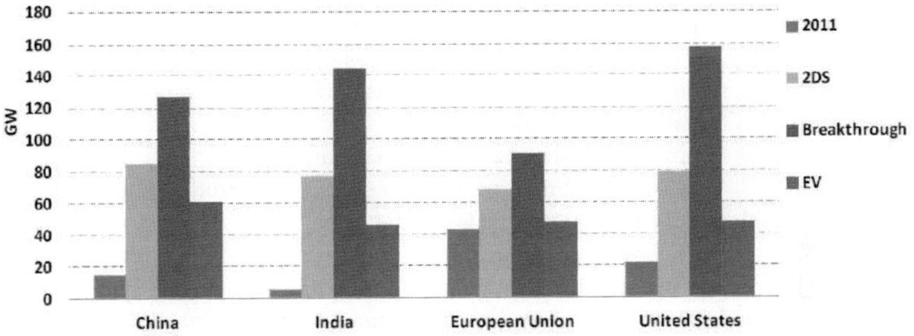

2DS: reference case; storage costs equal to current pumped hydro costs
Breakthrough: LCOE comes down to $90/MWh
EV: high EV penetration provides additional storage

Storage Technologies & Services Provided

Service	**Peak-shifting (arbitrage)**	**Energy smoothing and shaping**	**Ancillary services (e.g. frequency regulation)**
Timescale	hr.+	min. to hr.	sub-minute
Technology			
Pumped Hydro	*		
Compressed Air	*		
Flywheels			*
Lead-Acid			
Li-ion			
NaS			
Flow Batteries			
Hydrogen			

Solid box = can provide (* mans "can potentially provide"). Hashed box = potentially provide, Empty box = cannot provide
SOURCE: [DOE (2013b), SBC (2013)][1]

R&D Thrusts by Storage Technology

Technology	Primary R&D Challenges
Pumped Hydro	Cost of underground/seawater reservoirs to remove geographic dependence & mitigate environmental harms
Compressed Air	Adiabatic and isothermal processes to avoid use of gas to heat air upon decompression; artificial reservoirs to remove geo. dependence
Flywheels	Materials cost; flywheel stability
Lead-Acid	Improve cycle life and depth of discharge
Li-ion	Find lower-cost materials for the negative electrode; increase energy density & cycle life; confirm safety; scale-up from small, cylindrical to large-format, prismatic cells for energy storage applications
Molten Salt	Improve durability/ manufacturability of ion-conducting separator; improve corrosion resistance; alternative chemistries to reduce operating temperature, improve safety
Flow Batteries	Improve specific energy and cycle life; alternative chemistries to lower system cost
Hydrogen	Improve efficiency of oxygen reactions while reducing cost of oxygen electrodes

SOURCES: [DOE (2013b), SBC (2013)]

CONCLUSIONS

- Electricity storage can provide vital grid services needed for integrating renewable and nuclear generation.
- Worldwide deployment is still relatively small and dominated by pumped hydro, but that may change with R&D investment and electricity market reforms

RECOMMENDATIONS

1. Create a level playing field, allowing storage to participate in provision of all market services for which it is qualified.
2. Support deployment of mature storage technologies as widely as possible (especially pumped hydro), by characterizing resources to identify those with the lowest environmental impact and highest potential to help integrate base load and intermittent generation.
3. Continue to fund research, development, and demonstration projects across a wide range of advanced energy storage technologies, including grid-enabled electric vehicle systems.

DISPATCHING RENEWABLE AND NUCLEAR POWER

BRUCE STRAM
Element Markets
Houston, TX, USA

A MODERN MARVEL

- Electric power is extremely valuable.
  - Runs myriad equipment, provides light, etc.
  - Value greatly improves if it is readily available and reliable and high quality (voltage and frequency).
- The laws of physics: electric power flows virtually instantaneously on a grid according to least resistance.
- Managing that flow to provide value requires a complex grid.

THE GRID

- Economies of scale have dictated large generation facilities.
  - Location determined by transport logistics for fuel, environmental factors etc.
  - Location at demand centers often not appropriate.
- The grid connects generation with demand.
  - Heavy transmission lines connect to nodes which step down power for various consumers, heavy industrial to home owners.

Key Issues: Grid Operation

- Power demand cycles.
  - Anticipated (within limits) daily and seasonal changes.
  - Capacity to meet peak demand efficiently.
- Power demand may vary on very short notice.
- Power generation equipment requires periodic maintenance and may fail in use.

Keeping the Grid Operating

- Two primary tools.
  - Reserve Capacity over and above peak demand.
  - A diversity of generation capacity for economic efficiency and reliability.
- Traditionally.
  - Base load: coal, nuke, maybe combined cycle NG.
  - Intermediate: CCNG, maybe coal, as demand increases.
  - Peakers: NG, oil turbines.

Hydro can operate across all of these modes.

RUDIMENTARY OPERATIONS

- Coal and Nuclear units meet low demand at night.
- Additional units come on as demand grows in the morning.
- The NG turbines come on to meet peaking demand in the afternoon.
- Cost efficiency is served by minimizing short run cost per kWh, i.e., mostly fuel cost times heat rate.

Intermittency

- Traditional power sources: rated in MW.
- Renewable MW rated per natural conditions:
  - PV per solar intensity (insolation)
  - Wind turbines per wind velocity
- These don't perform to operator desires.
- The rest of the grid typically adapts to renewable production, zero fuel cost implies must dispatch.
- Intermittency, relative to rated capacity, leads to low load factors.

SYSTEM COSTS

- Stability, reserve margin, and grid connection for generation comes at a cost.
- Intermittent power sources (solar, wind) can disproportionately incur such costs.
  - More reserve margin
  - More spinning reserve
  - More other auxiliary services
  - More connection cost (facilities must be located where the sun shines best and wind blows)
  - Lower load factors on transmission lines

Allocation of System Costs

- Traditionally these costs have been passed on per fixed or per kWh charges to customers rather than being borne by generators.
- If and as intermittent sources become major supply sources, these costs become quite significant.
- This can be viewed as a hidden subsidy to renewables in addition to direct subsidies.

RENEWABLE AND CARBON GOALS

- Many nations, especially in Europe, and states in the US have set ambitious renewable and/or carbon reduction goals.
  - 30 to 50% power from renewables
  - Carbon emissions below 1990 or a fraction thereof

- Increasing renewable capacity to reduce carbon can cause production to exceed demand.
  - Ever more ambitious goals require more capacity
  - Germany and Denmark export excess power
  - Energy revenue to traditional producers can fall

RISING INCREMENTAL COSTS?

- Adding more renewable capacity (to further reduce carbon or reach RPS goals) means that capacity will not be utilized at peak production period.
- Capex is incrementally spread over fewer kWh.
- Not everyone can export power.

POWER PRICE VS RENEWABLE SHARE

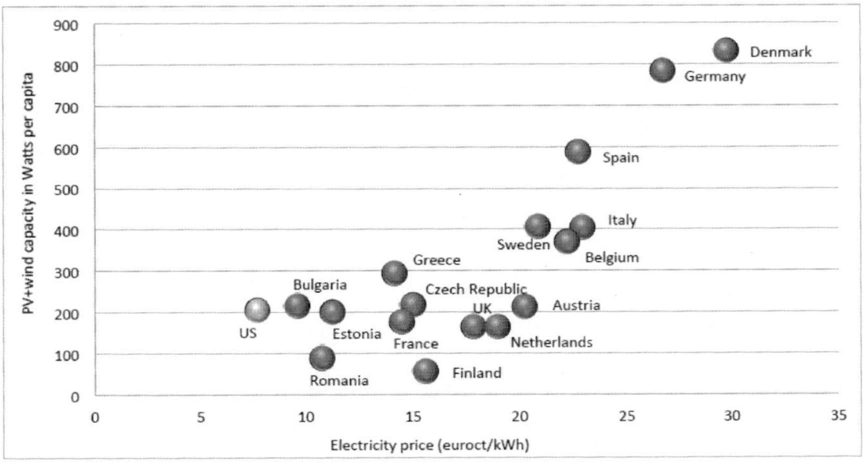

NUCLEAR AND THE GRID

- Operationally, nuclear plants are much like coal.
  - Preference for continual operation, base load
  - High load factors amortizes high cap costs over more kWh
  - High cap costs militate against cycling nukes even if technically feasible
- Large associated steam turbines help grid stability "momentum" to carry through glitches.
- May require redundant and robust grid connection to assure cooling pump operation at least with old designs.

BREAKTHROUGHS NEEDED

- Neither nuclear nor renewable can provide a carbon free power alternative with existing technology at costs close to what we've enjoyed.
- Low cost power storage would help both.
- More exotic carbon free power technologies.
  - Beaming from space?
  - The choices are slim.

ENVIRONMENTAL IMPACT OF NUCLEAR AND RENEWABLES

CHARLES MCCOMBIE
McCombie Consulting, Arius Association
Gipf-Oberfrick, Switzerland

MICHAEL JEFFERSON
ESCP Europe, Business School
London, UK

ENVIRONMENTAL IMPACTS: TOPICS

We will illustrate only some key and some lesser known facts

1. Resource Requirements (Materials; Water and Land Use)
2. Emissions (CO_2 Emissions; Toxic Emissions)
3. Health Effects
4. Accidents
5. Wastes
6. Public Attitudes
7. Additional Topical Issues (e.g., impacts on bird life, bats, threatened species)
8. Conclusions and Recommendations

LCA: AN APPROACH TO COMPARISON

- Within a Life Cycle Analysis (LCA) impacts, including costs, are assessed and summed throughout all stages.
- These types of analyses were extensively studied within the ExternE Project of the EU which ran through to 2005.
- The Ecoinvent database was produced in Switzerland. A large database was also assembled for the recent IPCC report on climate change.

CO_2 IS RISING!

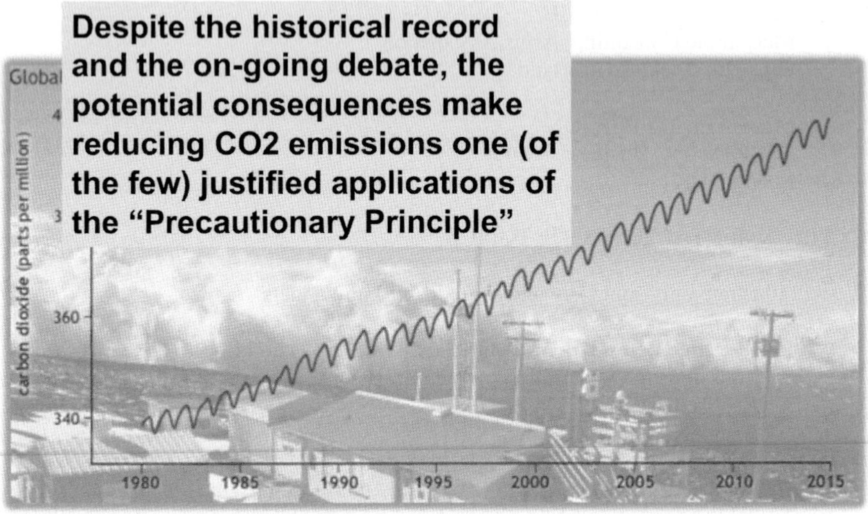

Global average carbon dioxide concentrations since 1980, with photo of Mauna Loa Observatory in background. Adapted from "State of the Climate in 2014".

"GREENHOUSE GASES"—THE GOOD AND THE BAD

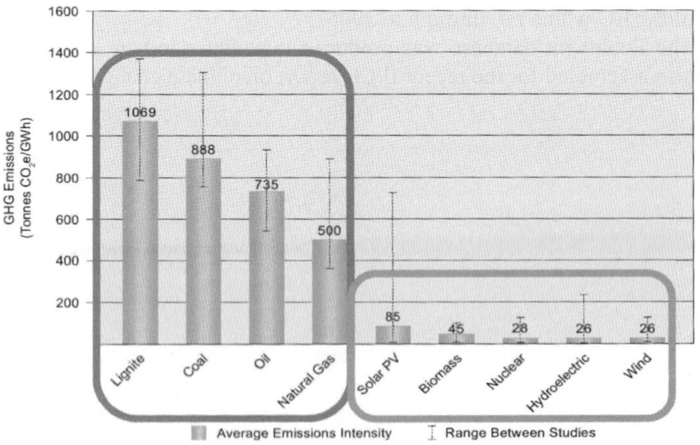

WNA website (82 studies)

MOJAVE PV PLANT

Desert Sunlight generates 550 megawatts; 8 million photovoltaic (PV) panels
Average high temperature in July: 40°C In January: 18°C. Average annual rainfall: 10 cm.

CADMIUM IN SOLAR CELLS

- Cadmium is very toxic, although CdTe may be less so.
- Desert Sunlight Solar Farm in the USA has 8 million solar panels, each with around 6g of Cd.
- This implies that the Cd inventory for the 550 MWe peak capacity plant is around 40t.
- Assuming a 25y plant lifetime and 50% availability, the toxic waste inventory per GWe year is around 6 t/GWe.y.
- For a 1000 MWe NPP, the long-lived wastes are ca 30 t/GWe.y of spent fuel or 1.4 t/GWe.y of HLW.

http://instituteforenergyresearch.org/

YOLL COMPARISON FOR GERMANY

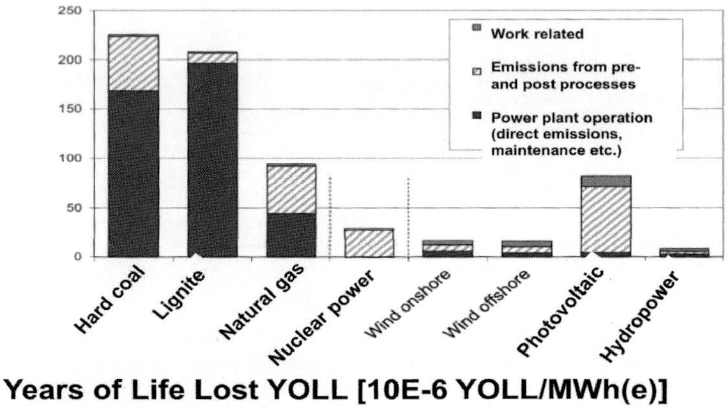

P. Preiss, S. Wissel, U. Fahl, R. Friedrich, A. Voß
http://www.ier.uni-stuttgart.de/publikationen/arbeitsberichte/downloads/Arbeitsbericht_11.pdf

Figure 1. Percentage of respondents having mainly or very favourable opinions or impressions of different energy sources for producing electricity

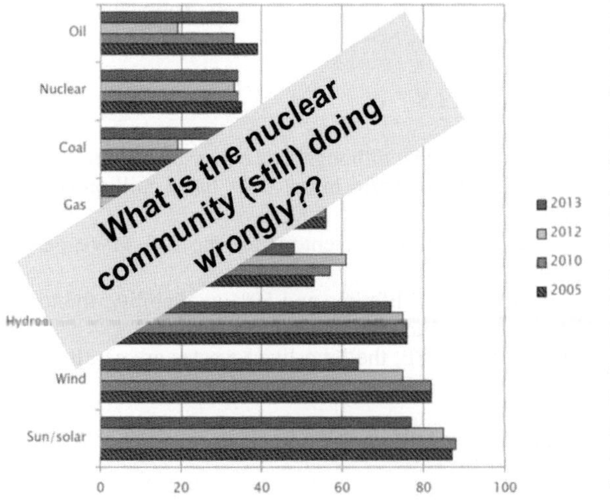

THE PUBLIC WORRIES WHEN THEY RECEIVE MIXED MESSAGES FROM THE SCIENTIFIC COMMUNITY

SELECTED CONCLUSIONS

- The strongest proponents of both nuclear and renewables tend to have an **overly optimistic view** of their favoured technology.
- Nuclear faces **special challenges** because of "radiation phobia" among the public and the environmental impacts of some renewable energy sources are underestimated by the public.
- Energy policies of many governments often appear to be based on **subjective judgements** of risks and selected vested interests.
- **Life cycle analyses of energy systems** are the most promising approach to developing objective criteria that can aid future decision making.

NUCLEAR POWER: STATUS REPORT AND FUTURE PROSPECTS

ROBERT J. BUDNITZ
Earth Sciences Division, Lawrence Berkeley National Laboratory
Berkeley, CA, USA

SCOPE

- Emphasis in the paper on issues of:
  - Safety (this talk will *only* discuss safety)
  - Physical security
  - Proliferation
  - Economics
  - Waste issues

REACTOR SAFETY

- Why is safety *the* major issue?
  - A major NPP accident is perceived as being *qualitatively different* from most other man-made accidents:
    - long-lived contamination
    - potential loss of life
  - However, the financial impacts are *not* qualitatively different!

WHAT DO WE KNOW ABOUT SAFETY?

- We know that the *frequency of a major core-damage accident*, on average, is about ~ few times 10^{-5}/year.
- We know that the *likelihood of a major release*, given above, is *a few percent*.
- So the frequency of a major release is about ~ 10^{-6}/year.
- With ~ 400 reactors worldwide, *one core-damage accident every several decades*.

TRENDS IN REACTOR SAFETY

- Safety is clearly improving worldwide.
- Evidence:
  - Initiating event frequencies have *dropped.*
  - Failure rates of equipment have *dropped.*
  - Human error rates have *dropped.*
  - Major *improvements in design* against the rare severe external events (winds, floods, earthquakes)
  - Major cross-fertilization of *shared experience*—much less isolation
- Is today's overall safety level "adequate"?
  - This is always an individual or societal judgment!

WHY CAN'T WE DO BETTER?

- The worldwide average of ~ few x 10^{-5}/year (core damage frequency) is controlled by:
  - How today's designs rely on *active systems*
  - How today's designs rely on certain vital *operator interventions*
  - How much "*margin*" exists between equipment capacities and what is needed in upset conditions.
- Achieving much better requires *radically* different approaches.

NEW DESIGN APPROACHES

- Large LWRs
  - More "passive" designs
  - Less equipment
  - Less reliance on operator actions
- Small (modular) LWR designs
  - Smaller → much more margin in key areas
    - (Thermal margins, slower evolutions in upsets, simplicity)
- Advanced concepts
  - Some incorporate *all* of these improved approaches!

"NEWCOMER" COUNTRIES?

- Major CULTURE issues:
  - Political culture (including an *independent* regulator)
  - Social culture (corruption!)
- Technical and financial infrastructure

A country without *all* of these features is unlikely to achieve adequate nuclear safety over the long run.

INTERNATIONAL INSTITUTIONS

- IAEA (the foremost)
- OECD/Nuclear Energy Agency
- Regional groupings
- International vendor (owner's) groups

RECOMMENDATIONS

- Technical recommendations:
  - SMRs have great promise. We need:
    - Technical R & D (and engineering development)
    - Development of a regulatory review capability
  - Training and infrastructure

    - Especially important in "newcomer" countries
- Policy and institutional recommendations for achieving adequate safety:
  - "Culture" issues are major barriers.
  - An independent regulatory agency is crucial.
  - Need to continually strengthen the international safety regime.

SESSION 3

ENERGY END-USE MANAGEMENT

THE CONTROL OF COMPLEX SYSTEMS: ENERGY IN CITIES

KIMBERLY GRAY
Department of Civil and Environmental Engineering, Northwestern University
Evanston, IL, USA

INTRODUCTION

- Networks: Energy, Cities, and the Control of Complex Systems, 12 – 14 May 2014.
- 26 physicists, engineers, transportation and operations researchers, network scientists, computer scientists, experts in energy, atmospheric science, statistics, policy makers, network economists.
- Examine applications of complex systems theory to energy systems, with special attention to rapidly growing urban populations.
- Science of Complex Networks promises enhanced stability and maximum efficiency, increased resilience, thereby facilitating the transition to and control of sustainable cities.

THEMES

- Energy Emergencies and Imperative of End-use management
- Cities offer golden opportunities; action is here
- Issues:
  - Energy Equity
  - Dependable, reliable service (supply matched to demand)
  - Impact
  - Grid systems—by very nature complex systems
  - Transportation systems—accessibility

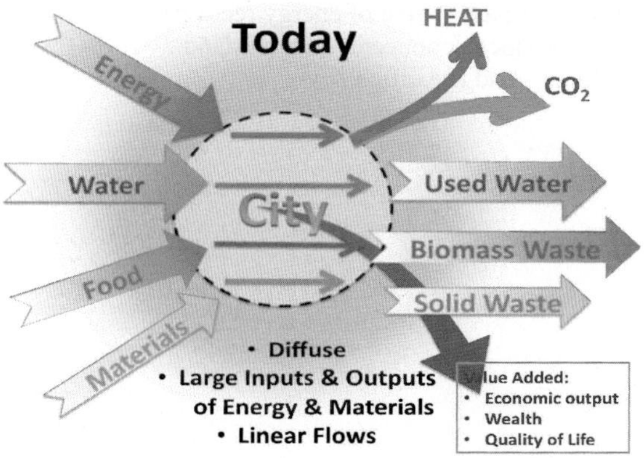

SOTRA NETWORK

From a set of 301 distinct bus stops, 203 were geolocated using OSM + googlemap. These stops represent 17 express bus routes, 67 regular bus routes (called the monbus), and 1 special route meant for trader with heavy goods (called marche bus).

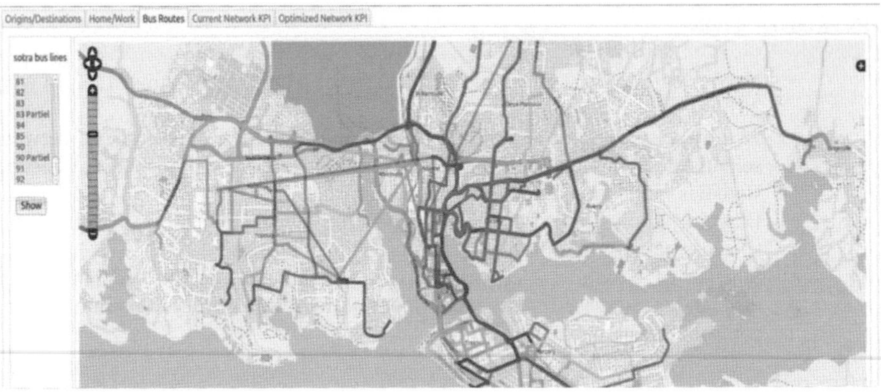

http://researcher.watson.ibm.com/researcher/view_group_subpage.php?id=

ANOTHER EXAMPLE: GRID SYSTEMS

- Transition to smart
- Support network of renewable sources
- Robust—microgrids, isolate
- Resilient—**Example, N. Germany**
- Couple to other complex systems:
  - Transportation system—electric vehicles, storage
  - Waste—biomass source of energy recovery
  - Land use—local, connected to hinterland
  - Buildings—end use efficiency

BRAESS' PARADOX IN COMPLEX NETWORKS?

Original Network

Add link:

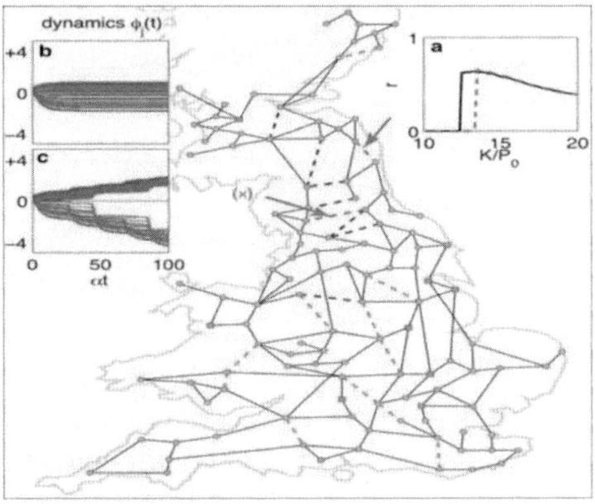

Generators: $P = 11P_0$
Consumers: $P = -P_0$
Links: $K = 13P_0$
Damping: $\alpha = P_0$

ANOTHER EXAMPLE: GRID SYSTEMS

- Transition to smart
- Support network of renewable sources
- Robust—microgrids, isolate
- Resilient
- Couple to other complex systems:
  - Transportation system—electric vehicles, storage
  - Waste—biomass source of energy recovery
  - Land use—local, connected to hinterland
  - Buildings—end use efficiency

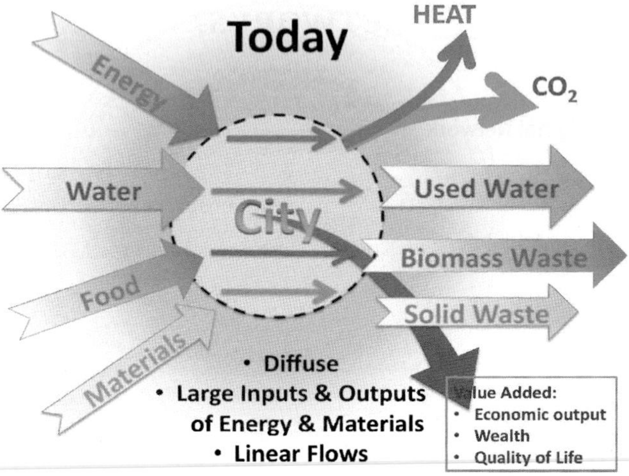
Today
HEAT
CO_2
Energy
Water
City
Used Water
Biomass Waste
Food
Solid Waste
Materials
• Diffuse
• Large Inputs & Outputs of Energy & Materials
• Linear Flows
Value Added:
• Economic output
• Wealth
• Quality of Life

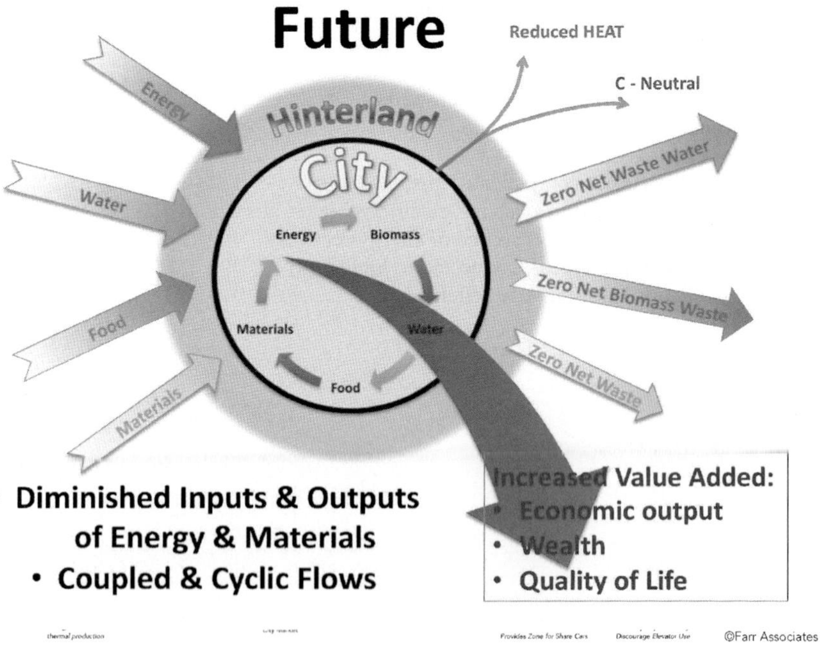
Future
Reduced HEAT
C - Neutral
Energy
Hinterland
City
Water
Zero Net Waste Water
Energy
Biomass
Zero Net Biomass Waste
Food
Materials
Water
Zero Net Waste
Food
Materials
• Diminished Inputs & Outputs of Energy & Materials
• Coupled & Cyclic Flows
Increased Value Added:
• Economic output
• Wealth
• Quality of Life
©Farr Associates

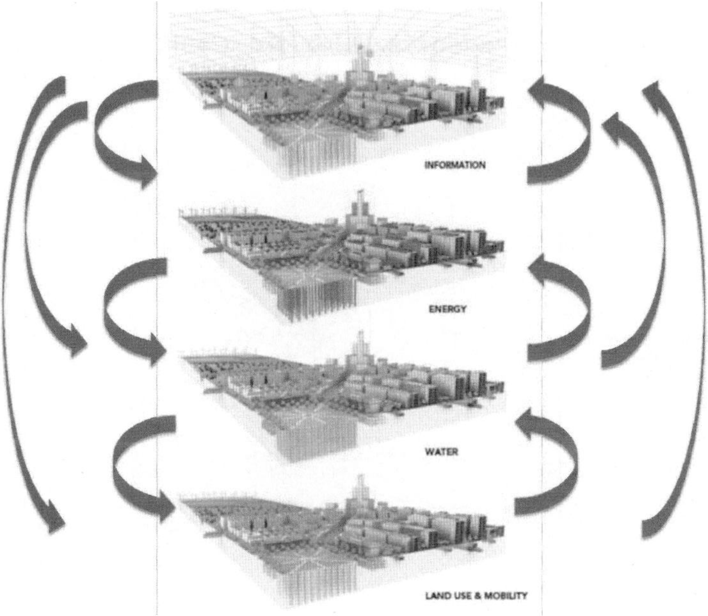

Unintended Consequences

- Hydrologic-Climate-Energy-Health Networks
- Simulate weather and atmospheric interactions & model effects of various technical innovations/ adaptation strategies over regions
- For instance – What are the effects of green roofs on climate, energy, health
- Unintended consequences – large reduction in precipitation

Georgescu, M. et al. (2014) *PNAS,* 111(8)2909-2914.

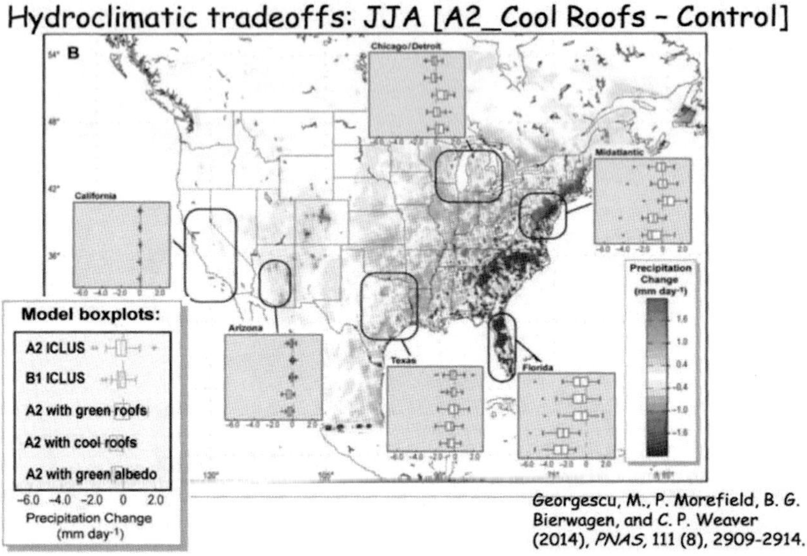

TODAY

- We have many of tools to improve efficiency of current system
- Integration of Urban Energy Systems
  - Shift to local, distributed
  - Proliferation of sensors
  - Handling of Big Data
  - Yet, technical capacity >>> institutional capacity

GOING FORWARD

- BUT—Need to go to next level of complex system thinking:
  - Interdependent Networks, Networks of Networks
  - Coupling Energy Networks to Water to Transportation to Waste to Materials, etc.
  - Greater security—just how secure, minor perturbations can propagate cascading failures.
  - And these networks are not the networks of today.

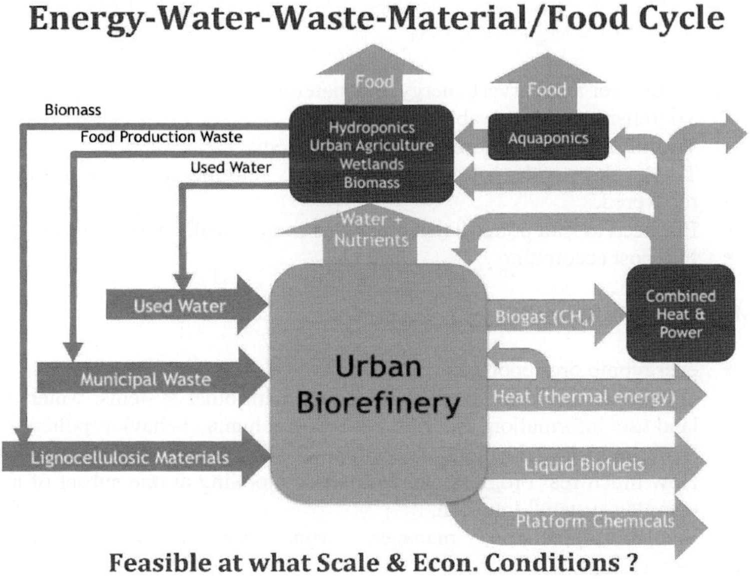

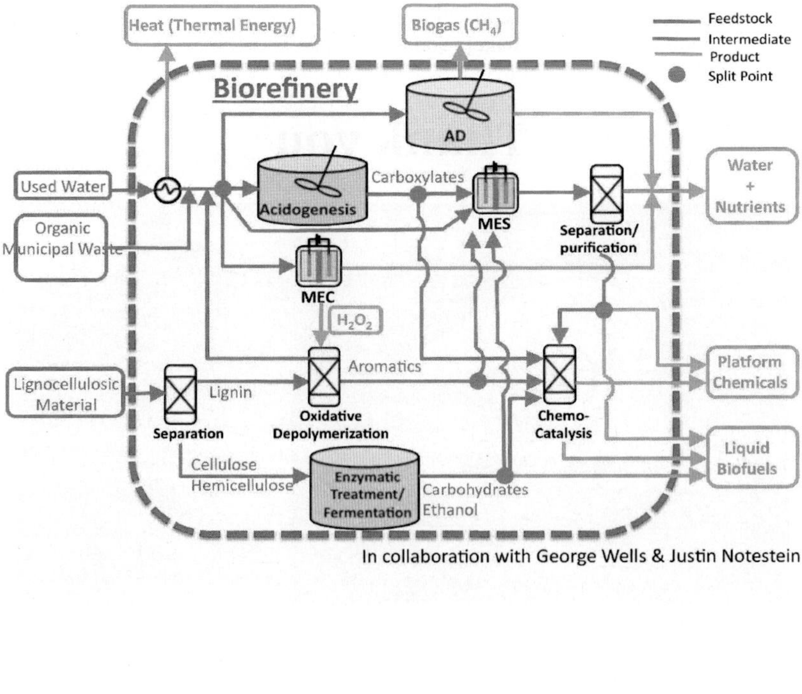

In collaboration with George Wells & Justin Notestein

EXPAND VIEW OF END-USE MANAGEMENT TO COUPLED SYSTEMS & CYCLES OF HUMAN METABOLISM

- To recover and harvest energy and materials.
- To transition to renewable energy.
- To shift to locally tailored distributed systems.
- To develop cycles of water-waste-food; reframe waste as resource to be recovered.
- But need to find policies that make this economically competitive.
- Full cost accounting.

SUMMARY WORKSHOP 2014

- energypmp.org/reports/
- Urban energy issues must be integrated with other systems: water, mobility, land use, information, security, governance, human behavior, policy.
- Better models and better data needed—feed on one another!
- How much real progress can be made by looking at one subset of a coupled complex system? Intuition.
- Involve operational managers from industry: government-industrial partnerships.
- How do we provide tools for decision makers?
- For example—global climate models.
  - Need predictions on local scale.
  - Uncertainty too large, blows up at more granular scale; the scale of interest to policy makers and to cities.

Thank you

POLICIES FOR AN ENERGY SYSTEM TRANSFORMATION

ROBERT N. SCHOCK
World Energy Council, Senior Advisor
London, UK

ABSTRACT

There is a planetary emergency in energy, and the management of energy end-use networks is, today, the foremost part of the problem as well as its solution. Fortunately the science of complex networks is now able to provide an understanding of how these systems respond to perturbations and eliminates the element of chance in designing and optimizing them. Urban areas are a very critical element of the larger concern and form the base on which to rest the solutions. With potential solutions in hand, engaging decision-makers on the understanding of energy networks and implementing proactive policies is not only possible, but likely.

INTRODUCTION

The search for supplies of primary energy over the past several hundred years has been eminently successful. In the past 100 years alone, primary energy supplies have increased by orders of magnitude and consist of many forms. Today mankind has available energy in diverse forms from petroleum, natural gas, coal, nuclear fission, hydroelectric, wind, solar (both photovoltaics and thermal), and biomass. These supplies (Figure 1) have kept up with mankind's growth in population, as well as in economic development (IEA, 2015; WEC, 2013; Means, 2014) and show every sign of continuing this trend.

Technical progress has made energy available that was once considered unavailable (Hubbert, 1956). This energy also comes from a wide area in the globe (BP, 2015) and Table 1. While there is still much energy poverty in the world, that is not the result of not having enough energy, or even of the price being too high, but rather non-energy related economic and social factors.

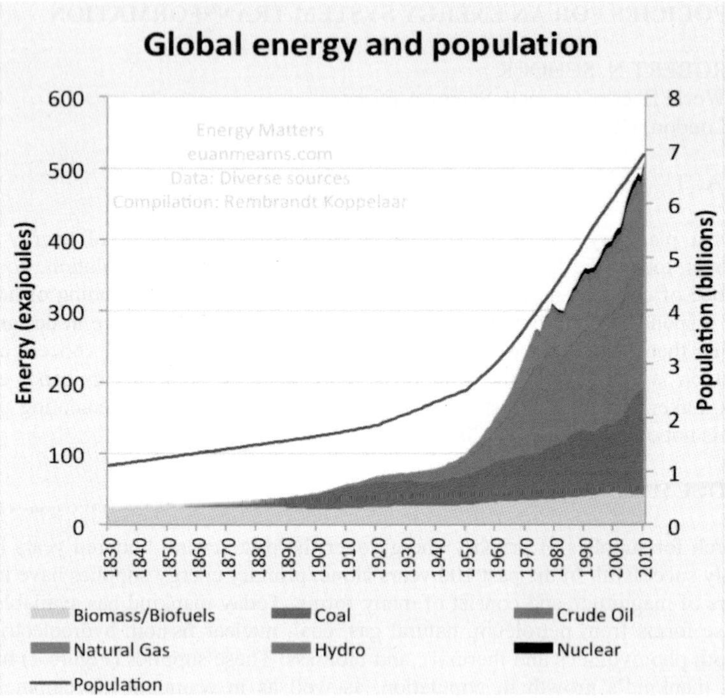

Fig. 1: Population and energy, 1830-2010 (Means, 2014)

Table 1. Global energy production by region. Note: Hydroelectricity, Biofuels, and Renewable Energy (Other) are consumption figures. Mtoe = million tonnes of oil equivalent. (BP, 2015).

2014	Oil (mtoe)	Gas (mtoe)	Coal (mtoe)	Hydro (mtoe)	Bio (mtoe)	Other (mtoe)
N. America	867	866	551	154	31	74
S. and C. America	391	158	65	155	20	22
Europe & Eurasia	834	902	442	196	12	124
Middle East	1340	541	1	5	0	0
Africa	392	182	152	28	0	3
Asia Pacific	397	478	2723	342	8	94

While attention to ways to utilize energy has taken place parallel to finding more primary energy, there is much more that could be done in energy end-use management, not only to make energy more widely available, but to make it less expensive. At present, the world wastes 60% of the primary energy utilized (Figure 2). Rather an achievable goal is to utilize at least 80% of primary energy. The economic stakes are enormous. Today, global investments in devices to utilize energy, alone are of the order of 300B U.S.$/yr. This is but a small part (10 percent) of the overall economic investment in energy end-use which is of the order of 3T U.S.$/yr. (GEA, 2012).

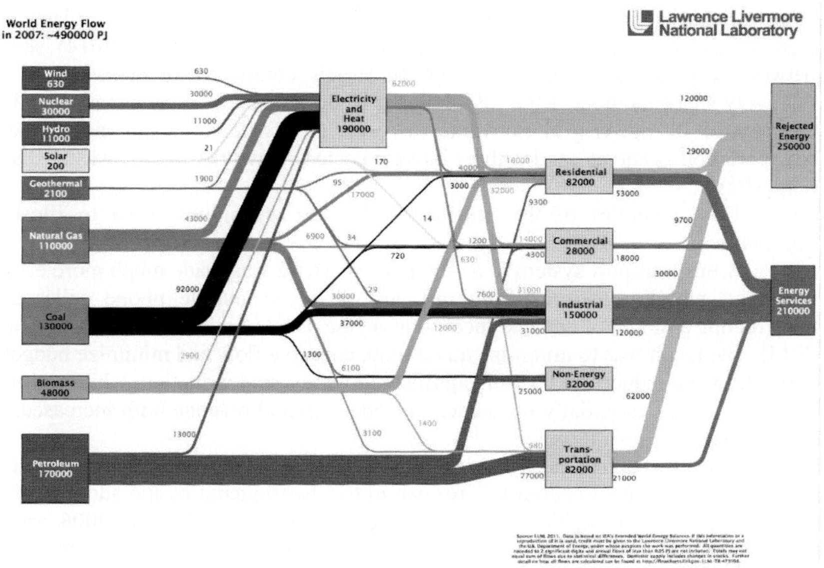

Fig. 2: World energy flow in 2007. (Lawrence Livermore National Laboratory, 2015). PJ = 10^{15} joules.

COMPLEX NETWORK THEORY

What is required is to take the holistic view of this very complex energy end-use system. But the system is so complex that it is often fractured into parts that have their own rules and regulations. Fortunately, complex networks involving energy end-use lend themselves to being designed for optimum efficiency and sustainability by using the rapidly developing physics and mathematics of capability of complex networks. Using basic physics to study the stability of complex networks, these can be applied to the networks which we live with and utilize every day. Related networks can also be controlled, thereby increasing the resilience of each and thus the entire larger system. Another problem in improving the management of energy end-use systems is that common-sense improvements are often in conflict with existing laws, codes, and regulations (Gray, 2014).

Network dynamics are typically non-linear and comprise a large number of dynamic variables. Systemic properties of a network can depend as much on the interactions between components as they do on the properties of the individual components (e.g., generators, distribution stations, households, etc.). The same connections that provide functionality to a network can also serve as conduits for the spread of failures and instabilities that would otherwise be limited to small areas. Optimization and control of network behavior in general requires ongoing systematic assessment of the state of the system, and in many cases forecasts. Determining the state of a network depends on the number and placement of measurements made in one or more variables across the network. Likewise, control depends on number, placement, and form of the control inputs. Recent research (Motter and Schock, 2014) shows that network response to perturbations can be largely controlled by interventions using a relatively small fraction of the total number of nodes.[1] These concepts are applicable to the control of power-grid networks, as well as efficient, clean, and cost-effective transportation systems and other interconnected networks (e.g., waste disposition networks).

Two examples of the utilization of these techniques serve to illustrate the possibilities.

A bus transport system in a city in West Africa was made much more efficient by analyzing mobility patterns from millions of riders' mobile phone calls, and then constructing a model of demand incorporating the data to optimize the system (Calabrese, 2014). The result was to minimize travel time, conserve flow and minimize budgets. Over 90 routes were reduced to 65 with significant journey time reduction, while waiting times for buses stayed essentially the same, and ridership and revenue both increased, a much more sustainable operation.

Models of actual electric grids with distributed power sources in the UK, Italy and Germany show that cascading failures can in fact be triggered by the addition of another link, the opposite of intuition or "common sense" (Witthaut, 2014). In 2006, after a very large cascading failure of the electric grid, the president of one large utility said, "We need more interconnections!" A model of the system showed that cascading failure can in fact be triggered by the addition of such a link. On the other hand dynamic stability can be shown to be achieved using tests of periodic fluctuations in load.

In practice, a network's response to perturbations can be largely controlled with a small fraction of nodes. Motter et al. (2013) shows that a full electric grid network composed of 678 nodes and comprising 1644 edges can be controlled with 170 nodes and comprises all possible edges.

Common sense (often human intuition) sometimes leads us to conclusions that are false. Everyone uses their intuition and experience to make decisions but the science of complex network theory allows us obviate the need to rely on intuition and to understand how parts of networks are really interacting, in theory avoiding unintended results. An example of this phenomenon is that we often believe that properties change monotonically as the number of available interactions is varied. Also that some structures of networks facilitate while others inhibit synchronization (harmony) of the network. Both of these beliefs are, in fact, false.

In practice, a network's response to perturbations (natural or man-made) can be largely controlled by a small fraction of nodes. And control interventions can be

---

[1] In the mathematics associated with networks (graph theory), nodes refer to objects (e.g., power stations in an electrical grid, storage facilities in a materials supply network). Links between nodes are referred to as edges (e.g., power transmission lines, supply transportation routes).

identified *a priori*. Furthermore, planning a network to achieve a desired outcome can be effective even when a stable state cannot be reached directly. But it is important that control in real time depends on efficient data collection and processing.

IMPLEMENTING ENERGY END-USE POLICIES

Finally, to implement effective control of any system, decision makers must be involved. Policymakers are confused when receiving disparate opinions and the result is no implementation. Competing energy technologies and systems or plans leave policymakers confused and the result of this confusion is usually to do nothing, or worse the overall system becomes chaotic. It is therefore important to send clear signals about what technologies are applicable and how they integrate with interdependent networks such as mobility, waste, water, or security. The result of careful use of complex network theory to provide solutions will be clarity on the results that can be achieved.

There are a number of key elements of successful policies, but these are among those most important:

Cost/benefit analysis for proposed actions (Paulson, 2014; Garman et al. 2014)

Elimination of less efficient technologies and enhanced research and development on those technologies that show promise to make a real difference in overall energy end-use efficiency. Plans to build human capacity to ensure expertise to implement and operate these new technologies.

Full-cost pricing of external factors and design of any subsidies to ensure they are consistent with goals and then terminate after effectiveness is shown.

In summary, there is a planetary emergency in energy, and the most critical path to stability is the management of energy end-use. Despite the complexity of this system, there are new tools available to both understand and build efficient networks, most notably the science of complex networks. With potential solution in hand, engaging decision-makers on the understanding of a system is not only possible, but likely.

Beyond energy end-use, related networks (e.g., water) can also be studied and organized and the results integrated.

LOOKING FORWARD

Having groups in place to study Planetary Emergencies, the Majorana Centre for Scientific Culture of the World Federation of Scientists is in a unique position to utilize network science for solutions to mitigate the threats already identified. Ultimately, this will allow the Centre to expand its studies into the connections and interactions between networks with which it is already involved (such as energy availability, security against threats, and limits of development, to name one possible connection), and to identify solutions to dealing with these and other interrelated planetary emergencies.

The science, taking the same traditional physics used in the understanding of the behavior of small systems such as the details of nuclear structure and behavior, applies it to the understanding of complex systems on a large scale. These systems can include any of the networks associated with the diverse threats mentioned above. It is now possible to envision the control of large networks with great precision in real time to achieve societal goals. Application to the management of these networks increases the likelihood of reliable and sustainable systems over those that traditionally rely on human intuition and which can be wrong. Related networks can now also be controlled thereby increasing their resilience and that of the entire larger system. Imagine the linked electric power, natural gas, water, telecommunication, and transportation networks, each with its own non-linear dynamics, time scales, and constraints. Add the existing political/policy constraints, which are geared to the 20th century, not the technologies of the 21st century, and implemented solutions are possible.

REFERENCES

1. BP (2015) *BP Statistical Review of World Energy 2015*. London.
2. Calabrese, F. (2014) "Understanding urban dynamics from digital traces." www.energypmp.org/reports.
3. Garman, D., K. Emanuel and B. Phillips (2014) "Breaking the Climate Deadlock." *Issues in Science and Technology*. XXXI No. 4:75-82.
4. GEA (2012) *Global Energy Assessment-Toward a Sustainable Future*. T.B. Johansson, A. Patwardhan, N. Nakicenovic and L. Gomex-Echeverri, eds., Cambridge U. Press, Cambridge UK.
5. Gray K.A. (2014) "Sustainable Cities." www.energypmp.org/reports.
6. Hubbert M.K. (1956). "Drilling and Production Practice." *American Petroleum Institute & Shell Development Co*. Publication No. 95, 9-22.
7. IEA (2015) *World Energy Outlook*. International Energy Agency. Paris, www.worldenergyoutlook.org.
8. Lawrence Livermore National Laboratory (2015) "Energy Flow Charts." www.flowcharts.llnl.gov.
9. Means, E (2014) "Energy and Mankind part 3." http://euanmearns.com/energy-and-mankind-part-3/.
10. Motter, A.E., S.A. Meyers, M. Anghel and T. Nishikawa (2013) "Spontaneous synchrony in power-grid networks." *Nature Physics*. 9:191-197.
11. Motter, A.E. and R.N. Schock (2014) "Networks: Energy, Cities and the Control of Complex Systems." www.energypmp.org/reports.
12. Paulson, H.M. (2014) "The Coming Climate Crash." *New York Times*. June 21, 2014.
13. WEC (2013) *World Energy Resources: 2013 Survey*. World Energy Council, London.
14. Witthaut, D. (2014) "Stability, Decentralization and Braess' Paradox in Oscillator Models for Power Grid Operation." www.energypmp.org/reports.

SESSION 4

IPCC: ISSUES AND ALTERNATIVES

GLOBAL WARMING AND THE IRRELEVANCE OF SCIENCE

RICHARD LINDZEN
Earth, Atmospheric and Planetary Sciences, Massachusetts Institute of Technology, Cambridge, USA

In many fields, governments have a monopoly on the support of scientific research. Ideally, they support the science because they believe objective research to be valuable. Unfortunately, as anticipated by Eisenhower in his farewell speech from January 17, 1961 (the one that also warned of the military-industrial complex), "Partly because of the huge costs involved, a government contract becomes virtually a substitute for intellectual curiosity." Under these circumstances, when the government wants a particular scientific outcome the ideal arrangement is vulnerable. However, as I hope to show, the problem is not simply bias. Rather, the powers that be invent the narrative independently of the views of even cooperating scientists. It is, in this sense, that the science becomes irrelevant. This was certainly the case in the first half of the twentieth century, where we just have to look at Lysenkoism[1] in the former Soviet Union, Social Darwinism, and Eugenics throughout the western world,[2] as well as, in the 1960s, the unfounded demonization of DDT.[3] Each phenomenon led to millions of deaths. And, in each case, the scientific community was essentially paralyzed, if not actually complicit.

Will climate catastrophism join this list? I fear so. The position of the policy world is clear. Here is President Obama's constant refrain:

> "Climate change is contributing to extreme weather, wildfires, and drought, and that rising temperatures can lead to more smog and more allergens in the air we breathe, meaning more kids are exposed to the triggers that can cause asthma attacks."

Pope Francis, President Hollande, and virtually all state leaders have chimed in with similar proclamations.

And yet, the whole proposition is without basis and highly implausible. The association with asthma that is regularly made by both Obama and Hillary Clinton is a good example of nonsense driven by focus groups who find this to be an effective scare theme. The other claims are no better. In the 70s, the scientific community regularly designated warm periods as 'climate optima'. That CO_2 was essential to plants and effectively a fertilizer was also widely understood. Thus, it was not surprising that the early environmental movement chose to promote fear of global cooling which, not surprisingly, was attributed to industrial emissions (most notably sulfates).[4] When, however, in the late 1970s, it was recognized that sulfates could be scrubbed, that the irreducible product of industrial emissions was CO_2, that CO_2 emissions were likely to warm rather than cool, and that there was an hypothetical process whereby this warming

[1] Medvedev, Zh.A. (1969) The Rise and Fall of T.D. Lysenko, 304pp, Columbia University Press.
[2] Lindzen, R.S. (1996) Science and politics: global warming and eugenics. in Risks, Costs, and Lives Saved, R. Hahn, editor, Oxford University Press, New York, 267pp (Chapter 5, 85-103).
[3] http://www.jpands.org/vol9no3/edwards.pdf.
[4] Ponte, L. (1976) The Cooling, Prentice-Hall, 306pp.

could be amplified (by what came to be known as the water vapor feedback[5]), this whole narrative was turned on its head. The hitherto optimal warming was now put forth as a consequence to be feared. President Carter's science adviser, Frank Press, had the National Research Council investigate the matter, leading to the famous Charney Report from 1979.[6] This report summarized the results of the primitive climate models of that period, and found that they had a range of sensitivities to doubling CO_2 of from 1.5 to 4.5 degrees Centigrade.[7] The report regarded such results as possible but attached little credibility to the models. The whole situation was eerily reminiscent of Orwell's *Animal Farm*, when 'four legs good, two legs bad' became 'four legs good, two legs better'. Repetition was the mechanism used to convince. So was the claim, already made by 1988 in *Newsweek* that, 'all scientists agreed.' The larger public thus had no reason to actually dig into the science. Indeed, the actual science had already become irrelevant. This new narrative depended not only on the allegation of consensus, but also on lineage. It was always pointed out that the greenhouse effect had already been identified in the early 19th Century by Tyndall,[8] later by Arrhenius,[9] and still later by Callendar.[10] While this was true, it was also the case that the effect was generally held to be of much less importance than changes in the general circulation related to transport. For example, in an important collection of papers from 1955 (Dynamics of Climate, Richard Pfeffer, editor, Pergamon Press) with contributions from the leading climate scientists of the period (Charney, Phillips, Lorenz, Eliassen, Smagorinsky, etc.), increasing CO_2 was barely mentioned, and the greenhouse effect was not mentioned at all. The model favored for global cooling alarm was the Budyko-Sellers[11] model which also focused on meridional heat transport. Only with the decision to push global warming alarm, did the greenhouse effect become central to the discussion of climate. Needless to add, consensus and lineage are not generally regarded as the backbones of science.

The implausibility or even outright silliness of global warming catastrophism (sometimes referred to as CAGW, catastrophic anthropogenic global warming) is so extensive that one hardly knows where to begin. It is crucial to emphasize catastrophism, because the situation is made even more incoherent by the intentional conflation of

---

[5] Manabe, S., and R.T. Wetherald (1975) "The effects of doubling the CO_2 concentration on the climate of a general circulation model." *J. Atmos. Sci.* 32:3. For definitions of feedback and discussions of water vapor see Lindzen, R.S. (1993) "Climate dynamics and global change." *Ann. Rev. Fl. Mech.*, 26:353-378.

[6] Carbon Dioxide and Climate: A Scientific Assessment: Report of an Ad Hoc Study Group on Carbon Dioxide and Climate Woods Hole, Massachusetts July 23-27, 1979 to the Climate Research Board Assembly of Mathematical and Physical Sciences National Research Council. This is commonly referred to as the Charney Report. Jule Charney was the chairman of the panel.

[7] The Intergovernmental Panel on Climate Change (IPCC) has not changed this range except for the fourth assessment which increased the lower bound to 2°C. The fifth assessment brought it back to 1.5°C while peculiarly but revealingly claiming that there was no basis for preferring one value to another.

[8] Tyndall, J. (1863) "On Radiation through the Earth's Atmosphere." *Philosophical Magazine.* ser. 4:25: 200-206.

[9] Arrhenius, S. (1896) "On the Influence of Carbonic Acid in the Air upon the Temperature of the Ground." *Philosophical Magazine* 41: 237-76.

[10] Callendar, G.S. (1938) "The Artificial Production of Carbon Dioxide and Its Influence on Climate." *Quarterly J. Royal Meteorological Society* 64: 223-40.

[11] A critical discussion of these models can be found in Lindzen, R.S. (1990) Dynamics in Atmospheric Physics, Cambridge University Press, New York, 310pp.

simple basic results that are widely agreed upon, but which have no catastrophic implications, with catastrophism itself.

Currently, there really is quite a lot of basic agreement within the climate science world: climate change exists; there has been warming since the Little Ice Age ended around the beginning of the 19th Century (well before emissions are regarded as contributing significantly); human emissions can contribute to climate change; levels of CO_2 in the atmosphere have been increasing. None of this is controversial and none of this actually implies alarm. However, in the policy world, as emerges from virtually any reading of the current political discourse and its attendant media coverage, the innocuous agreement is taken to be equivalent (with essentially no support from observations, theory or even models) to rampant catastrophism. There are numerous examples of the issuance of unalarming claims (regardless of their validity or lack thereof) that are interpreted as demanding immediate action. Perhaps the most striking example involves the iconic statement of the IPCC: *Most of the warming over the past 50 years is due to man.* Is this statement actually alarming? First, we are speaking of small changes. 0.25°C would be about 51% of the recent warming. Given the uncertainties in both the data and its analysis, this is barely distinguishable from zero. Evidence of this uncertainty is shown by the common adjustments made to the record.

Some charts from the weather page of the Boston Globe of March 12, 2013 (any other date would serve as well) illustrate how small the changes really are. In the attached figure we see the high and low temperatures for each day in the preceding month (black), the average high and low temperature for each date (dark gray) and the record high and low temperature for each date (light gray). The width of the black horizontal line corresponds to the change in the global mean temperature anomaly over the past 150 years.

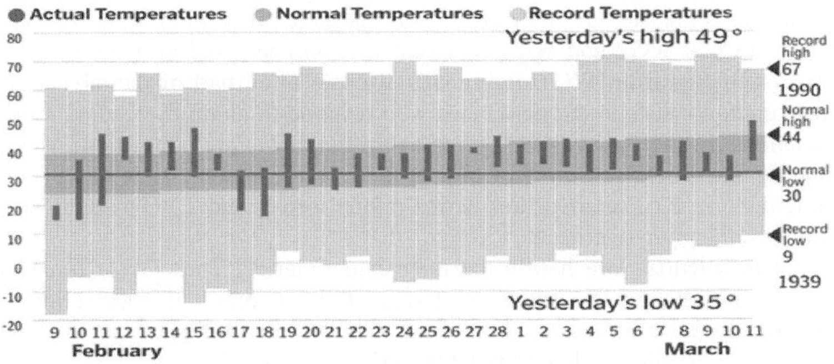

High and low temperatures result from the advection of air roughly along the path of the jet stream (and this path changes from day to day and year to year). Record breaking temperatures (regardless of the year that they occurred) correspond roughly to the warmest and coldest temperatures on the temperature map for March 11.

Second, the recent warming episode is not at all unprecedented. The almost identical episode from about 1919-1940 could not be attributed to man.

Third, the observed warming is completely consistent with low climate sensitivity. Alarm requires, for starters (and only for starters), high sensitivity. By sensitivity, we refer to how much warming we expect for each doubling of CO_2. High sensitivity is generally regarded as 3°C or more. If we were to assume that ALL warming over the past 50 years were due to added greenhouse gases, we would conclude that the sensitivity was about 1°C. How do models with much higher sensitivity manage to replicate the past 50 years? They do so by subtracting from the greenhouse warming essentially unknown aerosols which they then include as due to human emissions. However, in a recent paper from the Max Planck Institute, Stevens (2015)[12] finds that aerosols are limited and unable to compensate for the higher sensitivities. If man accounts for only 51% of the warming, then even modest warming becomes implausible.

Although it has become commonplace to fear warming, it is worth noting that the approximately 1°C warming since the 19th Century has been accompanied by the improvement of all indices of human welfare (including environmental quality).

Indeed, the very notion that climate is described by a single number that is forced by another single number, is itself a bit strange. For example, the force on a piston acting on a gas in a cylinder certainly does determine the pressure. However, as Budyko and Izrael[13] noted long ago, climate change is characterized by a relatively stable tropics and changes in the equator-to-pole temperature difference. This, crudely speaking, has to do with heat transport. Pursuing the analogy with the piston, would we really expect the flow through a pipe to depend on the mean pressure in the pipe rather than the gradient of pressure along the pipe?

Why then do scientists go along with this? The situation has been described by me earlier as consisting in an iron triangle.[14] At one vertex are the scientists who make meaningless or ambiguous statements. The scientific assessment of Working Group 1 of the IPCC is full of such statements. Then there is the second vertex: that of the advocates and media that 'translate' the statements into alarmist declarations. The advocates also include the IPCC's WG2 and WG3 that deal with impacts and mitigation by assuming worst case scenarios from WG1. Politicians also are often part of the advocacy efforts. The third vertex consists in the politicians who respond to alarm by feeding more money to the scientists in the first vertex. As far as the scientists are concerned, what's not to like? Should the scientist ever feel any guilt over the matter, it is assuaged by two irresistible factors: 1.) The advocates define public virtue; and 2.) His administrators are delighted with the grant overhead.

Of course, scientists are hardly the main beneficiaries. The current issue of global warming/climate change is extreme in terms of the number of special interests that opportunistically have strong motivations for believing in the claims of catastrophe despite the lack of evidence. In no particular order, there are the:

---

[12] Stevens, B.J. (2015) "Rethinking the lower bound on aerosol radiative forcing." *Journal of Climate*. 705 doi: 10.1175/JCLI-D-14-00656.1.

[13] Budyko, M.I., Izrael, Y.A. (1991) In Anthropogenic Climate Change, ed. M.I. Budyko, Y.A. Izrael, pp. 277-318. Tucson: Univ. Ariz. Press

[14] Lindzen, R.S. (2013) "Science in the Public Square." *Journal of American Physicians and Surgeons*. **18**: Fall issue.

- Leftist economists for whom global warming represents a supreme example of market failure (as well as a wonderful opportunity to suggest correctives),
- UN apparatchiks for whom global warming is the route to global governance,
- Third world dictators who see guilt over global warming as providing a convenient claim on aid (i.e., the transfer of wealth from the poor in rich countries to the wealthy in poor countries),
- Environmental activists who love any issue that has the capacity to frighten the gullible into making hefty contributions to their numerous NGOs,
- Crony capitalists who see the immense sums being made available for 'sustainable' energy,
- Government regulators for whom the control of a natural product of breathing is a dream come true,
- Newly minted billionaires who find the issue of 'saving the planet' appropriately suitable to their grandiose pretensions,
- Politicians who can fasten on to CAGW as a signature issue where they can act as demagogues without fear of contradiction from reality or complaint from the purported beneficiaries of their actions. (The wildly successful London run of *Yes, Prime Minister* dealt with this.)

All of the above special interests, quite naturally, join the chorus of advocates.

Strange as it may seem, even the fossil fuel industry is generally willing to go along. After all, they realize better than most, that there is no current replacement for fossil fuels. The closest possibilities, nuclear and hydro, are despised by the environmentalists. As long as fossil fuel companies have a level playing field, and can pass expenses to the consumers, they are satisfied. Given the nature of corporate overhead, the latter can even form a profit center.

In point of fact many of the foremost scientific supporters of alarm acknowledge the absence of a basis for catastrophism. Here are some remarks the presidents of the Royal Society (Martin Rees) and of the National Academy (Ralph Cicerone) published in the Financial Times.[15]

"Straightforward physics tells us that this rise is warming the planet. Calculations demonstrate that this effect is very likely responsible for the gradual warming observed over the past 30 years and that global temperatures will continue to rise—superimposing a warming on all the other effects that make climate fluctuate. Uncertainties in the future rate of this rise, stemming largely from the 'feedback' effects on water vapour and clouds, are topics of current research."

Rees and Cicerone are counting on the fact that most readers won't notice that the so-called 'uncertainties' are, in fact, the main issue; the straightforward physics is trivial.

They continue "Our academies will provide the scientific backdrop for the political and business leaders who must create effective policies to steer the world toward a low-carbon economy."

Clearly, despite the uncertainty, the policy is taken for granted.

---

[15] *Financial Times*. 9 April, 2010.

Here is an exchange from the BBC 4 interview of Ralph Cicerone on 13/07/2012. John Humphrys is the interviewer.

> John Humphrys: *You don't sound, if I can use this word, apocalyptic. I mean, you're not saying "If we don't do these things, we're going to go to hell in a handbasket, we're going to fry, in a few years".*
>
> Ralph Cicerone: *Well, there are people who are saying those things John.*
>
> Humphrys: *But not you.*
>
> Ralph Cicerone: *No. I don't think it's useful, I don't think it gets us anywhere, and we don't have that kind of evidence.*

The situation may have been best summarized by Mike Hulme, director of the Tyndall Centre at the University of East Anglia (a center of concern for global warming): "To state that climate change will be 'catastrophic' hides a cascade of value-laden assumptions which do not emerge from empirical or theoretical science."

Even Gavin Schmidt, Jim Hansen's successor as head of NASA's Goddard Institute of Space Studies, whose website, Realclimate.org, is a major defender of global warming, does not agree with claims of extremes: "General statements about extremes are almost nowhere to be found in the literature but seem to abound in the popular media…It's this popular perception that global warming means all extremes have to increase all the time, even though if anyone thinks about that for 10 seconds they realize that's nonsense."

Interestingly, basic meteorological theory tells us that extremes depend significantly on the temperature difference between the tropics and the poles—something that is expected to diminish in a warmer world.

On the other hand, there is quite a lot of 'science on demand.'

The well-established Medieval Warm Period is a problem for the narrative.

Michael Mann's Hockey Stick gets rid of the Medieval Warm Period.

The physics of moist convection requires that warming maximize in the tropical upper tropospheric troposphere, and models agree, but the data doesn't show this.

Ben Santer reworks the data to show the maximum.

Significant warming ended about 18 years ago, showing that CO_2 is not the major factor in climate.

Tommy Karl adjusts and rearranges the data to eliminate the pause.

Quite a few independent studies show that the outgoing radiation from the earth indicates low climate sensitivity.

Andy Dessler ignores the physical and mathematical constraints to claim the opposite (at a truly negligible significance level).

Antarctic sea ice is increasing.

Jim Hansen absurdly claims that this is what one should expect from global warming (which, however, has not been occurring for 18 years).

Basic dynamics of the atmosphere calls for reduced extremes and storminess in a warmer world.

John Holdren invents a cockamamie theory of tropospheric polar jets to claim that such an imaginary jet is destabilized with warming, leading to more and more extreme storminess.

As Pat Michaels showed,[16] there is a remarkable bias in publications. For articles in *Nature* and *Science* during the period July 1, 2005 through July 30, 2006, he found a total of 116 publications dealing with climate data. Of these, 84 were "worse", 10 were "better", and 22 were "neutral" with respect to earlier claims. The relative numbers for Science and Nature, respectively, were 34,50 (worse), 5,5 (better) and 9,13 (neutral). Assuming existing studies were equally likely to be better or worse, this result would have negligible likelihood. Of course, given Michaels' findings, it is almost certain that the existing studies were already biased—thus rendering likelihood almost infinitesimal.

In point of fact, the Climategate 1 and 2 email releases showed explicitly the breakdown in peer review.[17]

We have, thus far, ignored the 'impacts' industry where papers are published (and research is supported) attributing hundreds of things to the minimal warming that has occurred. The website http://whatreallyhappened.com/WRHARTICLES/globalwarming2.html lists some of these—ranging from acne to walrus stampedes to typhoid fever. Note that even in this extensive list, asthma is not mentioned.

Prof. Stefan Rahmstorf of the Potsdam Institute for Climate Impacts has reflected on the failure of his alarmist position to sway the world: "Sometimes I have this dream… I call the fire brigade. But they don't come because some mad person keeps telling them it's a false alarm. The situation is getting more and more desperate, but I can't convince the firemen to get going." Such nightmares over a few tenths of a degree seem a little exaggerated. One expects that a counsellor might be more effective than a fireman.

The take of political figures generally transcends the absurd. Senators McCain and Lieberman (Boston globe, February 13, 2007) offered the standard misreading of the

---

[16] Michaels, P. (2008)" Evidence for 'PUBLICATION BIAS' concerning global warming in *Science* and *Nature*," *Energy & Environment*, 19:287-301.

[17] For example: http://www.lavoisier.com.au/articles/greenhouse-science/climate-change/climategate-emails.pdf, http://wattsupwiththat.com/2012/01/06/250-plus-noteworthy-climategate-2-0-emails/.

IPCC's iconic statement: "The recent report by the Intergovernmental Panel on Climate Change concluded there is a greater than 90 percent chance that greenhouse gases released by human activities like burning oil in cars and coal in power plants are causing most of the observed global warming. This report puts the final nail in denial's coffin about the problem of global warming."

Of course, the IPCC wisely avoided making the claim that 51% of a small change in temperature constituted a 'problem.' This, they left to the politicians.

Secretary of State John Forbes Kerry goes much further in a lengthy speech delivered in Indonesia in February of 2015. Here are some selections:

> "...When I think about the array of global climate—of global threats—think about this: terrorism, epidemics, poverty, the proliferation of weapons of mass destruction—all challenges that know no borders—the reality is that climate change ranks right up there with every single one of them. And it is a challenge that I address in nearly every single country that I visit as Secretary of State, because President Obama and I believe it is urgent that we do so...
>
> ...it's compelling us to act. And let there be no doubt in anybody's mind that the science is absolutely certain...
>
> ...I know sometimes I can remember from when I was in high school and college, some aspects of science or physics can be tough—chemistry. But this is not tough. This is simple. Kids at the earliest age can understand this...(It should come as no surprise that Kerry proceeds to get literally everything wrong in his subsequent description of the science.)
>
> ...First and foremost, we should not allow a tiny minority of shoddy scientists and science and extreme ideologues to compete with scientific fact...
>
> ...This is not opinion. This is about facts. This is about science. The science is unequivocal. And those who refuse to believe it are simply burying their heads in the sand. Now, President Obama and I believe very deeply that we do not have time for a meeting anywhere of the Flat Earth Society..."

As usual, political figures improperly associate science as a source of unquestionable authority rather than a successful mode of inquiry.

Secretary Kerry's unsurprising lack of understanding as to what science is, is duplicated by Gina McCarthy (Head of U.S. EPA (which is spearheading America's War on Fossil Fuels) whose education consists in a B.S. in Anthropology from the University of Massachusetts, Boston Branch, and an M.S. in Environmental Health Engineering, Planning and Policy from Tufts University).

> "By now we all know that climate change is driven in large part by carbon pollution (typically conflating carbon with carbon dioxide) and it leads to more extreme heat, cold, storms, fires and floods."
>
> "We are way past any further discussion or debate. Scientists are as sure that humans are causing climate change as they are that cigarette smoke causes lung cancer. So, unless you want to debate that point, don't debate about climate change any longer because it is our moral responsibility to act. That responsibility right now is crystal clear. And that is why we have taken action."

"...the science has spoken on this. A low-carbon future is inevitable. We're sending exactly the right signals on what, at least EPA believes to be, a future of lower pollution that is essential for public health and the environment, that EPA's not just authorized but responsible to acknowledge and push towards."

Of course, some political figures skip any embarrassing pretenses concerning science and move directly to their agenda. Christiana Figueres, executive secretary of U.N.'s Framework Convention on Climate Change: "This is the first time in the history of mankind that we are setting ourselves the task of intentionally, within a defined period of time, to change the economic development model that has been reigning for at least 150 years, since the Industrial Revolution."

Ms. Figueres is not alone in taking this approach. Pope Francis' closest adviser castigated conservative climate change skeptics in the United States, blaming capitalism for their views. Speaking with journalists, Cardinal Oscar Rodríguez Maradiaga criticized certain "movements" in the United States that have preemptively come out in opposition to Francis's planned encyclical on climate change. "The ideology surrounding environmental issues is too tied to a capitalism that doesn't want to stop ruining the environment because they don't want to give up their profits."

It is difficult to know whether the statements of prominent political figures represent dishonesty, ignorance or both.

Ms. Figueres may be the most honest. No proposed measures will have any discernible impact on climate (regardless of one's view of the physics) unless one rolls back the industrial revolution everywhere and permanently—and even then significant impact on global climate is dubious. Of course, no country outside the western world would even consider this, though they are perfectly happy to endorse the efforts of the West to do so.

A constant feature of the public presentation of the issue is the exploitation of public ignorance. A large poster appearing in the Paris Metro showed the World Wildlife Fund's signature panda leading young people in mass demonstration (intentionally mimicking the storming of the Bastille) calling for the elimination of CO_2. Presumably these young people have never heard of photosynthesis and fail to realize that advanced forms of life would largely cease for levels of CO_2 less than about 150 ppmv.

In view of the very serious implications of proposed policies alleged to deal with global warming such as the restriction of access to electricity for the 1.3 billion human beings currently without such access, and the increased poverty for billions more with its obvious implications for health and longevity, etc., not to mention foregoing the well-established agricultural benefits of added CO_2,[18] a chemical essential to life as we know it rather than a pollutant (the US Navy regards levels of 5000 ppmv safe on nuclear submarines; ambient levels are currently 400 ppmv). It is clear that the issue of climate does constitute an emergency. However, as is so often the case, the emergency does not arise from science and technology, but rather from politics. It is worth examining whether science can play a role in the mitigation of this emergency. It is doubtful whether the

[18] Goklany, I.M. (2015) Carbon Dioxide, The Good News, Global Warming Policy Foundation, GWPF Report 18, 47 pp. Available from http://www.thegwpf.com/content/uploads/2015/10/benefits1.pdf.

answer will consist in research grants. However, science has much at stake. Its hard earned raison d'etre as our most effective tool for objective assessment is being squandered, and with it, the basis for public trust and support.

If we do nothing to stop this insanity, science will rightly be regarded as just another racket. This might just be more collateral damage than we can readily afford—attaining the deserved status of a planetary emergency.

THREE THINGS SCIENTISTS NEED TO KNOW ABOUT THE IPCC

DONNA LAFRAMBOISE
Journalist and a Senior Research Fellow at the Frontier Centre for Public Policy
Toronto, Canada

I am not a scientist. Rather, I'm an investigative journalist who has spent the past six years examining the climate debate. When I began my research, I heard a great deal about an organization called the Intergovernmental Panel on Climate Change, the IPCC. This is the most important climate body in the world because governments point to IPCC reports as the reason enormous amounts of money should be spent fighting climate change.

The IPCC is typically described as a paragon of scientific virtue, a model of transparency. We're told it is staffed by the world's top scientists and best experts, and that it has been awarded a Nobel Peace Prize. The IPCC, we are advised, is an organization whose conclusions can—and should—be trusted.

At the beginning, I believed all of that. I had no reason not to. But then I started to notice that many assertions about the IPCC are not, in fact, true. These, therefore, are the three things every scientist needs to know:

1. The IPCC is a political entity.
2. Scientists are not in charge.
3. The IPCC is a template that gets duplicated elsewhere. So even if your scientific specialty has nothing to do with climate you may, at some point, be invited to participate in an organization of this kind.

Let's start with point #1. **Is the IPCC about science or politics?** If you visit its website, you'll find the highly misleading claim that the IPCC is "a scientific body." But let's take another look at its name. It's called the Inter*governmental* Panel. The definition of 'intergovernmental' is "between two or more governments." There's only one way to become a member of the IPCC, and that is to be a UN-recognized government. Taiwan, with its 23 million people (four times the population of Norway), isn't recognized by the UN, and therefore cannot participate in the IPCC.

Now when governments work together they are not doing science—they're doing politics. There's nothing wrong with politics, but if a political body is trying to convince us that it's a scientific body, we should be wondering why. Back at the IPCC website, you'll notice the logos of two UN bodies in the top right corner—the World Meteorological Organization (WMO) and the United Nations Environment Programme (UNEP). Those two UN logos appear on every IPCC report, press release, and official speech. This is an organization consisting of governments, an organization firmly embedded in the UN structure.

What happens is that governments—not science academies—nominate individual scientists to help write IPCC reports. Thousands of individuals are nominated and hundreds are chosen to be lead authors. The public is then told that these are the world's finest scientific minds and that we should therefore trust these people's findings. Here's a news clipping from 1995. For 20 years, we've been told that the IPCC and the world's top scientists are one and the same.

Global warming is leading to climatic upheaval, say scientists

Experts have reached consensus after years of disagreement, reports Geoffrey Lean

GEOFFREY LEAN | Sunday 15 October 1995

SHARE TWEET SHARE REDDIT SHARE | Shares 0 PRINT

THE world's top scientists have at last reached agreement on global warming - and their message is just about as alarming as it could be for humanity. The world really is warming up, the scientists agree, and pollution is at least partly to blame.

VOICES

But here's a curious point. The IPCC seems strangely uninterested in the actual credentials of these individuals. We all know scientific expertise is rather specific. The website of the Ettore Majorana Centre for Scientific Culture, here in Erice, tells us "there are at least 10 recognized specialties" in subnuclear physics alone. And yet the IPCC never bothers to identify the expertise of its chosen personnel.

If you consult an IPCC-issued list of authors involved in its last report, you'll find that those who worked on a chapter about oceans came from Germany, Australia, Japan, Brazil and so forth. Why are we told that? Because the IPCC, being a UN body, cares deeply about *geographical* representation and *geopolitical* appearances. A journalist such as myself has to Google the names of each one of these authors in order to find out what it is that these people are experts in. The IPCC has a copy of all of their CVs, but it doesn't make those CVs public. Instead, it behaves as though a scientist's nationality is the critical piece of information.

Now let's take a look at point #2. **Scientists are not in charge.** The IPCC's latest report contains 60 chapters and totals more than 7,000 pages. Many good, sincere scientists toiled away on their own small portion of that enormous report. These people no doubt did their best to be honest and accurate. But here's the problem: almost no one will ever read that 7,000-page report. Journalists certainly won't; we don't have that kind of time. The same is true of politicians—the people who vote on new laws and determine how trillions of euros will get spent. The few occasions in which I've spoken to a politician about my book-length exposé of the IPCC, it has become clear their lives are way too busy. They don't have time to read 200 pages, never mind 7,000.

Everyone knows this. Which is why the IPCC also produces documents in the twenty to 30-page range bearing the title *Summary for Policymakers*. There's a Working Group 1 *Summary for Policymakers*, a Working Group 2 summary, and a Working Group 3 summary for different sections of that 7,000-page report. These are the documents that matter, the ones that actually get read. These are the documents that influence media coverage and shape new legislation.

Now if scientists were in charge at the IPCC, at the end of the process these summaries would be written up by a small group, released into the world, and we'd all read these scientists' unadorned words. But that's not what happens.

In fact, IPCC authors only *draft* these summaries. And then something incredible transpires. A big IPCC meeting takes place, attended by governments. Although some people in the room are scientists, the vast majority are diplomats, politicians, foreign affairs specialists, bureaucrats, and assorted other officials. These people then spend the next week *re-writing* the summary authored by scientists.

Every single paragraph in that draft gets projected onto large screens and discussed. The delegation from country X wants two sentences removed. The delegation from country Y insists that a new phrase be inserted. Graphs get added, tables get subtracted. And they keep talking about that one paragraph until all of the countries present are happy with it. Then it's taken down and the next paragraph is put up on the screen.

Even though there's a hard deadline that this process needs to finish by, these meetings aren't always well managed. So toward the end of the week, discussions often take place around the clock; they're still going on at 11 pm, and at 3 am. People who've been through this process say it becomes a bit of an endurance test. Who can stay awake? The bottom line is that this is a week of naked political horse trading that takes place behind closed doors. Journalists are not allowed to witness what goes on.

NEWS & MEDIA
UNITED NATIONS RADIO

TODAY'S NEWS | TODAY'S FEATURES | PROGRAMMES | REGIONS | ABOUT US | YOUR

2 Nov 2014

Science has spoken again on climate change: UN chief

LISTEN / DOWNLOAD

Science has spoken yet again on climate change and time is not on the world's side, according to the UN Secretary-General.

On Sunday, Ban Ki-moon was speaking in Copenhagen to launch the latest report of the Intergovernmental Panel on Climate Change (IPCC).

Compiled by hundreds of scientists from around the world, Mr Ban said the assessment is the most comprehensive appraisal of climate change ever undertaken.

"With this new Synthesis Report, science has spoken yet again with much more clarity and greater certainty. Citizens are increasingly restive—but also eager to

It's only after the diplomats have haggled over this *Summary* paragraph-by-paragraph that the final version gets officially released at a press conference. The world is then told that *science has spoken*. But what's just happened has nothing to do with science. Scientific truth is not determined in the dead of night by UN-level negotiations. On what planet would such an approach make scientific sense?

But the bad news doesn't stop there. There's actually a step in the IPCC process in which the original, lengthy report gets amended so that it conforms to the politically-negotiated *Summary*. I am not making this up.

Bloomberg.com

Food, Water Security Threatened by Warming, UN Panel Chief Says

By Alex Morales

Publication Delay

On Feb. 2, the panel said in its first of the four reports temperatures have risen by 0.76 degrees Celsius (1.37 Fahrenheit) since the 19th century, and will rise by another 1.1 to 6.4 degrees this century. It also said global warming is ``very likely'' caused by people, based on scientific evidence.

The panel has yet to release the full scientific report that forms the basis of the ``summary for policymakers'' issued last month, leading to charges that the process is politicized and the science is being amended to conform to the policy document. The process is ``transparent,'' and the report will likely come in the next two weeks, Pachauri said.

``The draft that was submitted to the plenary and what came out of it are not substantially different, but we necessarily have to ensure that the underlying report conforms to the refinements,'' Pachauri said, adding that scientists were present and able to comment at the meeting of government officials that negotiated the summary's wording.

``That's the strength of the IPCC: really it would mean that one particular country or a group of countries would have to bamboozle everybody else to bring about radical changes that don't have a basic rationale behind them,'' Pachauri said. Still ``this point is well taken, and in the next two reports you won't see this kind of delay,'' he said.

To contact the reporter on this story: Alex Morales in London at amorales2@bloomberg.net.

Here's a 2007 news clipping in which the IPCC's chairman explains to a reporter that the reason the *Summary* has been released before the full report, is because the IPCC needs to ensure that the report is consistent with the *Summary*. After the political meeting takes place, they go back and change the original report. In 2013, following one of these meetings, nine out of 14 chapters had to be revisited. In Chapter 11, more than 20 changes needed to be made.

As a journalist, I find this bizarre. If I spend months writing a long, investigative magazine article in which I've worked hard to describe matters accurately, there is no way on Earth it would be OK for the sponsors of the process—the magazine's advertisers—to then sit around and negotiate between themselves what should and should not be included in my magazine article. Absolutely out-of-the-question. But this is normal operating procedure at the IPCC.

So what's going on here? How do we explain this outlandish, perverse process? Well Rajendra Pachauri, who chaired the IPCC for 13 years, has said that its "main customer" was something called the UNFCCC. Jean-Pascal van Ypersele, who aspired to become the IPCC's new chairman, gave a presentation a few years ago that said the IPCC is "eager to continue serving the UNFCCC process." What does that mean?

The United Nations Framework Convention on Climate Change (UNFCCC) is an international treaty. If you're the UN, and you're trying to persuade nearly 200 different parties to act together, you need to start from somewhere. You need an established body of facts, a master document, a bible that everyone can use as a jumping-off point. IPCC reports, it turns out, serve that purpose. In other words, the IPCC's primary function is to enable a *political* instrument—a treaty between nation states.

Now maybe you're thinking to yourself: "What's the big deal? The IPCC's connected to a treaty. So what?" Well, why isn't this fact acknowledged when the IPCC spends five paragraphs telling us about itself on its website? Why is a treaty the IPCC exists to promote not even mentioned? Instead we're told about a *scientific* view, a *scientific* body, about thousands of *scientists* and *scientific* information. What's happening here is that science and scientists are being used. As window dressing. To obscure what's actually going on.

The UN recruits scientists to write a report. The UN then spends days re-writing the summaries—the only documents that really matter. For good measure, it goes back and tampers with the original report, too. And then, abracadabra, the UN points to the report and says: *Our treaty needs to be strengthened, science itself says so.* This is politics using the good name of science to accomplish its goals. This is not honest. And it's my job, as a journalist, to alert the public.

So now we come to my final point. **The 27-year-old IPCC is, in fact, a template**. It's part of a pattern in which the UN, again and again, exploits the good name of science. Between 2003 and 2008, the UN sponsored the International Assessment of Agricultural Science and Technology (IAAST). Described as an "IPCC for agriculture," this effort was led by Robert Watson, who had just wrapped up five years as IPCC chairman. In that instance, 400 scientists wrote the UN a big report.

In 2011, *Nature* reported that the UN wanted to establish an IPCC-like body on soil degradation. The article was entirely candid about the fact that such a body would assist a treaty known as the United Nations Convention to Combat Desertification.

And then there's the Intergovernmental Platform on Biodiversity and Ecosystem Services, *aka* the IPBES. In the words of the *Guardian* newspaper, this is an "IPCC for nature." Robert Watson is involved here, too. At the moment, he's a Vice Chair. And, what do you, know? This IPCC clone is linked to yet another UN treaty called the Convention on Biological Diversity.

Ladies and gentlemen: We live in a world in which the public is rightly suspicious of politicians, yet still holds scientists in high regard. Politicians are therefore eager to borrow the prestige of science, to camouflage their own agenda with a veneer of scientific authority.

At some point in your career, the UN may try to use you to further its goals. Be careful out there.

Visit http://wp.me/pSEKJ-5tZ *for a version of this presentation that contains all 27 slides, plus hyperlinks to source material.*

WHAT IS THE NON-GOVERNMENTAL INTERNATIONAL PANEL ON CLIMATE CHANGE (NIPCC) AND WHY DOES IT MATTER?

ROBERT CARTER
James Cook University
Townsville, Australia

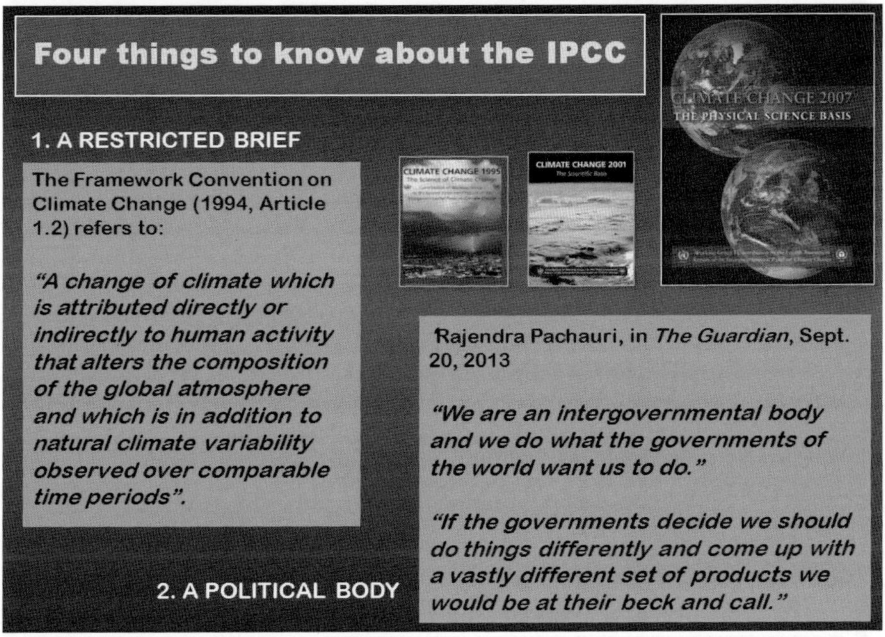

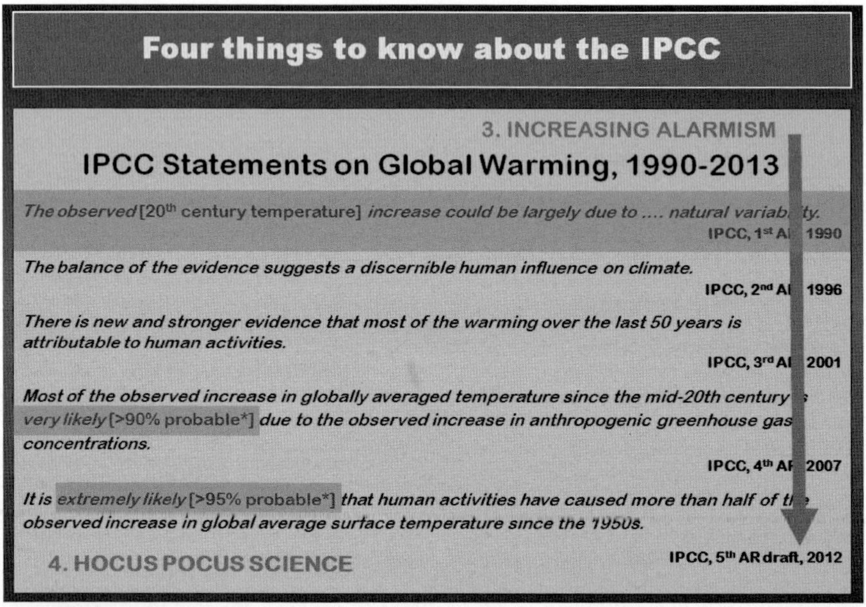

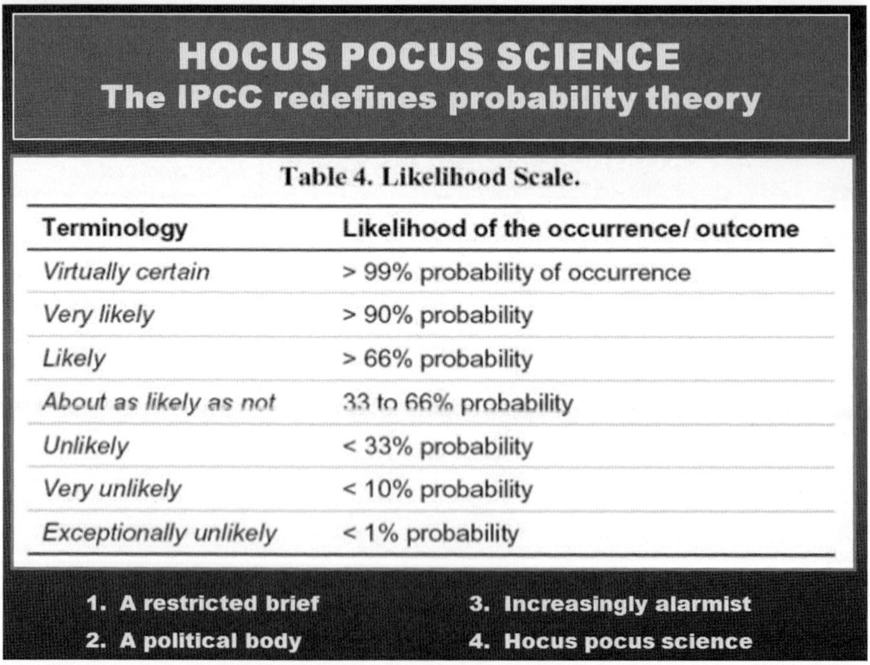

Terminology	Likelihood of the occurrence/ outcome
Virtually certain	> 99% probability of occurrence
Very likely	> 90% probability
Likely	> 66% probability
About as likely as not	33 to 66% probability
Unlikely	< 33% probability
Very unlikely	< 10% probability
Exceptionally unlikely	< 1% probability

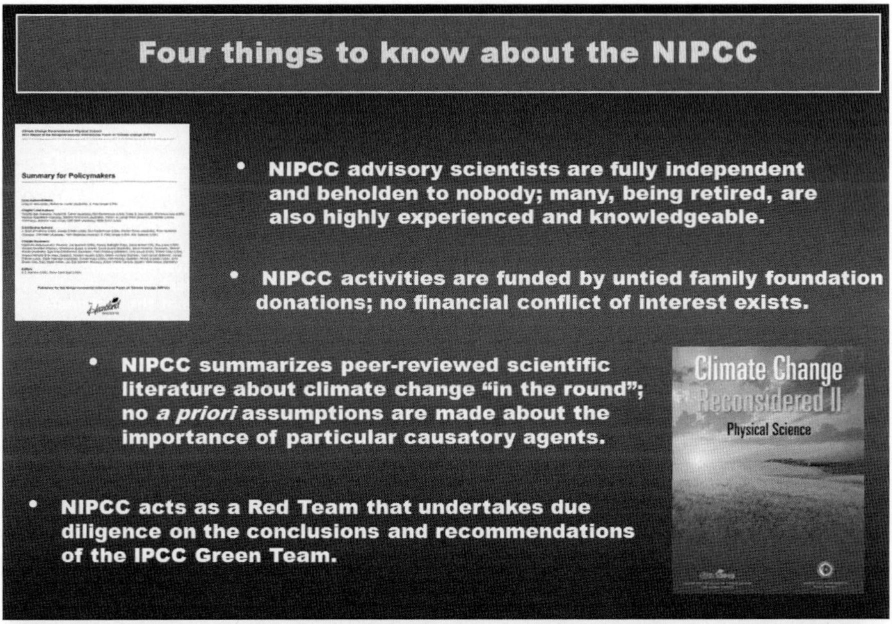
Four things to know about the NIPCC
Summary for Policymakers
• NIPCC advisory scientists are fully independent and beholden to nobody; many, being retired, are also highly experienced and knowledgeable.
• NIPCC activities are funded by untied family foundation donations; no financial conflict of interest exists.
• NIPCC summarizes peer-reviewed scientific literature about climate change "in the round"; no *a priori* assumptions are made about the importance of particular causatory agents.
Climate Change Reconsidered II
Physical Science
• NIPCC acts as a Red Team that undertakes due diligence on the conclusions and recommendations of the IPCC Green Team.

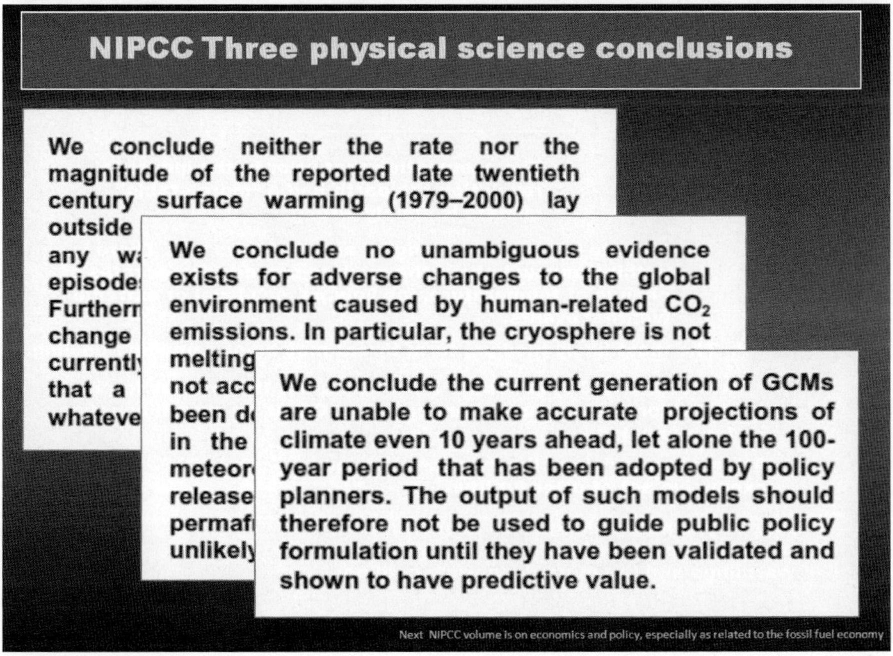
NIPCC Three physical science conclusions
We conclude neither the rate nor the magnitude of the reported late twentieth century surface warming (1979–2000) lay outside
any wa
episode
Furtherm
change
currently
that a
whateve
We conclude no unambiguous evidence exists for adverse changes to the global environment caused by human-related CO_2 emissions. In particular, the cryosphere is not melting
not acc
been d
in the
meteor
release
permaf
unlikely
We conclude the current generation of GCMs are unable to make accurate projections of climate even 10 years ahead, let alone the 100-year period that has been adopted by policy planners. The output of such models should therefore not be used to guide public policy formulation until they have been validated and shown to have predictive value.
Next NIPCC volume is on economics and policy, especially as related to the fossil fuel economy

cf. IPCC, NIPCC has 2 science volumes & 1 policy-economic volume
Climate Change Reconsidered II
Benefits and Costs of Fossil Fuels
NIPCC
Climate Change Reconsidered II
BENEFITS AND COSTS OF FOSSIL FUELS
The case for the huge benefit of fossil fuel technology as a means of improving mankind's lot continues to be led by a single small-sized American think tank.
In contrast, hydrocarbon industry leaders have signally failed to combat the dishonest demonization of their industry undertaken by eco-evangelists over the last 25 years.
COP 21, Paris, now being but a few months away, the time for continued pusillanimity is past.
NIPCC matters, then, because (i) it provides a balancing science bulwark to the IPCC's advice; and (ii) almost uniquely, it provides a defence against rampant (and some would say pathological) anti-capitalism.

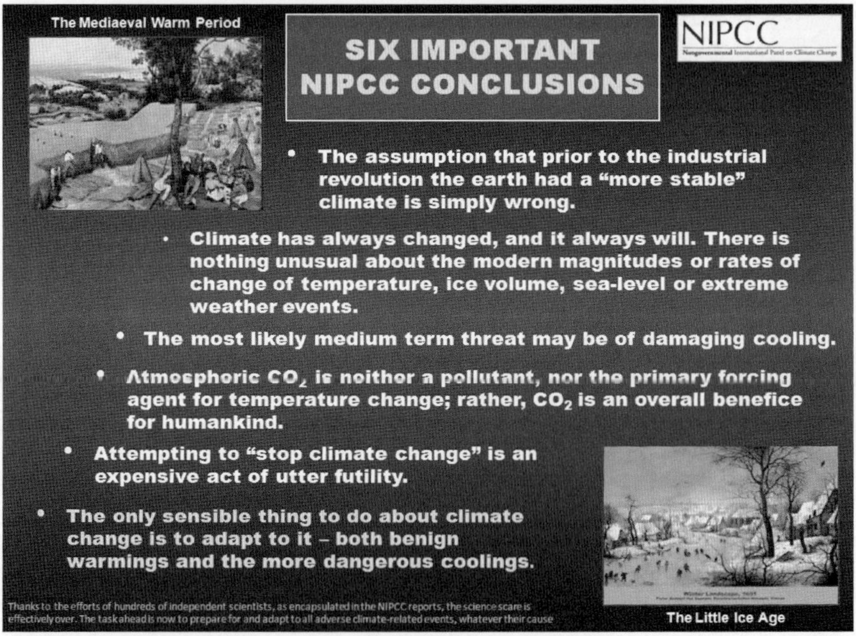
The Mediaeval Warm Period
SIX IMPORTANT NIPCC CONCLUSIONS
NIPCC
• The assumption that prior to the industrial revolution the earth had a "more stable" climate is simply wrong.
• Climate has always changed, and it always will. There is nothing unusual about the modern magnitudes or rates of change of temperature, ice volume, sea-level or extreme weather events.
• The most likely medium term threat may be of damaging cooling.
• Atmospheric CO_2 is neither a pollutant, nor the primary forcing agent for temperature change; rather, CO_2 is an overall benefice for humankind.
• Attempting to "stop climate change" is an expensive act of utter futility.
• The only sensible thing to do about climate change is to adapt to it – both benign warmings and the more dangerous coolings.
Thanks to the efforts of hundreds of independent scientists, as encapsulated in the NIPCC reports, the science scare is effectively over. The task ahead is now to prepare for and adapt to all adverse climate-related events, whatever their cause
The Little Ice Age

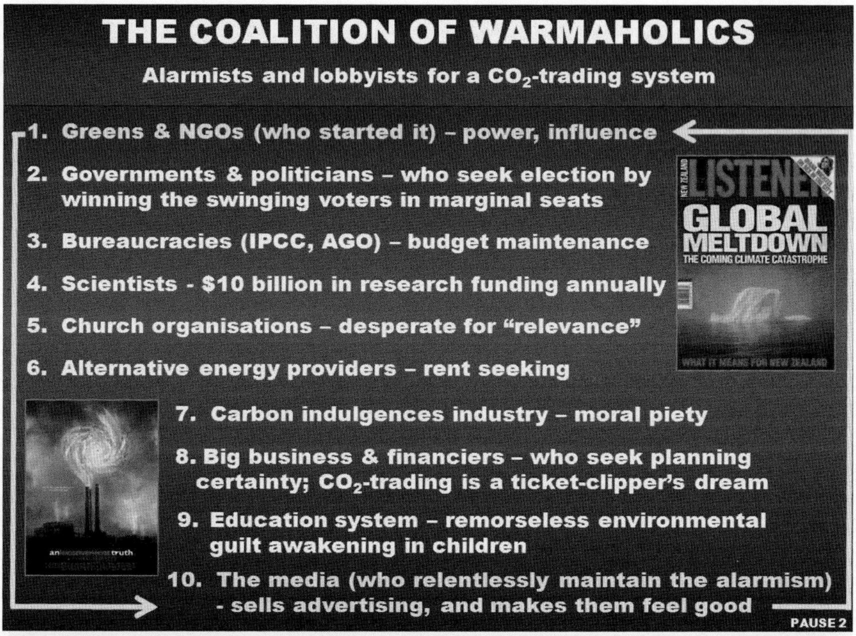
THE COALITION OF WARMAHOLICS
Alarmists and lobbyists for a CO_2-trading system
1. Greens & NGOs (who started it) – power, influence
2. Governments & politicians – who seek election by winning the swinging voters in marginal seats
3. Bureaucracies (IPCC, AGO) – budget maintenance
4. Scientists - $10 billion in research funding annually
5. Church organisations – desperate for "relevance"
6. Alternative energy providers – rent seeking
7. Carbon indulgences industry – moral piety
8. Big business & financiers – who seek planning certainty; CO_2-trading is a ticket-clipper's dream
9. Education system – remorseless environmental guilt awakening in children
10. The media (who relentlessly maintain the alarmism) - sells advertising, and makes them feel good
LISTENER
NEW ZEALAND
GLOBAL
MELTDOWN
THE COMING CLIMATE CATASTROPHE
WHAT IT MEANS FOR NEW ZEALAND
truth
PAUSE 2

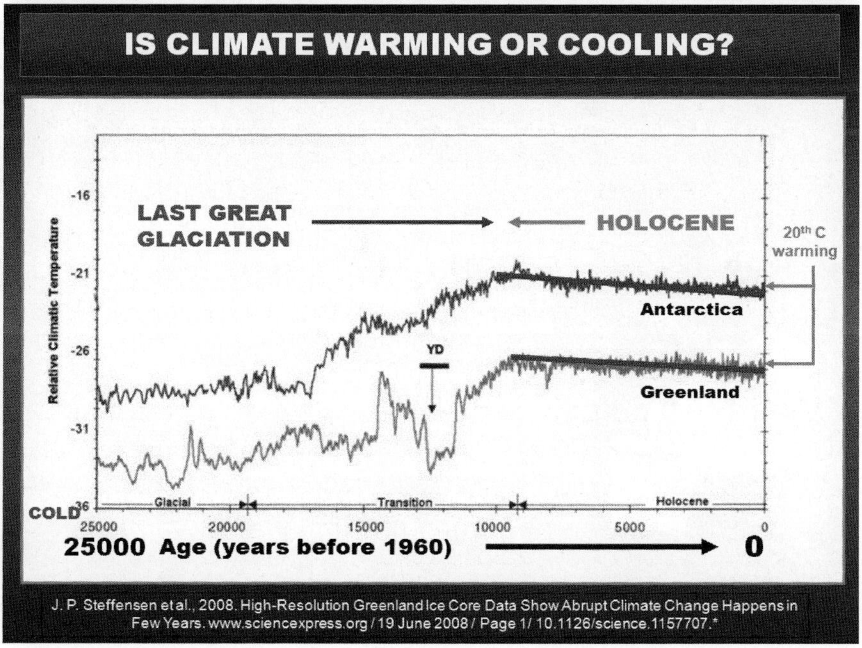
IS CLIMATE WARMING OR COOLING?
LAST GREAT GLACIATION
HOLOCENE
20th C warming
Antarctica
Greenland
YD
Relative Climatic Temperature
-16
-21
-26
-31
COLD -36
Glacial
Transition
Holocene
25000
20000
15000
10000
5000
0
25000 Age (years before 1960) 0
J. P. Steffensen et al., 2008. High-Resolution Greenland Ice Core Data Show Abrupt Climate Change Happens in Few Years. www.sciencexpress.org / 19 June 2008 / Page 1/ 10.1126/science.1157707.*

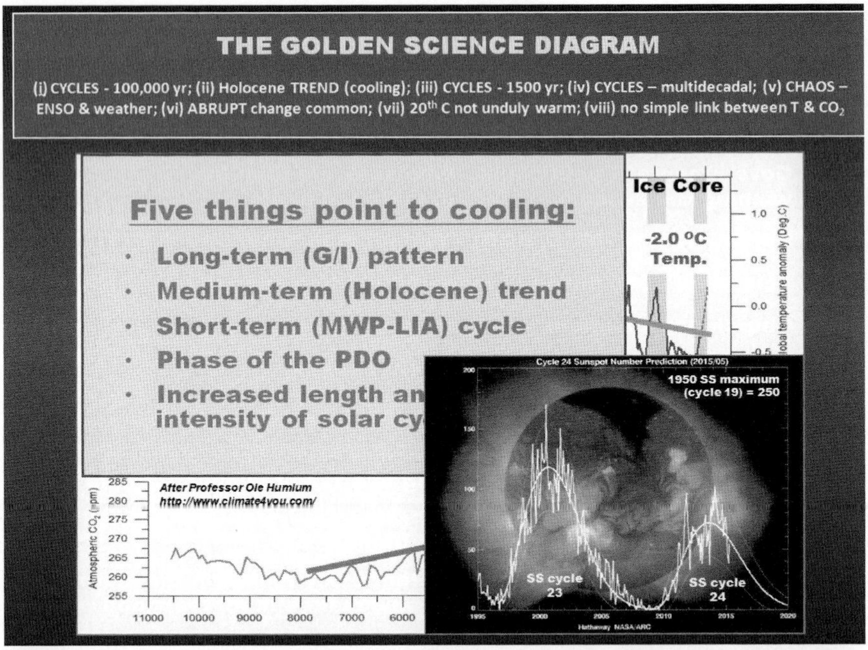
THE GOLDEN SCIENCE DIAGRAM
(i) CYCLES - 100,000 yr; (ii) Holocene TREND (cooling); (iii) CYCLES - 1500 yr; (iv) CYCLES – multidecadal; (v) CHAOS – ENSO & weather; (vi) ABRUPT change common; (vii) 20th C not unduly warm; (viii) no simple link between T & CO2
Five things point to cooling:
Long-term (G/I) pattern
Medium-term (Holocene) trend
Short-term (MWP-LIA) cycle
Phase of the PDO
Increased length an
intensity of solar cy
Ice Core
-2.0 °C
Temp.
Cycle 24 Sunspot Number Prediction (2015/05)
1950 SS maximum
(cycle 19) = 250
SS cycle
23
SS cycle
24
Hathaway NASA/ARC
After Professor Ole Humlum
http://www.climate4you.com/
Atmospheric CO2 (ppm)

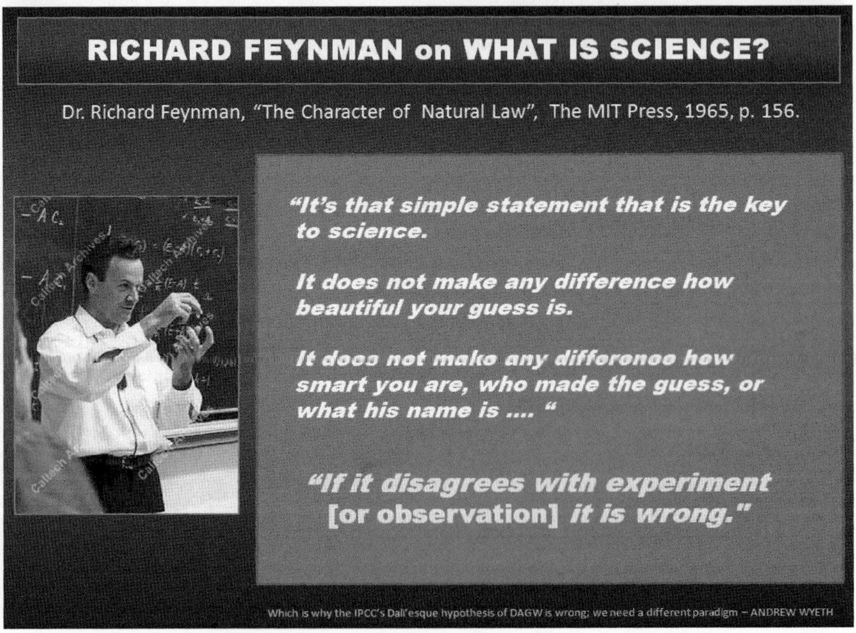
RICHARD FEYNMAN on WHAT IS SCIENCE?
Dr. Richard Feynman, "The Character of Natural Law", The MIT Press, 1965, p. 156.
"It's that simple statement that is the key to science.
It does not make any difference how beautiful your guess is.
It does not make any difference how smart you are, who made the guess, or what his name is .... "
"If it disagrees with experiment [or observation] it is wrong."
Which is why the IPCC's Dali'esque hypothesis of DAGW is wrong; we need a different paradigm – ANDREW WYETH

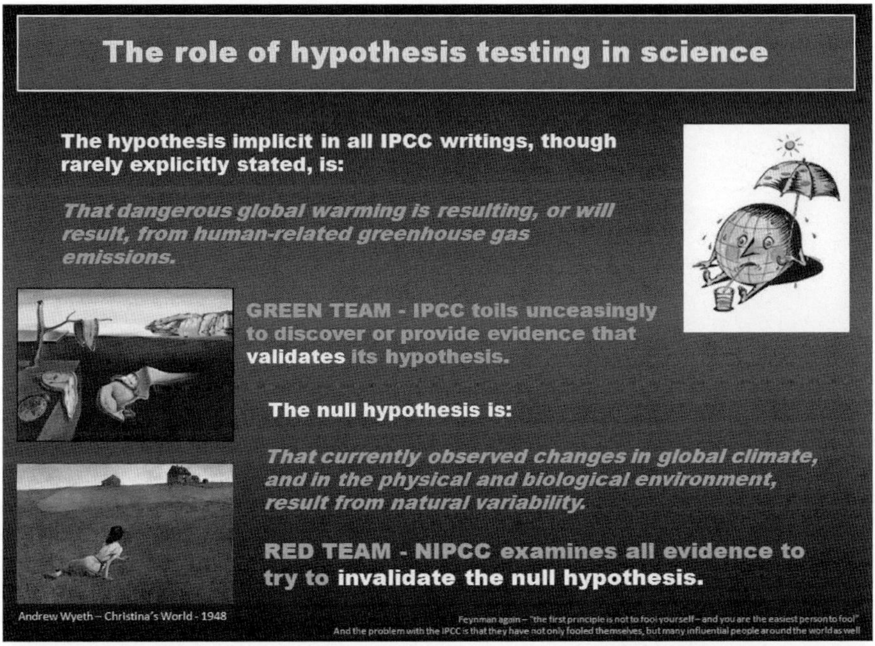
The role of hypothesis testing in science
The hypothesis implicit in all IPCC writings, though rarely explicitly stated, is:
That dangerous global warming is resulting, or will result, from human-related greenhouse gas emissions.
GREEN TEAM - IPCC toils unceasingly to discover or provide evidence that validates its hypothesis.
The null hypothesis is:
That currently observed changes in global climate, and in the physical and biological environment, result from natural variability.
RED TEAM - NIPCC examines all evidence to try to invalidate the null hypothesis.
Andrew Wyeth – Christina's World - 1948
Feynman again – "the first principle is not to fool yourself – and you are the easiest person to fool"
And the problem with the IPCC is that they have not only fooled themselves, but many influential people around the world as well

Christina is most surely a lead member of the climate RED TEAM
Andrew Wyeth – Christina's World – 1948 - MoMA

GUARDIAN ANGELS
Cristo Redentor
Paul Landowski & Heitor da Silva Costa - 1931
Rio de Janiero
40 m high
Angel of the North
Antony Gormley - 1998
Gateshead, England
There is another guardian angel that few have observed,
Though many have seen the site at which it is erected
54 x 20 m
We humans are a strange species – strive for impartiality, but prone to emotionalism – don't have to be religious to feel the power of Brazil's famous statue...

The 2000 Millennium Sculptures – Clemens Jöckle Woytek
Mt. Hakepa, Pitt Island, Chatham Islands, NZ
Climate Change Reconsidered II
NIPCC – The guardian of integrity in climate science
Team Red
ASTRONOMER
Questing
"They give thought to the creation of heaven and earth"
Ibn Rushd Averroes, 12th C
PHILOSOPHER
Wisdom
Integrity
NEWBORN
New ideas
Future hope
MASAI WARRIOR
Dignity
Strength
Millennium TV show in 2000 (ii) 10,000 km to S. America – SIZE ISN'T EVERYTHING

THE EU'S UNILATERAL CLIMATE POLICY AND FLAWED SCIENCE ADVICE

BENNY PEISER
The Global Warming Policy Foundation
London, UK

Since the fall of the Berlin Wall, the European Union has committed to unilateral efforts to tackle climate change. It has felt it a *duty* to set an example to the rest of the world through radical climate policy-making at home.

The EU's climate agenda was founded on three key assumptions that were widely embraced and promulgated by European leaders in the late 20th century:

> First, the belief that global warming was an urgent threat that needed to be prevented without delay and at all costs;
>
> Second, that the world was running out of fossil fuels, so that oil and gas prices would rise inexorably and renewable energy would become ever-more cost competitive;
>
> And third, that the world would agree on a legally binding carbon reduction target creating a ready-made market into which EU firms could sell their low-carbon knowhow.

All of these core beliefs turned out to be mistaken.

We first have to look back at the early 1990s and remember what the Intergovernmental Panel on Climate Change predicted at that time.

In its first assessment report published in 1990, the IPCC predicted that continuing carbon dioxide emissions would lead to a rise of the average global surface temperature, initially by 0.3°C per decade and accelerating thereafter.

> *"Based on current model results, we predict under the Business-as-Usual (Scenario A) emissions of greenhouse gases, a rate of increase of global mean temperature during the next century of about 0.3°C per decade…This will result in a likely increase in global mean temperature of about 1°C above the present value by 2025."*

It was this expectation of accelerating global warming that drove the EU, back in 1996, to adopt a global temperature target, saying that it should not exceed 2 degrees above the pre-industrial level.

They can hardly have guessed that in the decades that followed, instead of the predicted rapid warming, the global surface temperature would essentially come to a standstill. Instead of accelerated global warming there has been no rise in average global surface temperature for nearly 20 years.

The EU's assumptions about energy prices were equally mistaken.

With the onset of the shale revolution in the U.S. and the prospect of a global expansion of unconventional oil and gas exploration there now seems to be no lack of fossil fuels. In fact, we have sufficient conventional and unconventional energy to meet the world's needs for most of this century. Fossil fuel prices have fallen and look set to remain low for the foreseeable future.

Meanwhile, instead of making Europe more competitive, astronomical subsidies for renewables—nearly 1 *trillion* euros to date—have given the EU the world's highest energy prices and have helped turn it into an economic basket-case: the only region of the world where economic growth has been absent for years.

EU policymakers also naively assumed that Europe's main competitors would follow its shift from cheap fossil fuels to expensive green energy. This has not happened and, in reality, was never a realistic expectation in the first place.

Lastly, legally binding CO_2 cuts at the global level have proven to be impossible to agree on, and are highly unlikely to be forthcoming anytime soon. The benefits that the proposed global deal would bring to the EU remain a pipedream, and an increasingly desperate one at that.

THE COSTS

No other major economy has followed the EU's policy direction. And none will be foolish enough to do so in the future, now that the costs for the economy and the pain for consumers are obvious for all to see.

With electricity prices in Europe now more than double those in North America, Europe's remaining and struggling manufacturers are unable to compete and are increasingly relying on government subsidies. They are rapidly losing both sales and market share.

And the pain has only just begun. The International Energy Agency estimates that in 2035, EU electricity prices will still be around twice those in the USA. According to the International Monetary Fund, these inflated energy prices are going to cut further into the EU's share of global manufacturing.

The EU's unilateral climate policies have become an existential threat to its energy-intensive industries.

It is no surprise, then, that European politicians are increasingly concerned.

What is new, however, is that this concern and open criticism is no longer limited to climate skeptical MEPs or policymakers from central and eastern European member states. In recent months, it is increasingly coming from the very centre of old Europe.

The threat of rising energy prices and the loss in competitiveness was the central theme of the energy summit of EU heads of government in Brussels last May.

Antonio Tajani, the EU's former Industry Commissioner, warned last year that Europe's climate policies were pushing electricity costs to destructive levels:

> *"We face a systemic industrial massacre. We need a new energy policy. We have to stop pretending, because we can't sacrifice Europe's industry for climate goals that are not realistic, and are not being enforced worldwide."*

If someone like Mr. Tajani can say that Europe's industry is being sacrificed on the altar of unrealistic climate goals, other leaders should listen.

Tajani has not been alone in sounding the alarm over the costs of unilateral climate policy.

Gunther Oettinger, the EU's then energy commissioner, declared last September that the EU should not adopt new binding CO_2 targets unless all major emitters do likewise. He pointed out that if they were imposed unilaterally, the result of such targets would be the export of production and emissions outside the EU.

And since it is the developing nations whose CO_2 emissions are increasing rapidly, Oettinger rightly stated the obvious, namely that anyone who thinks that cutting Europe's CO_2 emissions would make any difference to global warming is "arrogant or stupid".

Oettinger pointed out that the EU was responsible for just over 10% of global CO_2 emissions today, a sum that would fall to 4.5% in 2030.

THE EU'S NEW CLIMATE POLICY

When confronted with these concerns and faced with the threat of a Polish veto, EU leaders agreed a new approach to climate policy.

Gone is the focus on unilateral target-setting and green idealism. Instead the EU is now offering a conditional pledge linked to an unconditional demand.

The pledge is a new target for CO_2 emissions, cutting them by 40% below the 1990 level by 2030. But critically, this pledge is conditional on the Paris agreement being legally binding for all countries.

The conditional nature of the EU's CO_2 pledge was clearly spelled out by the European Council in its *2030 Climate and Energy Policy Framework* agreement adopted last October.

It states that with regards to its 2030 pledge:

> *"The European Council will revert to this issue after the Paris Conference [and] will keep all the elements of the framework under review."*

In February, the European Commission followed up with a document that communicated the EU's red line for the Paris climate summit.

In this document called, *The Paris Protocol,* the Commission demands that all nations represented at the Paris summit should sign what it calls the "Paris Protocol" and adopt legally binding CO_2 emissions targets similar to and as a carry-over of the Kyoto Protocol.

According to EU's red line, the Paris Protocol must deliver:

> *"Legally binding mitigation commitments that put the world on track towards achieving the below 2°C objective... Mitigation commitments under the Protocol should be equally legally binding on all Parties."*

Yet the chances of such an agreement remain slim.

For a start, an international protocol or treaty with legally binding CO_2 emissions targets would have to be ratified by national parliaments. This is unlikely to happen in the U.S., given the Republican majority in both houses.

Republican Senators have already warned governments that Obama's climate pledge is non-binding and will be contested by the U.S. Senate.

Republican senators are not alone either. It was reported last year that many EU member states and at least five EU commissioners had been pushing for a reduced CO_2 target of 35%.

Moreover, Poland and other member states remain opposed to any new binding targets in the absence of a legally-binding UN agreement.

There are strong indications from Poland and other Visegrad nations that the 40% emissions pledge would be an unacceptable burden if other major economies did not follow suit.

Many former Warsaw Pact states continue to rely on Russian gas imports for their energy needs. Many fear that new unilateral CO_2 limits would make coal power generation unviable and consequently lead to an even greater dependence on Russia.

The prospect of increasing renewable energy sources is also an unappealing prospect. Governments in Eastern Europe look at Germany where, as a result of astronomical subsidies, domestic consumers are paying the second highest amount for electricity in the EU.

But it is the position of the developing nations that makes a binding treaty an unlikely outcome for Paris.

China and India have consistently made clear their view that CO_2 targets should only be binding for developed nations. In their view, and that of the other 24 members of the so-called Like-Minded Developing Countries group, developing nations should be exempt.

China and India are countering Western pressure to fall into line by demanding a legally binding compensation package of $100 billion per year, as promised by President Obama at the UN climate conference in Copenhagen in 2009.

They are resisting attempts by the U.S. and the EU to end the legal distinction between developing and developed nations, and it is likely that they will still be holding out in Paris at the end of the year.

They have also stated repeatedly that they will not allow any outside body to review their Intended Nationally Determined Contributions and reject any form of international review as a violation of their sovereignty.

In the face of this hard line from the developing world, it is hard to see any kind of legally binding targets being agreed in Paris.

WHAT HAPPENS IF THE EU'S CLIMATE STRATEGY FAILS IN PARIS?

Given that UN climate deals have to be agreed by consensus, it is almost certain that the EU's key demand will be rejected by most of the developing nations.

It is equally likely that all parties involved—the EU and the U.S. on the one hand and China and India on the other—will proclaim victory after the final Paris agreement is signed.

Nevertheless, there are considerable uncertainties about what will happen to the EU's conditional 2030 pledge in such circumstances.

The European Council would only be able to make its 40% emissions reduction pledge legally binding if all Member States accepted new, nationally binding CO_2 targets under a *Burden Sharing Mechanism*.

However, in the event that the Paris conference fails to make its nationally determined mitigation pledges legally binding, the EU should abandon or at least delay making its own 40% pledge legally binding.

Alternatively, EU leaders could simply agree to make the 40% pledge binding at EU, but not at national levels.

Such a soft exit strategy would emulate the EU's dodgy renewable energy target for 2030 which, while being called an EU-wide and binding target, does not oblige members states to adopt any legally binding renewables targets domestically.

And since the 27% EU-wide green energy target is unenforceable, no country will be punished for failing to meet any target.

This option would provide a logical and justifiable exit from unilateral self-sacrifice and would allow member states to decide their climate and energy policies domestically and back in line with their national interest.

SESSION 5

DYING OCEANS

THREATS TO COASTAL OCEAN SUSTAINABILITY: CURRENT STATUS AND FUTURE TRENDS IN DEAD ZONES

ROBERT DIAZ
Virginia Institute of Marine Science, College of Williams and Mary
Gloucester Point, VA, USA

RUTGER ROSENBERG
Department of Biological and Environmental Sciences, University of Gothenburg, Fiskebäckskil, Sweden

ABSTRACT

Development associated with human populations has led to the globalization of many environmental problems. The most serious of these problems in coastal seas are related to eutrophication and its impact on species abundance/composition and dissolved oxygen budgets. In open oceans the most serious problems are related to global climate change and its direct impact on temperature and pH, and indirect impacts on oxygen budgets. Development of low dissolved oxygen (hypoxic) areas in coastal areas is the most widespread deleterious consequence of eutrophication. The occurrence of dead zones is increasing, a trend that is consistent with the increase in human population and activities that result in increased fluxes of nutrients to coastal waters. Currently, most oxygen depletion events are seasonal, but trends toward longer periods and lower concentrations of oxygen are evident. Similarly, naturally occurring open ocean oxygen minimum zones are expanding into shallower water from the effects of global climate change. As a result, the future status of hypoxia and its consequences for the environment, society and economies will depend on a combination of climate change (primarily from warming, and altered patterns for wind, currents and precipitation) and land-use change (primarily from expanded human population, agriculture and nutrient loadings). The overall forecast is for all forms of hypoxia to worsen, with increased occurrence, frequency, intensity and duration. It is not all bad news, hypoxia in areas predicted to receive less rainfall may improve, and the consequences of regional eutrophication-induced hypoxia can and have been reversed with long-term and persistent efforts to manage and reduce nutrient and or organic matter loads, which has lead to the restoration of ecosystem services. Reversing the expansion of oxygen minimum zones will more difficult, requiring a global management approach to mitigate global warming trends. Hypoxia and dead zones represent a potentially enormous ecological and economic threat to global ecosystem services, which is in the billions of U.S. dollars annually.

KEY MESSAGES

- Between the 1950s and 1960s, eutrophication has emerged as a problem that threatens and degrades coastal ecosystems, alters fisheries, and impacts human health in many areas of the world. Hypoxia is one of the most acute symptoms of eutrophication.

- The global extent of eutrophication and its threats to human health and ecosystem services are just beginning appreciated, but much remains unknown relative to social and economic consequences. Virtually all of our information on hypoxia and its effects is from North America and Europe. We know very little about conditions in the most populated parts of the planet (India, Asia, and Indonesia) or oceanic islands (mostly in the Pacific).
- Close to 1000 areas around the world have been identified as experiencing the effects of eutrophication. Of these, over 600 have problems with hypoxia, but through nutrient and organic loading management about 75 systems can be classified as recovering.
- Threat of declining oxygen is more than of academic interest. It is one of many stressors that are a looming 'perfect storm' that threatens to seriously degrade ecosystem services we have come to depend upon. Annual losses in water quality and nutrient cycling services from hypoxia are likely in the billions of U.S. dollars.
- The importance of maintaining adequate levels of oxygen in our coastal and ocean systems is summarized by the motto of the American Lung Association: "when you can't breathe, nothing else matters".

Introduction

Virtually all ecosystem services providing food to humans relay on oxygen to support organism growth and production. In addition, important processes regulating water quality and nutrient cycling are controlled by oxygen. In the mid 20th century alarming trends of declining oxygen concentrations started to appear in both coastal and open ocean systems related to a combination of anthropogenic and natural factors. The majority of these trends have been linked to anthropogenic activities associated with an ever-expanding population. This expansion has led to extensive modification of landscapes at the expense of ecosystem function and services that we have come to rely upon, including pervasive effects on coastal primary production from excess nutrients to overfishing (Millennium Ecosystem Assessment, 2005). Long-term records of nutrient discharges provide compelling evidence of a rapid increase in the fertility of many coastal ecosystems starting in the 1960s. On a global basis, by 2050, coastal marine systems are expected to experience, from today's levels, at least a doubling in both nitrogen and phosphorus loading, with serious consequences to ecosystem structure and function (Foley et al. 2005; Gruber and Galloway, 2008).

The question asked by Foley et al. is: are land use activities degrading the global environment in ways that undermine ecosystem services, which in turn undermine human welfare? When it comes to dissolved oxygen the answer is yes. In marine ecosystems, oxygen depletion has become a major structuring force for communities and energy flows at global scales.

Eutrophication can be defined as an increasing rate of primary production and organic carbon accumulation in excess of what an ecosystem is normally adapted to processing. It is one part of a complex of multiple stressors that interact to shape and direct ecosystem level processes (Cloern, 2001). The visible ecosystem response to eutrophication is the excessive growth of algae and vegetation in coastal areas as primary production increases in direct response to nutrient enrichment. One of the most serious threat from this eutrophication is the reported decrease in oxygen in bottom waters created by the increased flux of organic matter to the sea bed, which can lead to "dead zones" (Diaz and Rosenberg, 2008). In the past this eutrophication-induced low oxygen or hypoxia was mostly associated with rivers, estuaries, and bays. But dead zones have now developed in continental sea, such as the Baltic Sea, Kattegat, Black Sea, Gulf of Mexico, and East China Sea.

What appear to be small changes in oxygen can have major consequences to animals living in an oxygen limited milieu. Physiologically, higher temperatures also increase metabolic requirements for oxygen and increase rates of microbial respirations and, therefore, oxygen consumption. For salmonid fishes, oxygen can become limiting at relatively high values and even air-saturation can be limiting at higher temperatures (Fry, 1971). Concentrations of dissolved oxygen below 2 to 3 mg O_2/L are a general threshold value for hypoxia for marine and estuarine organisms, and 5 to 6 mg O_2/l in freshwater. However, species and life stages differ greatly in their basic oxygen requirements and tolerances (Vaquer-Sonyer and Duarte, 2008).

The relatively low solubility of oxygen in water combined with two principal factors lead to the development of hypoxia and at times anoxia. These factors are water column stratification that isolates the bottom water from exchange with oxygen rich surface water and decomposition of organic matter in the isolated bottom water that reduces oxygen levels. Both factors must be at work for hypoxia to develop and persist in deeper waters.

While unintended fertilization of marine systems, mainly from excess nitrogen, has been linked to many ecosystem level changes associated with eutrophication, there are also natural processes that can lead to eutrophic-like conditions along continental margins and produce similar ecosystem responses. Coastal upwelling zones that occur along the western side of continents are highly productive but also can produce severe hypoxia from respiration of fluxed organic matter (<0.7 mg O_2/l). Oxygen minimum zones (OMZs) are another natural low oxygen phenomenon. They are persistent oxygen-depleted areas occurring primarily in the eastern Pacific Ocean, south Atlantic west of Africa, Arabian Sea and Bay of Bengal at intermediate depths about 200 to 1000 m (Helly and Levin, 2004). Where OMZs contact the bottom, the benthic fauna are adapted to oxygen concentrations as low as 0.1 mg O_2/l. This is in stark contrast to the faunal response to eutrophication-induced hypoxia in coastal and estuarine areas where oxygen concentrations of <1 mg O_2/l lead to mass mortality and major change in community structure (Levin et al. 2009).

THE HEART OF THE PROBLEM

Over-enrichment of our water primarily with nutrients, such as nitrogen and phosphorus from fertilizers, has emerged as one of the leading causes of water quality impairment. These nutrients have lead to excess production of organic matter that is called eutrophication. It is an indirect result of activities that support a rapidly expanding human population and directly related to agricultural practices, increased industrial activities, combustion of fossil fuels, and municipal sewage discharges, all of which contribute to the increased flow of nitrogen and phosphorus to terrestrial and aquatic environments. On a global basis, humans add more nitrogen to the land or ocean than is supplied by natural biological nitrogen fixation (Figure 1. Gruber and Galloway, 2008). Most of these nutrients end up in the sea and fuel eutrophication, which increasingly threatens the health of coastal ecosystems and fisheries, as well as human health.

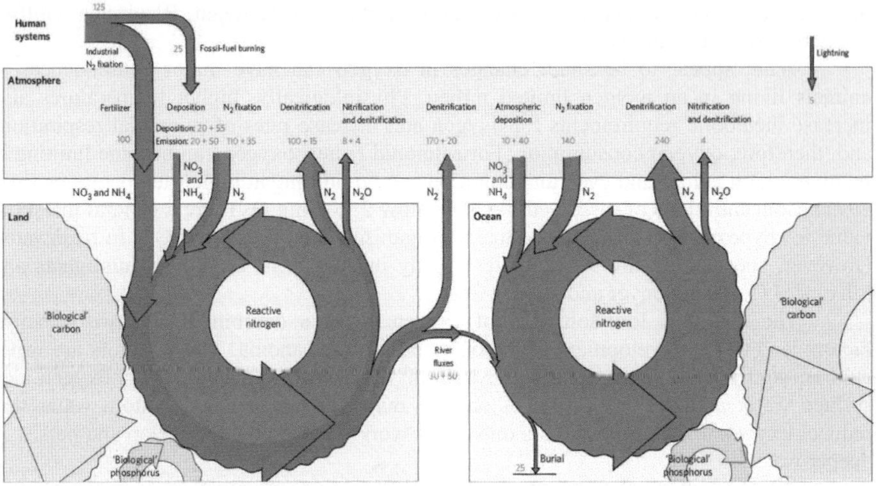

Fig. 1: Depiction of the global nitrogen cycle on land and in the ocean from Gruber and Galloway, (2008). Major processes that transform molecular nitrogen into reactive nitrogen are shown. There is also a tight coupling between the nitrogen cycles on land and in the ocean with those of carbon and phosphorus. Blue fluxes denote 'natural' (unperturbed) fluxes; orange fluxes denote anthropogenic perturbation. Numbers are in Tg N per year and are for the 1990s.

Virtually all the ocean's food provisioning ecosystem services for humans require oxygen to support organism growth and production. Oxygen is just absolutely necessary to sustain the life of all the fishes and invertebrates we have come to depend upon. By the early 1900s dissolved oxygen was a topic of interest in research and management, and by 1920s it was recognized that a lack of oxygen was a major hazard to fishes. But it was not obvious that dissolved oxygen would become critical in shallow coastal systems until the 1970s and 1980s when large areas of low dissolved oxygen started to appear with associated mass mortalities of invertebrate and fishes. From the middle of the 20th century to today, there have been drastic changes in dissolved oxygen concentrations and dynamics in marine coastal waters. Diaz and Rosenberg (1995) noted that no other environmental variable of such ecological importance to estuarine and coastal marine ecosystems as dissolved oxygen has changed so drastically, in such a short period of time.

GLOBAL PATTERNS IN HYPOXIA

Since the 1960s alarming trends of declining oxygen concentrations have immerged in both coastal areas and in the open oceans (see Diaz and Rosenberg, 2008; Keeling et al. 2010; Conley et al. 2011). Many of these trends have been linked to human activities. So, how did we get to where we are? In 2009 an interesting article proposed that there are planetary boundaries for a safe operating space for humanity (Rockström et al. 2009). While oxygen was not one of the nine boundaries discussed, it is influenced by many of the processes discussed (Table 1). The expanding size of our human population has led to three of the boundaries being crossed, which are climate change, rate of biodiversity loss, and the nitrogen cycle. Of these, alterations to the nitrogen cycle have the most direct consequences for dissolved oxygen followed by climate change. The more nutrients added to the sea the more organic matter will be produced, which will create a greater oxygen demand when it is decomposed potentially leading to more hypoxia.

Table 1. Planetary boundary processes (Rockström et al. 2009) and how exceeding them will affect oxygen.

Earth-system process	Parameters	Proposed boundary	Current status	Pre-industrial value	Consequences for oxygen
Climate change	(i) Atmospheric carbon dioxide concentration (parts per million by volume)	350	387	280	More carbon dioxide in water reduces oxygen concentration
	(ii) Change in radiative forcing (watts per metre squared)	1	1.5	0	Warmer water holds less oxygen
Rate of biodiversity loss	Extinction rate (number of species per million species per year)	10	>100	0.1-1	Lower oxygen will stress more species
Nitrogen and Phosphorus Cycle	(i) Amount of N2 removed from the atmosphere for human use (millions of tonnes per year)	35	121	0	More N and P entering coastal systems will increase primary production, which will in turn decompose and lower oxygen increasing hypoxia
	(ii) Quantity of P flowing into the oceans (millions of tonnes per year)	11	8.5-9.5	~1	
Stratospheric ozone depletion	Concentration of ozone (Dobson unit)	276	283	290	Unknown

Earth-system process	Parameters	Proposed boundary	Current status	Pre-industrial value	Consequences for oxygen
Ocean acidification	Global mean saturation state of aragonite in surface sea water	2.75	2.9	3.44	More acidic waters contain less oxygen
Global freshwater use	Consumption of freshwater by humans (km3 per year)	4,000	2,600	415	Reduced river flow would improve low oxygen conditions
Change in land use	Percentage of global land converted to cropland	15	11.7	Low	More cropland leads to more nutrient runoff, see Nitrogen and Phosphorus Cycle
Atmospheric aerosol loading	Overall particulate concentration in atmosphere on a regional basis		Not Set		Unknown
Chemical pollution	Amount of persistent organic pollutants, plastics, endocrine disrupters, heavy metals and nuclear waste in the global environment		Not set		Unknown

To bring this problem of low oxygen into focus, it has taken this unintended consequence of the green revolution less than 40 years to significantly alter the global nitrogen cycle on land and in our coastal systems. It was in the mid-1940s that the global use of industrial fertilizer surpassed all natural forms of fertilizer for growing corps. This laid the groundwork for the rapid rise in the areas affected by hypoxia in the 1970s and 1980s. This lag is the time it took for excess organic matter from primary production to buildup and overwhelm an ecosystem's assimilative capacity, to the detriment of higher trophic levels but to the benefit of microbes. By comparison it has taken over 100 years for the industrial revolution to significantly alter the global carbon cycle.

Prior to 1960, there were about 45 systems with reports of eutrophication-related hypoxia (Figure 2). During the 1960s, another 60 systems were added. The 1970s saw estuarine and coastal ecosystems around the world becoming over enriched with organic matter from expanding eutrophication and the number of oxygen-depleted ecosystems jumped to 180. In the 1980s many more systems reported hypoxia for the first time bringing the total to 330. An additional 170 hypoxic areas were reported in the 1990s. By the end of the 20th century, hypoxia had become a major, worldwide environmental problem with only a small fraction of systems (75) showing signs of improvement. At the end of the first decade of the 21st century another 190 sites reported, bringing the total to about 640. An additional over 400 coastal sites globally were identified as areas of concern that currently exhibit signs of eutrophication and are at risk of developing hypoxia (Diaz et al. 2010; Conley et al. 2011).

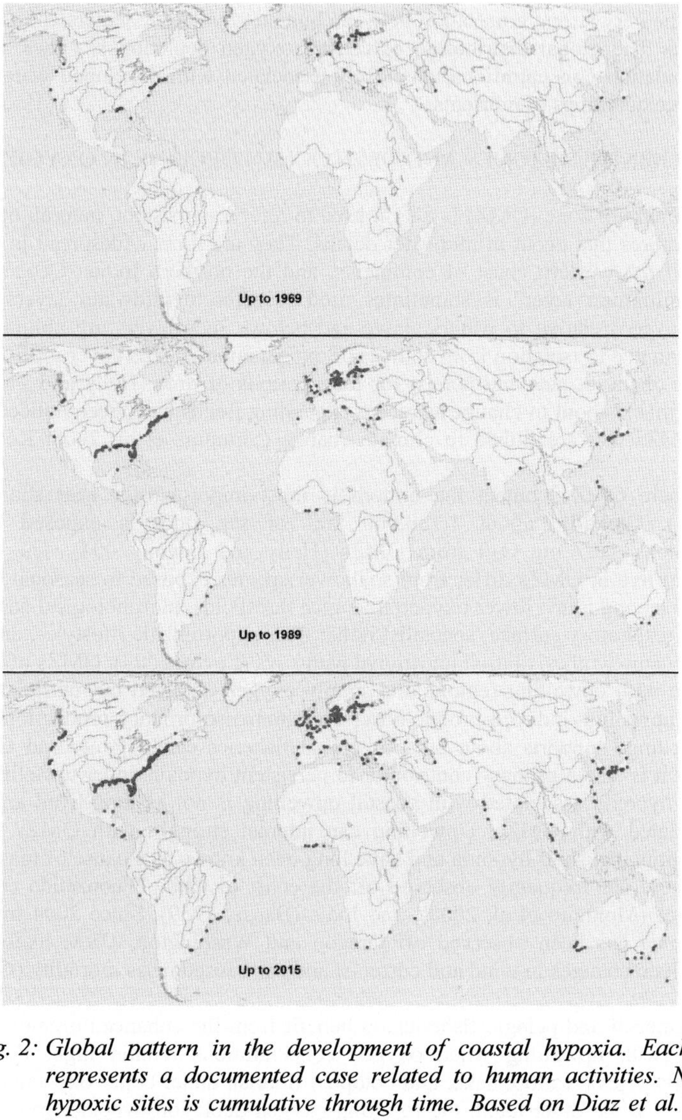

Fig. 2: Global pattern in the development of coastal hypoxia. Each red dot represents a documented case related to human activities. Number of hypoxic sites is cumulative through time. Based on Diaz et al. (2010) at http://www.wri.org/project/eutrophication.

There are signs of a slowing growth in hypoxic systems, mostly because North America and Europe are well studied and about completely reported. We predict that as more data become available on oxygen conditions in Asia, the Indopacific, and Pacific Islands, the number of dead zone will more than double to over 1,000. Our prediction is based on the very strong correlation between human population centers and the presence of hypoxia. The global distribution of coastal oxygen depletion is either centered on major population concentrations, or closely associated with developed watersheds that deliver large quantities of nutrients.

OXYGEN MINIMUM ZONES AND OPEN OCEAN DECLINE IN OXYGEN

Oxygen minimum zone (OMZ) is a term used to refer to persistent, natural, oceanic low-oxygen features that occur at midwater depths. They are most widespread in the eastern Pacific, off the western coast of continents, and the northern Indian Ocean. The term "oxygen minimum layer" is sometimes used to refer to midwater layers exhibiting reduced oxygen relative to waters above and below. Such layers are ubiquitous in the global ocean due to isolation from sources of oxygenation. OMZs cover million of square kilometers of the oceans and persist for long periods of time (Helly and Levin, 2004). However, this is about to change as recent expanding trends have been noticed in several Pacific OMZs that are linked to climate change (Stramma et al. 2008; Keeling et al. 2010).

Where OMZs contact the bottom is most important for near shore fisheries (Figure 3). Globally, about 1,150,000 km^2 of sea floor is exposed to oxygen concentrations <0.7 mg O_2/l from OMZs (Helly and Levin, 2004). The specialized species that live in OMZs differ from shallower species exposed to seasonal or episodic hypoxia in having much lower oxygen tolerance thresholds, morphological adaptations to maximize respiratory surface, specialist rather than opportunistic lifestyles, and potential to utilize chemosynthesis-based nutritional pathways (Levin, 2003). OMZs often occur in association with upwelling regions and supply oxygen depleted but nutrient rich water that fuels upwelling-induced productivity. This immense primary production associated with upwellings supports about 80% of global fisheries catch (Pauly and Christensen, 1995), but it is also a primary contributor to persistent hypoxia under upwelling regions. However, hypoxia associated with coastal upwelling is not as long-lived and stable as that associated with OMZs. Upwelling can interact strongly with OMZs to produce intense continental shelf hypoxia and have long been known as a cause of mass mortality events. These are frequently observed off the coast of Chile (Fuenzalida et al. 2009), West Africa (Monteiro et al. 2008), and India (Banse, 1959). Since 2004 intense inner shelf hypoxia has been observed off Oregon and Washington, USA, likely linked to climate driven changes in wind and currents, and has caused mass mortality (Grantham et al. 2004).

Demersal and pelagic fisheries do benefit from the enhanced organic production associated with upwelling when oxygen concentrations remain high. For example, demersal fisheries (scallop, hake, and octopus) flourish under better oxygenated El Niño conditions. On the other hand, the resulting oxygen depletion, whether from mid-ocean OMZs or coastal upwelling or their interaction, affects midwater plankton and pelagic organisms creating areas of low biodiversity, and hostile environmental conditions for many commercially valued fisheries resources (Ekau et al. 2009). Monsoon-driven upwelling along the western coast of India migrates onshore and offshore depending on winds and currents. When pushed onshore, there are declining catches of fishes and prawns and diminished demersal fisheries. Recent intensification in the severity of low oxygen associated with this upwelling zone is related to a combination of climate change and land derived nutrients, and will most likely further impact inshore fisheries (Naqvi et al. 2006).

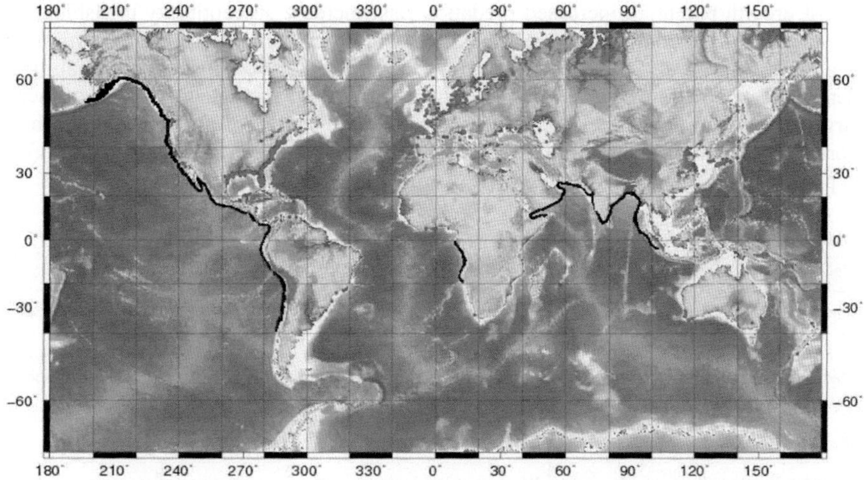

Fig. 3: World distribution of outer shelf areas threatened by hypoxia from upwelling or OMZs (black lines) and eutrophication-induced coastal hypoxia (Red dots). Based on Levin et al. (2009).

The upper depth limit of OMZs has major implications for fisheries. Expansion of OMZs toward the surface in the eastern tropical Pacific has limited the depth distribution of tropical pelagic marlins, sailfish, and tunas into a narrower surface layer of oxic water about 50 to 100 m thick (Prince and Goodyear, 2006). The high-performance physiology of these fishes leads to a relatively high hypoxic threshold making any reduction in oxygen problematic (Brill, 1996). Declining oxygen and expansion of the OMZ in the tropical northeast Atlantic Ocean toward the surface is also restricting usable habitat of billfishes and tunas. From 1960 to 2010 Stramma et al. (2012) found hypoxia-based habitat compression to decreasing their suitable habitat by 15%. The combination of shallowing of OMZs encroaching onto outer continental shelfs and increased coastal eutrophication-induced hypoxia will eventually reduce suitable habitat space for both pelagic and demersal fishes.

ENVIRONMENTAL CONSEQUENCES OF HYPOXIA

Relative to oxygen, we do not know where the boundaries are on catastrophe for setting a planetary boundary (Table 1). The amount, severity, and duration of hypoxia and anoxia must all be factored in. While there are at times spectacular events such as mass mortalities of sessile organisms, there is little information on population level effects of hypoxia and anoxia. We do know that when a dead zone forms, oxygen in the water is so low that fish, crabs, shrimp, and other marine life will swim away to areas of higher oxygen concentration. This is the origin of the term dead zone, a place where fishermen cannot find anything to catch. This escape must come at a cost to the individuals in terms of lost growth potential, increased predation risk, increase susceptibility to fishing gear, etc. (Table 2). A combination of other stressors from climate change, pollution, and habitat destruction also produce similar responses to dead zones that further complicates identifying population level responses to hypoxia. We do know that hypoxia influences harvest, both through effects on processes underlying production such as growth and mortality (including effects on juveniles before they are subject to fishing mortality) and on processes influencing catchability (e.g., emigration, avoidance behavior).

Table 2. Generalized response of populations to hypoxia and potential economic effect.

Factor	Result	Economy
Mortality	Loss of stock, may take years to recover	Lower landings, Increased time fishing
Reduced recruitment	Smaller populations, effect may be long lasting	Lower landings, Increased time fishing
Reduced growth	Smaller individuals	Lower individual value
Poor body condition	Weaker individuals	Lower value
Increased Migration	Energy resources diverted to movement	Smaller individuals, Increased time fishing
Aggregation	Exposure to increased risks of predation and exploitation	Less time fishing, Easier to catch
Altered behavior	More/Less susceptible to fishing gear	Increased or decreased catch-ability

Earliest account of hypoxia stressed systems are from European fjords, such as the Drammensfjord in Norway appears to have been persistently hypoxic and anoxic since at least the 1700s based on foramaniferan proxies (Alve, 1995). Even in this small fjord with extended residence time of deep water, historic anoxia has been made worse over the last two centuries by eutrophication. Rivers with population and industrial centers, such as the Mersey Estuary, England, where a combination of factors including hypoxia led to the elimination of salmon by the 1850s but is now recovered through concerted management efforts (Jones, 2006). In the U.S. the earliest account of ecological stress associated with hypoxia come from Mobile Bay, Alabama, where in the 1850s, hypoxic bottom water pushed by tides and wind into shallow water caused mobile organisms to migrate and concentrate at the water's edge. These events became known as "Jubilees" as it was easy for people to picked up the hypoxia stressed fish and crabs (May, 1973). In all cases, hypoxia caused the movement of mobile fauna, which has energetic consequences, and reduced or eliminated the trophic base for bottom feeding species, which has adverse consequences for a system's higher level energy flows (Baird et al. 2004).

There is a similarity of faunal response across systems to varying types of hypoxia that range from beneficial to mortality (Diaz and Rosenberg, 1995; Vaquer-Sonyer and Duarte, 2008). Consequences of low oxygen are often sublethal and do affect growth, immune response, and reproduction. Mobile fauna have to contend with two simultaneous problems, loss of habitat as they are forced to migrate into higher oxygen waters, and reduced or changed prey resources for feeding. Fauna that cannot move initiate a graded series of behaviors to survive and will eventually die as oxygen declines or extends through time. The result is a hypoxia-based habitat compression when hypoxia overlaps with essential habitat such as nursery areas, feeding grounds, or deeper, cooler refuge waters during the summer (Coutant, 1990). Hypoxic habitats avoided are lost to the ecosystem for varying lengths of time and are not productive for fisheries.

Small and large systems exposed to long periods of hypoxia and anoxia have lower annual secondary production with productivity a function of how quickly benthos can recruit and grow during periods when oxygen is normal. The extreme case would be the perennial hypoxic/anoxic areas of the Baltic Sea that cover about 70,000 km^2 where benthic invertebrate production is near zero. Under normal oxygen condition this area of the Baltic Sea should be producing about 1.3 million metric tons (mt) wet weight of potential benthic prey for bottom feeding predators. In Chesapeake Bay, which has about 3,500 km^2 of seasonal hypoxia that lasts about three months, about 75,000 mt wet weight of potential prey for fish and crab predators is lost or translocated. In the northern Gulf of Mexico severe seasonal hypoxia covers about 20,000 km^2 and leads to approximately a 210,000 mt wet weight loss of prey from the fisheries forage base (Diaz and Rosenberg, 2008). The question remains as to whether a system can recover the secondary production lost to hypoxia during periods of normal oxygen. The Chesapeake has about nine months to recover and the northern Gulf of Mexico six months.

The elimination of benthic prey and hypoxia-based habitat compression can have profound effects on ecosystem functions as organisms die and are decomposed by microbes. Up food chain energy transfer is inhibited in areas where hypoxia is severe as benthic resources are killed directly and mobile predators avoid the area (Figure 4). As mortality of benthos occurs, microbial activities quickly dominate energy flows (Baird et al. 2004). This energy diversion tends to occur in ecologically important places and at the most inopportune time for predator energy demands, and causes an overall reduction in an ecosystem's functional ability to transfer energy to higher trophic levels and renders the ecosystem potentially less resilient to other stressors (Diaz and Rosenberg, 2008). Systems reporting mass mortality provide primary examples of degradation in trophic structure.

Hypoxia has clear mortality effects on sessile, and at times mobile, organisms but its population level effects in coastal environments remain uncertain. Much of the evidence supporting negative effects from hypoxia comes from laboratory experiments, localized effects in nature, fish kills, and our intuition that lack of oxygen can lead to dire consequences (Rose et al. 2009). Conclusive evidence of wide-spread population level response to hypoxia is lacking. Quantifying the effects of hypoxia on fish populations, whether large or small, is critical for effective management of coastal ecosystems and for cost-effective and efficient design of remediation actions. The potential for interaction of direct and indirect effects, and subtle changes in vital rates (such as reproduction and recruitment) leading to population responses complicates field study and management, but does not excuse us from quantifying the population losses due to hypoxia. As coastal ecosystems continue to decline their capacity to deliver ecosystem services will also decline (Millennium Ecosystem Assessment, 2005). At some point the consequences of hypoxia will become obvious.

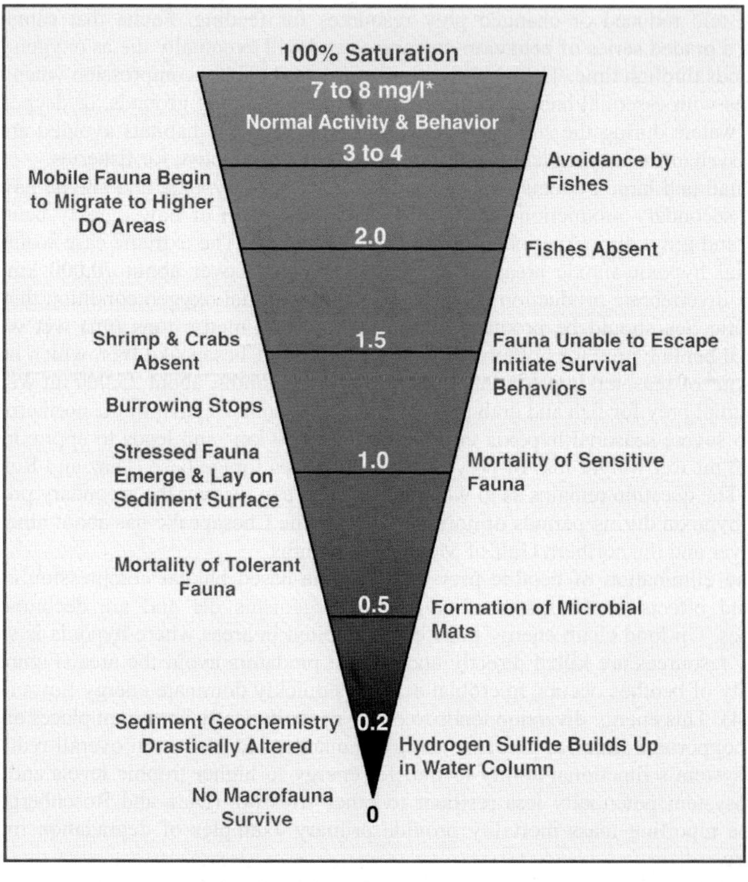

Fig. 4: Range of behavior and ecological impacts as dissolved oxygen levels drop from saturation to anoxia.

ECONOMIC CONSEQUENCES OF HYPOXIA

Economic effects attributable to hypoxia are subtle and difficult to quantify even when mass mortality events occur. However, from assessing ecological effects of hypoxia it is known that populations can experience a range of problems that at some point will negatively affect economic interests (Table 2). Much of the problem in assess economic consequences is related to the multiple stressors acting on targeted commercial populations (habitat degradation, over exploitation, pollution) and also factors that stress fisher's economics (aquaculture, imports, economic costs of fishing, fisheries regulations).

Because of their devastating effects on stocks and fishermen, it is possible to estimate effects of mass mortality events. Losses from hypoxia related mortality of oysters in Mobile Bay, USA, in the early 1970s was U.S.$500,000 (in 1970 dollars), but greater economic losses were associated with the declining stocks and poor recruitment of oysters (*Crassostrea virginica*) associated with recurring severe hypoxic (May, 1973). Estimated losses, both actual and to the resources, to marine related industries from the New York Bight hypoxic event in the summer of 1976 were over U.S.$570 million (Figley et al. 1979). Much of this loss was in surf clams (*Spisula solidissima*) that accounted for >U.S.$430 million. Factored by the area of hypoxia (987 km^2) the 1976 event cost about U.S.$580,000 per km^2 for resources and fisheries related activities and U.S.$165,000 per km^2 for just the resources lost.

Lack of identifiable economic effects in fisheries landing data does not imply that declines would not occur should conditions worsen. In the northern Gulf of Mexico brown shrimp landing appear to be inversely related to the area of hypoxia (O'Connor and Whitall, 2007). Whether this relationship will remain linear or transform into a catastrophe function at some critical point is not known (Kemp et al. 2009). Other large systems have suffered serious ecological and economic consequences from seasonal hypoxia; most notable are the Kattegat with localized loss of catch and recruitment failures of Norway Lobsters (*Nephrops norvegicus*) in the late 1980s, and northwest continental shelf of the Black Sea which suffered regional loss of bottom fishery species also in the 1980's (Karlsen et al. 2002; Mee 1992, 2006).

Economic valuation of losses from hypoxia seem small relative to the total value of fisheries, but the key point is that losses from hypoxia are measureable in economic terms. For example, the valuation of recreational fishing relative to hypoxia in the Patuxtant River, a tributary in Chesapeake Bay, showed that as oxygen declined to mild hypoxic levels total losses for striped bass (*Morone saxatilis*) fishing was about U.S.$10,000 with a net present value of about U.S.$200,000. If the same water quality was allowed to occur in the entire Chesapeake Bay, the net present value of the losses due to hypoxia would be >U.S.$145 million (Lipton and Hicks, 2003). A similar analysis of recreational fishing in northeast and middle Atlantic regions found that overall as oxygen declined capture rate of fish declined (Bricker et al. 2006). The finding of effects of hypoxia on recreational fishing for one species would lead one to believe that similar effects are likely being experienced by commercial fishers. This has been documented for the brown shrimp fisheries in North Carolina. Huang et al. (2010) found that hypoxia may reduce annual harvest by about 13%, valued at about 1.2 million dollars annually.

Experience with other hypoxic zones around the globe shows that both ecological and fisheries effects become progressively more severe as hypoxia worsens (Diaz and Rosenberg, 1995; Caddy, 1993). This would lead one to believe that at some point economic loses will become more obvious and costly. However, currently the direct connection of hypoxia to fisheries landings at large regional scales is weak, likely related to a number of factors that include confounding effects of eutrophication, overfishing, and compensatory mechanisms that alter or mask effects of hypoxia on landings.

Breitburg et al. (2009) found a hint of a connection with a possible decline in landings of benthic species in systems where ≥40% of the bottom area becomes hypoxic.

The most pervasive and measurable effects of hypoxia is on biogeochemical processes, which support ecosystem services of water quality and nutrient cycling (Millennium Ecosystem Assessment, 2005). Oxygen is a key regulator of these two services with hypoxia disrupting most biogeochemical processes. For example, the cycling of nitrogen is complex and dependent on the concentration of oxygen (Middleburg and Levin 2009). Following Costanza et al. (1997) approach to valuing global ecosystem services, which is problematic for many economists, at least an order of magnitude approximation of the impacts of hypoxia can be made. Costanza et al. (1997) estimated that marine ecosystems provide at least U.S.$21 trillion worth of services annually, with the majority of this total outside market valuation in contributions from waste treatment and nutrient cycling. Hypoxia covers about 4% (240,000 km^2) of all coastal and estuarine habitats annually (Diaz and Rosenberg, 2008). If we assume a complete loss in water quality and nutrient cycling services to hypoxia, the amount of loss is approximately U.S.$170 billion. This represents about 0.8% of all marine ecosystems services.

GLOBAL CHANGE AND HYPOXIA

Since the early 2000s many assessments of global environmental and resource health identify hypoxia as one of the factors threatening coastal and ocean life, for example the Millennium Ecosystem Assessment (2005). In addition, climate model predictions and observations reveal regional declines in oceanic dissolved oxygen linked to global warming (Matear and Hirst, 2003; Gilbert et al. 2009; Deutsch et al. 2011). To understand hypoxia and its effects on ecosystems requires several perspectives that start at a local level, move to regional, and finally to a global perspective. The most important scale is local for stressors like coastal development, nutrients, pollution, and eutrophication. This is the scale (<1 to 1,000 km^2) at which impacts are most pronounced and we have the most information. At regional scales (>1,000 to 1,000,000 km^2) it is a mix of influences from land and sea. This involves local land-based impacts and processes bleeding into regional seas, and large scale open ocean processes that are at the boundaries of regions, such as upwelling and thermocline depth. Much of the problem with hypoxia at local and regional levels can be directly tied to concentrations of human populations and agriculture, both of which have significantly altered the global nitrogen cycle (Figure 1). Global scales factors that influence oxygen and hypoxia are changes in circulation patterns, climate, temperature and pH.

Climate change, whether from global warming or from microclimate variation, will have consequences for eutrophication-related oxygen depletion that will progressively led to an onset of hypoxia earlier in the season and possibly extending it through time. Climate change may make systems more susceptible to development of hypoxia through its direct effects on water column stratification, precipitation patterns, and temperature. These effects will likely occur primarily thought warming, which will lead to increased water temperatures and subsequent decrease in oxygen solubility. Warmer surface waters will extend and enhance water column stratification, a key factor in the development of hypoxia. Warmer water will increase organism metabolism, which is the key process for lowering oxygen concentrations. In addition to warming, future climate predictions include large changes in precipitation patterns. If changes in precipitation lead to increased runoff to estuarine and coastal ecosystems, stratification and nutrient loads are likely to increase and worsen oxygen depletion. Conversely, if stratification decreases due to lower runoff or is disrupted by increased storm activity or intensity, the chances for oxygen depletion should decrease (Table 3).

Table 3. Influence of climate drivers on the extent and severity of hypoxia (Modified from Boesch et al. 2007).

Climate Driver	Direct Effect	Secondary Effect	Influence on Hypoxia
Increased temperature	More evaporation	Decreased stream flow	+
		Land-use & cover changes	+/-
	Less snow cover	More nitrogen retention	-
	Warmer water	Stronger stratification	+
		Higher metabolic rates	+
More precipitation	More stream flow	Stronger stratification	+
		More nutrient loading	+
	More extreme rainfall	Greater erosion of soil P	+
Less precipitation	Less stream flow	Weaker stratification	-
		Less nutrient loading	-
Higher sea level	Greater depth	Stronger stratification	+
		Greater bottom water volume	-
		Less hydraulic mixing	+
	Less tidal marsh	Diminished nutrient trapping	+
Summer winds and storms	Weaker, less water column mixing	More persistent stratification	+
	Stronger, more water column mixing	Less persistent stratification	-
	Shifting wind patterns	Weaker/stronger upwelling potential	+/-

Climate related changes in wind patterns are of great concern for coastal system as wind direction and strength influence the strength of upwelling/downwelling, which in turn affect stratification strength and delivery of less-oxygenated deepwater and nutrients into shallow coastal areas. Even relatively small changes in wind and current circulations could lead to large changes in the area of coastal seabed exposed to hypoxia. Changes in the pattern of upwelling on Pacific coast off the Oregon and Washington coasts due to shifts in winds that affected the California current systems appeared to be responsible for the recent development of severe hypoxia over a large area of the inner continental shelf (Grantham et al. 2004).

Thus the future status of hypoxia and its consequences for the environment, society and economies will depend on a combination of climate change (primarily from warming, and altered patterns for wind, currents and precipitation) and land-use change (primarily from expanded human population, agriculture and nutrient loadings). Our expanding population has led to intensification of agriculture/aquaculture and threatens local and regional coastal systems everywhere. Expanding energy demands associated with our population now drive climate change that threatens all ecosystems at all scales from local to global.

RESTORATION AND THE FUTURE OF HYPOXIA

Recognizing the negative consequences of hypoxia and related coastal eutrophication, many nations have recently made major socioeconomic commitments for reducing nutrient loads to the adjacent estuaries, bays and seas, upon which they depend. Worldwide, of the over 600 coastal hypoxia systems, about 75 have responded positively to remediation. All but one improved from management of point discharges. The northwest continental shelf of the Black Sea is the only exception, responding positively to reduction on nutrient runoff. Once the second largest anthropogenic hypoxic area on earth, it is now remediated through concerted efforts to reduce point discharges and runoff from agricultural lands (Mee, 2006; Langmead et al. 2008).

Much of the scientific interest and management concern about hypoxia includes a focus on the general problem of eutrophication caused by the over enrichment of coastal waters from a combination of sewage/industrial discharge and nutrient runoff. Nutrients generally increase biological production, while hypoxia acts in the opposite direction, reducing biomass and habitat quality. Overall, the combination of stressors associated with eutrophication has and continues to degrade our coastal systems. The future effects of hypoxia and nutrient enrichment on food webs and fisheries will be strongly influenced by the extent to which these two factors co-occur. Unless the leakage of nutrients from land-based sources to the sea can be reduced, the future for our estuarine and coastal resources looks bleak. Where applied, nutrient management has reversed the effects of hypoxia. But concerted effort in the future will be needed to allow more systems to recover, particularly for those systems affected primarily by land-runoff.

In addition to the complexity of interactions among physical, chemical, and biological processes, two key elements hinder a global assessment of the effects of hypoxia:

1. Lack of details on the timing of occurrence and area covered by hypoxia
2. Accessibility of data on various ecosystem components.

For hypoxia in Europe and North America much is known about its occurrence in coastal areas, including spatial and temporal patterns. Less is known from the other continents, which have most of our population, and Pacific islands. For all systems, less is known about long-term trends, factors controlling dissolved oxygen depletion and replenishment, impacts on ecological processes, and economic losses. The World Resources Institute eutrophication project has made a start by compiling data on hypoxic and eutrophic areas that are in danger of becoming hypoxic (Diaz et al. 2010). A similar effort was conducted by Conley et al. (2011) for the Baltic Sea. The conclusion of Conley et al. (2011) for the Baltic could easily be applied to the rest of the globe: "The Baltic Sea coastal zone displays an alarming trend with hypoxia steadily increasing with time since the 1950s effecting nutrient biogeochemical processes, ecosystem services, and coastal habitat." To formulate effective strategies for remediating coastal hypoxia it is essential to have an understanding of what the specific drivers are and what responses to expect with various remediation approaches. Some of these drivers are nationally controlled, while others lack defined ownership/responsibility. Making management of hypoxia including the control of drivers of eutrophication a transboundary issue.

ACKNOWLEDGEMENTS

This work is based on Managing Ocean Environments in a Changing Climate, Sustainability and Economic Perspectives edited by Kevin Noone, Ussif Rashid Sumaila, and Robert Diaz (Elsevier 2013, ISBN 978-0-12-407668-6). Supported was provided in part by grants from the Packard Foundation to the World Resources Institute (http://www.wri.org) and from the Okeanos Foundation (http://www.okeanos-foundation.org/) to the Stockholm Resilience Centre, Stockholm University.

REFERENCES

1. Alve, E. (1995) "Benthic foraminiferal distribution and recolonization of formerly anoxic environments in Dramrnensfjord, southern Norway." *Marine Micropaleontology.* 25:169-186.
2. Baird, D., Christian, R.R., Peterson, C.H. and Johnson, G.A. (2004) "Consequences of hypoxia on estuarine ecosystem function: Energy diversion from consumers to microbes." *Ecological Applications.* 14:805-822.
3. Banse, K. (1959) "On upwelling and bottom trawling off the Southwest coast off India." *Journal of the Marine Biological Association of India.* 1:33-49.
4. Boesch, D.F., Coles, V.J., Kimmel, D.G. and Miller, W.D. (2007) Ramifications of climate change for Chesapeake Bay hypoxia, in: Pew Center (ed.), Regional impacts of climate change, four case studies in the United States, Pew Center on Global Climate Change, Arlington, Virginia, pp. 57-70.
5. Breitburg, D.L., Hondorp, D.W., Davias, L.W. and Diaz, R.J. (2009) "Hypoxia, nitrogen and fisheries Integrating effects across local and global landscapes." *Annual Review Marine Science.* 1:329–350.
6. Bricker, S., Lipton, D., Mason, A., Dionne, M., Keeley, D., Krahforst, C., Latimer, J. and Pennock, J. (2006) Improving methods and indicators for evaluation coastal water eutrophication: a pilot study in the Gulf of Maine, NOAA Technical Memorandum NOS NCCOS 20, National Ocean Survey, Silver Springs, Maryland. 81 p.

7. Brill, R.W. (1996) "Selective advantages conferred by the high performance physiology of tunas, billfishes, and dolphin fish." *Comparative Biochemistry and Physiology.* 113:3–15.
8. Caddy, J.F. (1993) "Towards a comparative evaluation of human impacts on fishery ecosystems of enclosed and semi-enclosed seas." *Review of Fisheries Science.* 1:57–95.
9. Cloern, J.E. (2001) "Our evolving conceptual model of the coastal eutrophication problem." *Marine Ecology-Progress Series.* 210:223-253.
10. Conley, D.J., Carstensen, J., Aigars, J., Axe, P., Bonsdorff, E., Eremina, T., Haahti, B.-M., Humborg, C., Jonsson, P., Kotta, J., Lännegren, C., Larsson, U., Maximov, A., Rodriguez Medina, M., Lysiak-Pastuszak, E., Remeikaité-Nikiené, N., Walve, J., Wilhelms, S. and Zillén, L. (2011) "Hypoxia Is Increasing in the Coastal Zone of the Baltic Sea." *Environmental Science and Technology.* 45:6777-6783.
11. Costanza, R., d'Arge, R., de Groot, R., Farber, S., Grasso, M., Hannon, B., Naeem, S., Limburg, K., Paruelo, J., O'Neill, R.V., Raskin, R., Sutton, P. and van den Belt, M. (1997) "The value of the world's ecosystem services and natural capital." *Nature.* 387:253–260.
12. Coutant, C.C. (1985) "Striped bass, temperature, and dissolved oxygen: a speculative hypothesis for environmental risk." *Transactions of the American Fisheries Society.* 114:31–61.
13. Deutsch, C., Brix, H., Ito, T., Frenzel, H. and Thompson, L. (2011) "Climate-Forced Variability of Ocean Hypoxia." *Science.* 333:336-339.
14. Diaz, R.J. and Rosenberg, R. (1995) "Marine benthic hypoxia: A review of its ecological effects and the behavioural responses of benthic macrofauna." *Oceanography and Marine Biology, an Annual Review.* 33:245-303.
15. Diaz, R.J. and Rosenberg, R. (2008) "Spreading dead zones and consequences for marine ecosystems." *Science.* 321:926-928.
16. Diaz, R., Selman, M. and Chique, C. (2010) Global Eutrophic and Hypoxic Coastal Systems. World Resources Institute. Eutrophication and Hypoxia: Nutrient Pollution in Coastal Waters. http://www.wri.org/project/eutrophication
17. Ekau, W., Auel, H., PSystems. World Resources Instit(2009) "Impacts of hypoxia on the structure and processes in the pelagic community (zooplankton, macro-invertebrates and fish)." *Biogeosciences Discussion.* 6:5073-5144.
18. Figley, W., Pyle, B. and Halgren, B. (1979) Socioeconomic impacts. in: R.L. Swanson & C.J. Sindermann (eds.), Oxygen depletion and associated benthic moralities in New York Bight, 1976, NOAA Professional Paper 11, U.S. Government Printing Office, Washington, D.C., pp. 315-322.
19. Foley, J.A., DeFries, R., Asner, G.P., Barford, C., Bonan, G., Carpenter, S.R., Chapin, F.S., Coe, M.T., Daily, G.C., Gibbs, H.K., Helkowski, J.H., Holloway, T., Howard, E.A., Kucharik, C.J., Monfreda, C., Patz, J.A., Prentice, I.C., Ramankutty, N. and Snyder, P.K. (2005) "Global consequences of land use." *Science.* 309:570-574.
20. Fry, F.E.J. (1971). The effect of environmental factors on the physiology of fish. Academic Press, New York, NY.

21. Fuenzalida, R, Schneider, W., Graces-Vargas, J., Bravo, L. and Lange, C. (2009) "Vertical and horizontal extension of the oxygen minimum zone in the eastern South Pacific Ocean." *Deep-Sea Research Part II.* 56:992-1003.
22. Grantham B.A., Chan, F., Nielsen, K.J., Fox, D.S., Barth, J.A., Huyer, A., Lubchenco, J. and Menge, B.A. (2004) "Upwelling-driven nearshore hypoxia signals ecosystem and oceanographic changes in the northeast Pacific." *Nature.* 429:749-754.
23. Galloway, J.N., Townsend, A.R., Erisman, J.W., Bekunda, M., Cai, Z., Freney, J.R., Martinelli, L.A., Seitzinger, S.P. and Sutton, M.A. (2008) "Transformation of the nitrogen cycle Recent trends, questions, and potential solutions." *Science.* 320: 889–892.
24. Gilbert, D., Rabalais, N.N., Diaz, R.J. and Zhang, J. (2009) "Evidence for greater oxygen decline rates in the coastal ocean than in the open ocean." *Biogeosciences Discussion.* 6:9127–9160.
25. Gruber, N. and J.N. Galloway. (2008) "An earth-system perspective of the global nitrogen cycle." *Nature.* 451:293-296.
26. Huang, L., Smith, M.D. and Craig, J.K. (2010) "Quantifying the Economic Effects of Hypoxia on a Fishery for Brown Shrimp Farfantepenaeus aztecus." *Marine and Coastal Fisheries: Dynamics, Management, and Ecosystem Science.* 2:232-248.
27. Helly, J.J. and Levin, L.A. (2004) "Global distribution of naturally occurring marine hypoxia on continental margins." *Deep-Sea Research Part I.* 51:1159–1168.
28. Jones P.D. (2006) "Water quality and fisheries in the Mersey estuary, England: A historical perspective." *Marine Pollution Bulletin.* 53:144-154.
29. Karlson, K., Rosenberg, R. and Bonsdorff, E. (2002) "Temporal and spatial large-scale effects of eutrophication and oxygen deficiency on benthic fauna in Scandinavian and Baltic waters—a review." *Oceanography and Marine Biology: An Annual Review.* 40:427-489.
30. Keeling, R.F., Körtzinger, A., and Gruber, N. (2010) "Ocean deoxygenation in a warming world." *Annual Review of Marine Science.* 2:199-229.
31. Kemp, W.M., Testa, J.M., Conley, D.J., Gilbert, D. and Hagy, J.D. (2009) "Temporal responses of coastal hypoxia to nutrient loading and physical controls." *Biogeosciences.* 6:2985–3008.
32. Langmead O., McQuatters-Gollop, A., Mee, L.D., Friedrich, J., Gilbert, A.J., Gomoiu, M.-T., Jackson, E.L., Knudsen, S., Minicheva, G. and Todorova, V. (2009) "Recovery or decline of the northwestern Black Sea: a societal choice revealed by socio-ecological modelling." *Ecological Modelling.* 220:2927–2939.
33. Levin, L.A. (2003) "Oxygen minimum zone benthos adaptation and community response to hypoxia." *Oceanography and Marine Biology: an Annual Review.* 41:1–45.
34. Levin, L.A., Ekau, W., Gooday, A.J., Jorissen, F., Middelburg, J.J., Naqvi, S.W.A., Neira, C. Rabalais, N.N. and Zhang. J. (2009) "Effects of natural and human-induced hypoxia on coastal benthos." *Biogeosciences.* 6:2063–2098.
35. Lipton, D. and Hicks, R. (2003) "The cost of stress: low dissolved oxygen and economic benefits of recreational striped bass (*Morone saxatilis*) fishing in the

Patuxent River." *Estuaries.* 26:310-315.

36. Matear, R.J. and Hirst, A.C. (2003) "Long-term changes in dissolved oxygen concentrations in the ocean caused by protracted global warming." *Global Biogeochemical Cycles.* 17:1125, doi:10.1029/2002GB001997.
37. May, E. (1973) "Extensive oxygen depletion in Mobile Bay, Alabama." *Limnology and Oceanography.* 18: 353-366.
38. Mee, L. (2006) "Reviving dead zones." *Scientific American.* 295:78-85.
39. Mee, L.D. (1992) "The Black Sea in crisis: a need for concerted international action." *Ambio.* 21:278-286.
40. Middelburg, J. and Levin, L.A. (2009) "Coastal hypoxia and sediment biogeochemistry." *Biogeosciences.* 6:1273–1293.
41. Millennium Ecosystem Assessment. (2005) Ecosystems and Human Well-being: Synthesis. Island Press, Washington, DC. 155 p.
42. Monteiro, P., van der Plas, A., Melice, J.-L., and Florenchie, P. (2008) "Interannual hypoxia variability in a coastal upwelling system Ocean-shelf exchange, climate and ecosystem-state implications." *Deep-Sea Research Part I.* 55:435–450.
43. Naqvi, S.W.A., Naik, H., Pratihary, A., D'Souza, W., Narvekar, P.V., Jayakumar, D.A., Devol, A.H., Yoshinari, T., and Saino, T. (2006) "Coastal versus open-ocean denitrification in the Arabian Sea." *Biogeosciences.* 3:621-633.
44. O'Connor, T. and Whitall, D. (2007) "Linking hypoxia to shrimp catch in the northern Gulf of Mexico." *Marine Pollution Bulletin.* 54:460-63.
45. Pauly, D. and Christensen, V. (1995) "Primary production required to sustain global fisheries." *Nature.* 374:255–257.
46. Prince, E. D. and Goodyear, C. P. (2006) "Hypoxia-based habitat compression of tropical pelagic fishes." *Fisheries Oceanography.* 15:451-464.
47. Rockström, J., Steffen, W., Noone, K., Persson, Å., Chapin, F.S., Lambin, E.F., Lenton, T.M., Scheffer, M., Folke, C., Schellnhuber, H.J., Nykvist, B., de Wit, C.A., Hughes, T., van der Leeuw, S., Rodhe, H., Sörlin, S., Snyder, P.K., Costanza, R., Svedin, U., Falkenmark, M., Karlberg, L., Corell, R.W., Fabry, V.J., Hansen, J., Walker, B., Liverman, D., Richardson, K., Crutzen, K. and Foley, J.A. (2011) "A safe operating space for humanity." *Nature.* 461:472-476.
48. Rose, K.A., Adamack, A.T., Murphy, C.A., Sable, S.E., Kolesar, S.E., Craig, J.K., Breitburg, D.L., Thomas, P., Brouwer, M.H., Cerco, C.F., and Diamond, S. (2009) "Does hypoxia have population-level effects on coastal fish? Musings from the virtual world." *Journal of Experimental Marine Biology and Ecology.* 381:S188-S203.
49. Stramma, L., Johnson, G.C., Sprintall, J., and Mohrholz, V. (2008) "Expanding oxygen-minimum zones in the tropical oceans." *Science* 320:655-658.
50. Stramma, L., Prince, E.D., Schmidtko, S., Luo, J., Hoolihan, J.P., Visbeck, M., Wallace, D.W.R., Brandt, P. and Körtzinger, A. (2012) "Expansion of oxygen minimum zones may reduce available habitat for tropical pelagic fishes." *Nature Climate Change.* 2:33–37.
51. Vaquer-Sunyer, R. and Duarte, C.M. (2008) "Thresholds of hypoxia for marine biodiversity." *Proceedings of the National Academy of Sciences of the United States of America.* 105:15452–15457.

CHALLENGES TO OCEAN LIFE: FISHERIES, GLOBAL WARMING AND POLLUTION

DANIEL PAULY
UBC Fisheries Centre
Vancouver, BC, Canada

WHAT ARE THE MAJOR PROBLEMS FOR LIFE IN THE OCEAN NOW, AT THE DAWN OF THE 21ST CENTURY?

I would argue that they are, in decreasing order of importance:

1. Ever-expanding marine fisheries, whose impacts on marine life are still widely underestimated;
2. Global warming, whose effects on fisheries have already kicked in;
3. Pollution, whose impacts on marine life was long overestimated, and which now begins to live up to its reputation.

We won't deal with ocean acidification (too discouraging).

The first issue with fisheries is that we not only do not know precisely what they catch and where, but that some argue we don't need to know fisheries catches.

Yet, fisheries catches are one of our strongest interactions with the oceans.

COMMENT

Shale gas and oil will run out faster than champions predict p.307

Napoleon Chagnon sets the record straight p.310

A flattering life of impish physicist Freeman Dyson p.311

Iran needs international help to fight HIV epidemic p.314

Does catch reflect abundance?

Researchers are divided over the wisdom of using estimates of the amount of fish hauled in each year to assess the health of fisheries.

POINT

Yes, it is a crucial signal

The only data available for most fisheries are the weight of fish caught each year, insists Daniel Pauly.

In developed countries such as the United States, Australia and members of the European Union, many fisheries are monitored by fisheries scientists using expensive stock assessments. To infer the size of the fish populations being exploited, scientists use the age and size distributions of the fish caught; the results of scientific surveys carried out from research vessels; and information about growth and migration from tag and recapture studies. Yet the only data ▶

COUNTERPOINT

No, it is misleading

Many factors as well as abundance determine the hauls of fishermen, warn Ray Hilborn and Trevor A. Branch.

The major database on all the fisheries of the world is the *FAO Yearbook Fishery and Aquaculture Statistics*. This collates the amount (in weight) of haddock, bream, cod and more than 1,000 other species hauled in each year by fishermen, whether from commercial trawlers or canoes, using estimates sent in by officials from individual countries.

For the past few years, researchers have been conducting analyses ▶

21 FEBRUARY 2013 | VOL 494 | NATURE | 303

© 2013 Macmillan Publishers Limited. All rights reserved

EXTENT OF THE GEOGRAPHIC BIAS IN 'GLOBAL' STUDIES OF FISHERIES STATUS PUBLISHED BY *SCIENCE*

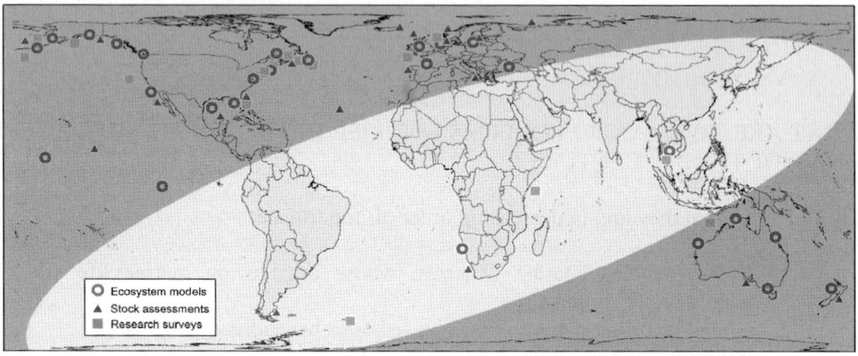

*Based on an article by Worm, Hilborn et al. (*Science. *2009)*

Official fisheries catches, as submitted by member countries to FAO are incomplete.

We addressed this by doing 'catch reconstructions', i.e., bottom-up re-estimation of total catches for all countries of the world, based on the principles that:

1. Every fishery casts a ''shadow' on the society in which it occurs, and
2. Zero is never a good estimate for a positive number that is not precisely known.

TOTAL RECONSTRUCTED CATCHES FOR WEST AFRICA: DOMESTIC

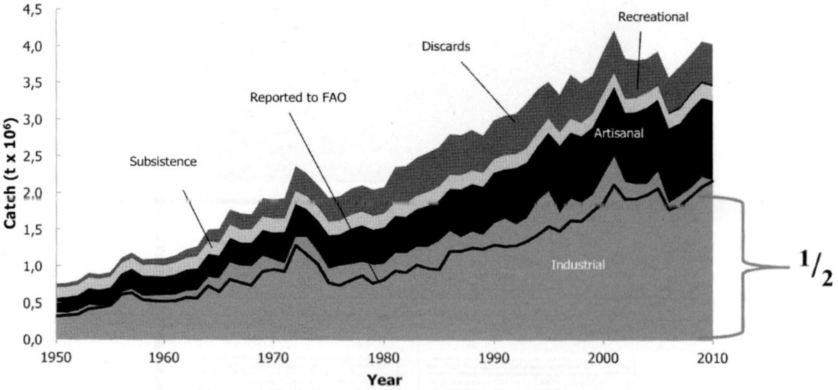

*Belhabib et al. (*Environmental Development. *in press)*

TOTAL RECONSTRUCTED CATCHES FOR WEST AFRICA: DOMESTIC AND FOREIGN

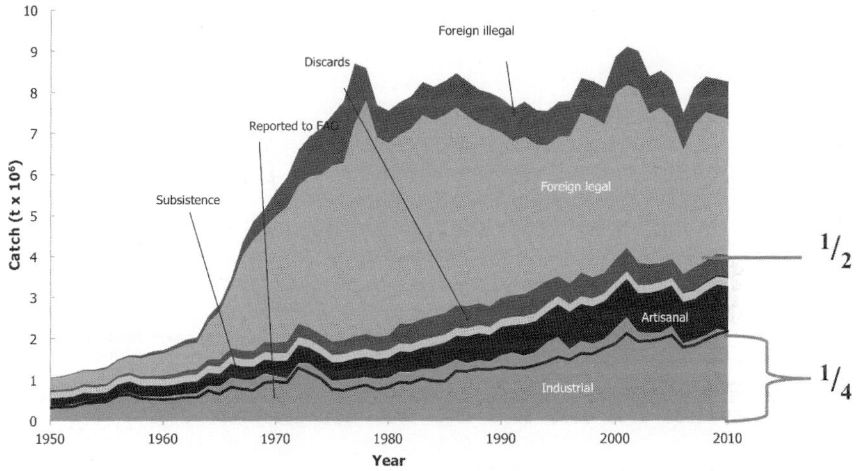

*Belhabib et al. (*Environmental Development. *in press)*

The reconstructions confirm that the world catch is declining; this trend is more marked that in the officially reported catch.

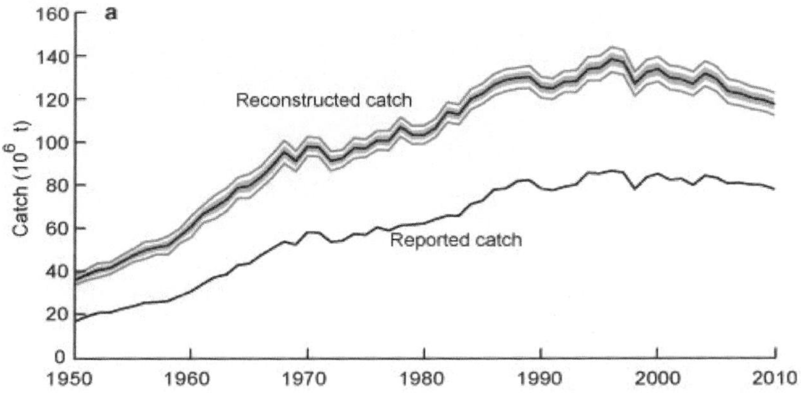

*Pauly and Zeller (*Nature Communication. *in review)*

This catch decline is not surprising, given the growth of 'effective' fishing effort (1950-2006).

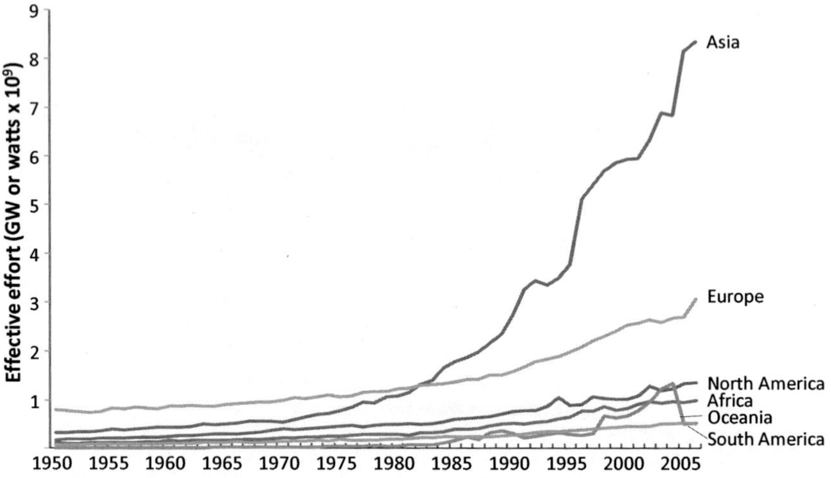

*Anticamara et al. (*Fisheries Research. *2011)*

This issue was long masked by fisheries expansion, here illustrated by mapping the catch of Spain in the 1950s…

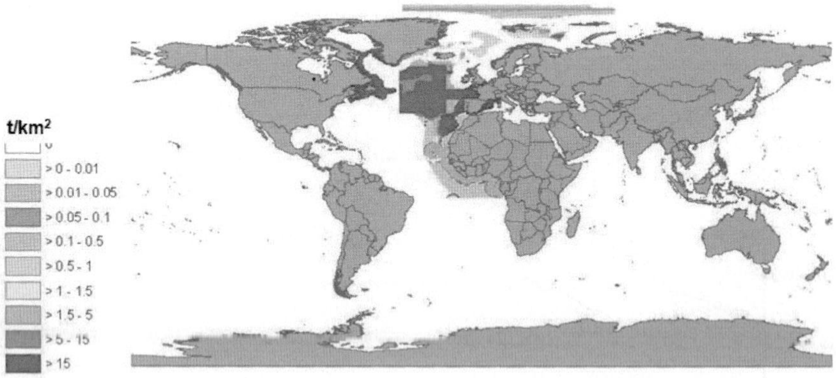

See www.searoundus.org

…and in the 2000-2004 (remember: Spain!)

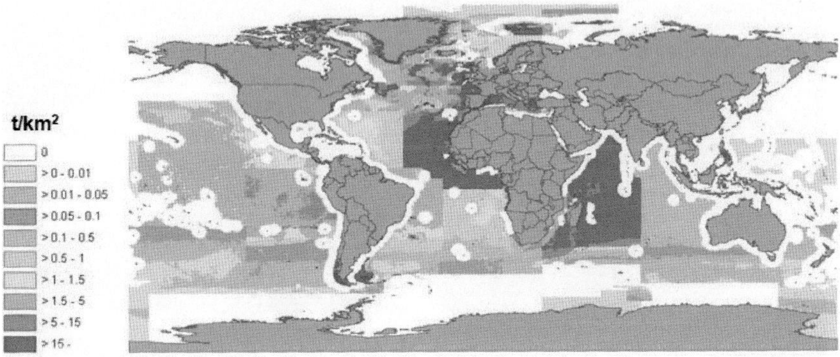

See www.searoundus.org

Now recall that ecosystem fluxes move up 'trophic pyramids,' and each species tends to have its own trophic level.

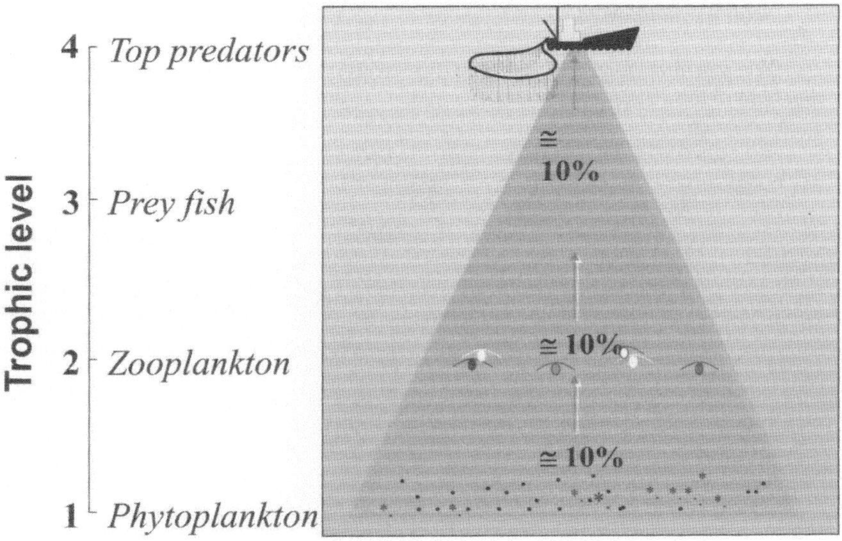

*Pauly and Christensen (*Nature. *1995)*

Now we know (from satellite data) the primary production of the ocean, which is usually high in coastal waters, and very low in the 5 central gyre of the oceans.

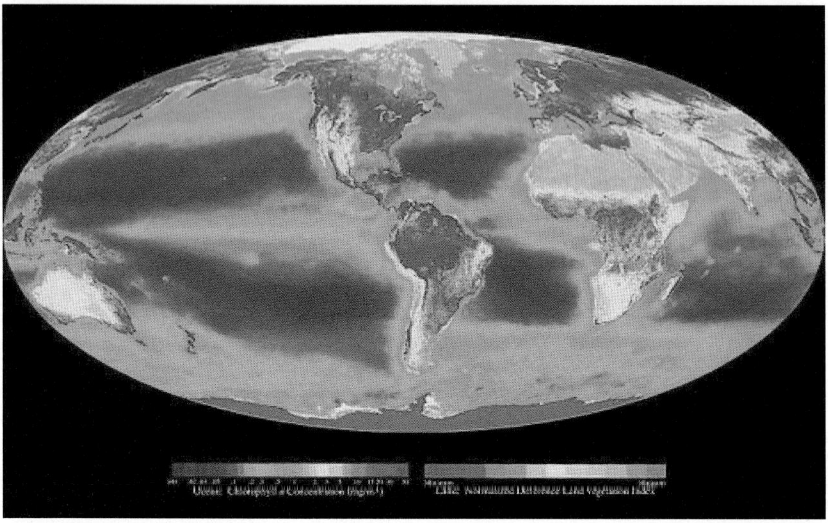

SeaWiFS data, NOAA

We can thus map the footprint (or 'seafoodprint') of fisheries onto the world ocean, here in the 1950s…

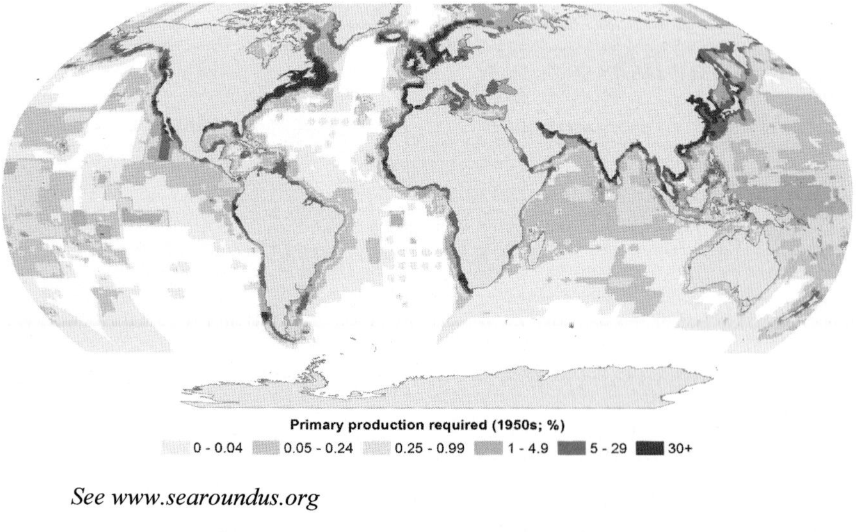

See www.searoundus.org

…and in the 2000s.

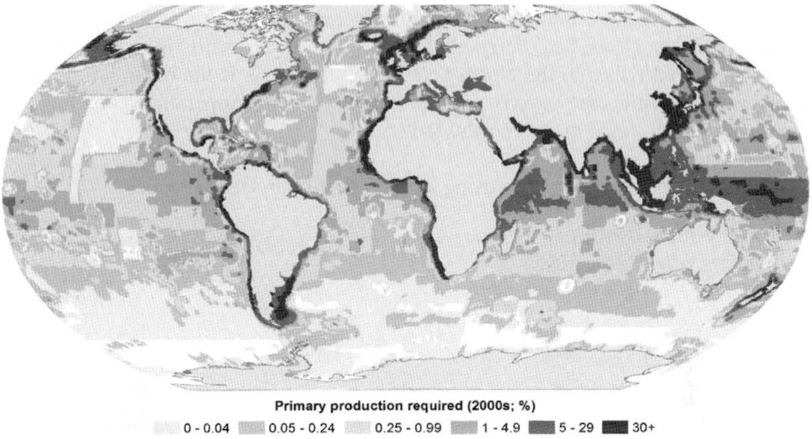

See www.searoundus.org

Observed climate-induced shifts in distribution ranges.

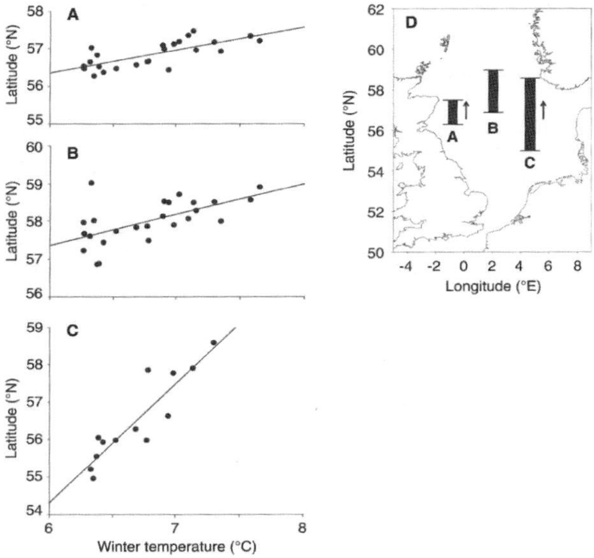

Poleward shifts in distribution ranges of marine species, e.g., in the North Sea (Perry et al. Science. *2005).*

Specifically, as ocean temperature increase, we would expect that many marine fishes and invertebrates would shift towards the pole to find cooler refuges. In fact, evidence of such climate-induced range-shifting has been observed in various regions. For example, in the North Sea, Perry *et al.* showed poleward shifts in mean distributional ranges of demersal fishes in the North Sea. Here, the y-axis is the mean latitudine of the species' ranges while the x-axis is the winter water temperature. As temperature increases, the mean range moves northward.

Simulating poleward shifts using temperature-abundance profiles.

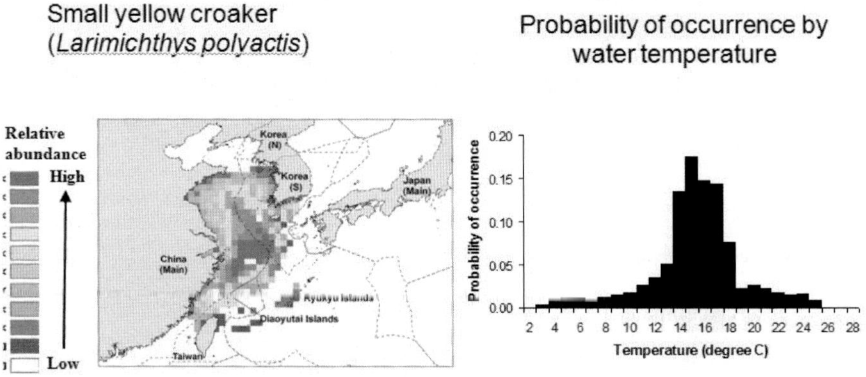

*Cheung et al. (*Marine Ecology Progress Series. *2008)*

The probability of occurrence by water temperature is inferred from overlaying predicted relative abundance map on map of sea water temperature. The resolution of the distribution maps is half degree lat by half degree long.

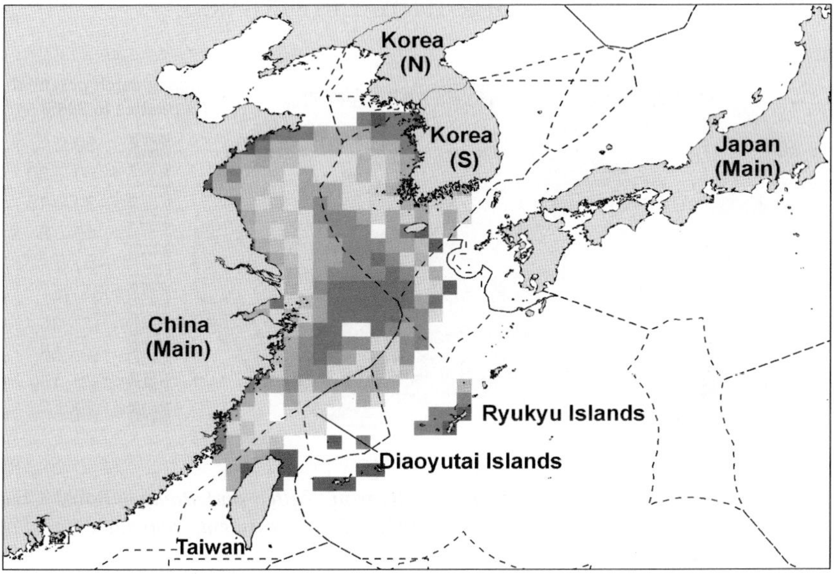

Small yellow croaker Year 0

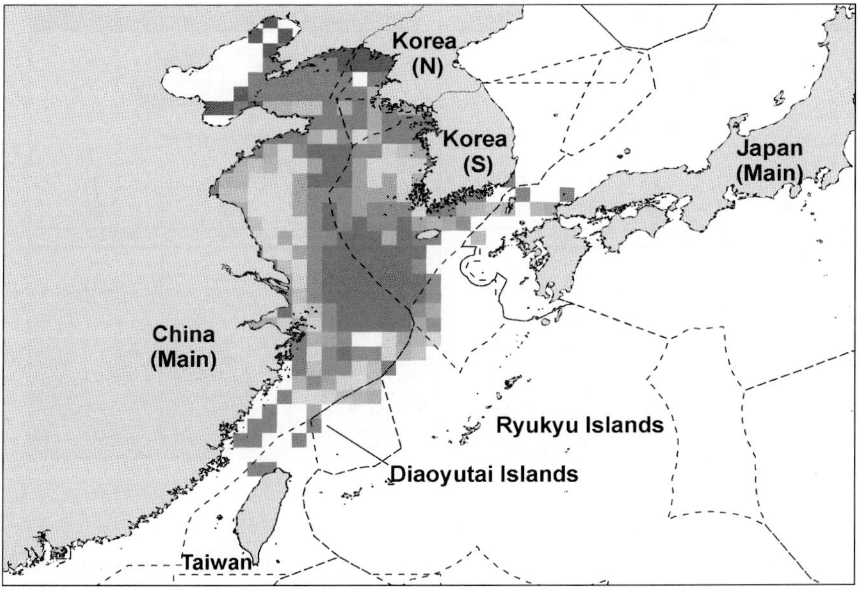

Small yellow croaker Year 30

PROJECTED CHANGE IN CATCH POTENTIAL IN 50 YEARS

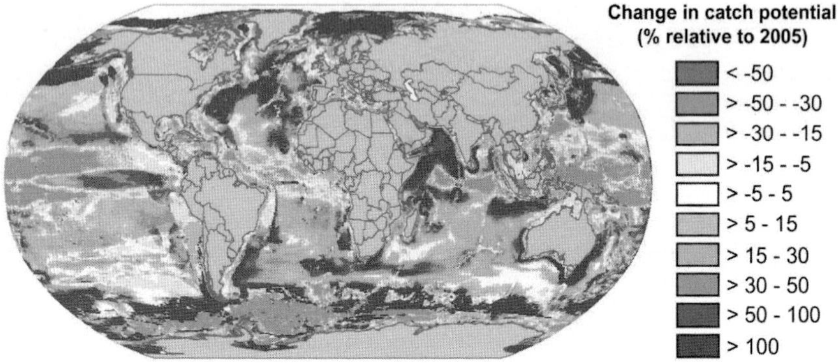

*Cheung, Lam, Kearney, Sarmiento, Watson, Zeller and Pauly (*Global Change Biology. *2009); see also IPCC, 5th Assessment, Summary for Policy Makers.*

IN SUMMARY

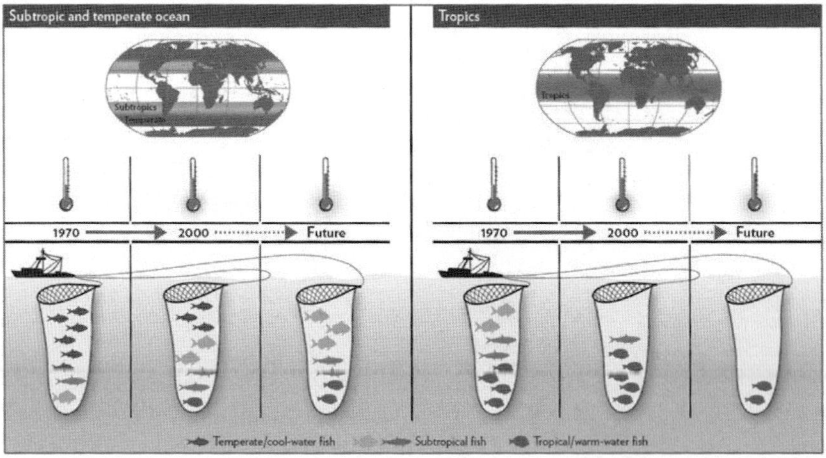

*Cheung, Watson and Pauly (*Nature. *2013)*

Marine pollutants take several forms; one of them are persistent Organic Pollutants (POPs), such as dioxin, whose land and ocean deposition we modeled as an input to food-web based models.

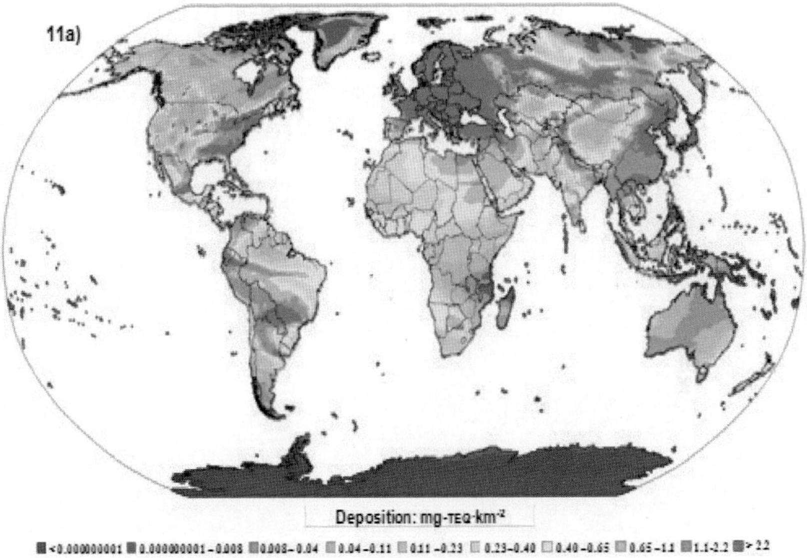

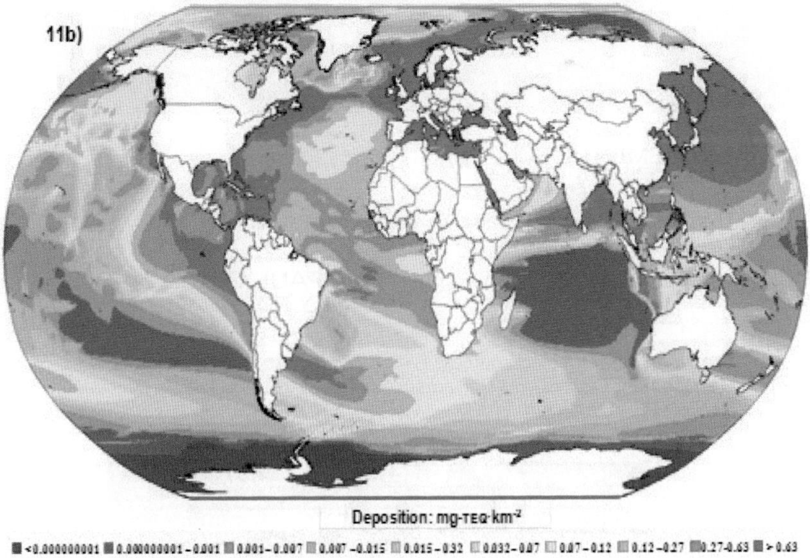

*Booth et al. (*Marine Pollution Bulletin*, 2013)*

Plastic pollution is an emerging problem, caused in part by fisheries, strongly affecting seabirds and marine mammals, but not strongly affecting fisheries.

ACKNOWLEDGEMENTS

- Thanks to The Pew Charitable Trusts and the Paul G. Allen Family Foundation for support

THE PEW CHARITABLE TRUSTS   THE PAUL G. ALLEN FAMILY FOUNDATION

- Thanks to other funding sources: FAO, US Western Pacific Fisheries Management Council, EU-Parliament, UNEP, BOBLME, MAVA Foundation, Rockefeller Foundation, WWF
- Thanks to all members of the *Sea Around Us,* past and present...

... sorry, I ran out of pictures....

and thanks to many other colleagues

visit us at **www.seaaroundus.org**

MARINE MAMMALS, SENTINELS OF OCEAN HEALTH

SAMUEL DOVER
Channel Islands Marine & Wildlife Institute
Santa Barbara, CA, USA

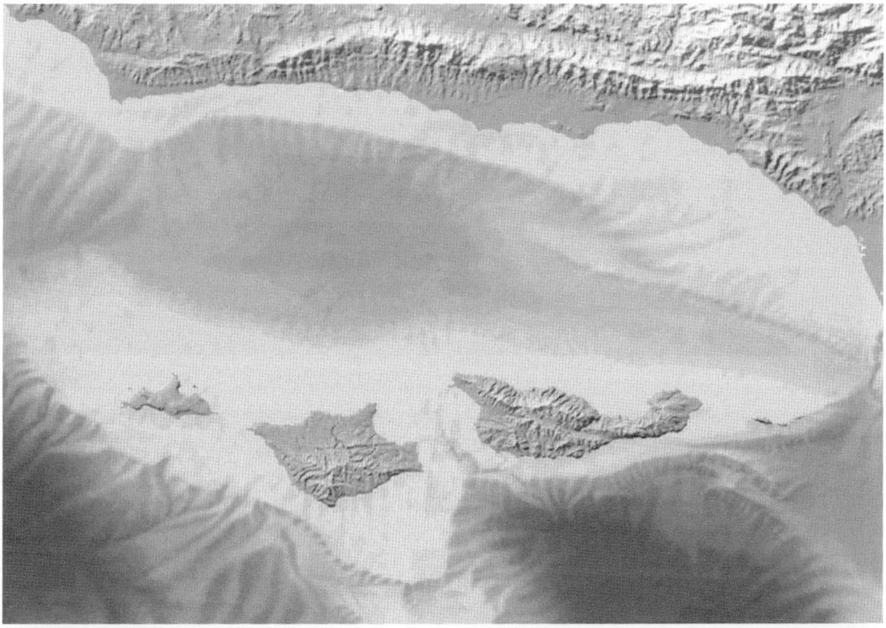

Why Marine Mammals?

* They reside in or depend on the Oceans
* Relatively long life-span
* Feed at high trophic level
* Have unique fat stores, depots for anthropogenic toxins
* Charismatic megafauna, observed and studied
* Diseases of marine mammals raise awareness of ocean health

Effects of Pollution On Marine Mammal Populations

Direct Effects

- Entanglements in Marine Debris
- Oil spills
- Toxic events
    - Acute vs. chronic

Trauma

- Ship/Boat strikes
- Habitat Destruction
- Human Interaction

Fin Whale Ship Strike

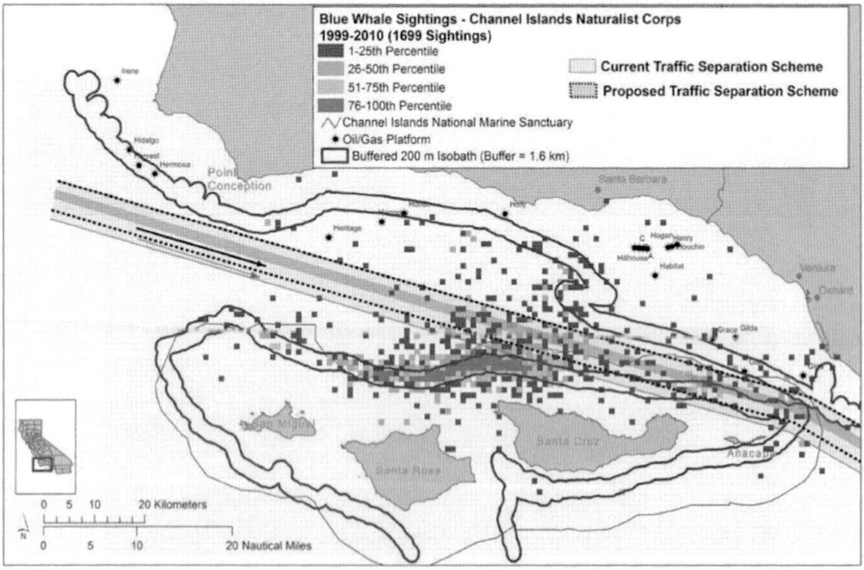
Blue Whale Sightings - Channel Islands Naturalist Corps
1999-2010 (1699 Sightings)
1-25th Percentile
26-50th Percentile
51-75th Percentile
76-100th Percentile
Channel Islands National Marine Sanctuary
Oil/Gas Platform
Buffered 200 m Isobath (Buffer = 1.6 km)
Current Traffic Separation Scheme
Proposed Traffic Separation Scheme
Point Conception
Santa Barbara
Ventura
Oxnard
Santa Cruz
Santa Rosa
0 5 10 20 Kilometers
0 5 10 20 Nautical Miles
N

One Solution to Ship Strikes: Move the Shipping Lanes

Santa Barbara Channel

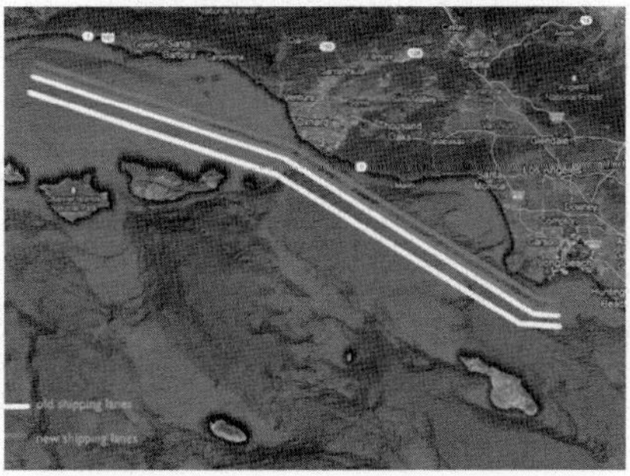

San Francisco Bay

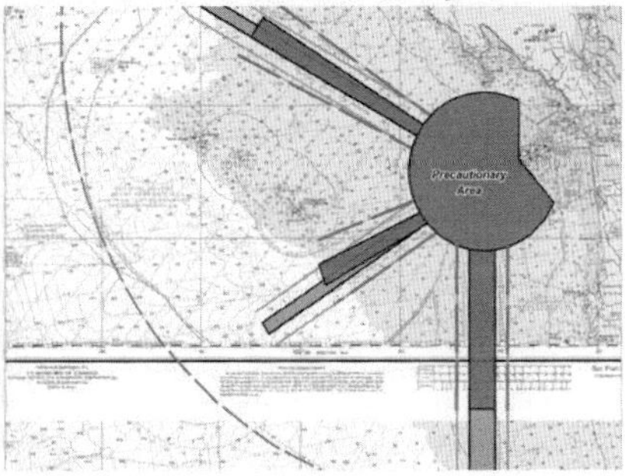

Effects of Pollution on Marine Mammals

Indirect effects

- Accumulation of toxins
  - Heavy metals
    - Mercury, Lead, Cadmium, etc.
  - Organochlorines
    - PCBs, PCDFs, PCDDs, Pesticides (DDT), etc.
- Immunosuppression and Pathogens
  - Viruses
  - Parasites
  - Bacteria
  - Other
- Harmful Algal Blooms
  - Increased in areas of nutrient rich water run-off
- Antibiotic Resistance
  - "One Health" approach

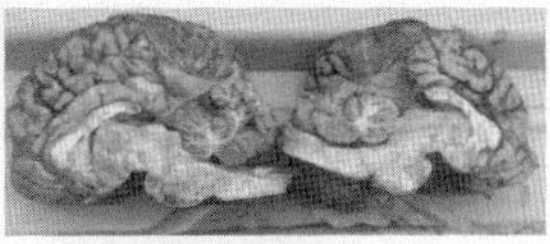

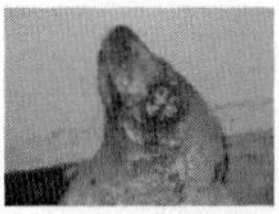

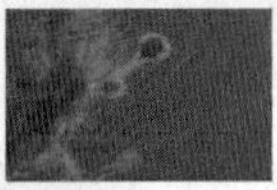

National Action Plan for Combating Antibiotic Resistance, March 2015

President's Council of Advisors on Science and Technology
World Health Assembly Resolution 67.25

Goals

1. Slow the Emergence of Resistant Bacteria and Prevent the Spread of Resistant Infections
2. Strengthen National One-Health Surveillance Efforts to Combat Resistance
3. Advance Development and Use of Rapid and Innovative Diagnostic Tests for Identification and Characterization of Resistant Bacteria
4. Accelerate Basic and Applied Research and Development for New Antibiotics, Other Therapeutics, and Vaccines
5. Improve International Collaboration and Capacities for Antibiotic-resistance Prevention, Surveillance, Control, and Antibiotic Research and Development

Major Threats to Resident Marine Mammals of the Mediterranean

* By-catch and driftnets
* Food resource depletion
* Pollution
* Maritime traffic
* Unregulated whale watching
* Underwater noise pollution
* Invasive species
* Coastal development

International Union for the Conservation of Nature

The IUCN Red List of Threatened Species™ is widely recognized as the most comprehensive, objective global approach for evaluating the conservation status of plant and animal species.

2005 Global Marine Species Assessment Group (GMSA)

As of March 2015, over 13,500 species have now been assessed under the IUCN Red List Categories and Criteria, over half of which (7,468) have been successfully peer-reviewed and now appear on the IUCN Red List of Threatened Species.

Thus far, approximately 11% of all marine species assessed have an elevated risk of extinction (listed on the IUCN Red List of Threatened Species as Critically Endangered, Endangered, or Vulnerable).

Resident Marine Mammals of the Mediterranean

* Common Bottlenose Dolphin *(Tursiops truncatus) V*
* Cuvier's Beaked Whale *(Ziphius cavirostris) ?*
* Fin Whale *(Balaenoptera physalus) V*
* Long-finned Pilot Whale *(Globicephala melas) ?*
* Risso's Dolphin *(Grampus griseus) ?*
* Sperm Whale *(Physeter macrocephalus) E*
* Striped Dolphin *(Stenella coeruleoalba) V*
* Short-beaked Common Dolphin *(Delphinus delphis) E*
* Mediterranean Monk Seal *(Monachus monachus) CE*

Monk Seals

Mediterranean *(Monachus monachus)*
- IUCN Critically endangered
- Less than 600 individuals remaining

Hawaiian *(Monachus schauinslandi)*
- Approximately 1100 individuals remaining

Caribbean *(Monachus tropicalis)*
- Last sighted in 1952
- Declared extinct in 2008

Distribution of Mediterranean Monk Seal

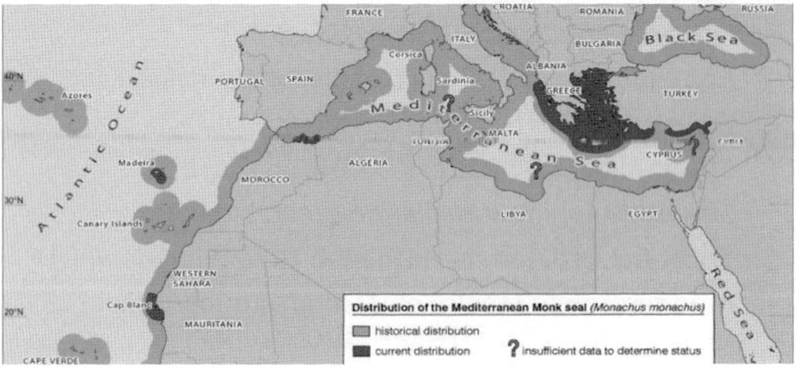

Mediterranean Monk Seal
Threats to Survival
Habitat Loss/Destruction
High infant mortality (50% after two months)
Human interaction
Deliberate vs Entanglement
Food shortages
Lack of International Coordination/Management
Disease Epidemics
Harmful Algae Blooms
1997 66% lost at Cape Blanc
Pollution
© Wild Wonders of Europe / Sá / nature

Questions?

SESSION 6

COUNTERING TERRORISM WITH TECHNOLOGY AND DATA

FRAGILITY OF THE ENERGY NETWORK

RICHARD LANZA
Dept. of Nuclear Science and Engineering, Massachusetts Institute of Technology, Cambridge, MA, USA

ABSTRACT

Modern cities are served by energy networks, especially electrical grids which are large, interconnected means for moving energy over wide geographic area. The peculiarities of electric grids lead to the potential of disruption over very wide geographical area. For example, electricity must be produced and consumed simultaneously and due to the interconnected nature, a change in one area of the grid can affect other areas, almost instantaneously; local and regional supply and demand must be balanced to avoid blackouts. Further, infrastructure components such as fuel supply, generating plants and transmission grids have limited capacity.[1] Attacks on the traditional electric grid and its components have been well recognized as a means for disruption. Currently, a combination of economic and environmental issues has led to a large increase in use of natural gas as a fuel for electric generation.[2,3] In some regions more than 50% of electrical energy is derived from natural gas. The natural gas network which supplies generating stations is thus a critical component of the electric network and though they are similar to electric grids in vulnerability, the natural gas network is far less flexible in operation and response to disruptions and thus may represent a significant security risk.[4] In this note, we give a brief overview of the gas network and comparison of its vulnerabilities especially as compared to the better studied electric grid. We find that a major difference in these systems is that whereas coal, oil and nuclear plants for example have significant amounts of on-site fuel, gas power plants typically have almost no reserve on site. In addition, electrical grids have well developed means for switching power between networks in short time, such capabilities do not typically exist in the gas network. Some means for mitigating the effects of attacks such as increased local storage capability will also be discussed.

ENERGY NETWORKS

There exists a vast amount of literature on the electrical grid but for the purpose of this note, we consider it to be a very simple system as shown in Figure 1. The fuel source is specifically included as this may be the point of vulnerability for natural gas fired generating stations but not for coal or nuclear generating stations. The generating stations do represent a significant and attractive target, but in the case of nuclear plants, they are likely to be well protected and present the terrorist with a difficult target but attacks are still a consideration. Transmission lines are of course spread out over large distance, often far from security forces and can easily be attacked. However, transmission lines fail during the normal operation of systems and means for rapid repair and switching of lines are well established. An especially vulnerable point are the large high voltage transformers used to connect the generator to the grid and also, to connect various parts of separate networks, and this will be discussed briefly later, especially with respect to the impact of simultaneous attacks.

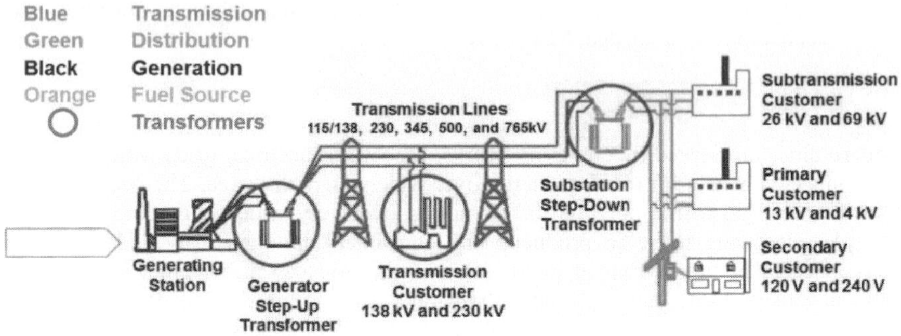

Fig. 1: Simplified electrical network.

NATURAL GAS NETWORKS

Natural gas typically is transported through the use of large pressurized pipe lines ranging from ~1 m diameter to as large as 1.2 m in diameter. Operating pressures are in the range of 60 to 125 bar (900–1800 psi, 6–12 MPa) and in most systems, compressors every 150–200 km. Most lines are either on the surface or buried to depths of 1–2 m, emerging from the ground at the compressor station. For redundancy, multiple lines may be run in parallel, with a series of lateral valves, usually located near the compressor stations, which allows bypassing of lines for repairs and maintenance. Costs of lines can be substantial, a rule of thumb for offshore lines ~\$25K-40K inch/km thus a 400-km, 48-in. line would cost ~\$480M.

The energy equivalent of natural gas varies somewhat, but for our purposes, we use 1 m^3 gas ~37 MJ and 1 cubic foot (cf) of gas ~1 MJ. Typical gas turbines used for electricity generation, can range in power to 500 MW. Depending on whether it is a conventional or combined cycle turbine, the efficiency for electric generation can be from 60 to 40%, averaging 50%. With these values, pipeline flows of 1 Mcf/day (10^6 cubic feet/day) results in 11.6 $MW_{thermal}$ or 5.8 $MW_{electric}$. Another unit of gas flow used is bcm/year, 10^9 m^3/year which is 0.6 $GW_{electric}$. Large gas lines in the U.S. may be 15000 Mcf / day or 90 GW and lines from Russia to Western Europe deliver ~150 GW of energy; gas lines are indeed the equivalent of very large electric lines.

While it is true that not all gas is used for electric generation, a significant and increasing amount of electric production is generated by gas. One notable exception is Germany, where electricity production using natural gas is *declining* while production using coal is *increasing* resulting in increased CO_2 production, largely driven by the economic realities of the German electric market. In the U.S., natural gas represents ~30% of electrical energy production and is anticipated to increase, mostly at the expense of coal.

ELECTRICAL NETWORKS COMPARED TO GAS

A recent report from the MIT Energy Initiative summarizes many of the salient difference between gas and electric systems:[3]

> "However, gas is different. The need for electric power systems to match supply and demand at every instant combined with the lack of gas storage at most generating plants means that short-run changes in the demand for gas from a gas-reliant electric power system closely mirror changes in the demand for its electricity. A pipeline outage generally causes electric generating capacity to go offline as well. While electric generators can buy gas in a fairly liquid market, spot prices of gas, like spot prices of electricity, can vary widely on a daily and even hourly basis. And, unlike a rail line, a pipeline's capacity is strictly limited; adding more capacity will generally require building another pipeline. Both pipelines and electric power systems offer interruptible contracts to large customers because it is sometimes necessary for both to shed load in order to maintain system integrity."

Gas pipeline networks have flow patterns in which successive branches of the network result in smaller pipes and do not have any easy and quick switch method. This is seen in Figure 2 (U.S.) and Figure 4 (Eastern Europe and FSU).[5] Note especially the decreasing sizes and capacity of pipelines as they branch out. Compare this to the electric grid in Figure 3 which has many crosslinks and connections between networks. Although this is a map of the U.S. system, similar observations can be made with the European energy network; gas networks tend to be linear, decreasing in density, while electric networks are denser and have many more points of interconnection.

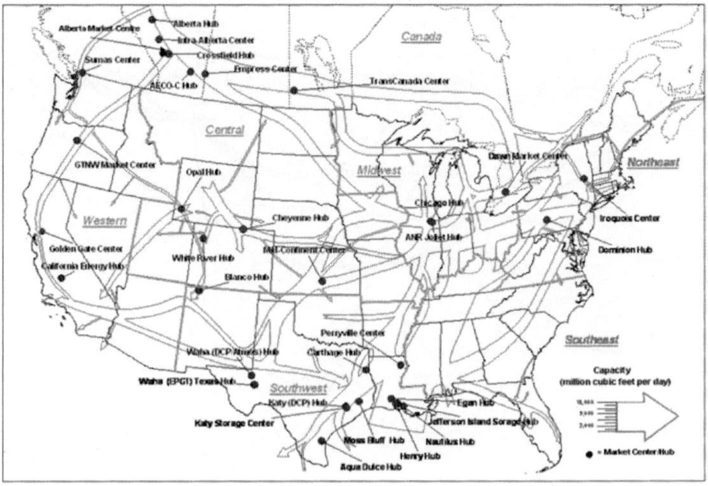

Fig. 2: U.S. gas network.

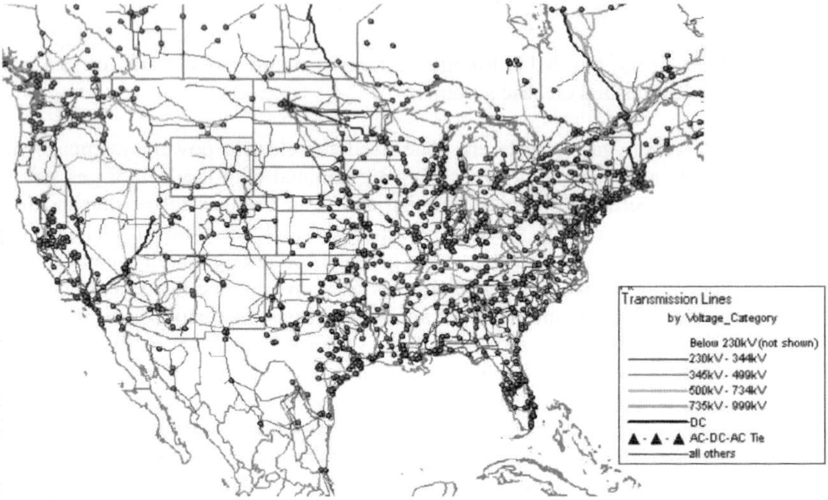

Fig. 3: U.S. electric network.

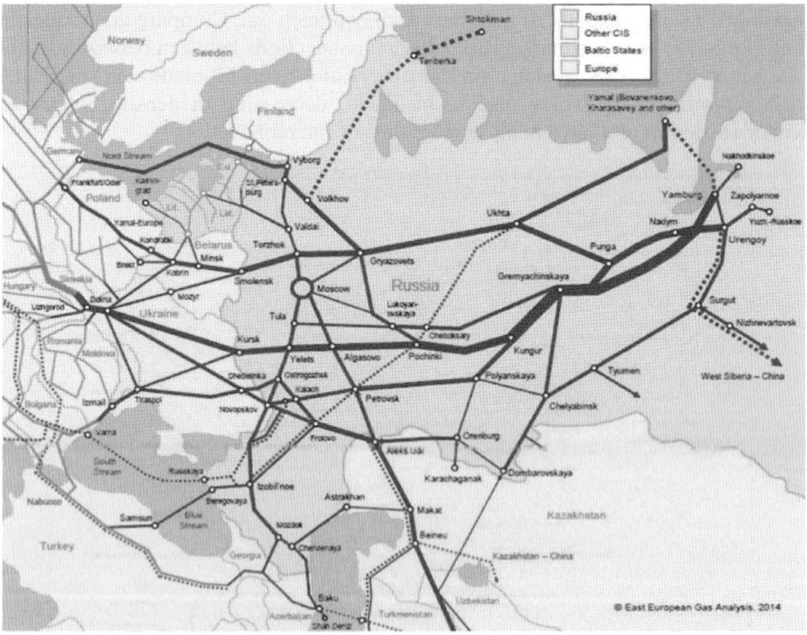

Fig. 4: Gas networks from FSU to Europe.

Table 1. Gas line capacity between FSU and Europe.[5]

Pipeline	Capacity		Destination of exports
Via Ukraine:	**bcm/year**	**GWE***	
Orenburg-Western border (Uzhgorod)	26	15.6	Slovakia, Czech, Austria, Germany, France, Switzerland, Slovenia, Italy
Urengoy-Uzhgorod	28	16.8	Slovakia, Czech, Austria, Germany, France, Switzerland, Slovenia, Italy
Yamburg-Western border (Uzhgorod)	26	15.6	Slovakia, Czech, Austria, Germany, France, Switzerland, Slovenia, Italy
Dolina-Uzhgorod - 2 lines	17	10.2	Slovakia, Czech, Austria, Germany, France, Switzerland, Slovenia, Italy
Komarno-Drozdowichi - 2 lines	5	3.0	Poland
Uzhgorod-Beregovo - 2 lines	13	7.8	Hungary, Serbia, Bosnia
Hust - Satu-Mare	2	1.2	Romania
Ananyev-Tiraspol'-Izmail & Shebelinka-Izmail - 3 lines	26	15.6	Romania, Bulgaria, Greece, Turkey, Macedonia
Total via Ukraine:	**142**	**85.2**	
Via Belarus:			
Yamal-Europe (Torzhok-Kondratki-Frankfurt/Oder)	33	19.8	Poland, Germany, Netherlands, Belgium, UK
Kobrin-Brest	5	3.0	Poland
Total via Belarus:	**38**	**22.8**	
St. Petersburg-Finland - 2 lines	6	3.6	Finland
Blue Stream (design capacity)	16	9.6	Turkey (possible to Greece, Macedonia)
Nord Stream (design capacity)	55	33.0	Germany, France, Czech and other
TOTAL EXISTING EXPORT CAPACITY:	**257**	**154.2**	
New Projects (Final Investment Decision Stage):			
South Stream	63	37.8	Bulgaria, Serbia, Greece, Italy and other
Other New Projects:			
Nord Stream-3 and -4	55	33.0	Germany, France, Czech, UK and other
Yamal-Europe-2	15	9.0	Poland, Slovakia, Hungary and other
Sub-total new capacity:	**133**	79.8	
TOTAL PLANNED EXPORT CAPACITY:	**390**	234.0	
Guaranteed contracted exports for 2020-2025	**158**	94.8	

***GWE calculated on basis of 50% conversion efficiency**

POINTS OF VULNERABILITY

We consider here the effect of interrupting the fuel supply for one or more generating stations. First, we can observe that in the case of a nuclear generating station, the time between re-fueling is relatively long, giving nuclear generators the ability to run for times from 6 to 9 months without refueling; coal generators typically have reserves of coal on site and can operate for periods of 30 to 60 days.

Natural gas powered generators typically do not have fuel storage on site and a failure of a pipeline soon shuts the generator (although it is possible to run some of these plants on oil, this is generally not available on site in most cases). A coordinated attack on several pipelines can simultaneously shut a large fraction of the electric supply and also residential heating. The results will be especially severe during a winter attack.

An example of an accidental failure in a 0.914 m diameter pipeline operating at 6.6 MPa is shown below. In this case, the proximate cause of the accident was traced to leakage in the lateral transfer valves which are the means for bypassing sections of pipelines.

Photo 2. Downstream line-break section of Line 100-2 (top of photo)

Photo 3. Upstream line-break section of Line 100-2

Fig. 4: Failure of 0.9 m diameter pipeline.

As with transmission lines, gas pipelines are long—often in remote areas—and are only minimally protected. However, the repair time for gas pipeline is considerably longer, especially if the attack is at one or more of the compressor stations in the line. A compressor station is a complex collection of compressors and coolers as well as valves for transferring gas flow to redundant lines as in Figure 5.

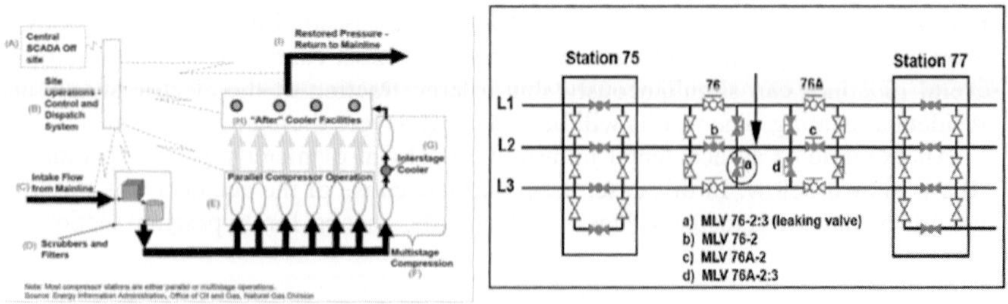

Fig. 5: Compressor schematic.

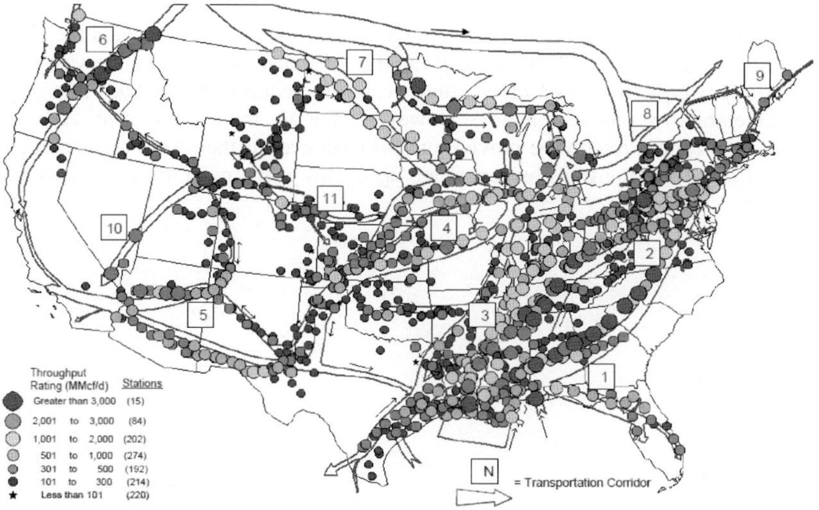

Fig. 6: Distribution of pipeline compressors by capacity in the U.S.

Replacement times for turbine compressors are long; times of many months to years, during which the capacity of a particular pipeline may be essentially zero. As a result of complaints from the public over noise, many of the natural gas powered compressors have been replaced with electrically powered compressors, linking the natural gas system even more tightly to the electric grid. Thus, unlike an attack on an isolated high voltage transmission line, repair times are long and, as we have seen earlier, switching to alternative pipelines is not an option.

An analogous situation exists with respect to large HV transformers which are used throughout the electrical grid to link grids and generating stations. As with compressors, repair times are very long and, due to the size of the units, transportation of fully assembled units is difficult. A recent article in the Wall Street Journal paraphrased a report from the U.S. Federal Energy Regulation Commission (FERC):

> "The U.S. could suffer a coast-to-coast blackout if saboteurs knocked out just nine of the country's 55,000 electric transmission substations on a scorching summer day, according to a previously unreported federal analysis. The study by the Federal Energy Regulatory Commission concluded that coordinated attacks in each of the nation's three separate electric systems could cause the entire power network to collapse, people familiar with the research said. A small number of the country's substations play an outsize role in keeping power flowing across large regions. The FERC analysis indicates that knocking out nine of those key substations could plunge the country into darkness for weeks, if not months."
>
> Rebecca Smith "Nation's Power Grid Vulnerable To Sabotage"
> *The Wall Street Journal* 13 March 2014.

The exact details of the particular substations were not made public, but a map of the U.S. grid shows the locations of both large transformers/substations and the tie points which link large networks. Only a small number of transformers carry the vast majority of electric power, and as with other transmission line equipment they are often isolated and vulnerable to physical attack. However, due to their large size and weight, replacements cannot readily be moved into position even if they were available. In fact, delivery times for large transformers range up to two years; most are essentially custom built units.

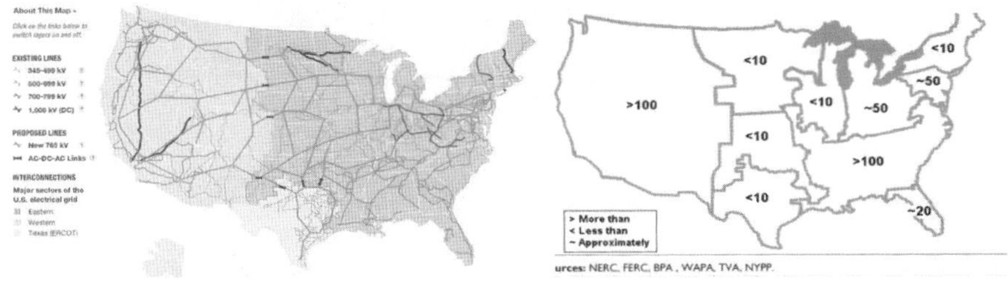

Fig. 7: Distribution of HV transformers by size.

SUMMARY AND SOME CONCLUSIONS

It appears that the target of opportunity for terrorists is more likely than not to be the peripheral parts of the energy network: gas supply lines and large transformers. Although these do not have the dramatic appeal of a (successful) attack on large facilities, they offer the advantage to the would-be terrorist of a rich number of potential targets and, if successful, the potential of creating wide spread disruption to a society with only a small investment.

Mitigating these attacks could be done in a number of ways. Obviously preventing attacks through surveillance and detection is the best way, but, given the dispersed energy network, a difficult one in practice. With respect to natural gas supply interruption, a fuel storage system at the generator or an energy storage system at the gas turbine could be considered. The energy storage approach is being considered for use in electric networks as way of dealing with the fluctuations on grids that may arise from the use of intermittent sources of energy such as wind or solar. If storage could also, mitigate the rapid loss of electric energy as the result of a terrorist act, then it may be more feasible. Transformers remain a problem due to their size and bulk. Programs are underway to design units which could be transportable by conventional means, perhaps broken up into smaller parts, and thus could be stored in reserve for emergencies.

REFERENCES

1. Future of the Electric Grid, MIT Energy Initiative, http://web.mit.edu/mitei/research/studies/the-electric-grid-2011.shtml.

2. "The Future of Natural Gas", MIT Energy Initiative Report (2011), http://mitei.mit.edu/system/files/NaturalGas_Report.pdf.
3. "Growing Concerns, Possible Solutions: The Interdependency of Natural Gas and Electricity Systems", MIT Energy Initiative Symposium Report (2013), http://mitei.mit.edu/system/files/2014-MITEI-Report-Growing-Concerns-Possible-Solutions.pdf.
4. N. Judson, "Interdependence of the Electricity Generation System and the Natural Gas System and Implications for Energy Security" Lincoln Laboratory Technical Report 1173(ESC-EN-HA-TR-2012-121) 15 May 2013.
5. East European Gas Analysis, http://www.eegas.com/.
6. For examples of this, see http://rrc-energy.com/gasmap or http://rrc-energy.com/electricitymap.
7. Pipeline Investigation Report PI1H0011, Transportation Safety Board of Canada 2011.
8. James Tobin, "Natural Gas Compressor Stations on the Interstate Pipeline Network: Developments since 1996", Energy Information Administration, Office of Oil and Gas, November 2007.
9. Rebecca Smith, "Nation's Power Grid Vulnerable To Sabotage", *Wall Street Journal* 13 March 2014.
10. U.S. Department of Energy, "Large Power Transformers and the U.S. Electric Grid", April 2014.
11. Paul W. Parfomak, "Physical Security of the U.S. Power Grid: High-Voltage Transformer Substations" U.S. Congressional Research Service Report R43604, June 17, 2014.
12. Rebecca Smith, "Grid Terror Attacks: U.S. Government is urged to take Steps for Protection", *Wall Street Journal* 6 July 2014 http://online.wsj.com/articles/grid-terror-attacks-u-s-government-is-urged-to-takes-steps-for-protection-1404672802.

MITIGATING HYBRID TERROR THREATS TO MEGA CRITICAL NATIONAL INFRASTRUCTURE

SALLY LEIVESLEY
Newrisk Limited London, World Federation of Scientists
London, UK

ABSTRACT

Cities are 'mega critical' national infrastructure (MCNI) which depend on energy and water systems. This paper identifies hybrid threats to the availability of energy and water and outlines a sustainable framework for mitigation which subsumes traditional resilience approaches. Sustainable systems maintain operational capability and can endure threats, changes in technology and environmental changes. Recommendations are made for proof of concept work to model a MCNI sustainability framework that prioritises security and control ahead of safety performance so that the energy and water systems remain operational under all-hazards and safety is delivered through systems that remain under control. Emerging risks are threats from terrorism, state actors and vulnerabilities to cyber-attacks on interconnected systems, and systems designed for convergence and integration of critical operations in smart cities, smart energy and water grids and intelligent structures.

INTRODUCTION

The criticality of water and energy infrastructure is a common factor within a number of the global threats emerging in 2014. The problem is one of complex systems in a world where the uncertainty of threats and consequences are being realised. A new method is therefore proposed for reducing catastrophic risks posed to water and energy, including the synergistic effects of loss of both resources to infrastructure such as nuclear power plants.

It is proposed that a new framework should be developed for national and city security, which is based on sustainability[1] and which subsumes resilience within the new framework by changing the concepts of patching, recovering systems, social processes and redundancy back-ups. The new model works through a design for sustainability and retrofit of legacy systems with measures that will increase endurance during impacts from multiple hazards.

A sustainability framework establishes:

1. A hierarchy of security over safety to provide controls that sustain operations that can deliver safety regardless of impacts. The concept applies generically to engineering design and physical, cyber and social processes.
2. A localised, distributed-threat monitoring process enabling isolation, workaround and continuity. Delivery can be through sensors, social environment monitoring, physical security or other cyber-space monitoring where information is reported on local failure points in critical components of a system. Examples would be sensors in sub stations of an energy grid, social

media monitoring of a crowd responding to a social incident that may become a riot or a CCTV camera identifying terrorist reconnaissance.

3. A process for horizon scanning of all hazards would exploit selected data from many locations: cyber, social, terrorism, nation conflict, human factors insiders or accidents and natural hazards. Governments use horizon scanning for policy to reduce uncertainty and integrate departmental inputs on emerging threats and developments to identify future conditions[2] but this has not readily translated into a method applied to energy and water systems.
4. A simple, integrated, 'system of systems' control of monitoring reports of critical functions can provide detection and early warning process and ensure that control of the system is retained for recovery. Exploitation of pre-set warnings from links to the few critical points of systems into an integrated strategic 'system of systems' would provide a pre-arranged catastrophic risk monitoring and management operation. It can handle legacy systems technology or wide ranging social activities—this needs proof of concept development.

It is hypothesised that the proposed sustainability framework overcomes both complexity and uncertainty by core consideration of critical functions and simple system of systems design to monitor the conditions of a system and deliver sustainability.

The objective is to reduce errors from assumptions inherent in probability analysis which does not reliably identify catastrophic failures. Probability analysis would however be retained within a sustainability framework as a method of managing general recurring systems failures (based on history and identifiable coupled events) i.e., below the catastrophic level.

Identification and management of catastrophic failures are the primary objectives of the sustainability framework and are managed through real time dynamic measures of the system and its environment and all the interfaces. Horizon scanning for emerging threats is a significant contribution along with systemic monitoring of the system's performance. The system of systems integrates key monitoring points or processes into a constant reporting to a system of system and cuts across many technologies that otherwise may solely be reporting to their own systems controller.

Real time monitoring of a system such as a nuclear power plant (NPP) or a nuclear fuel cycle or an energy distribution grid means all new risk factors are measured against the design strengths and vulnerabilities of the system and its environment. If a critical function in a system is open to failure from the factors being monitored then a system is at risk of a catastrophic event.

Recommendations for proof of concept development and modelling include the concepts of hybrid threat protection to cities, focusing on water and energy sustainability:

1. Proof of concept work by research and development of simulations of city critical functions using hybrid foreseeable threats: for example, terrorism, nation state conflict (including nuclear), cyber connectedness vulnerabilities, human factors such as insider threats, and natural hazards such as solar storms.

2. Stress testing to reduce the effects of unforeseeable threats at the catastrophic level—where the tests identify unexpected opportunities for pathways to critical functions. The stress testing quantifies whether risk controls will operate under complex scenarios where hybrid factors may come into action—for example, state sponsored terror and cyber-attacks on water and energy systems.

It is concluded that the new proposed framework will change the risk profile of city sustainability, especially MCNI such as smart cities at a time when emerging threats may otherwise cause considerable upheavals.

SUSTAINABILITY

The concept of sustainability re-visits the logic that has led to the evolution of concepts of resilience and safety in public policy and systems design.

Sustainability of physical and social systems, regardless of continuous attacks or threats, is a new concept with challenges for design, planning, policy and operational priorities. This subsumes the historical consideration of resilience which has assumptions based on safety, use of back-ups, cyber system 'patches' and many techniques in supply chain hardening to mitigate a catastrophic failure of critical functions. Resilience—regardless of many protection measures having been pre-established—was found to fail within four days in the Fukushima supply chain to global vehicle manufacturing following Japan's suffering of an earthquake, tsunami and NPP incident.[3] The concepts of back-up and supply chain do not contribute to catastrophic risk reduction as much as considerations of design for sustainability and a framework for security and control of the system which brings social and technological factors to bear on mitigation of the unexpected threats. Where a new design is not feasible and systems are already in operation, the introduction of a system of systems control with distributed monitoring acts as an 'add-on' to a sustainability framework.

The level of connectedness in the cyber environment has changed the risk profile of smart cities and many social processes. Therefore counter-intuitively, modern cities (critical to nation state performance) may be far more vulnerable to a range of catastrophic disturbances. The instability of cyberspace as a highly connected system means that in a sustainable framework, previous drivers such as safety would be secondary to the overall security of the system.[4] This is regardless of whether physical, social or cyber systems in cities are being assessed and it is an important differentiation to past assumptions because an ability to control any of these systems is essential to deliver safety and other outputs. Loss of control over a system through any number of pathways is the ultimate failure of a system under attack.

Control is therefore the top factor in sustainable systems design and without it safety cannot be properly addressed. This has been evident in SCADA[5] failures in systems caused by cyber threats such as Stuxnet and as-yet-undefined factors which may have caused the loss of systems control on board flight MH370 in March 2014 causing the airplane to be lost.

Currently there is preparation of a new IAEA Nuclear Security Series publication entitled *"Computer Security Controls for Instrumentation and Control Systems at*

Nuclear Facilities" which discusses computer security issues that should be addressed during the life cycle of digital Instrumentation & Control ("I&C") systems applied at nuclear facilities.[6] The potential damage to the control systems of NPPs including deception in instrumentation highlights the risk posed to cities proximate to plants or dependant on sustainable energy. In smart cities which have the greatest levels of connectedness and reliance on cyber systems, it is important that critical functions such as water and energy be sustained regardless of attacks on critical infrastructure (for example, in the instance of NPPs).

Once control is assumed to be at the top of the design hierarchy for physical, cyber or social processes then safety systems can be addressed and enabled because control of the system means that denial of access or deception operations or other events can be mitigated or their consequences reduced and the safety functions can perform as designed.

Below the top sustainable systems operation to deliver control, it is logical to place the resilience functions of preparation for and patching vulnerabilities, back-ups and testing (as applicable) and these remain as important contributors to the integrity of the system whether cyber, physical or social. The underlying driving assumption is that nations, cities, social processes and technology/infrastructure for both water and energy are not operating within calm and harmonious environments but may at any time have impacts on and changes to the status and integrity of the system—whether from terrorism, systems failures, cyber- attacks or natural disasters. The position of smart cities in contributing to national security through sustaining populations, economic production and international perceptions makes these cities MCNI in any nation.

EMERGING THREATS

Cities, particularly smart cities with high levels of sophisticated systems connections for integration and performance (cyber, social and physical) contribute to national productivity and as such are mega critical national infrastructure (MCNI). The term 'mega' refers to the fact that cities are a hub of many forms of CNI and act as unique drivers of the national economy and security for the population.

There are well-recognised threats to the supply and distribution of energy and water but emerging threats require the adoption of a wide horizon scanning process that informs the design, build, operation recovery and decommissioning. Some current threats are:

1. The 'Islamic State Caliphate' established in Iraq and Syria by the ISIS terrorist group, has taken control of energy and water supplies upon which Syria and Iraq are dependant. On August 3, 2014, for example, the Mosul Dam was seized giving ISIL a capacity to unleash a wall of water that would threaten many lives.[7]
2. Unsettled nuclear programmes in North Korea and Iran where a nuclear incident or conflict may affect regional and environmental security, including water resources.
3. Processes for water and energy production and distribution that are internet connected infrastructure carry vulnerabilities linked to cyber-space where the

high numbers of connections and efficiencies produced by integration of systems and convergence of data, along with wireless connections[8] create a plethora of points for unauthorised entry onto the network. Cyber-attacks or physical attacks on critical components of the system can have widespread consequences through the connections. Connections to cloud services[9] for efficient operation has provided additional levels of interconnection with other vulnerable systems and use of such systems is driven by business appetite and performance benefits along with a political will to create energy cost reductions for the public especially through smart grids for energy and water delivery to the public.

4. Social media that shapes public pressures on governments.
5. The lack of international standards and scientific monitoring of cyber space creates a catastrophic risk of uncontrollable variables as international agreements and laws traversing political and geographic boundaries are not available to monitor or restrict attacks on MCNI through cyberspace. Additionally codes of practice and education of city administrations for building regulation for structures housing or delivering MCNI (including communications) have not been developed for world-wide use. This means that horizon scanning of all hazards will have a risk matrix with cyber threats that are uncontrollable through law - this impacts on national security and public well-being.
6. Human factors leading to loss events include deterministic events whereby insiders in cyber operations act as unauthorised controllers of activities on the system and who may disrupt, deny or destroy critical system components by acting inside the security perimeters of an organisation. Accidental or negligent human factors are some of a number of factors known to contribute to opportunities for cyber space invasion by hackers and more serious organised criminal organisations and nation states.[10]

Appendix B provides examples of some of these threats to smart cities alongside assumptions underlying the proposed sustainable systems approach. Solutions to/mitigation against catastrophic risks are summarised in the Conclusions.

DISCUSSION

A common risk across the emerging threats is MCNI and this is recognised by national governments. One of the assumptions in a sustainable framework is that without knowing the catastrophic threats, it becomes possible to determine to a greater degree the vulnerabilities and to identify which controls are critical to maintaining control and sustaining the system's outputs.

At a practical level, city managers can develop a sustainable framework by bringing together critical functions such as water and energy in an integrated approach to risk management and applying broad horizon scanning on all hazards. Maintenance of city populations and national economies depends on water and energy which are, in turn, interdependent critical functions; therefore an initial focus on developing practical policy and procedures for sustainability will create a significant security protection for a city and its inhabitants.

The assumptions used in this analysis are that smart cities in comparison to less sophisticated urban areas are more vulnerable to energy and water disruption from deliberate or natural hazard incidents owing to a number of factors—including:

1. Interconnectedness of increasing inputs into the systems;
2. Interdependent systems between water and energy for production and distribution;
3. A lack of visibility of underlying systems and vulnerabilities in the highly integrated systems used for operating such as smart grids; and
4. Centralisation of functions, convergence of data and reliance on the internet and, at times, wireless transmission.

These risk factors have generally been recognised within systems engineering but they are also common to social interaction, political, physical security and many other processes that occur within cities.

Apart from energy, water is a top criticality for cities as populations will immediately search for water sources and move to new locations to find water or may refuse to remove themselves from danger if safe havens are without water.[11] Energy criticalities are a less immediate problem as whilst these may create population movements and disorder within a few days they may provide more time for intervention. The interdependency of energy production and water distribution (including clean water) is a risk factor underlining the criticality of access to both resources.

A sustainable approach to MCNI requires horizon scanning of multiple causes and multiple threats occurring simultaneously or in close proximity to create critical loss events which will reduce functionality and from which there may not be an easy recovery i.e., a catastrophic failure of water or an energy system or both.

A key to hybrid threat mitigation is the acknowledgement of critical failure points (such as those applied to energy and water) and then approaching the challenge of mitigation regardless of assumptions on future causation of failure. There may be hybrid causes leading to loss of access to energy or water supply in smart cities such as terrorism, state sponsored acts, catastrophic natural hazards such as solar storms[12] or the effects of technology and conflict such as IEMI[13] and HEMP.[14] The range of threats underlines the requirement for solutions based on a joined-up scientific effort across disciplines and pooling of solutions from international scientific capability.

ISIL AND OTHER TERROR THREATS

> *"The Islamic State of Iraq and the Levant [ISIL] declared itself an 'Islamic caliphate' on 29 June 2014, led by Caliph Ibrahim...ISIL spokesman Abu Muhammad al-Adnani said the caliphate will extend from the northern Syrian city of Aleppo to Diyala Province in Iraq. He described the establishment of the caliphate as 'the dream in all the Muslims' and 'the hope of all jihadists.' They...urged other radical Sunni groups to pledge their allegiance. ISIL announced that it should now be called 'The Islamic State' and declared its chief,*

Abu Bakr al-Baghdadi, as "the caliph" of the new state and "leader for Muslims everywhere..."[15]

Table 1 is a global snapshot of some of the MCNI terror attacks in June 2014 and it puts ISIL into context as a serious and immediate threat to the spread of terrorism to many countries owing to the clusters of foreign followers who have joined what is now the Islamic State - including young women urged to travel abroad to marry fighters and settle there. The leader Abu Bakr al-Baghdadi as Caliph of an 'Islamic State'[1] presents a new focal point for motivation and terror plots. A similar threat may exist in other countries such as parts of Afghanistan and Pakistan which are already linked in terror operations and flourishing within a fragile Afghan state.

ISIL has demonstrated a capability to capture and retain oil refineries and to threaten dams in Syria and Iraq. The Iraq Beiji refinery incident was one example evidenced in June 2014—this refinery incorporates a thermal power plant which serves Baghdad's power supply.[16] In Syria, ISIL was able to commence sale of product from some oilfields (using crude local production) with customers including the Syrian Government and for smuggling into Turkey.[17] Dams have also been used by ISIS as weapons and loss of control over dams is a strategic threat to Iraq. Reuters reported in April - '*Iraq insurgents use water as weapon after seizing dam*' when the ten Fallujah dam gates were closed flooding upstream to protect the insurgents against security forces. This affected irrigation and led to the decline of water levels in the Euphrates and *"electricity shortages in towns south of Baghdad, which rely on steam-powered generators that depend entirely on water levels...the power supply from Mussayab power station had decreased to 90 megawatts from 170 megawatts."*[18]

Table 1. June 2014 MCNI Terror Attacks- Examples.[19,20]

Iraq/ Syria	Caliphate announced removing state control from Syria and Iraq of land, cities, refineries, energy distribution, dams—ISIL
Pakistan	Jinnah International Airport, Karachi attack 5 hours, 8 June—militia style Tahreek-e-Taliban Pakistan (TTP) suicide and small arms, at least ten militants killed 23 people- similar to Mehran Naval Base attack 2011.
Yemen	Seiyun Airport attack 26 June, AQAP (Al-Qaeda in the Arabian Peninsula)
Afghanistan	Attacks on elections- 6 June Presidential candidate convoy; 14 June multiple attacks with IEDs, (Improvised Explosive Devices) small arms, rockets, mortars, grenades—150 attacks, 82 deaths—suspected Taliban militants.
Nigeria	IED close to Emab Plaza Abuja, 21 killed; 8 June military barracks attacked by suspected female suicide operative killing one person; 27 June VBIED (Vehicle Borne Improvised Explosive Device) outside mosque, Kano. Device of 13 charges in possibly improvised shaped charge formation. Boko Harum suspected.
Lebanon	Suicide bomber Beirut Hotel 25 June

Table 1 which is an extract of attacks in June from a database of some of the world's hotspots shows that MCNI of all types may be subject to terror attack. The lesson for sustainable infrastructure is that that any framework for sustainability should integrate

[1] http://www.aljazeera.com/news/middleeast/2014/07/islamic-state-caliph-lauds-iraq-rebellion-20147512574517772.html. Last accessed 20140715.

its horizon scanning information to apply across all critical functions. Water and energy are two functions of top importance to cities and national security.

Table 1 demonstrates that in June 2014 there were examples of significant threats to airports, military sites, shopping centres, places of worship, hotels and the infrastructure for elections (officials, voters, persons standing for office).

NUCLEAR THREATS

There are emerging threats of future use of nuclear weapons, attacks on nuclear weapons facilities and NPPs which could affect regional water and energy supplies along with population casualties and other consequences realised at the time of an incident. Water treatment failure and contamination of land and water resources (including seaways and seafood) would be a serious outcome if such threats eventuated. Some of the current threats are:

1. Israel's Dimona plant. *Militants from Hamas's Qassam Brigades said they had launched long-range M-75 rockets towards Dimona.*[21] The payload of these rockets which are a Palestinian variant of the Iranian Fajr-5 has been reported to be 80 kilograms.
2. Iran's nuclear programme facilities including the underground facility at Fordow and the NPP Bushehr, which if attacked with conventional weapons without warning could cause radiation release and contamination of seawater, fish stocks, water treatment units and agriculture. In the event of a more serious conflict (for example, if a nuclear weapon was used on an underground nuclear facility) the contamination may be more serious—although this would depend on the explosive performance of such weapons. Any high altitude burst of a nuclear weapon using HEMP (High Altitude Electromagnetic Effects) to disable communications and response to an attack may shadow across borders and waterways to affect electronic equipment that is not shielded outside the country under attack (depending on the yield of the weapon)
3. Iraq and Syrian university nuclear education programme resources have become an unexpected threat—Iraq has reported 40 kgs. of yellowcake now under control of ISIL: "*the takfiri terrorists also have taken some 40 kilograms of uranium compounds used for research at Mosul University*"[22] though this may be of commercial value for re-sale rather than for use in any form of contamination attack.
4. North Korea presents a threat to South Korea in the event of an attack on South Korean nuclear power plants or the use of a nuclear weapon as well as a threat to Japan which has recently been in the direction of mid-range rocket tests from North Korea. Japan has shut-down its nuclear facilities following Fukushima but there is a government objective to reopen facilities that have passed a safety case. Japanese nuclear containment and storage sites would be a problem if impacted with a North Korean weapon.

SMART CITY INTERCONNECTEDNESS

The trend for efficiency and political appetite means that smart grids in energy and water distribution are extending vulnerabilities down to the level of business and households. When systems, utilising cloud services and many different cyber systems configurations, link into a smart grid there is a two way risk posed to (i) the householders or business and, from that end point, (ii) back into the grid. The problem is one of business appetite for efficiency in distribution and for information analysis that brings more effective usage and this requires the use of internet connections for the operation of the grid. These may not be easily secured and the appetite for cheap and easily accessible cloud hosting services is driving business to expand and benefit from such efficiencies. Smart grids have strong political support for delivery of cheaper energy to the population.

Countries are responding with guidance and codes of practice to protect against some of the new threats entering MCNI through the fabric of structures supported by cyber systems. The IET (Institution of Engineering and Technology, previously known as the Institution of Electrical Engineers) in the UK has issued industry guidance on intelligent building interfaces with cyberspace and is due to issue a new code of practice for structures and cyber security.[23] Implementation of such guidance will mitigate some of the interconnectedness risks within design by systems engineers and software designers and the operational security designed by security managers. Insider threats and other human factors are now being considered within this new guidance including the high risks posed by mobile devices interacting with critical infrastructure.[24]

SOCIAL MEDIA

Social media has changed the speed of public pressure on governments as well as bringing benefits to the public. The drive towards participating in a social media driven activity may be as a result of similarity in beliefs, emotions, or interests in political or social issues or, alternatively, it may be driven by people who have objectives to learn, persuade, or project power and force.

However social media also introduces happenstance where connectedness creates new sets of information or access to communities of interest without active or deliberate intention by persons in a network. All of this contributes to uncertainty in social behaviour.

Social media is used by terror networks during crises to engage globally with sympathetic communities and to counteract government and/or security force statements to the media. Emotional ‘pull’ and the narrative ‘push’ are characteristic of the communications in these phases. ISIL used social media effectively in the build up to the announcement in June 2014 of the Caliphate on lands taken from Syria and Iraq. The global Caliphate’s intentions were clear in videos that were re-broadcasted by mainstream media of five young foreigners in Syria stating their intentions and recruiting people from their homelands with an appeal to those feeling guilty for not being involved in their fight and also to individuals who were depressed but might seek relief through joining them. This video was titled “*There is no life without jihad*”.[25] This was followed by a broadcast sermon from Abu Bakr al-Baghdadi on Friday 4th July, 2014 in the Great Mosque in Mosul which was a controlled and visually persuasive professionally produced video announcing he was Caliph Ibrahim and it contained a call for followers to use Ramadan as the time to attack 'for greater rewards'.

In September 2013 in the Westgate Mall attack in Nairobi the capacity of terrorists to use social media in real- time was demonstrated by the on-line activity throughout the incident. "*During this attack, Al-Shabaab created and used as many as six Twitter accounts, each time opening a new account very shortly after the active account was shut down for violating Twitter's Terms of Service....*"[26] A significant posting by them stated' "*...a great pleasure to have had you completely enthralled for more than 100 hours. What a wonderful audience you've been!*" Twitter account @xx was closed down.[27]

Although governments are using real time social media analysis to bring this medium into intelligence assessments, there has not been a successful technique for balancing and competing with the social media strategies adopted by terrorist and other anti-social groups, especially as government strategies focus primarily on blocking access to the accounts which is always a game of 'catch-up'.

UNGOVERNED CYBERSPACE

A strategic risk that is still unresolved by nations is the fact that cyberspace is ungoverned in a global context. Nations have attempted to create restrictions and internal laws but these handicap trade and operations of global corporations who then have to try to manage operations across conflicting requirements for communications and data.

The ungoverned cyberspace favours terrorists and organised crime. There is also a severe threat from nation states operating deceptively through proxies in order to avoid public contravention of international laws or detection in hostile acts of reconnaissance, cyber espionage or destruction of MCNI.

Examples of attacks that have occurred during low level conflicts are where one nation will experience loss of banking facilities, web sites representing governments and critical energy resources such as petrochemical resources. It was reported in the case of Saudi Aramco, October 23, 2012[28] that:

> "*A person with privileged access to the Saudi state-owned oil company's computers, unleashed a computer virus to initiate what is regarded as among the most destructive acts of computer sabotage on a company to date. The virus erased data on three-quarters of Aramco's corporate PCs—documents, spreadsheets, e-mails, files...the (U.S.) secretary of defense, Leon E. Panetta...cited the Aramco sabotage as 'a significant escalation of the cyber threat.'* Prior to that *'a computer virus that attacked Iranian oil companies and came to light in May. Iranian oil ministry officials have claimed that the Wiper software code forced them to cut Internet connections to their oil ministry, oil rigs and the Kharg Island oil terminal, a conduit for 80 percent of Iran's oil exports.*"

UNCERTAINTY AND COMPLEXITY

Inherent risk factors include the complexity of systems delivery in smart cities, growth in interconnectedness of energy sources into grids, political and business appetite for smart grids and connections to cloud services. These inherent risk factors create many layers of uncertainty, even before taking into account deterministic threats from terrorism, nation conflict or natural hazards.

A way to reduce difficulties associated with uncertainty and complexity is to consider a framework with a system of systems link retrofitted to CNI. It would run with underlying hypotheses of which factors indicate a catastrophic risk to the functioning of the specific infrastructure. A system of systems for catastrophic failure would use broad horizon scanning to identify threats, a dynamic analysis of the inherent systems weaknesses and weak interfaces.

The real-time operation means that understanding of residual risk of the system would dynamically change and indicate vulnerability to specific threats that would cause catastrophic failure. A system of systems side-steps a mass of threat data and chooses a few threads that indicate a critical function of the system may be at risk or an overwhelming threat may be in the environment. The constant analysis and updating of emerging threats provides this capability.

Assumptions are made in systems of systems of which combinations of factors will be important in predicting a catastrophic failure of the system and these assumptions also have to be dynamically updated. Although complex system of systems are provided for individual components of infrastructure there has not yet been development of catastrophic monitoring through an all-hazards system of system which looks for simple pathways between emerging hazards and vulnerabilities in the system where these may cause a failure of an irrecoverable part of the system i.e., a critical function. Some examples of key assumptions on threats and the systems of systems approaches are given in Appendix B.

It is recommended that because water and energy are critical and interdependent systems that are essential for sustaining cities, that a concept of broad horizon scanning and a simple system of systems approach be considered for new design in the security framework for the security of critical cities. The horizon scanning for all threats including conflict, terrorism, natural hazards, emerging cyber-attack processes, insider threats such as 'Snowden'[29] and other human factor failures.

This solution is relevant to complex systems and when there is an uncertainty of the attack surface and of the attack vectors. The objective is to minimise the residual risk by sustaining the operation of systems regardless of the source of the impact.

It requires a lifecycle design concept and should influence the underlying compilation of systems, human factors complexities, operating procedures, response procedures, catastrophic events, technology growth and decommissioning of structures or systems. Systems designers of cyber systems (whether hardware, software or networks) and policy makers in social, physical security or other systems can benefit from sustainability as a guiding concept as it widens the understanding of threats and creates different priorities for design so that security and maintenance of control will be paramount. This approach is inclusive of the dynamics of social media pressures which suddenly surface to impact on governments. So the policies in a sustainable framework would require social media analysis for public sentiments and a strategy to respond.

There is an uncertainty of complex systems supporting smart cities with water and energy which increases when hostile states may be assisting terror networks with arms, training, and tactics and targeting.

Hostile nation sponsors are limited by their own international agreements through the UN and regional ties but can subvert the intention of these agreements by support to terrorist networks (potentially in return for attacks on MCNI). Catastrophic events may happen which would not be possible within the less coordinated and less technically knowledgeable terror groups but once a hostile state supports the terror group there is a potential for a devastating success on MCNI. It is evident that weakened states are becoming playgrounds for multiple international interests by nations who use asymmetric groups to target and disrupt national leadership through successful attacks on MCNI thereby reducing quality of life and public perception of (and ultimately, the capability for) the government's capacity to provide security.

RECOMMENDATIONS

Proof of concept research is a way in which cities may test whether an integrated all-hazards approach within a framework for sustainability will deliver a reduction in the catastrophic risks to MCNI. Because water and energy are interlinked and are top priorities for city sustainability and key factors in potential instability of smart cities, it is recommended that energy and water sustainability be modelled together in establishing a proof of concept system.

It is further recommended that intervention by scientists in social media with trusted explanations and commentary (where practicable) may reduce the shock impacts currently seen in public panic, riots and influence of extreme groups on beliefs and behaviour in the society. This is important in times of crisis and especially crises such as water distribution disruption which has a fast effect on public stability and geographic movements.

Although there is a high uncertainty of risk factors cyberspace, the risk can be reduced if there is a move towards global governance. This would include drafting a discussion document on a governance framework for cyber space based on the IAEA model whereby:

1. A specialist technical international body is accepted by nations through an agreement
2. The recommendations, inspections and technical developments made through and/or undertaken by this body create governance and international reduction in frictions from different requirements in cyberspace operations.
3. It is to be noted that CERN has a profile that may fit the function of being an expert organisation and as this facility already has UN observer status[30] the acceptance by nations of its technical capability may be feasible if such an option were to be considered.

It is concluded that uncertainty and complexity in energy and water systems—whether technical or social—can be reduced by a framework for sustainability.

It is recommended that MCNI proof of concept research and stress testing be followed by evolution of codes of practice for systems design, initiatives in social media including scientific input at times of crisis, and a model for international governance of cyberspace.

The framework for sustainability would be a substantial change from the current processes of patching systems, patching social processes (such as laws on information intercept) and relying on 'back-ups' and testing supply chains. All of these assist general process performance but may not create sustainable systems operations under emerging threats. There would also be an attempt to tighten insider security which has occurred following the Snowden incident.

Sustainability in design and operation would not reduce the importance of probability analysis for risk assessments but would provide a primary structure encompassing the current activities. The design would focus on reducing the catastrophic level of failures through design of systems as well as providing policy and operational practices for sustainability in an uncertain and changeable environment.

Therefore, in light of the aforementioned discussion, a number of recommendations can be made (per Appendix A) for inclusion in a new Framework for City Sustainability for water and energy which should be developed and modelled, with a view to being adopted and applied in a manner which may serve to reduce and/or mitigate the risk of catastrophic events to MNCI.

APPENDIX A

RECOMMENDATIONS FOR A SUSTAINABILITY FRAMEWORK TO SECURE ENERGY AND WATER SUPPLY AND DISTRIBUTION FOR CITIES

- Design a city relevant system of systems to monitor threats to water and energy.
- Use broad horizon scanning functions.
- For smart cities critical to national security, design real time national horizon scanning with constant readjustment of the residual risk analysis on the national security risk register.
- Use fusion and exploitation of data drawn from Big Data sets (underlying hypotheses with runs for security threat identification); national intelligence inputs (police, security agencies, defence intelligence); and business intelligence including markets. Based on assumptions (see Appendix B) this would provide political, social, physical, cyber and communications pathways for reducing residual risk to MCNI of smart cities.
- A framework for control over national resources with water and energy recognised within the top risks for real time monitoring and risk reduction.
- Use of system of systems for integrated water/energy warnings and resilient core network[31] monitoring of distributed critical components such as energy sub-stations.
- Policy for this framework would sit within a national security strategy and be integrated with policies for all other MCNI.
- Lifecycle of design to decommissioning in cyber security requires capacity to manage changes in the system and its environment.[32] The IET in the UK has produced a guidance document on this 2013 and is producing a code of practice to outline lifecycle security management.[33]

- Horizon scanning based on all-hazards working directly with all interfaces in the cyber environment, physical security, political, social, human factor and other inputs to exploit key factors indicating a critical function risk that is emerging to threaten CNI.

APPENDIX B

This Appendix B outlines practical examples of the operation of a new sustainability framework based on assumptions that focus on system control rather than system resilience.

Assumptions underpinning a framework for sustainability and CNI threats are explained and related to solutions.

There are assumptions that can be tested and included in a system of systems approach to early identification of potential failure modes of water and energy in a smart city.

- Water/Energy interdependencies in smart cities are core factors in infrastructure sustainability.
- Cities as physical and social systems are a MCNI factor in national security—in a less developed form Mosul and other cities lost in Iraq to ISIS provided terrorist control over critical water and energy systems.
- Minor criticalities may lead to unexpected mega criticalities when there are connected systems and an escalation of consequences.
- All ancillary feeds would have, or be updated to have, a high level of security and same type of framework.
- Real time horizon scanning can replace risk update methods.
- Determination of residual risk of multiple risk factors in the environment—these risk factors are in interaction with security and safety controls at any point in time.
- Probability calculations based on historical data can be replaced by real time scanning to identify current and emerging risk profiles of MCNI.
- Connectedness of smart infrastructure in cities including water and energy are increasing pathways to cyber and physical attack and vulnerability to natural hazards from escalation of consequences.
- Security is the top of the hierarchy in design and operations—this places safety (which usually dominates design and operations decisions) into a secondary priority position. This hierarchy of security over safety would apply to design for cyber, physical and social systems and would put into effect a strategy for smart city sustainability. See technical discussion in NATO paper 2014, *'Novel Approaches to Cyber Defence'*.[34]
- Design for security by multidisciplinary teams is the top element of risk reduction and multidisciplinary analysis of all factors in horizon scanning creates broader risk matrix and options for security and safety.

Examples of Emerging Threats to Energy and Water for Smart Cities (as well as less developed cities)

- ISIL Caliphate and other conflict risks including proxy nation attacks.
- Nuclear Risks—solar storms, HEMP.
- Nuclear Risks—radiation contamination e.g., Media reports of city panic buying during shortage or contamination events. Tokyo was under pressure over several weeks following release of caesium and iodine from Fukushima explosions.[35] Fukushima went into dark controls from loss of power and without control systems safety could not be implemented as an emergency procedure to reduce explosion and radiation release. Chernobyl was a test of safety systems but it was loss of control that removed capability to institute safety systems.
- Technical risks—IEMI (Improvised Electro Magnetic Interference).
- Cyber interdependencies.
- Natural hazards.
- Human factors failures that are accidental.
- Physical connectedness of water and energy.
- Insider threats (e.g., Snowden) controlling cyber systems.

END NOTES

[1] Sustainability has many meanings and the definition within this conceptual paper sustainability is keeping an existing system operational and able to endure despite changes in the environment or impacts on the system. See Sandborn, P., and Myers, J., Designing Engineering Systems for Sustainability, *"The word 'sustain' has comes from the Latin sustenare meaning "to hold up" or to support, which has evolved to mean keeping something going or extending its duration.* http://www.enme.umd.edu/ESCML/Papers/SustainmentChapter.pdf.

[2] The UK Government has defined horizon scanning '*as an overall term for analysing the future: considering how emerging trends and developments might potentially affect current policy and practice. This helps policy makers in government to take a longer-term strategic approach'*, https://www.gov.uk/government/news/horizon-scanning-programme-a-new-approach-for-policy-making.

[3] Japan's earthquake, tsunami and Fukushima Daiichi plant incident created a supply chain crisis in some British vehicle manufacturing by 15 March, four days later. This was reported on the BBC and the author was interviewed on the severity and consequences on that date.

[4] @InProceedings{NATO-STO-SAS-106:2014, Kuntze, N. and Rudolph, C. and Leivesley, S. and Hanna, J. and Endicott-Popovsky, B., Novel Approaches to Cyber Defence: Strategic Analysis and Assessments in the Operating Environment with Case Study of Distributed Energy Systems SustainabilitySTO-MP-SAS-106 Analysis Support to Decision Making in Cyber Defence and Security , 2014, June NATO Science and Technology Organization.

[5] Supervisory Control and Data Access.

[6] http://www.iaea.org/NuclearPower/Meetings/2014/2014-09-03-09-05-TM-NPE.html. See also the information sheet prepared for this meeting- http://www.iaea.org/NuclearPower/Downloadable/Meetings/2014/2014-09-03-09-05-TM-NPE/TM47145_Garching_TM_Sept_2014_Info_Sheet_(Final).pdf.

[7] *Islamic State grabs Iraqi dam and oilfield in victory over Kurds*, http://in.reuters.com/article/2014/08/03/iraq-security-idINKBN0G305D20140803. Last accessed 20140803.

[8] IP-Enabled Sensor Networks, *International Journal of Distributor Sensor Networks*. http://www.hindawi.com/journals/ijdsn/si/452437/cfp/.

[9] A discussion on the difficulties for companies using cloud services and the limitations of encryption appears in a paper, *The Capabilities and Vulnerabilities of the Cloud*, McKelvey, N., and Houston-

Callaghan, E., *International Journal of Engineering and Technology* Volume 2 No. 6, June, 2012. http://iet-journals.org/archive/2012/june_vol_2_no_6/131471335299896.pdf. Last accessed 20140803.

[10] '*The 20 controls (and sub-controls) focus on various technical measures and activities, with the primary goal of helping organisations prioritise their efforts to defend against the current most common and damaging computer and network attacks. Outside of the technical realm, a comprehensive security program should also take into account many other areas of security, including overall policy, organisational structure, personnel issues and physical security.*' Critical Security Controls guidance, CPNI, Centre for Protection of National Infrastructure. http://www.cpni.gov.uk/advice/cyber/Critical-controls/.

[11] 20140725. An unusual example of the way in which water needs will shape behaviour emerged in an interview on media with a Palestinian woman told to leave her dwelling prior to bombing by Israel but who had returned with her children to face certain bombardment to destroy housing. She explained that the nominated location had no water for the children.

[12] Leivesley, S., Effects of solar storms (ElectroMagnetic Pulse EMP) on system critical control devices, CNI and solar storms: Integrating risk mitigation: Case examples - water and communications. http://conferences.theiet.org/solar-storms/programme/index.cfm.

[13] IEMI – Improvised electromagnetic interference.

[14] HEMP – High Altitude Electromagnetic Pulse.

[15] http://www.globalsecurity.org/military/world/para/aqi.htm. Last accessed 20140727

[16] http://wiki.openoil.net/index.php?title=Baiji_Refinery. Last accessed 20140715.

[17] http://www.businessweek.com/articles/2014-06-20/iraqs-isil-jihadists-want-an-oil-state-of-their-own-too, last accessed 20140715.

[18] http://www.reuters.com/article/2014/04/11/us-iraq-security-idUSBREA3A0Q020140411. Accessed 20140715.

[19] Examples drawn from data base material within TESS, News Briefs and Triton Afghanistan and Pakistan Supplement published June 2014, Allen-Vanguard International.

[20] Examples drawn from data base material within TESS, News Briefs and Triton Afghanistan and Pakistan Supplement published June 2014, Allen-Vanguard International.

[21] https://www.google.co.uk/?gws_rd=ssl#q=dimona+rocket+attack+gaza&tbm=nws. Last accessed 20140715.

[22] http://tehrantimes.com/oped/116970-iraq-wmd-redux-takfiri-terrorists-seize-chemicals-at-muthanna-and-uranium-at-mosul. Last accessed 20140715, also see- *Exclusive - Iraq tells U.N. that 'terrorist groups' seized nuclear material*, http://in.reuters.com/article/2014/07/09/us-iraq-security-nuclear-idINKBN0FE2L620140709.

[23] http://www.oxfordmartin.ox.ac.uk/cybersecurity/people/607, last accessed 20140803.

[24] In 2013 the IET with support from CPNI produced the first guidance on intelligent buildings connectedness risks and design management principles throughout the building lifecycle - Resilience and Cyber Security in the Built Environment (Author Hugh Boyes) is available online - http://www.cpni.gov.uk/documents/publications/2013/2013063-resilience_cyber_security_technology_built_environment.pdf?epslanguage=en-gb . The new code of practice Cyber Security in the Built Environment is in its final drafting process prior to publication by the IET (Author Hugh Boyes).

[25] http://www.nydailynews.com/news/world/australian-british-jihadists-urge-muslims-join-jihad-syria-iraq-article-1.1837418, last accessed 20140716.

[26] See http://allafrica.com/download/resource/main/main/idatcs/00071990:38c422c2c9053c64f0874 e4f6d2c55c1.pdf New York Police Department ANALYSIS OF AL-SHABAAB'S ATTACK AT THE WESTGATE MALL IN NAIROBI, KENYA.

[27] Triton Report for September, Allen Vanguard 2013.

[28] http://www.nytimes.com/2012/10/24/business/global/cyberattack-on-saudi-oil-firm-disquiets-us.html?pagewanted=all&_r=0 , last accessed 20140716.

[29] Edward Snowden: Leaks that exposed U.S. spy programme, http://www.bbc.co.uk/news/world-us-canada-23123964, last accessed 20140803.

[30] http://press.web.cern.ch/press-releases/2012/12/cern-granted-status-observer-united-nations-general-assembly, last accessed 20140727.

[31] Kuntze, N., Rudolph, C and Leivesley, S., Resilient Core Networks for Energy Distribution, Paper for IEEE, PES, July 27-31, 2014. Washington, DC.
[32] IET, 2013, http://www.cpni.gov.uk/documents/publications/2013/2013063-resilience_cyber_security_technology_built_environment.pdf?epslanguage=en-gb.
[33] IET, Draft Code in process, Cyber Security in the Built Environment, to be published by IET Standards.
[34] @InProceedings{NATO-STO-SAS-106:2014, Kuntze, N. and Rudolph, C. and Leivesley, S. and Hanna, J. and Endicott-Popovsky, B., Novel Approaches to Cyber Defence: Strategic Analysis and Assessments in the Operating Environment with Case Study of Distributed Energy Systems SustainabilitySTO-MP-SAS-106 Analysis Support to Decision Making in Cyber Defence and Security, 2014, June NATO Science and Technology Organization.
[35] http://enenews.com/japan-pm-fukushima-contamination-has-never-done-any-damage-to-tokyo-radioactive-water-at-plant-was-blocked-study-tokyo-was-contaminated-experts-radioactive-water-is-constantly-flowing-out. Last accessed 20140727.

COUNTERING (SEMI-) AUTONOMOUS TERRORISM

ANNETTE SOBEL
Critical Infrastructure Protection and Health Security Initiatives, Texas Tech University, Lubbock, TX, USA

INTRODUCTION

The paradigm of countering semi- or fully autonomous terrorism (CAT) has arisen out of lack of a unifying theory that may be translated from intelligence collection to actionable intelligence and effective countermeasures development. There is a range of operational scenarios that may be used to demonstrate sustainability and capacity of terrorist networks, alliances, and hubs; however, the failure and success of these organizational elements is non-linear and difficult to predict. This author believes the work of Tricia Bacon et al. (2015) may be relevant and translatable to real-world operational deployments and science and technology research investment.

DISCUSSION OF THE MODEL

The framework of understanding autonomous terrorism and strategies for disruption of effects-based operations is focused upon organizational learning theory and adaptation. According to Jackson et al. (2005) organizational learning is "...a process through which members of a group (systematically) acquire new knowledge that can be applied in strategic decision-making, tactical planning or design, and operational activities..." The success of organizational learning is predicated upon organizational adaptation, which refers to a group's "...ability to make adjustments as a result of environmental changes, goal-structure changes, or other changes..." (Fiol and Lyles, 1985), such as evolving mission requirements.

The potential real-world scenarios include: lone wolf; state supported; state-assisted; failure of state and opportunistic; exploitation; and naturally occurring (ideologically centered). The 'lone-wolf' attack has emerged as a predominant tactic, due to its dynamic, highly adaptive and responsive nature. Regardless, network-centric operations remain ubiquitous. For the purpose of this discussion, an alliance (network subtype) is defined as a formal or informal security and informational relationship among two or more terrorist entities involving some degree of coordination or synergy under certain conditions challenging mission success.

The concept of autonomous terrorism is predicated on the assumption of synergy of interrelationships. This concept does not imply full autonomy, with the exception of the lone wolf scenario. Other tactics require interdependencies for due to inadequate resources, skills, access, information, political landscape, and/or denial and deception.

Contextual operational considerations form the basis for the dynamics of (semi-) autonomous terrorism. Specifically, foreign capacity building and technology and human countermeasures, training and scaling for real world events, (largely opportunistic in nature), logistics, financing, and capability disruption and compromising counter-intelligence operations and information collection are contributing factors. These factors may all or individually drive the formation and dissipation of alliances and hubs.

Conventional wisdom dictates that initiation of alliances or hubs occurs with a convergence of common enemies or ideologies. This singular event may be ad hoc or post hoc events, and is non-linear, i.e., prediction is difficult, and analogous to weather prediction: precise but not accurate temporally or spatially. Hence, policy and long-term strategic intervention intended to interrupt or thwart alliance building is mostly ineffective. Alternative approaches to analyze and identify strategies to disrupt alliances are highly sought after.

As previously stated, dynamic alliances respond to common enemies and threats or ideological solidarity (i.e., a shared mental model or belief system). As a result, a unique identity may arise from this co-existence and resulting affinity. Identity affinity is dynamic, situation-dependent and continuously evolving to ensure survivability. Dynamic conditions render an alliance's longevity highly dependent upon trust within the organization, indigenous knowledge and resources, and the holistic threat environment. In summary, terrorist groups seek alliances that satisfy experiential, resource and adaptation capacity. Simplistically, the organization's needs may be categorized as resource-driven, training-dependent, and/or sanctuary-based.

OPERATIONAL SCENARIO CONSIDERATIONS

This author contends the most compelling reason to adopt a knowledge acquisition framework with characteristics of (semi-) autonomous terrorism is the 'internet of things' and the inherent vulnerabilities of a ubiquitous cyber infrastructure (UCI). UCI by its very nature excludes humans from some aspects of the decision-making the loop and constrain the human-computer interaction choke points into engineered systems. Hence, these constrained 'portals' present subtle anomaly detection cues. Since (semi-) autonomous terrorism relies heavily on rule-based decision-making, early warning features and subtle cues at critical chokepoints may be detected. This model emphasizes hybrid of mitigation strategies to include: big data analytics applied to vulnerability assessment, and human factors and social science techniques.

SUMMARY

The Permanent Monitoring Panel for Mitigation of Acts of Terrorism of the World Federation of Scientists is investigating a new path for strategic policy and science and technology investment to address the root case analysis of terrorist networks based on features of organizational structure and function. This approach appears to have merit based upon metrics of organizational survivability traceable to adaptation. This framework also offers great promise in the assessment of terrorism alliance vulnerabilities in megacities.

Critical Infrastructure presents ubiquitous vulnerabilities due to high reliance upon distributed engineering systems, notably, energy systems and distribution networks, big data analytics and cyber-infrastructure attack points. Examples of systems of interest include: nuclear power plants; intelligent micro-grids; distributed water systems; election voting platforms; hospital laboratory and electronic patient records (cloud-based, health informatics as a whole); and financial networks. These systems represent highly cyber-dependent critical infrastructure, representing important cases for data collection, simulation, assessment, and analysis of the validity and extensibility of this framework.

Validated operational scenarios (reviewed by PMP subject matter experts) emphasize superimposing a cyber attack on these engineered 'super' systems. This approach enables simultaneous, simulated network analysis at the human-centered and engineered system chokepoints, thus allowing for an accelerated analytic process and early 'triage' of actionable information. This approach is desirable due to its simultaneous scalability, and representation of multiple chokepoints and vulnerable avenues for attack. This paper describes the emerging science of big data analytics applied to counter-terrorism and non-proliferation, and is enhanced through use of a complex systems approach.

This paper describes an approach that analyzes multiple terrorist 'entity' intersections and nonlinear, measurable effects and features. The data employed is representative, but realistic from a deployability and scientific perspective, based upon current system capabilities. Notionally, this framework emphasizes an approach to risk mitigation that visualizes attack probabilities as semi- and fully autonomously driven by vulnerability, access and motivation. The author has further delineated the operational aspects of this approach to augment the initial validation completed by work of the team led by Dr. Tricia Bacon at American University.

REFERENCES

1. Andrews, K.R. (1980) The concept of corporate strategy, Homewood, IL: Irwin.
2. Argyris, C., & Schon, D.A. (1978) Organizational learning, Reading, MA: Addison-Wesley.
3. Bacon, T., (2015) Strange Bedfellows or Brothers-in-Arms: Why Terrorist Groups Ally, Unpublished Briefing, American University, Washington, D.C.
4. Barnard, C.I. (1938) Functions of the executive. Cambridge, MA: Harvard University.
5. Bartunek, J.M. (1984) "Changing interpretive schemes and organizational restructuring: The example of a religious order." *Administrative Science Quarterly*. 29:355-372.
6. C. Marlene Fiol and Marjorie A. Lyles (Oct., 1985) "Organizational Learning." *The Academy of Management Review*. Vol. 10, No. 4, pp. 803-813.
7. Jackson, B.A. et al. (2005) Aptitude for Destruction: Case studies of organizational learning in five terrorist groups. Rand Corp.

UNMANNED AERIAL VEHICLE: TOOL FOR FIRST RESPONDERS AND TERRORISTS[1]

ALAN MOORE
Research Institute, Von Braun Research Hall, University of Alabama in Huntsville, Huntsville, AL, USA

FRIEDRICH STEINHÄUSLER
Physics and Biophysics Division, University of Salzburg
Salzburg, Austria

INTRODUCTION

The concept of an unmanned aerial platform dates back to the mid-1800s when Austrians sent off unmanned, bomb-filled balloons as a way to attack Venice. The drones we see today had their beginning in the early 1900s.[2] The design of drones was accelerated during the course of World War I and II. Advances in technology, digital circuits and miniature components during the past two decades have further accelerated drone design, capability and utility. The conflicts in Iraq and Afghanistan demonstrated the utility of very small drones or unmanned aerial vehicles to conduct surveillance and attack operations. Small-unmanned platforms have been shown to be of commercial interest to those in the oil pipeline business, farmers, First Responders and others. The small unmanned platforms, by their very existence, can also be used by terrorists in what is discussed here as a hypothetical terrorist group (*Threat Cell)* to disrupt civilian and government activities.

To discuss possible applications of UAVs, this paper has been organized into two parts. Part One addresses the potential use and employment of very small Unmanned Aerial Vehicles (UAVs) or Unmanned Aerial Systems (UASs)[3] by a non-state actor or terrorist organization to accomplish a specific objective. Part Two of the paper highlights the advantages that can be realized by the use of UAVs by local First Responders in the assessment, establishment of situational awareness and implementation of a plan of action to mitigate an ongoing event. For ease of discussion the term UAV as used in this paper addresses both UAV and UAS.

UAVs come in a variety of sizes from a very small platform that would fit in one's hand (i.e., *Black Hornet*), to a very large platform that is capable of operations at high altitudes with a flight duration of several hours or even days (i.e., *Global Hawk*). Due to UAV system cost, size, operational and support considerations, it is the smaller UAVs that can be expected to provide the basis for a terrorist threat. These same size platforms can also provide the First Responders with platforms of high utility at relatively low cost. The discussion in Parts One and Two of this paper will be limited to what is

---

[1] The research leading to these results has received funding from the European Union Seventh Framework Programme (FP7/2007-2013) under grant agreement n∘261817, the BRIDGE project (www.bridgeproject.eu).
[2] Unmanned Aerial Vehicle, Wikipedia.
[3] UAVs are powered aerial vehicles sustained in flight by aerodynamic lift over most of their flight path and guided without an onboard crew. The term UAS (Unmanned Aircraft System) includes the aerial platform itself together with the ground control station, data link and support equipment.

generally categorized as Class 1 UAVs (Micro and Mini) as shown below. The small physical size, ease of assembly and employment, and relatively low cost are attributes that make the Micro and Mini UAVs ideal platforms for use by someone wanting to create an incident or disruption of an event. The same attributes make them prime candidates for use by First Responders.

Today there are over sixty nations in the world developing and producing one or more UAVs of various sizes, propulsion and guidance methodologies. National retail outlets are selling very capable Micro UAVs with a small ground control device and two cameras for under $400.00. Primary schools, colleges and Universities are teaching classes involving the design, construction and operation of both ground and aerial UAVs, many of which may include GPS guidance.

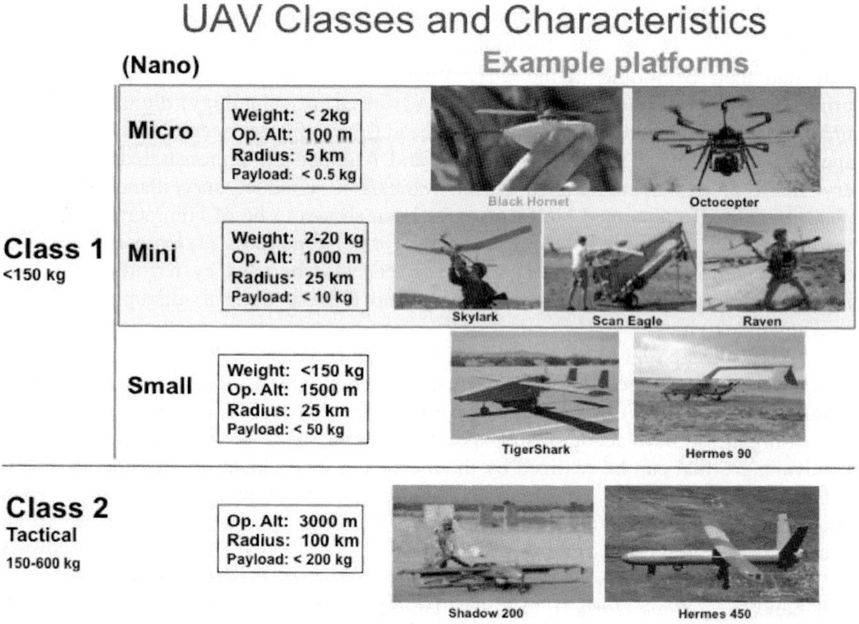

COMMERCIAL VERSUS TERRORIST UAV

In recent years there has been considerable interest and attention given to the utility of employing UAVs for commercial use. Much of this interest has resulted from the information and understanding gained by observing the employment of UAVs in Iraq and Afghanistan. As the military use of UAVs expanded over the past decade a number of nations and equipment suppliers within nations developed a multitude of different size and capability based UAVs. Due to platform weight and performance constraints, Micro and Mini UAVs are constructed with very similar materials and equipped with similar propulsion and guidance systems that add to the difficulty of distinguishing between a commercial use platform and a platform being employed for other objectives.

The expanding use of UAVs in domestic commercial applications and the pending approval by national aviation authorities for the use of UAVs in designated airspace is resulting in a blurring of the differences between commercial and military Micro and Mini UAVs. The expanded availability of Micro and Mini UAVs at local department stores results in capable platforms being available to anyone for very little money. These platforms with their integrated cameras and simple ground station can be easily modified to incorporate other payloads such as explosives and CBRN elements. Modifying the department store platforms to incorporate GPS guidance by high school and college students is known to have occurred in the United States. The expanded use and availability of commercial UAVs of various types, with a mix of components significantly helps to camouflage the threat.

The Micro and Mini UAVs for the most part require almost no logistics support base. As the platforms are small they can be easily disassembled, transported to the launch site and reassembled using hand tools. For the most part no launch system or geographical area is required for the employment of the UAV. This is particularly true for threat UAVs, as most missions will be "one-way".

A terrorist or a single person wanting to make a political statement or disrupt an event could easily obtain a Micro or Mini UAV, modify the payload and employ the system with little or no training from a small field, a roof top, apartment balcony or terrace, or outside of a car or small truck.

TYPE MISSION OBJECTIVE/PAYLOADS

The threat organization may want to make a political statement, undermine local government security, disrupt the national economy or create a diversion involving First Responders that would set the stage for an attack on the primary objective, i.e. bank robbery, kidnapping, or murder of a government official. A payload tailored to the desired effects to be achieved can be integrated into the platform. If the threat's objective is to create panic or disrupt a high profile international event, the Micro and Mini UAV provides an appropriate platform. The payload can be tailored to the effects desired. A number of different types of payloads can be integrated into the Micro and Mini UAV. The most common would be an explosive device that would create a measureable effect on the intended target. Other potential payloads can be chemical, biological or radiological dispersal devices, depending on the timing of the effects desired. For example, if a UAV dispersed a white flour substance or crashed into a group of people and dispensed this same white substance, panic would ensue. No one would be in the position to evaluate or categorize the substance for several hours. If such an attack took place at a high profile international event, the panic that ensued would be captured by the media and transmitted around the world in real time. A follow-up media story a week or ten days later on what the UAV was really carrying would probably never be broadcast. The threat UAV could be equipped with surveillance devices, but the probability of this application is considered small.

MISSION PERFORMANCE PERIMETERS

The wide range and capability of platforms in the Micro and Mini classes of UAVs enables the threat organization to tailor the platform, guidance, propulsion and the desired effects to be achieved to the specific mission. Different missions or flight durations may require different platforms. A review of the "UAV Classes and Characteristics" above indicates there a wide range of platform alternatives, provided the payload weight can be kept within the stated boundaries.

As noted earlier, most if not all threat missions will be one-way missions. This greatly reduces the probability of detection of the UAV operators as well as the need to identify a secure landing zone.

MISSION PLANNING

Mission planning for the threat employing UAVs takes some time and must consider a number of factors based on the proposed launch site and the location of the target. Obviously a flight route planned inside a built up city is more complex than outside a city where there are fewer tall buildings and communications towers. In selecting the launch site, the time of flight and anticipated weather conditions are of importance. Long flight times increase the probability of navigation errors and increased detection. Heavy rain or high winds will preclude the launch of a UAV, as it will be destroyed shortly after launch by the severe weather conditions.

Of key important to the threat UAV operator is survivability, at least up to and including mission launch. This may require the pre-positioning of the UAV components at the designated launch site well before the time of launch.

The requirement of the threat to launch a UAV during daylight or at night will directly impact the type of platform guidance. A short-range daytime launch would enable line-of-sight control of the platform by an iPad or small computer. Another option is the flying of pre-programmed waypoints using GPS. This guidance option would also be applicable for UAV night operations. Care must be taken to ensure the UAV control and data link frequencies, if required, will not be jammed by normal communications transmissions in the launch area, enroute to the target or in the target area.

THEORETICAL THREAT SCENARIO

The following example serves to integrate the various threat UAV employment considerations identified above into a single theoretical scenario that could take place today.

A high profile international event will be taking place in the country within the next six months. Construction of the venue is taking place at this time and will conclude within a couple of days of the initial event. The primary venue is a large open-air arena that is expected to seat 80,000 to 100,000 persons. A large four-sided television presentation screen will be suspended in the middle of the venue, projecting pictures of the events taking place within the venue. The initial event will take place at night and will be covered by worldwide media.

A terrorist cell (*Threat Cell*) has decided to disrupt the opening event to demonstrate their ability to undermine state security and embarrass the host nation. The secondary objective is to create panic among the participants resulting in the possible cancellation of the international event. It is the intent of the Threat Cell to survive the initial attack and escape to execute a similar type attack in the near future. To preclude pre-mission execution detection, the Threat Cell has prepositioned the UAV platforms and other required equipment weeks prior to the establishment of the security cordon around the venue. The aerial platforms have been placed in rented apartments. To enhance mission success the Threat Cell has selected and equipped three different UAVs that will be launched and will follow three different independent flight paths to arrive at the venue at approximately the same time.

The day of the initial international ceremony the local residents that make up the Threat Cell and have lived in the area for over a year, assemble the UAVs and check out the propulsion and guidance systems. The Threat Cell has determined their objectives can be achieved if they deploy a UAV payload of standard commercially available white baking flour, as it will take the authorities several hours to determine the composition and lethality of the payload. The attack will cause the participants to panic with the full impact of the terrorist event being broadcast locally on the large stadium screen and internationally by the media covering the event.

Today the advantage of employing UAVs in a terrorist mode rests with the offense. To defend against the Micro and Mini UAVs, many of which are identical to established in-service commercial platforms, is very difficult. The following are a number of issues that have to be addressed to counter a potential hostile UAV threat.

- Detection, classification and track.
- The development of engagement decision logic tailored to the event to be protected, taking into account potential collateral damage.
- The design, development and fielding of cost effective effectors to counter the UAV threat, while not creating unacceptable collateral damage if and when employed.
- The consideration of "open facility" versus "closed facility" venues for the conduct of high profile international events.

UAV AS TOOL FOR FIRST RESPONDERS

The Need for Advanced Situational Awareness

Situational Awareness in disaster management is the heightened consciousness or cognizance of currently developing events on a disastrous scene. Some considers situational awareness the single most important safety item at any emergency incident. First responders are often thrown into unfamiliar locations or situations, where a major emergency can be unfolding fast. Maintaining good situational awareness is crucial to their safety and the outcome of the situation. Knowledge of the events occurring in the surroundings of first responders operating on scene is essential, in particular a large-scale emergency situation to protect the first responders involved as well as the persons in the vicinity. Time pressure, inadequate and possibly contradicting information leading to confusion, unfamiliar and possibly hostile surroundings, multiple simultaneous hazardous incidents—all this impedes on the first responders′ capability to maintain quality situational awareness during an emergency.

Currently a significant gap exists between the need to obtain real-time information about an unfolding crisis, identification of the needs to master the crisis, the resources available and intact and that need to be brought in from the outside, and the actually provided time-delayed and frequently incomplete data on particular hazards as first responders approach the scene of a large scale incident.

It is important that situational awareness does not concentrate on generating an 'overview' of an incident but instead offers a dynamic combination of perspectives created, assimilated and filtered from a number of sources to best support those involved in the response effort. As such, situational awareness needs to assist the resource manager and the incident commander by addressing inter alia the issues: (1) Increased situational awareness during an emergency by supplying real-time information; (2) Filtering of information which distracts the first responder from maintaining quality situational awareness; (3) Maintaining quality situational awareness throughout the whole cycle of managing a crisis and its forever changing boundary conditions.

Optimal Conceptual Design

Situational awareness represents one of the most important components of proactive risk management. Ideally, a first responder on scene should be able to maintain constant situational awareness, updated in real-time by an uninterrupted flow of information. However, at present decisions on high risk and high stress operation are frequently based on fuzzy data associated with significant uncertainties, obtained from multiple sources of largely differing quality.

The optimal design of such a system needs to fill this gap by providing the incident commander with real-time data on the extent of the disaster and its consequences on the population and the environment. The optimal conceptual design for advanced situational awareness consists of a three component-system:

- *Unmanned aerial vehicle* (UAV) quadro-, hexa-, or octocopter (depending on mission and associated payload) as flying platform, equipped with visual camera (VIS), infrared camera (IR) and multiple sensors.
- *Expert System* for filtering and evaluating incoming data from UAV, its cameras and sensors.
- *2D/3D computer-based modelling* of structures or plume, using either data from UAV or from database incorporated into the Expert System.

The output from these three system components provides the following information:

(a) Extent and architectural details of physical damage of structures on scene;
(b) Number of victims outdoors on scene;
(c) Magnitude of fires on scene;
(d) Environmental hazards due to uncontrolled release of toxic substances;
(e) Propagation of toxic liquid and gaseous effluents as a function of time (e.g., plume dispersion).

(f) Comparison of actual environmental measurement data with exposure limits and recommendations.

The following sections describe the different components of such an optimal design concept for a UAV dedicated to deployment by First Responders.

UAV for First Responders (UAV-FR)

A UAV-FR is optionally under manual control or on a pre-programmed automatic flight pattern. Typically, such a flying platform houses a number of electric motors, a Global Positioning System (GPS), two different cameras (VIS and IR), on-board computer and several solid state sensors. Uninterrupted flight duration is mostly in the range of 20 to 30 minutes per battery pack with a 2.5 kg payload. If fitted with a GPS-device, this additional feature enables the device to follow a predefined flight pattern, using several predefined GPS-waypoints. Also, upon reaching the reserve capacities of the battery packs, the UAV-FR will automatically switch to homing mode, i.e., returning to the base station without the help of the ground-based operator or induce a soft emergency landing operation, if the homing mode for whatever reason is no longer possible to achieve.

Cameras and Sensors

Ideally a UAV-FR is equipped with two on-board lightweight devices: visual (VIS)-camera and infrared (IR)-camera. Both cameras are attached to a remotely controlled flexor, enabling the ground-based operator to change the camera-angle within a range of 90°. The horizontal angle can be adjusted using the onboard set of rotors.

The two cameras generate digital output signals that are fed into a data-processing unit aboard the UAV-FR, equipped with an ultra mini PC. The software installed on this onboard PC handles the incoming video stream as a live feed or as snapshots in preselected time intervals via a downlink transponder to the ground-based operator. The onboard-storage medium enables recording of the continuous flow of VIS-IR-image data for later use, and at the same time acts as a flight-recording system by storing flight-relevant parameters along with the GPS-coordinates onto the drive.

The UAV-FR system also houses environmental sensors. Such solid state sensors enable the monitoring of the various environmental parameters during the flight:

- Electronic sniffers, e.g., CO, CO_2, C_2H_6, CH_4, N_2O, O_2, H_2S, SO_2, NH_3, NO_2, Cl_2, ClO_2, PH_3, HCN, C_4H_{10} (butane), CH_4 (methane), C_3H_8 (propane);
- Environmental radiation, using a solid-state radiation detector for gamma- and high-energy beta-radiation.
- Aerosol number concentration measured with an optical particle counter
- Aerosol mass measurement for a first approximation of the characterisation of the aerosol.
- Standard meteorological data, such as ambient air temperature (T), relative humidity (rh), and atmospheric pressure (p).

The aerosol inventory gathered by the UAV-FR is helpful in determining a safe hovering-distance away from hazardous plume concentrations, which could potentially damage vital parts of the aviation elements (rotors, motors). All on-board sensors are connected to the on-board storage-management system to generate GPS-supported datasets of the relevant sensor input for either (a) immediate or (b) later use. Visualization of the data is done graphically on display units in order to facilitate fast recognition of hazard threshold levels.

Thereby, the system provides the incident commander on the ground with a visual presentation of:

- Video/photo (VIS) images supplemented with IR-images.
- Environmental data (gases, radiation, aerosol characteristics, T, rh, and p).

EXPERT System

An EXPERT System analyzes automatically incoming data from the UAV-FR. Incoming data is rated and combined with expert recommendations, such as issued by international organizations. The aim is to use data on environmental measurements collected by the UAV-FR for automatic pre-analysis and data reduction in order to assist in improving situation awareness. In addition to this numerical information such an EXPERT System recommends actions by the first responders, based on national legislation, international recommendation and international databases (e.g., on chemical toxicity).

This enables the user—automatically or upon individual demand—to compare the measured atmospheric data with information contained in different international databases. In case the measured data indicate hazardous substances, or exceed legal limits or recommended values, a visual and audible alarm is triggered. Furthermore, corrective actions (e.g., evacuation or sheltering) can be recommended. Ultimately, this information is passed on to the incident commander, assisting in the acceleration of the decision making process.

Examples for practical application on scene are listed below:

- *Infrared data* from the UAV-FR serve to locate zones that feature increased temperatures as compared to their surroundings. These zones are found by using the IR camera.
- *Data from the VIS camera* provide additional visual information, correlating the optical information with location details of the site by using GPS data.
- *Radiological data* from the UAV-FR determine alternatively the location of a radiological source, its activity or assess whether the source is dispersed into the air. Data pairs on activity value/GPS position enable the user to assess a radiological source.
- *Chemical data* from the UVA-FR provide information on the release of hazardous gases into the atmosphere. Different gas sensors can be installed on the flying platform.
- *Aerosol data* assist in the selection of adequate respiratory protective equipment for the First Responders by measuring the particle size distribution and determining the hazard potential of the aerosol.

Since all the above listed data sets are executed automatically, incoming (raw) data is reduced to its crucial content, processed, analyzed automatically and then passed on to the incident commander, providing decision support.

Modeling Module

Data from the flying platform and the EXPERT System are transmitted to the *Modelling Module (MM)*. The MM creates computer-based models of the incident site, and of the plume in case of an uncontrolled release of gases and aerosols from the incident site. Also, it is possible to draw on the pre-programmed generic models of reality-based structures created in advance (e.g., library of critical infrastructure (CI)). Thereby, the user has the option to either use in a first approach the most fitting CI library model as is, or subsequently approximate the CI generic model to the actual conditions on scene, using the visual data transmitted by the UAV-FR. Such a modelling module enables the user to assess the physical damage to buildings and the associated health hazards due to collapsing structures to people after an explosion and/or uncontrolled release of toxic substances.

Furthermore, the user can predict the dispersion of toxic gases and radioactive substances resulting from an uncontrolled release, causing hazardous plumes by using a 2D plume model.

Also, the number of injured persons and their type of injuries to be expected from an explosion, the number of radioactively contaminated persons as well as the number of dead can be estimated by using the UAV-FR data for modelling the incident.

SUMMARY

Today a threat organization could employ Micro or Mini UAVs with difference guidance schemes and payloads over a sizeable geographical area to further their cause. To defend against such a threat will require the development and fielding of sensors and effectors to counter the low, slow and small UAV platforms, in such as was to minimize collateral damage. The near-term gap in defensive capability places high profile international and national events at potential risk from UAV attack.

In case of UAVs used by first responders, the different components of the UAV-FR assist the incident commander in obtaining a more accurate, real-time update on the incident, strengthening the capabilities of managing resources and reducing the risk for the first responder on scene. Ultimately such a system enhances the efficiency of the management of a crisis.

TEXAS PROGRAMS AND TECHNOLOGIES FOR UAS

ROBERT DUNCAN
Vice President Research, Texas Tech University
Lubbock, TX, USA

LUIS CIFUENTES AND JERRY HENDRIX
Texas A&M University
Corpus Christi, TX, USA

BEI BEI REN, DAVID WEINDORF, STEVE PRESLEY, ANNIE SOBEL, KAY TINDLE, AND DARIO SOLIS
Texas Tech University
Lubbock, TX, USA

Lone Star UAS

LSUAS Headquartered at Texas A&M / Corpus Christi (TAMU-CC)

11 Test Ranges across S/SW Texas, both coastal and inland

Various COAs with specific operational and altitude restrictions

Padre Island COA: 517 NM^2, altitude up to 3,000 ft., since 2011

http://lsuasc.tamucc.edu

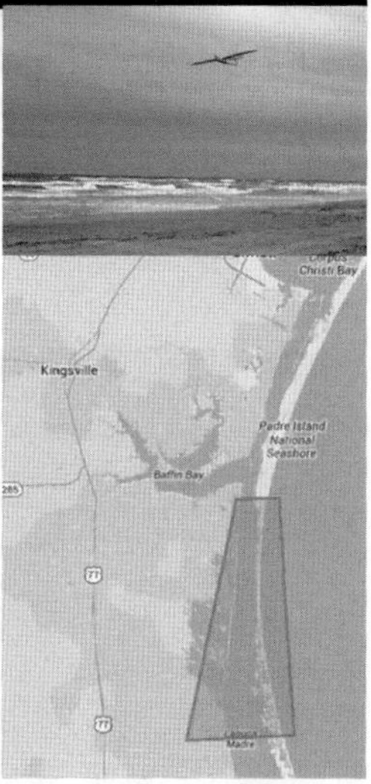

Lone Star UAS (Cont.)

Close support of multiple UAV / UAS operations

Many different UAVs and data capture applications

Mission Control Room →

Center for Autonomous Vehicles and Sensor Systems – Pegasus UAV

TEXAS TECH UNIVERSITY™

UAVs Research at Ren's Lab

Dr. Beibei (Helen) Ren
Assistant Professor
Department of Mechanical Engineering
Texas Tech University
Phone: +1 806-834-6692
Email: beibei.ren@ttu.edu

Research Overview at Ren's Lab (*Dynamic Systems & Control Lab)*

Control Theory

- *Robust Control*
- *Adaptive Control*
- *Neural Networks*
- *Boundary Control of Systems Modeled by Partial Differential Equations*
- *Real-time Optimization Using Extremum Seeking*

Control Application

- *Micro-electro-mechanical Systems (MEMS)*
- *Unmanned Aerial Vehicles (UAVs)*
- *Renewable Energy*

http://realinf.com/

Ren's Research Thrust on UAVs

Research Goal:

Pushing the boundaries of UAVs control in extreme weather and uncertain environment conditions to accomplish complex tasks

Challenges faced in real applications of UAVs and our solutions:

- Model uncertainties and external disturbances, e.g., large winds, or uncertain landing platforms etc.
  => Robust motion control and path planning algorithms
- Limited flight time => Solar energy powered UAVs
- Autonomous charging =>Wireless charging technique

Applications:

- Critical infrastructure health monitoring (e.g., wind turbine, oil/gas pipeline)
- Precision agriculture (e.g., 3D scanning of cotton plants)

Quadrotor motion control

Objective:

Precise position control in the presence of model uncertainties and external disturbance

Challenges:

- Robust control algorithm development for both attitude and position
- Precise position information acquisition and data fusion

Quadrotor Sensing Configuration

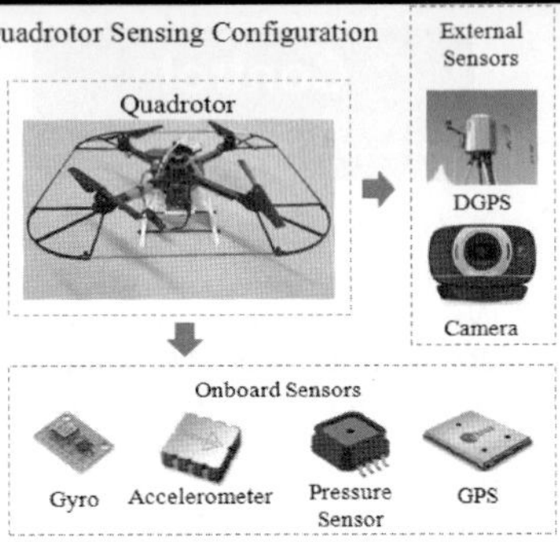

Quadrotor position control with a single external camera has been achieved at the current stage

- Logitech C615 HD Webcam
- Asctec Hummingbird Quadrotor

Quadrotor Control Scheme

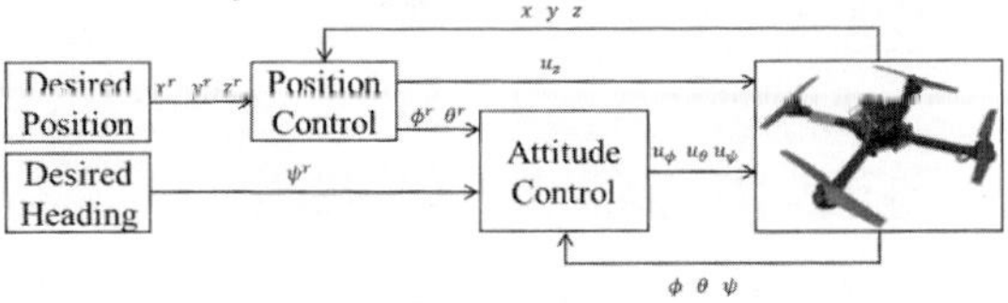

Solar energy powered UAV

Objective: to extend flight time

- Thin-film solar cell is adopted to reduce extra weight
- Continual power supply

Challenges:

- Maximum power capture from the sun light
- Effective deployment of thin-film solar cell

Image courtesy: http://www.gizmag.com

A boost DCDC converter based solar power system with maximum power point tracking (MPPT) algorithm was developed at the current stages

- 550mm x 680mm area solar panel
- 24V DC battery
- TMS320F28335 controller for MPPT
- 48W charging power was achieved

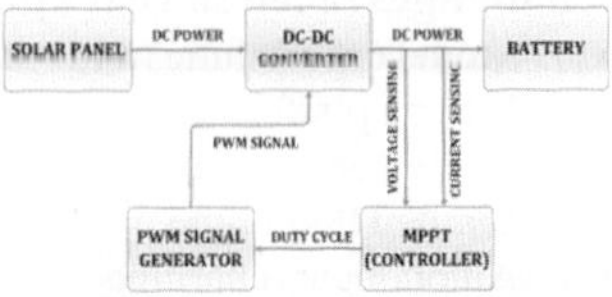

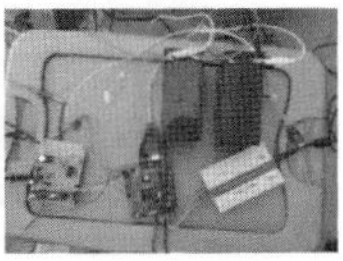

Wireless system for autonomous UAV charging

Objective : UAV wireless autonomous charging

- Without plugging in/out for battery charging
- Convenience and safety

Challenges:

- Autonomous landing control
- Autonomous charging control
- High efficiency

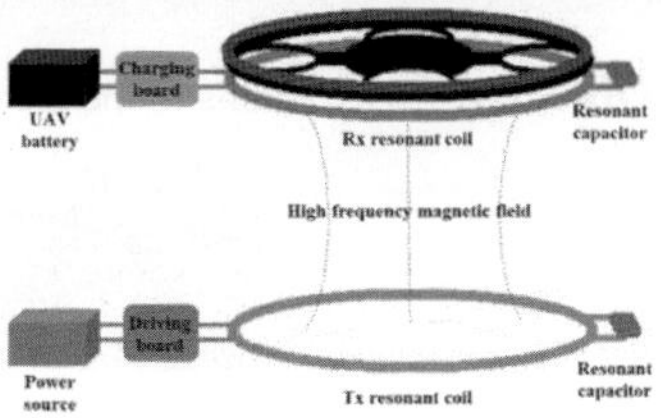

A wireless charging system was developed at the current stage

- 30cm diameter coils
- 89.4 kHz resonant frequency
- 40 W transferred power was achieved within 10 cm

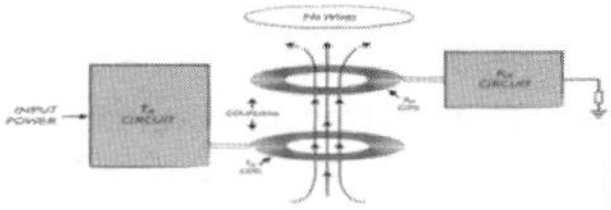

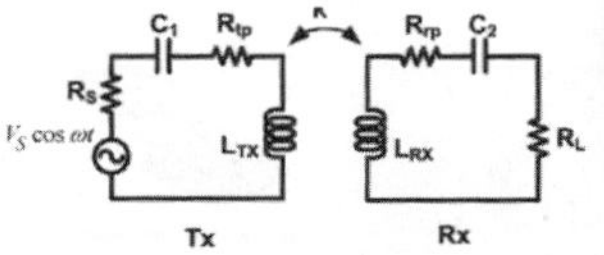

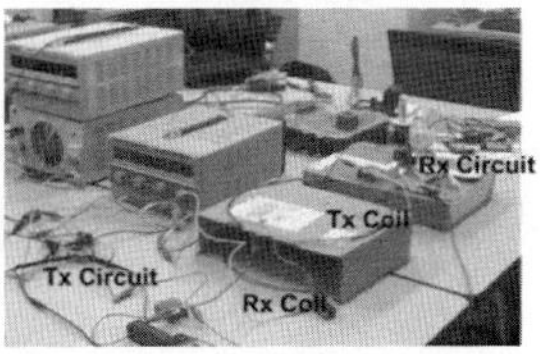

Critical infrastructure health monitoring (wind turbine & oil/gas pipeline)

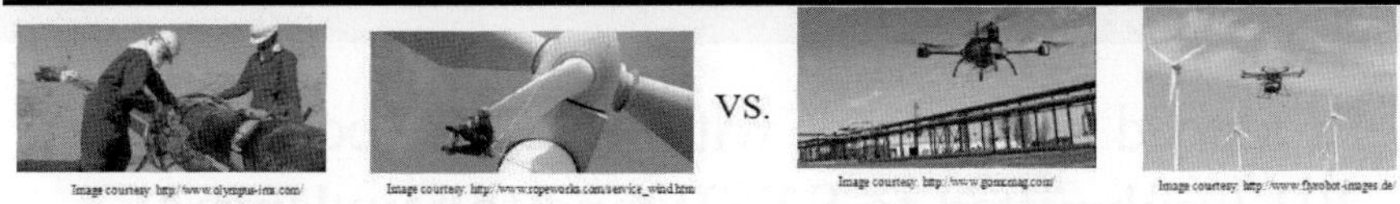

Human Inspection **UAV Inspection**

Objective: Inspect critical infrastructure damage/breakdown and prolong its life span

Approach: Using portable radar and camera to get the structure health information

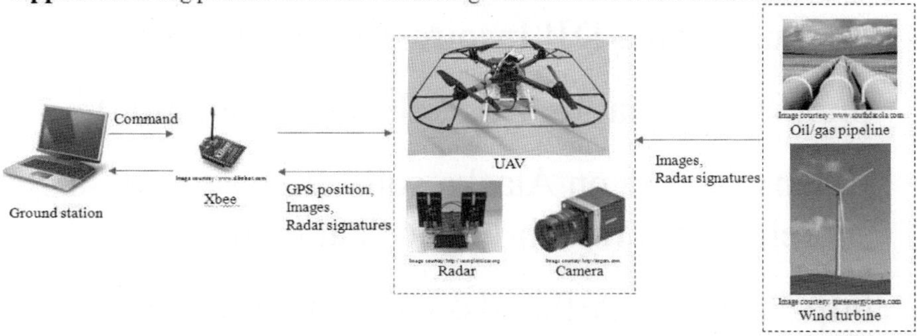

Precision agriculture – 3D scanning of cotton plants

Objective: 3D scanning of the cotton ball to determine the maturity of the plant

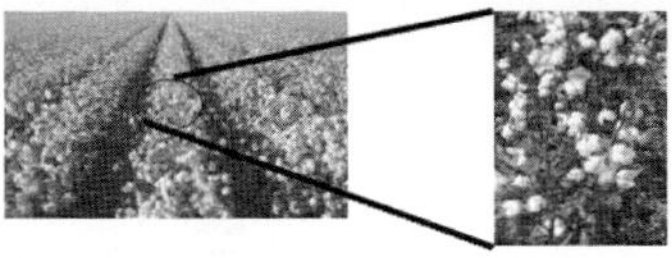

Approach:

- Develop a motion control and path planning scheme for UAV which is capable of communicating with a RGB-D sensor platform to scan the plant
- Develop a light weight RGB-D scanning platform which can compensate the motion and shakes due to the UAV motion and environmental causes (wind)

Between Earth and Sky / PBS Documentary

Produced at low-cost within Texas Tech. Airs in 2017, submitted to SXSW and to Sundance, David Weindorf, Executive Producer

First use of the DJI Inspire 1 UAV in Alaska

Shot in Cinema 4K (twice HDTV resolution)

Extensive data on Alaska soils analysis and shoreline loss (much larger than thought)

http://betweenearthandskyblog.tumblr.com

Erosion and Soils Remote Sensing in Alaska

Erosion and Soils Remote Sensing in Alaska

Homeland Security and Defense

- UAS for Bio-surveillance
- Prepositioning of first responders and assets (Example: Law Enforcement)
- Containment of biological threat
- Terrain mapping; optimization of ingress/egress routes
- Security monitoring of Biosafety Laboratories
- Support to training

SESSION 7

CYBER SECURITY ISSUES ROUND TABLE

FROM BAD TO WORSE: THE NEW DIMENSION OF CYBER THREAT

ROUND TABLE DISCUSSION

PANELISTS:

WILLIAM BARLETTA
U.S. Particle Accelerator School, Department of Physics, Massachusetts Institute of Technology, Cambridge, MA, USA

PAVAN DUGGAL
Associates Cyberlaw India
New Delhi, India

VLADIMIR BRITKOV
Institute for Systems Analysis, Russian Academy of Sciences
Moscow, Russia

MONA AL-ACHKAR JABBOUR
Faculty of Law, Lebanese University, Pan Arab Observatory for Cyber Security
Beirut, Lebanon

AXEL LEHMANN
Institute for Technical Computer Sciences, German Armed Forces University Munich
Neubiberg, Germany

HOWARD A. SCHMIDT
Ridge Schmidt Cyber
Washington, DC, USA

SESSION 8

SYSTEMATIC ERRORS IN CLIMATE MEASUREMENTS

ARE LABORATORY MEASUREMENTS OF CO_2 ADEQUATE TO PREDICT CLIMATE SENSITIVITY?

WILLIAM HAPPER
Cyrus Fogg Brackett Professor of Physics, Physics Department, Princeton University, Princeton, NJ, USA

CLIMATE MODELS DON'T WORK; RED IS OBSERVED; BARS ARE MODELS

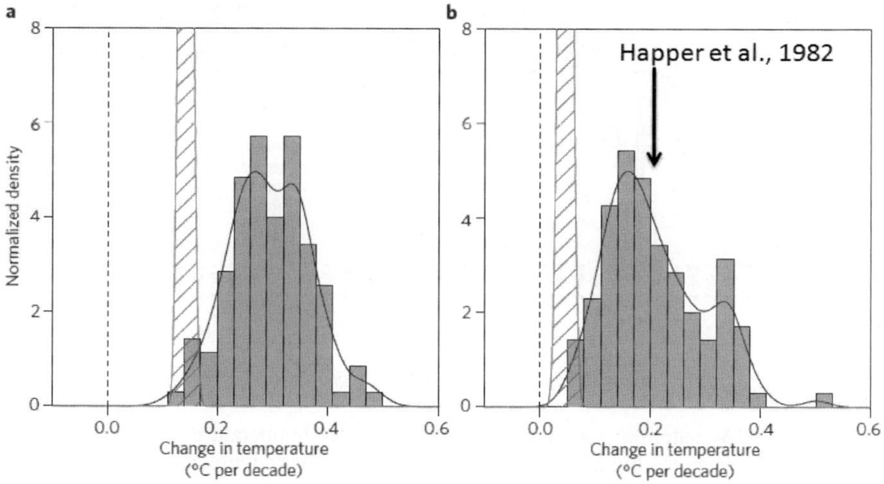

Fig. 1: Trends in global mean surface temperature. a.) 1993-2012. b.) 1998-2012. Histograms of observed trends (red hatching) are from 100 reconstructions of the HadCRUT4 dataset.[1] *Histograms of model trends (grey bars) are based on 117 simulations of the models, and black curves are smoothed versions of the model trends. The ranges of observed trends reflect observational uncertainty, whereas the ranges of model trends reflect forcing uncertainty, as well as differences in individual model responses to external forcings and uncertainty arising from internal climate variability.*

[1] *Nature Climate Change*. Vol. 3, p. 767, September 2013.

SATELLITE TEMPERATURE MEASUREMENTS OF LOWER ATMOSPHERE ALSO SHOW LITTLE WARMING

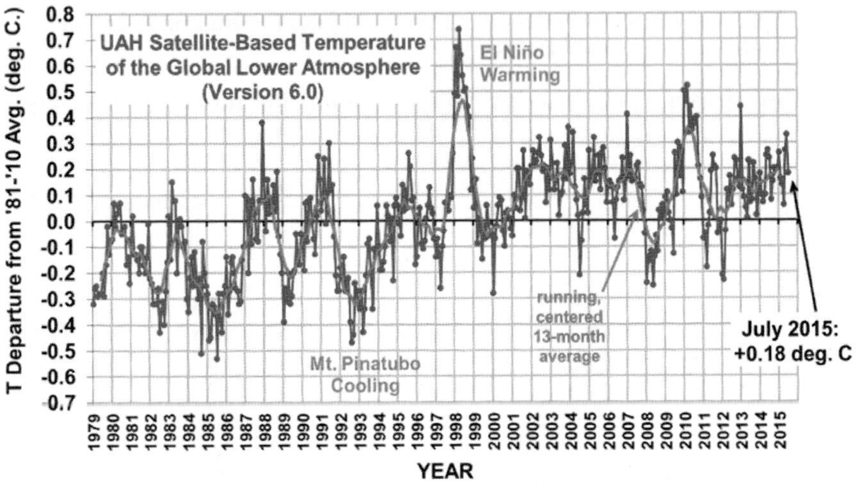

Fig. 2: http://www.drroyspencer.com/latest-global-temperatures/.

NASA satellite measurements show little temperature rise of the lower troposphere since the big El Nino of 1998. The atmospheric temperature is deduced from the microwave (millimeter wave) thermal emission of O_2 molecules at frequencies around 60 GHz. The temperature of the lower atmosphere is strongly affected by volcanic eruptions, when high-altitude aerosols reflect so much sunlight that they cool the entire earth. The El Nino/La Nina cycle causes large changes in the surface temperature of the tropical Pacific Ocean. The lower tropospheric temperatures measured by satellites should rise more rapidly than surface temperatures because of the latent heat released when water vapor of the rising air condenses into clouds.

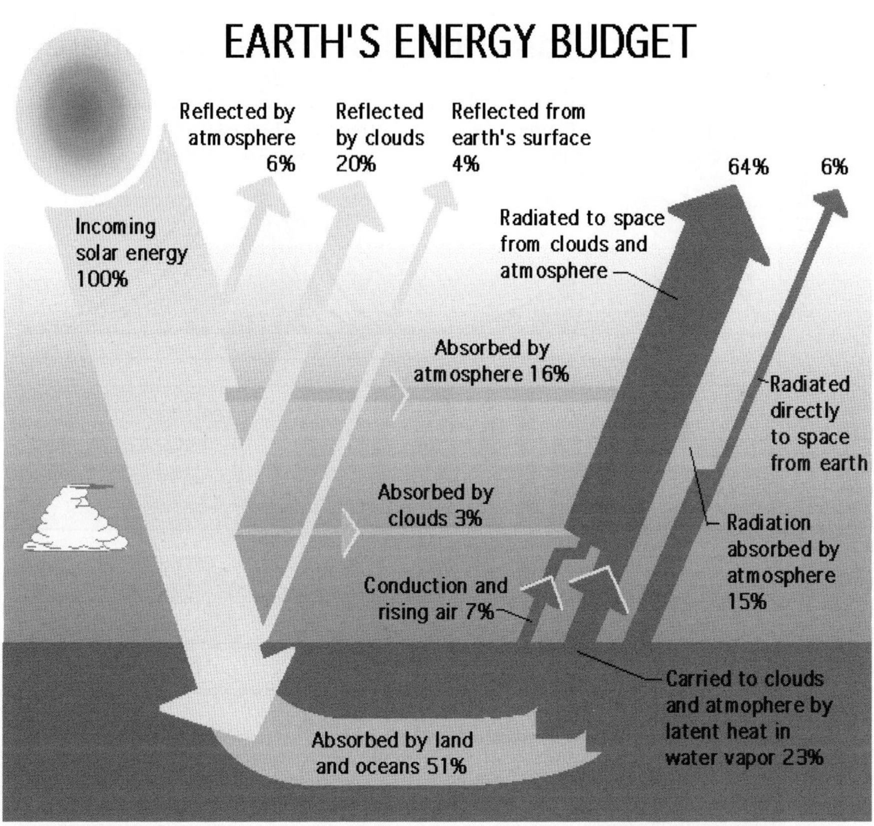

Fig. 3.

Aside from a very small amount of geothermal heat, the earth's surface is heated by sunlight with wavelengths of order half a micron, characteristic of the 5525K blackbody temperature of the sun's surface. The earth's surface is cooled mainly by convection, including cooling from the evaporation of water. The surface also cools by radiating directly to space in the thermal infrared window, with wavelengths of order 10 microns, characteristic of a blackbody with the 300K surface temperature of the earth. Most of the thermal radiation to space does not come from the surface but from greenhouse gases, predominantly H_20 and CO_2, and from clouds or other particulates in the atmosphere above the surface.

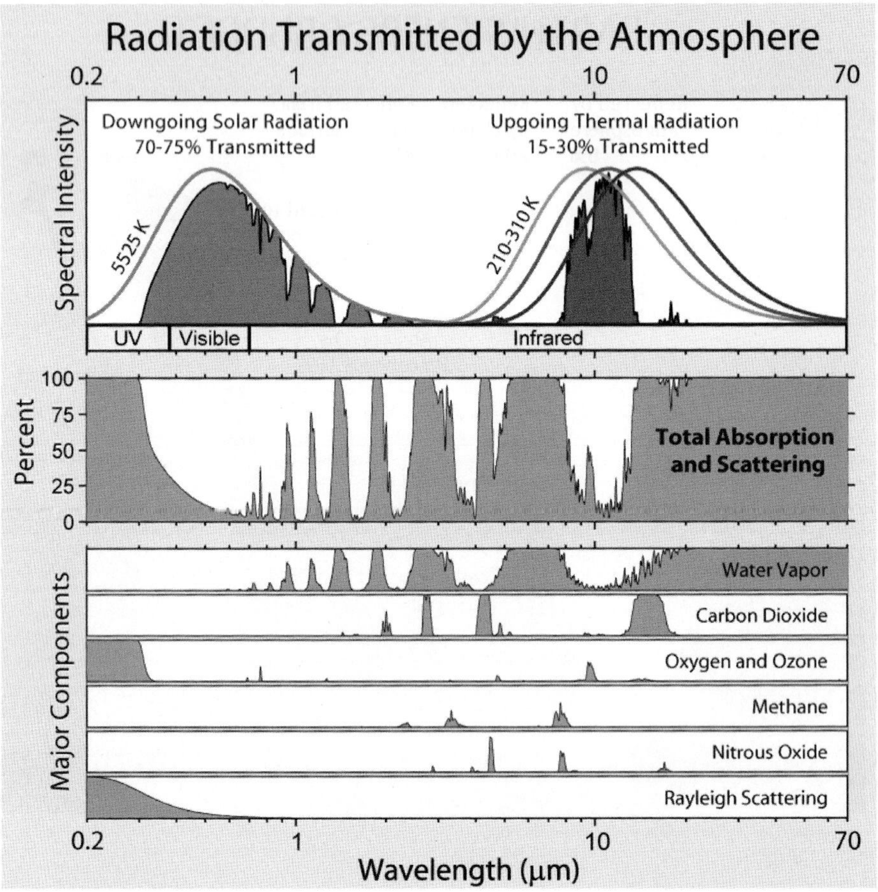

Fig. 4.

There is substantial absorption of both sunlight and thermal radiation by the atmosphere, even in the absence of clouds. The shortest wavelengths of sunlight are subject to Rayleigh scattering, which makes our blue sky. Overtones of the vibrational modes of water molecules absorb substantial amounts of near infrared sunlight. Molecules of H_2O, CO_2, O_2, O_3 and others absorb so strongly in the thermal infrared that even for clear skies, only a narrow band of radiation with wavelengths near 10 microns can escape from the surface to space with no further scattering. Even though strongly absorbed frequencies cannot be radiated directly to space from the ground, they can be radiated to space by molecules at a sufficiently high altitude that the absorption by the air above them is negligible. This high-altitude radiation is weaker than it would be at the surface, since the air is cooler at higher altitudes than at the surface, and the molecules therefore do not radiate as strongly as they would if they were closer to the surface and warmer.

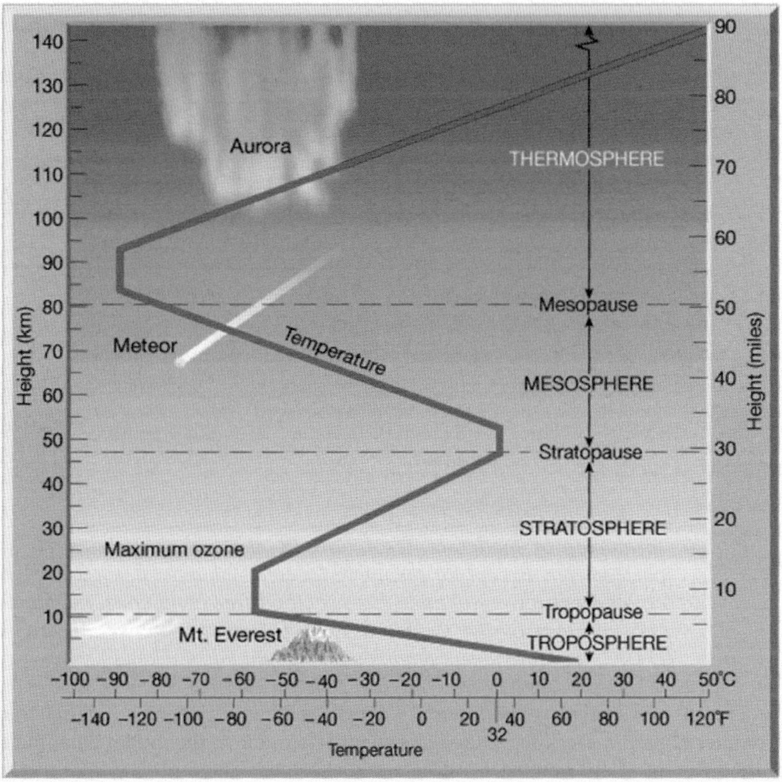

Fig. 5.

The temperature lapse rate of the atmosphere is an important factor that affects greenhouse warming. There is a rapid drop of temperature from the surface to the tropopause, where the convective mixing of the atmosphere stops and the stratosphere begins. Although the pressure and temperature drop rapidly with increasing altitude in the troposphere, the entropy per unit mass of air is nearly constant, since it can only be decreased by the relatively slow loss of energy by radiation. With no clouds, the surface can radiate directly to space in the thermal infrared window at wavelengths around 10 microns. For wavelengths outside the atmospheric window, heat is radiated to space by H_2O, CO_2 and other greenhouse molecules and by cloud tops. The more radiation to space that comes from the thermal emission of molecules or clouds at high, cold altitudes, as opposed to radiation from the warm surface or from the relatively warm lower troposphere, the more warming of the surface there will be.

EARTH'S THERMAL RADIATION MEASURED BY A SATELLITE

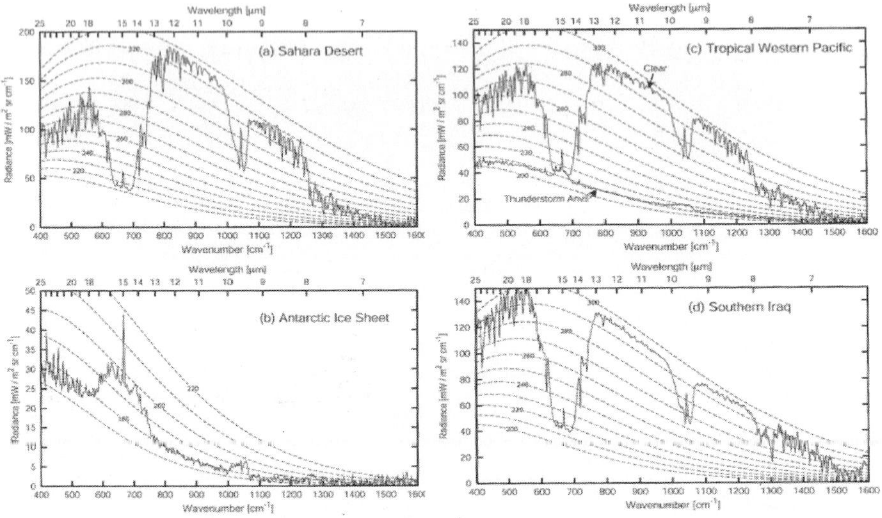

Fig. 6: Examples of moderate resolution IR spectra observed by a satellite spectrometer. Except for the curve labeled, "thunderstorm anvil" in panel (c), all spectra were obtained under cloud-free conditions. (Nimbus-4 IRIS data courtesy of the Goddard EOS Distributed Active Archive Center (DAAC) and instrument team leader Dr. Rudolf A. Hanel.)

These are actual experimental data, obtained from a Fourier-transform spectrometer aboard a satellite. Although generally similar to the model data, the spectra are quite different at different parts of the earth's surface. There is very little water vapor over Southern Iraq which was in dry subsiding air from the Hadley cell at the time of this measurement. There is an extreme temperature inversion over the Antarctic ice sheet, where the surface temperature is only 180K. Over Antarctica, the CO_2 and O_2 bands show up in emission from the warmer atmosphere above the ice sheet. The atmosphere over Antarctica is so cold that it contains practically no water vapor. Finally note the near complete absence of infrared emission from the top of the thunderstorm anvil over the Tropical Western Pacific, the site of the celebrated "warm pool" that is involved in El Nino's and La Nina's. The suppression of infrared radiation by high, cold cloud tops, especially cirrus clouds, is a potent global warming mechanism. However, for low clouds, the increase in visible albedo is a potent cooling mechanism.

CO_2 VIBRATIONAL MODES

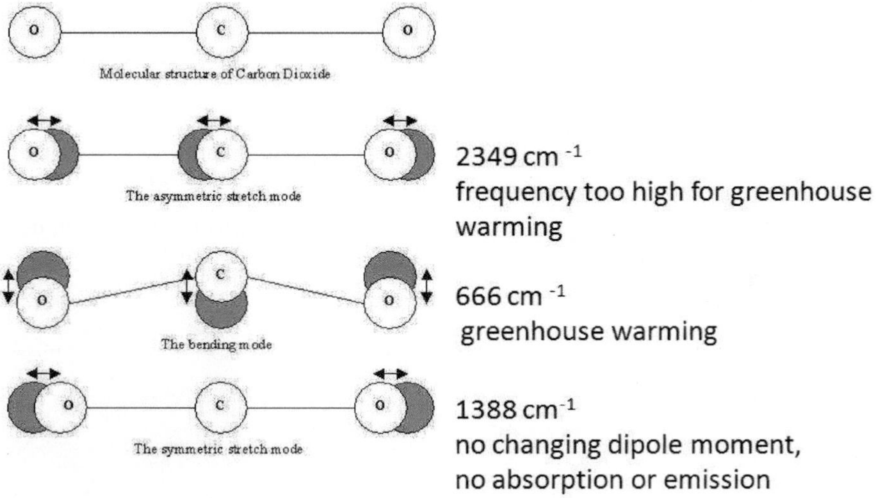

Fig. 7.

The CO_2 molecule has three vibrational modes: the infrared active bending mode and asymmetric stretch mode, and symmetric stretch mode, which is not infrared active. The resonance frequencies of the asymmetric stretch mode lie outside the thermal emission spectrum of the earth and also outside of most of the visible spectrum of the sun, so the asymmetric stretch mode contributes little to global warming. The frequency of the symmetric stretch mode is nearly twice the frequency of the bending mode, so there can be strong mixing of the excited bending vibration with the symmetric stretch mode. This mixing, often called Fermi resonance, since it was first explained by Enrico Fermi, makes CO_2 a more potent greenhouse gas.

Each of the strongly negative O atoms of the CO_2 molecule has a net negative charge, leaving the C atom in the middle positively charged with about half the elementary charge (the charge of a proton). It is the accelerations of these charges during the vibration and rotation of the CO_2 molecule that allows it to absorb and emit radiation. Those vibrations and rotations that produce changing electric dipole moments radiate most efficiently, very much like the whip antenna an a automobile, which also transmits and receives electric dipole radiation.

FERMI RESONANCES

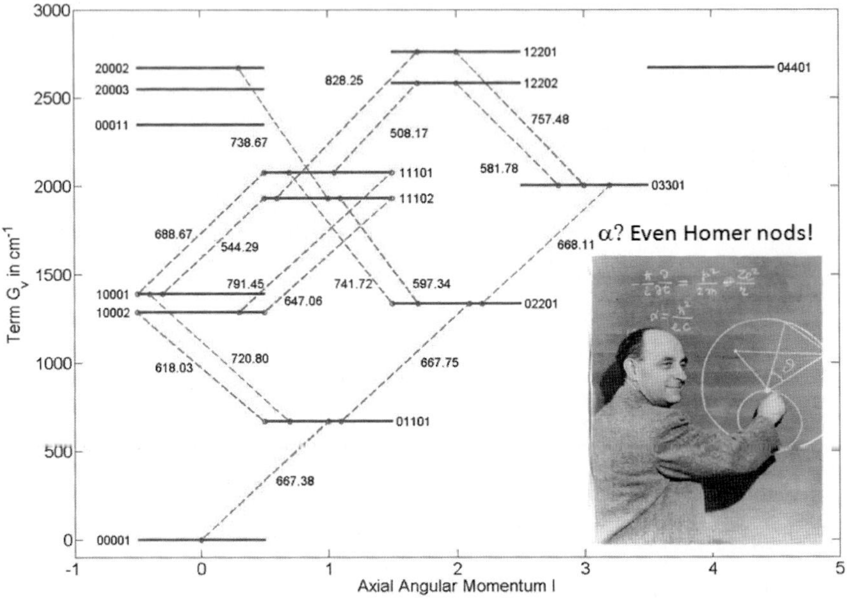

Fig. 8.

The low-lying vibrational levels of the CO_2 molecule for the most abundant isotopomer $^{16}O^{12}C^{16}O$, or 626. Note the splitting of the higher bending modes by Fermi resonance interactions with the symmetric stretch mode, and with the asymmetric stretch mode. The transition frequencies for a non-rotating molecule are indicated in cm^{-1}. The states are labeled with the HITRAN code, for which the integers, reading from left to right, mean: 1) v_1 = quanta of symmetric stretch vibration; 2) v_2 = quanta of bending-mode vibration; 3) l = quanta of "vibrational" angular momentum around the axis of symmetry; 4) v_3 = quanta of asymmetric stretch vibration; 5) r = ranking index for multiples split by Fermi mixing.

ROTATIONAL SIDEBANDS

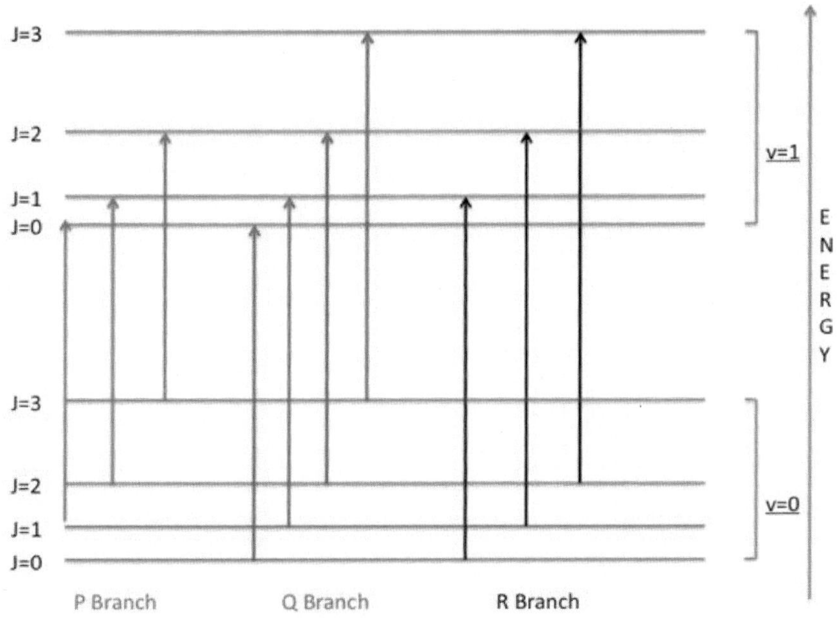

Fig. 9: Schematic diagram of P, Q, and R branch transitions.

Each possible vibrational transition frequency is split into many hundreds of lines (sidebands) by the rotation of the molecule. The total angular momentum J of the molecule is a good quantum number. If J' is the rotational quantum number of the upper vibrational state and J is the rotational quantum number of the lower vibrational state, transitions for which $J' = J - 1$ are called P transitions, transitions with $J' = J$ are called Q transitions, and transitions with $J' = J + 1$ are called R transitions. For the bending mode of the CO_2 molecule, the frequencies of the Q transitions, which corresponds to rotation of the molecule around the axis of transverse vibration, have little dependence on the rotational quantum number J of the lower state, are nearly the same as for a non-rotating molecule. The P and R transitions have lower and higher frequencies, respectively, than the non rotating molecule, and the frequencies depend strongly on the rotational quantum number J of the lower state.

This sketch shows transitions between two vibrational states with axial angular momentum $L = 0$, for example, for the asymmetric stretch state of the CO_2 molecules where the electric dipole moment is along the symmetry axis. For the bending mode of CO_2, that contributes to global warming, the electric dipole moment is perpendicular to the symmetry axis, and the axial angular momentum quantum numbers, L' and L, of the upper and lower states must differ by one unit, that is $L' = L + 1$ or $L' = L - 1$, as illustrated on the previous slide. If the O atoms are identical, the states L = 0 and odd J are missing because of exchange symmetry. States with nonzero L and non-identical O atoms are narrow parity doublets.

PHYSICS OF RADIATION

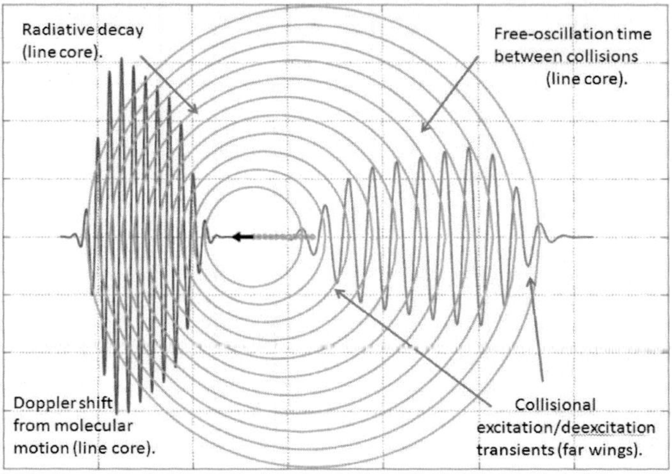

Fig. 10.

ATTENUATION COEFFICIENT (E-FOLDINGS/LENGTH)

$$\kappa = N\sum_{eg}\sigma_{eg} = N\sum_{eg} S_{eg}G_{eg} \longleftarrow \textbf{Line shape}$$

N is the number density of CO_2 molecules, and σ_{eg} is the cross section of the transition from a lower state g to an upper state e; G_{eg} is the line shape function.

Line strength (in cm)

$$S_{eg} = \frac{8\pi^3 \nu_{eg}|D_{eg}|^2 e^{-E_g/kT}\left(1 - e^{-hc\nu_{eg}/kT}\right)}{hcZ}$$

ν_{eg} = frequency (in cm^{-1}) of the transition, D_{eg} = the electric dipole matrix element, E_g = lower-state energy, T = absolute temperature, k = Boltzmann's constant, h = Planck's constant, c = speed of light, T = absolute temperature.

Partition function

$$Z = \sum_j e^{-E_j/kT}$$

The important line strength factor S_{eg} is a bit like a temperature-dependent "oscillator strength." It takes into account both absorption and stimulated emission of molecules that are in thermal equilibrium at a temperature T. Since it is not possible to calculate the matrix element D_{eg} of the electric dipole moment with sufficient accuracy for them to be of use in modeling the greenhouse effect, line strength factors are measured and tabulated, often at a reference temperature of 296 K, as in the HITRAN data base.

To get absorption cross sections it is necessary to have line shape factors G_{eg} as well as line strength factors. The line strength factors S_{eg} have been determined very well by experiment. The line shape factors G_{eg} have been determined experimentally to be very nearly Lorentzian (or Doppler-broadened Lorentizian = Voigt profiles) near line centers, but very little is known experimentally about the far wings of the profiles.

THOUSANDS OF LINES! LINE STRENGTHS VARY BY 10 ORDERS OF MAGNITUDE

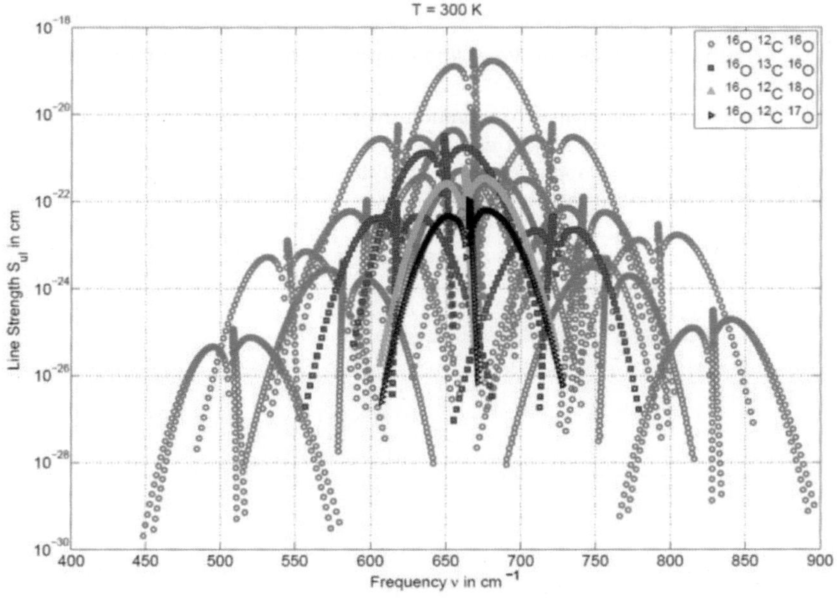

Fig. 11.

Normalized line shapes

$$\int_0^\infty G_{eg} d\nu = 1$$

A Lorentzian line shape

$$G_{eg} = \frac{\mu_{eg}/\pi}{\mu_{eg}^2 + (\nu - \nu_{eg})^2}$$

μ_{eg} = broadening; ν = frequency; ν_{eg} = resonance.

A Voigt line shape

$$G_{eg} = \frac{\mu_{eg}}{\pi}\sqrt{\frac{m}{2\pi kT}} \int_{-\infty}^{\infty} \frac{e^{-mv^2/2kT} dv}{\mu_{eg}^2 + (\nu - \nu_{eg}[1 + v/c])^2}$$

Neither Lorentzian nor Voigt line shapes are correct in the far wings!

The intrinsic line shapes of vibration-rotation transitions of the CO_2 molecule in the earth's atmosphere are almost completely determined by collisions, except for the high stratosphere where Doppler broadening of the lines is no longer negligible, and one should use a Voigt profile near line center.

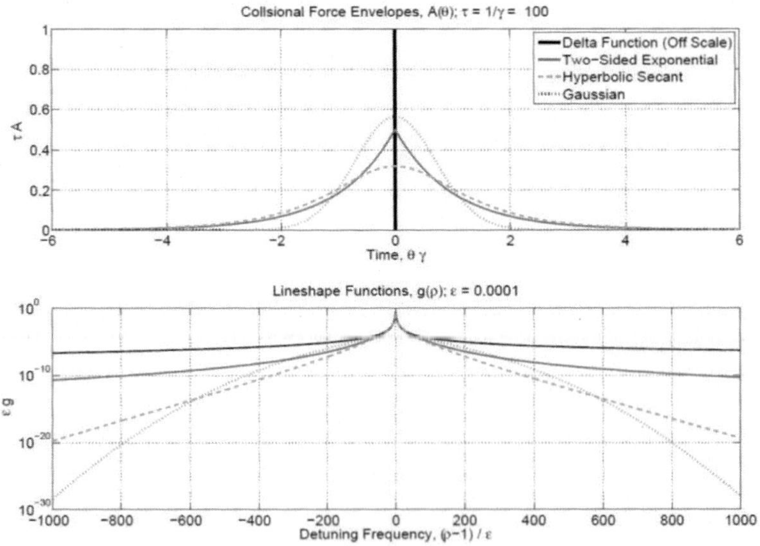

Fig. 12.

Schwartzschild-Milne Equation:
Downwelling Flux at the Surface

$$J(0) = \int_0^\infty \kappa(z) B(z) e^{-\rho(z)} dz$$

K. Schwartzschild

E. A. Milne

Optical Depth from Surface to Altitude z

$$\rho(z) = \int_0^z \kappa(z') dz'$$

M. Planck

Planck Brightness

$$B = \frac{2hc^2\nu^3}{e^x - 1}, \quad \text{where} \quad x = \frac{hc\nu}{kT}$$

Fig. 13.

The basic radiative transfer equations describe the net effect of spontaneous and stimulated emission and of absorption by molecules. Quite similar equations describe the gain of lasers, where the distribution of the molecules over the quantized energy states is far from thermal equilibrium and one can get gain.

ALTITUDE DEPENDENCE OF CROSS SECTIONS COMPLICATES LINE-BY-LINE CALCULATIONS OF RADIATIVE FLUX!

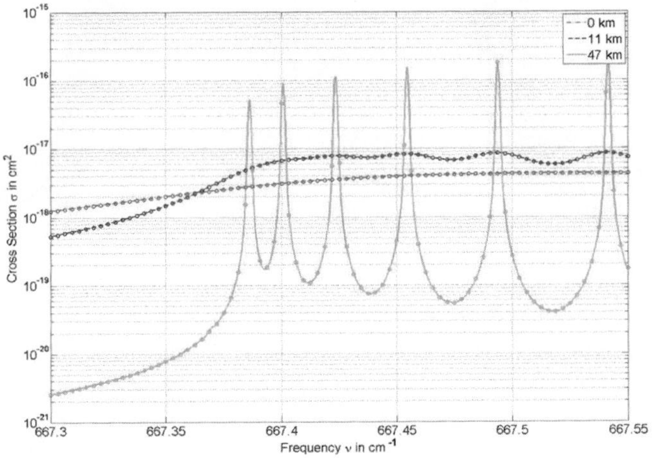

Fig. 14.

This is a fine-scale plot of the absorption cross section for CO_2 near the very center of the 667 band. The resonances are for Q-branch (J' – J = 0) transitions from rotational states with angular momentum quantum numbers J = 0, 2, 4,...of the ground vibrational level to the first excited vibrational level of the the most abundant isotopomer $^{16}O^{12}C^{16}O$ or 626. Because the O nuclei are identical bosons, the rotational levels with odd J are missing. Pressure broadening at the earth's surface smears out the resonance lines, but they are well resolved at the top of the stratosphere, 47 km, where the pressure is only a few millibar. The continuous lines are for Voigt profiles, and the points are Voigt profiles with suppressed wings. Wing suppression makes no difference at the center of the absorption band.

INCREASING CO_2 CONCENTRATION C BY 1000 ONLY INCREASES DOWNWELLING BY 10!

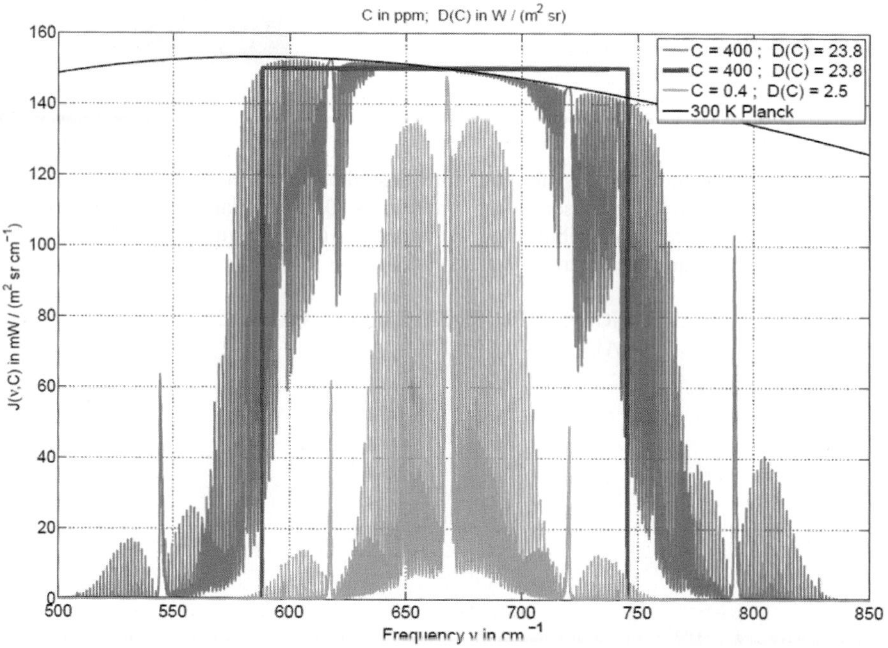

Fig. 15.

LOGARITHMIC DEPENDENCE OF DOWNWELLING IR ON CO_2 CONCENTRATION

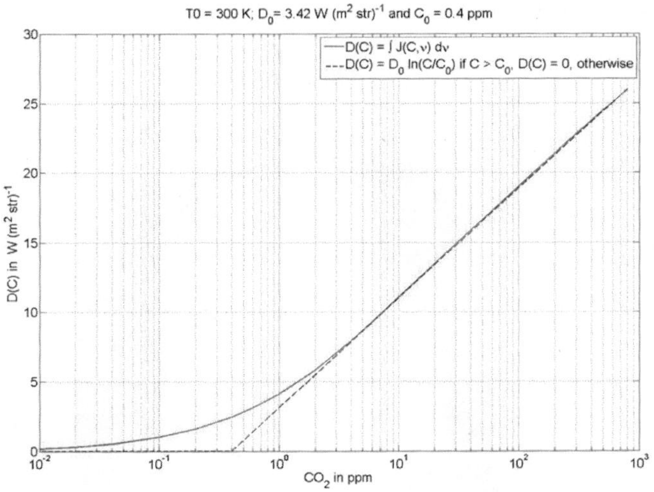

Fig. 16.

FAR WINGS OF INTENSE CORES LINES GIVE TOO MUCH RADIATIVE FORCING FOR PURE VOIGT PROFILES!

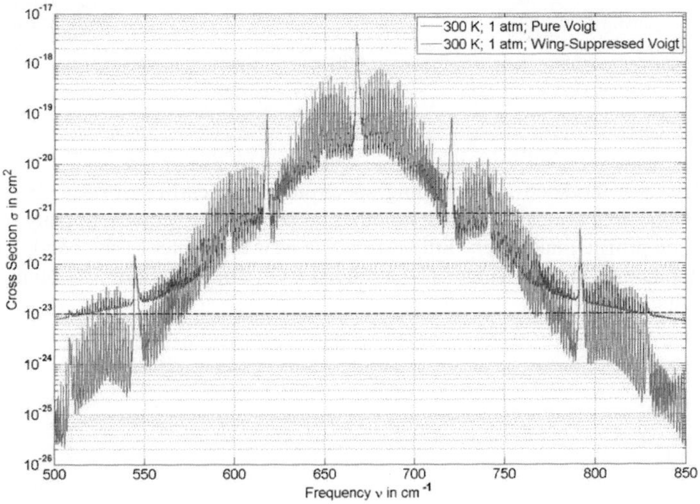

Fig. 17.

Assuming Voigt line shapes greatly overestimates the net CO_2 cross section in the wings of the band. Cross sections with value between 10^{-23} and 10^{-21} cm^2 are responsible for almost all of the greenhouse warming from additional CO_2. Current column density of CO_2 is 8.6 x 10^{21} cm^{-2}.

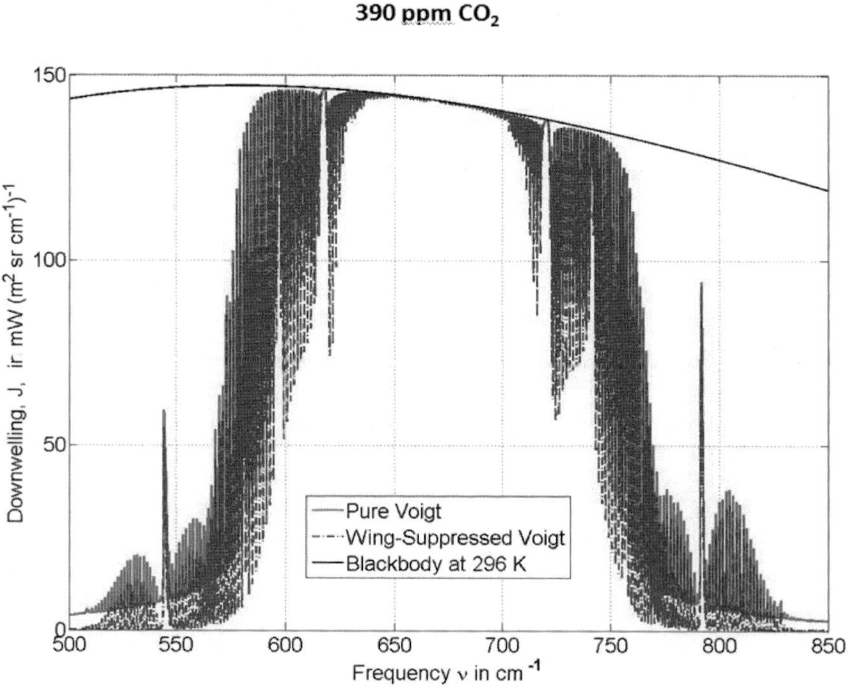

Fig. 18.

This is a modeled downwelling spectrum at the earth's surface. It comes from CO_2 in a cloud-free atmosphere with no other greenhouse molecules, notably no $H_2 0$. The assumed line profile makes very little difference near the center of the band, but it makes a big difference in the far wings. The assumed CO_2 concentration, 390 ppm, is close to today's value.

NEGLECTING LINE PEDESTALS GIVES OVERESTIMATE OF RADIATIVE FORCING BY ABOUT 40%

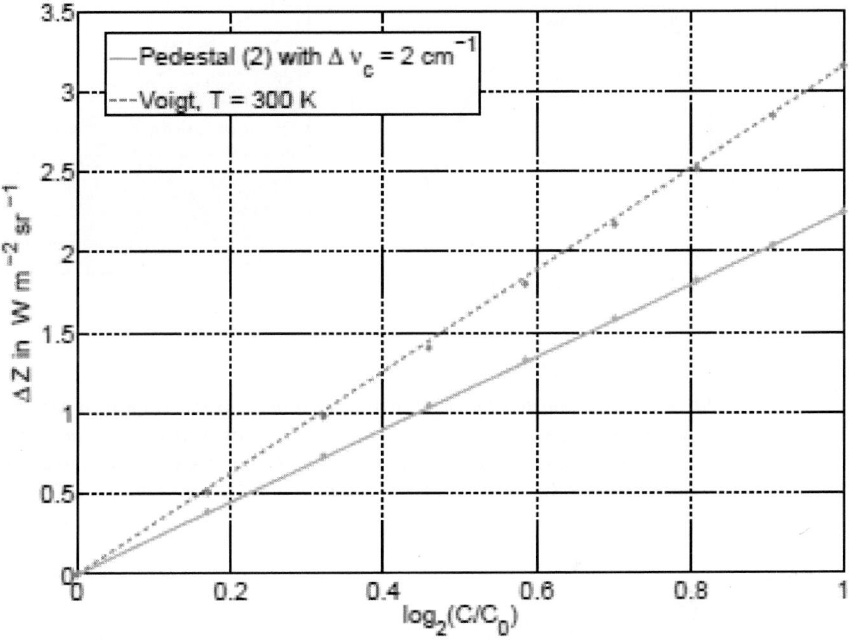

Fig. 19.

VOIGT LINE SHAPES DON'T WORK IN FAR WINGS!

The inadequacies of the Voigt profile for far-wings of the CO_2 band have already been demonstrated experimentally many times. A recent example is the balloon observations cited in the recent book by Hartmann et al. (2008).

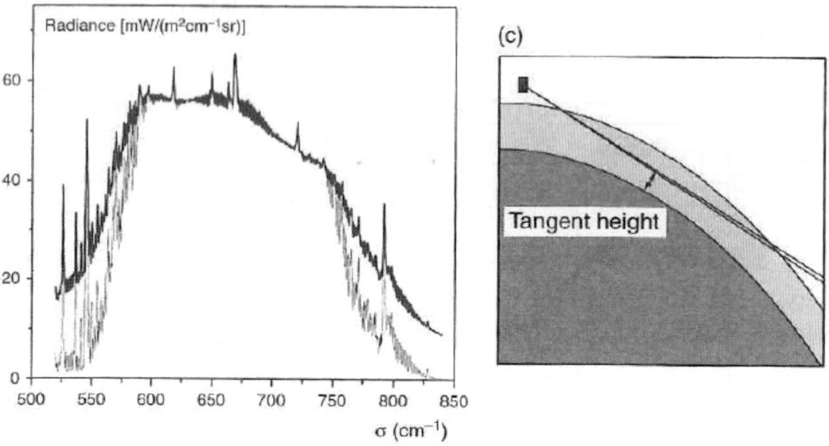

Fig. 20: Radiance emitted by the Earth atmosphere in the region of the ν_2 CO_2 band for a 1 cm^{-1} resolution. The thin line gives measured values obtained by a balloon-borne instrument at 40 km altitude looking down to a tangent height of 10 km. The thick line corresponds to predictions using purely Voigt line shapes.[2]

SUMMARY

- The cause for the striking over prediction of warming by climate models is not known.
- Neglect of natural variability, solar influences, exaggerated feedbacks, etc., are possibilities.
- Use of incorrect far-wing line shapes can lead to 40% overestimates of CO_2 radiative forcing.
- Line strengths, pressure-broadening, level energies and statistical weights have been determined by laboratory measurements.
- Far-wing properties of line shape have been ignored or guessed. They need to be measured—in the atmosphere.

[2] Hartmann, Boulet and Robert (2008) *Collisional Effects on Molecular Spectra,* Elsevier.

SYSTEMATIC ERROR IN CLIMATE MEASUREMENTS: THE SURFACE AIR TEMPERATURE RECORD

PATRICK FRANK
SLAC, Stanford University
Palo Alto, CA, USA

ABSTRACT

Systematic measurement errors of surface air and sea surface temperature (SST) sensors are surveyed. Field-calibrations reveal that the traditional Cotton Regional Shelter (Stevenson screen) and the modern Maximum-Minimum Temperature Sensor (MMTS) shield suffer daily average 1σ systematic measurement errors of ±0.44°C or ±0.32°C, respectively, stemming chiefly from solar and albedo irradiance and insufficient wind-speed. Marine field calibrations of bucket or engine cooling-water intake thermometers revealed typical SST measurement errors of 1σ = ±0.6°C, with some data sets exhibiting ±1°C errors. These systematic measurement errors are not normally distributed, are not known to be reduced by averaging, and must thus enter into the global average of surface air temperatures. Modern floating buoys exhibit proximate SST error differences of ±0.16°C. These known systematic errors combine to produce an estimated lower limit uncertainty of 1σ = ±0.5°C in the global average of surface air temperatures prior to 1980, descending to about ±0.36°C by 2010 with the gradual introduction of modern instrumentation. At the 95% confidence interval, the rate or magnitude of the global rise in surface air temperature since 1850 is unknowable.

INTRODUCTION: SYSTEMATIC ERROR

Systematic error enters into experimental or observational results through uncontrolled and often cryptic deterministic processes.[1] These can be as simple as a consistent operator error. More typically, error emerges from instrumental inaccuracy or an uncontrolled experimental variable. Instrumental inaccuracy arises from malfunction or lack of calibration. Uncontrolled variables can impact the magnitude of a measurement and/or change the course of an experiment. Figure 1 shows the impact of an uncontrolled variable, taken from the author's published work.[2,3]

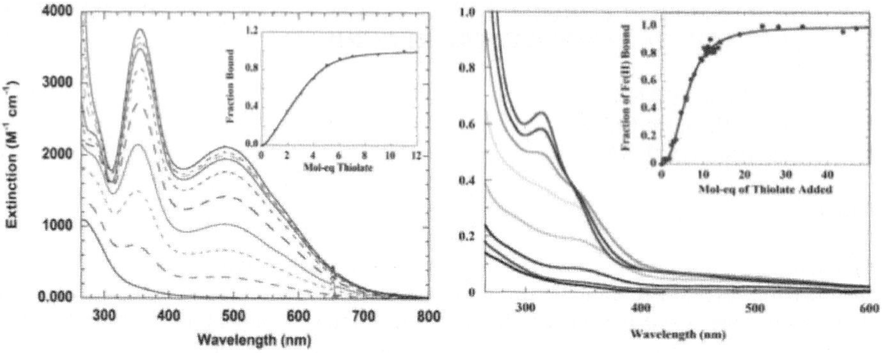

Fig. 1: Left, titration of dissolved ferrous iron under conditions that allowed an unplanned trace of air to enter the experiment. Inset: the incorrect data precisely follow equilibrium thermodynamics. Right, the same experiment but with the appropriately strict exclusion of air. Inset: the correct data reflect distinctly different thermodynamics.

Figure 1 shows that the inadvertent entry of a trace of air was enough to completely change the course of the experiment. Nevertheless, the erroneous data display coherent behavior, and follow a trajectory completely consistent with equilibrium thermodynamics. To all appearances, the experiment was physically valid. In isolation, the data are convincing. However, they are completely wrong because the intruded air chemically modified the iron.

Figure 1 exemplifies the danger of systematic error. Contaminated experimental or observational results can look and behave just like good data, and can rigorously follow valid physical theory. Without care, such data invite erroneous conclusions.

By its nature, systematic error is difficult to detect and remove. Methods of elimination include careful instrumental calibration under conditions identical to the observation or experiment. Methodologically independent experiments that access the same phenomena provide a check on the validity of results. Careful attention to these practices is standard in the physical sciences.

The recent development of a new and highly accurate atomic clock illustrates the extreme care physicists take to eliminate systematic error. Critical to achievement of its 10^{-18} second accuracy, was removal of systematic error produced by black-body radiation from the instrument itself.[4]

As a final word, systematic error does not average away with repeated measurements. Repetition can even increase error. When systematic error cannot be eliminated and is known to be present, uncertainty statements must be reported along with the data. In graphical presentations of measurement or calculational data, systematic error is represented using uncertainty bars.[1]

SYSTEMATIC ERROR IN SURFACE TEMPERATURE MEASUREMENTS

Land Surface Air Temperature

Fig. 2: Alpine Plaine Morte Glacier, Switzerland, showing the air temperature sensor calibration experiment carried out by Huwald, et al., in 2007 and 2008.[5] Photo credit: Bou-Zeid, Martinet, Huwald, Couach, 2.2006 EPFL-ENAC.

During most of the 20th century, land surface air temperatures were measured using a liquid-in-glass (LiG) thermometer housed in a box-like louvered shield (Stevenson screen or Cotton Regional Shelter (CRS)).[6,7] After about 1985, thermistors or platinum resistance thermometers (PRT) housed in an unaspirated cylindrical plastic shield replaced the CRS/LiG sensors in Europe, the Anglo-Pacific countries, and the U.S. Beginning in 2000, the U.S. Climate Research Network deployed sensors consisting of a trio of PRTs in an aspirated shield.[6,8-10] An aspirated shield includes a small fan or impeller that ventilates the interior of the shield with outside air.

Unaspirated sensors rely on prevailing wind for ventilation. Solar radiance can heat the sensor shield, warming the interior atmosphere around the sensor. In the winter, the albedo of a snow-covered surface can also produce a warm bias.[11] Significant systematic measurement error occurs when air movement is less than 5 m/sec.[10,12]

In 2007 and 2008 calibration experiments carried out on the Plaine Morte Glacier (Figure 2) tested the field accuracy of the RM Young PRT housed in an unaspirated louvered shield, situated over a snow-covered surface. In a laboratory setting, the RM Young sensor is capable of ±0.1°C accuracy. Field accuracy was determined by comparison with air temperatures measured using a sonic anemometer, which takes advantage of the impact of temperature on the speed of sound in air and is insensitive to irradiance and wind-speed.

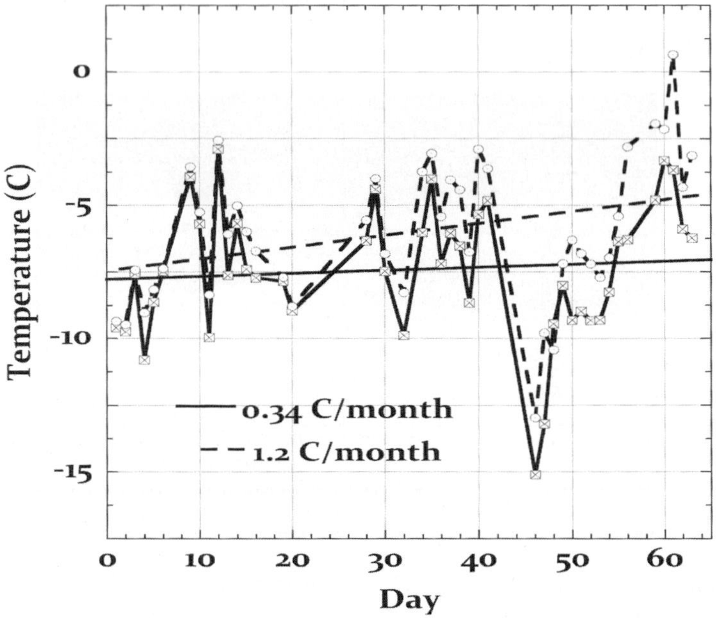

Fig. 3: Temperature trends recorded simultaneously on Plaine Morte Glacier during February-April 2007. (—), Sonic anemometer, and; (- -), RM Young PRT probe.

Figure 3 shows that under identical environmental conditions, the RM Young probe recorded significantly warmer winter air temperatures than the sonic anemometer. The slope of the RM Young temperature trend is also more than 3× greater. Referenced against a common mean, the RM Young error would enter a spurious warming trend into a global temperature average. The larger significance of this result is that the RM Young probe is very similar in design and response to the more advanced temperature probes in use world-wide since about 1985.

Figure 4 shows a histogram of the systematic temperature error exhibited by the RM Young probe.

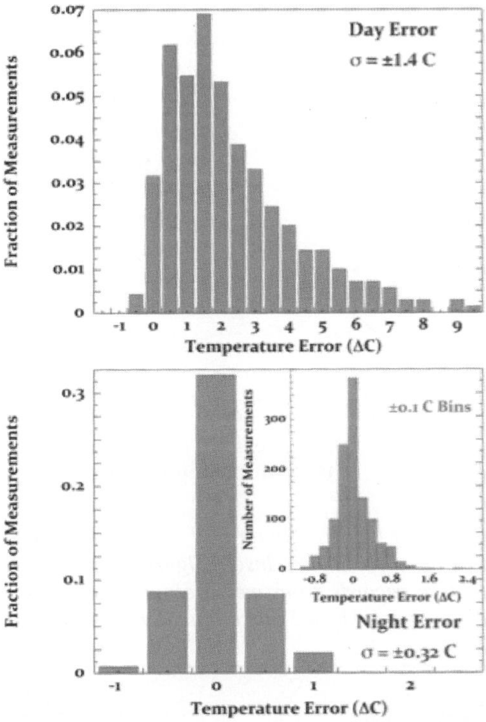

Fig. 4: RM Young probe systematic error on Plaine Morte Glacier. Top: day time error, averaging 2.0±1.4℃. Bottom, night-time error, averaging 0.03±0.32℃. Inset: asymmetry of the night error becomes clear at higher resolution.

The RM Young systematic errors mean that, absent an independent calibration instrument, any given daily mean temperature has an associated 1σ uncertainty of 1±1.4°C. Figure 4 shows this uncertainty is neither randomly distributed nor constant and cannot be removed by averaging individual measurements. Subtraction of the average bias will not remove the non-normal 1σ uncertainty. Entry of the RM Young station temperature record into a global average will carry that average error along with it.

Before inclusion in a global average, temperature series from individual meteorological stations are subjected to statistical tests for data quality.[13] Air temperatures are known to show correlation R = 0.5 over distances of about 1200 km.[14,15] The first quality control test for any given station record includes a statistical check for correlation with temperature series among near-by stations. Figure 5 shows that the RM Young error-contaminated temperature series will pass this most basic quality control test. Further, the erroneous RM Young record will pass every single statistical test used for the quality control of meteorological station temperature records worldwide.[16,17]

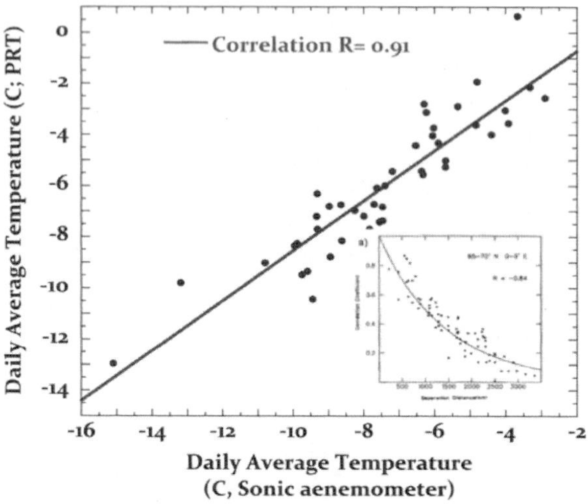

Fig. 5: Correlation of the RM Young PRT temperature measurements with those of the sonic anemometer. Inset: Figure 1a from[14] showing correlation of temperature records from meteorological stations in the terrestrial 65-70°N, 0-5°E grid. The 0.5 correlation length is 1.4×10^3 km.

Fig. 6: Calibration experiment at the University of Nebraska, Lincoln (ref.,[12] Figure 1); E, MMTS shield; F, CRS shield; G, the aspirated RM Young reference.

Figure 6 shows the screen-type calibration experiment at the University of Nebraska, Lincoln. Each screen contained the identical HMP45C PRT sensor.[12] The calibration reference temperatures were provided by an aspirated RM Young PRT probe, rated as accurate to <±0.2°C below 1100 Wm^{-2} solar irradiance.

These independent calibration experiments tested the impact of a variety of commonly used screens on the fidelity of air temperature measurements from PRT probes.[11,12,18] Screens included the traditional Cotton Regional Shelter (CRS, Stevenson screen), and the MMTS screen now in common use in the U.S. Historical Climate Network, among others.

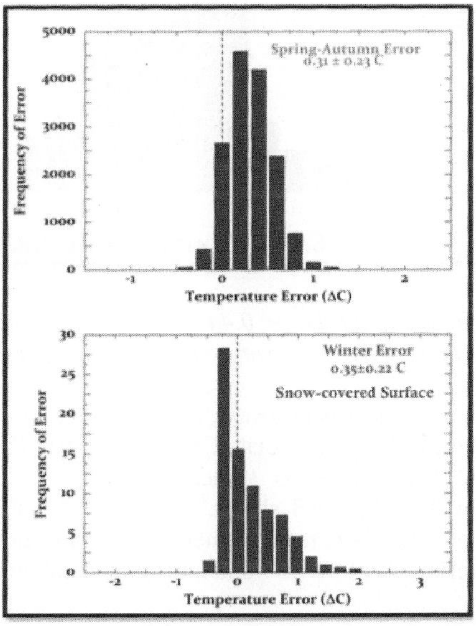

Fig. 7: Average systematic measurement error of an HMP45C PRT probe within an MMTS shelter over a grass (top) or snow-covered (bottom) surface.[11,12]

Figure 7, top, shows the average systematic measurement error an MMTS shield imposed on a PRT temperature probe, found during the calibration experiment displayed in Figure 6.[12] Figure 7, bottom, shows the results of an independent PRT/MMTS calibration over a snow-covered surface.[11] The average annual systematic uncertainty produced by the MMTS shield can be estimated from these data as, $1\sigma = 0.32\pm0.23$°C. The skewed warm-bias distribution of error over snow is similar in magnitude to that produced by the unaspirated RM Young shield in the Plaine Morte experiment (Figure 4).

Figure 8 shows the average systematic measurement error produced by a PRT probe inside a traditional CRS shield.[12]

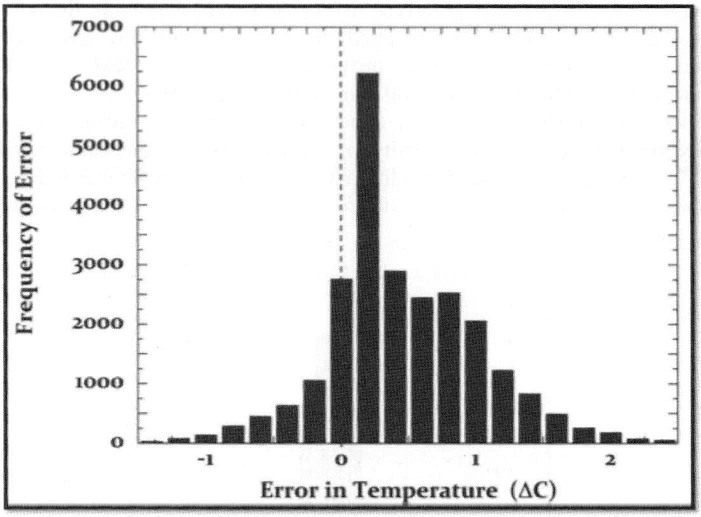

Fig. 8: Average day-night 1σ = 0.44 ± 0.41°C systematic measurement error produced by a PRT temperature probe within a traditional CRS shelter.

The warm bias in the data is apparent, as is the non-normal distribution of error. The systematic uncertainty found associated with the CRS shelter was 1σ = 0.44 ± 0.41°C. The HMP45C PRT probe is at least as accurate as the traditional LiG thermometers housed within the CRS shield.[19,20] The PRT/CRS experiment may then estimate a lower limit of systematic measurement uncertainty present in the land-surface temperature record covering all of the 19th and most of the 20th century.

Sea-Surface Temperature

Although considerable effort has been expended to understand sea-surface temperatures (SSTs),[21-28] there have been very few field calibration experiments of sea-surface temperature sensors. The history and methodological evolution of SST measurements are too complex to review here, however. Bucket- and steamship engine cooling-water intake thermometers provided the bulk of early and mid-20th century SST measurements. Sensors mounted on drifting and moored buoys have come into increasing use since about 1980, and now dominate SST measurements.[29] Attention is therefore focused on calibration studies of these instruments.

The series of experiments reported by Charles Brooks in 1926 are by far the most comprehensive field calibrations of bucket and engine-intake thermometer SST measurements carried out by any individual scientist.[30] Figure 9 presents representative examples of the described systematic error in bucket and engine intake SSTs.

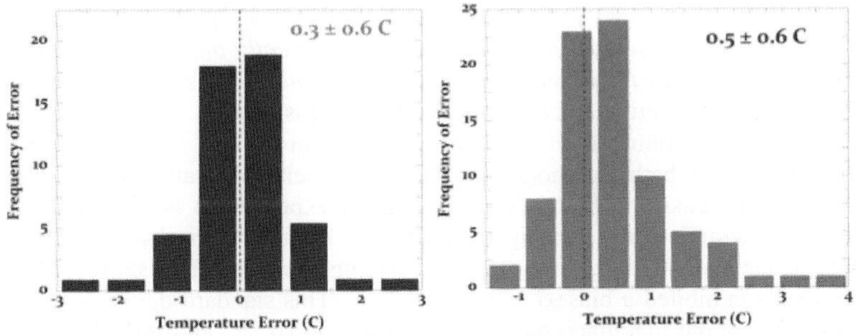

Fig. 9: Systematic measurement error in one set of engine-intake (left) and bucket (right) sea-surface temperatures reported by Brooks.[30]

In 1963 Saur reported the largest field calibration experiment of engine-intake thermometers, carried out by volunteers aboard twelve U.S. military transport ships engaged off the U.S. central Pacific coast.[31] The experiment included 6826 pairs of observations. Figure 10 shows the experimental results from one voyage of one ship.

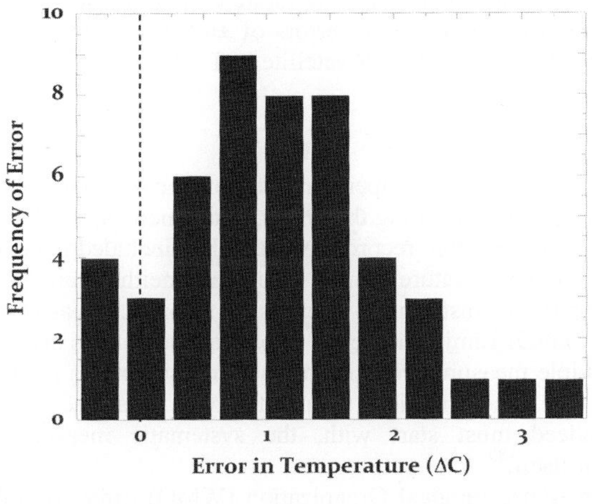

Fig. 10: Systematic error in recorded engine intake temperatures aboard one military transport ship operating June-July, 1959. The mean systematic bias and uncertainty represented by these data are, $1\sigma = 0.9\pm0.6$°C.

Saur reported Figure 10 as, "*a typical distribution of the differences*" reported from the various ships. The ±0.6°C uncertainty about the mean systematic error is comparable to the values reported by Brooks, shown in Figure 9. Saur concluded his report by noting that, "*The average bias of reported sea water temperatures as compared to sea surface temperatures, with 95 percent confidence limits, is estimated to be 1.2±0.6°F* [0.67±0.33°C] *on the basis of a sample of 12 ships. The standard deviation of differences* [between ships] *is estimated to be 1.6°F* [0.9°C]. *Thus, without improved quality control the sea temperature data reported currently and in the past are for the most part adequate only for general climatological studies.* [bracketed comments added]" Saur's caution is instructive, but has apparently been mislaid.

Significant contributions to the SST record have been obtained using measurements from bathythermograph (BT) and expendable bathythermograph (XBT) instruments.[32] Extensive BT and XBT calibration experiments have revealed multiple sources of systematic error, principally stemming from both mechanical problems and calibration lacunae.[33-35] Relative to a reversing thermometer standard, field BT measurements exhibited a bias±σ = 0.34±0.43°C.[35] This standard deviation is more than twice as large as the manufacturer-stated accuracy of ±0.2°C and reflects the impact of uncontrolled field variables.

The SST sensors in deployed floating and moored buoys were never field-calibrated during the 20th century, allowing no general estimate of systematic measurement error. However, Emery estimated a 1σ = ±0.3°C error by comparison of SSTs from floating buoys co-located to within 5 km. [28] SST measurements separated by less than 10 km are considered coincident. A similar ±0.26°C buoy error magnitude was found relative to SSTs retrieved from the Advanced Along-Track Scanning Radiometer (AATSR) satellite.[36] The error distributions were non-normal. More recently, Argo buoys field calibrated against very accurate CTD (conductivity-temperature-depth) measurements exhibited average RMS errors of ±0.56°C,[37] similar in magnitude to the reported average ±0.58°C buoy-AMSR satellite SST difference.[38]

DISCUSSION

Until recently,[39,40] systematic temperature sensor measurement errors were neither mentioned in reports communicating the origin, assessment, and calculation of the global averaged surface air temperature record, nor were they included in error analysis.[15,16,39-46] However, systematic temperature sensor errors are neither randomly distributed nor constant in time, space, or instrument. There is no theoretical reason to expect that these errors follow the Central Limit Theorem,[47,48] or that such errors are reduced or removed by averaging multiple measurements; even when measurements number in the millions. A complete inventory of contributions to uncertainty in the surface air temperature record must include, indeed must start with, the systematic measurement error of the temperature sensor itself.[39]

The World Meteorological Organization (WMO) offers useful advice regarding systematic error.[20]

> "*Section 1.6.4.2.3 Estimating the true value—additional remarks.*
>
> "*In practice, observations contain both random and systematic errors. In every case, the observed mean value has to be corrected for the systematic error insofar as it is known. When doing this, the estimate of the true value remains inaccurate because of the random errors as indicated by the expressions and because of any unknown component of the systematic error. Limits should be set to the uncertainty of the systematic error and should be added to those for random*

errors to obtain the overall uncertainty. However, unless the uncertainty of the systematic error can be expressed in probability terms and combined suitably with the random error, the level of confidence is not known. It is desirable, therefore, that the systematic error be fully determined."

Thus far, in production of the global averaged surface air temperature record, the WMO advice concerning systematic error has been followed primarily in the breach. Systematic sensor error in air and sea-surface temperature measurements has been woefully under-explored and field calibrations are few. Nevertheless, the reported cases make it clear that the surface air temperature record is contaminated with a very significant level of systematic measurement error. The non-normality of systematic error means that subtracting an average bias will not reduce the measurement uncertainty about the global temperature mean.

Further, the magnitude of the systematic error bias in surface air temperature and SST measurements is apparently as variable in time and space as the magnitude of the standard deviation of systematic uncertainty about the mean error bias. I.e., the mean systematic bias error was 2°C over snow on the Plaine Morte Glacier, Switzerland, but was 0.4°C over snow at Lincoln, Nebraska. Similar differences accrue to the engine-intake systematic error means reported by Brooks and Saur. Therefore, removing an estimate of mean bias will always leave the magnitude ambiguity of the residual mean bias uncertainty. In any complete evaluation of error, the residual uncertainty in mean bias will combine with the 1σ standard deviation of measurement uncertainty into the uncertainty total.

A complete evaluation of systematic error is far beyond any scope of the analysis presented here. However, to the extent that the above errors are representative, a set of estimated uncertainty bars due to systematic error can be calculated (Figure 11).

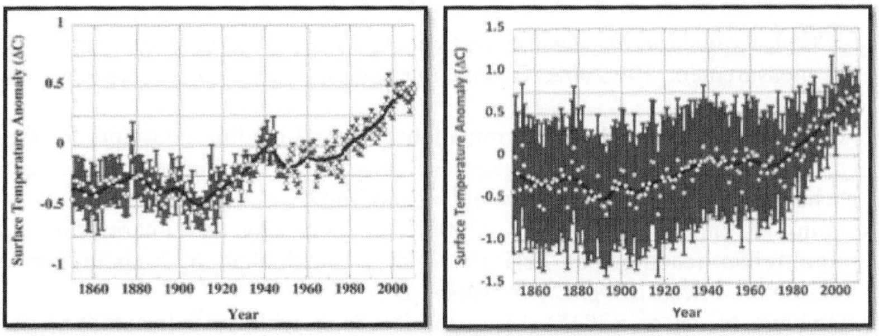

Fig. 11: The 2010 global average surface air temperature record obtained from website of the Climate Research Unit (CRU), University of East Anglia, UK. http://www.cru.uea.ac.uk/cru/data/temperature/. Left, error bars following the description provided at the CRU website. Right, error bars reflecting the uncertainty width due to estimated systematic sensor measurement errors within the land and sea surface records. See the text for further discussion.

The uncertainty bars in Figure 11 (right) reflect a 0.7:0.3 SST:land surface ratio of systematic errors. Combined in quadrature, bucket and engine-intake errors constitute the SST uncertainty prior to 1990. Over the same time interval the systematic error of the PRT/CRS sensor,[39,49] constituted the uncertainty in land-surface temperatures. Floating buoys made a partial contribution (0.25 fraction) to the uncertainty in SST between 1980-1990. After 1990 uncertainty bars are further steadily reduced, reflecting the increasing contribution and smaller errors of MMTS (land) and floating buoy (SS) sensors.

Figure 11 (right) is very likely a more accurate representation of the state of knowledge than is Figure 11 (left), concerning the rate or magnitude of change in the global averaged surface air temperature since 1850. As the revised uncertainty bars represent non-normal systematic error, the air temperature mean loses any status as the most probable value.

Systematic measurement error produced by the instrument itself constitutes a lower limit of uncertainty. The uncertainty estimate developed herein shows that the rate or magnitude of change in global air temperature since 1850 cannot be known within ±1°C prior to 1980 or within ±0.6°C after 1990, at the 95% confidence interval. No support can be found in the surface air temperature record for an emergency concerning climate.

ACKNOWLEDGEMENTS

The author gratefully acknowledges Prof. Hendrik Huwald and Dr. Marc Parlange, Ecole Polytechnique Federale de Lausanne (EPFL), CH-1015 Lausanne, Switzerland, for generously providing the Plaine Morte sensor calibration data entering into Figure 3, Figure 4, and Figure 5. This work was carried out without any external funding.

REFERENCES

1. JCGM, Evaluation of measurement data—Guide to the expression of uncertainty in measurement 100:2008, Bureau International des Poids et Mesures: Sevres, France.
2. Frank, P., et al. (2001) "Determination of ligand binding constants for the iron-molybdenum cofactor of nitrogenase: monomers, multimers, and cooperative behavior." *J. Biol. Inorg. Chem.* 6(7): p. 683-697.
3. Frank, P. and K.O. Hodgson (2005) "Cooperativity and intermediates in the equilibrium reactions of Fe(II,III) with ethanethiolate in N-methylformamide solution." *J. Biol. Inorg. Chem.* 10(4): p. 373-382.
4. Hinkley, N., et al. (2013) "An Atomic Clock with 10^{-18} Instability." *Science*. 341(p. 1215-1218.
5. Huwald, H., et al. (2009) "Albedo effect on radiative errors in air temperature measurements." *Water Resorces Res.* 45(p. W08431; 1-13.
6. Parker, D.E., et al. (1994) "Interdecadal changes of surface temperature since the late nineteenth century." *J. Geophys. Res.* 99(D7): p. 14373-14399.
7. Quayle, R.G., et al. (1991) "Effects of Recent Thermometer Changes in the Cooperative Station Network." *Bull. Amer. Met. Soc.* 72(11): p. 1718-1723; doi: 10.1175/1520-0477(1991)072<1718:EORTCI>2.0.CO;2.

8. Hubbard, K.G., X. Lin, and C.B. Baker (2005) "On the USCRN Temperature system." *J. Atmos. Ocean. Technol.* 22(p. 1095-1101.
9. van der Meulen, J.P. and T. Brandsma(2008) "Thermometer screen intercomparison in De Bilt (The Netherlands), Part I: Understanding the weather-dependent temperature differences)." *International Journal of Climatology*. 28(3): p. 371-387.
10. Barnett, A., D.B. Hatton, and D.W. Jones (1998) Recent Changes in Thermometer Screen Design and Their Impact in Instruments and Observing Methods WMO Report No. 66, J. Kruus, Editor. World Meteorlogical Organization: Geneva.
11. Lin, X., K.G. Hubbard, and C.B. Baker (2005) "Surface Air Temperature Records Biased by Snow-Covered Surface." *Int. J. Climatol.* 25(p. 1223-1236; doi: 10.1002/joc.1184.
12. Hubbard, K.G. and X. Lin (2002) "Realtime data filtering models for air temperature measurements." *Geophys. Res. Lett.* 29(10): p. 1425 1-4; doi: 10.1029/2001GL013191.
13. Menne, M.J. and C.N. Williams (2009) "Homogenization of Temperature Series via Pairwise Comparisons." *J. Climate*. 22(7): p. 1700-1717.
14. Briffa, K.R. and P.D. Jones (1993) "Global surface air temperature variations during the twentieth century: Part 2 , implications for large-scale high-frequency palaeoclimatic studies." *The Holocene*. 3(1): p. 77-88.
15. Hansen, J. and S. Lebedeff (1987) "Global Trends of Measured Surface Air Temperature." *J. Geophys. Res.* 92(D11): p. 13345-13372.
16. Brohan, P., et al. (2006) "Uncertainty estimates in regional and global observed temperature changes: A new data set from 1850." *J. Geophys. Res.* 111(p. D12106 1-21; doi:10.1029/2005JD006548; http://www.cru.uea.ac.uk/cru/info/warming/.
17. Karl, T.R., et al. (1989) "The Recent Climate Record: What it Can and Cannot Tell Us." *Rev. Geophys.* 27(3): p. 405-430.
18. Hubbard, K.G., X. Lin, and E.A. Walter-Shea, (2001) "The Effectiveness of the ASOS, MMTS, Gill, and CRS Air Temperature Radiation Shields." *J. Atmos. Oceanic Technol.* 18(6): p. 851-864.
19. MacHattie, L.B. (1965) "Radiation Screens for Air Temperature Measurement." *Ecology*. 46(4): p. 533-538.
20. Rüedi, I. (2006) WMO Guide to Meteorological Instruments and Methods of Observation: WMO-8 Part I: Measurement of Meteorological Variables, 7th Ed., Chapter 1. World Meteorological Organization: Geneva.
21. Berry, D.I. and E.C. Kent (2011) "Air–Sea fluxes from ICOADS: the construction of a new gridded dataset with uncertainty estimates." *International Journal of Climatology*. p. 987-1001.
22. Challenor, P.G. and D.J.T. Carter (1994) "On the Accuracy of Monthly Means." *J. Atmos. Oceanic Technol.* 11(5): p. 1425-1430.
23. Kent, E.C. and D.I. Berry (2005) "Quantifying random measurement errors in Voluntary Observing Ships' meteorological observations." *Int. J. Climatol.* 25(7): p. 843-856; doi: 10.1002/joc.1167.
24. Kent, E.C. and P.G. Challenor (2006) "Toward Estimating Climatic Trends in SST. Part II: Random Errors." *Journal of Atmospheric and Oceanic Technology*. 23(3): p. 476-486.

25. Kent, E.C., et al. (1993) "The Accuracy of Voluntary Observing Ships' Meteorological Observations-Results of the VSOP-NA." *J. Atmos. Oceanic Technol.* 10(4): p. 591-608.
26. Rayner, N.A., et al. (2003) "Global analyses of sea surface temperature, sea ice, and night marine air temperature since the late nineteenth century." *Journal of Geophysical Research-Atmospheres*. 108(D14).
27. Emery, W.J. and D. Baldwin (1999) "In situ calibration of satellite sea surface temperature." in Geoscience and Remote Sensing Symposium, 1999. *IGARSS '99 Proceedings*. IEEE 1999 International.
28. Emery, W.J., et al. (2001) "Accuracy of in situ sea surface temperatures used to calibrate infrared satellite measurements." *J. Geophys. Res.* 106(C2): p. 2387-2405.
29. Woodruff, S.D., et al. (2007) The Evolving SST Record from ICOADS, in Climate Variability and Extremes during the Past 100 Years, S. Brönnimann, et al. eds, 2007, Springer: Netherlands, pp. 65-83.
30. Brooks, C.F. (1926) "Observing Water-Surface Temperatures at Sea." *Monthly Weather Review*. 54(6): p. 241-253.
31. Saur, J.F.T. (1963) "A Study of the Quality of Sea Water Temperatures Reported in Logs of Ships' Weather Observations." *J. Appl. Meteorol.* 2(3): p. 417-425.
32. Barnett, T.P. (1984) "Long-Term Trends in Surface Temperature over the Oceans." *Monthly Weather Review*. 112(2): p. 303-312.
33. Anderson, E.R. (1980) Expendable bathythermograph (XBT) accuracy studies; NOSC TR 550 1980, Naval Ocean Systems Center: San Diego, CA. p. 201.
34. Bralove, A.L. and E.I. Williams Jr., A Study of the Errors of the Bathythermograph 1952, National Scientific Laboratories, Inc.: Washington, DC.
35. Hazelworth, J.B., Quantitative Analysis of Some Bathythermograph Errors 1966, U.S. Naval Oceanographic Office Washington DC.
36. Kennedy, J.J., R.O. Smith, and N.A. Rayner (2012) "Using AATSR data to assess the quality of in situ sea-surface temperature observations for climate studies." *Remote Sensing of Environment*. 116(0): p. 79-92.
37. Hadfield, R.E., et al. (2007) "On the accuracy of North Atlantic temperature and heat storage fields from Argo." *J. Geophys. Res.: Oceans*. 112(C1): p. C01009.
38. Castro, S.L., G.A. Wick, and W.J. Emery (2012) "Evaluation of the relative performance of sea surface temperature measurements from different types of drifting and moored buoys using satellite-derived reference products." *J. Geophys. Res.: Oceans*. 117(C2): p. C02029.
39. Frank, P. (2010) "Uncertainty in the Global Average Surface Air Temperature Index: A Representative Lower Limit." *Energy & Environment*. 21(8): p. 969-989.
40. Frank, P. (2011) "Imposed and Neglected Uncertainty in the Global Average Surface Air Temperature Index." *Energy & Environment*. 22(4): p. 407-424.
41. Hansen, J., et al. (1999) "GISS analysis of surface temperature change." *J. Geophys. Res*. 104(D24): p. 30997–31022.
42. Hansen, J., et al. (2010) "Global Surface Temperature Change." *Rev. Geophys.* 48(4): p. RG4004 1-29.
43. Jones, P.D., et al. (1999) "Surface Air Temperature and its Changes Over the Past 150 Years." *Rev. Geophys*. 37(2): p. 173-199.

44. Jones, P.D. and T.M.L. Wigley (1992) Corrections to pre-1941 SST measurements for studies of long-term changes in SSTs, in Proc. Int. COADS Workshop, H.F. Diaz, K. Wolter, and S.D. Woodruff, Editors. 1NOAA Environmental Research Laboratories: Boulder, CO. p. 227–237.
45. Jones, P.D. and T.M.L. Wigley (2010) "Estimation of global temperature trends: what's important and what isn't." *Climatic Change*. 100(1): p. 59-69.
46. Jones, P.D., T.M.L. Wigley, and P.B. Wright (1986) "Global temperature variations between 1861 and 1984." *Nature*. 322(6078): p. 430-434.
47. Emery, W.J. and R.E. Thomson (2004) Data Analysis Methods in Physical Oceanography. 2nd ed. Amsterdam: Elsevier.
48. Frank, P., Negligence (2015) "Non-Science, and Consensus Climatology." *Energy & Environmen*t. 26(3): p. 391-416.
49. Folland, C.K., et al. (2001) "Global Temperature Change and its Uncertainties Since 1861." *Geophys. Res. Lett*. 28(13): p. 2621-2624.

SUN-CLIMATE CONNECTION: FROM BOUNDARY DATA TO CLIMATIC RESPONSES

WILLIE SOON
Harvard's Center for Astrophysics
Harvard, MA, USA

Themes

(1) The multi-dimensional problem in obtaining absolute Total Solar Irradiance (TSI) measurements

(2) Avoid being a slave to the false paradigm of greenhouse gas forcings and feedbacks by the IPCC and others

(3) Present this topic in context of General Circulation Model (GCM) representations of physical, chemical and biological processes—consider only the model simulation of absolute climatology

So, how well can the 36 climate models in Paleoclimate Modeling Intercomparison Project (PMIP) simulate mid-Holocene Climate Optimum temperatures in China?

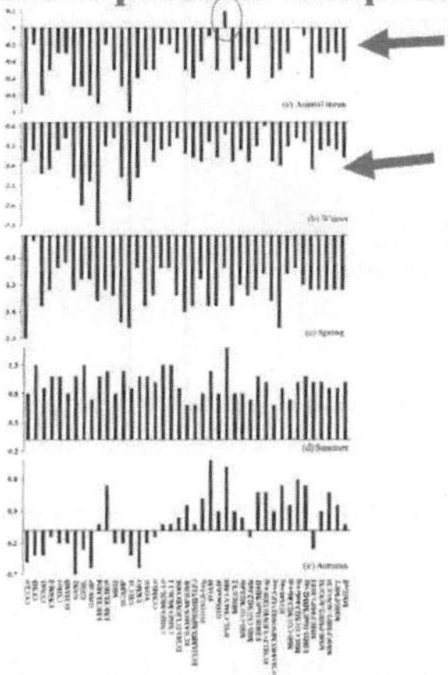

35 out of 36 and 36 out of 36 PMIP models simulated cooler annual-mean and winter temperatures, respectively, during Mid-Holocene (6000 years ago) than present

Jiang et al. (2012) Journal of Climate, vol. 25, 4135-4153

So, how well can the 36 climate models in Paleoclimate Modeling Intercomparison Project (PMIP) simulate mid-Holocene Climate Optimum temperatures in China?

"[A] few studies suggested that winter warming was stronger than the annual mean during the mid-Holocene. The existence of *Ceratopteris* [subtropical species] at Baiyangdian [of North China Plain] and Tancheng during the mid-Holocene ... implied that winter temperature was about 6 K warmer than it is today ... The discovery of nuts from Helicia plants dating back to mid-Holocene at Baohuashan indicated a 6.3 K warming in January ... Stable forest in the Qinghai Lake area during the mid-Holocene implied a January warming of 8 K ... Pollen records from Bayanchagan, Hidden, Hulun, and Ren Co Lakes also indicated a strong warming in winter ... Using the proxy data before the 1990s, winter warming was estimated to be 70-100% stronger than that of the annual mean [January mid-Holocene temperatures up to 10 to 11 K warmer than present], particularly in Eastern China." (p. 4147)

Jiang et al. (2012) Journal of Climate, vol. 25, 4135-4153

Standard IPCC Storyline: DOES NOT change in the AR5

Warming Post-1980 Cannot Be Explained Without CO_2

IPCC AR5 (2013) Chapter 10, FAQ 10.1 Figure 1 (Final Draft, June 7, 2013 version)

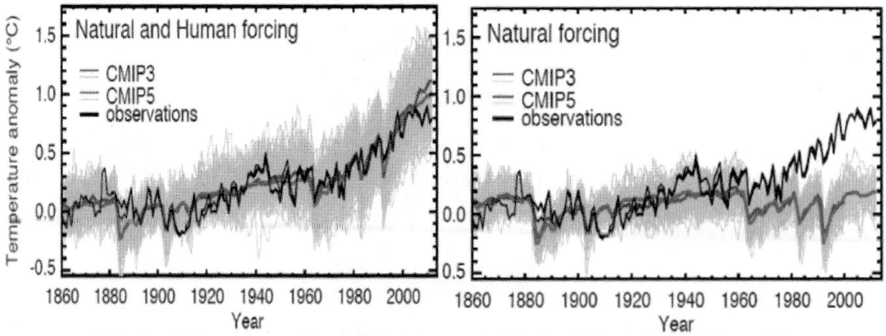

147 simulations from 17 models **52 simulations from 17 models**

Beware of the normalization present in the graphical representation in these IPCC charts!

Reality: Does this looks like reasonable agreement between observation and model calculations of Earth temperature?

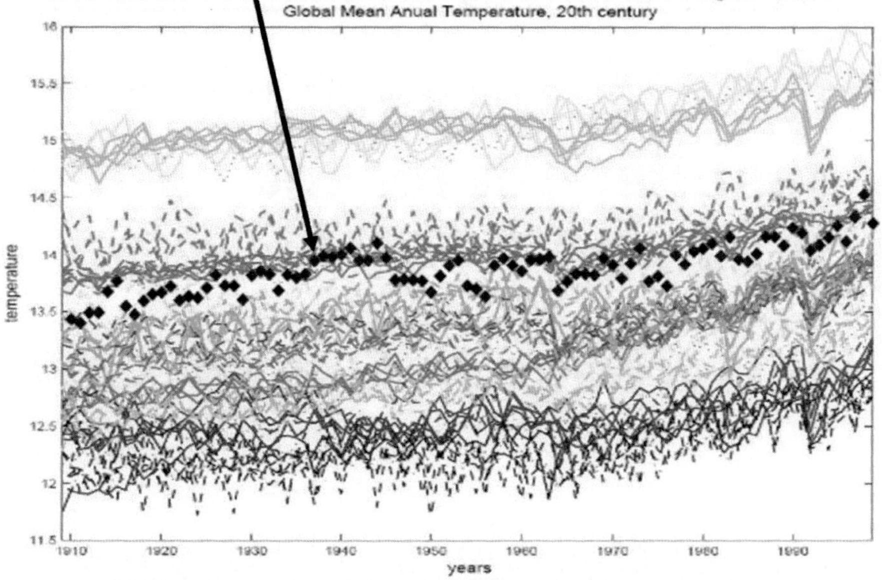

Important public science-education-policy questions

If CO_2 control on the climate is so self-apparent and scientifically correct, then why the graphical "confusion"?

Will climate scientists stand up and condemn the IPCC+USGCRP+USEPA?

Key conclusions on climate models

(1) GCMs may get the right results, but for the wrong reasons!

(2) Garbage In Gospel Out!

Key conclusions on climate models

(3) GCMs hide the key physics – Absolute TSI calibration problem

p. 8-32: "However, the few tenths of a percent bias in the absolute TSI value has a minimal consequences for climate simulations ..."

Untrue! Allowing the freedom of setting any TSI value hides the fundamental problem in getting the correct mean climatology of the Earth.

Thirteen instruments with 13 TSI values: How do you correctly normalize them all?

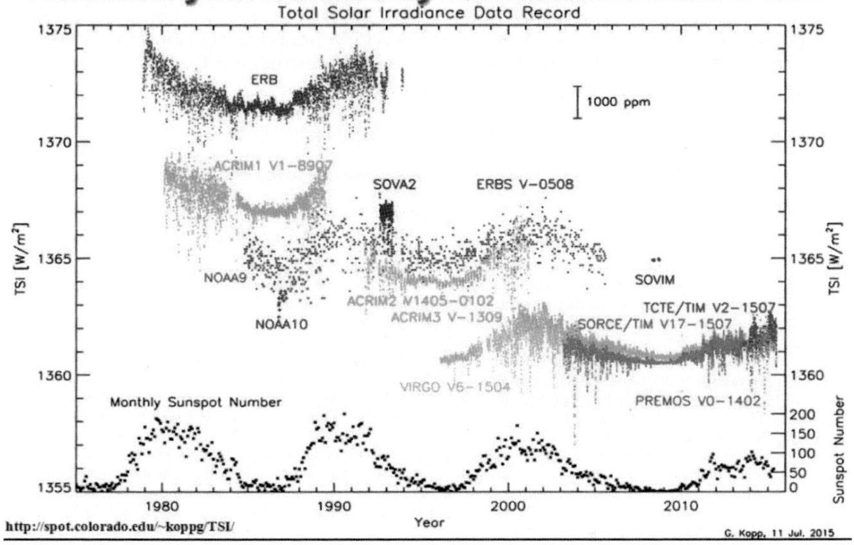

How well can we measure the Earth's radiation and energy budgets?

The best measurements of Earth's radiation budget by CERES has an imbalance of 6.5 W/m²

Toward Optimal Closure of the Earth's Top-of-Atmosphere Radiation Budget

NORMAN G. LOEB,* BRUCE A. WIELICKI,* DAVID R. DOELLING,* G. LOUIS SMITH,+
DENNIS F. KEYES,# SEIJI KATO,* NATIVIDAD MANALO-SMITH,# AND TAKMENG WONG*

*NASA Langley Research Center, Hampton, Virginia
+National Institute of Aerospace, Hampton, Virginia
#Science Systems and Applications, Inc., Hampton, Virginia

(Manuscript received 14 May 2008, in final form 18 July 2008)

ABSTRACT

Despite recent improvements in satellite instrument calibration and the algorithms used to determine reflected solar (SW) and emitted thermal (LW) top-of-atmosphere (TOA) radiative fluxes, a sizeable imbalance persists in the average global net radiation at the TOA from satellite observations. This imbalance is problematic in applications that use earth radiation budget (ERB) data for climate model evaluation, estimate the earth's annual global mean energy budget, and in studies that infer meridional heat transports. This study provides a detailed error analysis of TOA fluxes based on the latest generation of Clouds and the Earth's Radiant Energy System (CERES) gridded monthly mean data products [the monthly TOA/surface averages geostationary (SRBAVG-GEO)] and uses an objective constrainment algorithm to adjust SW and LW TOA fluxes within their range of uncertainty to remove the inconsistency between average global net TOA flux and heat storage in the earth–atmosphere system. The 5-yr global mean CERES net flux from the standard CERES product is 6.5 W m^{-2}, much larger than the best estimate of 0.85 W m^{-2} based on observed ocean heat content data and model simulations. The major sources of uncertainty in the CERES estimate are from instrument calibration (4.2 W m^{-2}) and the assumed value for total solar irradiance (1 W m^{-2}). After adjustment, the global mean CERES SW TOA flux is 99.5 W m^{-2}, corresponding to an albedo of 0.293, and the global mean LW TOA flux is 239.6 W m^{-2}. These values differ markedly from previously published adjusted global means based on the ERB Experiment in which the global mean SW TOA flux is 107 W m^{-2} and the LW TOA flux is 234 W m^{-2}.

Loeb et al. (2009) Journal of Climate, vol. 22, 748-766

Toward Optimal Closure of the Earth's Top-of-Atmosphere Radiation Budget

NORMAN G. LOEB,* BRUCE A. WIELICKI,* DAVID R. DOELLING,* G. LOUIS SMITH,+
DENNIS F. KEYES,# SEIJI KATO,* NATIVIDAD MANALO-SMITH,# AND TAKMENG WONG*

*NASA Langley Research Center, Hampton, Virginia
+National Institute of Aerospace, Hampton, Virginia
#Science Systems and Applications, Inc., Hampton, Virginia

"The 5-yr global mean CERES net flux from the standard CERES product is 6.5 W/m^2, much larger than the best estimate of 0.85 W/m^2 based on observed ocean heat content data and model simulations. The major source of uncertainty in the CERES estimate are from instrument calibration (4.2 W/m^2) and the assumed value for total solar irradiance (1 W/m^2)."

Loeb et al. (2009) Journal of Climate, vol. 22, 748-766

"Heat" is hiding in the deep, abyssal ocean

"Some basic elements of the sampling problem are compiled in Table 2. About 52% of the ocean lies below 2000 m and about 18% below 3600 m. By defining a volume as having been 'probed' if at least [one station] existed within a roughly 60x60 km^2 box in the interval 1992-2011 … [A]bout 1/3 (11% of total volume) of water below 2000 m was sampled … Of the 18% lying below 3600 m, about 17% was measured. … [M]any papers assume no significant changes take place in the deep ocean over the historical period … The history of exploration suggests, however, that blank places on the map have either been assumed to be without any interesting features and dropped from further discussion, or at the other extreme, filled with 'dragons' invoked to explain strange reports (G. de Jode, 1578)." —
Wunsch and Heimbach (2014)

The best measurements of Earth's radiation budget by CERES has an imbalance of 6.5 W/m²

TABLE 1. Global mean clear- and all-sky SW, LW, and net TOA radiative fluxes, solar irradiance, and CRE for satellite-based data products (units in W m^{-2}).

		CERES				
Product name	ERBE S-4	ES-4 Ed2_rev1	SRBAVG-nonGEO Ed2D_rev1	SRBAVG-GEO Ed2D_rev1	GEWEX SRB Version 2.86	ISCCP FD
Time period	02/85 – 01/89			03/00 – 02/2005		
Solar irradiance	341.3	341.3	341.3	341.3	341.8	341.5
LW (All sky)	235.2	239.0	237.7	237.1	240.4	235.8
SW (All Sky)	101.2	98.3	96.6	97.7	101.7	105.2
Net (All Sky)	4.9	4.0	7.0	6.5	−0.3	0.5
LW (Clear Sky)	264.9	266.6	266.4	264.1	268.1	262.3
SW (Clear Sky)	53.6	49.3	51.2	51.1	54.5	54.2
Net (Clear Sky)	22.8	25.4	23.7	26.2	19.2	25.0
LW CRE	29.7	27.6	28.7	27.0	27.7	26.5
SW CRE	−47.6	−49.0	−45.4	−46.6	−47.2	−51.0
NET CRE	−17.9	−21.4	−16.7	−19.7	−19.5	−24.5

Loeb et al. (2009) Journal of Climate, vol. 22, 748-766

Atmos. Chem. Phys., 11, 13421–13449, 2011
www.atmos-chem-phys.net/11/13421/2011/
doi:10.5194/acp-11-13421-2011

Atmospheric Chemistry and Physics

Earth's energy imbalance and implications

J. Hansen[1,2], M. Sato[1,2], P. Kharecha[1,2], and K. von Schuckmann[3]
[1]NASA Goddard Institute for Space Studies, New York, NY 10025, USA
[2]Columbia University Earth Institute, New York, NY 10027, USA
[3]Centre National de la Recherche Scientifique, LOCEAN Paris, hosted by Ifremer, Brest, France

Received: 2 September 2011 – Published in Atmos. Chem. Phys. Discuss.: 29 September 2011
Revised: 30 November 2011 – Accepted: 7 December 2011 – Published: 22 December 2011

How does he know this?

The precision achieved by the most advanced generation of radiation budget satellites is indicated by the planetary energy imbalance measured by the ongoing CERES (Clouds and the Earth's Radiant Energy System) instrument (Loeb et al., 2009), which finds a measured 5-yr-mean imbalance of 6.5 W m^{-2} (Loeb et al., 2009). Because this result is implausible, instrumentation calibration factors were introduced to reduce the imbalance to the imbalance suggested by climate models, 0.85 W m^{-2} (Loeb et al., 2009).

Beware of the highly misleading post-2009 paper by CERES PI: The budget imbalance of 6.5 W/m² is hidden!

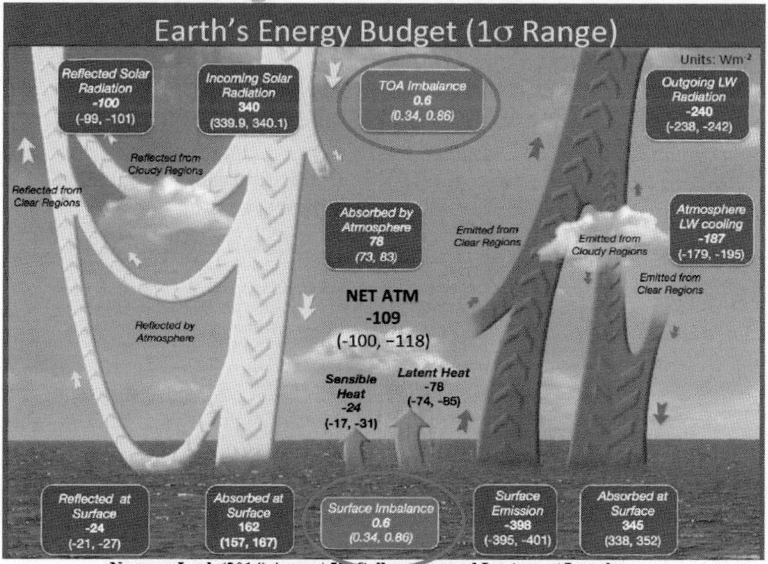

Norman Loeb (2014) August 5's Colloquium and Lecture at Langley

The view from the now very popular 4-D assimilation reanalysis:

Quarterly Journal of the Royal Meteorological Society Q. J. R. Meteorol. Soc. 137: 553–597, April 2011 A

RMetS
Royal Meteorological Society

The ERA-Interim reanalysis: configuration and performance of the data assimilation system

D. P. Dee[a*], S. M. Uppala[a], A. J. Simmons[a], P. Berrisford[a], P. Poli[a], S. Kobayashi[b], U. Andrae[c], M. A. Balmaseda[a], G. Balsamo[a], P. Bauer[a], P. Bechtold[a], A. C. M. Beljaars[a], L. van de Berg[d], J. Bidlot[a], N. Bormann[a], C. Delsol[a], R. Dragani[a], M. Fuentes[a], A. J. Geer[a], L. Haimberger[e], S. B. Healy[a], H. Hersbach[a], E. V. Hólm[a], L. Isaksen[a], P. Kållberg[c], M. Köhler[a], M. Matricardi[a], A. P. McNally[a], B. M. Monge-Sanz[f], J.-J. Morcrette[a], B.-K. Park[g], C. Peubey[a], P. de Rosnay[a], C. Tavolato[e], J.-N. Thépaut[a] and F. Vitart[a]

[a] *European Centre for Medium-Range Weather Forecasts, Reading, UK*
[b] *Japan Meteorological Agency, Tokyo, Japan*
[c] *Swedish Meteorological and Hydrological Institute, Norrköping, Sweden*
[d] *European Organisation for the Exploitation of Meteorological Satellites, Darmstadt, Germany*
[e] *Department of Meteorology and Geophysics, University of Vienna, Austria*
[f] *School of Earth and Environment, University of Leeds, UK*
[g] *Korea Meteorological Administration, Seoul, Korea*
*Correspondence to: D. P. Dee, ECMWF, Shinfield Park, Reading RG2 9AX, UK. E-mail: dick.dee@ecmwf.int

The view from the now very popular 4-D assimilation reanalysis:

3.1.6. *Incoming solar radiation*

For solar irradiance, ERA-Interim uses a constant value of 1370 W m^{-2} throughout, i.e. no account is taken of the solar cycle. Variations due to the varying distance between the Earth and the Sun are incorporated as described in Paltridge and Platt (1976).

Due to a programming error in the calculation of incident solar radiation as a function of solar zenith angle, the global solar radiation in ERA-Interim is overestimated by about 2 W m^{-2}. This error, which has been present in the ECMWF forecast model for a very long time, has also affected ERA-40. Model simulations have shown that the systematic impact of this error on the model climate is confined to the upper stratosphere, which is warmed by approximately 1 K globally.

Earth's Energy Budget: The view from reanalysis assimilation

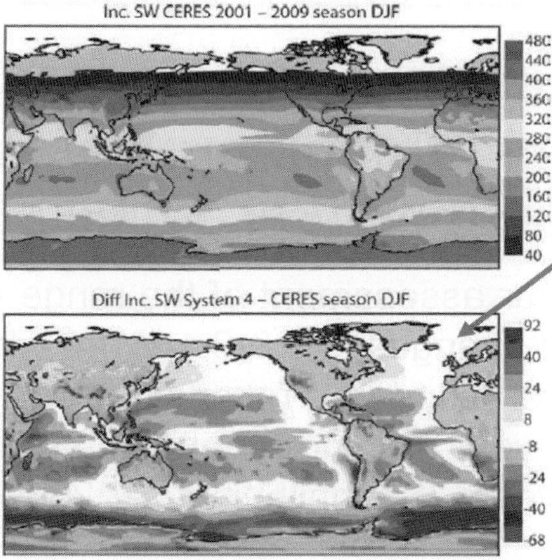

Regional differences in the observed solar radiation by CERES and the SW from current ERA's System 4 forecasting system model are much larger than 6.5 W/m²

Dee et al. (2014) Bulletin of the American Meteorological Society

Key conclusions on climate models

(4) Dangerous ignorance: TSI reconstruction problems in Steinhilber et al. (2009, 2012)

p. 8-33: "Our assessment of the range of RF from TSI changes is 0.0-0.1 W/m^2 which covers several updated reconstructions using the same 7-year running mean past-to-present minima years (Steinhilber et al., 2009 …"

A careful look into the most popular reconstructed TSI values and results by Steinhilber et al. (2009 and 2012)

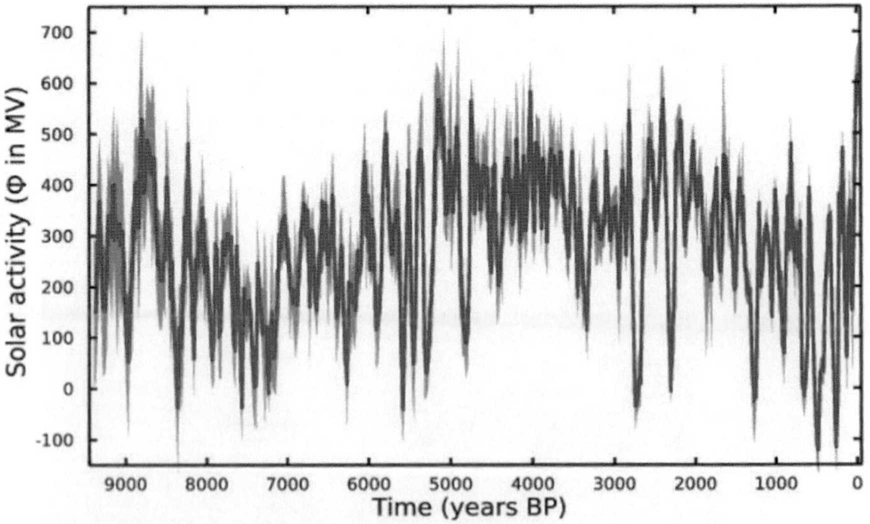

Steinhilber et al. (2012) PNAS, vol. 109, 5967-5971; (2009) GRL, vol. 36, 2009GL040142

A careful look into the most popular reconstructed TSI values and results by Steinhilber et al. (2009 and 2012)

Highly questionable TSI-B_r correlation and its physical relevance

3. Reconstruction of Total Solar Irradiance

[18] After converting ϕ into B_r, TSI is calculated using the TSI-B_r relationship of *Fröhlich* [2009]

$$\begin{aligned} \mathrm{TSI} = {} & (1364.64 \pm 0.40)\,\mathrm{Wm^{-2}} \\ & + (0.38 \pm 0.17)\,\mathrm{Wm^{-2}\,nT^{-1}}\,B_r. \end{aligned} \tag{4}$$

Steinhilber et al. (2012) PNAS, vol. 109, 5967-5971; (2009) GRL, vol. 36, 2009GL040142

A careful look into the most popular reconstructed TSI values and results by Steinhilber et al. (2009 and 2012)

Highly questionable TSI-B_r correlation and its physical relevance

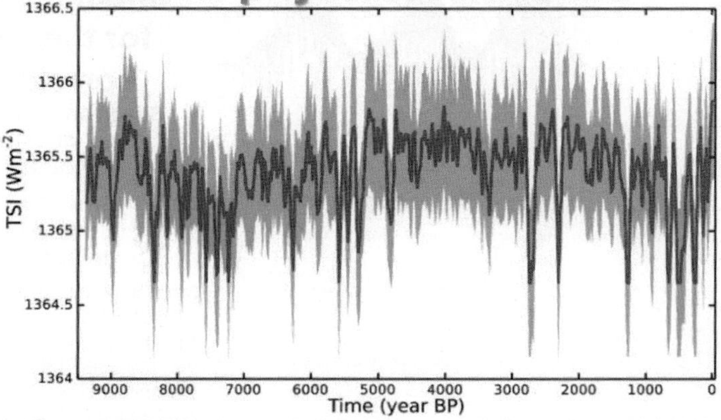

Steinhilber et al. (2012) PNAS, vol. 109, 5967-5971; (2009) GRL, vol. 36, 2009GL040142

The IPCC has split its pants!

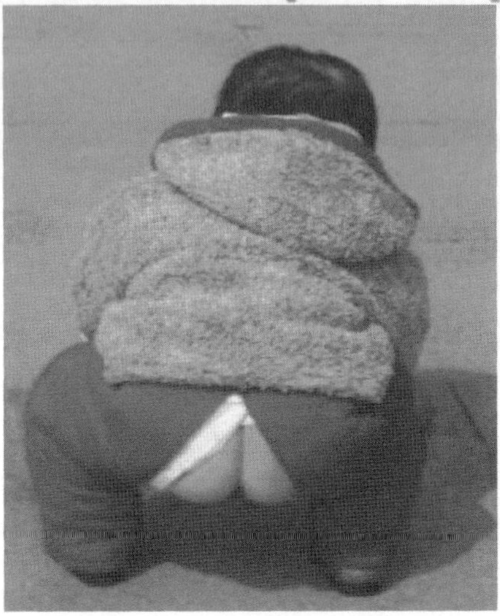

A careful look into the most popular reconstructed TSI values and results by Steinhilber et al. (2009 and 2012)

Highly questionable TSI-B_r correlation and its physical relevance

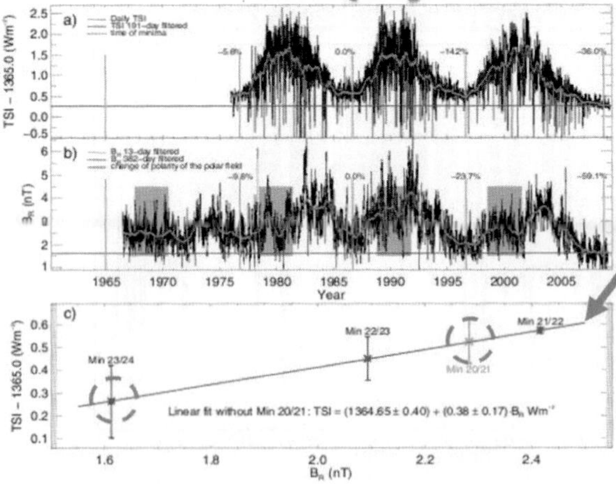

Shocking basis for the TSI-B_r correlation:

Based on only 2 points plus 2 questionable points!!!

Two points that should NOT be included in the fit

Frohlich (2009) Astronomy and Astrophysics, vol. 501, L27-L30
Steinhilber et al. (2012) PNAS, vol. 109, 5967-5971
Steinhilber et al. (2009) GRL, vol. 36, 2009GL040142

The schizophrenic TSI-B_r correlations from Frohlich's many papers?

Why do the values of TSI and B_R keep changing ?

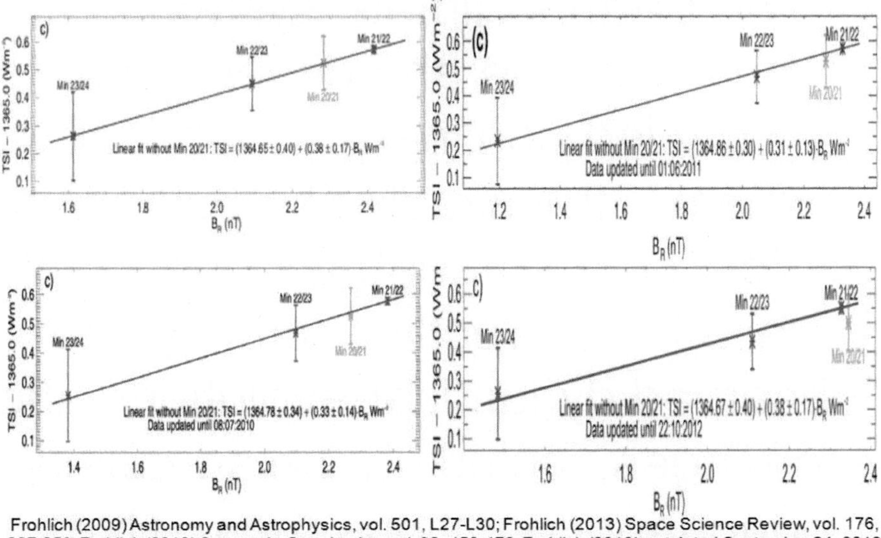

Frohlich (2009) Astronomy and Astrophysics, vol. 501, L27-L30; Frohlich (2013) Space Science Review, vol. 176, 237-252; Frohlich (2012) Surveys in Geophysics, vol. 33, 453-473; Frohlich (2013)-ppt dated September 24, 2013

Frohlich's latest file dated May 3, 2013

Five different versions in five years!

Note all four points now are fit for the latest TSI-B_r correlation

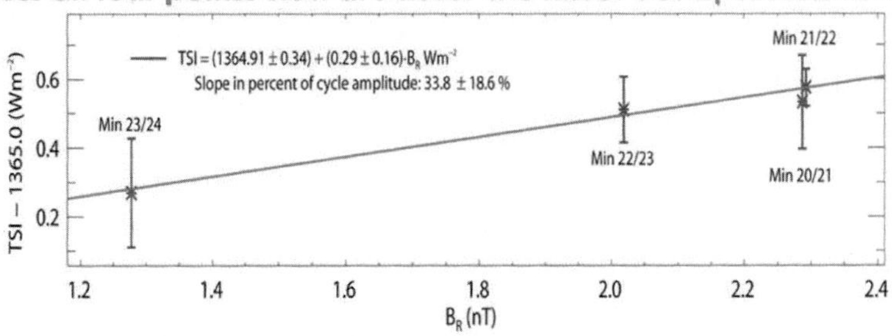

One can hardly accept such a weak result (based strictly on a correlation of two points!!!) let alone argue that changes in the Sun's TSI are small and thus have minimal or no influence on global climate!

Conclusion:

The Earth's climate system is a multi-variable and dimensional, coupled nonlinear dynamical system – the simplistic description as forcing and feedback has simply failed in correctly capturing the observed dynamical evolutions and changes, including the so-called boundary conditions themselves.

The most current climate models used in the IPCC reports do not conserve mass!

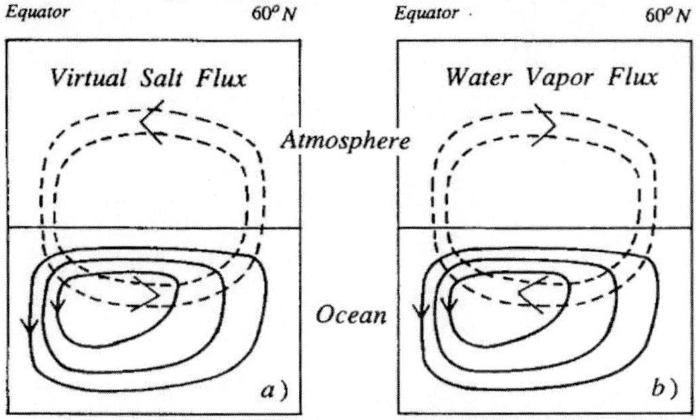

FIG. 3. (a) Meridional circulations in an atmosphere–ocean coupled model. Old models based on virtual salt flux. (b) New model based on natural boundary condition of plain water flux.

A "Dragon" is hiding in the deep, abyssal ocean

A "Dragon" is hiding in the deep, abyssal ocean

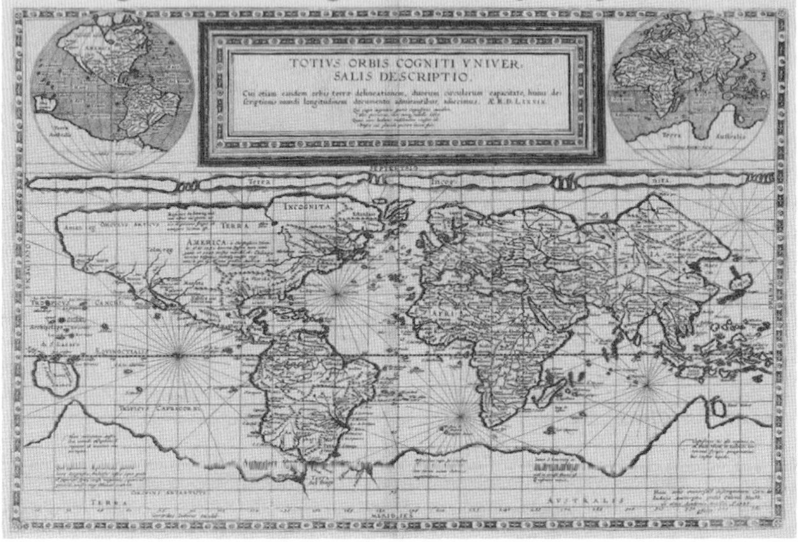

A "Dragon" is hiding in the deep, abyssal ocean

James Marshall & Marie-Louise Osborn Collection/Beinecke Rare Book & Manuscript Lib.

A prototype wetsuit — one of several innovations presented in the 1590s manuscript MS Osborn a8.

DYNAMICAL EVIDENCE FOR CAUSALITY BETWEEN GALACTIC COSMIC RAYS AND GLOBAL TEMPERATURE

ANASTASIOS TSONIS
Department of Mathematical Sciences, University of Wisconsin-Milwaukee
Milwaukee, WI, USA

ETHAN R. DEYLE AND GEORGE SUGIHARA
Scripps Institution of Oceanography, University of California San Diego, La Jolla, CA, USA

ROBERT M. MAY
Department of Zoology, University of Oxford, Oxford UK

KYLE L. SWANSON AND JOSHUA D. VERBETEN
Department of Mathematical Sciences, University of Wisconsin-Milwaukee, Milwaukee, WI, USA

GELI WANG
LAGEO Institute for Atmospheric Physics, Chinese Academy of Sciences, Beijing, CHINA

THE DYNAMICS

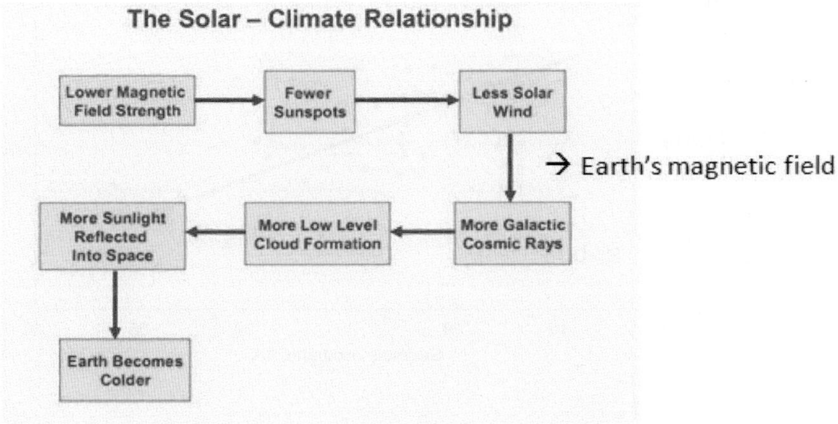

At its simplest, the relationship between the solar magnetic field strength and the Earth's climate is this: lower magnetic field strength means few sunspots, fewer sunspots means less solar wind, less solar wind means more galactic cosmic rays, more galactic cosmic rays means more low level cloud formation, more low level clouds means more sunlight reflected back into space, which in turn means less heating of the Earth's surface and atmosphere.

GCR AND SUNSPOTS

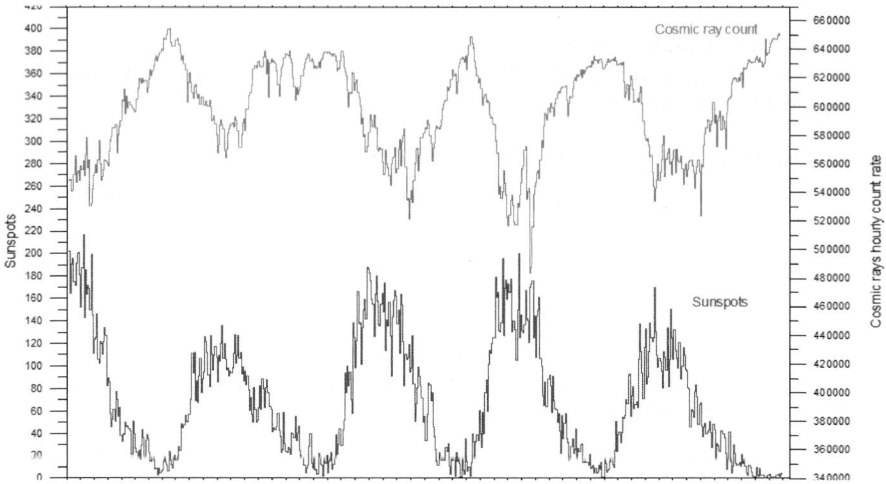

GCR AND AA INDEX

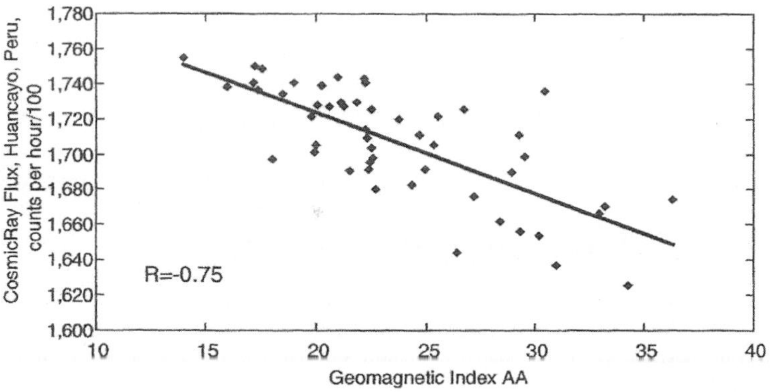

CORRELATION BETWEEN CLOUD COVER AND CR

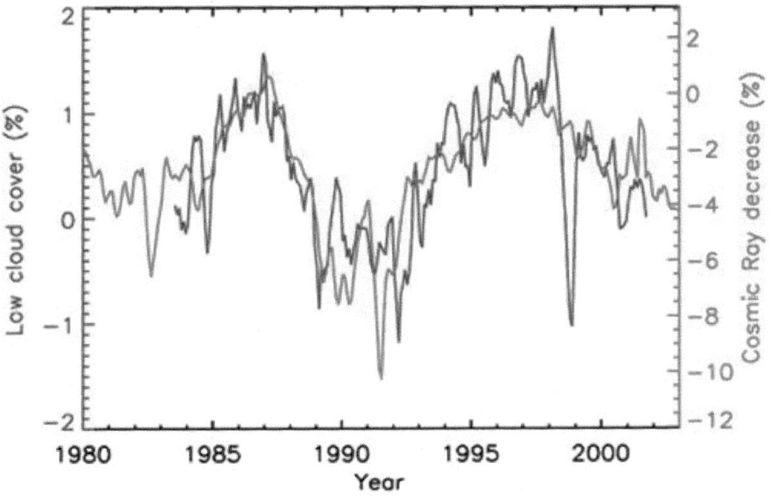

LORENZ SYSTEM

$$dx / dt = -ax+ay$$

$$dy/ dt = -xz+bx-y$$

$$dz/ dt = xy-cz$$

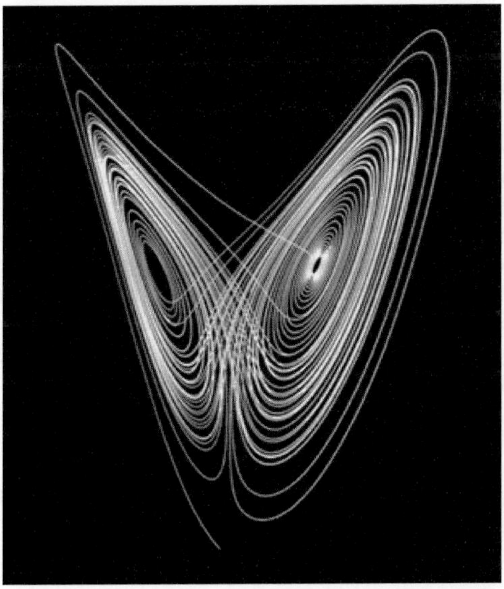

TAKEN'S THEOREM

Convergent cross mapping (CCM) tests for correspondence between shadow manifolds.

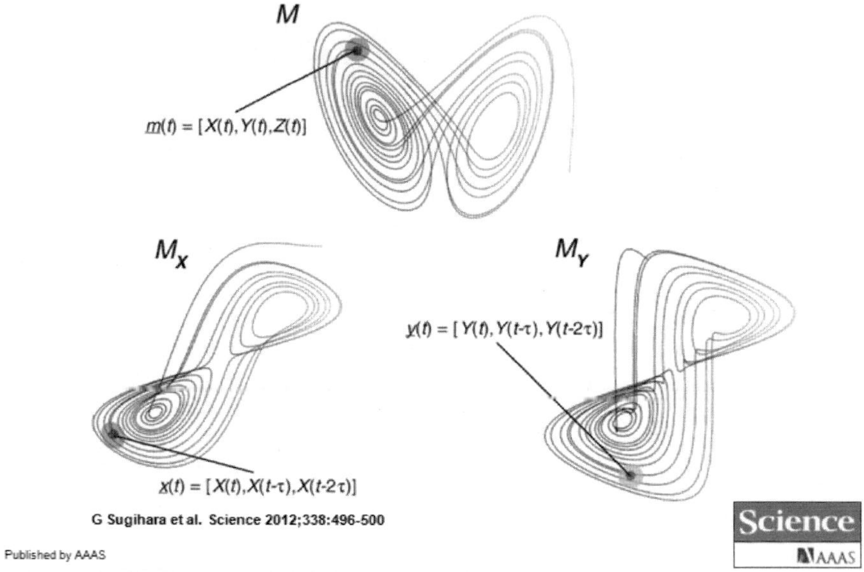

Convergent cross mapping (CCM) tests for correspondence between shadow manifolds. This example based on the canonical Lorenz system (a coupled system in X, Y, and Z; eq. S7 without V) shows the attractor manifold for the original system (M) and two shadow manifolds, M_X and M_Y, constructed using lagged-coordinate embeddings of X and Y, respectively (lag = τ). Because X and Y are dynamically coupled, points that are nearby on M_X (e.g., within the red ellipse) will correspond temporally to points that are nearby on M_Y (e.g., within the green circle). That is, the points inside the red ellipse and green circle will have corresponding time indices (values for t). This enables us to estimate states across manifolds using Y to estimate the state of X and vice versa using nearest neighbors (*3*). With longer time series, the shadow manifolds become denser and the neighborhoods (ellipses of nearest neighbors) shrink, allowing more precise cross-map estimates (see movies S1 to S3).

Main point

When X causes Y, information about X is imprinted in Y. We say Y cross maps X and prediction of X, based on nearby neighbors of a terminal point of a time series of Y, is good and its accuracy should increase as the Y "library" increases in size.

COSMIC RAYS VS. GLOBAL TEMPERATURE

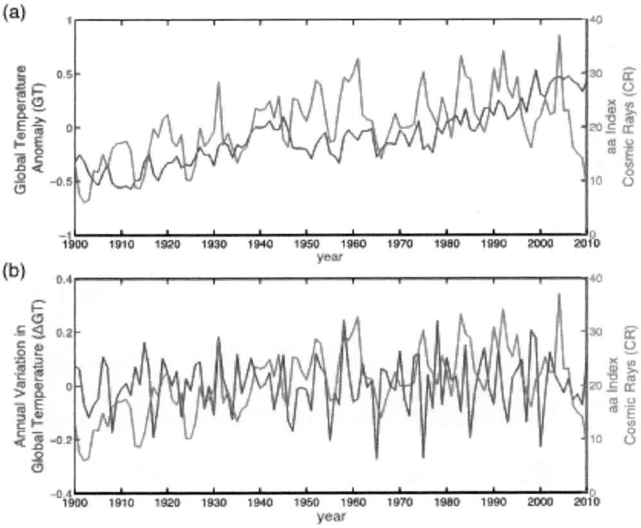

EVIDENCE FOR DYNAMICS

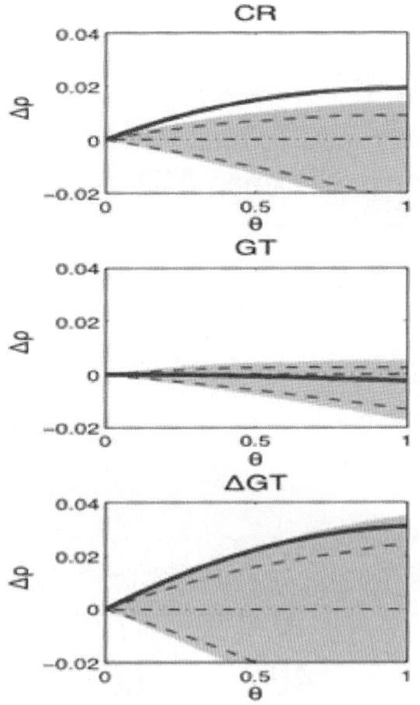

CCM (D=5) AND SURROGATES 1

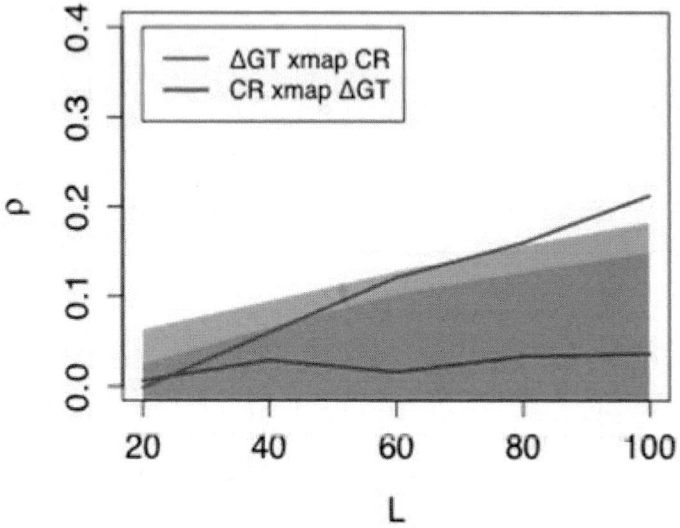

CCM (D=5) AND SURROGATES 2

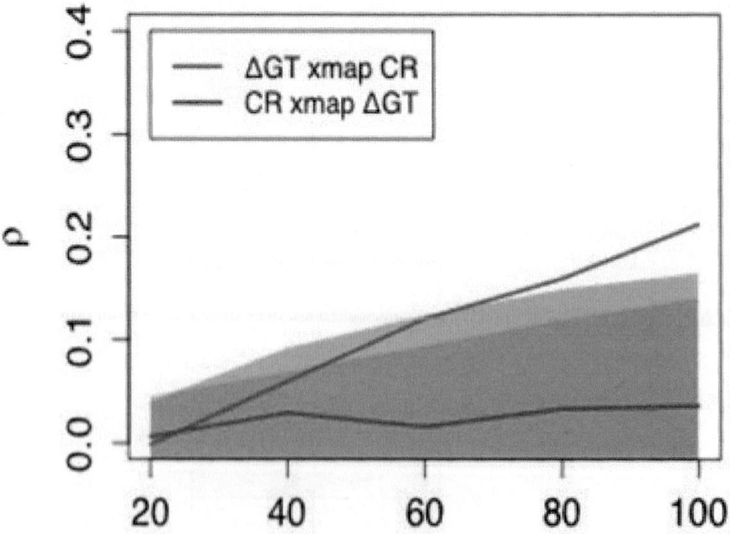

CCM AND SURROGATES 1

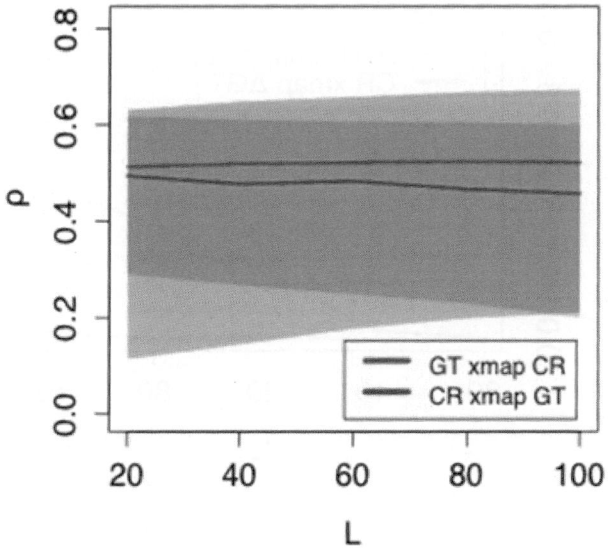

CCM BETWEEN CR AND CCSM4 NCAR MODEL

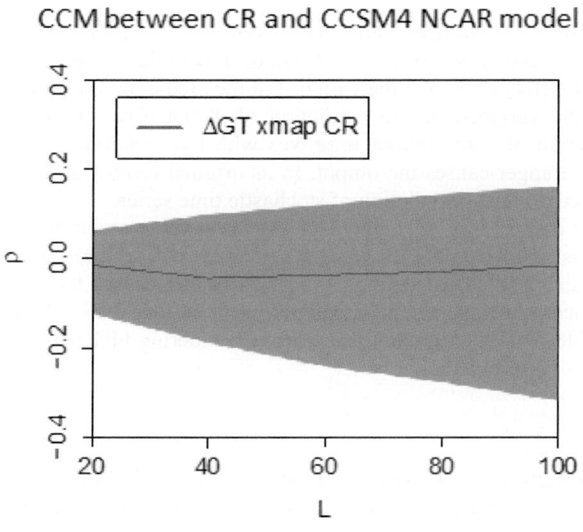

CCM EMBEDDING DIMENSION 8

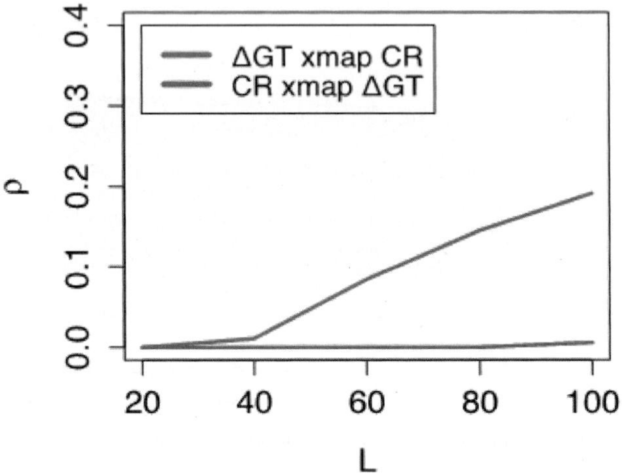

We have confirmed these results with actual CR data measured at the surface (Colorado data), but due to limited data the statistical significance (85%) is not as high as with the proxy data (95%).

GRANGER CAUSALITY

- According to Granger causality, given two simultaneously recorded time series X_t and Y_t where $t = 1, N$ denotes sampling times, we say that Y has causal influence on X if the variance of the prediction error of X given Y is less than the variance of the prediction of X not given Y. This means that if prediction of some output improves with the addition of an input, then the input Granger-causes the output. In its original formulation Granger causality is based on linear prediction of stochastic time series.
- In our case if X is ΔGT and Y is CR the p value assuming an AR-1 (AR-2) process is 0.82 (0.97). If X is CR and Y is ΔGT, the respective p values are 0.81 and 0.64. Thus neither X nor Y Granger-causes the other. In fact, the variance explained in the prediction error is less than 1% regardless which variable is used to predict the other. Considering higher order AR processes does not improve these results.

CONCLUSIONS

- Our results indicate weak to moderate coupling between cosmic rays and year-to-year changes of global temperature. They resonate with the physical and chemical evidence emerging from laboratory studies suggesting, in theory, a dynamical link between galactic cosmic rays and global temperature. In the name of completeness, it should be noted that they indicate the presence of a significant non-traditional forcing in the climate system, which should not be ignored. This is another interesting piece of the puzzle in our understanding of factors influencing climate variability.

SESSION 9

ENERGY FOR THE POOR

ENERGY FOR THE RURAL POOR: A SOLVABLE HUMAN PLANETARY EMERGENCY

WILLIAM BARLETTA
U.S. Particle Accelerator School, Department of Physics, Massachusetts Institute of Technology, Cambridge, MA, USA

WHAT IS ENERGY POVERTY?

Energy poverty can be defined as the lack of sufficient and safe modern energy for basic needs. A minimum consumption of commercial energy is equivalent to one ton of oil (TOE = 11.6 MWh) per capita per year. This definition is consistent across all countries with acceptable standards of living i.e., Human Development Index (HDI) ≥0.8.

The essential appliances for basic human needs in rural area are two light bulbs, a fan, a small fridge, a basic television or computer and a small pump for subsistence level irrigation. This list implies that electricity should cost <$0.10/kWh. Moreover, electricity is not the only energy need. Modern fuels to replace wood charcoal and dung and improved stoves are also needed.

WHY IS ENERGY POVERTY IMPORTANT?

Energy is the wellspring of human development and a basic contributor to human health and well-being. The critical importance of energy supplies as view by every nation-state is demonstrated by the role of oil in international politics.

Of its many forms, electricity is the most important form of modern energy for development. It can be easily and efficiently converted to heat, light and mechanical power for end use through appliances. Therefore, sufficient access to electricity must be a focus of global efforts to alleviate poverty and its related social ills.

WHERE ARE THE ENERGY POOR?

For those who live in the industrialized world personal access to electrical power must seem virtually limitless. In fact for industrialized nations, the population's level of access to energy is frequently limited only by individuals' interest in consumption. According to the World Energy Council, in the U.S. and Canada the residential per capita energy use exceeds 4.5 MWh/yr while in Germany the figure is only ~1.7 MWh/yr. Residential electricity use in France is double that in Italy.

In contrast meaningful access, that is, electricity consumption for basic needs, is beyond the reach of the poorest 4 billion. The average residential use in India is only 131 kWh/yr and in Nigeria it is only 74 kWh/yr. In fact, ~20% of the world's population (~80% rural) has no access to electricity for heating, cooking and lighting. (The percentage of the population with no access is in China ~ 1%, while in India it is ~ 33% and in Sub-Saharan Africa it is ~72%, i.e. ~600M. people).

A plot of energy usage has been given by Berkeley's Laboratory Institute for Globally Transformative Technologies (Figure 1).[1]

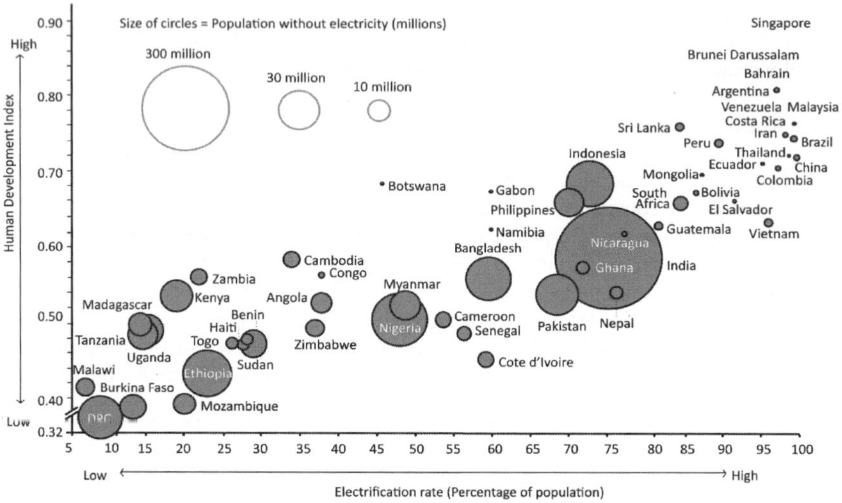

Fig. 1: Where is energy poverty? (From reference 1).

WHAT ARE THE EFFECTS OF ENERGY POVERTY?

Countries with lower electrification rates are clearly worse off in terms of human development. Health facilities, food security, education, and sustainability are all severely limited by the lack of electricity. Access to clean water and sanitation is constrained without effective pumping capacity.

Health is severely impacted by the household use of primitive solid fuels as they lead to a high incidence of pulmonary diseases. In addition, energy poverty limits human rights in countries having large populations without access to electricity. This is an issue of social justice. In those countries, the most severe burdens of energy poverty (including ill health) fall disproportionally on women.

WHY IS ENERGY POVERTY A NON-LINEAR PROBLEM?

Despite having the most acute need for cheap energy, the poor must resort to dirtier and less efficient forms of energy with a higher unit price. For example, according to the World Bank the unit cost of lighting with kerosene is ~70x higher than the most expensive thermal generation of electricity. Light from a candle costs 150x more.

[1] "*The 50 Most Transformative Scientific & Technological Breakthroughs Required for Sustainable Global Development*" by LBNL Institute for Global Transformative Technologies *https://ligtt.org/50-breakthroughs*, (2014). Herein after, "50 Breakthroughs."

Unfortunately, a study by the UN Development Agency study[2] finds that modern fuels and improved stoves to meet most basic cooking needs are out of the reach of the majority of people in rural areas of developing countries.

WHAT WOULD IT COST TO ALLEVIATE ENERGY POVERTY?

Roughly 80 million additional people need reliable energy access every year. The problem is not only large; it is growing rapidly. The International Energy Agency estimates[3] that to provide universal modern energy access by 2030, the global investment would have to average $48 B/year with the majority being focused on sub-Saharan Africa. Of this amount $18 B/yr is needed for grid extension and $26B/yr to construct off-grid systems.

WHAT IS BEING DONE?

For countries of the Organisation for Economic Co-operation and Development (OECD) energy security and mitigating the impact on climate change dominate the global energy agenda. In contrast, the present annual global investment in energy poverty is ~$9B with the primary emphasis being incremental change—for short of the amount needed to alleviate energy poverty in the next 20 years. The estimate of present efforts as compared with what is needed can be seen in Figure 2 taken from the IEA report.[3]

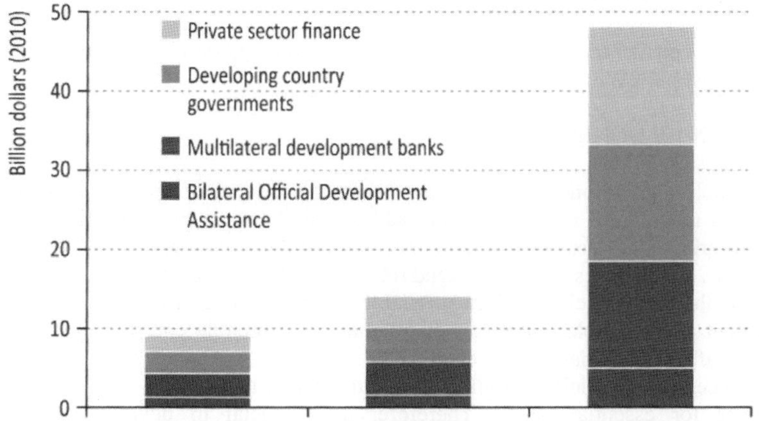

Fig. 2: Annual average investment in modern energy access by source of financing.

[2] Energy for a Sustainable Future from the UNDP, www.un.org/.../AGECCsummaryreport%5B1%5D, (2010).

[3] *World Energy Outlook: Energy for All* from the IEA, http://www.worldenergyoutlook.org/resources/energydevelopment/energyforallfinancingaccessforthepoor/.

IS THIS ONLY A QUESTION OF MONEY?

Although efforts at incremental change using stand alone energy supply need to be continued, incremental advances will not do the job. It is unrealistic to assure that the level of global investment needed will quintuple to the amounts estimated to do the job using present commercial technology. Therefore, industrialized nations must lower the cost of providing universal access to electricity by making large investments in energy science aimed at technological breakthroughs in energy supply, energy storage and electrical energy distribution. Success in these endeavors also will provide large economic benefits to industrialized nations. While electricity is not the only energy need of the rural poor; it is an essential first start.

WHAT ARE COST TARGETS?

The upper segment of low-income users (making $3-5/day) can afford to pay $0.10/kWh for reliable electricity, while the affordable electricity budget of the poorest four billion people is ~$8/month. Either the cost of electricity must be highly subsidized by the governments of developing countries or the actual costs must be reduced dramatically. Presently estimates of actual costs (depending on the source of generation) are as follows:

- The cost of generation is $0.20-0.60/kWh;
- With currently available storage, this cost doubles.
- Distribution adds another $0.05-0.15/kWh.

To meet the economic capabilities of the poor, modern technology must drive down the cost per kWh.

WHAT CAN BE DONE?

Unfortunately stand-alone systems cannot sustain the level of appliance usage that is intrinsic to meaningful access.[4] Instead, addressing shortfalls by expanding robust off-grid mini-grids could provide an additional ~29 million people annually with access for basic needs. These are essential to extend reliable electricity access to rural areas.

As the most efficient and most reliable source of energy for the rural poor will be intermittent (for example, via solar photovoltaics), affordable energy storage, which is efficient, safe, easy to maintain, is essential.

Access to electricity is useful only to the extent that people can use it to operate appliances for essential needs. Therefore, it is crucial to develop super-efficient equipment and appliances that allow electricity to deliver greater benefits within a given budget. SEAD, the Super-efficient Equipment and Appliance Development Initiative,[5] is aimed at this goal.

---

[4] "50 Breakthroughs".

[5] SEAD is a voluntary collaboration among governments to promote the manufacture, purchase, and use of energy-efficient appliances, lighting, and equipment worldwide. http://www.superefficient.org/.

WHY IS SCIENCE IMPORTANT TO THIS PROBLEM?

Energy technology breakthroughs can improve access to electricity to levels essential for human needs:

1. Radically cheaper solar PV for affordable off-grid electricity generation. The technology has the potential to be widely available, easiest to operate and maintain, highly modular, dependable, and have the highest potential for cost-reduction. Realizing the full promise of solar energy requires multi-scale solutions (from the nanoscale to kilometer scale) and cross-scale integration. The underpinnings are discovery science and accelerated materials discovery that drive new paradigms in component assembly and integration. These enable production scale-up and cross-scale integration that from a basis for further discovery science, etc.
2. Rechargeable battery storage technology to enable affordable and reliable renewable electricity in off-grid mini-grids in the cost range of \$0.10/kWh. It should be highly modular, robust, durable, compact and easy to transport or build. The challenge for battery storage is depicted in Figure 3.[6]

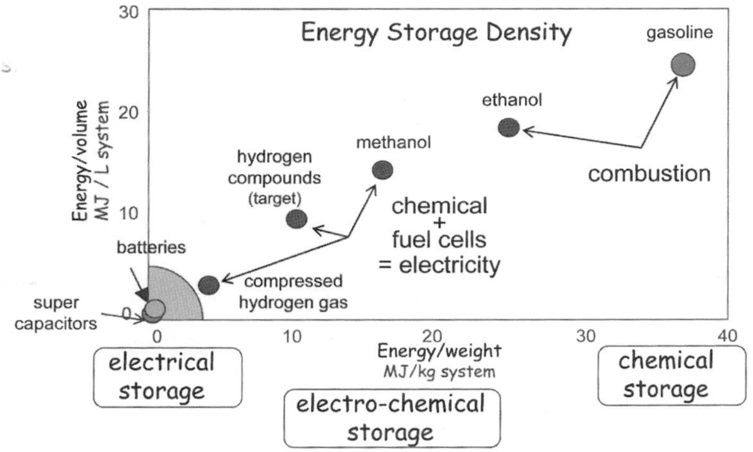

Fig. 3: The challenge for advanced battery technology.

3. Super-efficiency, cheaper appliances using new materials. Super-efficiency is loosely defined as achieving large efficiency gains of 50% or more in essential appliances.

---

[6] George Crabtree, Private communication.

We have a moral imperative to repair the world.[7] Alleviating energy poverty is a planetary emergency. Our inability to complete the job is no excuse for our not starting in earnest.

ACKNOWLEDGEMENTS

I extend my thanks to my Berkeley colleagues, Prof. Ashok Gadgil and Dr. Shashi Buluswar for their essential help in developing this presentation. George Crabtree has provided key information on the prospects for developing highly improved batteries.

[7] The Hebrew expression is tikkun olam (to repair the world), a practice that helps to avoid social disharmony.

SESSION 10

TERRORIST MOTIVATIONS

WOMEN IN THE ISLAMIC STATE

LYDIA WILSON
Centre for the Resolution of Intractable Conflict, Harris Manchester College
Oxford, UK

This was actually taken in the Kurdish region of Iraq—I'm sure the Islamic State wouldn't appreciate the face showing!

Less attention than males going to join the Islamic State.

Beginning to change with more press attention, with articles that fall into two categories: one misguided and one based on work being done at ICSR/ISD.

Misguided still in assumptions that these young people, male and female, are on the fringes of society in some way; drop outs, loners, perhaps petty criminals and involved with drugs. This is not the profile of most Western women we have seen make it there or who have been stopped. Another common assumption about the women in particular is that they are escaping a conservative background with no access to men to find a husband— "jihadi bride".

The other half all go back to ISD/ICSR research—all. A recent example was an article in the NYT just a few days ago about the "Bethnal Green Three" (again, all based on this work in King's and the ISD).

BETHNAL GREEN THREE

Top students, happy families, popular: what pushed them to go to the Islamic State?

PUSH FACTORS

Complex and variable, but common factors include:

- Feeling isolated from the mainstream culture.
- Anger at perceived Islamophobia, inside and outside the country.
- Anger/sadness at foreign policy in the Muslim World, both interventionist and standing by.

PULL FACTORS

- Romantic ideas.
- Finding a stronger Muslim identity, with a strong sense of belonging and sisterhood.
- Personal messages asking for them as individuals.
- **Most of all**: the historic opportunity to build a utopia on earth and fulfill God's instructions.

We have seen this idealism many times through history, with communism springing to mind.

ROMANCE OF A JIHADI

Women are being actively recruited with many roles open to them. The Islamic State needs female doctors, teachers, nurses, and even religious police. The Al-Khansaa brigade is a female religious police brigade, as a response to an attack on a checkpoint by men dressed in bruqas—the Islamic State saw the need of people who could check women. This soon developed into a force that ensured sharia law was being adhered to and also to deal out the punishments. (Interestingly, the punishments have driven some women away from the State altogether: we have two witness reports from women who made it to Turkey who say they couldn't stomach the violence. Though that's not to say this is a female phenomenon: men have also left due to the violence, and the BBC recently showed a member of the Al-Khansaa brigade who wasn't bothered by the floggings she carried out at all, saying that if they had broken God's law, someone had to do the appropriate punishments.)

COMPARISONS: COMMUNISM

And I have met it in the communist group the PKK [Slide of Shorosh] and other places in Kurdistan, with Westerners going to the region to support the fledgling state in the Kurdish part of Syria.

COMPARISONS: PKK

OK, this is a man, but I've heard about women joining up, and I have to say—I've been tempted myself: the camaraderie, the conviction, the simplicity and rigor of life. There are other comparisons I am looking at: the Dutch women who went to join FARC and fought in Columbia, and the "Arrow Girls", the equivalent of the "Arrow Boys", formed in response to women.

NEXT STEPS

- More work on Arabic language sources.
- A comparison with other ideologies that have drawn commitment from people in other countries.
- **Direct reports**: the Australian family's slaves who escaped to tell the tale.

ZEYNAB, 14, FROM AUSTRALIA

This is Zeynab Shourouf's twitter profile photo. Zeynab is very active on social media, exhorting sisters outside the caliphate to join as soon as they can; praising her beloved husband; praising the Islamic State and life there.

There are many fascinating pieces to the Australian's family's story, but for my purposes, their case study has provided a way to assess the online activity of Zeynab. For the Yazidi slaves that her father obtained managed to escape and have spoken about Zeynab's life, reporting that she was desperate to go back to Australia, begging her father to allow her. It is vital for understanding the trajectory of these idealistic young women. (It would also be invaluable of course in adding to the picture of life under the Islamic State.)

LESSONS

- We have to take these women seriously, and not dismiss them as "brainwashed adolescents".
- It is vital to understanding any gaps between the face they give to the world and the reality.
- Only from this more nuanced understanding can we begin to come up with counter-narratives and techniques for prevention and deradicalisation.

We took communists who defected to the USSR seriously. A pscyhologist in the group who sadly couldn't be with us this year, Baruch Fischoff, who feels that there is a gap in general in developmental psychology, as a result of not treating adolescents in general more seriously. He would like to know in what ways these are and are not within the normal range of developmental processes, so this will hopefully be another strand to the research.

The Islamic State has issued guidelines in posting on social media, and many families have described a feeling that there are people standing behind their children or siblings telling them what to say—or at least sometimes.

I have plenty of ideas for counter narratives and work with young people in the West, but there is no point sharing them in public until we have much more evidence concerning the individuals, the general trends in adolescent development, the nature of the spread of idealism, and of how this idealism in particular spreads and appeals.

INSTRUCTIONS FROM ISIL

- Act offensive rather than defensive—say, "You are agents of Iran, rather than we are not agents of Iran".
- Keep spreading the words of truth even if 'haters' dismiss it.
- Work on exposing the lies of leaders/presidents and religious clerks.
- Create original posts rather than re-tweeting.

FEARS ABOUT THE IRANIAN DEAL AND THE POTENTIAL FOR LOOMING ARMS RACE IN THE MIDDLE EAST

RICHARD DAVIS
Davis Energy
Washington DC, USA

HIGHLIGHTS OF RESEARCH CONDUCTED IN SUPPORT OF THE PMPT—MOTIVATIONS GROUP

1. Willingness to commit costly sacrifices for values.
2. Cognitive Structure of Moral Decision Making Leading to Terrorism.
3. Processes of radicalization and de-radicalization—development of a deterrence, prevention and de-radicalization framework.
4. Mutual Influence of Moral Values, Mental Models and Social Dynamics of Intergroup Conflict.
5. Dynamics of Values within Armed Groups (tension between values and responsibilities to host populations).
6. Theoretical Development of the 'Devoted-Actor' from real-world conflict environments.
   - Violation of transitivity in preferences.
   - Rule-bound logic of moral appropriateness.
7. Others—See www.artisinternational.org.

RESPONSES TO IRAN NUCLEAR DEAL

- Nuclear Deal Signed 14 July 2015.
- China, France, Germany, Russia, Iran and the EU have adopted or will adopt.
- United States—Congress must vote on ratification of deal before September 12.
  - A Majority in U.S. Congress is opposed and will vote to not adopt deal.
  - President is expected to veto Congress.
  - According to House Leadership, the support for veto over-ride is growing as un-decided members join the opposition.
  - Only 100 of 2500 vetoes have been overcome by 2/3 Congress (2 of 34 national security based vetoes).
- Saudi Arabia, Egypt, Turkey and other Gulf States are declaring intent on becoming nuclear power as a result of Iran's program.

Note: we currently have an attitudinal survey running in Iran on perceptions to the nuclear deal.

OPPOSITION TO DEAL IS FEAR-BASED

Fear Based Arguments include:

- Iran has declared intent of destroying another country.

- Iran's active support of Hezbollah, Hamas and the Houthis—the deal provides further resources for Iran to support these groups.
- Iran wants to export the Shia revolution against Sunni Arab interests.
- Iran will cheat in the deal and sanctions 'snap-back' will be ineffective due to the complexities.
- Deal gives Iran too much with insufficient enforcement mechanisms.
- In 10-15 years Iran will be permitted to have the bomb and will be able to act with impunity in the region.

An arms race in the Middle East will result.

REGIONAL REACTION TO DEAL

- Saudi Arabia declared intent to pursue 'nuclear capabilities', which goes beyond the 'Pakistan Relationship'.
  - Privately—Saudis expect Iran to cheat: In response to Q: 'if Iran cheats, will Saudi violate the NPT?' A: 'Yes. Everyone will'.
  - Publicly—they are saying they will buy deterrence.
- Egypt, Turkey and other Gulf States are expressing fear and a need to have a 'balance-of-power' in the region (code for interest in related nuclear capabilities).

OUR SCIENTIFIC ENDEAVORS

- Mutually Assured Destruction is still dominant framework for 'balance of power' related to nuclear, which suggests regional proliferation may result. New theoretical pathway needs to be explored and developed.
- Accepting the fears as *real (not just political rhetoric)* and understanding ways to address and reduce those fears with cultural sensitivity.
- Enhanced technical and policy pathways for nations [including Iran] to go nuclear for energy and medicine without dual use weaponization interests or capabilities.
- Developing population based understandings of the ways to enhance popular support against weaponization.

CURRENT RESEARCH (ARTIS AND OXFORD)

- Work directly with the leaders and populations in the Middle East.
- Dissecting power-politics from values based interests in nuclear capabilities. Understanding related aspirations and fears that drive interests.
- Identification of pathways for leaders and populations to resist nuclear weaponization capabilities.
- Pathways for acceptance of non-dual use of nuclear technology.
- Alternative theoretical framework to Mutually Assured Destruction.

DEVELOPMENTS IN THE PSYCHOLOGY OF RELIGIOUS FUNDAMENTALISM AND ITS RELATION TO POLITICALLY-MOTIVATED VIOLENCE

JOHN ALDERDICE
House of Lords
London, UK

WHY FOCUS ON MOTIVATIONS?

- The range and variability of vulnerabilities to attack is too great.
- If we can better understand the motivation for attack, it may be more possible to undermine the wish to attack.
- The 'rational actor' model of understanding of motivations has failed.
- Attempts to use military force to 'mitigate' the effects of terrorism have made the situation worse—Afghanistan—Iraq—Libya.
- The United Nations as an instrument of the will of the international community is deeply split.

WHAT HAVE WE LEARNT ABOUT MOTIVATIONS?

- Where there is serious on-going violence, we always find a group of people who feel humiliated and disrespected.
- They also have a perception of profound injustice and 'unfairness'.
- The usual 'democratic' methods of redress have been frustrated and unavailable to resolve these feelings, adding to the anger and rage.
- Motivation is therefore essentially emotional—indeed the words come from same root; it starts with how they feel not what they think.
- The change in the way the large group **thinks** is a **result.**
- All is exacerbated once action is taken in expression of the rage.

WHAT CHANGES ABOUT THE WAY OF THINKING?

- The sense of group coherence increases and significance of individual differences reduces.
- Eventually my identity becomes fused with that of the large group—the mosaic and complexity of identities and thinking dissolves.
- My own best social and economic interests are no longer critical drivers in decision-making.
- 'Sacred Values' become the key drivers, and the 'grammar' or 'syntax' of thinking changes—logical, rational, best-interest judgement is lost—fundamentalism is a **modality** of 'large group' thinking.

FUSION

- We all have a sense of identification with various groups of which we feel a part and it contributes to our complex identity.
- We call it 'fusion' when a single social identity becomes the central feature of our identity, with an exclusive level of significance.
- If another group member is threatened, I am threatened.
- I am prepared to make very costly and even total sacrifices for the group—my well-being, my family, my life, my children.
- The group/community is not bonded by kinship but by shared powerful emotional experiences.

<u>Testing for Fusion</u>

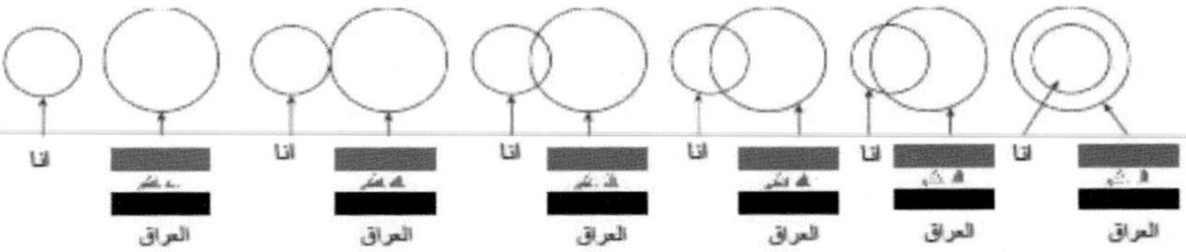

SACRED VALUES

- Primary group identity is bounded by 'Sacred Values'—not necessarily religious beliefs but transcendental ideologies (Ghandi and salt).
- When people become 'fused' with their group (because of feelings of threat/humiliation/unfairness) rational thinking dissolves.
- Sacred Values thinking is different—processed in the brain as duties bound by rules rather than utilitarian cost/benefit judgements.
- Offering material benefits to compromise 'Sacred Values' often backfires and produces rage and the potential for violence.
- Symbolic interactions can engage with Sacred Values thinking.

THE DANGERS OF RATIONAL JUDGEMENT

In Alan Greenspan's testimony to Congress, 23 October 2008, he confessed that he had been shocked that in the face of the global economic crisis the markets did not operate as he had expected. When pressed he went further and said:

> *"... an ideology is a conceptual framework with the way people deal with reality. Everyone has one. You have to—to exist, you need an ideology. The question is whether it is accurate or not....and what I'm saying to you is, yes, I found a flaw. I don't know how significant or permanent it is, but I've been very distressed by that fact."*

The flaw was that he believes that free markets would operate rationally so that...

"... the self-interests of organizations, specifically banks and others were such as they were best capable of protecting their own shareholders and their equity in the firms."

In other words, Greenspan believed (with some evidence over years of relative stability) that the banks and commercial companies were rational (and moral) actors, and he was shocked to discover during the global economic crisis that they were not. Treating them as rational actors resulted in economic catastrophe.

A THEORY OF FUSION, SACRED VALUES AND IRRATIONALITY

- In neurological development the simple, predictable, and reflex is inhibited to enable the evolution of complex, adaptable and voluntary.
- Damage leads to loss of higher functions, resulting in re-emergence of primitive reflexes, and gradual reparative redevelopment.
- Individual mental development is similar—trauma results in a loss (usually partial) of function and reappearance of 'primitive' function.
- Do large groups also show loss of more developed thinking and judgement, reappearance of the earlier and more primitive function?
- Can we engage with a different 'grammar' and encourage 'repair'?

LESSONS LEARNT ABOUT MOTIVATION

- Motivation is driven by emotion—how we feel.
- How we feel **then** affects how we think.
- Under threat individual identity becomes 'Fused' with group identity.
- The way of thinking changes to 'Sacred Values'—fundamentalism is about the way of thinking, more than the content of the thinking.
- Responding with socio-economic thinking not only does not work—it is counter-productive.
- We are learning about a different mode of large group functioning.
- Need to differentiate the small group terrorism and Daesh insurgency.

SESSION 11

EMERGING RISK OF ENDOCRINE DISRUPTING HERBICIDE

EMERGING RISK OF GLYPHOSATE

PETERSON MYERS
Environmental Health Sciences
Charlottesville, VA, USA

J. Rockström et al. (2009) "A safe operating space for humanity." *Nature*. 461:472–475. doi:10.1038/461472a; pmid:19779433

IN THE BEGINNING...

Glyphosate-based herbicides entered commerce in 1970s.

- Deemed 'safer than table salt' by producers.
- Based on mechanism of action.
- Largely unpublished data from manufacturer.
- Using toxicology not reflecting today's science.

TODAY...

- Most widely used pesticide in history.
- Initial assumptions about safety now challenged.
- Data on different mechanisms of action.
- IARC: 'probable human carcinogen'.

THIS PLENARY:

- Prof. Charles Benbrook on history and extent of use.
- Prof. Michael Antoniou on toxicological understanding.
- Patrick Holden, CBE, on the future of agriculture without glyphosate.

SOURCES AND MECHANISMS OF HEALTH RISKS FROM GLYPHOSATE-BASED HERBICIDES

MICHAEL ANTONIOU
King's College London School of Medicine, Department of Medical and Molecular Genetics, London, UK

ROBIN MESNAGE
King's College London School of Medicine, Department of Medical and Molecular Genetics, London, UK

Earth Open Source: free download

www.earthopensource.org

Chapter 4 on glyphosate-based herbicide toxicity

Accepted Manuscript

Potential toxic effects of glyphosate and its commercial formulations below regulatory limits

R. Mesnage, N. Defarge, J. Spiroux de Vendômois, G.E. Séralini

PII: S0278-6915(15)30034-X
DOI: 10.1016/j.fct.2015.08.012
Reference: FCT 8369

To appear in: *Food and Chemical Toxicology*

Industry & regulators say glyphosate/Roundup is safe!

Basis of Claims for Glyphosate Safety

High permitted acceptable daily intake (ADI) for glyphosate due to perceived:

- **Low toxicity based solely on EPSPS enzyme inhibition (shikimate biochemical pathway present in plants & some bacteria)**
- **Rapid excretion via urine / faeces**
  No bioaccumulation
- **[Assumption: urine levels will always be higher than those in blood, breast milk, organs]**

Current Acceptable Daily Intake (ADI) / chronic Reference Dose (cRfD)

- EU/Aus: 0.3mg/kg/day
- China/Russia: 1mg/kg/day
- USA: 1.75mg/kg/day

➢ ***Are these exposures really safe?***

Environmental & Analytical Toxicology

Antoniou et al., J Environ Anal Toxicol 2012, S:4
http://dx.doi.org/10.4172/2161-0525.S4-006

Review Article Open Access

Teratogenic Effects of Glyphosate-Based Herbicides: Divergence of Regulatory Decisions from Scientific Evidence

M Antoniou[1], MEM Habib[2], CV Howard[3], RC Jennings[4], C Leifert[5], RO Nodari[6], CJ Robinson[7]* and J Fagan[8]*

[1]Head, Gene Expression and Therapy Group, Department of Medical and Molecular Genetics, King's College London School of Medicine, UK
[2]Professor of entomology, former director, Institute of Biology, UNICAMP, and former provost of extension and community affairs, UNICAMP, São Paulo, Brazil
[3]Professor, Centre for Molecular Biosciences, University of Ulster, Northern Ireland
[4]Affiliated research scholar, Department of History and Philosophy of Science, University of Cambridge, UK
[5]Research development professor for ecological agriculture at the University of Newcastle, UK. Interests: director and trustee of the Stockbridge Technology Centre Ltd (STC), UK
[6]Professor, Center for Agricultural Sciences (department of plant science), Federal University of Santa Catarina, Brazil
[7]Research director, Earth Open Source, London, UK. Interests: editor, GM Watch, UK
[8]Director, Earth Open Source. Interests: employed at a GMO testing and certification company

Abstract

The publication of a study in 2010, showing that a glyphosate herbicide formulation and glyphosate alone caused malformations in the embryos of *Xenopus laevis* and chickens through disruption of the retinoic acid signalling pathway, caused scientific and regulatory controversy. Debate centred on the effects of the production and consumption of genetically modified Roundup Ready® soy, which is engineered to tolerate applications of glyphosate herbicide. The study, along with others indicating teratogenic and reproductive effects from glyphosate herbicide exposure, was rebutted by the German Federal Office for Consumer Protection and Food Safety, BVL, as well as in industry-sponsored papers. These rebuttals relied partly on unpublished industry-sponsored studies commissioned for regulatory purposes, which, it was claimed, showed that glyphosate is not a teratogen or reproductive toxin.

However, examination of the German authorities' draft assessment report on the industry studies, which underlies glyphosate's EU authorisation, revealed further evidence of glyphosate's teratogenicity. Many of the malformations found were of the type defined in the scientific literature as associated with retinoic acid teratogenesis. Nevertheless, the German and EU authorities minimized these findings in their assessment and set a potentially unsafe acceptable daily intake (ADI) level for glyphosate. This paper reviews the evidence on the teratogenicity and reproductive toxicity of glyphosate herbicides and concludes that a new and transparent risk assessment needs to be conducted. The new risk assessment must take into account all the data on the toxicity of glyphosate and its commercial formulations, including data generated by independent scientists and published in the peer-reviewed scientific literature, as well as the industry-sponsored studies.

Our conclusion: ADI should be at least:
3-fold lower based on industry data
12-fold lower based on academic data

Glyphosate: multiple toxic effects

- Antibiotic
- Nutrient metal chelator; e.g. Mn, Mg, Zn, Co
- DNA mutagen
- **Endocrine (hormone) disruptive chemical (EDC)**

Glyphosate-based herbicides

MONSANTO imagine

Backgrounder
Summary of Human Risk Assessment and Safety Evaluation on Glyphosate and Roundup® Herbicide
Downloaded 17 June 2014

"Under present and expected conditions of use, Roundup herbicide does not pose a health risk to humans." *"Roundup is placed in U.S. EPA's least toxic category (IV) for acute oral, dermal and inhalation toxicity. Thus, the Roundup formulation is considered to be practically nontoxic by all these routes of exposure. ...*

Roundup herbicide, like glyphosate, has very low acute toxicity, which means very high exposure is required to cause an adverse effect.

Glyphosate does not adversely affect reproduction or development.

There is no evidence ? endocrine disruption.

Environmental Health Sciences

Endocrine Disruptive Chemicals

- Can work at very low doses
- Non-linear dose response
- Events start in the womb
- Regulators world-wide ignore EDC potential in toxicity evaluation of chemicals

Evidence for Glyphosate/Roundup as an EDC?

Glyphosate/Roundup have EDC Capability:
Retinoic Acid System

Roundup and glyphosate caused malformations in chicken and frog embryos at doses far below (~96ppm) those used in agricultural spraying (Paganelli, A. et al. 2010):

- Neuronal and cranial development defects
- Toxicity via interference with the retinoic acid signalling pathway

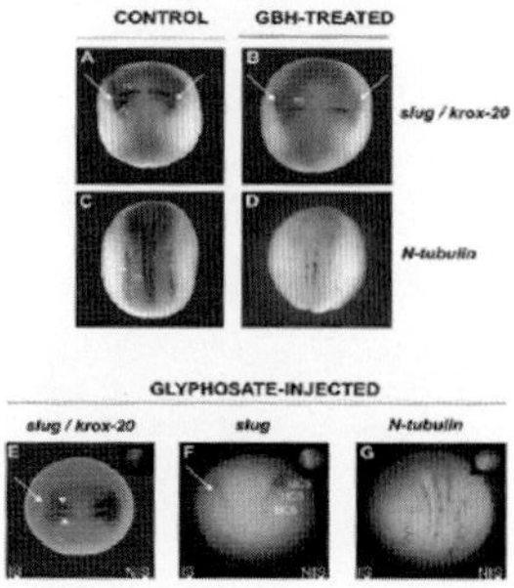

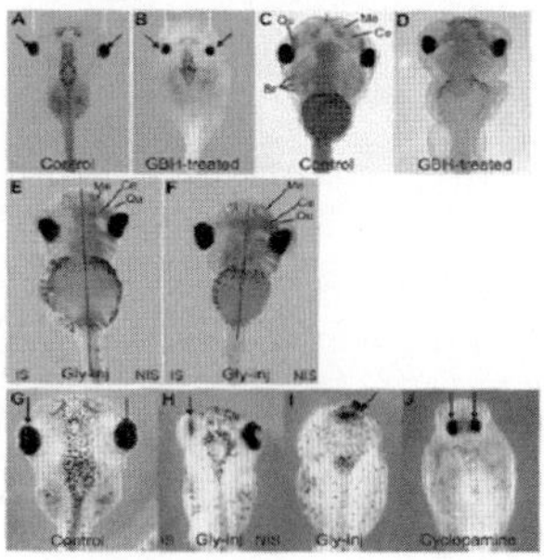

Glyphosate/Roundup have EDC Capability:
Estrogen System

Roundup formulation at 0.00023% glyphosate dilution (Hokanson, R. et al. 2007):

- **Replaced and worked synergistically with estrogen to promote human breast cancer cell growth**
- **Dysregulation of large numbers of genes**

Glyphosate mimics estrogen in promoting human cell cancer cell growth at extremely low (starting at 1.69ng/L or 1.69ppt) concentrations **(Thongprakaisang, S. et al. 2013):**

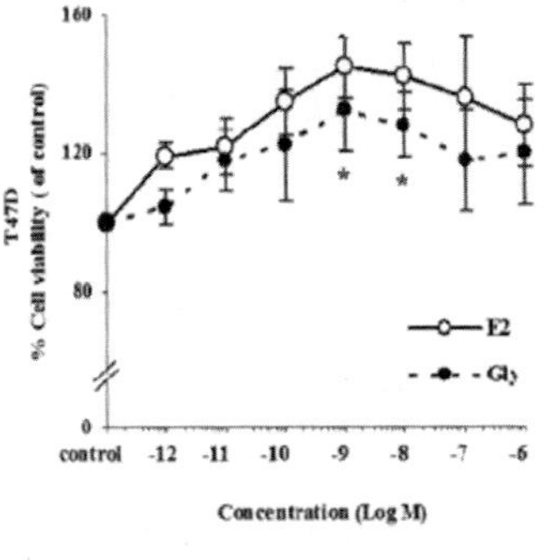

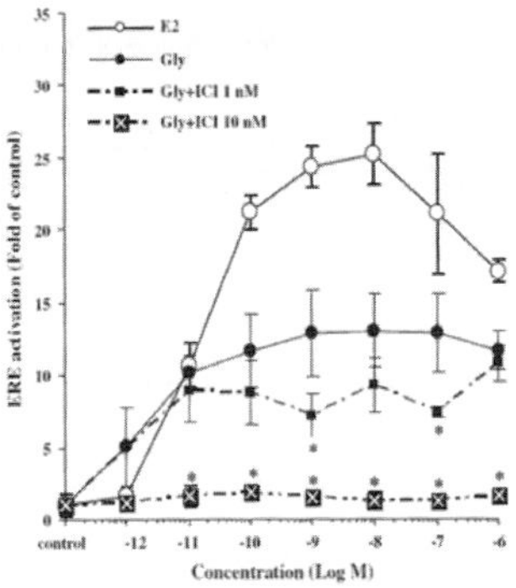

Glyphosate Formulations: "Adjuvants"

Pesticide formulations contain "adjuvants" to allow "active ingredient" to work.

Adjuvants:

- Mostly chemically modified tallowenes (POEA)
- Claimed to be "inert"

Increasing evidence shows some adjuvant components *more* toxic than active ingredient:

Mesnage et al. 2012

How much glyphosate in general human population?

Table 1 Glyphosate concentrations in human urine samples (mean and maximum values) and resulting estimates of previous exposure, compared to ADI or AOEL

References	Analytical method, LOD/LOQ	Participants	Urine concentrations [μg/l]		Estimated exposure or systemic dose [μg/kg bw]		Percentage of ADI or AOEL [%]
			Mean	Maximum	Mean	Maximum	
Acquavella et al. (2004)	HPLC following ion exchange LOD 1 μg/L	**48** male farmers from Minnesota and South Carolina (USA), **their spouses** and **79** children	**3.2**	**233** (farmer) **29** (child)*	**0.11** (systemic dose)	**8.3** (systemic dose)	**8.3** % of AOEL (maximum value), ca. **0.1** % of AOEL (mean value)
Curwin et al. (2007)	Immunoassay (fluorescent microbeads) LOD 0.9 μg/L	**48** women, **47** men, **117** children from "farm" and "non-farm" households in Iowa	**1.1–2.7** (in different groups)	**18** ("farm child")*	**0.5** (dietary exposure highest mean) **0.1** (systemic dose highest mean)		**0.1** % of ADI
Mesnage et al. (2012)	HPLC-MS LOD 1 μg/L LOQ 2 μg/L	**1** farmer, his wife and 3 children, presumably Europe	n.a. (only single values available)	**9.5** (farmer) **2** (child)*	**0.33** (systemic dose)		**<0.4** % of AOEL
Hoppe (2013)	GC–MS/MS following derivatisation LOQ 0.15 μg/L	**182** citizens from 18 European countries	**0.21**	**1.82**		**0.3–0.4** (dietary exposure)	**<0.1** % of ADI
Markard (2014)	GC–MS/MS (presumably) LOQ 0.15 μg/L	**40** male and female German students	**n.a.** (22 samples above LOQ)	**0.65**		**0.13** (dietary exposure)	**≪0.1** % of ADI
Krüger et al. (2014)	ELISA partly validated against GC–MS LOD/LOQ not given	**>300** (mostly from Germany)	**≤2**	**5**		**0.83** (dietary exposure)	**<0.2** % of ADI
Honeycutt and Rowlands (2014)	ELISA LOQ 7.5 μg/L	**35** women, men and children from USA	**n.a.** (13 samples above LOQ)	**18.8**		**3.3** (dietary exposure) **0.66** (systemic dose)	**<0.7** % of ADI, **<0.7** % of AOEL

n.a. not applicable

* For children, comparisons to reference values were not performed since age, body weight and urine volume were not known

Niemann et al. 2015

Environmental & Analytical
Toxicology

Krüger et al., J Environ Anal Toxicol 2013, 3:5
http://dx.doi.org/10.4172/2161-0525.1000186

Research Article Open Access

Field Investigations of Glyphosate in Urine of Danish Dairy Cows

Monika Krüger[1], Wieland Schrödl[1], Jürgen Neuhaus[1] and Awad Ali Shehata[1,2,3*]

[1]Institute of Bacteriology and Mycology, Veterinary Faculty, University of Leipzig, An den Tierkliniken 29, D-04103 Leipzig, Germany
[2]Avian and Rabbit Diseases Department, Faculty of Veterinary Medicine, Minoufiya University, Egypt
[3]Albrecht Daniel Thaer-Institute of Agronomy, University Leipzig, Leipzig, Germany

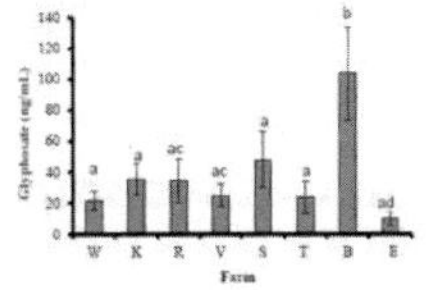

Figure 1: Glyphosate in urine of Danish cows. Significant glyphosate differences were seen between different farms (ab: p< 0.001, cd: p< 0.001) using t-test.

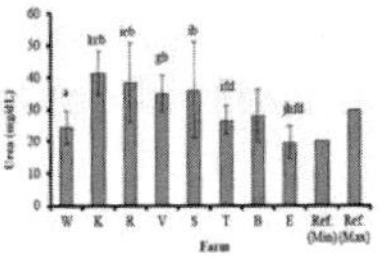

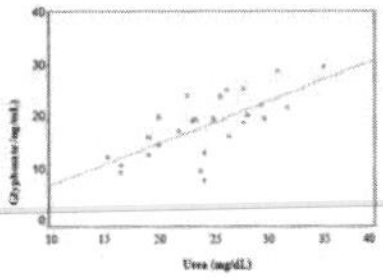

".. increased level of the enzymes GLDH, GOT, and CK, and the lipid profile marker cholesterol in all cows as well as high urea levels from cows at half of the farms point to livers, kidneys and muscles damage."

Levels detected to date in humans are below regulatory set limits but are they safe?

Need for long term laboratory animal toxicity feeding studies at ADI / cRfD levels

Séralini et al. Environmental Sciences Europe 2014, 26:14
http://www.enveurope.com/content/26/1/14

Environmental Sciences Europe
a SpringerOpen Journal

RESEARCH **Open Access**

Republished study: long-term toxicity of a Roundup herbicide and a Roundup-tolerant genetically modified maize

Gilles-Eric Séralini[1*], Emilie Clair[1], Robin Mesnage[1], Steeve Gress[1], Nicolas Defarge[1], Manuela Malatesta[2], Didier Hennequin[3] and Joël Spiroux de Vendômois[1]

1. Toxicity study extending Monsanto's 90-day feeding study

2. Only laboratory animal feeding study looking at long-term health effects of complete Roundup herbicide formulation

Study of toxic effects of Roundup herbicide & GMO NK603 corn in adult male and female rats under a daily regimen for 24 consecutive months

↓

60 Rats
Roundup
(10 Males / 10 Females)
(3 doses)

↓

Doses of Roundup: half of agricultural dilution
400mg/L
0.05ug/L – daily intake 4ng/kg bw/day

Summary of Findings from lowest dose R treatment group

- **Increased incidence of liver and kidney pathology**

- Testosterone/estrogen imbalance suggesting endocrine disrupting effects
- Pituitary dysfunction in females

- **Unexpected increased trend in tumour incidence:**
- ➢ Roundup in females (mammary tumours)
- ➢ Statistically significant vs controls in lowest Roundup treatment group

- ➢ Females: euthanized due to large mammary tumours.

Roundup-associated kidney toxicity in farm workers

Int. J. Environ. Res. Public Health **2014**, *11*, 2125-2147; doi:10.3390/ijerph110202125

OPEN ACCESS

International Journal of
Environmental Research and Public Health
ISSN 1660-4601
www.mdpi.com/journal/ijerph

Hypothesis

Glyphosate, Hard Water and Nephrotoxic Metals: Are They the Culprits Behind the Epidemic of Chronic Kidney Disease of Unknown Etiology in Sri Lanka?

Channa Jayasumana [1,2,*], Sarath Gunatilake [2,†] and Priyantha Senanayake [3,†]

OPEN ACCESS Freely available online — PLOS | NEGLECTED TROPICAL DISEASES

Viewpoints

Chronic Kidney Disease Epidemic in Central America: Urgent Public Health Action Is Needed amid Causal Uncertainty

Pedro Ordunez*, Ramón Martinez, Ludovic Reveiz, Evelina Chapman, Carla Saenz, Agnes Soares da Silva, Francisco Becerra

Pan American Health Organization, Washington, D.C., United States of America

Carcinogenicity of tetrachlorvinphos, parathion, malathion, diazinon, and glyphosate

WHO International Agency for Research on Cancer

Lancet Oncol 2015
Published Online
March 20, 2015

	Activity (current status)	Evidence in humans (cancer sites)	Evidence in animals	Mechanistic evidence	Classification*
Tetrachlorvinphos	Insecticide (restricted in the EU and for most uses in the USA)	Inadequate	Sufficient	..	2B
Parathion	Insecticide (restricted in the USA and EU)	Inadequate	Sufficient	..	2B
Malathion	Insecticide (currently used; high production volume chemical)	Limited (non-Hodgkin lymphoma, prostate)	Sufficient	Genotoxicity, oxidative stress, inflammation, receptor-mediated effects, and cell proliferation or death	2A†
Diazinon	Insecticide (restricted in the USA and EU)	Limited (non-Hodgkin lymphoma, leukaemia, lung)	Limited	Genotoxicity and oxidative stress	2A†
Glyphosate	Herbicide (currently used; highest global production volume herbicide)	Limited (non-Hodgkin lymphoma)	Sufficient	Genotoxicity and oxidative stress	2A†

EU=European Union. *See the International Agency for Research on Cancer (IARC) preamble for explanation of classification system (amended January, 2006). †The 2A classification of diazinon was based on limited evidence of carcinogenicity in humans and experimental animals, and strong mechanistic evidence; for malathion and glyphosate, the mechanistic evidence provided independent support of the 2A classification based on evidence of carcinogenicity in humans and experimental animals.

Table: IARC classification of some organophosphate pesticides

Summary of Seralini Study Findings
Low dose Roundup toxicity

Unexpected low dose toxicity from Roundup (0.05ug/L glyphosate equivalent):

- **Half level permitted in drinking water in EU (0.1ug/L)**
- **14,000 times lower than that permitted in drinking water in USA (700ug/L)**
- **20,000 times lower than Australian Drinking Water Guidelines (2011) [1mg/L]**

- **Vastly below ADI /cRfD**
  55,000-145,000 times below EU/AUS levels
  300,000-850,000 times below US levels

- **Acting as EDC?**

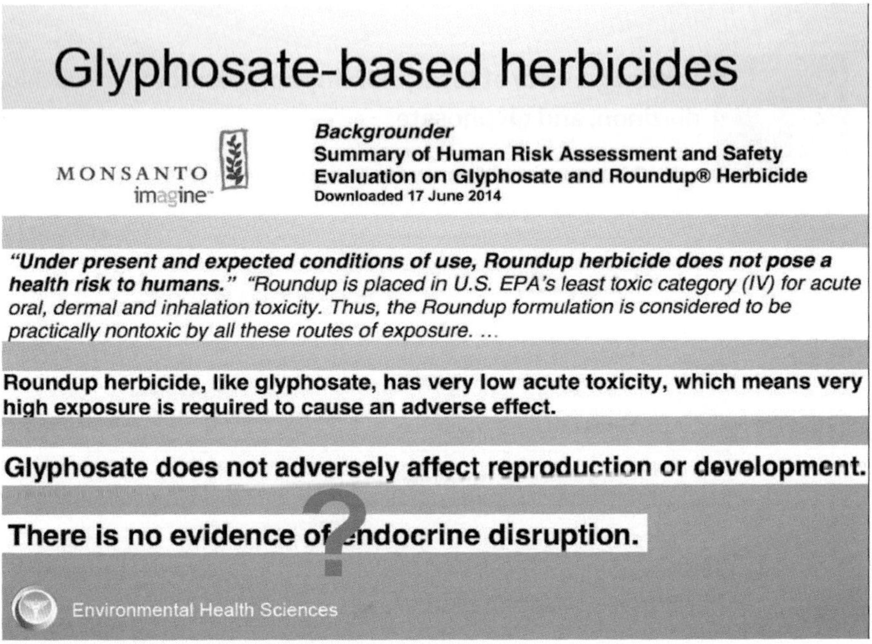

Current Acceptable Daily Intake (ADI) / chronic Reference Dose (cRfD)

- EU/Aus: 0.3mg/kg/day
- China/Russia: 1mg/kg/day
- USA: 1.75mg/kg/day

➢ ***Are these exposures really safe?***

Glyphosate/Roundup as an EDC

Huge implications:

- ❑ For health
- ❑ For policy
- ❑ For GMO crops

What can be done?

More toxicity tests are needed with appropriate study design

Glyphosate and its commercial formulations have to be tested at an environmental level

Because glyphosate is a potential endocrine disruptor, future studies should incorporate testing principles from endocrinology (hormone dosages)

Future studies of laboratory animals should use designs that examine the full lifespan of the experimental animal, including a prenatal period

The lack of data on toxicity is not a proof of safety

Glyphosate-based herbicides cannot be considered as safe without being tested

Limited toxicokinetics studies for glyphosate, but .....

Single very high dose analysis in rats shows:

- **Ready uptake from the gastrointestinal tract from aqueous solutions (23-40%)**
- **Easy diffusion into tissues**
- **Conversion to AMPA**

[Brewster et al. 1991; Anadon et al. 2009]

- However note:
- **Impossible to determine bioaccumulation from single dose exposures**
- **No ADI / cRfD tested, especially on long term basis**
- **No long-term testing of complete Roundup formulation**

Industry study: rats fed commercialised Roundup tolerant GM NK603 corn for 90 days

Signs of toxic effects on **liver** and **kidney** function. (Hammond et al. 2004; see de Vendomois et al. 2009).

Academic studies: mice fed commercialised Roundup tolerant GM soya

- **Mice fed GM soy:** disturbed **liver**, pancreas and testes function (Malatesta et al. 2002; Malatesta et al. 2003; Vecchio et al. 2004).

- **Mice fed GM soy over 24 months:** more acute signs of ageing in the **liver** (Malatesta et al. 2008):

Is GM process or Roundup residues in feed causing these effects?

Glyphosate/Roundup have EDC Capability:

Glyphosate herbicide altered hormone levels in female catfish, decreased egg viability and harmful to reproduction (Soso, A.B/ et al. 2007).

Roundup disrupted production of the steroid hormone progesterone in mouse cells (Walsh, L.P. et al. 2000).

Glyphosate herbicide was a potent EDC in rats, causing disturbances in reproductive development after exposure during puberty (Romano, R.M. et al. 2010).

In vitro experiments in human cells, glyphosate herbicides at 0.5 ppm prevented the action of androgens, the masculinizing hormones and disrupted the action and formation of estrogens, the feminizing hormones at (Gasnier, C. et al. 2009).

Roundup altered testicular function after 8 days of exposure at a concentration of 0.5% causing: increase in abnormal sperm formation at 2, 3, and 4 months after exposure, change gene expression in sperm cells; could alter the balance of the sex hormones androgen and estrogen; negative impact on sperm quality (Cassault-Meyer, E. et al. 2014).

Detection of Glyphosate Residues in Animals and Humans

Krüger, M. et al. *J Environ Anal Toxicol.* 2014, **4**:2

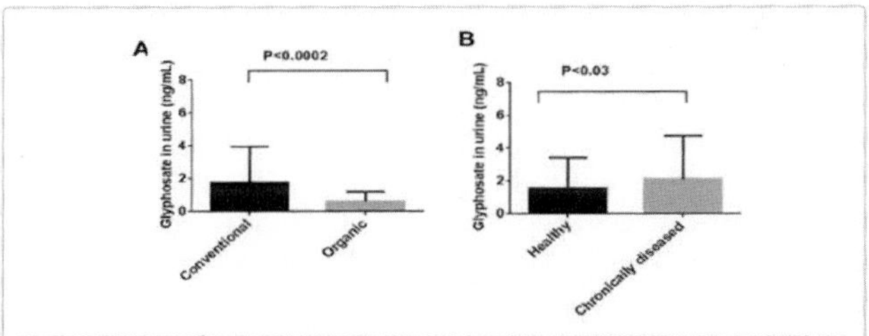

Figure 3: Glyphosate in humans. A) Comparison of glyphosate excretion with urine of humans with conventional (N=99) and predominantly organic (N=41) feeding. B) Glyphosate in healthy (N=102) and chronically (N=199) diseased humans.

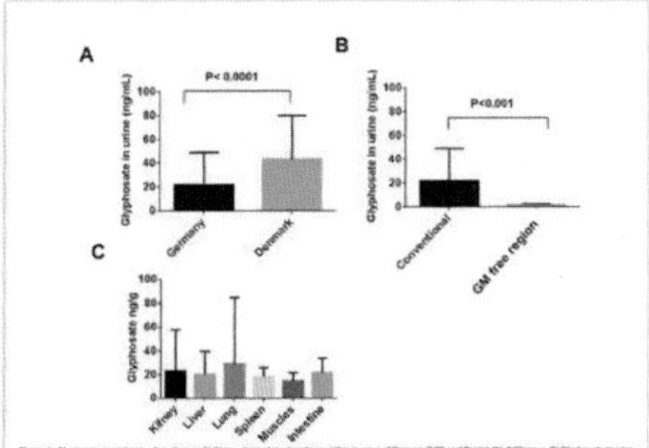

Figure 1: Glyphosate excretion in urine of cows. A) Comparison of glyphosate excretion in urine of German (343) and Danish (N=242) cows. B) Glyphosate in urine of conventional (N=343) and organic (N=32) livestock husbandry in Germany. C: Glyphosate accumulation in different organs.

GLOBAL TRENDS IN THE USE OF GLYPHOSATE-BASED HERBICIDES

CHARLES BENBROOK
Center for Sustaining Agriculture and Natural Resources, Washington State University, Enterprise, OR, USA

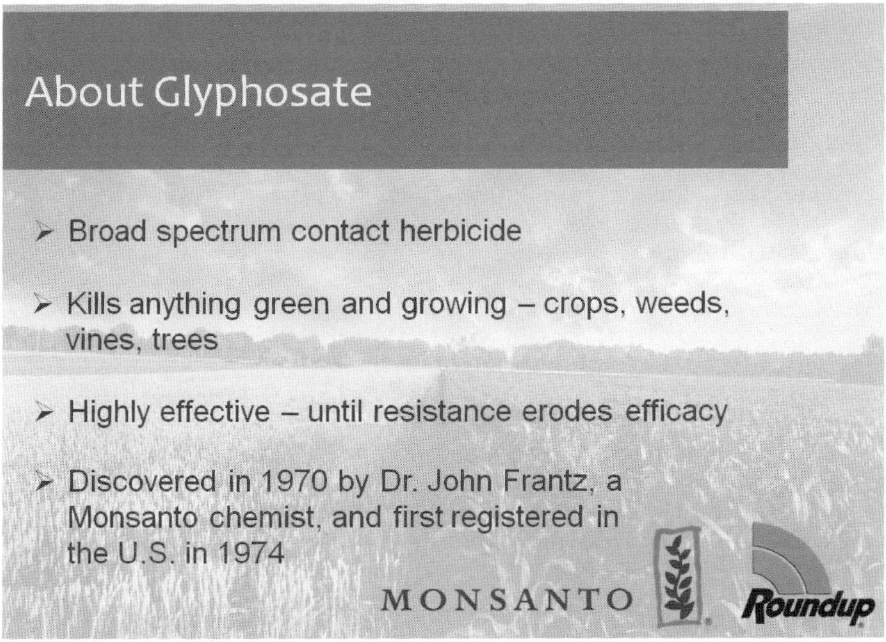

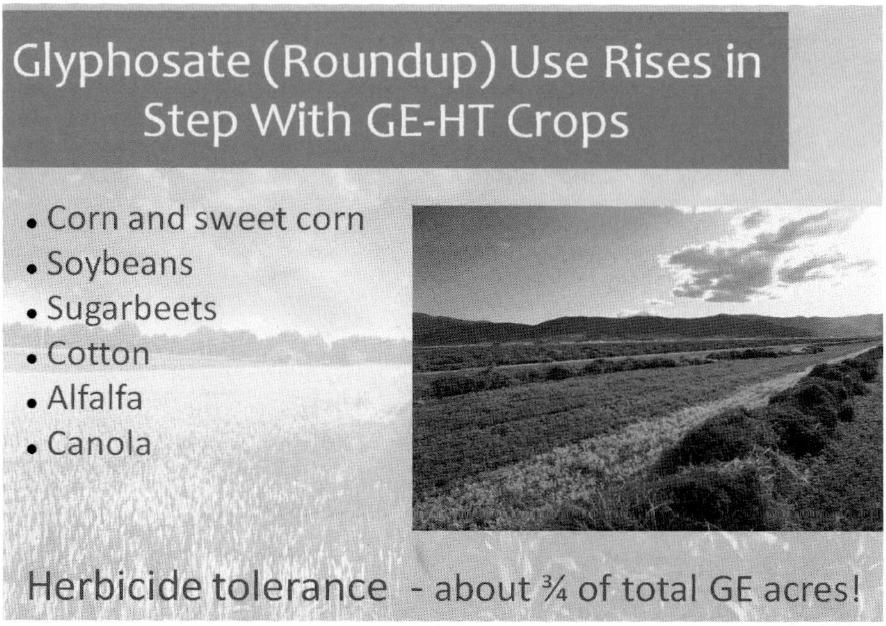
Glyphosate (Roundup) Use Rises in Step With GE-HT Crops
• Corn and sweet corn
• Soybeans
• Sugarbeets
• Cotton
• Alfalfa
• Canola
Herbicide tolerance - about ¾ of total GE acres!

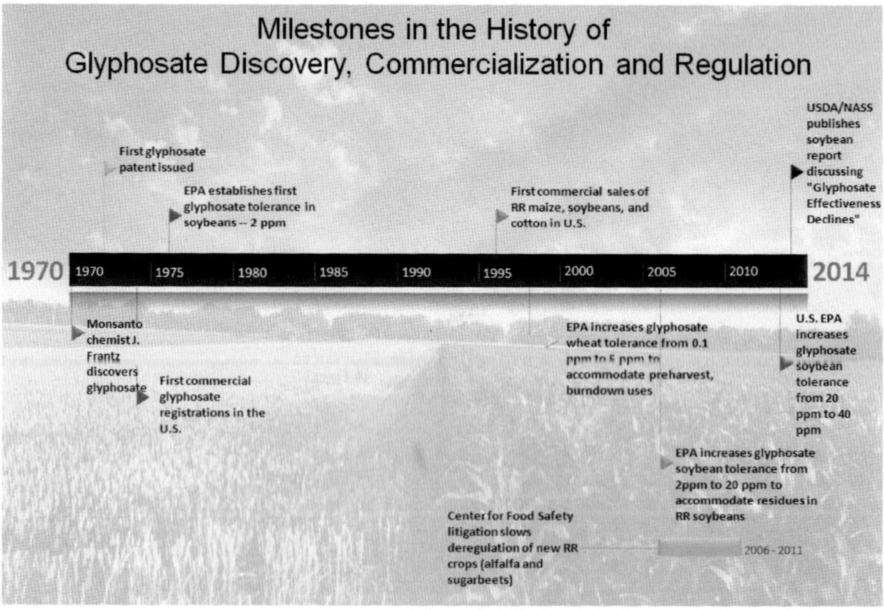
Milestones in the History of Glyphosate Discovery, Commercialization and Regulation
1970
1970
1975
1980
1985
1990
1995
2000
2005
2010
2014
First glyphosate patent issued
EPA establishes first glyphosate tolerance in soybeans -- 2 ppm
First commercial sales of RR maize, soybeans, and cotton in U.S.
USDA/NASS publishes soybean report discussing "Glyphosate Effectiveness Declines"
Monsanto chemist J. Frantz discovers glyphosate
First commercial glyphosate registrations in the U.S.
EPA increases glyphosate wheat tolerance from 0.1 ppm to 5 ppm to accommodate preharvest, burndown uses
U.S. EPA increases glyphosate soybean tolerance from 20 ppm to 40 ppm
EPA increases glyphosate soybean tolerance from 2ppm to 20 ppm to accommodate residues in RR soybeans
Center for Food Safety litigation slows deregulation of new RR crops (alfalfa and sugarbeets)
2006 - 2011

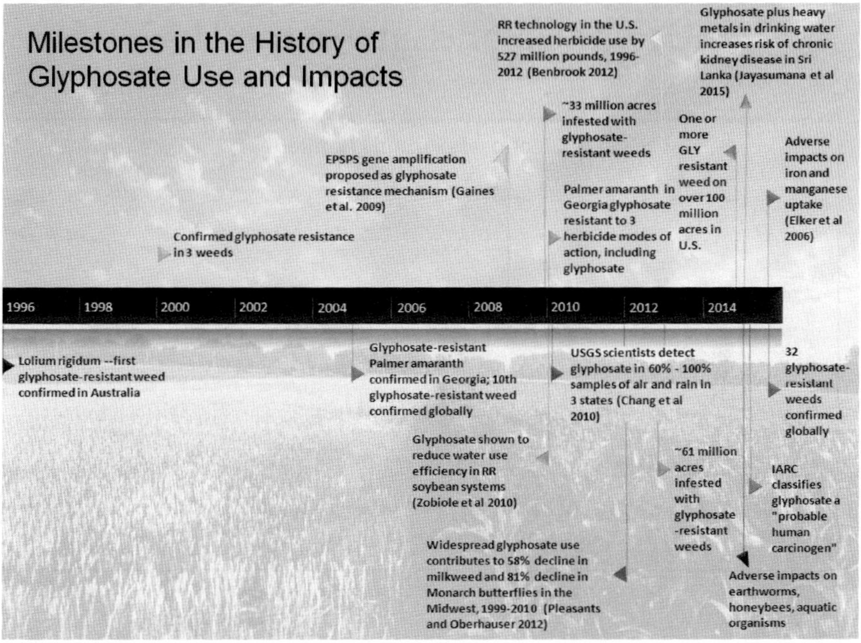

Milestones in the History of Glyphosate Use and Impacts
RR technology in the U.S. increased herbicide use by 527 million pounds, 1996-2012 (Benbrook 2012)
Glyphosate plus heavy metals in drinking water increases risk of chronic kidney disease in Sri Lanka (Jayasumana et al 2015)
~33 million acres infested with glyphosate-resistant weeds
One or more GLY resistant weed on over 100 million acres in U.S.
Adverse impacts on iron and manganese uptake (Elker et al 2006)
EPSPS gene amplification proposed as glyphosate resistance mechanism (Gaines et al. 2009)
Palmer amaranth in Georgia glyphosate resistant to 3 herbicide modes of action, including glyphosate
Confirmed glyphosate resistance in 3 weeds
1996
1998
2000
2002
2004
2006
2008
2010
2012
2014
Lolium rigidum --first glyphosate-resistant weed confirmed in Australia
Glyphosate-resistant Palmer amaranth confirmed in Georgia; 10th glyphosate-resistant weed confirmed globally
USGS scientists detect glyphosate in 60% - 100% samples of air and rain in 3 states (Chang et al 2010)
32 glyphosate-resistant weeds confirmed globally
Glyphosate shown to reduce water use efficiency in RR soybean systems (Zobiole et al 2010)
~61 million acres infested with glyphosate-resistant weeds
IARC classifies glyphosate a "probable human carcinogen"
Widespread glyphosate use contributes to 58% decline in milkweed and 81% decline in Monarch butterflies in the Midwest, 1999-2010 (Pleasants and Oberhauser 2012)
Adverse impacts on earthworms, honeybees, aquatic organisms

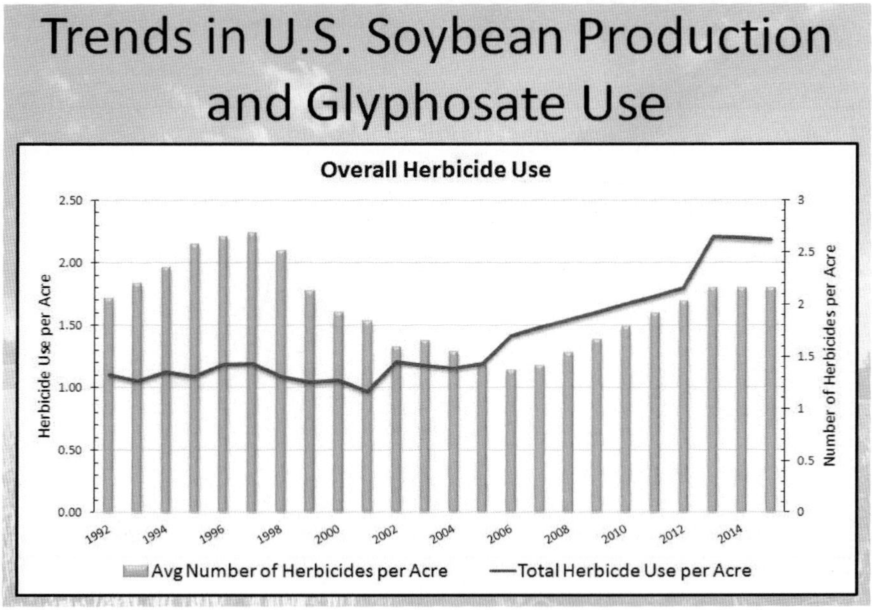

Trends in U.S. Soybean Production and Glyphosate Use
Overall Herbicide Use
Herbicide Use per Acre
Number of Herbicides per Acre
2.50
2.00
1.50
1.00
0.50
0.00
3
2.5
2
1.5
1
0.5
0
1992
1994
1996
1998
2000
2002
2004
2006
2008
2010
2012
2014
Avg Number of Herbicides per Acre
Total Herbicde Use per Acre

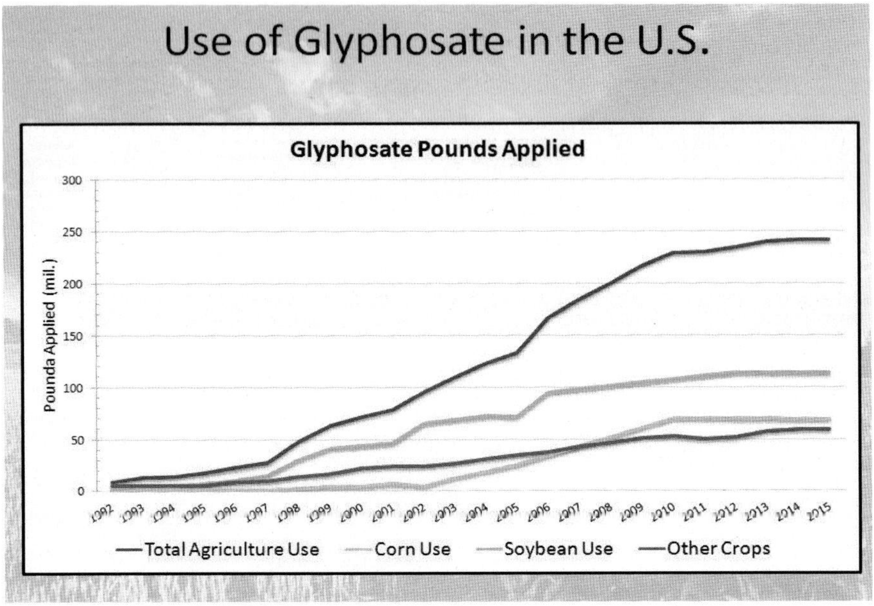
Use of Glyphosate in the U.S.
Glyphosate Pounds Applied
Pounda Applied (mil.)
300
250
200
150
100
50
0
1992 1993 1994 1995 1996 1997 1998 1999 2000 2001 2002 2003 2004 2005 2006 2007 2008 2009 2010 2011 2012 2013 2014 2015
Total Agriculture Use
Corn Use
Soybean Use
Other Crops

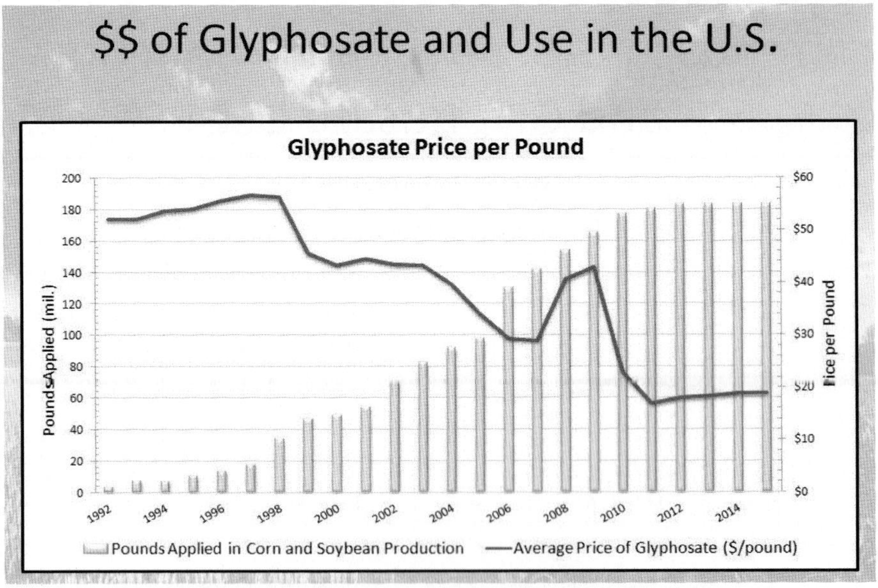
$$ of Glyphosate and Use in the U.S.
Glyphosate Price per Pound
Pounds Applied (mil.)
200
180
160
140
120
100
80
60
40
20
0
$60
$50
$40
$30
$20
$10
$0
1992 1994 1996 1998 2000 2002 2004 2006 2008 2010 2012 2014
Pounds Applied in Corn and Soybean Production
Average Price of Glyphosate ($/pound)

Consequences of Glyphosate Overuse in the U.S.

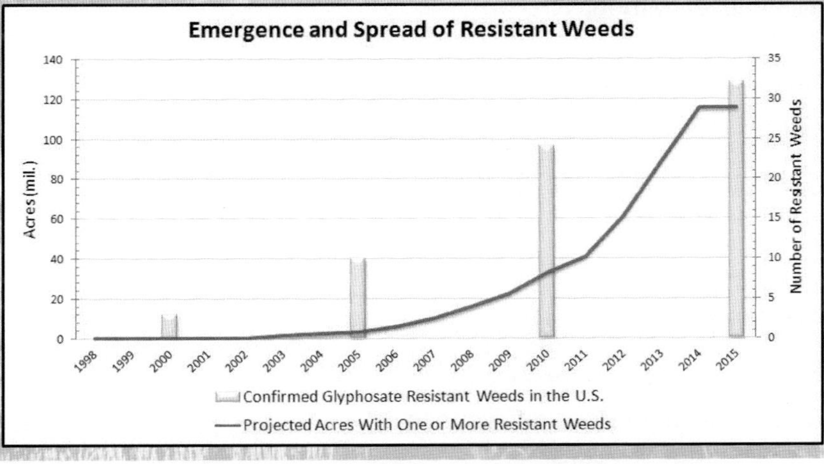

How Have GE-HT Crops Impacted Pesticide Use in the U.S.?

Simulation model uses widely accepted USDA data are available on:

- Acres planted to corn, soybeans, and cotton
- Percent crop acres planted and not planted to major GE traits
- Herbicide and glyphosate use rates per acre
- Use rates of insecticides displaced by *Bt* corn and cotton

USDA United States Department of Agriculture
National Agricultural Statistic Service

From 1996-2011 in the U.S. --

Herbicide use on RR crops ↑ 527 million pounds compared to likely level without RR technology

Insecticide use on *Bt* corn/cotton ↓ 123 million lbs

Overall pesticide use ↑ 404 million pounds

~7% increase in overall pesticide use since 1996

~20% increase in 2011 and progressively larger differences in subsequent years

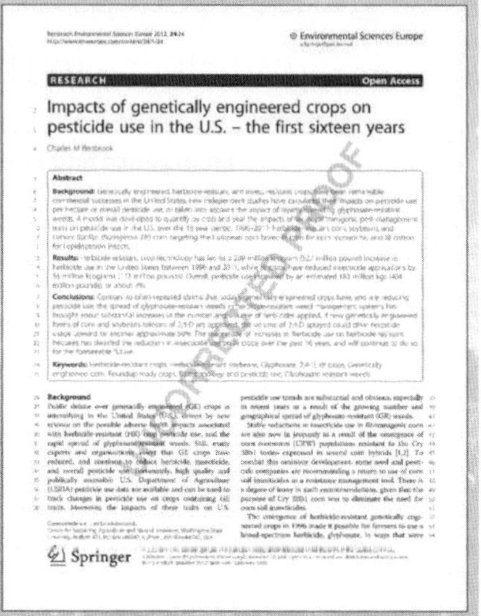
RESEARCH Open Access

Impacts of genetically engineered crops on pesticide use in the U.S. – the first sixteen years

Springer

Key Results: 2011 Herbicide Rates

Each acre planted to a HT variety required substantially more herbicides than acres not planted to HT crops:

- 0.73 pounds per acre more in the case of soybeans
- 0.41 pound/acre corn
- 0.86 pound/acre cotton

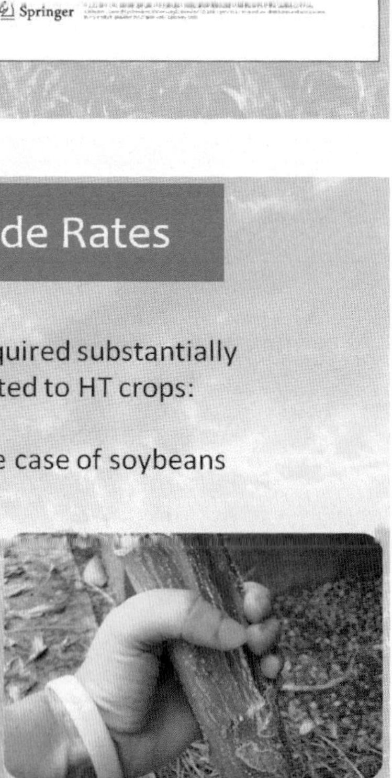

C. Benbrook. 2012 "Impacts of Genetically Engineered Crops on Pesticide Use in the U.S. – The First Sixteen Years," *Environmental Sciences Europe*

Agricultural vs. Non-Ag Use

Agricultural and Non-Agricultural Use of Glyphosate in the United States: 1974-2014

	1974	1982	1990	1995	2000	2005	2010	2012	2014
Glyphosate Use (kg)									
Agricultural	363	2,268	3,357	12,474	35,720	71,441	104,398	106,488	112,973
Non-Agricultural	272	1,270	2,404	5,670	8,958	9,299	9,299	9,299	9,299
Total	635	3,538	5,761	18,144	44,679	80,739	113,697	115,787	122,272
Glyphosate Use (lb)									
Agricultural	800	5,000	7,400	27,500	78,750	157,500	230,159	234,766	249,063
Non-Agricultural	600	2,800	5,300	12,500	19,750	20,500	20,500	20,500	20,500
Total	1,400	7,800	12,700	40,000	98,500	178,000	250,659	255,266	269,563
Share Agricultural	57.1%	64.1%	58.3%	68.8%	79.9%	88.5%	91.8%	92.0%	92.4%
Share Non-Agricultural	42.9%	35.9%	41.7%	31.3%	20.1%	11.5%	8.2%	8.0%	7.6%

Agricultural vs. Non-Ag Use

Global Agricultural and Non-Agricultural Use of Glyphosate: 1994 - 2014

	1994	1995	2000	2005	2010	2012	2014
Glyphosate Use (kg)							
Agricultural	42,868	51,078	169,350	356,403	545,062	616,331	658,170
Non-Agricultural	13,428	16,000	42,472	46,389	48,700	53,819	54,173
Total	56,296	67,078	211,821	402,792	593,762	670,150	712,343
Glyphosate Use (lb)							
Agricultural	94,507	112,607	373,348	785,726	1,201,643	1,358,764	1,451,002
Non-Agricultural	29,604	35,274	93,633	102,269	107,364	118,649	119,430
Total	124,111	147,880	466,982	887,995	1,309,008	1,477,413	1,570,432
Share Agricultural	76%	76%	80%	88%	92%	92%	92%
Share Non-Agricultural	24%	24%	20%	12%	8%	8%	8%

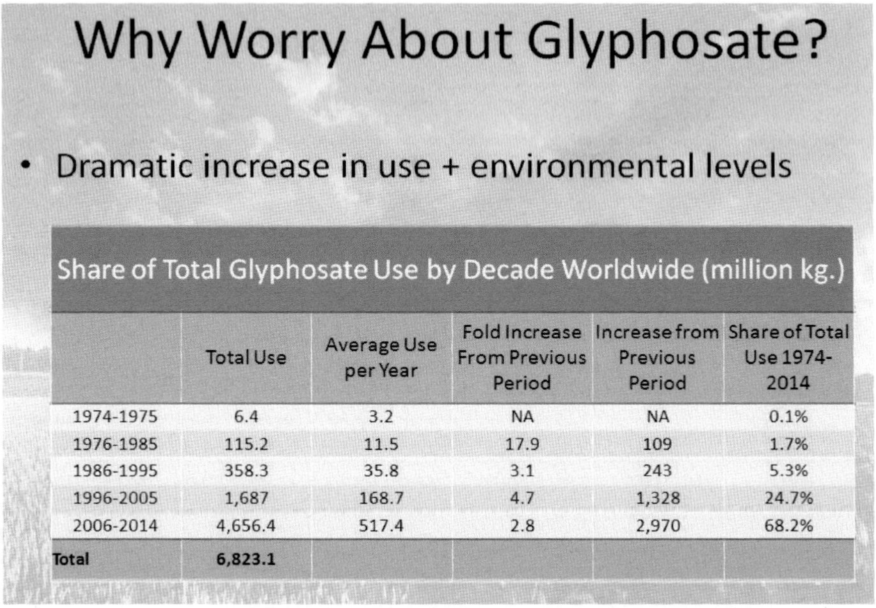
Why Worry About Glyphosate?
• Dramatic increase in use + environmental levels
Share of Total Glyphosate Use by Decade Worldwide (million kg.)

	Total Use	Average Use per Year	Fold Increase From Previous Period	Increase from Previous Period	Share of Total Use 1974-2014
1974-1975	6.4	3.2	NA	NA	0.1%
1976-1985	115.2	11.5	17.9	109	1.7%
1986-1995	358.3	35.8	3.1	243	5.3%
1996-2005	1,687	168.7	4.7	1,328	24.7%
2006-2014	4,656.4	517.4	2.8	2,970	68.2%
Total	**6,823.1**				

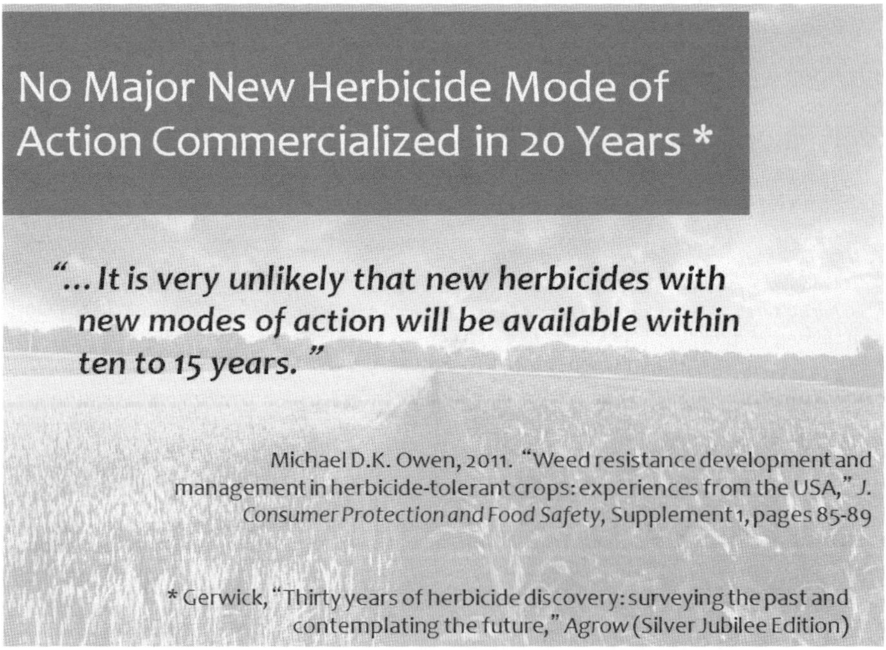
No Major New Herbicide Mode of Action Commercialized in 20 Years *
"... It is very unlikely that new herbicides with new modes of action will be available within ten to 15 years."
Michael D.K. Owen, 2011. "Weed resistance development and management in herbicide-tolerant crops: experiences from the USA," J. Consumer Protection and Food Safety, Supplement 1, pages 85-89
* Gerwick, "Thirty years of herbicide discovery: surveying the past and contemplating the future," Agrow (Silver Jubilee Edition)

Ominous Trends in Plant Health

Declining plant health triggered by changes in genetics, planting densities, and crop management during the GE crop era

2014 – 18% corn was treated with fungicide (USDA - NASS data)

- ✓ Less than 1% of corn acres were treated with fungicides in all previous NASS surveys

RESPONDING TO RESISTANT WEEDS

Farmers must diversify the tactics embedded in their weed management systems – integrate "many little hammers"

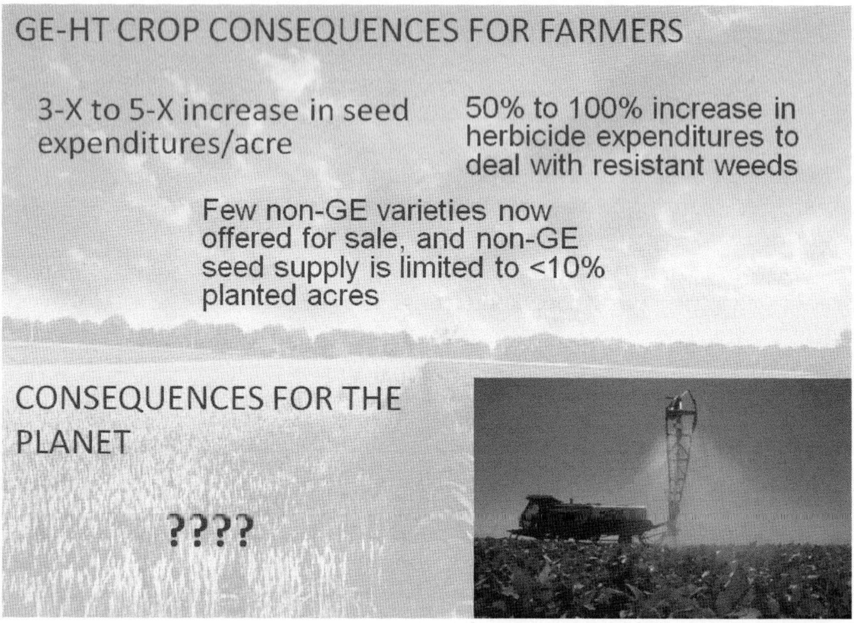
GE-HT CROP CONSEQUENCES FOR FARMERS
3-X to 5-X increase in seed expenditures/acre
50% to 100% increase in herbicide expenditures to deal with resistant weeds
Few non-GE varieties now offered for sale, and non-GE seed supply is limited to <10% planted acres
CONSEQUENCES FOR THE PLANET
????

FARMING WITHOUT GLYPHOSATE

PATRICK HOLDEN
Sustainable Food Trust
Bristol, UK

In this paper I will:

- Outline the history of the development of this herbicide;
- Detail current dependency on the use of glyphosate;
- Discuss the issue of vested interests;
- Describe how farming systems rely on glyphosate
- Describe the limitations of no-till and min-till as alternative farming systems;
- Explain why mixed-farming systems offer the best models for farming without glyphosate;
- Describe the key characteristics of an alternative farming system.

Most agriculture, both in the UK and globally, has changed beyond recognition since I started farming over forty years ago. A high proportion of arable crops are now grown in ways that are only possible because of the use of glyphosate and these would scarcely be recognisable to a farmer from the early 1970s. Take glyphosate away and these production methods would no longer be viable, or in many cases, even possible. Would this be a disaster for global agriculture or actually its salvation? Let's start by considering the history and current use of glyphosate.

HISTORY OF GLYPHOSATE

Glyphosate was first used in the U.S. in 1971 and in the UK in 1976. The glyphosate molecule was actually first synthesised in 1950 by a Swiss chemist, Henri Martin, working for the pharmaceutical company Cilag. Its herbicidal properties, however, were not discovered for another 20 years. Cilag was bought by Johnson and Johnson in 1959, and eventually the research samples created in 1950 were acquired by Monsanto, which at the time was trying to develop water-softening agents. One of the chemists at Monsanto tested a range of compounds for their herbicidal activity on perennial weeds and glyphosate was found to be the most effective of these. Monsanto began to synthesise glyphosate in May 1970.[1] Speedy field trials were carried out and it was patented in March 1971 as the well-known herbicide Roundup®.

Farmers soon began to prefer glyphosate to other broad-spectrum herbicides in use at this time, such as paraquat, because in addition to killing the green growth above ground it also killed the roots underground, preventing regrowth of grass and perennial weeds. However, because glyphosate is not a selective weed killer, it could only be used before crops were sown or after they were harvested; selective weed killers were still needed on the growing crops. Use of Roundup, as well as pesticides in general, rose dramatically throughout the 1970s and 1980s. A slight dip in the use of all pesticides in the mid-1980s has been linked to the introduction of Integrated Pest Management techniques,[2] but total usage levels in the 1990s continued to increase at a phenomenal rate.

In 1991, Monsanto's patent for Roundup® expired—although several related patents didn't expire in until 2000—leading to a drop in the price of glyphosate, as generic versions came onto the market. Other agrochemical companies including Dow, Syngenta and NuFarm developed their own versions, but Monsanto has continued to dominate the market. In part, it did this by developing genetically modified (Roundup Ready) crop varieties, which are resistant to glyphosate, thereby allowing farmers to spray their crops with herbicides after sowing (post-emergence) as well. Roundup is also widely used to kill crop regrowth and accelerate crop ripening. Such crops can be used for human and animal consumption, though they are rejected by seed merchants and maltsters because the practice damages the germination ability and leads to deformed plants.

USE OF GLYPHOSATE TODAY: GLOBAL

In 2012, the global glyphosate market was valued at U.S.$5.46 billion and mainstream analysts expect it to grow to U.S.$8.79 billion by 2019.[3] To put this into perspective, the global value of pesticides as a whole was estimated to be U.S.$51.9 billion in 2013, and projected to rise to U.S.$75.9 billion by 2019. In summary, glyphosate accounts for about 12% of the value of global pesticide sales.

A total of 719,000 tonnes of glyphosate were used worldwide in 2012, an increase of more than 10% from the 650,000 tonnes used in 2011. Given that global pesticide use was 2.3 million tonnes in 2013,[4] glyphosate alone accounts for almost a third of total global pesticide use.

Analysts also expect demand for the herbicide to continue to grow as emerging economies in Asia and Latin America increasingly embrace genetically modified (GM) crops, and due to the increasing popularity of no-till farming systems, both of which rely on glyphosate to control weed growth. GM crops currently account for just under half of total glyphosate use. It is expected that demand for glyphosate will also continue to increase in countries that already grow GM soybean, maize, cotton and oilseed rape, such as in Argentina, South Africa, India and China. With minimum tillage (min-till) already well established, sales trends in Europe will be closely linked to the extent to which EU countries do or do not start growing GM crops.

Currently, more than 30% of current glyphosate demand comes from the Asia Pacific region, followed by North America and Latin America. It is the most widely used herbicide in the United States.

USE OF GLYPHOSATE TODAY: UK

Glyphosate is also the most widely used agricultural herbicide in the UK. In 2014, almost 1,800 tonnes of glyphosate were used on arable crops, accounting for 25% of all herbicide use (over 120 herbicides are in active use in the UK but glyphosate is by far the most commonly used).[5] More than two-thirds of use is in crop systems growing wheat and oilseed rape. The remaining one-third is used mostly in the production of winter and spring barley, potatoes, oats, beans, sugar beet, peas, linseed, triticale, and rye.

Glyphosate use has increased 32% in only two years. In 2014, just over 2 million hectares of arable land were sprayed with glyphosate, almost half of all the arable land in the UK.[6]

Glyphosate is also one of the most widely used herbicides for soft fruit, with 2,432kg sprayed on 4,240 hectares in 2012, an increase of 44% since 2012.[7] 12,309kg of glyphosate were also applied to 22,050 hectares of orchards in 2012.[8] It was the second most commonly used herbicide applied to grassland (to kill the grass before reseeding or conversion to arable cropping) and fodder crops, and the sixth most widely applied 'active substance' on outdoor vegetable crops.

Friends of the Earth, has estimated that if GM glyphosate resistant crops are allowed to be grown within the EU, overall glyphosate use could increase by 800%.[9]

VESTED INTERESTS

Given the huge dependence on glyphosate by current farming systems, thc changes that will be necessary for farmers to manage without it and the profits it makes for the agrochemical industry, it's not difficult to see why opposition to a ban on glyphosate will be fierce and prolonged.

In March 2015, the WHO's International Agency for Research on Cancer (IARC) published an evaluation of the carcinogenicity of five organophosphate pesticides and concluded that glyphosate was 'probably carcinogenic to humans'.[10] This was mostly based on "convincing evidence that glyphosate also can cause cancer in laboratory animals" as well as some evidence that agricultural exposure to glyphosate in humans could cause cancer. Monsanto reacted strongly to these findings, accusing the IARC of having an agenda and 'cherry picking' the evidence. Philip Miller, Monsanto's vice president for global regulatory affairs, was quoted as saying that the conclusion was "starkly at odds with every credible scientific body that has examined glyphosate safety".

Yet, a carcinogenicity study submitted to the U.S. EPA as long ago as 1981 gave an early indication of a problem, finding an increase in testicular tumours in rats at the highest dose tested. In addition, a study published in the British Journal of Cancer in 1998,[11] and by the American Cancer Society in 1999,[12] both found strong statistical associations between occupational exposure to glyphosate and the development of Non-Hodgkin's Lymphoma.

It has therefore taken 17 years, but as a result of the IARC's evaluation, the U.S. Environmental Protection Agency (EPA) has said it will re-evaluate glyphosate as well as a new herbicide which combines glyphosate with 2,4-D, another herbicide in use since the 1940s which has also been linked to increased cancer risk. But given the lobbying power of Monsanto and the enormous reliance on glyphosate for the vast majority of farmers, it seems inevitable that the EPA will drag its feet.

Re-evaluation in Europe is similarly uncertain. Roundup was last re-approved in 2002, but the scheduled 2012 re-evaluation was postponed until 2015 apparently because the European Commission and the European Food Safety Authority were overworked. Germany is the EU rapporteur member state for glyphosate and submitted its glyphosate renewal assessment report in January 2014 concluding that, 'the available data do not show carcinogenic or mutagenic properties of glyphosate'. As such, Germany not only recommended re-approval but also an increase in the acceptable daily intake from 0.3 to

0.5 mg per kg body weight per day. This has, however, been widely criticised amid claims of industry influence in the review process.[13]

Researchers[14,15] have also drawn together large amounts of evidence indicating a wide range of possible additional problems. These relate to both long term soil and human health issues linked to the use of glyphosate and Roundup, yet none of these associations have so far been taken seriously by regulators.

CURRENT USES OF GLYPHOSATE

While I have obviously been aware of these changes, in my own farming life they have largely passed me by. I began farming organically in 1973, so using Roundup was never an option for me. Since the productivity of crops and animals on my 120 hectare mixed organic farm in West Wales has steadily increased over the last 40 years I have never felt the need to use glyphosate, or any other herbicides. This plus first-hand knowledge of many similar farms gives me the confidence to say that it is perfectly possible to farm productively without the use of this uniquely broad-spectrum herbicide.

Before we can consider how to farm without the use of glyphosate, we need to understand the different uses of glyphosate by food producers and consider what, if any benefits they confer:

1. Min-till/no till cultivation on non-Roundup Ready crops—Roundup is used to kill all growing plants, both weeds and volunteer crops before planting and after harvest. This allows the next crop to be established with little or no cultivation.

2. Min-till/no-till of Roundup Ready crops—as in number 1 above, but can be sprayed onto the growing crop to control weeds allowing Roundup to replace all other herbicides except on the growing areas of land now becoming infested with weeds that are resistant to Roundup. Crops can be established even more quickly since spraying can be delayed until after the crop has emerged.

3. Conversion of grassland to arable crop production—this allows grass to be killed off completely making ploughing of grassland quicker, easier and more effective.

4. Desiccation of crops before harvesting to accelerate ripening—Roundup is not actually a crop desiccant (other products such as Reglone are licensed for such use) but it is frequently applied to crops (GM and non-GM alike) shortly before harvest to kill off any remaining weeds and green crop ‘regrowth’, to make harvesting easier.

5. Control of grass and weeds amongst perennial crops, such as apple orchards—Roundup is widely used to control weeds around fruit trees and between other perennial crops such as strawberries. It is also used to control weeds on non-cropped areas.

NO TILLAGE AND MINIMUM TILLAGE FARMING SYSTEMS

As can be seen, both with GM and non-GM crop varieties, the main use of glyphosate is associated with no tillage (no-till) and the similar min-till, which includes some shallow cultivation. These avoid ploughing with the crop generally established directly into the stubble or other residues of the previous crop.

Such methods actually fit with the long-term aims of early members of the organic farming movement, such as Frank Newman Turner,[16] to grow crops without inverting the soil by ploughing, except for the reliance on chemical weed control. And they have been widely advocated by agronomists and many environmentalists based on claims, initially backed by some seemingly credible research, which indicated that they reduced greenhouse gas losses from soil. For many years claims were made that no-till systems reduced nitrous oxide emissions form soils. However, further research eventually demonstrated that it generally increases them.[17] Advocates then turned their attention to arguing that soil carbon levels increased under no-till and min-till systems, something which was widely accepted and led to many claims that moving all crop production to reduced tillage systems could have huge greenhouse gas mitigation benefits. Yet, here too it has now been shown that these claims were misleading. A comprehensive review of the evidence by scientists from the UK's leading Rothamsted Research found that even where there was a small increase in soil carbon in the top few inches of soil there was always a loss of carbon from soil lower down resulting in no overall gain in most of the studies examined.[18]

Attention then focused on claimed yield increases associated with no-till systems especially on dry lands. However, while there is still a consensus that no-till can help to conserve moisture in regions with very low rainfall, a major review of over 600 peer-reviewed studies found that on average such methods produced crop yields 6-9% lower than comparative production using tillage.[19,20]

We can conclude from this that moving away from the use of Roundup to support no-till systems will have little if any negative impacts in relation to greenhouse gas emissions and may actually bring some benefits, as well as increasing crop yields. However, there is another more significant obstacle to overcome, and this relates to the problem of specialisation.

WHY MIXED-FARMING IS FUNDAMENTAL

The vast majority of farmers have moved away from more traditional mixed farming systems in recent decades, specialising either in all arable crop production or all livestock production. Even farms that still have both livestock and crops rarely integrate these enterprises, growing crops on all suitable fields and keeping grazing animals only on land unsuitable for crop production. This specialisation has gone hand in hand with a rapid process of farm size expansion. Large farms that can afford the most modern machinery continue to expand while small producers are generally unable to compete.

Along with this we have seen a substantial reduction in the number of workers employed and a reduction in the number of crops grown. Where rotations do exist they are generally limited to very large areas of two alternating crops, such as corn and soy in the mid-west of the U.S., or wheat and oilseed rape on most of the best arable farmland in the UK. Taking the last example, two crops of wheat are generally followed by one of oilseed rape in a three-year rotation. Yields are highest when winter varieties are established in the autumn but this means the farm needs to be able to plant large areas of crops in a very short time. This is especially the case with oilseed rape since the best yields come from crops sown in late August or early September, immediately after the wheat has been harvested. No-till systems allow farmers to sow straight into the stubble with the drill and then spray the crop with Roundup as weeds emerge later.

Moving away from no-till systems, which will be necessary in many situations if glyphosate use is banned. However, this will be strongly resisted by farmers who are financially locked into such systems due to heavy borrowing for the expensive machinery required for no-till agriculture.

A DIFFERENT APPROACH

The ideal alternative approach is one that will allow the greatest number of production-related problems to be resolved simultaneously. Without glyphosate there will be increased need for ploughing and soil cultivation, but to bring maximum benefits it will also need more complex crop rotations, which include:

- Grass and grazing livestock, or prolonged use of green manure crops, to build soil carbon to increase soil stability and water retention;
- Legumes, especially forage legumes, to fix atmospheric nitrogen naturally;
- Deeper rooting crops and species to reduce the effect of compaction on sub-soils, store atmospheric carbon at deeper levels and make crops less drought prone;
- A sufficient range of plants to provide continuity of nectar for pollinating insects throughout the growing season;
- The return of crop residues, livestock manure, uncontaminated sewage and/or green waste, each appropriately composted.

The use of grass and forage legumes such as clover and lucerne (alfalfa) grown in rotation with crops can greatly reduce weed burdens and crop diseases, leading to zero or reduced need for herbicides and fungicides. Reducing the frequency with which crops like oilseed rape are grown can also greatly reduce the risk of pest attack and the need for insecticides.

However, many other herbicides are still available to provide weed control in the absence of Roundup. Farmers will simply need to rely on the advice of a good agronomist, as they did in the past, to identify which weeds are present and which herbicides will be most appropriate.

Moving away from Roundup should also provide additional stimulus for other chemical-free methods of weed control. Scientists in Denmark have found that weed populations can be reduced by more than 70% and yields increased by over 40% where crops are sown in a grid pattern at higher than normal densities instead of in traditional rows.[21] Many other methods for reducing weed completion are also available, some of them pioneered by organic farmers. One of these involves inter-cropping with legumes. This brings additional benefits in the form of increased nectar sources for pollinating insects, reduced nitrogen fertiliser applications, reduced soil run-off over winter and improved river and drinking water quality.[22]

To make the necessary changes will, of course, require significant determination on the part of governments, farmers, educationalists, agronomists and others and as such it is likely to meet with significant initial resistance. The first reaction of many farming organisations will be to seek an extension to any deadline placed on a glyphosate ban and hope that some agrochemical company will come up with a new product to replace Roundup that has similar benefits but none of the problems.

While it is likely that the search is already on for such a replacement chemical, perhaps the good news here in the UK at least is that an increasing number of farmers are being forced by circumstances, rather than a Roundup ban, to make some of the necessary changes. Pernicious arable weeds, such as blackgrass, have become a problem on land that has received so many applications of herbicides, including glyphosate, over the past decades that this weed is now resistant to all these products and they can no longer be controlled by chemicals. That's helpful and hopeful because knowledge of these problems and the more sustainable solutions already available to us will start to spread.

There is also growing concern in the UK that some of our best arable soils are losing carbon, and suffering from soil erosion as a result. This is particularly significant because crops grown in soils with higher organic carbon levels are better able to resist both drought and disease, both of which are expected to increase due to climate change. A recent report by the UK's Committee on Climate Change predicts that if current trends are allowed to continue, crop yields will fall,[23] and most farmland will eventually be unable to grow crops at all. While this is a dreadful situation, growing awareness of problems like this and, for example the dramatic decline in pollinating insects, could be just what is needed. These problems could force a rethink of current agricultural systems and lead to a redesign that will be fit to face the future challenges of food production.

REFERENCES

1. V.K. Nandula (ed.) (2010) *Glyphosate Resistance in Crops and Weeds: History, Development, and Management* (New Jersey: John Wiley & Sons).
2. USDA (2014) *Pesticide Use in U.S. Agriculture: 21 Selected Crops, 1960-2008* (United States Department of Agriculture Economic Research Service). Available at http://tinyurl.com/q3ac5h8, accessed August 13, 2015.
3. Transparency Market Research (2014) Global Glyphosate Market Expected to Reach U.S.$8.79 Billion in 2019. Available at http://tinyurl.com/nwar62d, accessed August 13, 2015.
4. PR Newswire (2014) Global Pesticides Market—Segmented by Type, Application Area and Geography. Available at http://tinyurl.com/nqxgwks, accessed August 13, 2015.
5. Fera (2014) *Pesticide Usage Survey Report 263—Arable Crops in the United Kingdom 2014* (Fera—Land Use & Sustainability Team). Available at http://tinyurl.com/pk3v54p, accessed August 13, 2015.
6. Ibid.; DEFRA (2015) Agriculture in the United Kingdom Data Sets. Available at http://tinyurl.com/czjlm6d, accessed August 12, 2015.

7. Fera (2013) *Pesticide Usage Survey Report 251—Soft Fruit in the United Kingdom 2012* (Fera—Land Use & Sustainability Team). Available at http://tinyurl.com/o2nt2yx, accessed August 13, 2015.
8. Fera (2013) *Pesticide Usage Survey Report 252—Orchards in the United Kingdom 2012* (Fera—Land Use & Sustainability Team). Available at http://tinyurl.com/pqwmctv, accessed August 13, 2015.
9. FoE (2013) *Introducing Glyphosate, the World's Biggest Selling Herbicide* (Friends of the Earth Europe). Available at http://tinyurl.com/pr47rov, accessed August 13, 2015.
10. WHO IARC (2015) *IARC Monographs Volume 112: Evaluation of Five Organophosphate Insecticides and Herbicides* (WHO IARC). Available at http://tinyurl.com/psn8dps, accessed August 13, 2015.
11. Nordström et al. (1998) "Occupational exposures, animal exposure and smoking as risk factors for hairy cell leukemia evaluated in a case control study." *British Journal of Cancer* **77**:2048-2052.
12. Hardell and Eriksson (1999) "A Case-Control Study of Non-Hodgkin Lymphoma and Exposure to Pesticides." *Cancer* **85**:1353-1360. Available at http://www.beyondpesticides.org/documents/acs-nhlymphoma-1999.pdf.
13. Corporate Europe Observatory (2015) The Glyphosate Saga and "Independent Scientific Advice" according to Germany, the UK and France. Available at http://corporateeurope.org/food-and-agriculture/2015/04/glyphosate-saga-independent-scientific-advice-according-germany-uk accessed August 16 2015.
14. Cox, C. (1998) "Responding to a Chemical Goliath." *Journal of Pesticide Reform* **18**:2-17.
15. Seneff, S. (2013) "Glyphosate: The Elephant in the Room." Slide presentation available at http://people.csail.mit.edu/seneff/WAPF_Slides_2012/Offsite_Seneff_Handout.pdf.
16. Newman-Turner, F. (1956) Fertility Farming, Chapter 4 Farming Without Plough or Chemicals, Faber and Faber. Available at http://journeytoforever.org/farm_library/turner/turner4.html.
17. Smith et al. (2008) "Greenhouse gas mitigation in agriculture." *Philosophical Transactions of the Royal Society B* **363**:789-813. Available at http://www.ncbi.nlm.nih.gov/pmc/articles/PMC2610110/pdf/rstb20072184.pdf.
18. Powlson et al. (2014) "Limited potential of no-till for climate change mitigation." *Nature Climate Change* 4:678-683. Available at http://www.nature.com/nclimate/journal/v4/n8/full/nclimate2292.html.
19. Pittelcow et al. (2015) "Productivity limits and potentials of the principles of conservation agriculture." *Nature* **517**:365-368.
20. UC Davis (2014) "No-till agriculture may not bring hoped-for boost in global crop yields, study finds." Available at http://news.ucdavis.edu/search/news_detail.lasso?id=11062.
21. Department of Plant and Environmental Sciences, University of Copenhagen (2015) "Crops can do their own weed control." Available at http://plen.ku.dk/english/news/2015/crops/. Accessed 16 August 2015.
22. Bilalis et al. (2010) "Weed-suppressive effects of maize-legume intercropping in organic farming." *International Journal of Pest Management* **56**:173-178.

23. Committee on Climate Change (2015) 'Reducing Emissions and Preparing for Climate Change: 2015 Progress Report to Parliament', Available at: http://tinyurl.com/owlkhfx.

SESSION 12

MECHANISMS OF SUSCEPTIBILITY/RESISTANCE TO HUMAN PATHOLOGIES

MECHANISMS OF SUSCEPTIBILITY TO INFECTIONS: THE HUMAN IMMUNODEFICIENCY VIRUS (HIV) PARADIGM

ISHWAR GILADA
President of AIDS Society of India
Chennai, India

TRUPTI GILADA
Seattle, USA

PRAPTI GILADA
Mumbai, India

MUMBAI…

- Bombay has no bombs and is a harbour not a bay.
- Churchgate has neither church nor gate—a railway station.
- No king ever stayed at Kings Circle.
- Nor did Queen Victoria stay at Victoria Terminus station.
- Nor is there any princess at Princess Street.
- There are no marines or sailors at Marine Lines.
- The Mahalaxmi temple is at Haji Ali not at Mahalaxmi station.
- Trams—used to terminate at Kings Circle, not Dadar Tram Terminus TT.
- Breach Candy is not a sweetmeat market, but there is a Hospital.
- You will not find lady fingers at Bhendi Bazar.
- Funny Bombay...aka Mumbai!

MY INDIA!!!

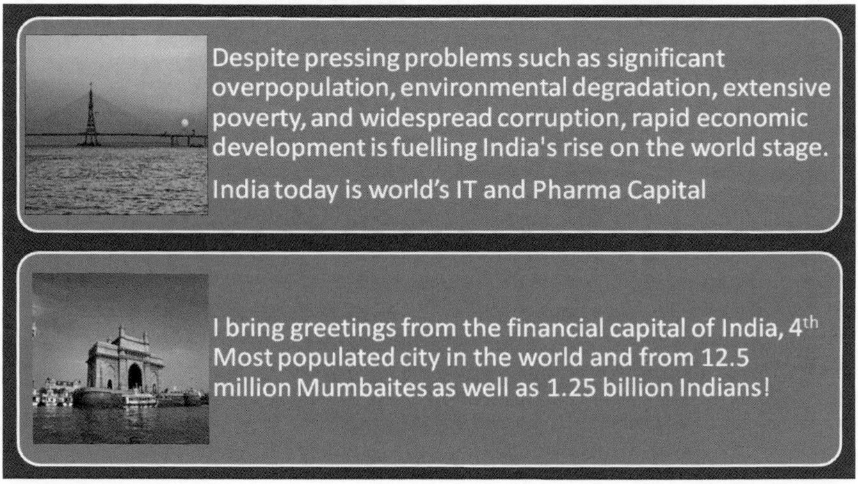

CHAIN OF INFECTION

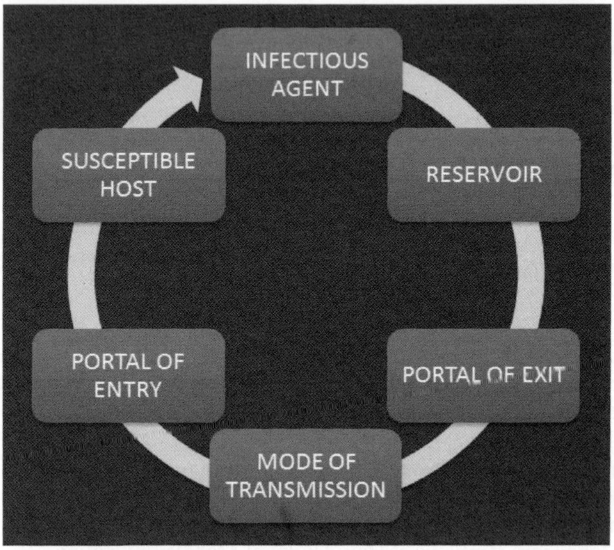

HIV-1 SUBTYPES: GLOBAL DISTRIBUTION

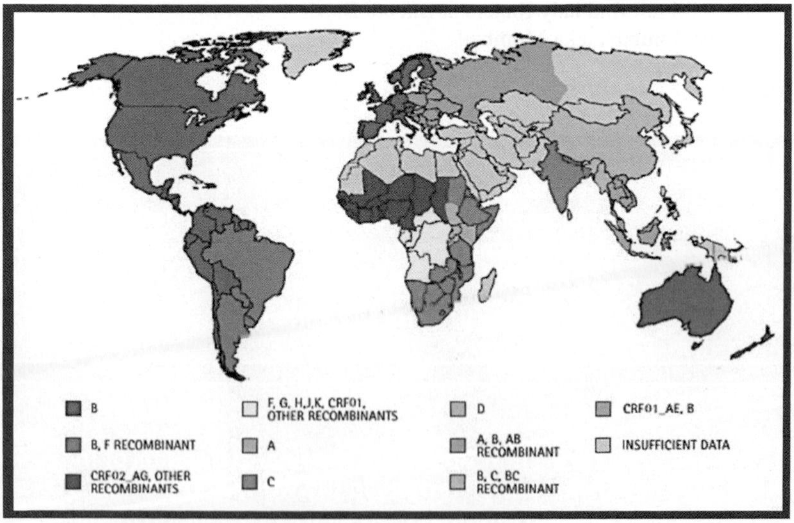

Source: IAVI Report, published by the International AIDS Vaccine Initiative, 2003.

SUBTYPES IN INDIA

- HIV-1 (97%), subtype C continues to predominate though A & B are also reported infrequently.
- HIV-1 recombinants—A/C and B/C are infrequent.
- HIV-2 (1-2%) subtype A is more common in India.
- HIV-1 + 2 (1-2%), dual infection, better than HIV-1.

Two distinct patterns of HIV epidemic in India

- Sex work driven in rest of India.
- Needle-sharing driven in north-east though sexual mode making its presence felt now!
- Overall more than 90% infections—sexual route.

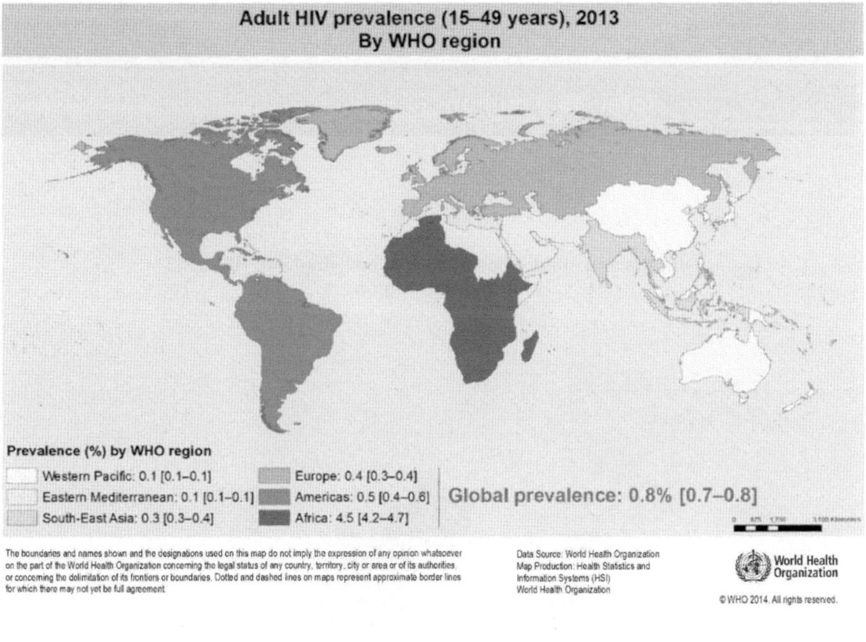

ADULTS AND CHILDREN ESTIMATED TO BE LIVING WITH HIV: 2014

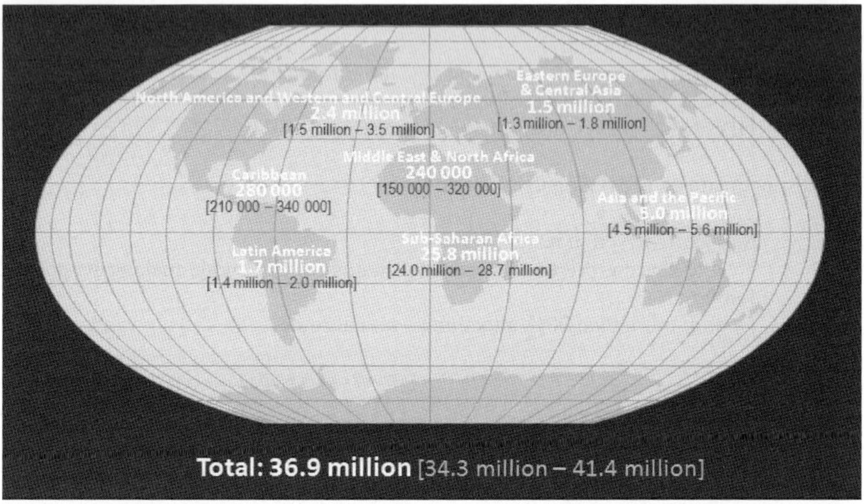

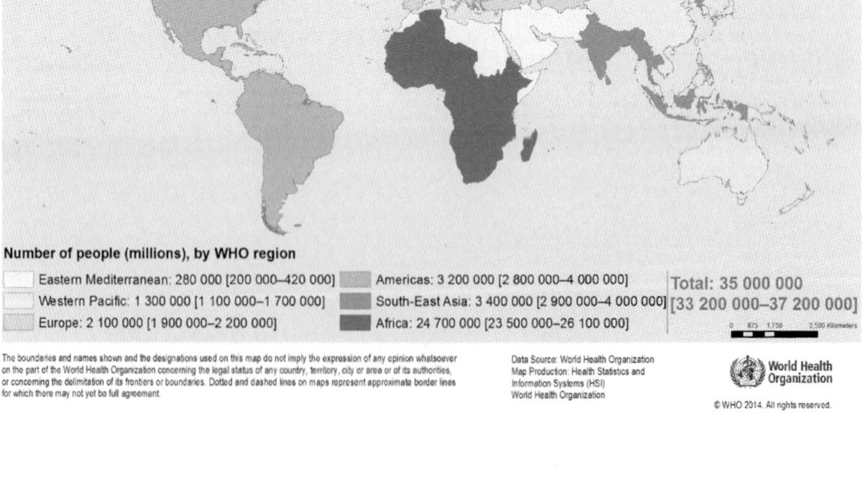

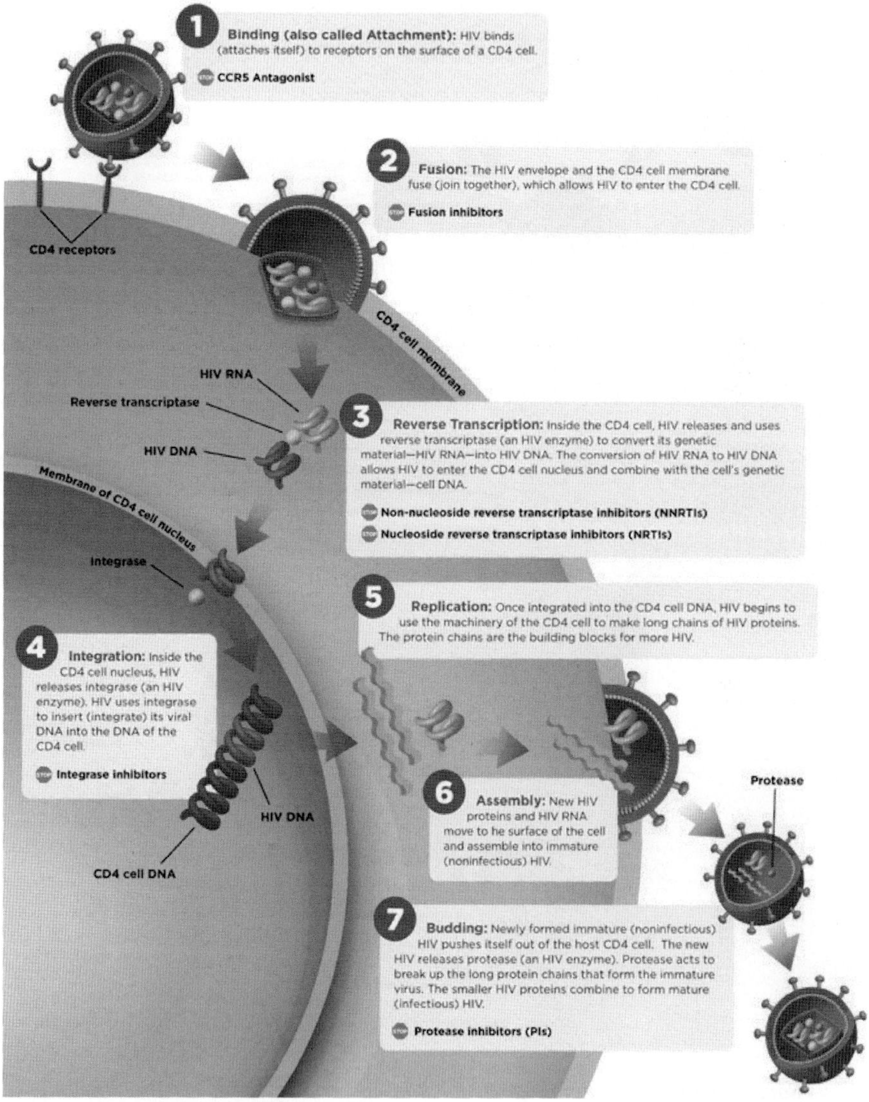

Source: https://aidsinfo.nih.gov/education-materials/fact-sheets/19/73/the-hiv-life-cycle.

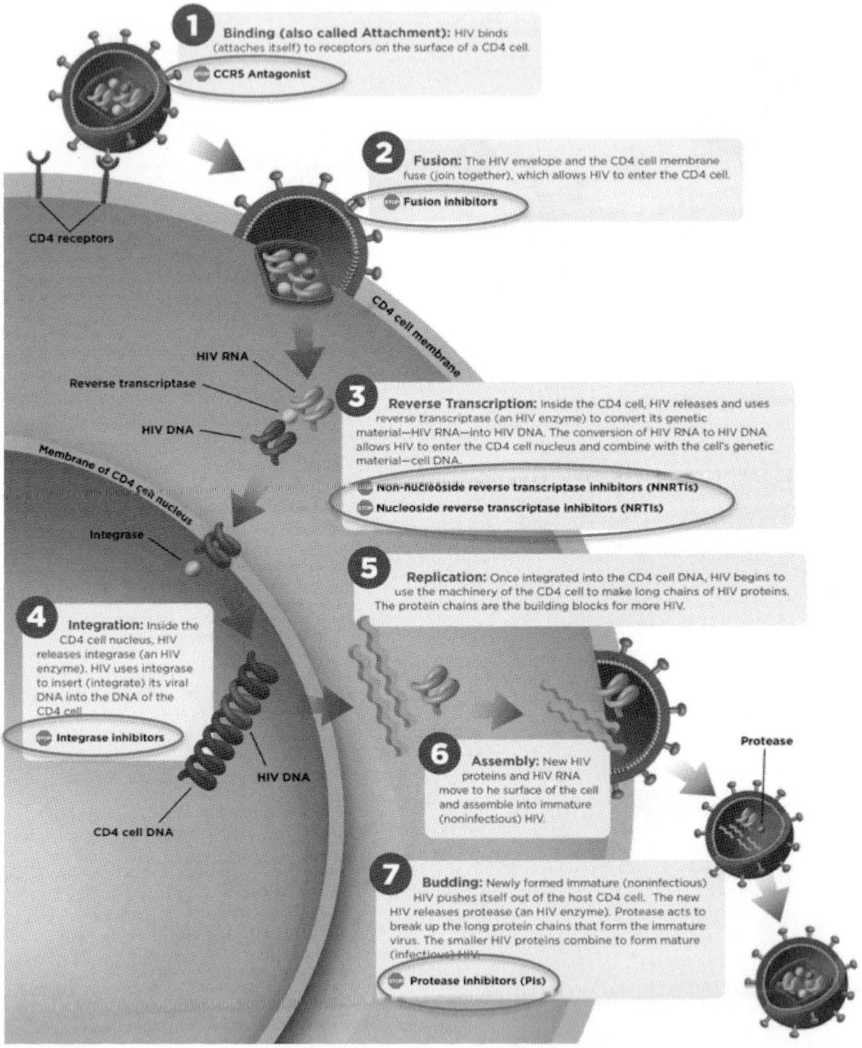

Source: https://aidsinfo.nih.gov/education-materials/fact-sheets/19/73/the-hiv-life-cycle.

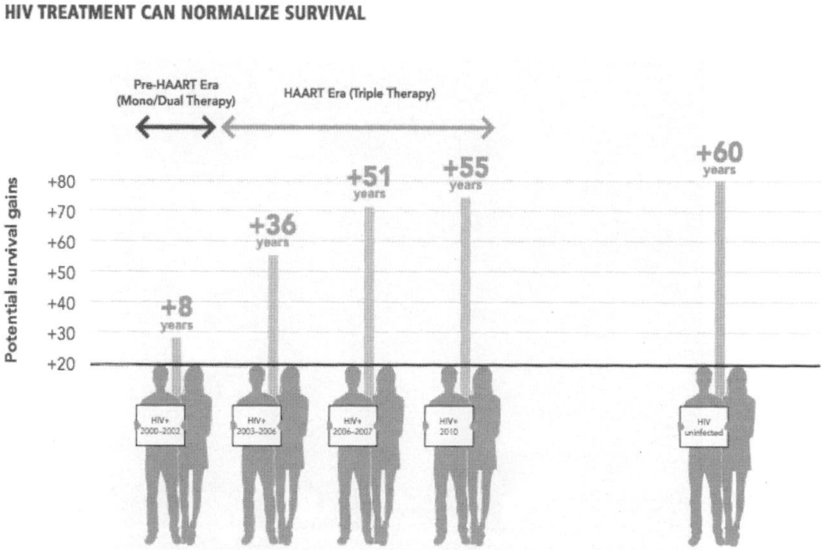

Expected impact of HIV treatment in survival of a 20 years old person living with HIV in a high income setting (different periods)

Source: Samji H et al., PLoS ONE, 2013.

COMPARISON OF SCENARIO FOR DRUG RESISTANCE TESTING

- DHHS
  Before initiation of ART:
  - TDR in 6-16% of HIV-infected patients.
  - In absence of t/t, resistance mutations may decline over time, become undetectable by current assays, but may persist and cause t/t failure when ART is started.
  - Identification of resistance mutations may optimize t/t outcomes.
  - Resistance testing (genotype) for all at entry to care.
  - Recommended for all pregnant women.
- India
  - TDR is < 5%.
  - Almost all patients receive simplified version I line ART.
  - We are unable to monitor virologically so far.
  - Irrational prescriptions are on a significant decline.
  - All pregnant women exposed to SD-NVP to be provided r/LPV based regimen.

CHAIN OF INFECTION

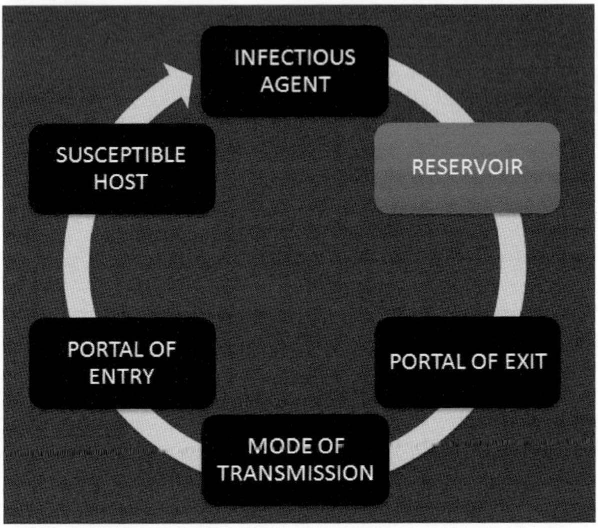

RESERVOIR

- Chromaffin Cells in Intestine
- Glial Cells in Brain
- Alveolar Spaces in Lungs
- Langerhans' cell in Skin
- Bone Marrow

NATURAL COURSE OF HIV INFECTION

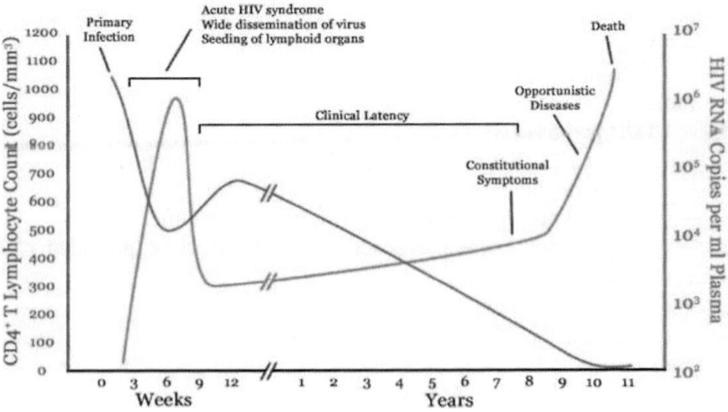

TASP: TREATMENT AS PREVENTION

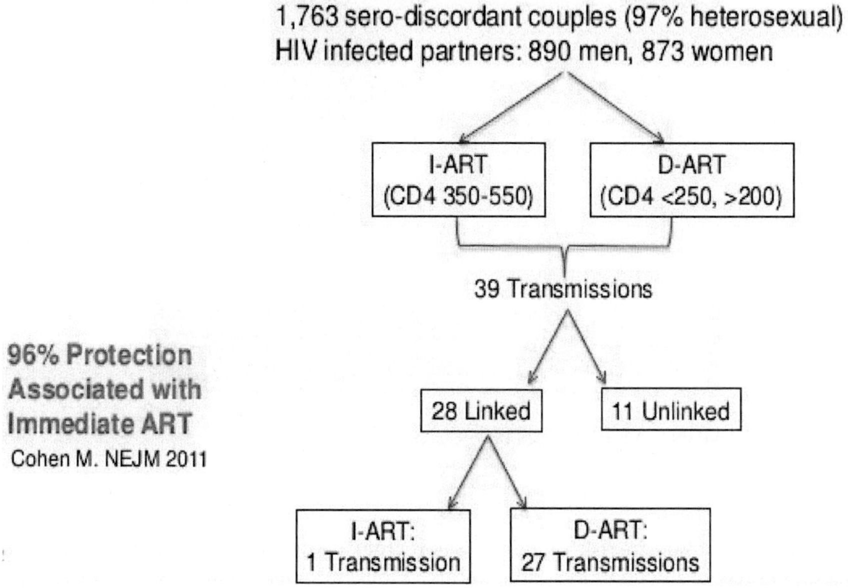

HPTN 052: MULTIVARIATE ANALYSIS OF FACTORS ASSOCIATED WITH LINKED TRANSMISSIONS

Variable	**HR**	**95% CI**
Treatment, immediate vs. delayed	0.04	0.01-0.28
Baseline CD4+ count, per 100 cells/mm^3 decrease	1.24	1.00-1.54
Baseline HIV-1 RNA, per 1 $\log_{10}$ copies/mL increase	2.85	1.51-5.41
Baseline condom use, 100% vs. < 100%	0.33	0.12-0.91
Sex of infected partner, male vs. female	0.73	0.33-1.65

- 61% of transmissions occurred from infected patient with CD4 + cell count > 350 cells/mm^3.
- All transmissions occurred prior to starting ART.
- 82% of transmissions occurred in African patients.

Cohen, M.S. et al. IAS (2011) Abstract MOAX0102.
Cohen, M.S. et al. *N Engl J Med.* (2011); [Epub ahead of print].

TASP: TREATMENT AS PREVENTION

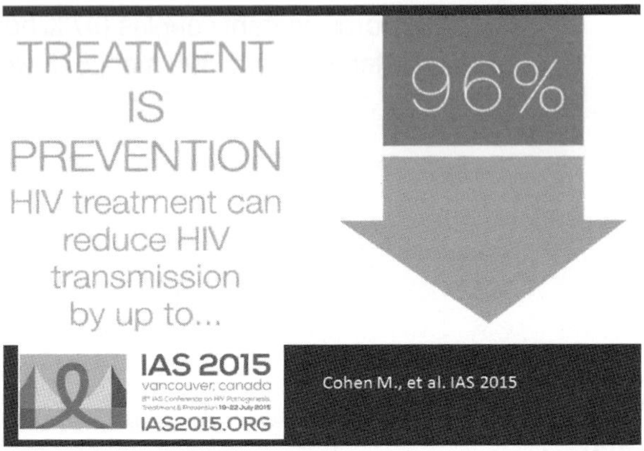

TASP: TREATMENT AS PREVENTION

HPTN052

93% reduction in risk of transmission from partner (linked transmission). 69% reduction in risk of infection including unlinked transmission.

- No index-to-partner (linked) HIV transmissions were observed when the index participant was stably suppressed on ART.
- ART is highly effective for prevention of HIV sexual transmission.
- Durable benefit of early ART in HIV prevention among HIV-discordant couples.

HIV PREVENTION IN CARE AND TREATMENT

- Due to increasing availability of ART, many PLHAs are accessing health care settings and clinics, providing an opportunity to reach a large number of infected persons with prevention messages and interventions.
- Health care providers in HIV clinic settings meet with patients regularly so can deliver consistent, targeted prevention messages/strategies during routine visits.
- Providers are considered authority figures and trusted sources of health information.

BENEFIT OF TasP

Individual Level

- As a risk-reduction strategy.

- Decreased risky sexual behaviors after ART initiation (Venkatesh et al. *AIDS* 2011; Jean et al. IAS 2013).
- Benefit of early ART (ART-CC Cohort Study 2009, NA-ACCORD 2009).

Population Level

- Reduces HIV infections in a population.
- Impact of early ART among a population with **diverse sexual partnerships** (both stable/non-stable partners, both sero-discordant/sero-concordant partners).

ART FOR TREATMENT AND PREVENTION

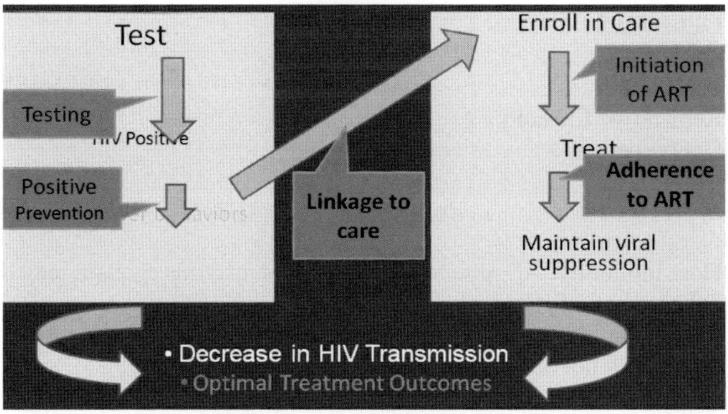

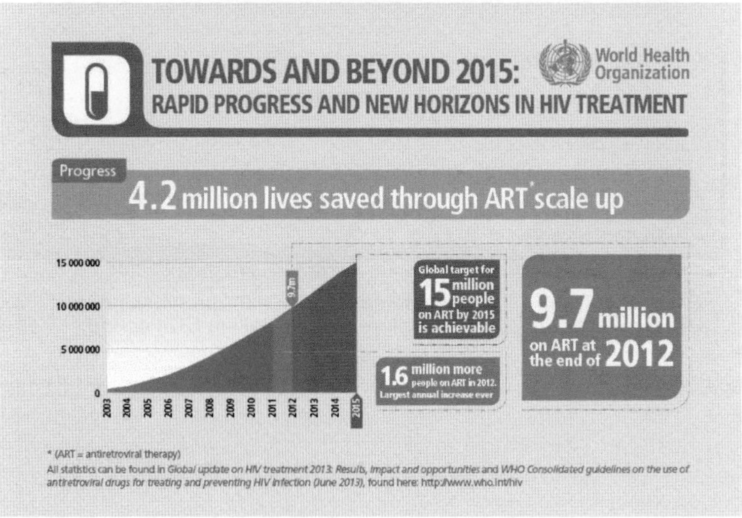

WHAT IS A “TEST AND TREAT” STRATEGY?

Improving access to HIV testing, care, support, and treatment.

Objectives

1. Increase number of PLHIV who know their HIV status.
2. Ensure that those who test positive are linked to care.
3. Increase number of PLHIV who are on treatment.
4. Support people who are on ART with regular clinical monitoring, management of side-effects, adherence and risk-reduction counseling, diagnosis and t/t of STIs.

VIRAL PROPERTIES

- Viral Load
- Replicative Capacity
- Genotype

HOST PROPERTIES

- Host-immunity—despite having same type of virus spouses may differ.
- Other genetic factors.
- Long-term non progressors: we have HIV+ve children, now in 20-25 age group, still not on ART.

TRANSMISSION EFFICIENCY

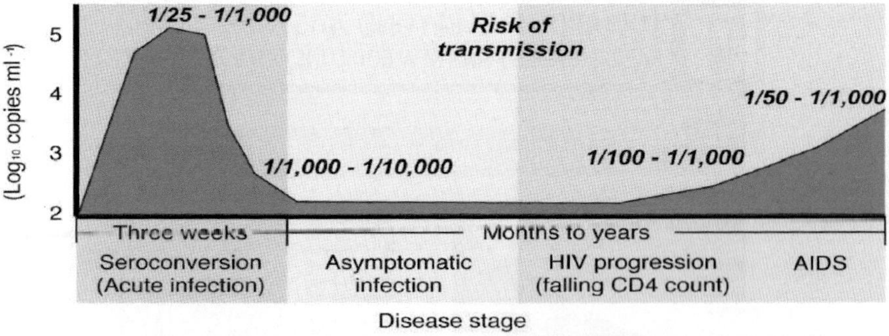

Source: Galvin, S.R. & Cohen, M.S. (2004) The role of sexually transmitted diseases in HIV infection. Nature Reviews Microbiology, 2(1).

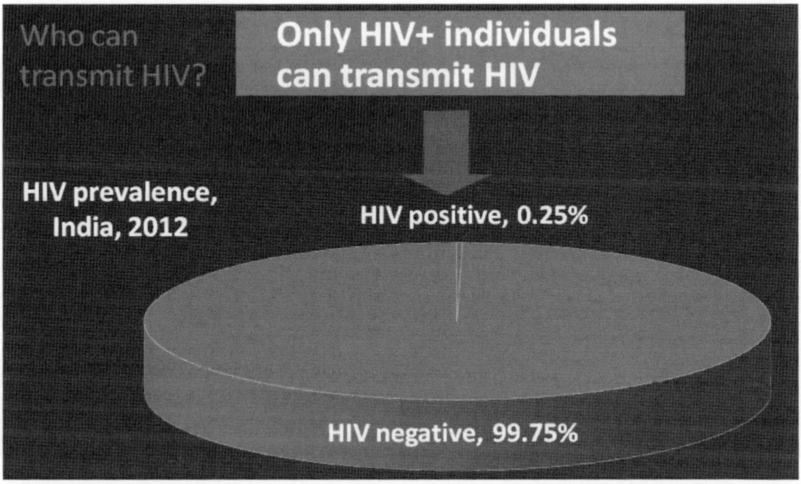

Focusing on ~2.4million HIV-infected people rather than ~1197 million uninfected is more practical and cost-efficient, targeted prevention approach.

CHAIN OF INFECTION

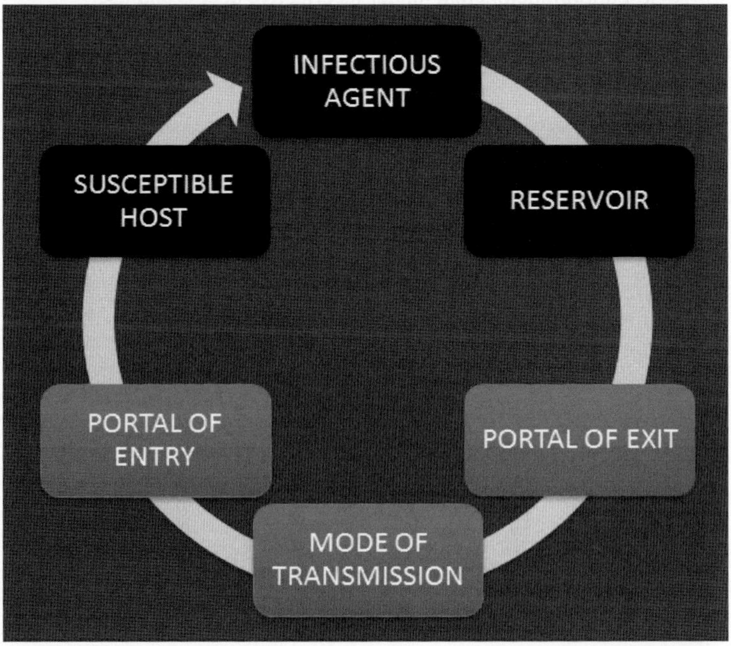

ESTIMATED NEW HIV INFECTIONS BY ROUTE OF TRANSMISSION, 2010

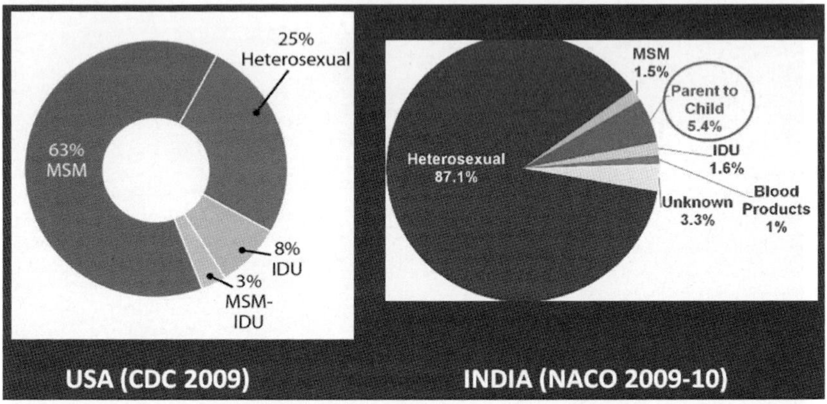

CONDOMS

- Male and female condoms: only devices that reduce the transmission of HIV/STIs and prevent unintended pregnancy.
- Programmes promoting condom must address stigma and gender-based and socio-cultural factors that hinder its effective access and use.
- Adequate investment in and further scale up of condom promotion is required.

Source: UNFPA, WHO and UNAIDS Statement. July 2015.

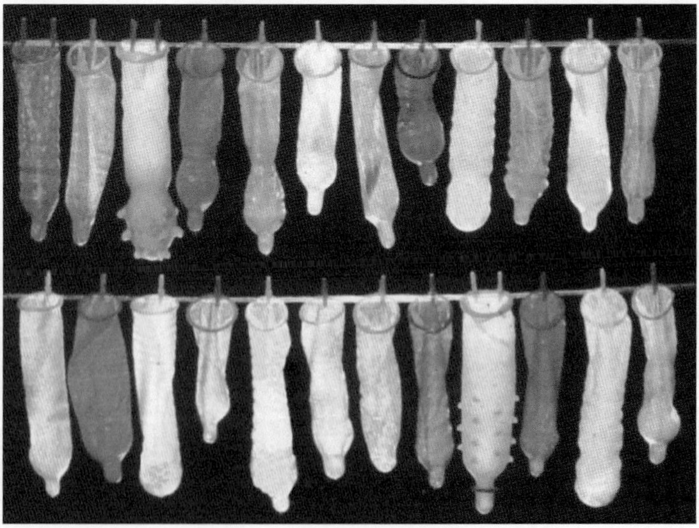

SAHELI PROJECT: PEER EDUCATION MODEL FOR SEX WORKERS

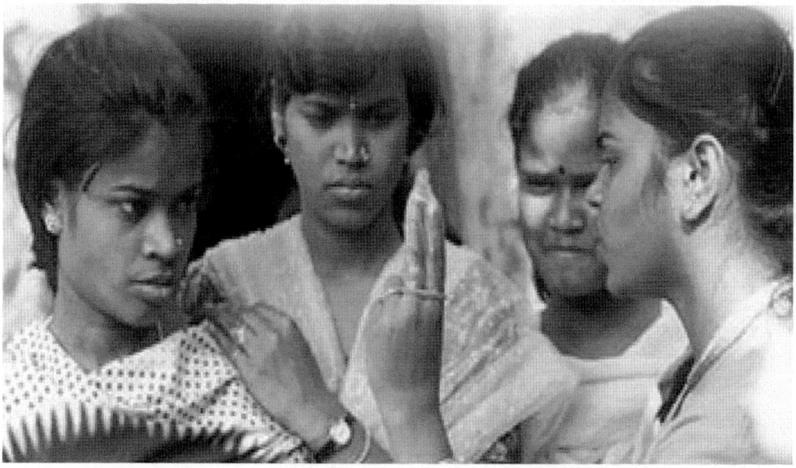

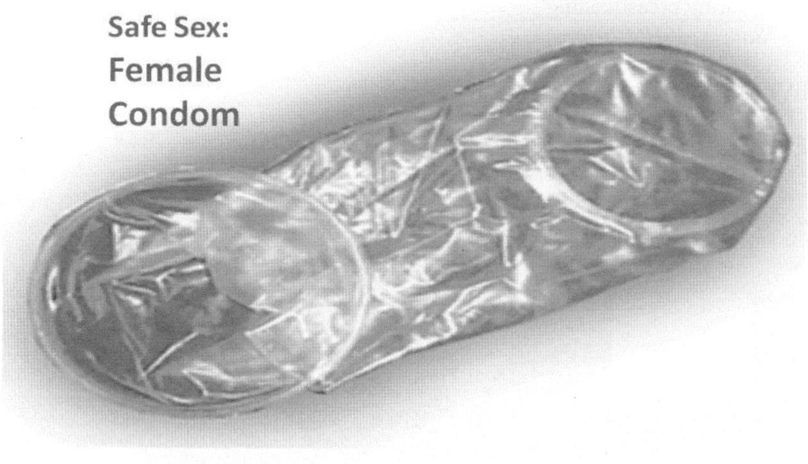

MICROBICIDES

- Compounds that can be applied inside the vagina or rectum to protect against STIs including HIV.
- At present, an effective microbicide is not available, Tenofovir jelly ~ 50% efficacy.

- Would greatly empower women to protect themselves and their partners; women can easily control and do not require the cooperation, consent or even knowledge of the partner.
- Recent cost-benefit analysis conducted (London School of Hygiene and Tropical Medicine) indicated that the introduction in 73 lower-income countries of a microbicide which reduced the risk of infection by 40%, at 30% coverage, would avert approximately 6 million HIV infections over 3 years in men, women and children.

MOTHER TO CHILD TRANSMISSION

- Number of children born annually with HIV has almost halved since 2009: down from 400,000 in 2009 to 240,000 in 2013.
- Every year, globally, an estimated 1.4 million women living with HIV become pregnant.
- Risk of MTCT without intervention: 15-45%.

Critical Period	**Rate**
During pregnancy	5-10%
During labor and delivery	10-20%
During breastfeeding	10-20%

Source: WHO 2015

PREVENTION OF MTCT

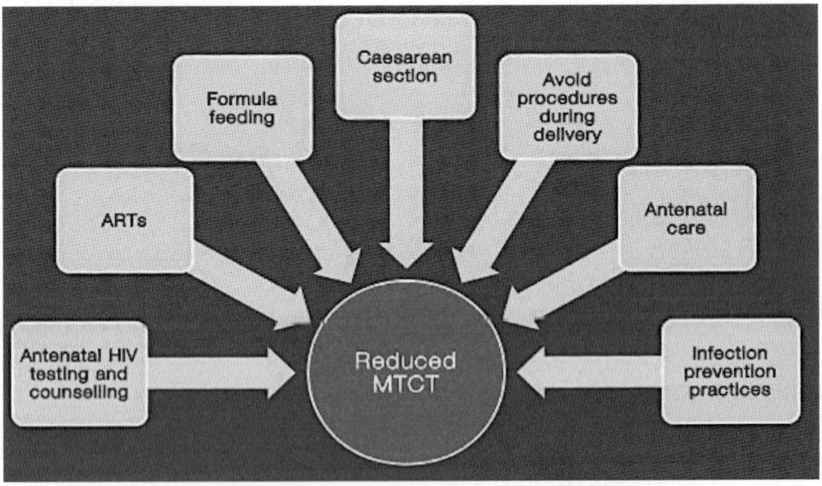

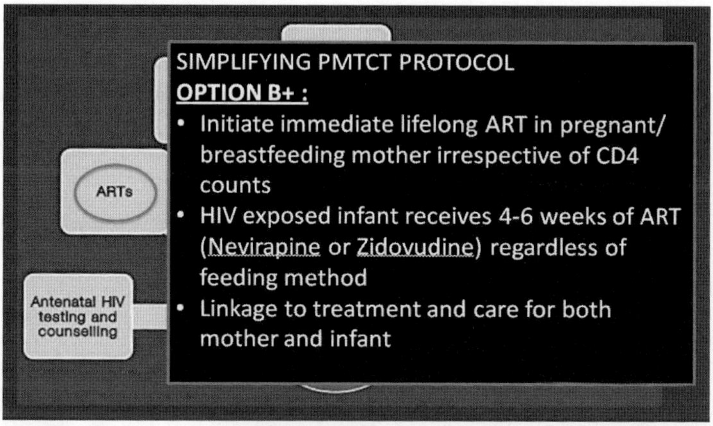

Today, Cuba became the first country to be validated by the World Health Organization (WHO) to have eliminated mother-to-child transmission of HIV and syphilis. The achievement marks a major progress in global public health. "Eliminating transmission of a virus is one of the greatest public health achievements possible," said Dr. Margaret Chan, WHO Director-General. "This is a major victory in our long fight against HIV and sexually transmitted infections, and an important step towards having an AIDS-free generation."

BLOOD AND BLOOD PRODUCTS

- Blood/organ donations: tested for HIV antibodies and HIV RNA.
- RNA test detects HIV at an earlier stage than HIV antibody.
- IDUs:
  - Globally, around 16 million IDUs, 3 million of them are PLHIV.
  - On average, 10% new HIV infections are caused by IDU and in parts of Eastern Europe and Central Asia, over 80% of all HIV infections is related to drug use.

Harm Reduction: evidence-based approach to HIV prevention, t/t and care for IDUs; comprehensive package, which includes needle-syringe programmes and opioid substitution therapy.

COMMUNITY OUTBREAK OF HIV-1 INFECTION LINKED TO INJECTION DRUG USE OF OXYMORPHONE—INDIANA, 2015

- J. Duwve et al. "Community outbreak of HIV infection linked to injection drug use of oxymorphone—Indiana, 2015."
- M. Spiller et al. "Network analysis of a contact network from an investigation of a community outbreak of HIV infection linked to injection drug use of oxymorphone—Indiana, 2015."

CUMULATIVE HIV INFECTIONS DIAGNOSED, SCOTT COUNTY, INDIANA THROUGH JUNE 14, 2015 (N=170)

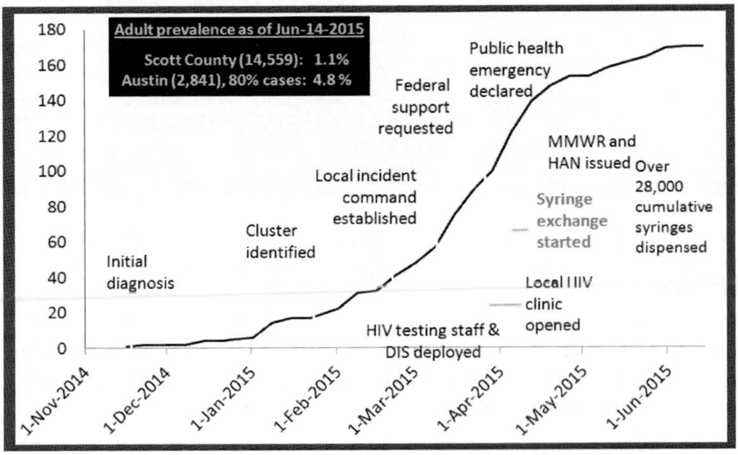

In terms of response time—some considerations…first few cases were not linked until mid-January. One DIS working the cases—and all STD cases, too. Reporting procedures are onerous—HIV reporting is treated differently than all other infectious disease reporting. Information is still considered more sensitive, more protected, so reporting is delayed. At end of January, number of cases/contacts mushroomed and became very rapidly unmanageable. Siloed databases made it difficult to review cases comprehensively. Lack of resources on the ground-building the system from the ground up. Laws didn't initially support syringe exchange—and there was resistance at the local level as well—lots of layers of response needed.

PEP (POST-EXPOSURE PROPHYLAXIS)

- ART (28 days) after a single high-risk event to prevent HIV.
- Occupational or non-occupational PEP.
- If started soon after exposure, PEP can reduce the risk of HIV infection by over 80%. Adherence to a full 28-day course of ART is critical to effectiveness of the intervention.
- Recent evidence shows PEP uptake has been insufficient: only 57% of the people who initiated PEP have completed the full course and rates were even lower at 40% for victims of sexual assault.

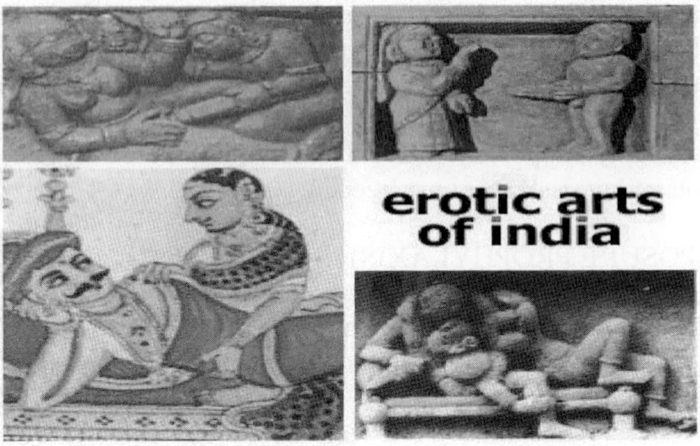

"Many Postures with One, Better than One with Many"—PHO.

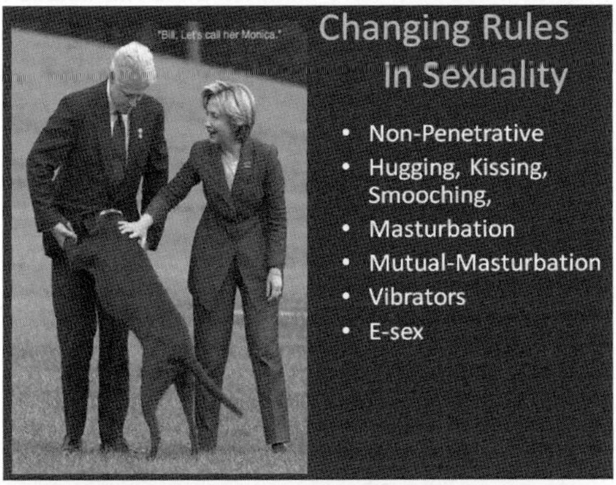

HIV PREVENTION TOOLS—INFECTION RATES

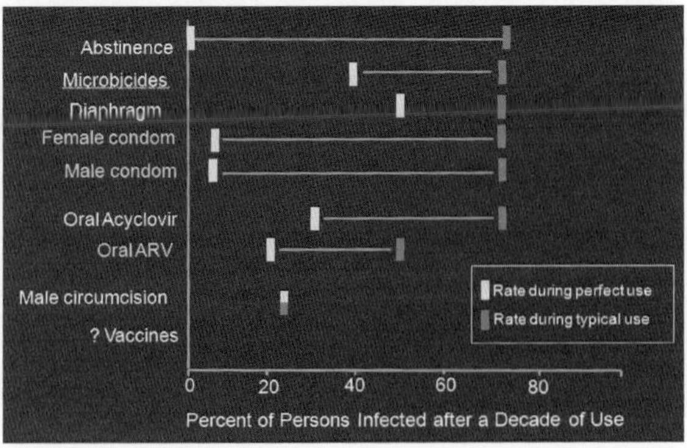

CHAIN OF INFECTION

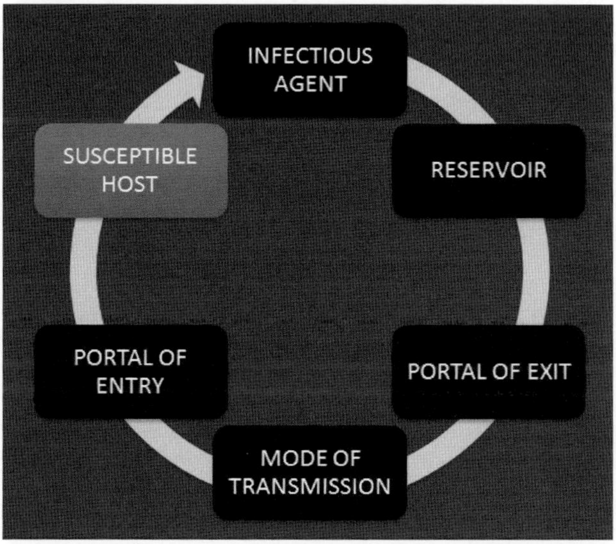

PREP: PRE-EXPOSURE PROPHYLAXIS

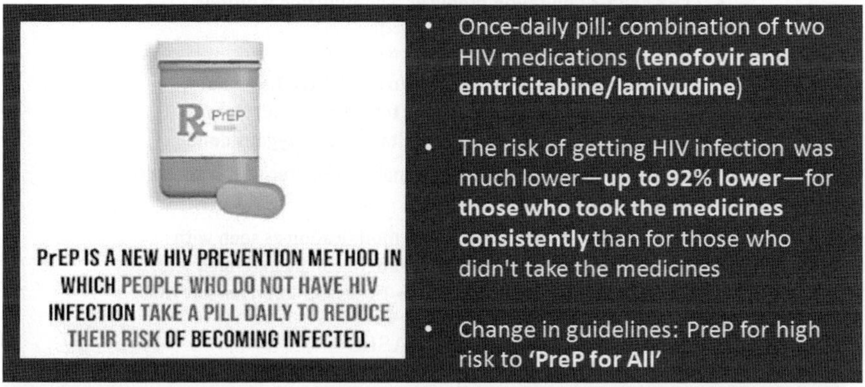

Grant, R. et al. (2010) "Pre-exposure Chemoprophylaxis for HIV Prevention in Men Who Have Sex with Men." *N. Engl. J Med.*

Thigpen, M. et al. (2012) "Antiretroviral Pre-exposure Prophylaxis for Heterosexual HIV Transmission in Botswana." *N. Engl. J Med.*

Baeten, J. et al. (2012) "Antiretroviral Prophylaxis for HIV-1 Prevention among Heterosexual Men and Women." *N. Engl. J Med.*

CCR5 Δ32 MUTATION

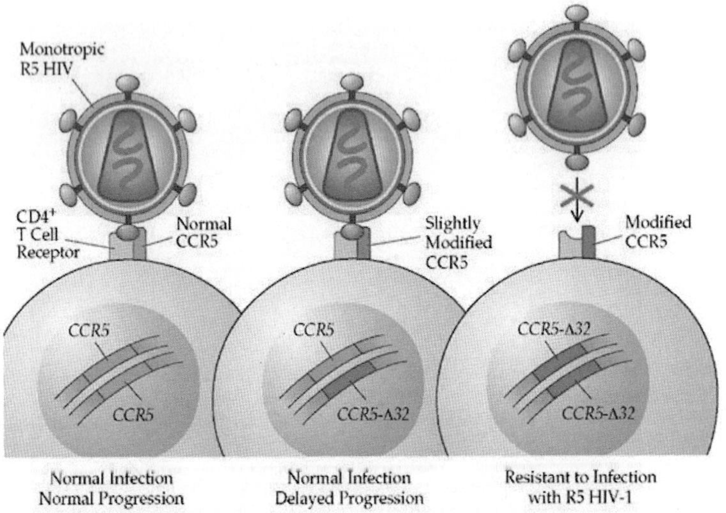

- **HIV (R5 virus) targets CD4 T-cell by binding to CCR5, one of the major co-receptor for HIV entry**
- **CCR5 delta-32 mutation produces a nonfunctional protein**
  - Homozygotes are resistant to HIV infection
  - Heterozygotes have slower disease progression
- **The "Berlin Patient" is HIV-free w/o HAART for 3.5 years following hematopoietic stem cell transplant (HSC) from an allogeneic, HLA matched, CCR5 delta-32 homozygous donor**
- **Zinc Finger Nuclease technology enables the precise disruption of the CCR5 gene**
- **The therapeutic potential of CCR5 modification as seen with natural mutations and the Berlin Patient can be extended with ZFN disruption of autologous CD4+ T cells (SB-728-T) in any HIV**

OTHER SEXUALLY TRANSMITTED DISEASES

<u>**INCREASED SUSCEPTIBILITY:**</u>

2-5 times more likely than uninfected individuals to acquire HIV infection

<u>**INCREASED INFECTIOUSNESS**</u>:

If also HIV-infected, more likely to transmit HIV through sexual contact than other HIV-infected persons

- Similar risk behaviours: target similar populations for expanded screening programs.
- Early detection and treatment of curable STDs: major, explicit component of comprehensive HIV prevention programs.

Wasserheit et al. 1992.

CIRCUMCISION

- Male circumcision reduces the risk of heterosexually acquired HIV infection in men by approximately 60%.
- Mixed results in men who have sex with men.
- Safe when provided by well trained health professionals in properly equipped setting.

Weiss, H.A. et al. (2000) "Male circumcision and risk of HIV infection in sub-Saharan Africa: a systematic review and meta-analysis." *AIDS.*

Siegfried, N. et al. (2003) "Male circumcision for prevention of heterosexual acquisition of HIV in men." Cochrane Database Syst Rev.2003

Kreiss, J.K. (1993) "The association between circumcision status and human immunodeficiency virus infection among homosexual men." *J Infect Dis*.

Buchbinder, S.P. et al. (2005) "Sexual risk and lack of circumcision associated with HIV seroconversion in MSM in the United States." *J Acquir Immune Defic Syndr*.

NON-SURGICAL CIRCUMCISION

- PrePex: medical device developed to facilitate non-surgical voluntary medical male circumcision of adults as part of a comprehensive HIV prevention strategy in Sub-Saharan Africa.
- Provides only partial protection.

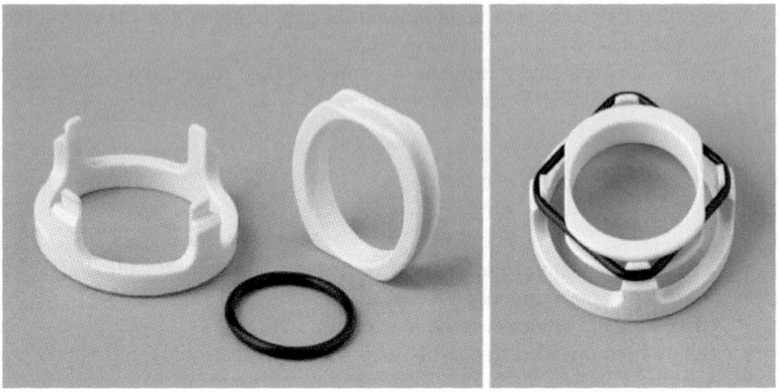

VACCINES and CURE

- Preventive that protect individuals who do not have or
- Therapeutic effect for persons who have or later contract HIV/AIDS.
- The most challenging area in the field of HIV.
- Thai AIDS Vaccine RV144 trial after ~70-odd failed trials had brought short lived celebrations.
- Phase-III Alvac (SanofiPasteur); AIDSVax (VaxGen) failed miserably in 2003 in USA, Uganda, Thailand.
- Globally, ~95 candidate preventive and therapeutic AIDS vaccines over 25 years costing millions, failed.
- NARI's Phase-I vaccine trial in India failed in 2008 mysteriously, failed recombinant adeno-associated viral vector-rAAV tried in 30 volunteers in Pune.

CURE RESEARCH

- Identify and characterize replication-competent HIV reservoirs throughout the body.
- Determine the effects of early ART, chemotherapy and transplantation, latency reversal, immune modulation, cell and gene therapies, and other interventions on the HIV reservoir.
- Develop assays and model systems, including nonhuman primate and humanized mouse models, to quantify and study HIV latency.
- Identify which reservoirs serve as the source of viral rebound when ART is stopped.
- Determine the role of HIV integration site and homeostatic proliferation on virus elimination strategies.
- Investigate strategies that lead to purging of latent reservoir and a cure for HIV or control of viral rebound and sustained remission.

FUNCTIONAL CURE

- Researchers' lead by Dr. Deborah Persaud from Johns Hopkins identified case of 'functional' HIV cure in HIV+ child on ART from second day of birth.
- Child-off ART for a year, and although DNA of HIV has been detected at very low levels in the child's cells, the virus is not reproducing.
- Past research, long term observation and clinical experience shows that the HIV-1 from MTCT is less virulent, resulting in many long-term survivors (non-progressors) among MTCT acquired HIV infected children than adults infected sexually.

HIV CONTROLLERS

- People who spontaneously suppress the virus.
- Two categories: Elite and Viremic.
  - Elite Controllers: at least 3 undetectable VL for >12 months without HAART.
  - Viremic controllers: a VL threshold of no more than 2,000 c/ml.
- They have high CD4 counts.
- CD4 count increases with HAART.
- May become back-bone for future research.

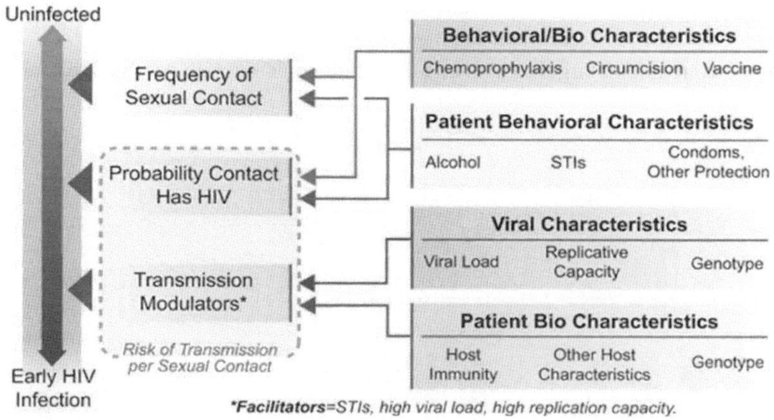

Prevention Issue. Reducing Transmission Risk after Exposure.

INFECTIOUSNESS: IN SIMPLE TERMS

- Disease Triangle: Host, Parasite, Environment.
- HIV Type and Sub types.
- Co-Infections, especially other STIs.
- Type of Sexual acts: traumatic/non-traumatic.

- Duration of Sexual activity.
- Frequency of Sexual acts.
- Protection used and its efficacy.
- Risk mapping of STI.

Police and Gang co-exist.

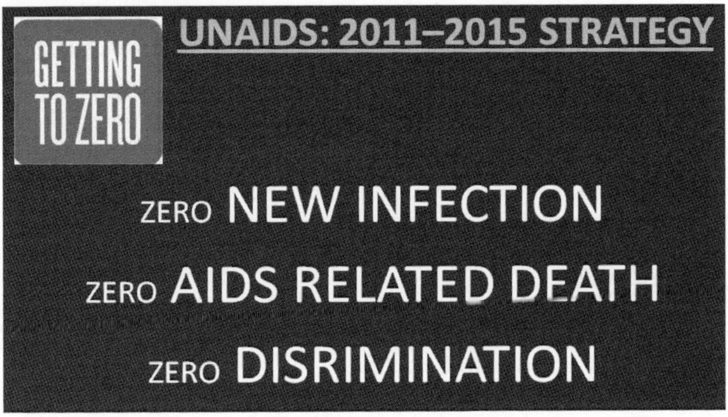

90–90–90 BY 2020

An ambitious treatment target to help end the AIDS epidemic (UNAIDS 2015).

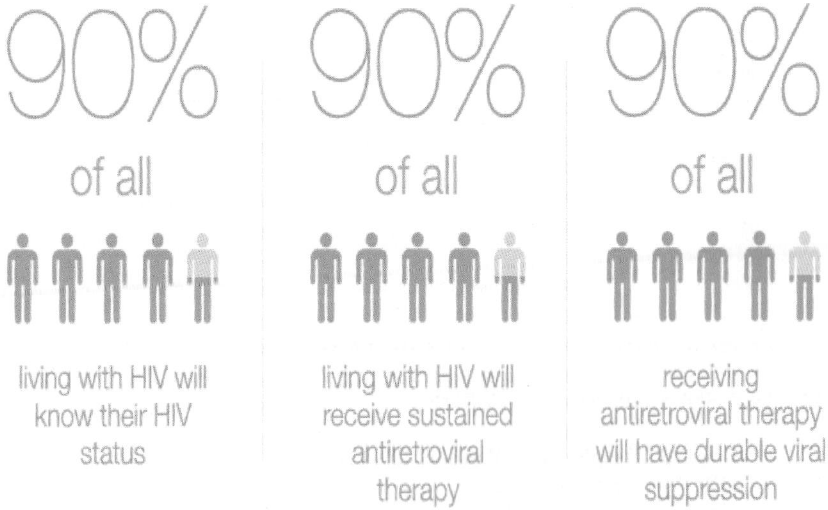

TOTAL ANTIRETROVIRAL TREATMENT COVERAGE BY REGION IN 2013

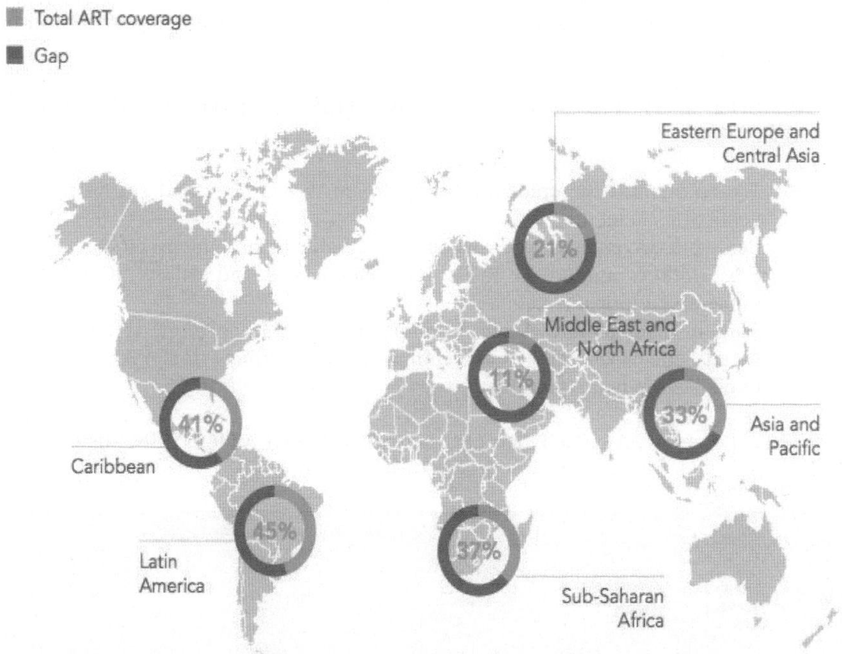

HIV IN INDIA: MILESTONES

- 1986 – First indigenous AIDS case & HIV infection.
- 1989 – HIV testing of blood units made mandatory in Mumbai and Maharashtra state.
- 1991 – First reports of HIV-2 in India.
- 1992 – NACP launched.
- 1989 – HIV testing of blood mandatory in India.
- 2001 – Prevention of mother-to-child transmission programme launched.
- 2004 – Free ART programme launched.
- 2010 – Free secondline ART added in programme.

WHY DID OUR EPIDEMIC NOT PROGRESS AFRICAN WAY?

- Perhaps the projections were overestimated.
- Perhaps the transmission circles in Africa are both males and females, in India only males.

- Perhaps our epidemic had low level drivers.
- Perhaps our targeted approach for preventive interventions was most appropriate.
- Primary Health Care much better than Africa.
- Also that we could implement positive prevention including ART roll-out well.

THE REALITY WHY THE EPIDEMIC IS BETTER CONTROLLED IN INDIA IS PERHAPS

- We recognised the epidemic being driven through sex and sex work in six years of its detection;
- Sex work driven epidemic was only observable in large metro cities;
- Sex workers started to insist on condom use and follow it among all vulnerable groups;
- Largely inadequately credited group for the stable epidemic is FSWs, Peer-based education and intervention—*Saheli Project* of Peoples Health Organisation being pioneer.

OVERALL HIV PREVALENCE RATES BY SUB-POPULATIONS IN 2008-2009 & 2012. *NACO*

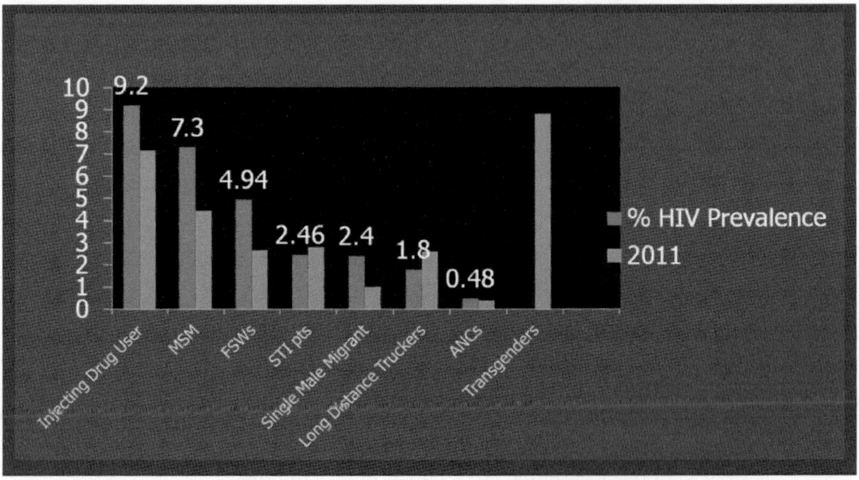

KEY CHALLENGES IN CONTROLLING HIV AMONG MSM

- Lower Coverage.
- Reaching married MSM.
- Lack of knowledge and sensitivity towards managing STIs among MSM.
- High level of social stigma.

A COMPLEX ISSUE: AWARE VS. UNAWARE

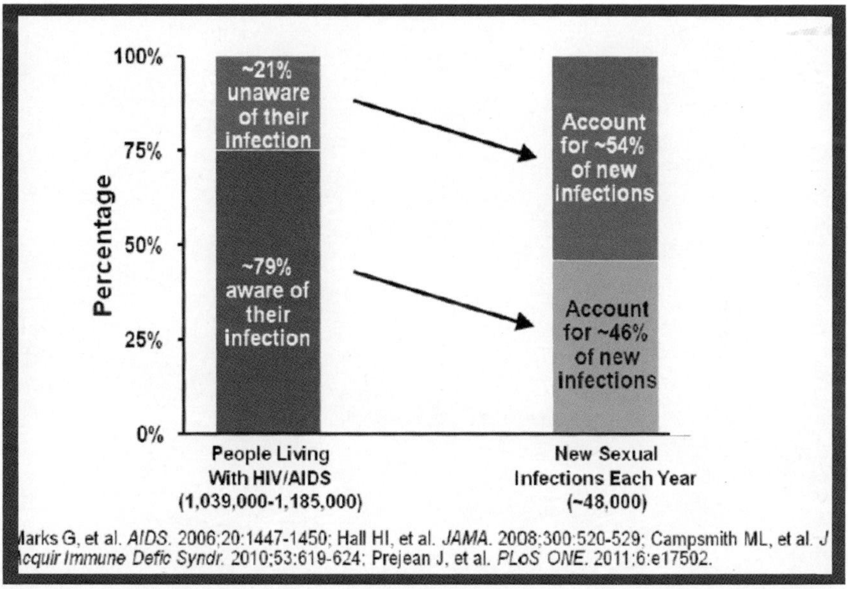

FIRST-LINE SCALE UP…

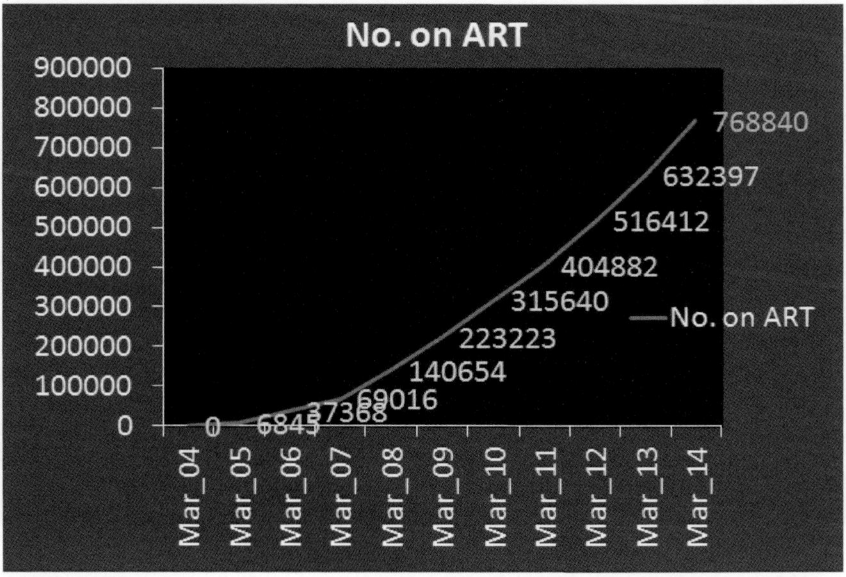

NO. OF PATIENTS ON 2ND LINE ART ~ A TREND 2008-14

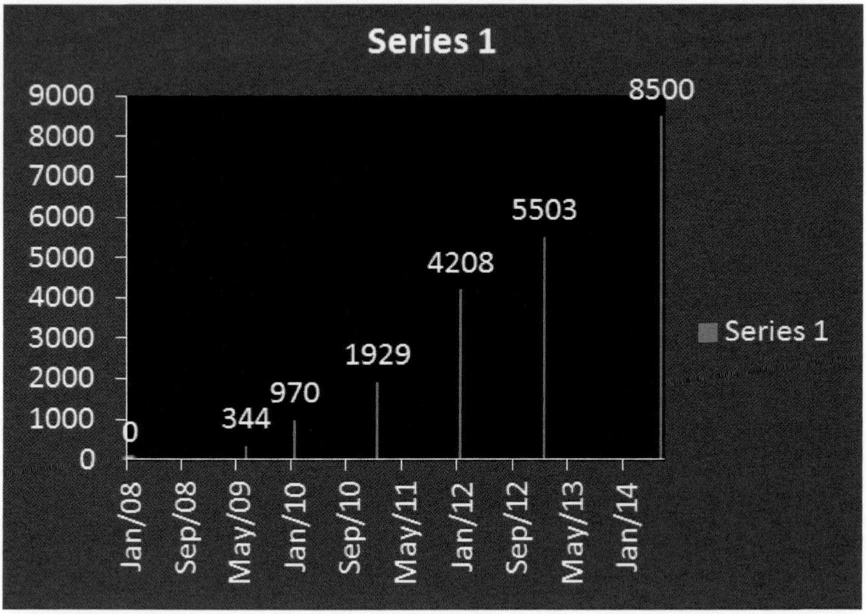

THE REALITY IS…

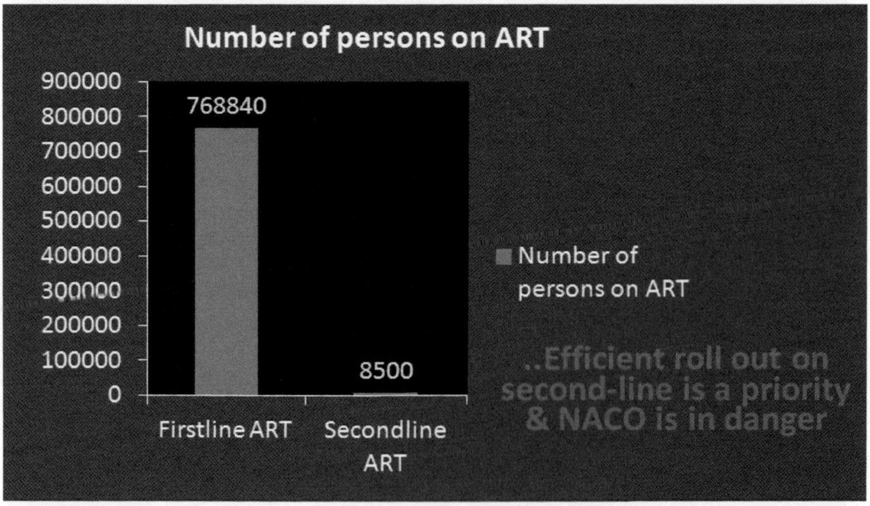

PATIENTS ON ART IN INDIA. *NACO*

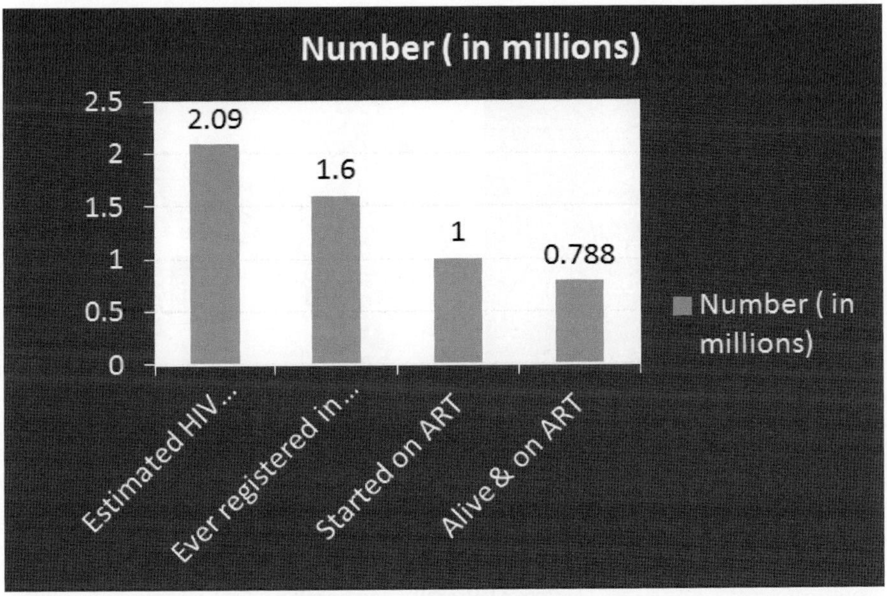

CATEGORIES OF DISTRICTS

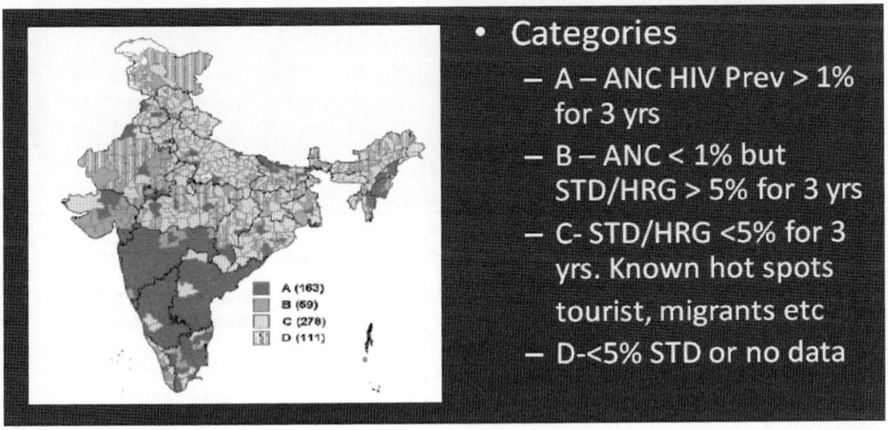

LARGE DIFFERENCES IN PREVALENCE AMONG STATES
Claeson, M. et al. 2008

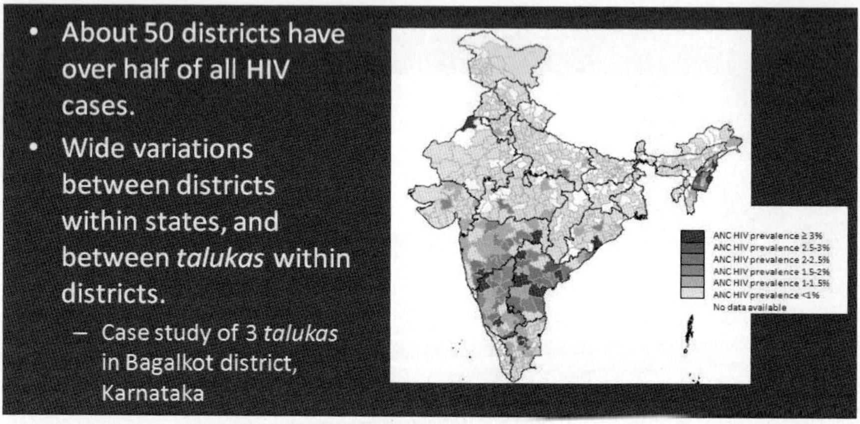

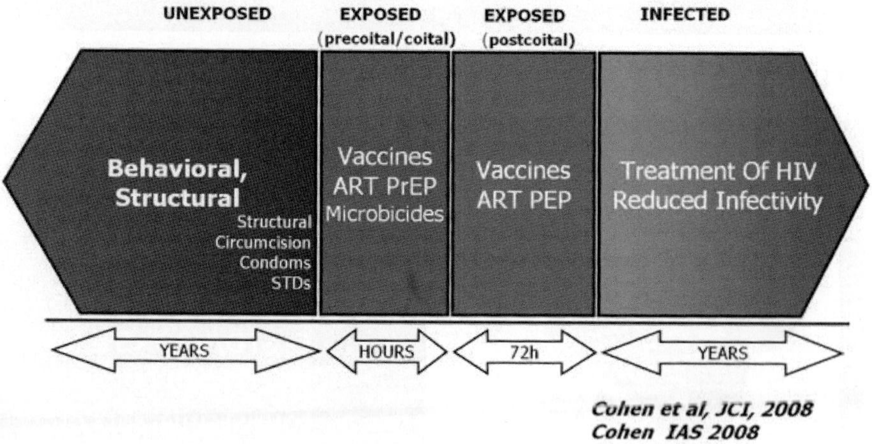

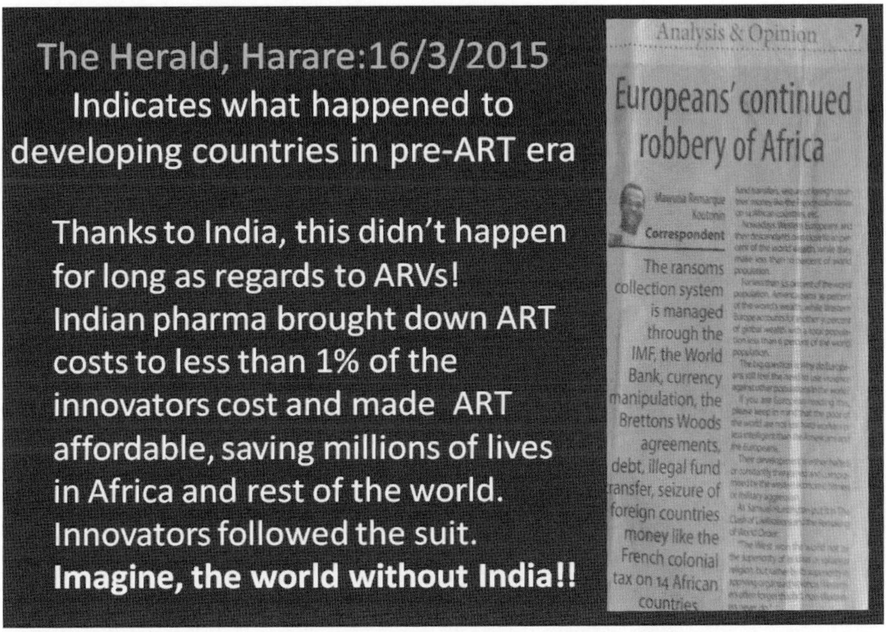

ADVANTAGE INDIA! GENERIC COMPETITION BRINGS PRICES DOWN *AND THE REASON—WHY U.S. PRESIDENT OBAMA WAS FURIOUS WITH INDIA/INDIANS?*

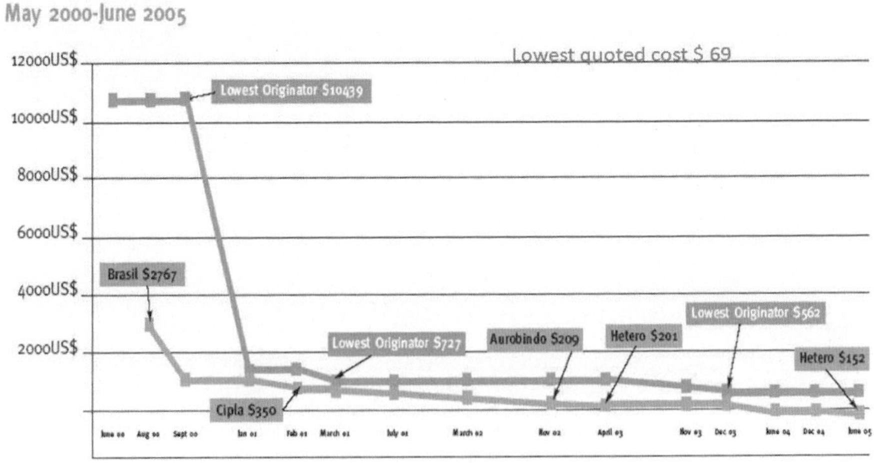

Sample of ARV triple-combination: stavudine (d4T) + lamivudine (3TC) + nevirapine (NVP). Lowest world prices per patient per year. Generic competition has shown to be the most effective means of lowering drug prices. During the last four years, originator companies have often responded to generic competition.

LESSONS FOR THE FUTURE

- No other disease has triggered such a fast and quality research and development like HIV.
- Enough known on pathogenesis, susceptibility, epidemiology.
- Abundant clinical material, but basic research facilities are not at par with the west → scope for research collaboration.
- Continuously evolving knowledge necessitates a periodical review of care continuum, academic update on latest info.
- Multi-sectoral collaboration among planners, care givers, pharma and researchers is crucial.
- The future is pregnant with hope and endless possibilities.

But we still have miles to go before we sleep!

THANKS

- Staff at Unison Medicare, other professional colleagues.
- My patients—who have been a great source of learning/inspiration, their family members.
- NACO, WHO, UNAIDS, MSF for data/graphics.
- Several scientists, clinicians, researchers, epidemiologists with whom I interact regularly.
- Organising Committee of 48th Session of the International Seminars on Planetary Emergencies.
- Especially Dr. A. Zichichi, Dr. Franco Buonaguro, Dr. Claude Manoli, Ms. Deirdre Sanso.
- All of you for your presence and attention.

ETIOPATHOGENIC MECHANISMS OF IMMUNE-RELATED CANCERS

SAM MBULAITEYE
National Institutes of Health/NCI/DCEG, Infections and Immunoepidemiology Branch, Bethesda, MD, USA

TODAY'S PRESENTATION

- Mechanisms of cancer development.
- Burkitt Lymphoma as model.
  - To understand cancer mechanisms.
  - To bridge the promise of science to the poor.
- Conclusions.

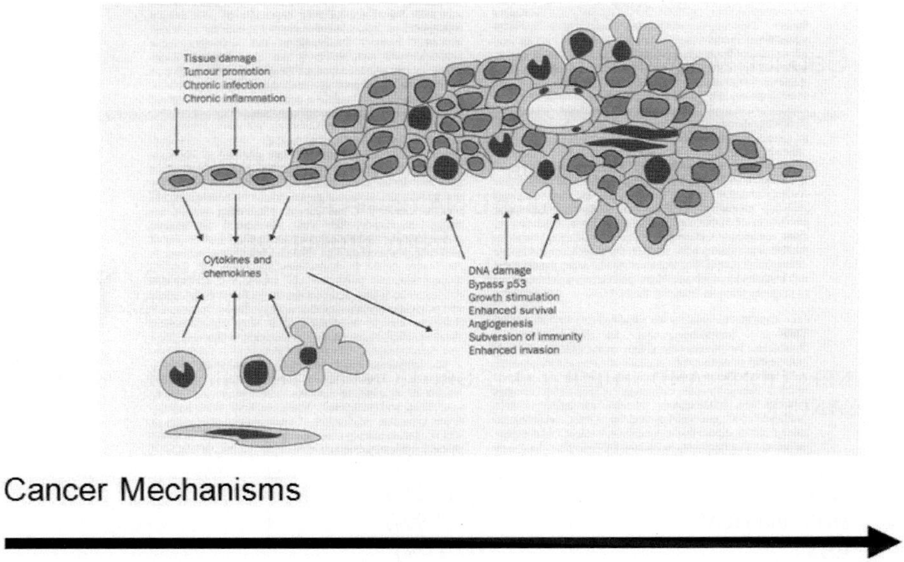

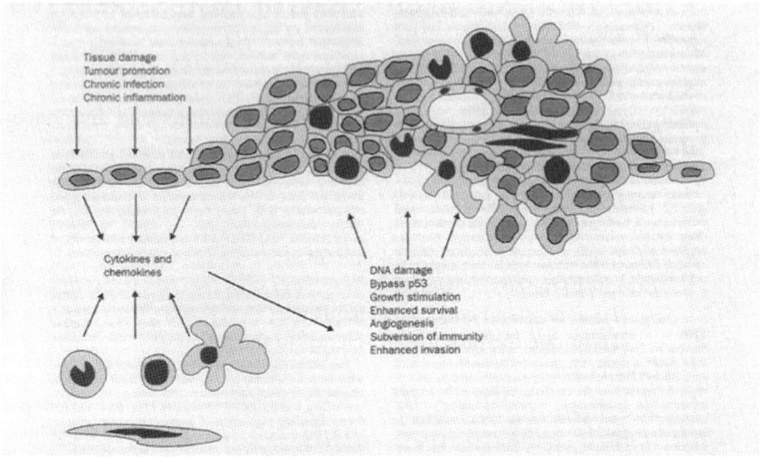

Cancer Mechanisms

Infections, Chemicals & Genetic susceptibility
Cellular DNA damage
Early precursors
Progressive clones
Metastases

INFECTIONS AND THEIR CONTRIBUTION TO THE GLOBAL BURDEN OF CANCER

H. pylori	(5.5%)
HPV	(5.2%)
HBV and HCV	(4.9%)
EBV	(1%)
HIV and KSHV	(0.9%)
Parasites, HTLV-1	(~0.1%)

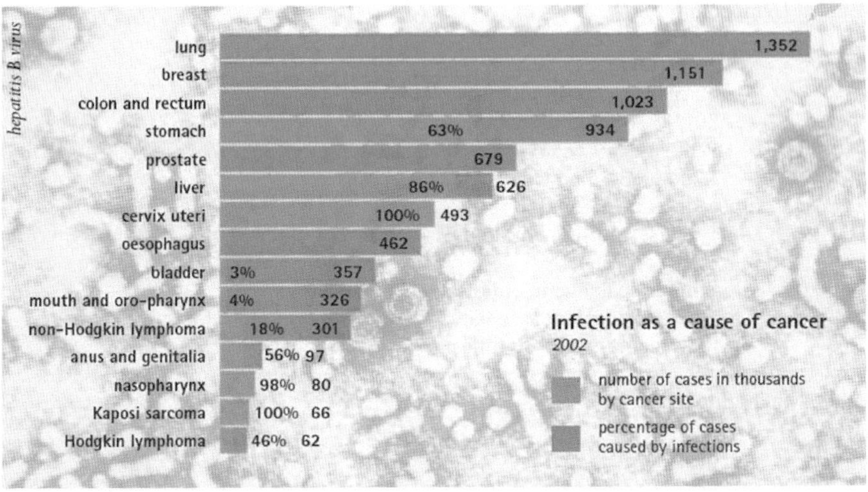

Infectious agents account for ~18% of all cancers and 25% of cancers in the developing world.

Burkitt lymphoma

218 THE BRITISH JOURNAL OF SURGERY

RAPER, F. P. (1956), Brit. J. Urol., 28, 436.
RAVAULT, P. P., PAPILLAN, J., PINET, F., and JACQUOT, F. J. (1956), Lyon méd., 88, 453.
ROSS, J. C., and TINCKLER, L. F. (1958), Brit. J. Surg., 46, 58.
SAMSON, P. (1945), J. thorac. Surg., 14, 330.
SANTY, P., GALY, P., GONIN, A., MARION, P., PAPILLON, J., and PINET, F. (1957), Pr. méd., 65, 307.
SCANNELL, J. G., and SHAW, R. S. (1954), J. thorac. Surg., 28, 163.
TUBBS, O. S. (1946), Thorax, 1, 247.
VAUX, D. M. (1938), J. Path. Bact., 46, 441.
WALSH, G. C., NORTON, G. I., BAIRD, M. M., and ROBERTSON, R. (1957), Canad. med. Ass. J., 76, 292.
ZEMAN, F. D. (1945), J. thorac. Surg., 14, 330.

A SARCOMA INVOLVING THE JAWS IN AFRICAN CHILDREN

By DENIS BURKITT
FROM THE DEPARTMENT OF SURGERY, MAKERERE COLLEGE MEDICAL SCHOOL, AND MULAGO HOSPITAL, KAMPALA, UGANDA

MALIGNANT tumours of the jaws in children, primary or secondary, are generally regarded as rare. A sarcoma involving the jaws in African children has recently come to be recognized at Mulago Hospital as a distinctive clinical condition and certainly the commonest malignancy of childhood.

Thirty-eight patients with this sarcoma in the jaws have been seen during the past 7 years; 32 of them were seen at Mulago Hospital and 6 at district hospitals. The tumour was diagnosed clinically in a further 8 children, but these have not been included in this series owing to lack of histological confirmation.

Fig. 246.—Showing age distribution in 38 cases.

Records of only 3 cases of this type of jaw sarcoma in children have been traced in the literature (Christiansen, 1938; Salmon and Darlington, 1944; Burford, Ackerman, and Robinson, 1944). Gelfand (1957) published an illustration of a sarcoma of the jaw in an African child without clinical details.

GEOGRAPHICAL DISTRIBUTION

Patients have not been limited to any particular area in Uganda, and have represented 11 different tribes. This sarcoma has also been observed in Kenya (Clifford, 1958), Tanganyika (Morris, 1958; Blackman, 1958), Nigeria (Thomas, 1958), the Belgian Congo (Thijs, 1958), and Southern Rhodesia (Gelfand, 1957). Patients with this syndrome have not yet been recognized in Johannesburg (Oettlé, 1958), Khartoum (Taylor, 1958), Lusaka (Buck, 1958), or Lourenço Marques (Prates, 1958).

CLINICAL FEATURES

These patients were from 2 to 14 years of age, 30 being between the ages of 2 and 7 years (Fig. 246).

In most cases the tumour started in the region of the alveolar process of a maxilla (Fig. 247) or the mandible (Fig. 249). Loosening of the deciduous molars was often the first symptom, the teeth in the involved area soon becoming embedded in tumour tissue only, and losing their insertion in bone. The next stage was irregular displacement of the teeth prior to their falling out. The tumour grew rapidly, grossly distorting the face. In only one patient (Case 20) did it ulcerate through the skin. Œdema of the eyelids and chemosis of the conjunctivæ indicated invasion of the orbit, and if the patient survived the eye became proptosed and finally destroyed. Less commonly the tumour presented as a swelling high in the maxilla with early invasion of the orbit (Fig. 248). Pain was not usually as severe as would have been expected from the appearance of the tumour. Within two or three months of onset of symptoms their relatives removed the majority of the children from hospital in a moribund condition.

Fig. 247.—Sarcoma involving the left maxilla and arising in relation to the teeth. Case 1, taken 3 months after onset of symptoms.

Unless secondary infection occurred, which was not usual, the regional lymph-nodes were not

Burkitt Lymphoma

- African (endemic).
  - 40-50 cases/ million African children.
  - 50-75% of all childhood cancers.
  - Peak age 6-9 years.
  - Male predominance.
- Sporadic (non-African).
  - 1-2 cases/million children.
  - 1% of all cancers.
  - Similar peak age in children; occurs in adults.
  - HIV or transplant associated.
  - Risk increased 20-60 times.

1963: Malaria linked to BL

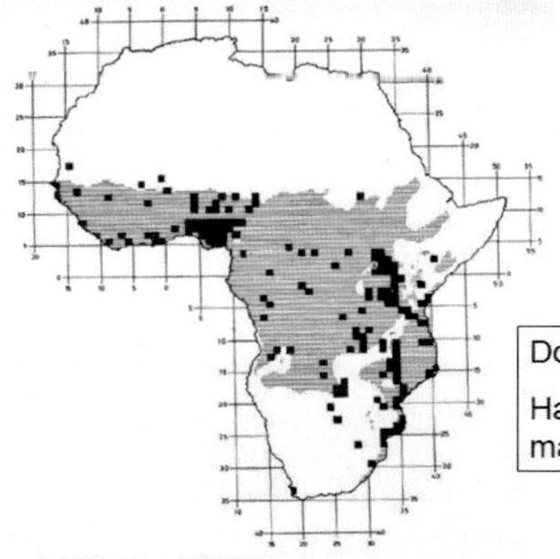

Dots – Burkitt lymphoma cases

Hatched lines – distribution of *Pf* malaria

A.J. Haddow, *EMJ* 1963

1964: Epstein-Barr Virus discovered in BL

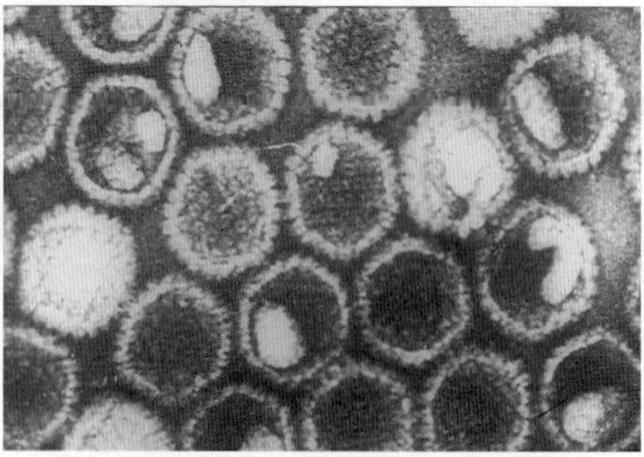

- First human virus linked to cancer.
- Pirated human genes discovered—linked to malignant transformation.
- EBV linked to multiple cancers—nasopharynx, stomach, lymphomas.
- EBV linked to INFECTIOUS MONO.
- Other cancer causing infections discovered.

1972: BL molecular abnormality of BL identified as *c-MYC* translocation into immunoglobulin loci

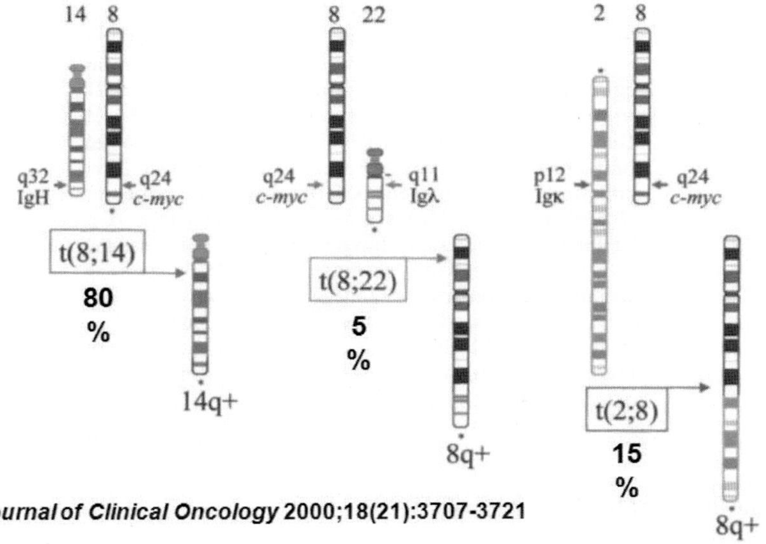

Journal of Clinical Oncology **2000;18(21):3707-3721**

1982: *c-MYC* translocation occur in both African and non-African BL, but at different locations

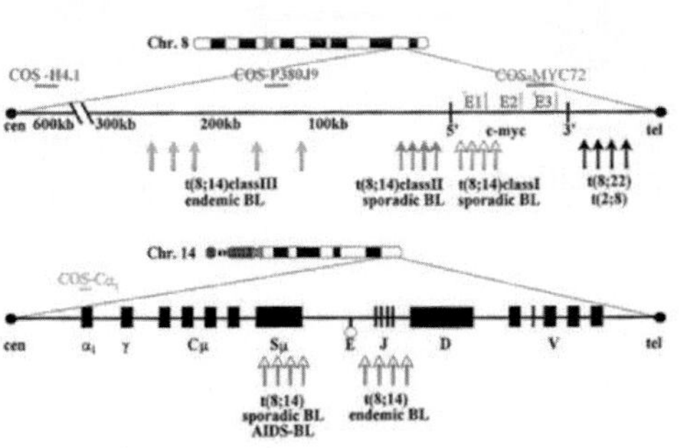

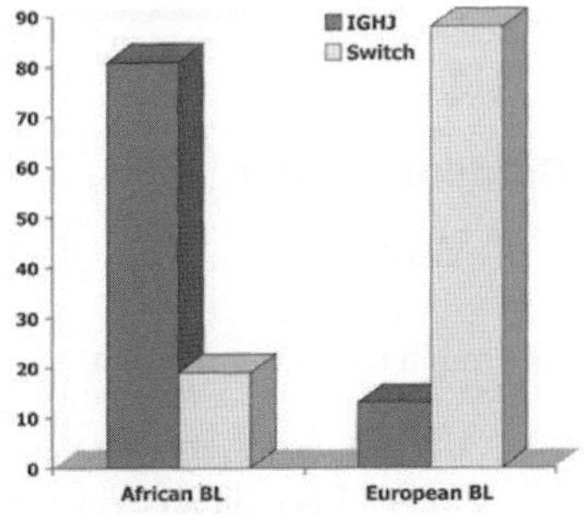

1965: BL cure reported

CURE OF BURKITT'S LYMPHOMA
Ten-year Follow-up of 157 Ugandan Patients

JOHN L. ZIEGLER IAN T. MAGRATH
CHARLES L. M. OLWENY

Division of Cancer Treatment, National Cancer Institute, Bethesda, Maryland 20205, U.S.A.; and Uganda Cancer Institute, Kampala, Uganda

Summary 192 Ugandan patients with Burkitt's lymphoma were treated with various regimens containing high-dose cyclophosphamide. 21 patients died during induction therapy, and 14 died after partial responses. Of 157 patients who responded completely to treatment, 16 were lost to follow-up (10 in remission when last seen). Of 141 evaluable responders, 69 died and 72 (51%) are alive and disease-free. 31 of the long-term survivors have relapsed at least once and have been successfully retreated. Overall, 34 surviving patients had central-nervous-system involvement, also successfully treated. It is concluded that Burkitt's lymphoma is curable in at least 50% of patients, and that relapse and central-nervous-system involvement are not incompatible with long survival.

THE LANCET, NOVEMBER 3, 1979

Natural history of Burkitt Lymphoma

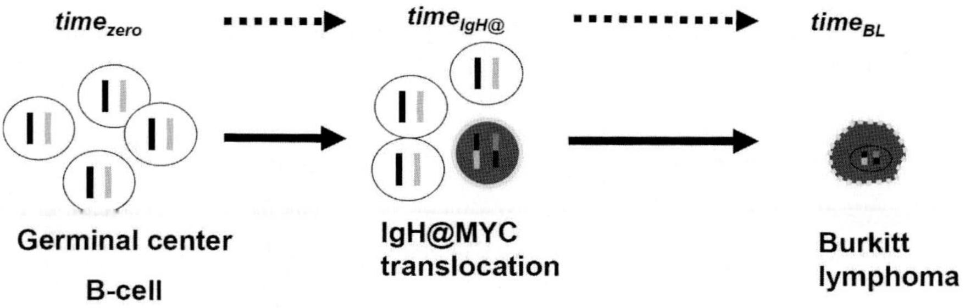

…but much remains unknown.

- Pathogens (*Pf* malaria and EBV).
  - Lack direct evidence for causation.
  - Don't completely understand.

- Strength of association between pathogen and BL risk.
- If specific pathogen proteins influence BL risk.
- If pathogen genetic diversity influences BL risk.

<u>Malaria antibodies higher in BL cases than controls</u>

Uganda: 126 cases, 70 controls

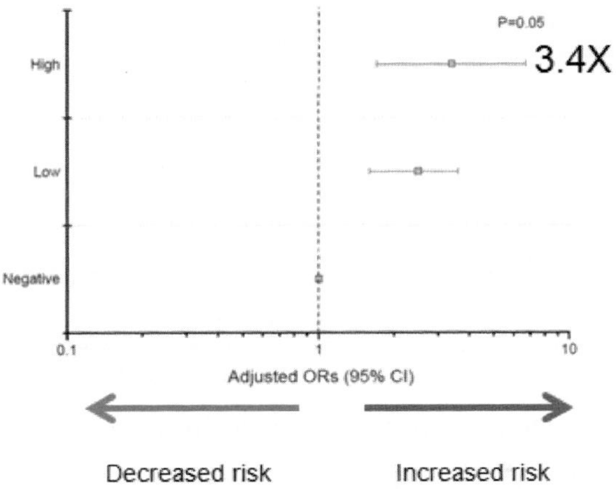

Carpenter et al. *IJC* 2008

<u>Malaria antibodies higher in BL cases than controls</u>

Uganda: 126 cases, 70 controls

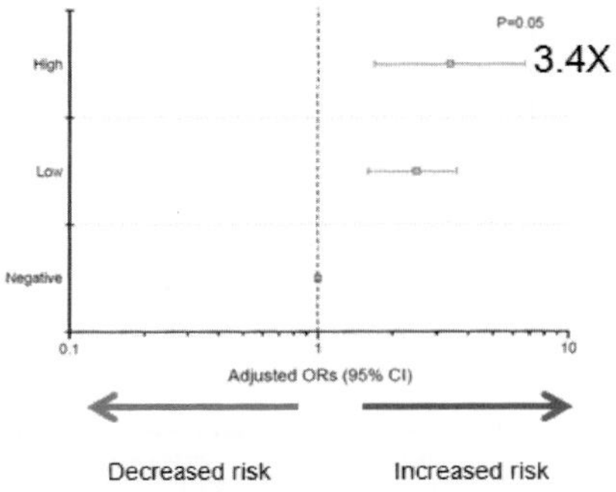

Carpenter et al. *IJC* 2008

Malawi: 129 cases, 90 controls

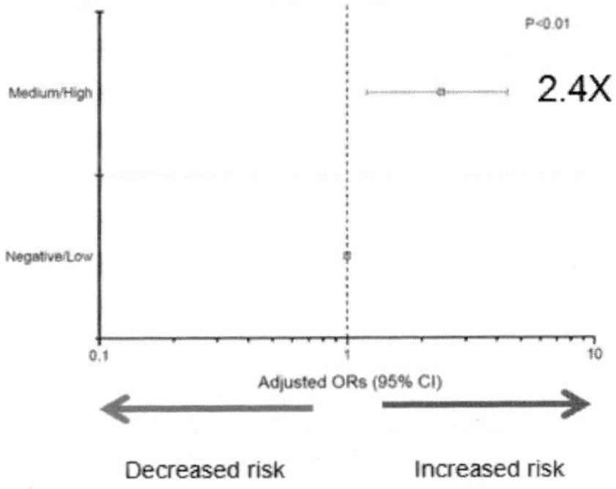

Mutalima et al. *PLOS One* 2008

Malaria antibodies lower in BL cases than controls

NCI Ghana Burkitt Study: 350 cases, 384 controls

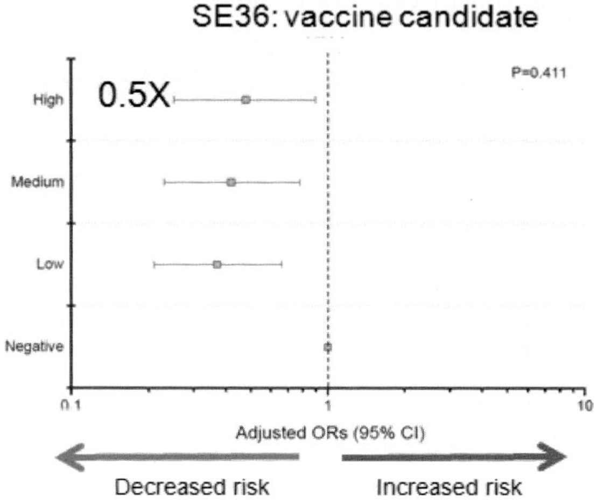

Guech-Ongey et al. *Int. J Cancer* 2011

Aka et al. *Blood* 2013

Malaria antibodies both risk and protective for BL

NCI Ghana Burkitt Study: 350 cases, 384 controls

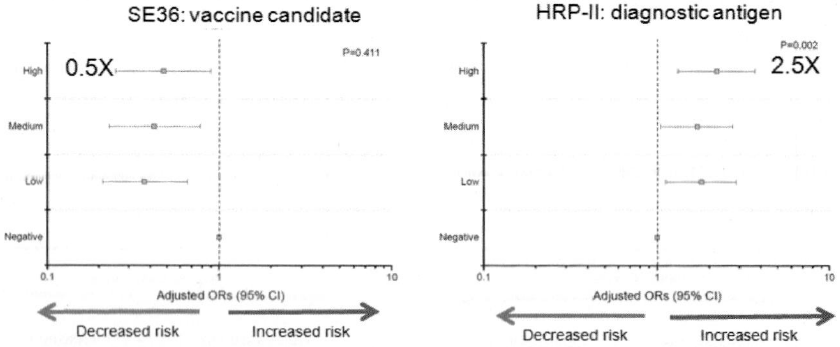

Guech-Ongey et al. *Int. J Cancer* 2011

Aka et al. *Blood* 2013

Risk of BL related to malaria profile

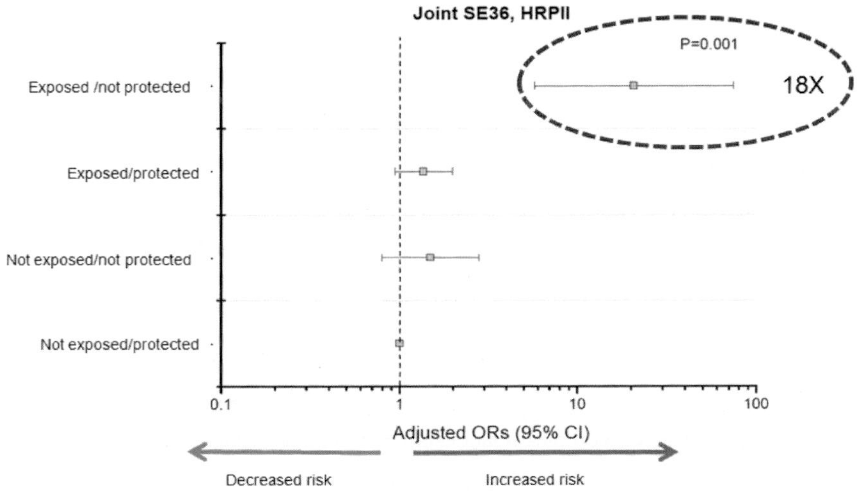

Guech-Ongey et al. *Int. J Cancer* 2011

Aka et al. *Blood* 2013

Malaria profile is *mostly* similar in children who get or don't get BL

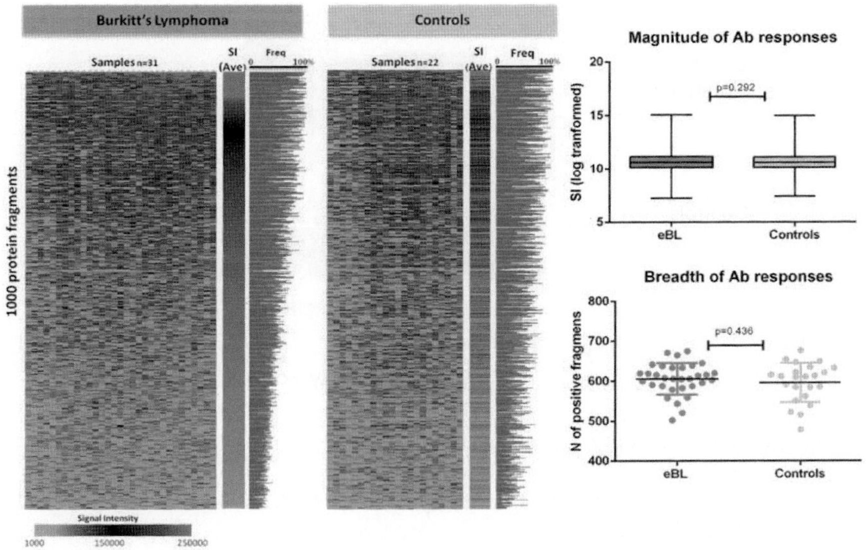

Six *Pf* peptides highly predictive of BL in proof-of-concept study

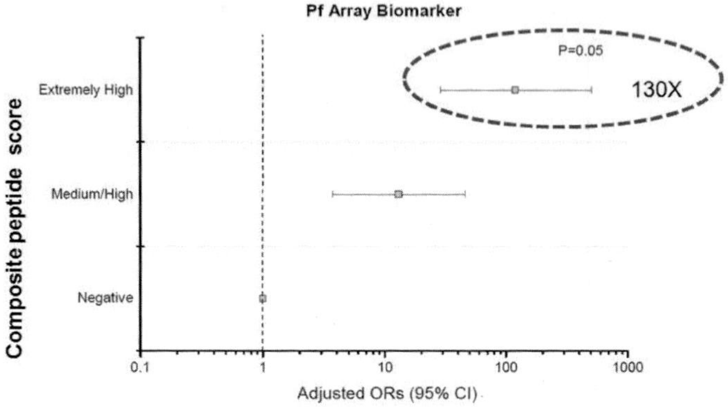

Antibody profiles suggest complex relationship between pathogen, host and BL

- Pathogen diversity drives BL
- Host genetic diversity predisposes to BL

Pf genetic diversity scores higher in BL cases than controls

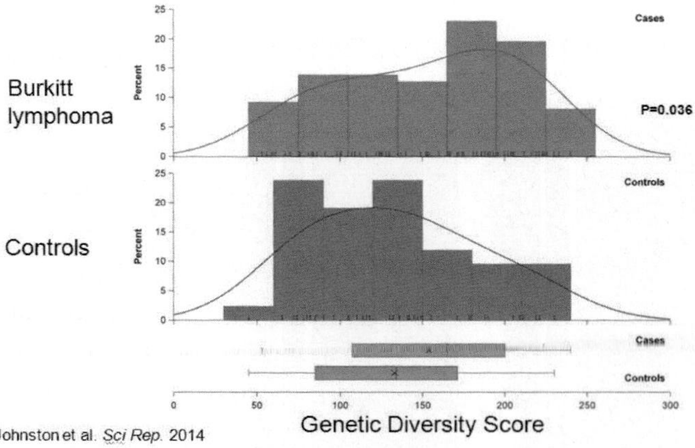

Johnston et al. *Sci Rep.* 2014

SUMMARY OF MALARIA STUDIES

- *Pf* is associated with Burkitt Lymphoma (risk is 2-130 times elevated in susceptible children).
- Limitation.

- Relevant *Pf* proteins not known.
- Therefore, malaria-related technologies cannot be used for early detection, treatment of early BL, or vaccination to prevent BL.

... molecular targets of BL remain unknown.

- Germ line predisposing factors.
- Somatic drivers of BL development.

TREATMENT OF BURKITT LYMPHOMA

- Most common childhood cancer in Africa.
- Curable.

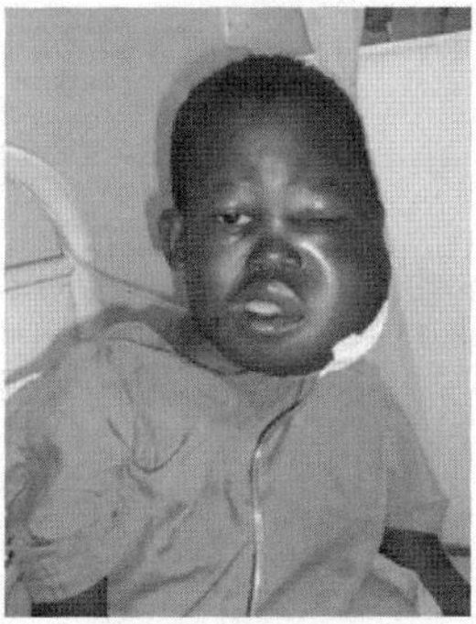

Before treatment

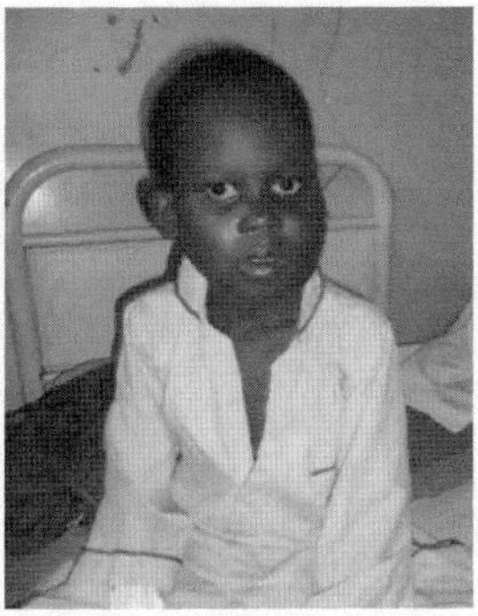

After treatment

BL in Ghana

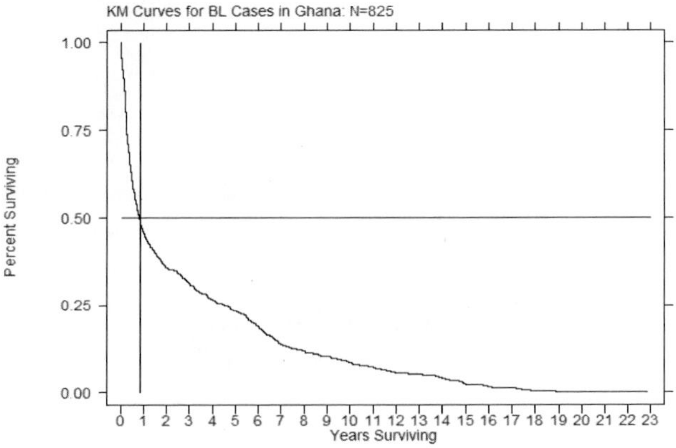

2012. Targeted therapies in BL

LETTER

Burkitt lymphoma pathogenesis and therapeutic targets from structural and functional genomics

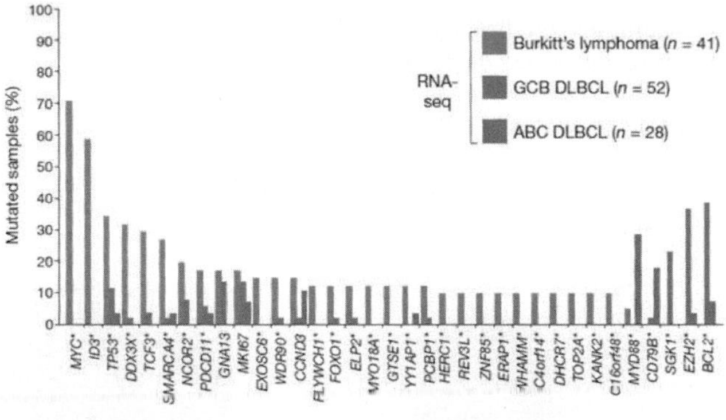

Figure 1 | Recurrently mutated genes in aggressive lymphomas determined by RNA-seq. Shown are genes that were recurrently mutated in BL based on RNA-seq analysis (≥4 out of 41 samples), as well as representative genes known to be recurrently mutated in DLBCL. Asterisks indicate differentially mutated genes ($P < 0.05$; Supplementary Table 9).

Targeted therapies in BL

LETTER

Burkitt lymphoma pathogenesis and therapeutic targets from structural and functional genomics

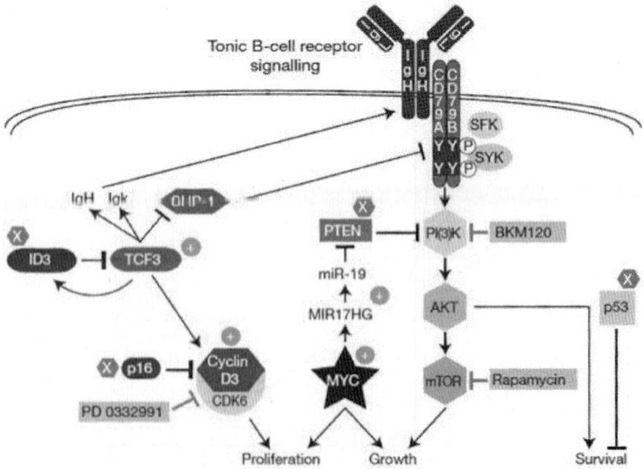

TCF3 and ID3 mutation frequencies in lymphoid cancers suggest therapeutic targets

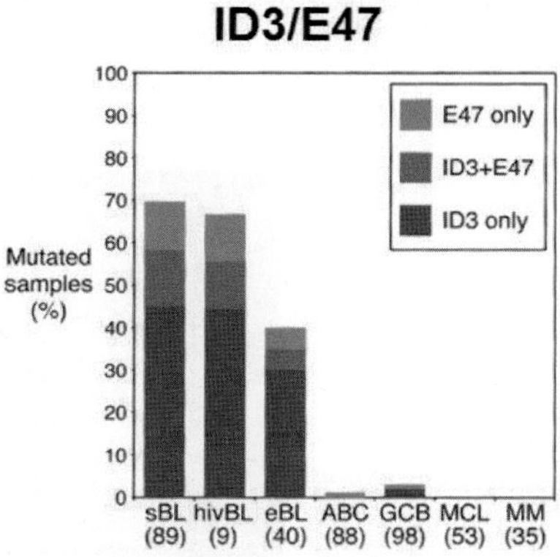

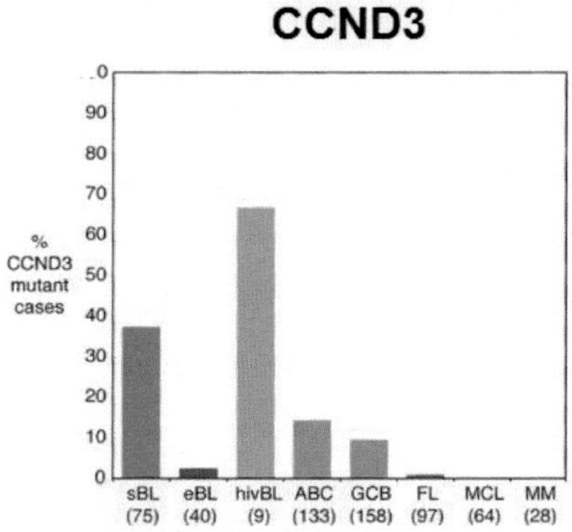

Schmitz et al. *Nature* 2012

2011 EMBLEM Study

- Epidemiology of Burkitt Lymphoma in East African Children and Minors (EMBLEM)
- Multicentre study in 6 regions in East Africa
- Regions are holoendemic malaria areas

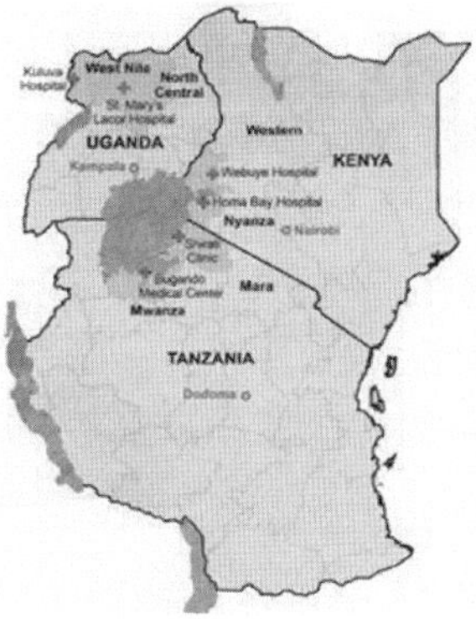

EMBLEM
High Quality, High Impact Research

NCI PIs: K. Bhatia, J.J. Goedert, R. Pfeiffer, S.J. Chanock

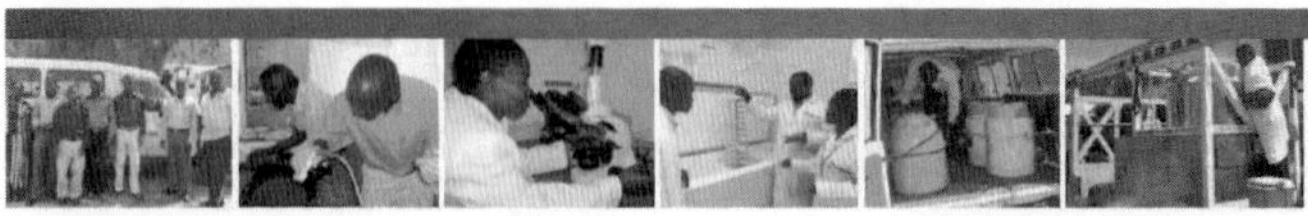

EMBLEM
High Quality, High Impact Research

Epidemiology of Burkitt Lymphoma in East-African Children and Minors

EMBLEM staff enrolling a study participant in Uganda

www.emblem.cancer.gov

Status since 2010

Research Capacity

Technical training for:

- Research Project Managers
- Health Researchers
  - Doctors
  - Laboratory technicians
  - Nurses

Laboratory equipment

- Bio-safety cabinets
- Light and fluorescence microscopes
- Ultra-low temperature freezers
- QBC/Drucker hematology instruments
- Bench-top centrifuges
- Liquid nitrogen shippers and tanks
- Portable ultrasound units and core biopsy guns
- CO2 incubators
- Rotary microtomes

Transport

- Research vehicles

Research

- Ethical approval in 4 countries
- Established six active field sites
- Spotted over 1,600 cases
- Randomly selected 300 control villages
- Collected over 30,000 biological samples

Mentoring/leadership

- Principal investigator training
- Study standardization meetings (weekly calls, bi-annual conferences)
- EMBLEM scientific updates (Gulu, Uganda - 2010; Mwanza, Tanzania, 2012; Kisumu, Kenya, 2014)
- International scientific meetings participation
  - African Organisation for Research and Training in Cancer – 2011, 2013, 2015
  - International Lymphoma Epidemiology Consortium 2013

Collaboration

National Institutes of Health Partners

National Institute of Allergy and Infectious Diseases
(ethics, logistics, magnesium studies)

Fogarty International Center
(Principal investigator training)

Division of Cancer Control and Population Sciences
(consortium support)

Center for Cancer Genomics
(BL genome sequencing)

Center for Information Technology
(ultrasound fixation assisted micro tissue processor)

Non-NIH Partner Projects

Ohio State University (BL diagnosis)

Partec/Sysmex (malaria diagnosis)

Fred Hutchinson Cancer Research Center
(molecular staging of BL)

NIH National Cancer Institute

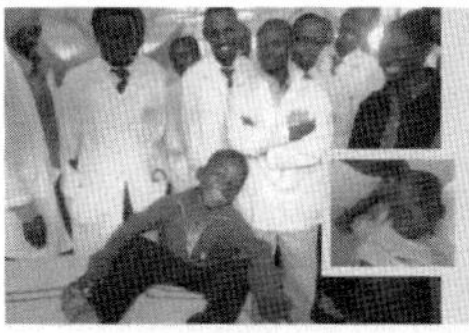

Burkitt Lymphoma (BL)

BL is the most common childhood cancer in Africa. Malaria, although both treatable and preventable, is the most widespread childhood infection. Understanding the role of malaria, Epstein-Barr virus, and genetics in BL will answer important scientific questions about the cause and biology of the disease. This information will be important for surveillance, diagnosis, clinical care, and prevention.

What is this study about?

EMBLEM is a multi-country and multi-year case-control study of childhood BL sponsored by the National Institutes of Health/National Cancer Institute (NIH/NCI) with study sites in rural areas of Northern Uganda, Northern Tanzania, and Western Kenya, where malaria endemicity is high year-round.

EMBLEM aims to:

- Enroll 1,500 cases and 3,000 controls and
  - Collect data about home and life experiences (including malaria exposures)
  - Collect biological samples

EMBLEM will evaluate:

- Genetic risk factors of BL
- Role of infections in BL

Work Flow

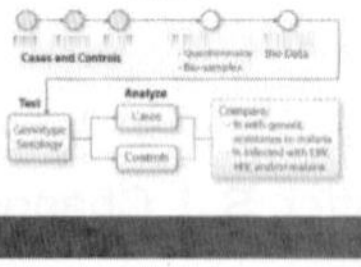

EMBLEM is a model for:

- Conducting networked research
- Building capacity for high quality research in Africa
- Mentoring the next generation of African scientists
- Establishing collaborative projects

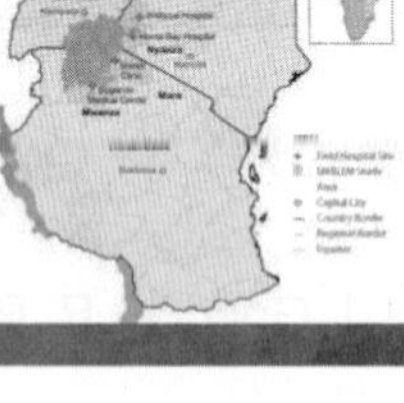

FIELD SITES

Uganda

St. Mary's Hospital Lacor

St. Mary's Hospital Lacor, a 480-bed non-profit charitable hospital, the main EMBLEM case-control enrollment site in north-central Uganda, is located 6 km outside Gulu. The local co-principal investigator is Dr. Martin D. Ogwang, Medical Director.

Kuluva Hospital

Hospital, located 12 km from the town of Arua, is the smaller Uganda enrollment site. It has a reputation of being the best Arua hospital with a long history of diagnosing and treating BL. The local co-principal investigator is Dr. Patrick Kerchan.

Uganda Virus Research Institute (UVRI)

UVRI, in Entebbe, was established in 1914 as the Yellow Fever Institute. It is a premier virology research center in Uganda and serves as the long-term EMBLEM biospecimen storage/international shipment site. In addition, the UVRI-Science and Ethics Committee provides ethical oversight on behalf of the Uganda National Council for Science and Technology. The local co-principal investigator is Dr. Steven Reynolds.

Makerere University, Department of Medical Microbiology

Professor Moses Joloba and Mr. Samuel Kirimunda at Makerere University are the EMBLEM study monitors and provide technical laboratory support for EMBLEM.

Tanzania

Bugando Medical Centre (BMC)

BMC is a 900-bed referral center in Mwanza. It is the main EMBLEM case-control enrollment site in northern Tanzania. The co-principal investigator is Dr. Nestory Masalu, Head-Oncology Department at BMC.

Shirati

Shirati is the smaller site, located in North Mara. It houses an office, laboratory, and treatment unit for BL. The co-principal investigator is Dr. Esther Kawira, medical director of the Shirati Health, Education and Development (SHED) Foundation. The SHED Foundation provides administrative and logistical support for the EMBLEM study in Tanzania.

National Institute of Medical Research (NIMR)

NIMR provides ethical oversight for the EMBLEM study in Tanzania. The Mwanza campus provides long-term storage and assistance with shipping research samples.

Kenya

Homa Bay District Hospital

Homa Bay District Hospital is located in Homa Bay County, in Nyanza Province with a bed capacity of about 250 beds. It is the main EMBLEM case-control enrollment site in Kenya.

Webuye District Hospital

Webuye District Hospital is located in Bungoma County, in Western Province with a bed capacity of 220 beds. It is the smaller case-control site in western Kenya.

Moi University

Moi University Teaching and Referral Hospital (MTRH), established in 1990, is the second national referral hospital in Eldoret, Kenya. The Research and Sponsored Project Office (RSPO) at Moi University provides administrative support for EMBLEM in Kenya. Moi University Institutional Research Committee exercises ethical oversight for the study on behalf of the Kenya National Council of Science and Technology. The local co-principal investigator is Dr. Constance N. Tenge, in the School of Medicine.

Kenya Medical Research Institute (KEMRI)

KEMRI is the national body responsible for carrying out health research in Kenya and will provide assistance with permanent storage and shipment of research samples for EMBLEM in Kenya. The KEMRI co-principal investigator is Dr. Peter Odada Sumba.

BL is Curable

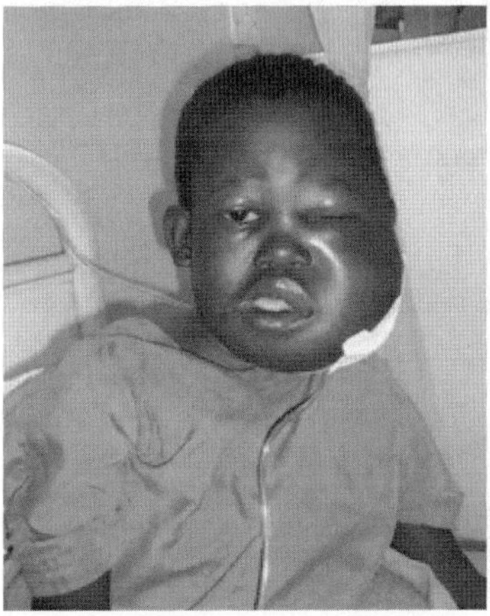

Before treatment

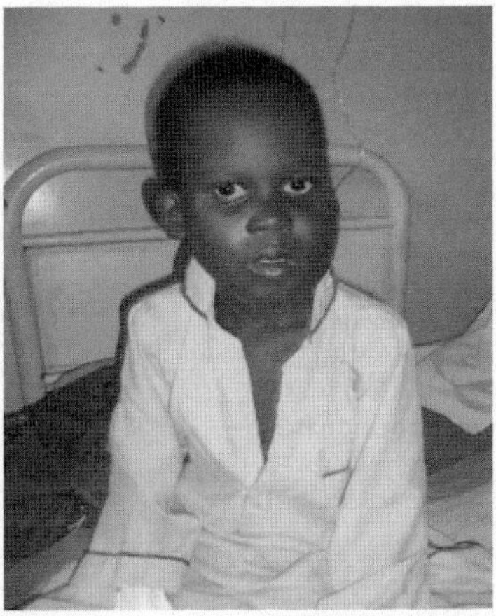

After treatment

2010

Viewpoint

African Burkitt's lymphoma: could collaboration with HIV-1 and malaria programmes reduce the high mortality rate?

Sam M Mbulaiteye, Ambrose O Talisuna, Martin D Ogwang, F Ellis McKenzie, John L Ziegler, Donald M Parkin

Lancet 2010; 375: 1661-63

2302 Cranberry Terrace, Silver Spring, MD, USA (S M Mbulaiteye MD); Uganda Malaria Surveillance Programme, Kampala, Uganda (A O Talisuna PhD); Department of Surgery, St Mary's Hospital, Lacor, Gulu, Uganda (M D Ogwang MMed); Fogarty International Center, NIH, Bethesda, MD, USA (F E McKenzie PhD); Global Health Sciences, University

Burkitt's lymphoma is an aggressive lymphoma that is common in sub-Saharan Africa but rare elsewhere.[1] First described in African children by Denis Burkitt,[2] the uneven distribution of Burkitt's lymphoma attracted worldwide interest and led to seminal discoveries. For example, Epstein-Barr virus, the first virus to be associated with a human cancer, chromosomal translocations, and elucidation of molecular aberrations in *MYC*, *Rb*, and *p53* genes—now regarded as central to the biology of cancer—is closely linked to Burkitt's lymphoma.[3,4] The striking response of this disease to chemotherapy led to more successful outcomes,[5,6] and 90% of children with children per year[12]—varies from $37 500 to $187 500 (0·02–0·08% of Uganda's health budget of $246 million).[13] Nevertheless, children with Burkitt's lymphoma—even those few who make it to the medical system—continue to die needlessly because proportionately fewer of them get life-saving drugs nowadays than they did in the 1960s and 1970s.

Data about the burden, distribution, and high mortality rate of Burkitt's lymphoma in Africa, as well as about availability of affordable curative treatments are sparse. Cancer registration in Africa is underdeveloped; many countries have no registries, and registries that do exist

ACKNOWLEDGEMENTS

Special thanks to
Robert J. Biggar
Joseph F. Fraumeni, Jr.
John L. Ziegler

National Cancer Institute
Benjamin Emmanuel
Peter Aka
Paul Levine
Ruth Pfeiffer
James Goedert
Kishor Bhatia
Louis M. Staudt
Stephen J. Chanock

St. Mary's Hospital Lacor, Gulu, Uganda
Martin D. Ogwang

Kampala Cancer Registry, Kampala, Uganda
Henry Wabinga

Shirati Hospital, Tanzania
Esther Kawira
Josiah Mogatti

Korle Bu University, Accra, Ghana
Janet Neequaye

Noguchi Memorial Institute, Accra, Ghana
Francis Nkrumah

University of Osaka, Japan
Masonari Yagi
Nirianne M Palacpac
Takahiro Tougan
Toshihiro Horii

University of York, U.K.
Robert Newton
Thomas W Johnston
Nora Mutalima

George Washington University, USA
Jeffrey Bethony
Maria Candida Vila
Amar Jariwala
Maria V. Periago

Queensland Institute of Medical Research, Brisbane, Australia
Denise Doolan
Carla Proietti
Angela Trieu
Lutz Krause
Jason Mulvenna

DECIPHERING NEURODEGENERATION: WHAT DO PHENOTYPES, MOLECULAR BIOLOGY AND GENETICS TELL US ABOUT PATHOGENESIS

EMANUELE BURATTI
International Centre for Genetic Engineering and Biotechnology
Trieste, Italy

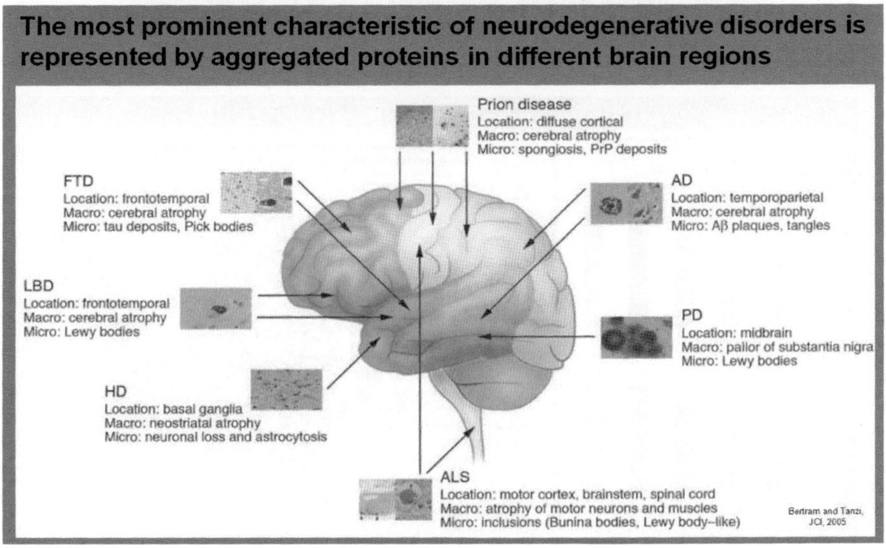

Why are neurodegenerative processes so insidious?

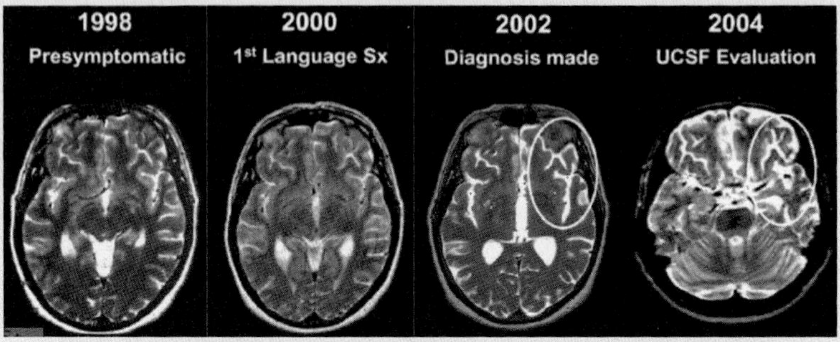

Seeley et al. *Brain*, 2008

What is the burden of neurodegenerative diseases in our aging population?

Europe has a rapidly aging population. Currently, 16% of the European population is over 65, and this figure is expected to reach 25% by 2030.

Amongst these disorders, the dementias are responsible for the greatest burden of disease, with Alzheimer's disease and related disorders affecting over **7 million people in Europe**,

This figure is expected to **double every 20 years** as the population ages.

Existing treatments are very limited and only treat the symptoms, not the causes

It currently **costs approximately €130 billion per annum** to care for people with dementia across Europe, highlighting age-related neurodegenerative disease as one of the leading medical and societal challenges faced by EU society.

JPND research

The burden is becoming excessive also for advanced health systems such as the USA:

Figure 1 Projected number of Medicare beneficiaries with Parkinson disease and Alzheimer disease in millions (2010–2030)

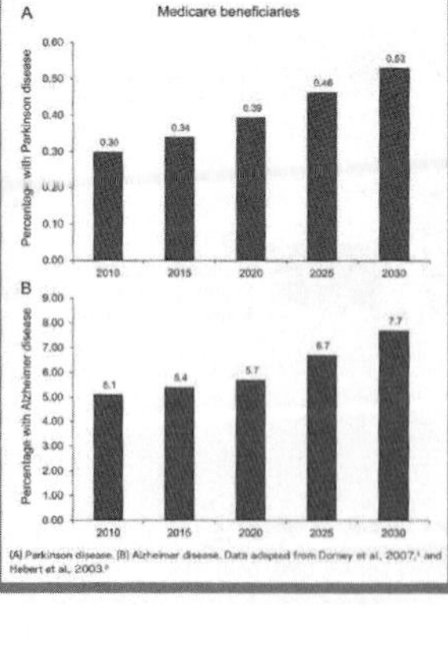

(A) Parkinson disease. (B) Alzheimer disease. Data adapted from Dorsey et al, 2007,[1] and Hebert et al, 2003.[8]

Figure 2 Proportion of Medicare beneficiaries (2002–2005) with an incident diagnosis of Parkinson disease who do not receive care from a neurologist within the first 4 years of diagnosis

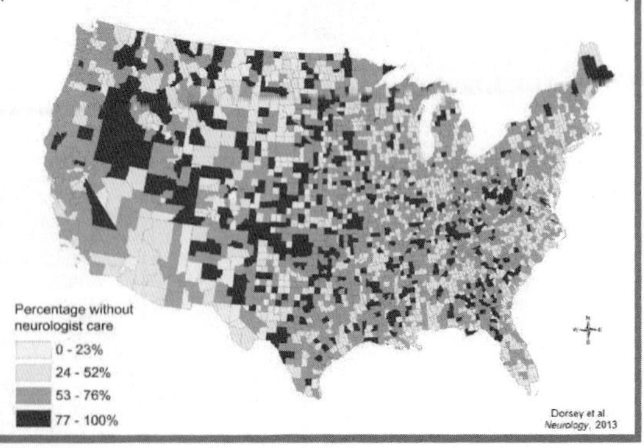

Dorsey et al *Neurology*, 2013

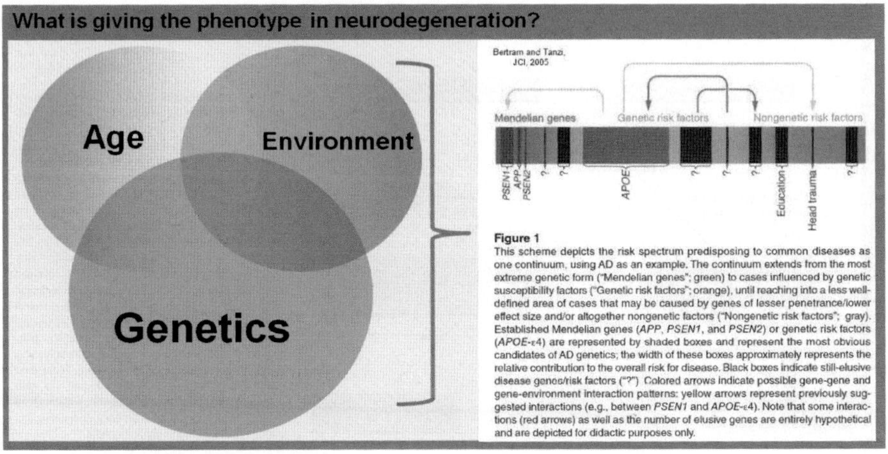
What is giving the phenotype in neurodegeneration?
Age
Environment
Genetics
Bertram and Tanzi, JCI, 2005
Mendelian genes
Genetic risk factors
Nongenetic risk factors
PSEN1
APP
PSEN2
?
?
APOE
?
?
?
Education
Head trauma
?
Figure 1
This scheme depicts the risk spectrum predisposing to common diseases as one continuum, using AD as an example. The continuum extends from the most extreme genetic form ("Mendelian genes"; green) to cases influenced by genetic susceptibility factors ("Genetic risk factors"; orange), until reaching into a less well-defined area of cases that may be caused by genes of lesser penetrance/lower effect size and/or altogether nongenetic factors ("Nongenetic risk factors"; gray). Established Mendelian genes (APP, PSEN1, and PSEN2) or genetic risk factors (APOE-ε4) are represented by shaded boxes and represent the most obvious candidates of AD genetics; the width of these boxes approximately represents the relative contribution to the overall risk for disease. Black boxes indicate still-elusive disease genes/risk factors ("?") Colored arrows indicate possible gene-gene and gene-environment interaction patterns: yellow arrows represent previously suggested interactions (e.g., between PSEN1 and APOE-ε4). Note that some interactions (red arrows) as well as the number of elusive genes are entirely hypothetical and are depicted for didactic purposes only.

GENES

Classical strategy to find disease genes:

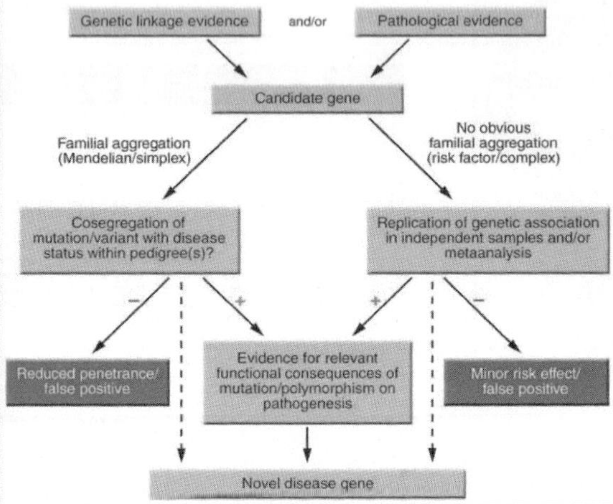

Figure 2
Flow chart of current strategies used to identify novel disease genes. This scheme outlines strategies for identifying mutations and/or polymorphisms causing or predisposing to disease. Candidate genes are chosen based on genetic linkage data and/or known or hypothesized pathobiological relevance to disease mechanisms. This procedure is referred to as the "candidate gene approach." An alternative and inherently similar strategy is based on the detection of formerly unknown genes/proteins according to genetic linkage data and is referred to as "positional cloning." Dashed lines indicate "shortcuts" allowing the definition of a novel disease gene based on the genetic evidence alone, e.g., *APOE*-ε4 in AD, of which the precise functional consequences remain unknown despite an established genetic role. Note that there are examples of genes/mutations with reduced penetrance or minor risk effects (red boxes) within bona fide disease genes (e.g., certain mutations in *PSEN1* in AD).

Bertram and Tanzi
JCI, 2005

Starting from TDP-43, many genes have now been associated with the ALS-FTD spectrum:

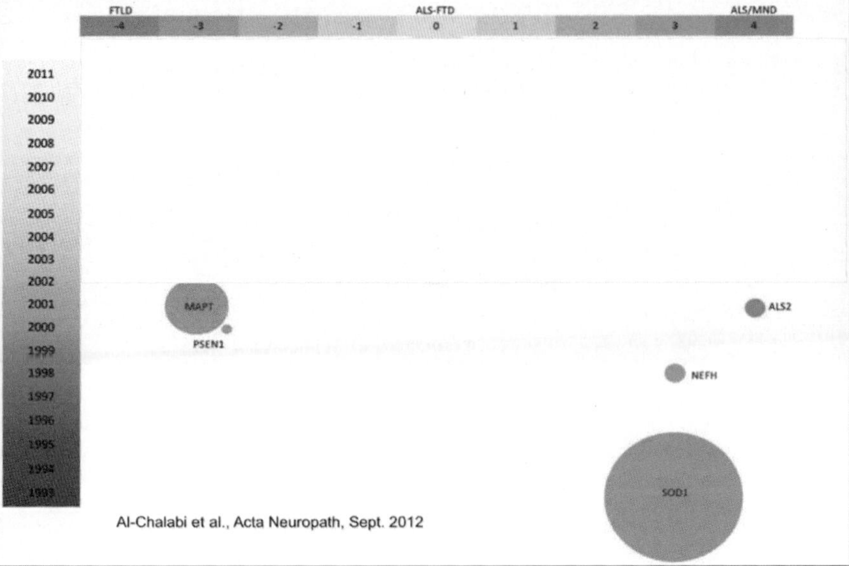

Al-Chalabi et al., Acta Neuropath, Sept. 2012

New strategies. Genome-wide association studies (GWAS) are an approach that involves rapidly scanning markers across the complete sets of DNA, or genomes, of many people to find genetic variations associated with a particular disease.

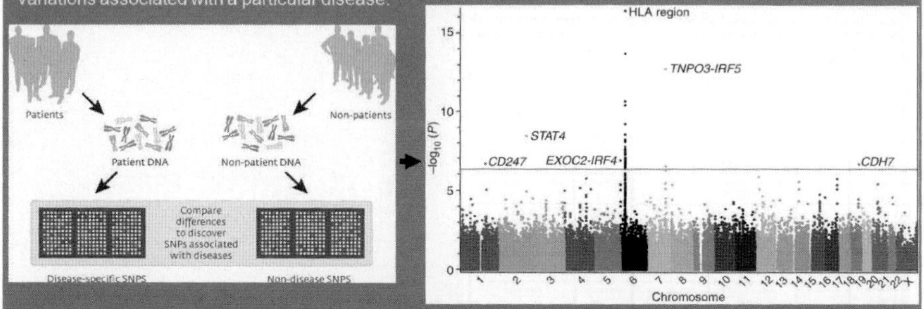

DNA from patients and controls is then purified from the blood or cells, placed on tiny chips and scanned on automated machines for selected markers of genetic variation, which are called single nucleotide polymorphisms, or SNPs.

Starting from TDP-43, many genes have now been associated with the ALS-FTD spectrum:

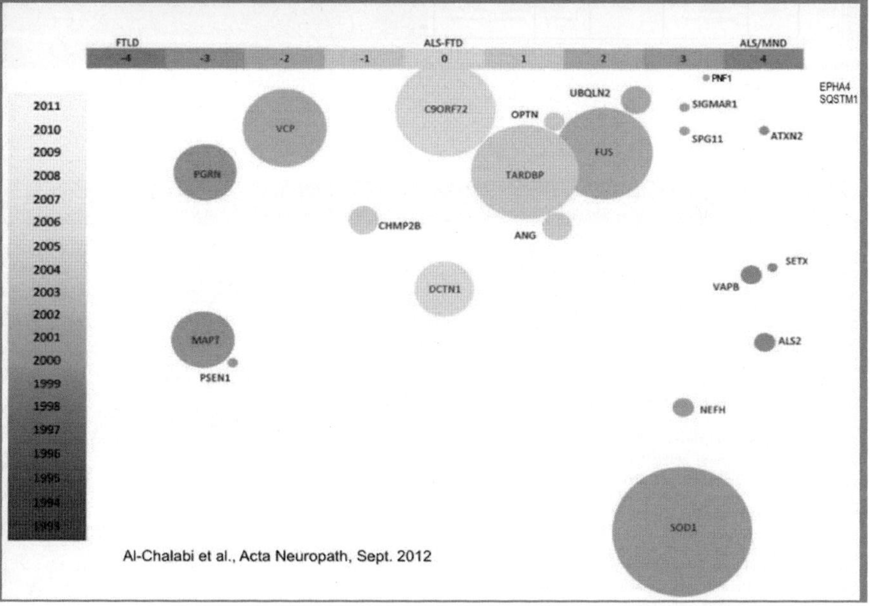

Al-Chalabi et al., Acta Neuropath, Sept. 2012

Knowing the genes involved also allows to prioritize research in metabolic areas that are more commonly associated with a specific disease:

Table 1
A summary of genetic, clinical and brain histopathology data together with the possible target of mutations in ALS and/or FTD (the black boxes capture the suggested continuum in all three categories).

Genetics of ALS and/or FTD			Continuum based on:		
Gene	Frequency in familial cases	Type of mutations	Clinical presentation	Brain pathology[a]	Likely pathological effect
SOD1	~20%	Mainly missense	ALS	SOD1/p62	Toxic aggregation
FUS	~5%	Mainly missense, & in-frame small deletions/insertions	ALS	FUS/p62	DNA/RNA metabolism
TARDBP (TDP43)	~3%	Mainly missense, & one truncating	ALS	TDP43/p62	DNA/RNA metabolism
C9orf72	~30%	G_4C_2-repeat expansion	ALS, FTD	TDP43/p62, p62/repeat-dipeptides, UBQLN2	Toxic RNA (?) Toxic aggregation(?) Low C9orf72 expression (?)
VCP	Rare	Missense	FTD, ALS, IBMPFD	TDP43/p62	Autophagy
SQSTM1 (p62)	~3%	Missense and nonsense	FTD, ALS, PDB	TDP43/p62	Autophagy
OPTN	Rare	Missense and nonsense (haploinsufficiency)	ALS/FTD, glaucoma PDB (by GWAS)	TDP43/p62	Autophagy
UBQLN2	Rare	Missense	ALS, FTD, SP, MS	TDP43/p62, UBQLN2, FUS, OPTN	Autophagy
GRN	~10%	Nonsense (haploinsufficiency)	FTD, CLN11	TDP43/p62	Autophagy/lysosomal pathway
CHMP2B	Rare	C-terminal truncation of the CHMP2B	FTD	p62	Autophagy/lysosomal pathway
MAPT	~10%	Missense and splicing of exon 10	FTD	Abnormal tau filaments (tangles)	Toxic aggregation (defect in neuronal cytoskeleton)

Hardy and Rogaeva, 2013

RNA metabolism

Autophagic processes

Interestingly, the recent discovery of RNA metabolism as a key player in neuronal death has confirmed the many links between neurodegeneration and cancer:

A wide number of epidemiological studies have suggested an association between cancer and neurodegenerative diseases. For example, a reduced risk of almost all types of cancer has been reported among individuals with Parkinson's disease (PD) (Driver et al. 2007; Inzelberg and Jankovic 2007; Moller et al. 1995). Similarly, it has been shown that Alzheimer's disease (AD) is associated with a decreased risk of cancer while patients with cancer have lower incidence of AD (Musicco et al. 2013; Roe et al. 2005). In addition, a significantly reduced risk of amyotrophic lateral sclerosis (ALS) has been observed amongst prostate cancer survivors (Freedman et al. 2013). On the other hand, a positive correlation between certain types of cancers and neurodegeneration has also been reported. Malignant melanoma and skin carcinoma occur more frequently in the PD population as compared to controls (Inzelberg and Israeli-Korn 2009; Olsen et al. 2006) while a significantly elevated risk of ALS amongst survivors of melanoma and tongue cancer has been observed (Freedman et al. 2013).

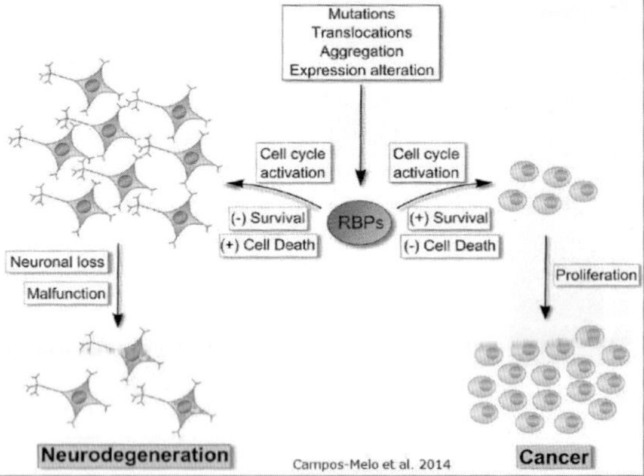

Campos-Melo et al. 2014

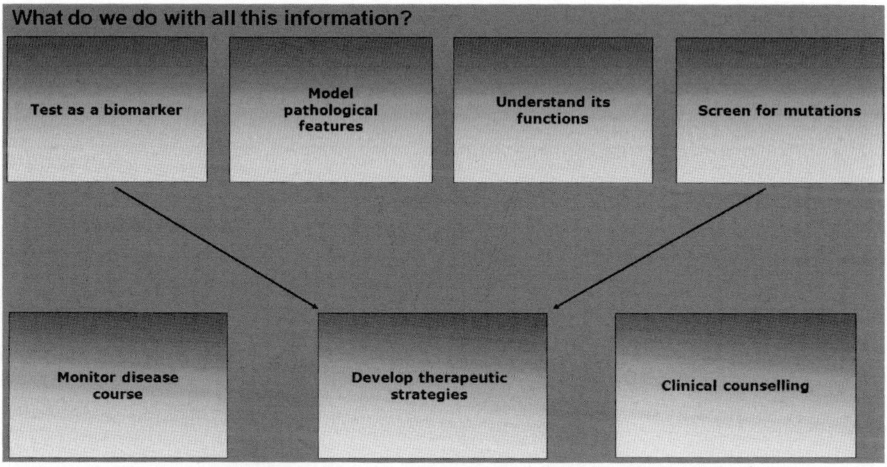

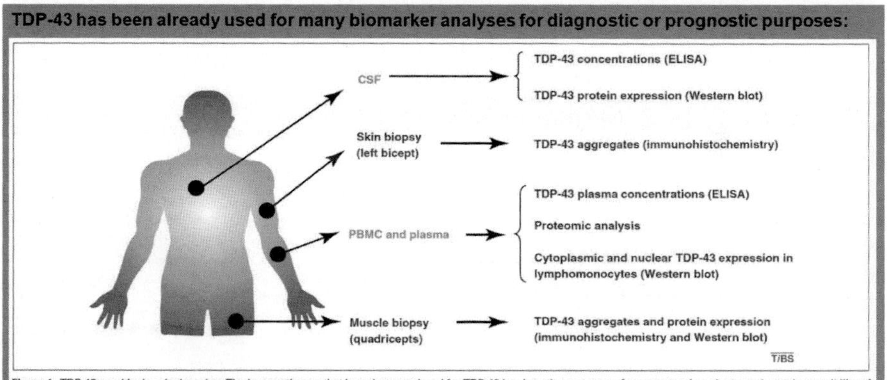

Figure 4. TDP-43 as a biochemical marker. The human tissues that have been analyzed for TDP-43 levels or the presence of aggregates, in order to evaluate the possibility of using TDP-43 as a biomarker to detect disease onset and progression, are shown. The most promising sources of TDP-43 material to be used for biomarker studies are highlighted in red, and include peripheral blood mononuclear cells (PBMCs) and cerebrospinal fluid (CSF).

TDP-43 likes to bind UG-repeated motifs in a sequence specific manner using both RRMs:

- RRM1 specifically recognizes one GU repeat and RRM2 recognizes also one UG repeat plus additional nucleotides that are not specifically recognized. Thus the emerging binding consensus sequence is 5'-N(N)GUGN(N)UGN-3'.
- A unique feature of the TDP-43-RNA complex is the reversed binding of the UG-rich RNA with the 5'end being bound on RRM1 rather than RRM2 as thus far always seen in tandem RRMs.
- TDP-43 binding to UG-rich RNA also showed a very clear correlation between binding affinity and inhibitory splicing function for interaction on RRM1, but not for RRM2 in which alanine mutations in the recognition sites on RRM2 has only little impact on the overall RNA binding affinity of TDP-43. Nevertheless, these interactions are of functional importance and required for the splicing function of TDP-43. Thus the role of RRM2 could be to direct the UG-rich RNA path for productive TDP-43 dimerization on pre-mRNA target sites.

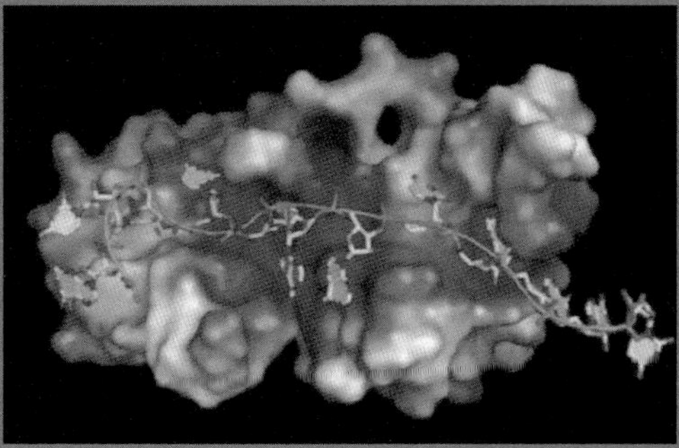

Lukavsky et al. 2013

Developing a loss of function model in flies:

A

B

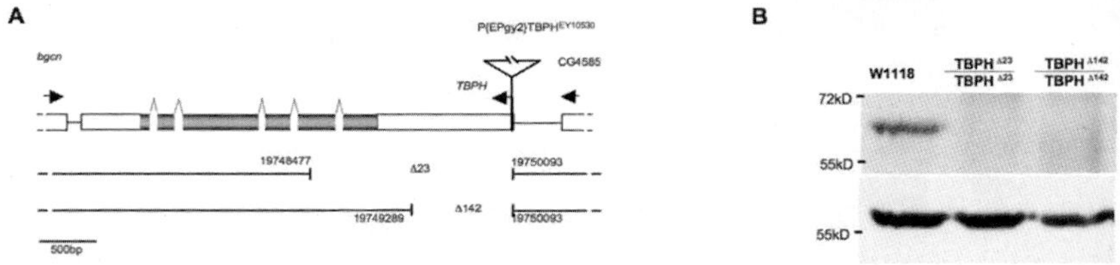

C

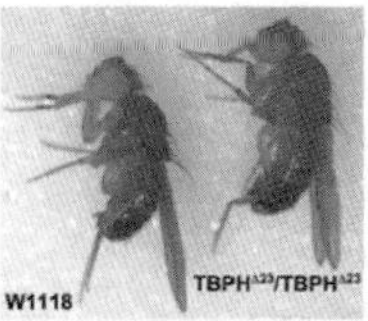

Eclosion Analysis

	Pupa n	Eclosed flies	Flies with motility defects
$TBPH^{\Delta 23}/CyO^{GFP}$	795	81%	0%
$TBPH^{\Delta 142}/CyO^{GFP}$	669	72%	0%
$TBPH^{\Delta 23}/TBPH^{\Delta 23}$	722	34%	100%
$TBPH^{\Delta 142}/TBPH^{\Delta 142}$	358	25%	100%
OregonR	361	89%	0%

D

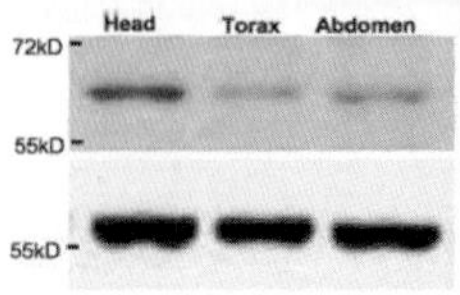

Flies apparently develop normally but cannot get out of the pupal cage without external help and show a reduced life span.

Feiguin et al. 2009

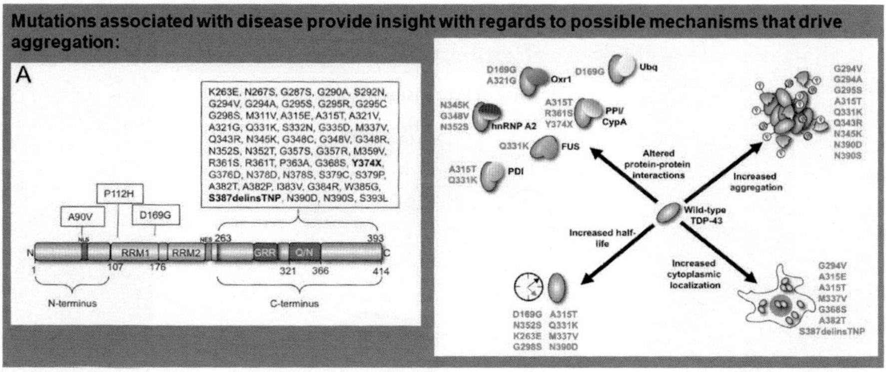
Mutations associated with disease provide insight with regards to possible mechanisms that drive aggregation:
A
K263E, N267S, G287S, G290A, S292N, G294V, G294A, G295S, G295R, G295C G298S, M311V, A315E, A315T, A321V, A321G, Q331K, S332N, G335D, M337V, Q343R, N345K, G348C, G348V, G348R, N352S, N352T, G357S, G357R, M359V, R361S, R361T, P363A, G368S, Y374X, G376D, N378D, N378S, S379C, S379P, A382T, A382P, I383V, G384R, W385G, S387delinsTNP, N390D, N390S, S393L
P112H
A90V
D169G
NLS
NES
263
393
N
RRM1
RRM2
GRR
Q/N
C
1
107
176
321
366
414
N-terminus
C-terminus
D169G A321G
Oxr1
D169G
Ubq
N345K G348V N352S
hnRNP A2
A315T R361S Y374X
PPI/ CypA
Q331K
FUS
A315T Q331K
PDI
Altered protein-protein interactions
Increased aggregation
G294V G294A G295S A315T Q331K Q343R N345K N390D N390S
Wild-type TDP-43
Increased half-life
Increased cytoplasmic localization
D169G A315T N352S Q331K K263E M337V G298S N390D
G294V A315E A315T M337V G368S A382T S387delinsTNP

AGING

As aging progresses, the body proteostasis network becomes less efficient which translates in greater chances of leading to misfolded protein accumulation, neuronal loss, etc.

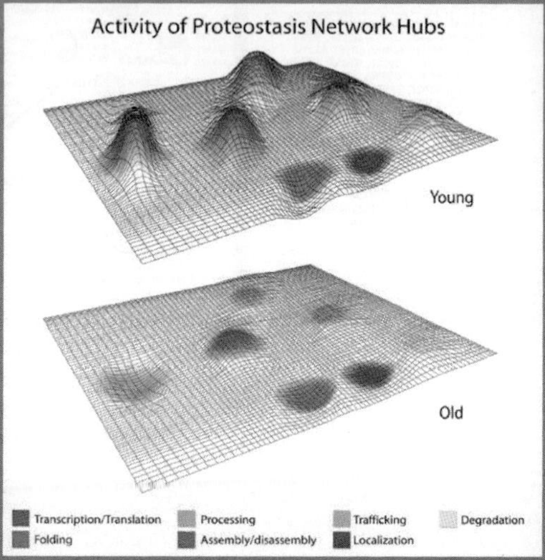

However, simple aging DOES NOT equate with neurodegeneration.

Analysis of alternative splicing associated with aging and neurodegeneration in the human brain

James R. Tollervey,[1] Zhen Wang,[1] Tibor Hortobágyi,[2] Joshua T. Witten,[1] Kathi Zarnack,[3] Melis Kayikci,[1] Tyson A. Clark,[4] Anthony C. Schweitzer,[4] Gregor Rot,[5] Tomaž Curk,[5] Blaž Zupan,[5] Boris Rogelj,[2] Christopher E. Shaw,[2] and Jernej Ule[1,6]

[1]MRC Laboratory of Molecular Biology, Hills Road, Cambridge CB2 0QH, United Kingdom; [2]MRC Centre for Neurodegeneration Research, King's College London, Institute of Psychiatry, De Crespigny Park, London SE5 8AF, United Kingdom; [3]EMBL–European Bioinformatics Institute, Wellcome Trust Genome Campus, Hinxton, Cambridge CB10 1SD, United Kingdom; [4]Expression Research, Affymetrix, Inc., Santa Clara, California 95051, USA; [5]Faculty of Computer and Information Science, University of Ljubljana, Tržaška 25, SI-1000 Ljubljana, Slovenia

Age is the most important risk factor for neurodegeneration; however, the effects of aging and neurodegeneration on gene expression in the human brain have most often been studied separately. Here, we analyzed changes in transcript levels and alternative splicing in the temporal cortex of individuals of different ages who were cognitively normal, affected by frontotemporal lobar degeneration (FTLD), or affected by Alzheimer's disease (AD). We identified age-related splicing changes in cognitively normal individuals and found that these were present also in 95% of individuals with FTLD or AD, independent of their age. These changes were consistent with increased polypyrimidine tract binding protein (PTB)–dependent splicing activity. We also identified disease-specific splicing changes that were present in individuals with FTLD or AD, but not in cognitively normal individuals. These changes were consistent with the decreased neuro-oncological ventral antigen (NOVA)–dependent splicing regulation, and the decreased nuclear abundance of NOVA proteins. As expected, a dramatic down-regulation of neuronal genes was associated with disease, whereas a modest down-regulation of glial and neuronal genes was associated with aging. Whereas our data indicated that the age-related splicing changes are regulated independently of transcript-level changes, these two regulatory mechanisms affected expression of genes with similar functions, including metabolism and DNA repair. In conclusion, the alternative splicing changes identified in this study provide a new link between aging and neurodegeneration.

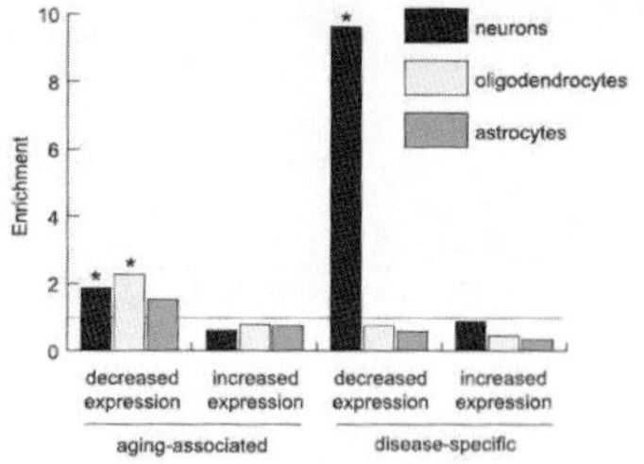

Figure 4. Analysis of cell-type-specific gene expression. The y-axis shows the enrichment of cell-type-specific genes (Cahoy et al. 2008) (see Methods) among the genes with increased or decreased age-related or disease-specific changes in transcript levels. (*) Significant difference in proportion of cell-type-specific genes (P-value < 0.0001, χ^2 test with Yates correction).

Neuroprotective actions such as use of antioxidants from food, exercise, etc. can slow down the aging process and thus represent an obstacle to neurodegeneration

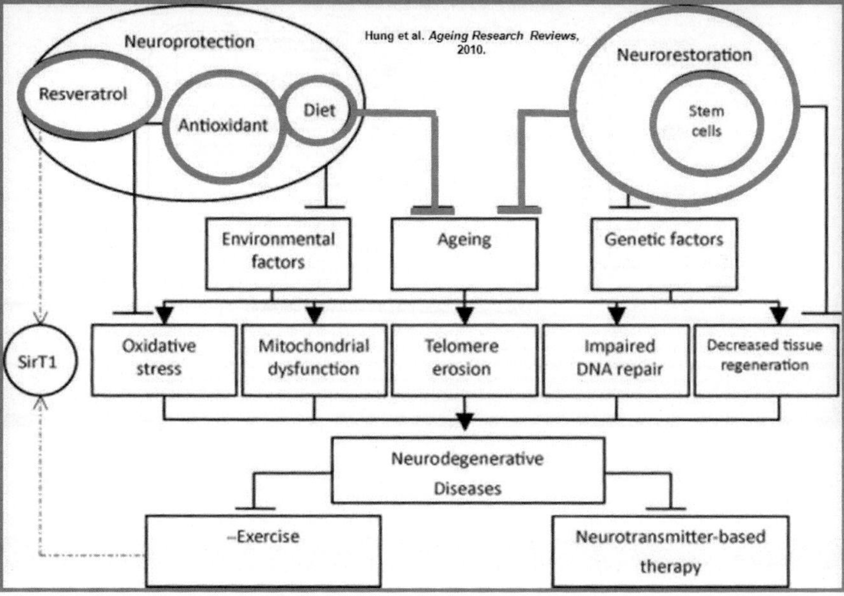

ENVIRONMENT

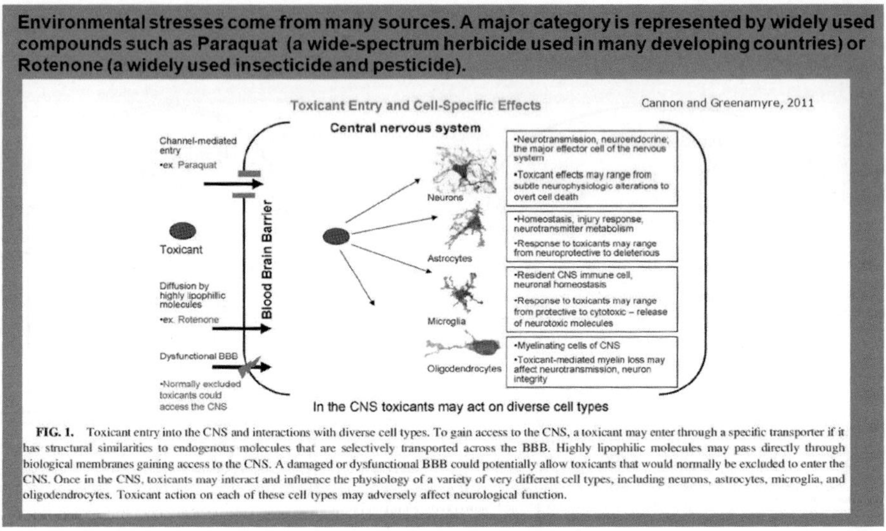

FIG. 1. Toxicant entry into the CNS and interactions with diverse cell types. To gain access to the CNS, a toxicant may enter through a specific transporter if it has structural similarities to endogenous molecules that are selectively transported across the BBB. Highly lipophilic molecules may pass directly through biological membranes gaining access to the CNS. A damaged or dysfunctional BBB could potentially allow toxicants that would normally be excluded to enter the CNS. Once in the CNS, toxicants may interact and influence the physiology of a variety of very different cell types, including neurons, astrocytes, microglia, and oligodendrocytes. Toxicant action on each of these cell types may adversely affect neurological function.

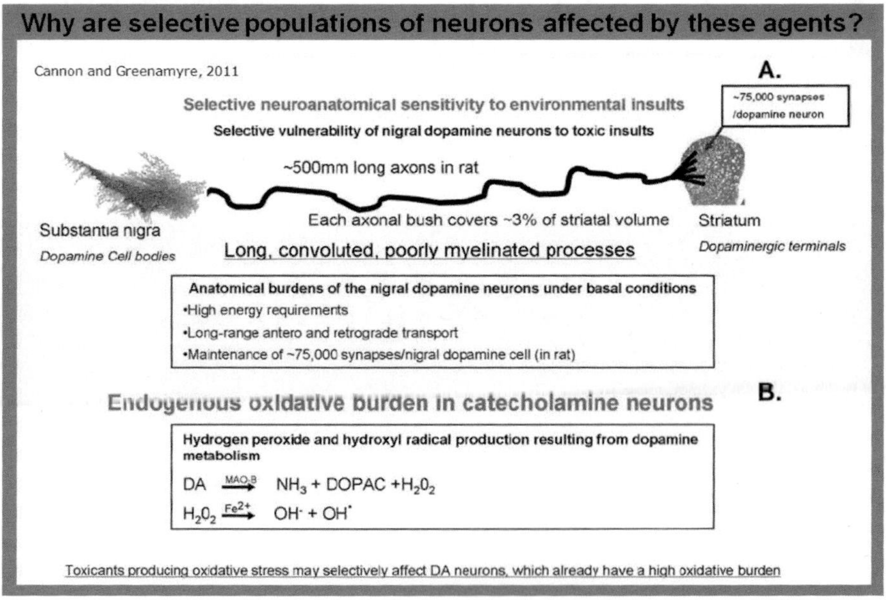

Traumatic brain injury (TBI) is also associated with neurodegeneration.

6 *P.M. Washington et al. / Experimental Neurology xxx (2015) xxx–xxx*

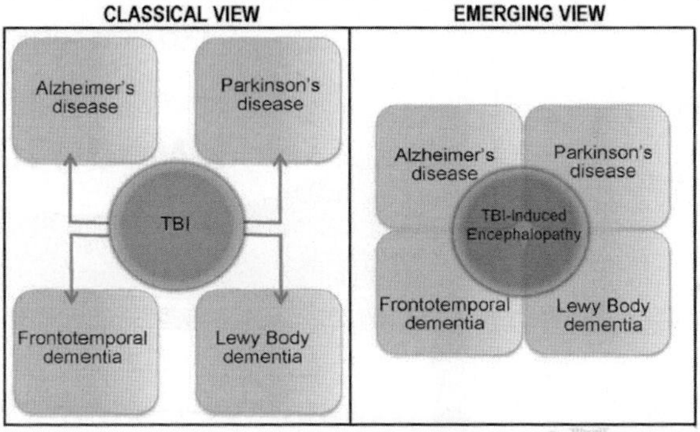

Fig. 2. The traditional view of the chronic effects of TBI classifies brain injury as a risk factor for the precipitation of individual neurodegenerative diseases. The emerging view supported in this review is that TBI-induced neurodegenerative disease, or traumatic encephalopathy, is a spectrum disorder that shares clinical and neuropathological hallmarks with other neurodegenerative disorders.

Conclusions:

- **Neurodegenerative diseases represent complex disorders that involve our genes, ageing process, and the environment.**

- **All these three factors influence each other. This can explain the huge spectrum of phenotypes that is observed for all neurodegenerative diseases in different individuals.**

- **In recent times, the weight given to genetic factors has been steadily increasing. Knowing the genes involved is very important from a number of reasons:**

  - **Diagnostic, because it allows to look for the presence of mutations or risk alleles in the population.**

  - **Prognostic, as measuring the expression level can be used as a biomarkers to predict disease course and evaluate the efficacy of novel therapeutic approaches.**

  - **Therapeutic, because it has opened new areas to research in fields that have not been previously associated with neurodegenerative processes such as RNA metabolism. In addition, it has allowed the development of new animal and cellular models that are expected to help the development of novel therapeutic approaches.**

ACKNOWLEDGMENTS:

PROJECTS SUPPORTED BY JPND

JPND research

RiMod-FTD

Risk and Modifying factors in Fronto-Temporal Dementia

AriSLA

FONDATION THIERRY LATRAN

Under the aegis of Fondation de France

SESSION 13

COSMIC OBJECTS UPDATE

ASTEROID AND COMET NUCLEUS IMPACT RISK ASSESSMENT

WALTER F. HUEBNER
Southwest Research Institute
San Antonio, TX, USA

PAUL. A. BRADLEY
Los Alamos National Laboratory
Los Alamos, NM, USA

ABSTRACT

How does the cost of countermeasures to avert an impact of an asteroid or comet nucleus on Earth compare with expected fatalities, injuries, economic and material losses (including infrastructure), and historic and cultural losses? Our objective is to assess the potential for casualties, injuries, and material and economic losses resulting from possible impacts by comparing the energetics of such impacts with the total energies dissipated during events of major natural disasters and compare these losses with the costs of implementing procedures to avoid such collisions. The emphasis of natural disasters is placed here on earthquakes because their energetics and consequences are similar but better understood and databases for fatalities and material and economic losses have been gathered by the insurance industry and are available for comparison. We also make comparisons with underground nuclear tests for which the energy release is well known. If the kinetic energy of a 100 m asteroid (often referred to as a "city killer"), which is smaller than the planned lower detection limit for asteroids, were completely converted to seismic energy in a collision with Earth, it would be equivalent to an earthquake of about magnitude 8 on the Richter scale, but its seismic effects depend on energy containment and partitioning. Extensive damage is caused by air blast (and tsunami for an impact at sea). Well over 100,000 people have died since 1920 in *each* of six earthquakes of magnitude 7.0 or larger. In the Haiyuan earthquake (magnitude about 8) in a sparsely populated area of rural central China (about 50 people per square kilometer) 200,000 to 273,000 people died, and the estimated time for recurrence of such an event is about 800 to 1600 years; not too different from the impact probability of a 100 m asteroid. Losses from injuries were over $1 billion in *each* of the seven most damaging earthquakes since 1994 and uninsured losses were as high as $210 billion in just *one* of these earthquakes. The goal here is to reduce or even avoid risks for fatalities and losses from impacts through knowledge-based actions. The results can be used to compare expected losses with the costs of countermeasures. Costs for countermeasures are based on NASA's Deep Impact mission that released an impactor at high speed into the nucleus of comet 9P/Tempel in 2005. The cost of that mission was about $330 million. Depending on the size and orbital parameters of a PHO, several countermeasure missions may have to be launched, but for multiple missions the cost of each individual mission will be less. *Readiness is the most important aspect for countermeasures to be successful.*

1. INTRODUCTION

The potential number of fatalities, injuries, and damage to infrastructure are of fundamental concerns when considering impacts of asteroids or nuclei of comets (here referred to as Earth impactors, but collectively also called potentially hazardous objects, PHOs) on Earth. Even for small objects, casualties and damage can be enormous. In the Tunguska event of 30 June 1908, about 2000 km^2 of forest were destroyed, even though the object exploded high in the atmosphere and apparently no major parts of it reached the ground intact. The object was estimated to be 50 to 60 m, but may have been only 40 m in mean diameter according to newer estimates (Boslough, private communication). Fortunately, the area was virtually uninhabited. However, as we shall discuss below, fatalities can be very large even in sparsely populated rural areas.

In order to better understand and put the disastrous consequences of PHO impacts on a firmer basis, we compare them to major earthquakes of similar energy. Another parallel between earthquakes and PHO impacts is that the time of occurrence cannot be predicted, but we know with certainty that they will occur. In earthquakes caused by the release of stored energy deep underground most of the energy goes into seismic waves. In an Earth impact the kinetic energy of a PHO is partitioned into excavating crater material, air blast, dust production, and heating that can lead to chemical reactions in the atmosphere, in addition to seismic waves. Comparisons with underground (nuclear) tests (UGTs) for which the energy yield is known, is another useful comparison. Again, the depth of the nuclear explosion is important for energy partitioning. In impacts at sea, the kinetic energy goes into sea bottom cratering, impact generated wave formation, air blast, melting of sediments on the ocean floor, and steam production. In airbursts, almost all the energy goes into airblast, shock formation, and radiation.

Some of the non-seismic consequences of a PHO impact may be more benign than an earthquake, while others, e.g., air blast, strong winds with entrained debris, dust, smoke, and chemical pollutants, may be more severe. In the absence of more detailed data, we will assume the consequences to be the same as if all of the impact energy went into seismic waves.

1.1. Impact velocity

The impact energy of a PHO depends on its impact velocity after passing through the atmosphere. As an object enters the atmosphere it is decelerated by air friction, which causes heating and ablation from its surface as discussed by Chyba et al. (1993) (after correcting a misprint)

$$m\frac{dv}{dt} = -\frac{1}{2}C_D\rho_a A_i v^2 + mg\sin\theta\,, \tag{1}$$

where $m = m(z)$ is the mass of the object at altitude z, $v = v(z)$ its velocity, $A_i = A_i(z)$ its cross sectional area in the direction of its motion, $\rho_a = \rho_a(z)$ the density of the atmosphere, t the time, C_D the coefficient for aerodynamic drag, g Earth's gravitational constant, and θ the angle of its trajectory with respect to the horizontal. For a PHO moving at hypervelocity through the atmosphere the last term in Eq. (1) is not important for estimating its final velocity. A significant amount of an Earth impactor's kinetic energy is dissipated and converted to heat during its passage through the atmosphere. The energy and mass loss from heat transport and ablation is given by

$$Q\frac{dm}{dz} = -\frac{1}{2}C_{\mathrm{H}}\rho_{\mathrm{a}}A_{\mathrm{i}}v^2\,, \tag{2}$$

where Q is the heat of ablation and C_{H} is the heat transfer coefficient. Ablation changes the cross sectional area A_{i} and the differences in pressures from the leading to the trailing surfaces can exceed the material strength of the projectile. Numerical solutions, even for an idealized spheroidal shape of a non-tumbling asteroid are difficult, at best. However, a general estimate can be obtained by assuming a cylindrical object of radius R with its axis along the trajectory. The hydrodynamic deformation of this cylindrical object is given by

$$R\frac{d^2R}{dz^2} \approx -\frac{1}{2}\,C_{\mathrm{D}}\frac{\rho_{\mathrm{a}}}{\rho_{\mathrm{i}}}\,, \tag{3}$$

where ρ_{i} is the material density of the impactor. These coupled differential equations are usually solved numerically using an Eulerian shock-physics code (e.g., CTH) with adaptive mesh refinement to sufficiently resolve the entry dynamics while still spanning the entire atmospheric region of interest [see, e.g., Mac Low and Zahnle (1994) and Boslough and Crawford (2008)].

Heat transport, ablation, and deformation depend strongly on material properties of the PHO (e.g., ice for a comet nucleus vs. more refractory materials for an asteroid) and its physical structure and material strength. Chemical heterogeneity combined with structural boundaries change heat flow and local heat storage. These properties when associated with low material strengths lead to disruption, even explosive disruption, of some asteroids. Disruption is more common with small objects, where the entire body can fragment. An increase of the cross section in the direction of motion leads to additional aerodynamic drag, shock, ablation, and internal stresses. Small asteroids, such as the Tunguska asteroid, will then explode, dissipating much of their energy in the atmosphere and only fragments may reach the ground. Thus, fractional mass loss is more important for a small object than for a large one and small projectiles are more likely to disintegrate or explode at the peak of their energy release rates.

For larger objects, for which the fractional mass loss is small, the impact velocity can be obtained from Eq. (1), neglecting the term for gravitational attraction

$$\begin{aligned} m\frac{dv}{dt} &= -m\frac{dv}{dz}\frac{dz}{dt} = -mv\sin\theta\frac{dv}{dz} \\ &= -\frac{1}{2}C_{\mathrm{D}}\rho_{\mathrm{a}}A_{\mathrm{i}}v^2\,. \end{aligned} \tag{4}$$

Solving for v, assuming a straight line trajectory and an exponential atmosphere with scale height H gives

$$v_0 = v_\infty exp \frac{-C_D \rho_0 A_i H sin\ \theta}{2m}, \tag{5}$$

where v_∞ is the PHO's velocity at entry into the atmosphere, v_0 its impact velocity on the ground, and ρ_0 the atmospheric density at ground level. For a spherical impactor of radius r

$$v_0 = v_\infty exp \frac{-3C_D \rho_0 H sin\ \theta}{8r\rho_i}. \tag{6}$$

Assuming approximate values for $C_D \approx 1, H \approx 8500\ \text{m}, \rho_0 = 1.2\ \text{kg/m}^3$, $sin\ \theta \approx 1$, and using the average density for all asteroids listed on the website http://NEOData.space.swri.edu $\rho_i \approx 2000\ \text{kg/m}^3$, we obtain $v_0 \approx v_\infty e^{-1.91/r} \approx 0.96 v_\infty$. for a 100 m diameter PHA (potentially hazardous asteroid) and $v_0 \approx 0.91 v_\infty$ for a 40 m (r = 20 m) PHA (the lower size limit for the Tunguska object). The difference in a PHO's energy at the top of the atmosphere (proportional to v_∞^2) and at impact on the ground (proportional to v_0^2), see, e.g., Sect. 2, is the energy dissipated in shock formation in the atmosphere, heat production, thermal radiation, electromagnetic pulse (EMP) effects, sublimation of material (phase transition from solid to gas), and overcoming internal strains that lead to disruption of the object. Thus, smaller PHAs, such as the Tunguska object, experience a larger fraction of energy dissipation during transit of the atmosphere and therefore tend to explosively disrupt before reaching the ground. This is borne out from the simultaneous solution of Eqs. (1) – (3) as carried out for example by Chyba et al. (1993).

1.2. Earthquake energy

We will now discuss the parallels and similarities in effects based on energy input and dissipation in the ground of major earthquakes and impacts by PHOs. The total energy released during an earthquake depends on the elastic and rheological material properties and the stored stress around the fault region. This total energy is partitioned into mechanical energy of fracturing and moving rocks, frictional heat, and elastic seismic waves. It is the energy in seismic waves, E_S, that is the major source for physical failures of infrastructures and the cause of associated casualties and economic losses. The *magnitude* of an earthquake is a number that characterizes the relative size or amount of seismic wave energy released by such an event. It is usually based on measurements of the maximum ground motion recorded by a seismograph (sometimes for a particular wave type and frequency) and is corrected for the decay of amplitudes with distance from the epicenter and depth of the source associated with geometric spreading and attenuation during wave propagation. Earthquakes originating near the Earth's surface are more relevant for comparison with PHO impacts, however, this information is not always readily available.

Several different methods and scales for categorizing earthquakes have evolved. We will summarize some of these briefly. In the press earthquakes are commonly classified by (1) the dimensionless Richter moment magnitude scale, M_W (W stands for mechanical work), developed by Richter and Gutenberg in 1935, or scientifically and technically by (2) the more modern *seismic moment magnitude*. The seismic wave energy, E_S, dissipated in an earthquake can then be obtained from the Gutenberg-Richter magnitude-energy relation (see, Gutenberg and Richter, 1942, 1956a, 1956b)

$$\log_{10} E_S = 1.5M_W + 4.8 , \tag{7}$$

where E_S is in units of J.

For scientific studies, the *seismic moment* is the measure to use since it has fewer limitations than magnitudes, which often reach a maximum value, called magnitude saturation. The seismic moment, M_0, is a measure of the *total* amount of energy that is transformed during an earthquake. It combines the average area, A, of the rupture and the offset of the fault, D, with a measure of the strength of the rocks, i.e., the shear modulus, μ,

$$M_0 = \mu AD . \tag{8}$$

Usually the seismic moment is measured directly from seismograms, since the size of the very long-period waves generated by an earthquake is proportional to the seismic moment. Typical values for the shear modulus are $\mu = 32$ GPa in the crust and $\mu = 75$ GPa in the mantle. In SI units, M_0 has dimension N m.

To determine the total seismic energy radiated from an earthquake one would have to integrate the energy released at all frequencies over the entire focal sphere, a conceptual sphere around the source at the beginning of the event. The spectrum of the average radiation over the focal sphere can be approximated by a constant level at low frequencies, which is proportional to the moment, M_0, and a uniform decrease with increasing frequency above some corner frequency, F_c; thus, the seismic energy is a function of both M_0 and F_c. For a given moment, the radiated energy will increase as F_c increases. The corner frequency defines the maximum slip velocity.

If one compares two earthquakes, one with a "slow creep-like deformation" (Kanamori, 1972), the other with a more typical rupture velocity near the shear wave velocity, much more energy would be radiated from the second earthquake because of its rich high frequency spectrum corresponding to the larger corner frequency than from the first of the two events.

Since the corner frequency varies systematically with the moment one can approximate the radiated seismic energy. For the average earthquake, the seismic wave energy, E_S, moment, M_0, and moment magnitude, M_W, are related (Kanamori, 1977)

$$M_0 = 2{\times}10^4 E_S , \tag{9}$$

with M_0 in N m and E_S in J. Combining with Eq. (7) gives

$$M_W = \frac{2}{3}(\log_{10} M_0 - 9.1) , \tag{10}$$

where M_w is dimensionless. The constant values in the equation are chosen to achieve consistency with the magnitude values produced by earlier scales, such as the *local magnitude* and the *surface wave magnitude*, M_S, both referred to as the "Richter" scale by news media. The surface wave magnitude M_S is determined from surface waves (typical period of 20 s) and can be measured over global distances. Most nuclear tests did not produce large earthquakes and their seismic signals were detectable over distances of order 1000 km or less. In effect this changes the values in Eq. (7) as less of the total available energy goes into seismic waves. In this case, it is more appropriate to use the body wave magnitude m_b(P), which is determined from local P waves (typical period of 1 s). The nuclear test earthquake versus yield scale uses m_b(P), and these can be related by the formula: $M_S = 1.059\, m_b(\mathrm{P}) - 0.6624$ (Devi and Kalita, 2013).

2. IMPACT ENERGY OF AN ASTEROID OR A COMET NUCLEUS

The kinetic energy of a spherical impactor at the time of collision with the Earth is

$$E_\mathrm{i} = \frac{1}{2}\frac{4\pi}{3} r^3 \rho_\mathrm{i} v_0^2 , \tag{11}$$

where r is the mean radius of the impactor, ρ_i is its material density, and v_0 is its speed relative to the Earth at the time of impact [see Eq. (6)]. Using the energy equivalent of 1 Mt of TNT = 4.18×10^{15} J, Eq. (11) becomes

$$E_\mathrm{i} = 6.26{\times}10^{-11} d^3 \rho_\mathrm{i} v_0^2 , \tag{11a}$$

where E_i is measured in Mt of TNT, d is the diameter of the PHO in m, ρ_i its density in kg/m^3, and v_0 is its impact speed in km/s.

In Figure 1 we use rounded numbers for densities and velocities to permit easy scaling. For a typical value for the density of an asteroid we assume ρ_i = 2000 kg/m^3 and for the nucleus of a comet ρ_i = 500 kg/m^3. Typical speeds of a PHO near and relative to the Earth are v_∞ = 17 km/s for asteroids and prograde comets. However, retrograde comets may have much higher speeds, e.g., Comet Halley had a speed of about v_∞ = 70 km/s with respect to the Earth. The actual speed of a PHO depends on its orbital parameters. Here we will use v_∞ = 10 km/s for asteroids and prograde comets, and v_∞ = 40 km/s for retrograde comets.

Over 10,600 near Earth asteroids (NEAs) and over 90 near Earth comets (NECs) have been discovered. Currently 1,458 NEAs are classified as potentially hazardous asteroids (PHAs) and 154 PHAs and 867 NEAs have an absolute brightness magnitude of 17.75 or brighter. This means that they are at least 1 km in mean diameter.

Using Eq. (6) and nominal values mentioned above, a single (non-binary) 100 m diameter asteroid with atmospheric entry speed of 10 km/s will have an impact energy equivalent to about E_i = 11.5 Mt of TNT as shown in Fig, 1. This is equivalent to the energy of over 700 Hiroshima bombs. If this energy were converted to seismic energy it would cause an earthquake of magnitude 8. Its total energy at the time of entry into the atmosphere is $E_\infty = (v_\infty/v_0)^2 \times 11.5$ Mt = 12.5 Mt of TNT (see Sect. 1.1). For a 100 m prograde comet nucleus with the same entry speed into the atmosphere the energy is equivalent to E_∞ = 3.1 Mt of TNT. For a 100 m retrograde comet nucleus with a nominal speed of v_∞ = 40 km/s the energy equivalent before atmospheric entry is E_∞ = 50 Mt of TNT.

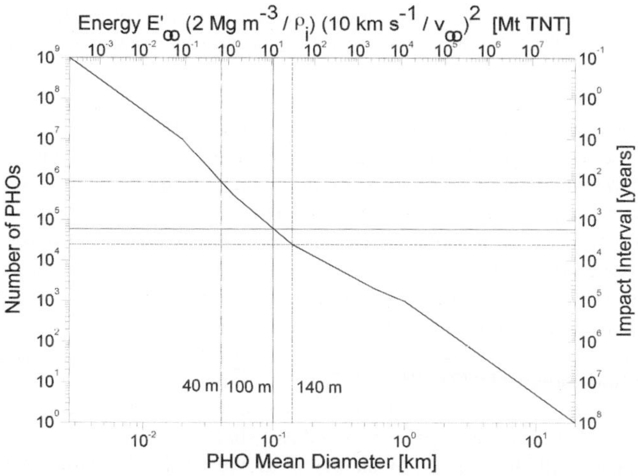

Fig. 1: Number of PHOs larger than a given size and their impact intervals vs. PHO size. The PHO size is also related to the energy at the time of atmospheric entry $E_\infty = E'_\infty\ \rho_i/2\ (v_\infty/10)^2$, with E_∞ in Mt of TNT, ρ_i in Mg/m^3 and v_∞ in km/s. The impact energy on the ground can then be obtained from $E_i = (v_0/v_\infty)^2\ E_\infty$, where v_0 is given by Eq. (6). PHO's of three different mean diameters are highlighted: 40 m (Tungaska object), dotted lines; 100 m (our nominal object), solid lines; and 140 m (minimum size of planned detectability) dashed lines. The deviation of the curve from a straight line is thought to be caused by the transition from monolithic to rubble pile structure of asteroids (Harris, 2007).

The Chelyabinsk, Russia, meteor of 15 February, 2013, was about 20 m in mean diameter. Objects of this type and size my plunge into Earth's atmosphere on average about every 30 years, ten times more often than previously thought.

3. DAMAGE ESTIMATES FROM AN IMPACT

In this section, we describe the various damage mechanisms from the impact of a small asteroid or comet nucleus with Earth. Depending on their structure and composition, the smallest objects will break up during their travel through the atmosphere. The Chelyabinsk and Tunguska objects are well-known examples of this. In this case, most of the damage will come from the overpressure of the blast wave generated by the object, as described in Sect. 3.1. Airbursts can also generate modest earthquakes, as described in Sect. 3.2. If the object survives its travel through the atmosphere, then it will impact and create a crater (see Sect. 3.3), an earthquake (see Sect. 3.2), a blast wave, and in some cases, a fireball. Because the Earth's surface area is about 70% water, there is a significant chance that a small asteroid can impact the ocean. In this case, an impact generated wave can reach the shore, along with the other damage mechanisms from a land impact, depending on the size of the impactor and whether the water depth is significantly greater than the impactor size or not. We discuss this problem in Sect. 3.4.

3.1. Blast damage from an air burst

Although the Tunguska event eventually showed that damage from an asteroid exploding in the atmosphere could be substantial, the Chelyabinsk event of 15 February 2013 showed that an airburst of an even smaller object could still be quite damaging. The fireball from the explosion with an estimated explosive yield of about 400 to 500 kt equivalent of TNT produced a blast wave that shattered windows on many buildings in Chelyabinsk. Over 1000 people were injured from flying glass and one building partially collapsed (Durda, 2013). The asteroid broke up between 45 and 30 km altitude based on the images of the fireball and other data. The thinness of the atmosphere at this altitude renders the blast less effective; we deduce a blast efficiency of about 0.8 from Table 3.68 of Glasstone and Dolan (1977). We used the example on page 108 and Fig. 3.72 of Glasstone and Dolan and determined an overpressure of about 6.9 kPa. Table 5.145 of Glasstone and Dolan indicates that glass windows shatter at peak overpressures of 3.5 to 7 kPa, consistent with the Chelyabinsk event. If we apply the same formulae to the Tunguska event with an explosive force of 5 Mt equivalent of TNT, the peak overpressure is about 20 kPa. Table 5.145 and the text of Chapter 5 of Glasstone and Dolan indicate that at this overpressure most buildings will be damaged. Figure 5.146 shows that a coniferous forest would be severely damaged out to a distance of about 15 - 17 km. This compares well with the estimated 2000 to 2200 km^2 of the actual area destroyed in the Tunguska region. If the Tunguska event had occurred over a city, the overpressure of 50 kPa and wind of 90 km s^{-1} at a distance of 10 km would blow down wooden structures (most houses), shatter glass windows, and knock down some trees. Even at a distance of 20 km, glass windows would shatter from the overpressure. In summary, the airburst damage from an asteroid exploding in air will be essentially the same as from an equivalent nuclear explosion, except that there will be no radioactive fallout from the asteroid explosion.

An asteroid exploding in the air or on the ground will produce a shockwave and hence an earthquake that can cause damage to buildings and structures. We start by analyzing earthquakes generated by underground nuclear tests (UGTs), which are close analogues to earthquakes generated by asteroid impacts. Figure 2 shows the body wave earthquake magnitude m_b(P) versus the quoted yield of many underground nuclear tests (Nuttli, 1986; Rodean, 1972; von Seggern and Lambert, 1972; Rodean, 1971; Ryall and Savage, 1969). We also show data from the Sedan test, which was at a shallow depth to generate the largest possible crater, and the earthquake recorded from the Chelyabinsk meteorite of 15 February 2013. The Sedan event was equivalent to an earthquake of magnitude m_b(P) = 4.75 (Sedan Event, 2014), which is 0.75 magnitude less than a comparable UGT (a factor of six less shaking), while the earthquake from the Chelyabinsk airburst (Durda, 2013) was 1.8 magnitudes less damaging in earthquake magnitude than a comparable UGT (a factor of about 60 less shaking). Underground nuclear tests were buried deep enough that all of the blast energy went into vaporizing the rock surrounding the device and into ground motion. Shallow burial events, such as Sedan or an asteroid impact, will have a significant amount of blast energy going into moving rock and even escaping as radiation. The 0.75 magnitude reduction reflects the reduced coupling of energy into ground motion. Finally, an airburst has the downward shock wave coupling to the ground, which is inefficient at producing ground motion.

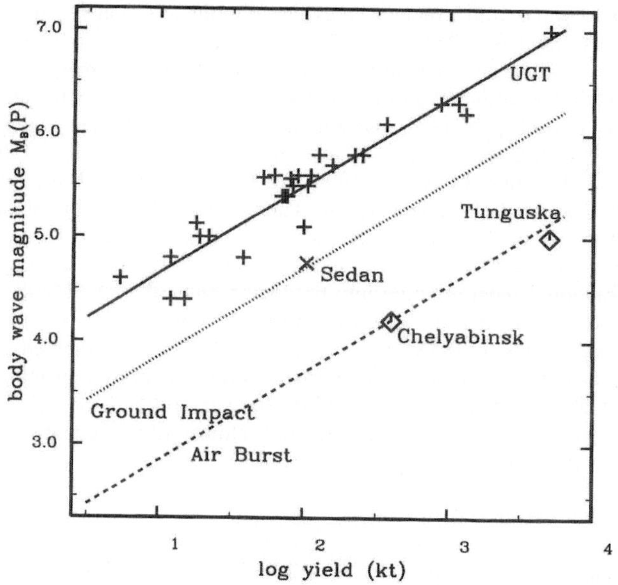

Fig. 2: Underground nuclear test (UGT) body wave earthquake magnitude, $m_b(P)$ (Nuttli, 1986), versus explosive yield (log kt) [solid line, "+" symbol, from DOE/NV—209-Rev 15 (2000)]. Also shown are the body wave earthquake magnitude, $m_b(P)$, of the shallow burial Sedan nuclear test and $m_b(P)$ from the explosion of the Chelyabinsk and Tunguska meteors. The earthquake magnitude, $m_b(P)$, is reduced by 0.75 for a surface burst (dotted line, "×" symbol) and by 1.8 for an airburst (dashed line, diamond symbols) compared to a comparable UGT.

If we make the assumption that a ground impact from a meteorite scales in energy yield as the underground test data, the body wave earthquake magnitude, $m_b(P)$, scales as the dotted line ("×" symbol) in Fig. 2. The fitting formula for the scaled impact earthquake magnitude is $m_b(P) = 0.853 \log (E_i) + 3.03$, where E_i is the impact energy in kt of TNT. [The fitting formula to the UGT data is $m_b(P) = 0.853 \log (E_i) + 3.79$ (plotted in Fig. 2 as the solid line, "+" symbols) and has an R^2 of 0.85.] The implied impact earthquake magnitudes are 6.3 for a 1 Mt impact, 7.2 for a 10 Mt impact, and 7.5 for a 20 Mt impact. The magnitudes of impact-generated earthquakes are therefore quite a bit less than that of the most devastating earthquakes (see Table 1, next section). There will still be considerable damage to buildings and structures that are not earthquake resistant. (Our formula for UGT yield fitting is comparable to that of Nuttli [$m_b(P) = 0.833 \log (E_i) +$

3.96] and Richards and Zhongliang (2011) [$m_b(P) = 0.75 \log (E_i) + 4.05$]. The fit for earthquake magnitude $m_b(P)$ given here in the range from 100 kt to 100 Mt is from about 0 to less than 0.3 larger than that of Richards and Zhongliang (2011) and about 0.1 smaller than that of Nuttli (1986).

The blast wave from a meteor breaking up in the atmosphere can also generate an earthquake (this effect is called "airslap"). This was observed for the Chelyabinsk and Tunguska explosions. The Chelyabinsk explosion generated a magnitude $m_b(P) = 4.2$ earthquake, while the earthquake from the Tunguska event is estimated at $m_b(P) = 5.0$. If we scale the impact generated earthquake magnitude formula downward by 1.1 magnitudes to match Chelyabinsk, we have: $m_b(P) = 0.853 \log (E_i) + 1.99$. If we use a yield of 5 Mt for Tunguska (Vasilyev, 1998), the formula estimated earthquake is $m_b(P) = 5.1$, which is close to the estimate of 5.0. Tunguska shows that the airslap earthquake from a multi-megaton blast is not that great. Such an object will create much more damage from the overpressure and blast induced wind, as will be described next.

The fraction of energy going into producing an earthquake versus other forms of damage can be calculated from Eqs. (9) and (10). A magnitude 5 earthquake has the equivalent energy release of 0.5 kt of TNT, which is 10^{-4} of the estimated 5 Mt blast from the Tunguska airburst. We obtain a similar fraction of energy going into an earthquake from the Chelyabinsk airburst. For both airbursts, very little of the asteroid reached the ground, so the impact energy is negligible. Thus, almost all of the burst energy goes into the shock wave in an airburst.

3.2. Impact crater damage

Meteor Crater in Arizona is a typical impact crater from a relatively small asteroid. Recent estimates suggest that an approximately 25 meter diameter iron meteor impacted with a force equivalent to about 10 Mt of TNT (Melosh and Collins, 2005). The resultant crater is 1.2 km in diameter. Assuming the asteroid reached the surface intact; a 500 kt explosion would create a crater about 500 m in diameter, while a 5 Mt explosion (approximately Tunguska size) would create a crater 1 km in diameter. Indeed, the lack of a crater at the Tunguska site made people realize that the object had to have exploded in the atmosphere. Finally, at 50 Mt, the crater would be about 2.2 km across, but ejecta would be sprayed out over tens of kilometers.

Using the impact effect estimator of Collins et al. (2005), we estimate the effect of a 100 m diameter stony asteroid hitting the city of Houston, Texas, (city population 2.2 million; metro area population 6.3 million). We chose typical parameters of 3000 kg/m^3 for the rock density, an initial impact velocity of 20 km s^{-1}, and an impact angle of 45°. A stony object 100 m in diameter is predicted to break up high in the atmosphere, but the fragments stay together enough to impact the ground at about 8 km s^{-1} with a blast energy equivalent to 11 Mt of TNT (this is nearly the same as the "Mike" nuclear test of 1951). The impact generated earthquake magnitude, $m_b(P)$ would be between 5.3 and 6.5. The fragment-generated crater would be about 1.5 km, and would eliminate the downtown area. The damage from the blast wave and flying debris would extend over many kilometers. Within 5 km of the epicenter, the destruction would be nearly total. The overpressure would be 130 kPa and impact generated winds would be about 200 m s^{-1}. Most buildings would collapse and many bridges would be damaged. Cars and trucks would be blown over. Between 5 and 10 km, buildings and other structures would be severely damaged by the peak overpressure (30 kPa) and wind of 70 m s^{-1} (250 km/h). The windows of houses and office buildings would be shattered and most of the trees would be blown down. This damage zone would extend to Loop 610, a circular highway about 10 km from the city center. At 25 km (the location of an outer circular highway called the Sam Houston Tollway), many windows would be shattered by the peak overpressure (9 kPa) and wind of 14 m s^{-1}. In short, the main part of a city the size of Houston would be destroyed, although the outlying areas would feel only minor effects.

In the case of a ground impact, about half of the energy goes into excavating a crater and hurling out ejected material, while the other half of the energy goes into the blast. About 0.1% of the energy goes into the earthquake, another 1% or less of the energy goes into melting material in the crater, and a negligible fraction goes into the fireball (if it occurs). The results are similar for a water impact, although some impact energy will be used to convert water into steam. The fraction of energy going into an earthquake (~10^{-3} for a ground impact and 10^{-4} for an airburst) is consistent with the commonly accepted fraction from experiments (Schultz and Gault 1975).

3.3. Impact-generated tsunami

Because 70% of the Earth's surface is ocean and many large cities lie on or near coasts, the greatest danger from an asteroid impact is from an impact-generated wave. Many people use formulae from Glasstone and Dolan (1977), Hills et al. (1994), and Collins et al. (2005) to compute the "tsunami" from an asteroid impact. However, these references assume the most of the impact energy goes into creating a wave, while more recent simulations suggest a different picture. Gisler et al. (2011) and Gisler (2013) show through simulations two major differences for small objects impacting water. Gisler et al. (2011) show that a 100 m diameter object hitting deep water produces a short wavelength wave that dissipates its energy relatively quickly. This implies that an asteroid hitting the ocean far from land will have negligible impact on the coastlines. Gisler et al. (2011) imply that a 100 meter object impacting 1000 km from shore would create no wave, while Eq. (2) of Hills et al. (1994) applies for an impact into deep water (water depth much greater than the impactor size)

$$h = \frac{1}{2} h_W = \frac{4500}{r} E_i^{1/2} , \tag{12}$$

where h_W is the peak-to-valley amplitude of the tsunami wave in meters, E_i is the energy yield of the impactor in gigatons of TNT, and r is the distance from the impact point in km. A 100 m asteroid would produce 11 Mt of yield and Hills et al. (1994) would predict a wave of 0.5 m, which would not cause much damage.

The big difference comes if an asteroid strikes near shore, and the water depth is less than 1 km deep. If we assume a water depth of 100 meters and the impact is 12 km from shore, then Eq. (4) of Hills et al. (1994) would imply a tsunami of 4 meters peak-to-trough amplitude. Gisler (2013) modeled the impact of a 200 m diameter body hitting water 100 m deep underlain by 100 m of sediments. Gisler's simulation shows that a large quantity of sediments will melt from the heat of the impact. These refrozen, glassy sediment particles and water will be carried by hurricane force winds for tens of km away from the impact site. These same hurricane force winds will interact with the water's surface to push forward a "storm surge" onto the shore that would be several meters high. Thus, the impact of a small asteroid near shore would produce a one-two-three punch consisting of: 1) a modest earthquake of approximately magnitude 5 that would arrive about 3 seconds after impact, 2) a debris-laden blast wave of at least 40 m s^{-1} along with an overpressure blast wave of 20 kPa that will shatter windows and "sandblast" everything with warm, wet glassy debris about 30 seconds after impact, and finally 3) the wind-generated impact wave several meters high would arrive several minutes later. Since many major cities lie near a coast on a shallow coastal plain, the effects of a several meter high impact generated wave can be devastating, especially after all of the structures have been "sandblasted" by hurricane force winds with glassy debris.

4. VULNERABILITY ASSESSMENT

Munich Reinsurance's Geo Risks Research unit developed a unique archive and database of natural catastrophes and natural disasters (http://www.munichre.com/de/group/default.aspx). In the following we make use of this database, as summarized in Tables 1 and 2, for the assessment and comparison of Earth impacts by asteroids and comet nuclei

Table 1. Earthquakes with the highest number of fatalities since 1900. (Selected data extracted with permission from Munich Reinsurance NatCatSERVICE database and USGS Earthquake Hazards Program.)

Date	Event	M_W	Total Losses [million $]	Fatalities	Injured Displaced
28 Dec. 1908	Messina Earthquake & Tsunami, Italy	7.2	116	86,000	
16 Dec. 1920	Haiyuan Earthquake, China	7.8-8.5	25	200,000-273,400	
1 Sep. 1923	Great Kanto Earthquake, Japan	7.9	2,800	142,800	
25 Dec. 1932	Changma Earthquake, China	7.6		77,000	
5 Oct. 1948	Ashgabat Earthquake, Turkmenistan	7.3		110,000	
31 May 1970	Ancash Earthquake & Huayco,* Peru	7.9	550	67,000	150,000
28 Jul. 1976	Tangshan Earthquake, China	7.5-7.8	5,600	242,769	799,000
20 Jun. 1990	Western Iran Earthquake	7.4		40,000-50,000	60,000 400,000
26 Dec. 2004	Sumatra–Andaman Earthquake & Tsunami, South & Southeast Asia	9.0-9.1	10,000	227,898	Unknown 1,700,000
8 Oct. 2005	Kashmir Earthquake, Pakistan, India, & Afghanistan	7.6	5,200	86,000-88,000	69,000
12 May 2008	Wenchuan Earthquake, Sichuan, China	7.9-8.0	85,000-87,587	84,000	374,177
12 Jan. 2010	Port-au-Prince Earthquake, Haiti	7.0	8,000	222,570-316,000	300,000 1,300,000

*Huayco is an Incan word used to describe the landslides that are a chronic threat to residents of steep Andean valleys.

The deadliest recorded earthquake was in Shensi, China, on 23 January, 1556, with an estimated $M_W \approx 8$ and about 830,000 deaths. Other deadly earthquakes were in 1138 in Aleppo, Syria, with 230,000 fatalities and in 856 in Damghan, Iran, with 200,000 people killed. The Haiyuan area is a rural area with a population density of about 50 people per square kilometer; but it had one of the highest death tolls from the earthquake of 1920. The total population of Haiyauna was about 360,000. Nearly all the houses collapsed in the small cities of Longde and Huining. Weiqi et al. (1987) estimate that the recurrence interval for a similar event would be about 800 to 1600 years; not too different from the impact probability of a 100 m asteroid.

The worst modern European earthquake destroyed about 90% of buildings in Messina and Reggio di Calabria. An 8 m high tsunami followed the earthquake, causing many deaths. In terms of geologic timescales, tsunamis are frequent catastrophes in the Mediterranean Sea (Maramai et al. 2007). In the Port-au-Prince earthquake in 2010, which is not only the most recent but also the weakest (M_W = 7.0) of the earthquakes listed in Table 1, 97,294 houses were destroyed and 188,383 damaged in the Port-au-Prince area and in much of southern Haiti.

The advent of disaster training and earthquake-resistant buildings in modern times has reduced thc number of fatalities from most of the recent earthquakes, but damage to buildings, infrastructure, and loss of economic output have made many recent earthquakes very expensive disasters. As shown in Table 2, even earthquakes of modest magnitudes (M_W = 6) have the potential to cause billions of dollars in losses in economically developed areas. Only one earthquake (the Sumatra–Andaman earthquake and Indian Ocean tsunami) appears on both lists. This earthquake and associated tsunami affected virtually the entire Indian Ocean basin. We note that a 100 m asteroid could cause large-scale loss of life and billions in damage if it hit the water near a populated area.

Table 2. The ten earthquakes with the highest insured losses. (Selected data extracted from Munich Reinsurance NatCatSERVICE database, with permission)

Date	Event	M_W	Country	Total Losses (million U.S.$)	Insured Losses (million U.S.$)	Fatalities
11 Mar. 2011	Tohoku Earthquake & Tsunami	9.0	Japan	210,000	35,000 – 40,000	15,840
17 Jan. 1994	Northridge Earthquake	6.8	USA	44,000	15,300	61
22 Feb. 2011	Christchurch Earthquake	6.1	New Zealand	16,000	13,000	181
27 Feb. 2010	Chile Earthquake & Tsunami	8.8	Chile	30,000	8,000	520
4 Sep. 2010	Canterbury Earthquake	7.0	New Zealand	6,500	5,000	
17 Jan. 1995	Hanshin Earthquake	6.9	Japan	100,000	3,000	6,430
26 Dec. 2004	Sumatra–Andaman Earthquake & Indian Ocean Tsunami	9.0	South & Southeas t Asia	10,000	1,000	220,000
17 Oct. 1989	Loma Prieta Earthquake	6.2	USA	10,000	960	68
13 Jun. 2011	Christchurch Earthquake	5.9	New Zealand	2,000	800	1
23 Oct. 2004	Chuetsu Earthquake	6.6	Japan	28,000	760	46

5. PREPAREDNESS AND EFFECTIVE APPROACHES TO RISK REDUCTION AND AVOIDANCE

Evacuation of the population of a potentially affected area has been suggested as a possible solution for an impending asteroid impact. While an orderly evacuation may save lives from the direct consequences of an impact, and is certainly a recommendation to be implemented, it does not address potential fatalities and injuries caused by a hectic exodus. Evacuation routes and the essential necessities, e.g., transportation requirements and fuel supplies, for an evacuation must also be planned well ahead of the impact time. This can be a particularly complex problem on islands, peninsulas, or mountainous regions. But most of all, evacuation leaves the infrastructure of modern societies completely exposed. As Table 2 shows, the total losses can be in the hundreds of billions of dollars and even some insured losses are in the tens of billions of dollars. In addition to the losses incurred from rebuilding infrastructures and replacing material properties, there is the loss of economic output if the impact affects one or more major cities. This loss could be even greater than the property losses.

When considering asteroid and comet nucleus collisions with Earth, impact *mitigation* is frequently discussed. This implies alleviation or lessening of effects. While this is admirable, here we focus on *countermeasures*. Clearly, we do not mean to just *lessen* the effects of a collision of a PHO with Earth, we want to *avoid* such a collision if at all possible!

5.1. Consequences of the Chelyabinsk Event

Shortly after and prompted by the Chelyabinsk event of February 15, 2013, the House Space Science Committee (HSSC) of the USA held a hearing on March 19, 2013 (see http://www.youtube.com/watch?v=XoMiJtQpIvg), in which the Director of the Office of Science and Technology Policy of the President, John P. Holden, the Commander of the U. S. Air Force Space Command, Gen. William L. Shelton, and the Administrator of NASA, Charles F. Bolden, Jr., testified. The testimony was mostly about finding, characterizing, and cataloging PHAs and PHCs, and plans for building telescopes to expand the ability to detect PHOs as small as 140 m. It must be remarked that this size is not a definitive number of a mean diameter. It is arrived at from the brightness of these objects when illuminated by the Sun. The brightness depends primarily on the size of the object and its albedo. Since the albedo is not known for each object, an average value is used. Thus, the actual size of an object depends on the deviations of its albedo from the average value.

The primary purpose of finding, characterizing, and cataloging PHOs is to initiate defensive measures and to launch countermeasures. After the primary presentation of the testimony, HSSC members were given the opportunity to ask the three testifiers questions. Congressman Bill Posey, from Florida, asked one of the most important and relevant questions. He said: "…what would we do if you detected, even a small one [PHA]—like the one that detonated in Russia [Chelyabinsk] headed for New York City in three weeks? What would we do?" The three testifiers pointed fingers at each other in an attempt to decide who should answer. Finally, Mr. Bolden said: "I would pray." Clearly, no plan is in place.

As already mentioned by Huebner et al. (2009, 2010) and Huebner (2011a,b), the preferred way to avoid Earth impacts is to deflect a PHO. Here we use "deflect" in a broad sense that includes all actions on a PHO's orbital motion that lead to avoiding a collision with Earth. One such action is to reduce the orbital speed of a PHO so that it arrives at the Earth orbit after the Earth has past the predicted collision point in its orbit. If the PHO is small, and "small" still needs to be defined more accurately, a kinetic hypervelocity impactor is the preferred technique (Walker and Chocron, 2011). For medium and large PHOs and for PHOs with short warning and response times a nuclear standoff explosion that ablates the surface and by its reactive force imparts momentum that changes the orbit is the preferred method.

Sophisticated radiation and neutron transport codes can be used to model effects from nuclear standoff explosions (Plesko et al. 2011). While radiation from such an explosion may be deposited in a PHO to a depth up to about 1 cm, neutrons can penetrate much deeper. Both, radiation and neutrons will heat the material and cause it to vaporize thereby imparting momentum to the object during the ablation process.

5.2. The Deep Impact Mission

The Deep Impact Mission was a NASA space probe launched on January 12, 2005, on a Delta II rocket to a comet nucleus. Its cruising speed with respect to the Earth was about 2.86 km/s. The mission was designed to study the interior composition of comet 9P/Tempel by releasing a 370 kg impactor unto its nucleus. About 173 days (less than 6 months) after launch, on July 4, 2005, the impactor successfully collided with the comet's nucleus. The impact excavated material from the interior of the nucleus, allowing analysis and images of the impact crater to be recorded.

The mission consisted of two parts, a flyby spacecraft and an impactor. Instruments on the Deep Impact spacecraft observed and captured images on approach, when the impactor hit, and during the flyby. The Deep Impact Spacecraft performed an evasive maneuver to fly past Comet 9P/Tempel at a safe distance of 500 km shortly after the impact.

The 370 kg impactor was 1 m in diameter and 0.8 m tall. It used an autonomous navigation computer, cameras, and a propulsion system to guide itself toward a suitable impact point. To reduce the chance of the impactor's materials being mistaken for cometary components, mission scientists built the impactor using materials unlikely to occur naturally in a comet; for example, about half its mass was copper. The impactor was completely vaporized during the impact.

The impactor struck the nucleus with a speed of about 10.3 km/s within 30 seconds of the predicted impact time. Most of the closing speed was delivered by the orbital motion of the comet nucleus. The impact energy was about 2×10^{10} J, equivalent to the explosive energy of almost 4.7 tons of TNT. The direct momentum change, Δp, caused by the impact is about 3.8×10^6 kg m s^{-1}. The mass of comet 9P/Tempel is very uncertain. Its size is about $14 \times 4 \times 4$ km. Assuming a density of 500 kg/m^3, its estimated mass might be about 6×10^{13} kg. Thus, with the impact in a direction opposite to the comet's motion, the comet's velocity change (without the momentum enhancement would be about $\Delta v = \Delta p/M = -\ 6.3 \times 10^{-8}$ m s^{-1}. If the entire spacecraft including the impactor with a total mass just over 1000 kg had collided with the comet nucleus, the velocity change would be about $\Delta v = - 1.7 \times 10^{-7}$ m s^{-1}, still much too small to effectively divert a nucleus of this size even if ten impacts of this type were executed.

5.3. Collision avoidance through momentum transfer to a PHO

A 100 m diameter comet nucleus has a mass of only about 2.6×10^8 kg and the Deep Impact experiment would have caused a velocity change of about 1.5×10^{-2} m s^{-1} from the impactor alone and nearly 4×10^{-2} m s^{-1} from a collision of the combined spacecraft with impactor. The actual value may be somewhat higher because of the reaction from the momentum of material excavated and ejected from the crater produced by the impact. If we assume a momentum enhancement factor of 1.5, then the velocity change may be as large as 6×10^{-2} m s^{-1}.

The orbital velocity of the PHO at time t is

$$v(t) = \left[\frac{2M_{\odot}G}{r(t)}\left(1 - \frac{r(t)}{2a}\right)\right]^{1/2} \tag{13}$$

where $M_{\odot} = 1.989 \times 10^{30}$ kg is the mass of the Sun, $G = 6.674 \times 10^{-11}$ m^3 kg^{-1} s^{-2} is the universal gravitational constant, and

$$r(t) = \frac{q(1+e)}{1 + e\cos\theta}, \tag{14}$$

is the heliocentric distance of the PHO. In Eq. (14), which is the equation for a conic section, e is the eccentricity, θ is the angle between polar axis in the direction of perihelion of the PHO and the radius vector of the position of the PHO, as defined in Fig. 3, and $q = a(1 - e)$ is the perihelion distance. Combining these equations and expressing distance in astronomical units, Eq. (13) becomes

$$v(t) = 4.2 \times 10^4 \left(\frac{1}{r(t)} - \frac{1}{2a}\right)^{1/2}, \tag{13a}$$

where 1 AU = 1.496×10^{11} m and the velocity $v(t)$ is in m/s.

In Fig. 3, we show a comet whose nucleus is a PHO that missed the Earth.

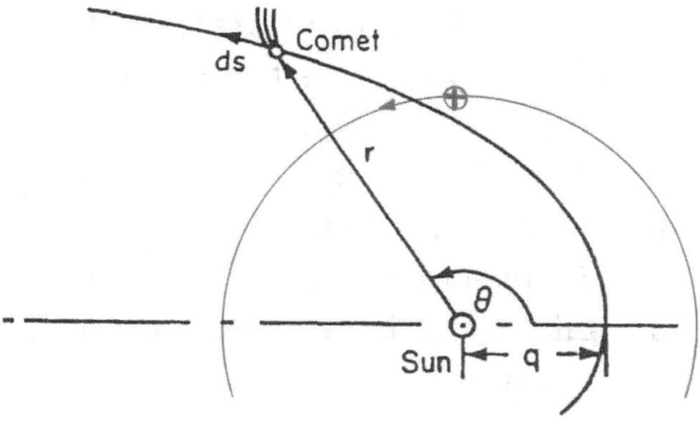

Fig. 3: Orbital elements: r heliocentric distance of comet, q perihelion distance of comet, θ angle between polar axis in direction of perihelion and radius vector of comet position. Earth and orbit of Earth at 1 AU around the Sun are shown in red.

As a simple example, we will assume that the orbital motion of a PHO is approximately at a right angle to the orbital motion of the Earth before a glancing impact (see Fig. 4a). Since the Earth's orbital speed is about 30 km/s, and the Earth's diameter is about 12750 km, delaying a PHO by about 430 s will cause it to miss the Earth. In order to delay a PHO's arrival time at Earth orbit by more than 430 s the encounter of the impact-imparting spacecraft must be at a distance s from Earth at a time t_i such that

$$\int_{t_i}^{t_1} \frac{ds}{v(t)} - \int_{t_i}^{t_1} \frac{ds}{v'(t)} = 430\ s, \tag{15}$$

where $v(t)$ and $v'(t)$ are, respectively, the PHO's velocities relative to Earth before and after the impact induced momentum transfer by the spacecraft collision. The time of momentum transfer is t_i and the time of the predicted impact with Earth (before the momentum transfer occurred) is t_1.

Assuming both $v(t) = v$ and $v'(t) = v - \Delta v$ to be constant in time, Eq. (15) can be approximated by $s/v - s/(v - \Delta v) = -(\Delta v/v)\, s/v = 430$ s, or $s = 430 \times v^2/\Delta v = 430 \times Mv^2/\Delta p$. This corresponds to a time prior to the expected Earth impact of $t_1 = s/v = 430 \times Mv/\Delta p = 430 \times v/\Delta v = 1.1 \times 10^8$ s, or about 3.5 years without a momentum enhancement factor. A momentum enhancement of 1.5 would reduce this time to about 2.3 years. Momentum enhancement can reduce the time needed to launch countermeasures and they should be investigated urgently.

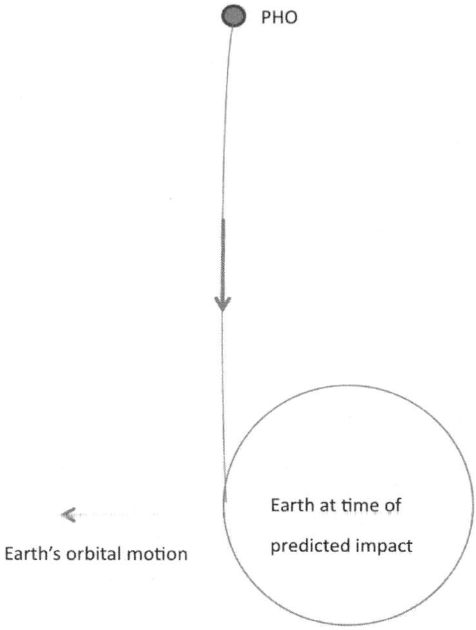

Fig. 4a: A simple example for collision avoidance when the orbit of the PHO is nearly perpendicular to the Earth's orbit.

If a PHO in an approximately perpendicular orbit to Earth's orbit is about to collide with Earth on the leading side of Earth's orbital motion, i.e. in the early morning hours, then it can be made to miss the Earth by delaying the PHO by at least 430 s (see Fig. 4b). In Fig. 5 we show the needed velocity change of a PHO to miss the Earth under these conditions. The assumed velocity of the PHO before momentum transfer is v = 10 km s^{-1}. The curve shown is without a momentum enhancement factor.

If the expected collision is not on the leading side of Earth's orbital motion, then an appropriately shorter time interval than 430 s and a corresponding smaller velocity reduction of the PHO will suffice. For non-perpendicular orbits the time delays of the PHO will be larger.

Instead of launching a specifically designed impactor from a spacecraft, the entire space craft may be used as a hypervelocity impactor. This would significantly increase the effective mass and momentum transfer. It may also be desirable to have several kinetic impactor spacecraft missions to impact the threatening PHO and ensure the intended deflection.

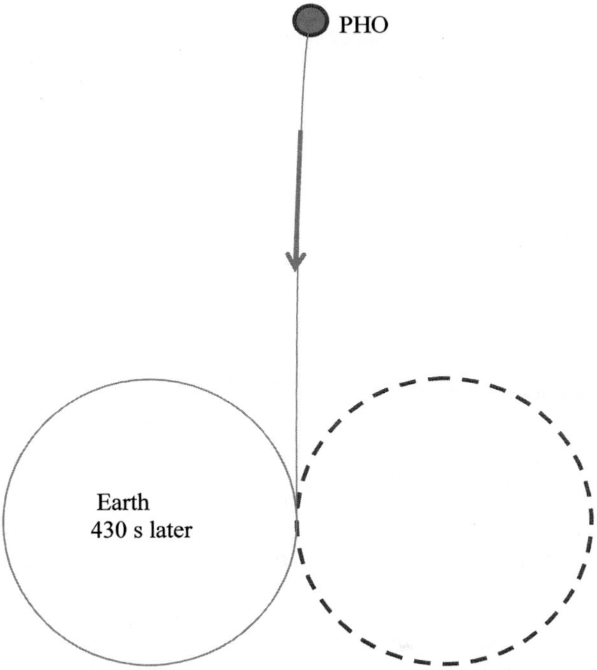

Fig. 4b: Since the Earth's orbital speed is about 30 km/s, and the Earth's diameter is about 12750 km, delaying a PHO by about 430 s (about 7 minutes) will cause it to miss the Earth.

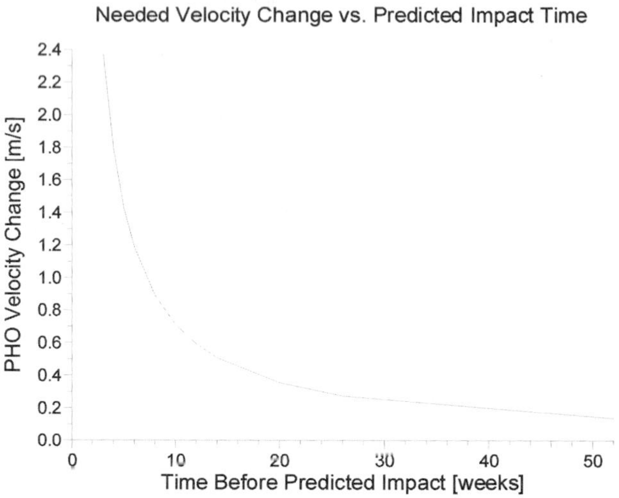

Fig. 5: It takes Earth about 430 s to move one Earth diameter in its orbit. The needed velocity change to delay a PHO's arrival by 430 s at Earth orbit is $\Delta v = -430 \times v/t_l$, where $v = 10\ km\ s^{-1}$ is he assumed velocity of the PHO relative to Earth and t_l is the time prior to the expected Earth impact when momentum Δp is transferred to the PHO. The corresponding momentum change of the PHO is $\Delta p = M\,\Delta v$.

6. MISSION COST

We base cost estimates to divert a PHO from impacting Earth on the Deep Impact mission. The total cost of NASA's Deep Impact mission was about $330 million. Using nearly identical multiple spacecraft will save on development costs and further savings will result from limiting the number of observer spacecraft to one or two. The proposed cost is trivially small when compared to the number of fatalities and losses than can be expected as listed in Tables 1 and 2.

7. CONCLUSIONS

The energy deposited by a 100 m mean diameter asteroid, a "city killer", is equivalent to an earthquake of up to magnitude $M_W = 8$. It has an impact energy equivalent to more than 700 Hiroshima bombs. It would be unconscionable, if not criminal, if we did not do everything possible to avoid PHO impacts. This Seminar, which is dedicated to the avoidance of nuclear war, must support these efforts. A 100 m asteroid is smaller than the planned lower detection limit of 140 m and is equivalent to less than 40% in mass of a

140 m object. The ability to detect and catalog about 90% of 140 m PHOs is still several decades from now.

By analogy with earthquakes, we have shown that asteroid impacts are not only a danger in high density population centers, but can cause large fatalities in sparsely populated areas because of ground shaking, in coastal areas because of tsunamis, and in mountainous regions because of huaycos. We note that the Haiyuan Earthquake in 1920 was in a sparsely populated area of rural central China, where the population density is only about 50 people per square kilometer. Still, about 270,000 people died. In addition to fatalities, PHO impacts can be the cause for large insured as well as uninsured economic losses and destruction of infrastructure ranging into hundreds of billion dollars, not to mention historic and cultural treasures.

Based on the cost of NASA's Deep Impact mission to the nucleus of comet 9P/Tempel of $330 million, developing countermeasures to avert a collision of an asteroid or comet nucleus with Earth is less than thc tangible losses, material and economic losses, and historic and cultural losses, not to mention the losses of lives that one can expect from an Earth impact. This is true even if multiple missions to counter a threatening PHO are required. *Readiness is the most important aspect for countermeasures to be successful.* Once, even a small threatening object is discovered it will be too late to prepare for countermeasures and small PHOs are the most abundant and therefore most likely to strike the Earth. We encourage investigation of momentum transfer enhancement factors.

Of the many natural disasters such as hurricanes, tornados, typhoons, earthquakes, and volcanic activity, all unpredictable, there are no countermeasures. Impacts by asteroids and comet nuclei have occurred ever since the formation of the planetary system four and a half billion years ago and they will continue for billions of years. The time and place of their occurrence is also unpredictable, but the fact that they will occur is absolutely certain, but we have developed the capabilities for countermeasures to avoid such occurrences. We have not yet developed the commitment to implement the development of countermeasures. We challenge this organization to take a leading role, politically as well as scientifically and technically, in implementing, guiding, and developing countermeasures.

There are many disastrous challenges humanity faces including communicable diseases, with disastrous consequences. These grow and advance over a period of time and thus can be controlled with a little preparation. Countermeasures against PHO impacts require time for preparation to be ready when the time comes to implement them. This time is now! We must do whatever needs to be done.

REFERENCES

1. Borovicka, J., Spurny P., Brown, P.G. et al. (2013) "The trajectory, structure and origin of the Chelyabinsk asteroidal impactor." *Nature,* **503**, 235-237.
2. Boslough, M.B.E., Crawford, D.A. (2008) "Low-altitude airbursts and the impact threat." *Int. J. Impact Eng.* **35**, 1441-1448.
3. Brown, P.G., Assink, J.D., Astiz, L. et al. (2013) "A 500–kiloton airburst of Chelyabinsk and an enhanced hazard from small impactors." *Nature,* **503**, 238-241.

4. Chapman, C.R. (2004) "The hazard of near-Earth Asteroid impacts on Earth." *Earth and Planetary Science Letters*, **222**, 1-15.
5. Chyba, C.F., Thomas, P.J., Zahnle, K.J. (1993) "The 1908 Tunguska explosion: Atmospheric disruption of a stony asteroid." *Nature* **361**, 40-44.
6. Collins, G.S., Melosh, H.J., Marcus, R.A. (2005) "Earth impact effects program: A web-based computer program for calculating the regional environmental consequences of a meteoroid impact on Earth." *Meteorit. Planet. Sci.,* **40**, 817-840.
7. Devi, A., Kalita, S. (2013) "Applicability of surface magnitude wave magnitude scale, M_S, and body wave magnitude scale, m_b, in the study of seismicity of Northeast India and its adjoining region." *IOSR Jour. Appl. Geol. Geophys.* **1**, 53-63.
8. DOE/NV—209-REV 15 (2000) United States Nuclear Tests July 1945 through September 1992, prepared by the United States Department of Energy, Nevada Operations Office, Las Vegas.
9. Durda, D.D. (2013) "The Chelyabinsk Super-Meteor." *Sky and Telescope* **125**, 24-31.
10. Gisler, G.R. (2011) "Calculation of the impact of a small asteroid on a continental shelf." *Proc. 2011 Planet. Defense Conf.,* Bucharest.
11. Gisler, G.R., Weaver, R.P., and Gittings, M. (2011) *Pure Appl. Geophys.* **168**, 1187-1198.
12. Glasstone, S., Dolan, P. J. (1977) "The Effects of Nuclear Weapons," U.S. Government Printing Office, Washington, D.C. Also: http://handle.dtic.mil/100.2/ADA087568.
13. Gutenberg, B., Richter, C.F. (1942) "Earthquake magnitude, intensity, energy, and acceleration." *Bull. Seism. Soc. Am.* **32**, 163-192.
14. Gutenberg, B., Richter, C.F. (1956a) "Earthquake magnitude, intensity, energy, and acceleration." *Bull. Seism. Soc. Am.* **46**, 105-145.
15. Gutenberg, B., Richter, C.F. (1956b) "Magnitude and energy of earthquakes." *Ann. Geofis*. **9**, 1-15.
16. Harris, A. W. (2007) "An Update of the Population of NEAs and Impact Risk." *Bull. Am. Astron. Soc*. **39**, 511.
17. Hills, J.G. (2003) Coastal Inundations by Tsunami from Large Asteroid and Comet Impacts, LA-UR-03-6279, Available from National Technical Information Service, Springfield, VA.
18. Hills, J.G. Goda, M.P. (1992) Airblast Damage from Small Asteroids, in Proceedings from the ECA Intercept Conference, LA UR 92-2049, Available from National Technical Information Service, Springfield, VA.
19. Hills, J.G. Goda, M.P. (2001) Tsunami from Asteroid and Comet Impacts, LA-UR-01-4764, Available from National Technical Information Service, Springfield, VA.
20. Hills, J.G., Nemchinov, I.V., Popov, S.P., Teterev, A.V. (1994) "Tsunami Generated by Small Asteroid Impacts." in *Hazards Due to Comets and Asteroids*, T. Gehrels, (ed.), Univ. Arizona Press, Tucson, AZ, pp. 721-778.

21. Huebner, W.F. (2011a) "Countermeasures to avoid asteroid and comet nucleus impacts on Earth: A summary of progress." in *International Seminars on Nuclear War and Planetary Emergencies.* R. Ragaini (ed.) pp. 381-396.
22. Huebner, W.F. (2011b) "PMP report for cosmic objects." in *International Seminars on Nuclear War and Planetary Emergencies.* R. Ragaini (ed.) pp. 583-588.
23. Huebner, W.F., Johnson, L.N., Boice, D.C., Bradley, P., Chocron, S., Ghosh, A., Giguere, P.T., Goldstein, R., Guzik, J.A., Keady, J.J., Mukherjee, J., Patrick, W., Plesko, C., Walker, J.D., Wohletz, K. (2009) "A comprehensive program for countermeasures against Potentially Hazardous Objects (PHOs)." *Solar Syst. Res.* **43**, 348-356.
24. Huebner, W.F., Boice, D.C., Bradley, P., Chocron, S., Clement, R., Ghosh, A., Giguere, P.T., Goldstein, R., Guzik, J.A., Johnson, L.N., Keady, J.J., Mukherjee, J., Patrick, W., Plesko, C., Tapley, M., Walker, J.D., Weaver, R P., Wohletz, K. (2010) "The engagement space for countermeasures against Potentially Hazardous Objects." in *Protecting the Earth Against Collisions with Asteroids and Comet Nuclei.* A Finkelstein, W. Huebner, V. Shor (eds.).
25. Kanamori, H. (1972) "Mechanism of Tsunami Earthquakes." *Phys. Earth Planet. Interiors*, **6**, 346-359.
26. Kanamori, H. (1977) "The energy release in great earthquakes." *J. Geophys. Res.* **82**, 2981-2987.
27. Kharif, C., Pelinovsky, E. (2005) "Asteroid Impact Tsunamis." *C. R. Physyque*. **6**, 361-366.
28. Mac Low, M.-M., Zahnle, K. (1994) "Explosion of comet Shoemaker Levy 9 on entry into the Jovian atmosphere." *Astrophys. J.* **434**, L33-L36.
29. Maramai, A., Graziuani, L., Tinti, S. (2007) "Investigation on tsunami effects in the central Adriatic Sea during the last century." *Nat. Hazards Earth Syst. Sci.* **7**, 15-19.
30. Melosh, H.J., Collins, G.S. (2005) "Planetary science: Meteor Crater formed by low-velocity impact." *Nature* **434**, 157.
31. Munich Reinsurance Geo Risks Research, http://www.munichre.com/de/group/default.aspx.
32. Nuttli, O.W. (1986) "Yield estimates of Nevada test site explosions obtained from seismic *Lg* waves." *J. Geophys. Res*. **91**, B2137-2151.
33. Plesko, C., S., Weaver, R.P., Bradley, P.A. Huebner, W.F. (2011) "Looking before we leap: An ongoing, quantitative investigation of asteroid and comet impact hazard mitigation." in *International Seminars on Nuclear War and Planetary Emergencies.* R. Ragaini (ed.) pp. 413-428.
34. Richards, P.G., Zhongliang, W. (2011) "Seismic monitoring of nuclear explosions," in *Encyclopedia of Solid Earth Physics*, 2nd Edition (Ed.: H.K. Gupta) Springer Dordrecht, Netherlands, p. 1144-1156.
35. Rodean, H.C. (1971) Nuclear-Explosion Seismology, (Livermore, DA: Univ. of California, U.S. AEC, Div. of Technical Information).
36. Rodean, H.C. (1972) "An energy approach to seismic coupling analysis." *J. Geophys. Res*. **77**, 3129-3145.

37. Ryall, A., Savage, W.U. (1969) "A comparison of seismological effects for the Nevada Underground Test Boxcar with natural earthquakes in the Nevada Region." *J. Geophys. Res.* **74**, 4281-4289.
38. Sedan Event (2014) http://www.nv.doe.gov/library/publications/newsviews/sedan.aspx.
39. Schultz, P.H., Gault, D.E. (1975) "Seismic effects from major basin formation on the Moon and Mercury." *The Moon,* **12**, 159-177.
40. USGS Earthquakes Hazard Program, http://earthquake.usgs.gov/earthquakes/world/most_destructive.php
41. Vasilyev, N. V. (1998) "The Tunguska meteorite problem today." *Planet. Space Sci.*, **46**, 129-150.
42. Von Seggern, D.H., Lambert, D.G. (1972) Analysis of Teleseismic Data for the Nuclear Explosion Milrow, Seismic data laboratory report #258, Available from National Technical Information Service, Springfield, VA.
43. Walker, J.D., Chocron, S. (2011) "Momentum enhancement from kinetic impactors and conventional Explosives." in *International Seminars on Nuclear War and Planetary Emergencies.* R. Ragaini (ed.) pp. 397-412.
44. Ward, S.N., Asphaug, E. (2000) "Asteroid Impact Tsunami. A Probabalistic Hazard Assessment." *Icarus* **145**, 64-78.
45. Weiqi, Z., Decheng, J., Peizhen, Z., Molnar, P., Burchfield, B.C., Qidong, D., Yipeng, W., Fangmin, S. (1987) "Displacement along the Haiyuan fault associated with the great 1920 Haiyuan, China, earthquake." *Bull. Seismol. Soc. Am.* **77**, 117-131.

ASTEROID IMPACT RISK

PAUL A. BRADLEY
Los Alamos National Laboratory
Los Alamos, NM, USA

WALTER F. HUEBNER
Southwest Research Institute
San Antonio, TX, USA

ABSTRACT

Of the major natural disasters, asteroid impacts stand out as being that one that can extinguish civilization in a single stroke and is possible to prevent with current technology. Most impact consequence discussions dealt with the rare asteroids of order 1 km and larger. However, smaller asteroids with diameters of 100 m or smaller are much more common and the larger asteroids can devastate a several thousand km^2 area. The size distribution of asteroids and their interaction with Earth's atmosphere imply that 99% of all asteroid impacts will results in airbursts, most of which will have little effect on the ground. As the Tunguska and Chelyabinsk impacts show, an airburst over a major city would be devastating. We find that a 100 m size asteroid (often referred to as a "city killer") can destroy much of a major city by the combination of impact cratering, blast wave, and earthquakes. If the impact is in the water, then devastating tsunamis can result. We note that while the latest planned NASA survey will discover many objects 100 m or smaller when completed several decades from now, the survey completion limit is set to the larger size of 140 m mean diameter.

1. INTRODUCTION

Asteroids strike the Earth on a constant basis. Most of these impacts are by small objects a few meters across and they break up harmlessly in the atmosphere. When an asteroid exceeds about 100 meters in diameter, it is possible for the asteroid to survive the transit through the atmosphere and reach the surface. Objects a few tens of meters across can explode in the atmosphere with enough energy to cause destructive airbursts over populated areas. The most recent was the Chelyabinsk event on 15 February 2013, and the 400 to 500 kt blast shattered windows over a wide area. The shattered glass injured about 1000 people and caused the partial collapse of a factory. The object was estimated to be about 20 m in mean diameter according to Brown et al. (2013). The Tunguska event of 30 June 1908 was also an airburst, and it flattened about 2000 km^2 of Siberian forest. Fortunately, the area was virtually uninhabited, so there were few fatalities. One complication is that both these objects hit without warning. Indeed, almost all of the asteroids that have hit Earth in recent times were not detected before they hit. Fortunately, a new NASA survey is planned to find at least 95% of all potentially hazardous objects (PHOs) 140 meters across and larger. This survey will find a large number of objects smaller than 140 m, but many will remain undetected and hence a threat. However, this survey may not be completed for another 40 years.

A PHO impact can cause damage in several ways. The first and most likely is the airblast from an object exploding in the atmosphere. If the PHO is large enough, it can reach the Earth's surface more or less intact and cause damage via the cratering event and earthquakes. However, 70% of the Earth's surface is covered by water, so a water impact is possible, which can also create a devastating tsunami. Finally impacts in a mountainous region can cause landslides. In order to better understand and put the disastrous consequences of PHO impacts on a firmer basis, we describe the consequences of hypothetical events of each type over populated areas.

The rest of the paper is organized as follows. We start in Section 2 by describing the risk of a PHO impact, which recent studies suggest is greater than previously appreciated. Section 3 describes the consequences of several hypothetical impact scenarios, including airbursts, water impacts, and ground impacts. Finally, we close with a summary and conclusions in Section 4.

2. THE RISK OF A PHO IMPACT AS A FUNCTION OF SIZE

Various surveys have been used to estimate the likelihood that a PHO of a given size will impact Earth. This determines the impact probability. We can determine the likelihood of damage to a place by determining the amount of damage that an object of a given size will cause if it hits Earth.

We use a time interval of 100 years for our impact likelihood estimates. Brown et al. (2013) and the B612 foundation results (2014) suggest that a 2 to 6 meter object will hit every eight months on average, a 6 to 10 meter object will hit every 2 to 4 years, a 20 meter object will hit about every 30 to 50 years, a 50 meter object would hit about every 200 years, and a 100 meter object would hit every 300 to 500 years. Thus, in a given century, there is less than a 1% chance of an object bigger than 40 to 50 meters in diameter will hit Earth. We will show in the next section that almost all asteroids smaller than 50 meters will break up during entry into the atmosphere and create an airburst. The bottom line is that over 99% of the asteroid impacts within a given century will be airbursts. Although fragments of asteroids up to several meters in size can hit the ground, most of the damage from the airburst will come from the overpressure and wind of the blast. Of the 26 objects hitting the atmosphere from infrasound data between 2000 and 2013, (B612 foundation 2014), only the Chelyabinsk impactor caused significant damage. Several of the other objects (Sudan, California, and Argentina) did have meteorites reach the ground, but they caused minimal damage. Seventeen of these 26 objects exploded over the ocean and no trace of the impactor could be found. The 17 out of 26 impactors (i.e., 65% fraction of impactors) exploding over the ocean closely matches the 70% of Earth's surface that is covered by water.

We have just shown that the odds of an object large enough to reach the ground sufficiently intact to create a crater or significant impact wave is less than 1% per century. Then one must ask how likely is it that a major city would be hit by an asteroid? The surface area of the Earth (including oceans) is 5.1×10^{8} km^2, while a major city and suburbs might have a radius of 25 km, for an area of 2000 km^2. Thus, a given major city covers only about 4×10^{-6} of the Earth's surface area. There are at least several hundred major cities, so we can multiply the above number by approximately 1000, so that the major cities cover at most 0.004 of the Earth's surface area. If an asteroid big enough to reach the ground is likely to hit once every 500 years, then the odds of a major city being hit within a given century is less than 8×10^{-6} chance in a given year. Although rare (magnitude 9 earthquakes happen more often), the damage to a major city can be catastrophic and more importantly, it can in principle be prevented if there is adequate warning and a deflection system in place. We note that technological development and cost are not limiting factors. We will touch on this topic again at the end.

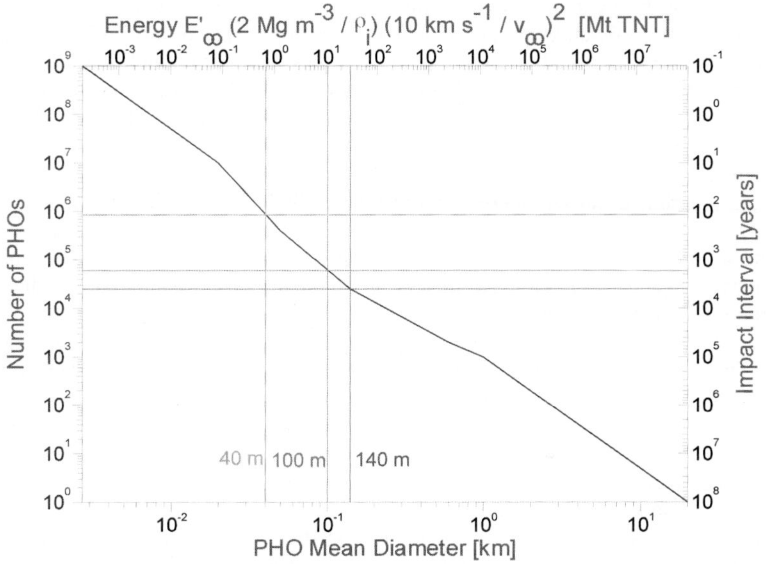

Fig. 1: Number of PHOs larger than a given size and their impact intervals vs. PHO size. The PHO size is also related to the energy at the time of atmospheric entry $E_\infty = E'_\infty\, \rho_i/2\, (v_\infty/10)^2$, with E_∞ in Mt of TNT, ρ_i in Mg/m^3 and v_∞ in km/s. The impact energy on the ground can then be obtained from $E_i = (v_0/v_\infty)^2 E_\infty$, where v_0 is a function of v_∞ and a decaying exponential that depends on the PHO trajectory through the atmosphere. PHO's of three different mean diameters are highlighted: 40 m (Tungaska object) is shown in green; 100 m (our nominal object) is shown in red; and 140 m (minimum size of planned detectability) is shown in blue. The deviation of the curve from a straight line is thought to be caused by the transition from monolithic (small asteroids) to rubble pile (larger asteroids) structure of asteroids (from Harris, 2007).

3. DAMAGE ESTIMATES FROM AN IMPACT

In this section, we describe the various damage mechanisms from the impact of a small asteroid or comet nucleus with Earth. Depending on their structure and composition, the smallest objects will break up during their travel through the atmosphere. The Chelyabinsk and Tunguska objects are well-known examples of this. In this case, most of the damage will come from the overpressure of the blast wave generated by the object, as described in section 3.1. If the object survives its travel through the atmosphere, then it will impact and cause damage from creating a crater, a blast wave, a possible fireball, and an earthquake (see section 3.2). Because the Earth's surface area is about 70% water, there is a significant chance that a small asteroid can impact the ocean. In this case, an impact wave can be generated, depending on the size of the impactor and whether the water depth is significantly greater than the impactor's size or not. We discuss this problem in section 3.3. Finally, the energy of the blast is great enough that the units are kilotons of TNT (1 kt = 4.18×10^{12} J) or megatons of TNT (1 Mt = 4.18×10^{15} J). For reference, the first nuclear test, Trinity (fielded 16 July 1945) had a yield of 21 kt.

3.1. Airburst damage

As mentioned earlier, there is more than a 99% chance that an asteroid hitting Earth will disintegrate in an airburst. In addition to the frequency of an asteroid of a given size hitting Earth, Figure 1 also shows the impact energy of the asteroid when it hits the atmosphere. Objects less than 10 meters in diameter will disintegrate high in the atmosphere with blast energies less than 100 kt and have little or no effect on the ground. An example of such an object is 2008 TC3, which was discovered about 19 hours before impact. The asteroid was estimated to be about 4 meters in diameter and the airburst had about a 1 to 2 kt yield. About 10 kg of meteorites reached the ground, causing no damage. A 10 meter object would explode with an energy of 20 to 40 kt, but would be high enough to not affect structures on the ground. The Chelyabinsk impactor had an estimated diameter of 20 meters. As was well reported in the news (Durda, 2013, Brown et al. 2013), this object disintegrated in the atmosphere with a blast energy of 400 to 500 kt of TNT, and caused about 1000 injuries from shattered glass. The Tunguska impactor was a "once-a-century" impact and the 3 to 5 Mt blast leveled about 2000 km^2 of forest. The isolation of Siberia meant there were few casualties. Both the Chelyabinsk and Tunguska airbursts produced earthquakes (magnitude 4.2 and 5, respectively). Kanamori (1977) relates the earthquake magnitude to seismic wave energy, and the two earthquakes had 0.03 kt (Chelyabinsk) and 0.5 kt (Tunguska) of seismic energy, or 0.01% of the airburst energy. This value is consistent with the commonly accepted fraction from experiments (Schultz and Gault, 1975). Both asteroids produced, at most, small fragments reaching the ground, so that a negligible amount of energy went into the kinetic energy of impactors. Thus, almost all of the energy (>99%) went into the blast, which is exactly what Collins et al. (2005) predict.

However, if the Tunguska impactor had detonated over a major city, such as Mumbai, India the result would be quite different. Mumbai has a population of 12.5 million and an average population density of 20,500 people km^{-2}. If we assume the asteroid is a 50 meter stone (a little bigger than Tunguska) with a density of 2000 kg m^{-3} impacting at 20 km s^{-1} and a 45° angle, the object would break up and detonate about 11 km above ground with an energy of 6 Mt. At a distance of 5 km from the epicenter, the blast would create an overpressure of ~15 kPa (2 psi) and a blast wind of up to 32 m s^{-1}. This kind of blast would damage wooden structures and shatter glass. Over 1.5 million people would likely be affected within this range. Even at a 10 km distance, the overpressure would be about 12 kPa (2 psi), with a wind of 27 m s^{-1}, resulting in heavy damage to wooden and poorly constructed buildings. This damage radius would encompass up to half of the population of Mumbai. At 15 km from the epicenter, the blast overpressure is 8 kPa (1.5 psi) and the 24 m s^{-1} wind would shatter many glass windows. In addition to the blast damage, the shockwave would also couple to the ground and cause an "airslap" earthquake of magnitude 5. In short, a city like Mumbai would suffer great destruction (and many dead) within the central parts, and even the suburbs would be a mess of broken glass and injured people. We note that the blast pattern would not be circular, but would likely have a "butterfly" shape (Boslough, 2013) with the "wings" extending perpendicular to the impactor's trajectory.

3.2. Impact crater damage

To reach the surface more or less intact, the object must either be an iron asteroid over 20 meters in diameter or a stony asteroid over 20 meters in diameter coming in at a nearly vertical trajectory in the atmosphere. If a stony asteroid arrives at a 45° angle and 20 km s^{-1}, then it needs to be over 80 meters in diameter to be intact enough at the ground to create a crater. Meteor Crater in Arizona is a typical impact crater from a relatively small asteroid. Recent estimates suggest that an approximately 25 meter diameter iron meteor impacted with a force equivalent to about 10 Mt of TNT (Melosh and Collins, 2005). The resultant crater is 1.2 km in diameter and 170 meters deep. In the case of an impact, about half of the energy goes into excavating a crater and hurling out ejected material, while the other half of the energy goes into the blast. About 0.1% of the energy goes into generating an earthquake, 1% or less of the energy goes into melting material in the crater, and a negligible fraction goes into the fireball (if it occurs). The results are similar for a water impact, although some impact energy will be used to convert water into steam.

We use the impact effect estimator of Collins et al. (2005) to estimate the effect of a 100 m diameter stony asteroid hitting the city of Houston, TX (city population 2.2 million; metro area population 6.3 million). We chose typical parameters of 2000 kg m^{-3} for the rock density, an initial impact velocity of 20 km s^{-1}, and an impact angle of 45°. A stony object 100 m in diameter is predicted to break up high in the atmosphere, but the fragments stay together enough to impact the ground at about 2.6 km s^{-1} with a blast energy of 50 Mt. The fragment-generated crater would be about 0.5 km in diameter, and would eliminate the downtown area of Houston. The damage from the blast wave and flying debris would extend for many kilometers. Within 5 km of the epicenter, the destruction would be nearly total. The overpressure is 370 kPa and the supersonic impact generated winds would be about 430 m s^{-1}. Almost all buildings would collapse and many bridges would collapse. Cars, trucks, and trees would be blown over. Between 5 and 10 km, buildings and other structures would be severely damaged by the peak overpressure (90 kPa) and wind of 160 m s^{-1} (350 mph). Many buildings would collapse, almost all houses would be destroyed, and most of the trees would be blown down. This damage zone would extend to Loop 610, a circular highway about 10 km from city center. At 25 km (the location of an outer circular highway called the Sam Houston Tollway), many windows would be shattered by the peak overpressure (16 kPa) and wind of 36 m s^{-1}. Finally, there would be an impact-generated earthquake with a magnitude between 5.3 and 6.5. In short, the main part of a city the size of Houston would be destroyed, and even the outlying areas would suffer significant damage.

Although rare, a 100 m iron asteroid would cause far greater damage because of the greater mass and the fact that the object would stay intact until impact. We assume an initial density of 6000 kg m^{-3}, an impact velocity of 20 km s^{-1}, and an impact angle of 45° on Houston, TX. The object will hit at 18 km s^{-1}, and create a crater 2.7 km diameter by 600 m deep. The blast energy would be 100 Mt, which is double the size of the largest nuclear explosion. In this case, the blast is big enough to create a fireball that would cause fires in clothing, trees, and grass out to 15 km, and it would be 40 times brighter than the Sun. At this distance, the airblast is 50 kPa (7.1 psi) with a wind of 100 m s^{-1} (220 mph). Many buildings and structures would be destroyed. Any buildings left standing would have all the windows blown out and all the trees would be blown over. Even at 50 km, the fireball would be 5 times brighter than the Sun. The overpressure of 8 kPa would be sufficient to shatter glass windows. Finally, there would be a magnitude 6 earthquake. In places with earthquake-resistant buildings, this would case minimal damage. Ordinary buildings and houses would suffer significant damage out to at least 20 km. Glass windows would be shattered out to at least 50 km away. An iron asteroid hitting the center of Houston would all but remove the city from the map.

3.3 Impact-generated tsunami

Because 70% of the Earth's surface is water and many large cities are near the coast, the greatest danger from an asteroid impact is from an impact-generated wave. Many people use formulae from Glasstone and Dolan (1977), Hills et al. (1994), and Collins, Melosh, and Marcus (2005) to compute the "tsunami" from an asteroid impact. However, these references assume the most of the impact energy goes into creating a wave, while more recent simulations suggest a different picture. Gisler, Weaver, and Gittings (2011) and Gisler (2013) show through simulations two major differences for small objects hitting the water. Gisler et al. (2011) show that a 100 meter diameter object hitting deep water produces a short wavelength wave that dissipates its energy relatively quickly. This implies that an asteroid hitting the ocean far from land will have relatively little impact on the coastlines. The simulations of Gisler et al. (2011) imply that a 100 meter object hitting 1000 km from shore would create no wave, while Eq. 2 of Hills et al. (1994) applies for an impact into deep water (water depth much greater than the impactor size)

$$h = \frac{1}{2} h_W = 4.5\, meters \left(\frac{1000\, km}{r}\right)\left(\frac{Y}{gigaton}\right)^{0.5}$$

where h_W is the peak-to-valley amplitude of the tsunami wave, Y is the explosive yield of the impactor in gigatons, and r is the distance from the impact point in km. A 100 m asteroid would produce 11 Mt of yield and Hills et al. (1994) would predict a wave of 0.5 m, which would not cause much damage.

The big difference comes if an asteroid strikes near shore, and the water depth is less than 1 km deep. If we assume a water depth of 100 meters and the impact is 12 km from shore, then Eq. 4 of Hills et al. (1994) would imply a tsunami of 4 meters peak-to-trough amplitude. Gisler (2013) modeled the impact of a 200 meter diameter body hitting water 100 m deep underlain by 100 m of sediments. Gisler's simulation shows that the sediments will melt from the hat of the impact and that refrozen, glassy sediment particles and water will be carried by hurricane force winds for tens of km away from the impact site. These same hurricane force winds will interact with the water's surface to push forward a "storm surge" onto the shore that would be several meters high. Thus, the impact of a small asteroid near shore would produce a one-two-three punch consisting of: 1) a modest earthquake of approximately magnitude 5 would arrive about 3 seconds after impact, 2) a debris-laden blast wave of at least 40 m s^{-1} and an overpressure of at least 20 kPa that will shatter windows and "sandblast" everything with warm, wet glassy debris about 30 seconds after impact, and finally 3) the wind-generated impact wave would arrive several minutes later and be several meters high.

We use Naples, Italy as a specific example for our damage estimates. The Bay of Naples is about 25 km in diameter and has an approximate depth of 150 meters. Once again, we use a 100 m stony asteroid (density 2000 kg m^{-3}) with an initial impact velocity of 20 km s^{-1}, and an impact angle of 45°. The object is predicted to break up high in the atmosphere, but the fragments stay together enough to impact the water at 8 km s^{-1}. The splash will send water over a wide area and the impactor will create a crater in the bottom of the bay that is 0.8 km in diameter and 0.17 km deep. The earthquake would be magnitude 5 and would arrive about 3 s after impact. This would get the attention of anyone that did not see the impact. The airblast would create an overpressure of 15 kPa (2.3 psi) that will shatter many glass windows about 45 seconds after impact. The 35 m s^{-1} blast wave would be especially devastating, since the impact would melt sediments on the floor of the bay, which would then solidify as they become entrained in the wind. The effect would be to "sandblast" everything that the hurricane force wind touches. Additional damage would come from the wind generated impact wave, which would arrive about 6.5 minutes after impact and be between 4 and 6 m high. This would inundate much of the coastline, especially on the north and east sides of the bay. We note for reference that the Straits of Messina earthquake in 1908 generated a tsunami that was 12 meters high.

4. SUMMARY AND CONCLUSIONS

Although the initial NASA Spaceguard survey has found over 90% of the PHOs greater than 1 km mean diameter and has determined that none of these objects has a chance of hitting Earth for over a century, there can still be regional devastation from the impact of an object smaller than 1 km. The new NASA survey has the goal of finding at least 90% of all PHOs over 140 m in diameter (several decades away at current detection rates), which is still more than 40% larger in diameter (and nearly three times as massive and energetic) than the objects we discuss, which are 100 m across or smaller. We choose 100 m as the upper limit, because these objects are predicted to hit Earth every few hundred years on average, which means there is a modest chance of one impacting Earth this century. The population of asteroids increases dramatically with decreasing size to where objects about 5 m across hit Earth at least once a year, with no harm done on the ground. Larger asteroids hit less often, but 99% of these will still result in airbursts, most of which will have little effect on the ground. However, as the Tunguska and Chelyabinsk impacts show, an airburst over a major city from an asteroid 20 m or more across would be devastating. A 100 m size asteroid (often referred to as a "city killer") can destroy much of a major city by the combination of impact cratering, blast wave, and earthquakes. If the impact is in the water, then devastating impact generated waves and "sandblasting" from windborne impact melted material can result. This latter effect can extend for many kilometers, which is important, since a major fraction of Earth's population lives near the coast.

The real question is what to do about the danger from these impacts? Although rare, the damage caused by an asteroid impact in or near a major city would be devastating to property, cause much loss of life, and disrupt the economy of that region. To prevent loss of life and possibly minimize property damage, early detection of these smaller asteroids is crucial. If there is enough lead-time, a pre-prepared deflection mission could be sent to deflect or disrupt the incoming threat. Huebner and Bradley (2014) discuss the tradeoffs between the cost of such a mission versus the cost of the damage from an impact. International agreements need to be made far ahead of any preparation and fielding of an "on alert" deflection system. We note that the technology to build telescopes to make the survey go faster, launch a rocket on a deflection mission, and have a launch system "on alert" has already been demonstrated. Compared to the global economy, the billions of dollars for such an effort is economically feasible. The challenge is having a global agreement concerning sharing responsibility and cost. Among the issues for international agreements is that the uncertainty in the trajectory of the asteroid creates a long (thousands of km), narrow (tens of km) "risk corridor" that will span many countries. One has to address subjects such as who pays for the system, who is responsible if the deflection effort fails or the deflected object goes off-target, and so on. The long, narrow risk corridor also complicates civil defense measures, such as evacuation. Waiting allows the asteroid trajectory to be refined and shrink the risk corridor, but then there is less time for any evacuation or deflection effort. The real keys to have options are 1) the availability of surveys to detect asteroids roughly 100 meters in diameter with enough warning time to make deflection (and/or evacuation) possible and 2) the availability of a "ready-to-launch" deflection system, backed up with international agreements, to make a deflection option possible. (Publication LA-UR-14-25643)

REFERENCES

1. B612 foundation impact video, at https://b612foundation.org/portfolio/impact-video (4/20/2014).
2. Borovicka, J., Spurny P., and Brown, P.G. et al. (2013) "The trajectory, structure and origin of the Chelyabinsk asteroidal impactor." *Nature*. 503:235-237.
3. Boslough, M.B.E. (2014) "Airburst Warning and Response." *Acta Astronautica*. in press.
4. Brown, P.G., Assink, J.D., and Astiz, L. et al. (2013) "A 500–kiloton airburst of Chelyabinsk and an enhanced hazard from small impactors." *Nature*. 503:238-241.
5. Chyba, C.F., Thomas, P.J., and Zahnle, K.J. (1993) "The 1908 Tunguska explosion: Atmospheric disruption of a stony asteroid." *Nature*. 361:40-44.

6. Collins, G.S., Melosh, H.J., and Marcus, R.A. (2005) "Earth Impact Effects Program: A web-based computer program for calculating the regional environmental consequences of a meteoroid impact on Earth." *Meteoritics and Planetary Science.* 40:817-840.
7. Durda, D.D. (2013) "The Chelyabinsk Super-Meteor." *Sky and Telescope.* 125: 24-31.
8. Gisler, G.R. (2011) "Calculation of the Impact of a Small Asteroid on a Continental Shelf", Proc. 2011 Planetary Defense Conference, Bucharest.
9. Gisler, G.R., Weaver, R.P., and Gittings, M. (2011) *Pure Appl. Geophys.* 168: 1187-1198.
10. Glasstone, S., Dolan, P.J. (1977) "The Effects of Nuclear Weapons," U.S. Government Printing Office, Washington, D.C. Also: http://handle.dtic.mil/100.2/ADA087568.
11. Harris, A.W. (2007) "An Update of the Population of NEAs and Impact Risk." *Bull. Am. Astron. Soc.* 39:511, (2007).
12. Hills, J.G., Nemchinov, I.V., Popov, S.P., and Teterev, A.V. (1994) Tsunami Generated by Small Asteroid Impacts, in *Hazards Due to Comets and Asteroids*, T. Gehrels, (ed.), Univ. Arizona Press, Tucson, AZ, pp. 721-778, (1994).
13. Huebner, W.F., and Bradley, P. (2014) "Asteroid and Comet Nucleus Impact Risk Assessment" in *preparation.*
14. Kanamori, H. (1977) "The energy release in great earthquakes." *J. Geophys. Res.* 82, 2981-2987.
15. Kharif, C., and Pelinovsky, E. (2005) "Asteroid Impact Tsunamis." *C.R. Physyque.* 6:361-366.
16. Maramai, A., Graziuani, L., and Tinti, S. (2007) "Investigation on tsunami effects in the central Adriatic Sea during the last century." *Nat. Hazards Earth Syst. Sci.* 7:15-19.
17. Melosh, H.J., and Collins, G.S. (2005) "Meteor Crater formed by low-velocity impact." *Nature.* 434:57.
18. Schultz, P.H., and Gault, D.E. (1975) "Seismic effects from major basin formation on the Moon and Mercury." *The Moon.* 12:159-177.
19. Vasilyev, N.V. (1998) "The Tunguska meteorite problem today." *Planet. Space Sci.* 46:129-150.

PLANETARY DEFENCE: FROM TALKING TO ACTION

ANATOLY ZAITSEV
Planetary Defense Center
Moscow, Russia

MIKHAIL A. KOZLOV AND VALERY I. LARIONOV
Extreme Situations Research Center LTD
Moscow, Russia

NIKOLAI A. MAKHUTOV
Working Group "Risk and Safety" under the President of the RAS
Moscow, Russia

VLADIMIR A. PUCHKOV AND ALEKSANDER A. TARANOV
EMERCOM of Russia
Moscow, Russia

VIKTOR A. SHOR
Institute of Applied Astronomy of the RAS
Moscow, Russia

ABSTRACT

Results of development of the International Planetary Defence System (PDS) named "Citadel" components are presented. It is shown, that modern rocket-space and nuclear technologies allow creating a fast reaction defence against dangerous celestial bodies (DCBs) up to several hundred meters in size. If the prevention of a DCB collision with the Earth is impossible, it is necessary to take a set of measures for minimisation of the space disaster damage. As an example, we present results of consequences simulation of two hypothetical asteroid falls—on the territory of the Moscow region and Sicily, obtained by using a special software system.

INTRODUCTION

One of the most serious natural hazards for the mankind is the threat of disastrous collisions of asteroids and comet nuclei with the Earth.[1] This was proven again by the Chelyabinsk meteorite fall. It showed that the world lacks not only means of defence from dangerous celestial bodies, but also effective detection means, which guarantee at least early detection of such objects and timely warning about the approaching threat. It becomes clear that with the present high density of the population and abundance of potentially dangerous technogenic objects (nuclear, chemical, etc.) even a small celestial body fall can lead to serious loss of life, and to material and ecological damage.

Since such events can occur any time, it would be unjustifiable carelessness not to take protective measures against this threat. It is obvious that the time has come for practical implementation of the numerous research and development on this issue that started more than a quarter of a century ago in the USSR[2] and continues now in Russia and in other countries. The research showed that all necessary technologies for creation of a PDS against asteroid and comet hazard already exist in the world. However, the creation requires consolidation of efforts of the entire international community. Documents and resolutions of many national and international organisations also call for urgent measures.

This article is a summary of results of our work on the problem of defence against DCBs. They are focused on two main areas. The first area covers the development of means and methods of active defence for deflection or destruction of DCBs. The second is the development of means and methods of passive defence for reducing damage in case the implementation of active defence is impossible.

DEVELOPMENT OF MEANS AND METHODS OF ACTIVE DEFENCE FROM DCBs

A possible variant of building a PDS against DCBs is proposed in the Concept of creating the International PDS "Citadel" developed by the Planetary Defence Centre (PDC) in cooperation with organisations of Russia, Ukraine and Kazakhstan.[3] In accordance with this concept, the system "Citadel" should include two echelons (levels):

- Echelon for short-term reaction (ESTR) "Citadel-1"
- Echelon for long-term reaction (ELTR) "Citadel-2".

In addition, composition of the ESTR may include auxiliary (backup) services:

- Services of regional defence on the basis of anti-missile weapons;
- Services for forecasting areas and consequences of celestial body falls.

The need for creation of a fast reaction echelon is caused by the lack of ability of early detection of all dangerous celestial bodies. Most of these bodies are small, that makes them difficult to detect. For example, 99.9% of the near-Earth asteroids (NEAs) are from tens to hundreds meters in size. The detection of such objects is most likely on approaching the Earth trajectories and only few days-months before the collision. Therefore, an ESTR of constant readiness is needed for defence against these space threats.

As shown in Zaitsev et al. (2010),[4] this echelon can be built within 5-7 years by using modern technologies. This will defend our planet from asteroid and partly comet hazard. If necessary, an ELTR can be formed on the basis of this echelon quickly enough to counter large celestial bodies discovered many years before the collision with the Earth. This will require mobilising potential of all states, which possess rocket-space and nuclear means. Organisation of defence from a hypothetical asteroid 2015 PDC can illustrate how this can happen. This asteroid was suggested to participants of the International Conference PDC 2015 for developing measures of defence against it. It was assumed that this asteroid was discovered on 13 April 2015, and its collision with the Earth may take place on 3 September 2022.

The first task to be addressed after discovery of such a potentially hazardous asteroid is to specify its trajectory and physical characteristics. The most effective way of solving this problem is to launch a reconnaissance spacecraft (SC) to the asteroid. Reconnaissance SCs of the small class are designed for studying DCBs. Latest achievements of the world space technologies are planned to be used for its development. This will allow to create a SC with minimum dimensions and mass characteristics (mass of about 100-200 kg).[4]

On approaching a DCB, the reconnaissance SC will use detection equipment and provide input data required for building an engineering model of the asteroid. Based on the model, a scheme of the interception and the action on the DCB will be developed.

If the ESTR was established before the moment of the 2015 PDC detection, the reconnaissance SC could be launched without waiting for the optimum launch date. For example, estimates indicate that if launch vehicles such as Proton, Ariane 5 and a number of others are used, the reconnaissance SC can be launched 1 July 2015 and will approach the asteroid 30 December 2016. Thus, if the collision with the Earth is confirmed we will have significant time reserve to make best decisions for defence of the Earth against the 2015 PDC.

Different technologies are studied for action on DCBs. Nuclear and kinetic means are most efficient for short-term interception.[5-7] They allow for destruction of asteroids of about 500 m in diameter and to deflect larger celestial bodies by using modern launch vehicles.

The optimum time for launching an intercepting SC is at the end of 2018. Characteristic energy C3 of these SCs will be 15-20 km^2/s^2. This gives the possibility to launch a spacecraft weighing about 4000 kg by Proton launch vehicle, Atlas V and a number of others.[8] Mass of the nuclear device will be about 1,000 kg,[4] and its capacity—1-2 Mt.[5]

Depending on the type of nuclear explosion (standoff, surface or subsurface), the momentum transferred to the asteroid may be from 10^9 to 10^{13} kg m/s for 1 Mt.[5] If the 2015 PDC asteroid diameter is 500 m and its density—2.7 kg/m^3, its mass equals to 1.8 10^{11} kg. In this case, the velocity impulse acquired by asteroid surface ranges from 5 mm/s to 0.5 m/s, (standoff explosion), 0.5 m/s (surface explosion), from 5 to 50 m/s (subsurface explosion).

The type of explosion will be selected not only to provide the needed velocity impulse of the asteroid, but also to avoid its destruction. For this purpose, several consecutive standoff explosions can be used providing a “soft” impulse on the asteroid.

Figure 1 shows a possible version of the intercepting flight to the 2015 PDC asteroid. Probable launching date can be 5 October 2018, and the arrival date—26 April 2020.

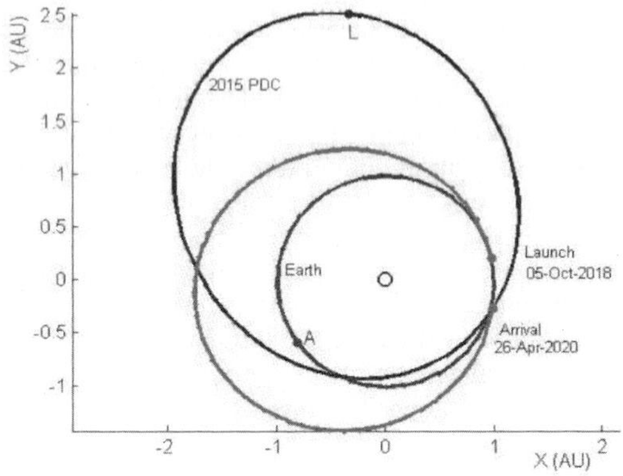

Fig. 1: Trajectory of the intercepting SC flight.

At approaching the DCB, the intercepting module separates from the spacecraft and moves to the DCB. The orbital module continues its motion and observes the interception process.

The asteroid flight outside the capture radius of the Earth must be ensured by selecting required velocity impulse to it. In case of 2015 PDC, the capture radius will be about 9000 km. Therefore, taking into account some margin it is necessary to ensure deviation of about 15,000-20,000 km from the centre of the Earth. Calculations show that if the impact to the asteroid is rendered at the end of April of 2020, the velocity change must be about 50 mm/s. Thus, near-surface nuclear explosion will be enough for this purpose.

It is important to realise, that any change of the orbit of a dangerous body in order to eliminate an immediate threat, can lead to resonant returns (or non-resonant approaches) or even a collision in the future. This potential danger has not been investigated in the present work.

Estimates show possibility of the defence from such as 2015 PDC asteroids with the lead time of 5-7 years.

Results of the research allow making a conclusion, that Russian and foreign technologies approve starting practical development of the International PDS against the asteroid-comet hazard. The PDS "Citadel" project can become the first one in this area.

DEVELOPMENT OF METHODS AND MEANS OF PASSIVE DEFENCE FROM DCBs

Passive methods and means of defence from DCBs will be auxiliary (reserve) components of the PDS. They are needed to support main PDS means in case of:

- Threat of a DCB's big (tens meter size) fragments fall after the DCB was destructed near the Earth;

- Short time reserve before a DCB collision with the Earth that prevents using main PDS means for the interception.

Since the PDS has not yet been created, the minimisation of damage from possible collisions becomes increasingly important. For mitigating the damage, it is necessary to determine in advance, location, time and explosive power of the collision, and affecting factors and their consequences for the population and territories. For this purpose, a Forecasting Service of areas and consequences of DCBs falls should be created.[9,10]

Under orders of the Ministry of Emergency Situations (EMERCOM) of Russia, the PDC together with A.A. Blagonravov Mechanical Engineering Research Institute, Institute of Applied Astronomy RAS, Institute of Geosphere Dynamics of RAS, Russian Federal Nuclear Center (RFNC–VNIITF) and Extreme Situations Research Center LTD has developed principles of this service establishment. Its most important element is the special software system. It combines various software that simulates catastrophic scenarios associated with the asteroid and comet falls. It has to ensure:

- Determination of conditions of a celestial body entry in the Earth atmosphere and its characteristics;
- Calculation of its motion in the atmosphere and the place of the hitting (explosion);
- Assessment of explosion power in the atmosphere, on the surface and in the water;
- Assessment of expected consequences for the surface infrastructure and biosphere.

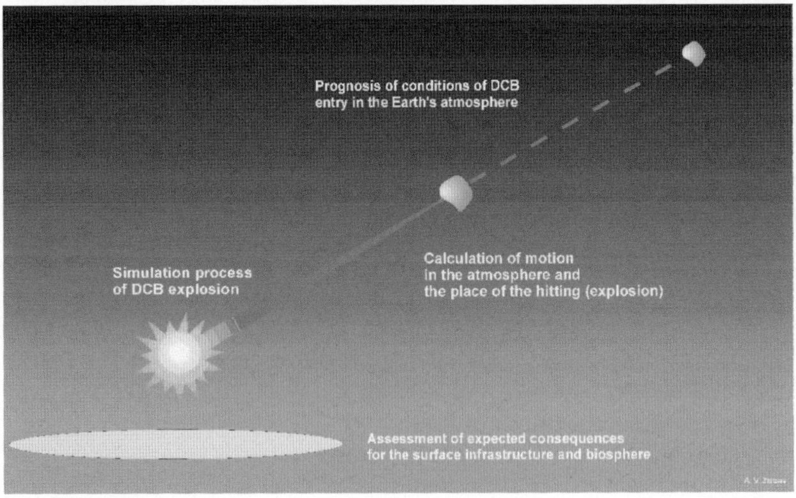

Fig. 2: Stages of forecast of celestial bodies' falls.

The asteroids and comets threat belongs to natural factors that can be discovered in advance. This makes possible for the EMERCOM of Russia to undertake a number of measures on minimising the potential losses (i.e., evacuation of people, material and cultural assets, and hazardous materials and facilities). It can be done if place and time of explosion are estimated with sufficient accuracy.

Realisation of these measures is more efficient through broad international cooperation. It is reasonable to have two regional segments of international Defence Service—Eurasian and American, each having its Forecast Centres. The Service should include Space-Based Observation Segment, optical and radar ground instruments, regional forecast centres and communication channels between all components of the Service. The suggested Service might become an important component of the future PDS.

Results of quantitative assessments of vulnerability and possible damage to the ground infrastructure and the Earth's biosphere from affecting factors of a hypothetical asteroids falling are presented below. They were obtained through a test simulation on the special software system. As an example, we present consequences of two hypothetical asteroids falls on the territory of the Moscow region and Sicily. Parameters of the hypothetical asteroid, falling to the Moscow region's territory:

- Diameter—250 m
- Density—3,000 kg/m3
- Velocity—14.9 m/s
- Angle of incidence—45°.

The final part of the DCB movement in the atmosphere can be completed by an explosion above the Earth surface (like the Tunguska explosion), above the land or the water area. The explosion will be accompanied by a wide range of damage factors—a shock wave, thermal surface radiation, seismic effects, and change of relief, tsunami and others. The special software system covers two main effects—the shock wave and seismicity.

An example of calculation of the asteroid impact on the Earth surface in the territory of Moscow region is presented below. The calculation results showed that the radius of the seismic field intensity from 10 to 6 points will make about 200 km (Figure 3). Thus, the loss of people from the air shock wave will be 3-4 orders less than the loss of people from the seismic effect of the impact.

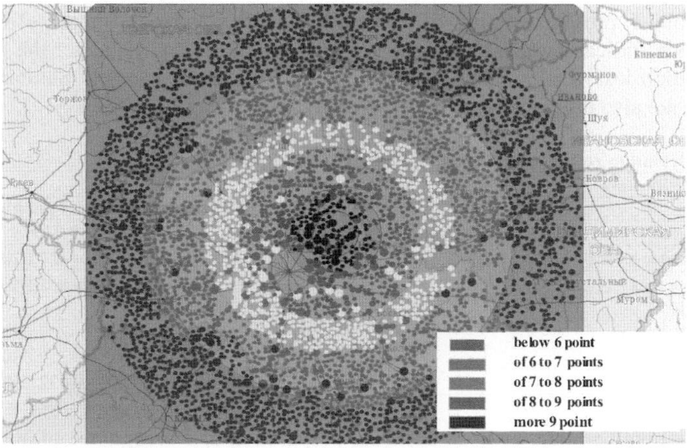

Fig. 3: Seismic field after the asteroid fall on the territory of Moscow region.

The assessment of possible consequences—size of affected territory, scales of destructions and human losses was carried out by simulation of the most important damaging factors of the DCBs explosion. For this purpose, an interface of impact models and pattern damage (destruction/damage) function was carried out. The use of this methodological approach allows estimating effects of DCB's explosion to the population and territories. Results of the possible consequences evaluation for the examined case of asteroid fall are shown in Figure 4. It presents data on the degree of settlements destruction at different distances from the asteroid impact point.

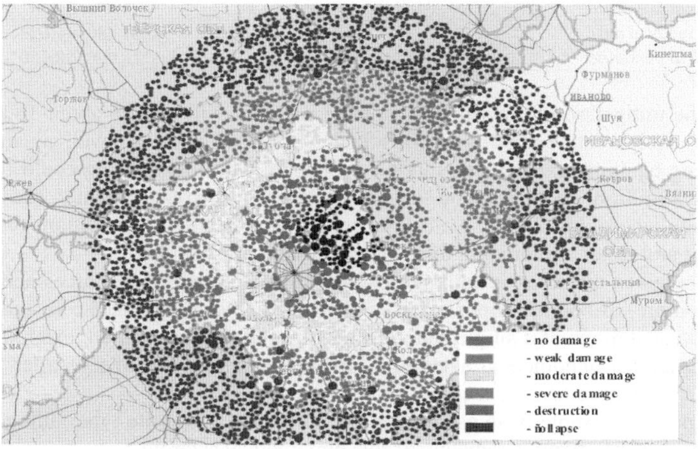

Fig. 4: Consequences of the asteroid fall on the territory of Moscow region.

Table 1 presents results of calculation of the asteroid impact effects, which include main indicators characterising scope of casualties and damage: the structure of the population lesion including irrevocable, sanitary and total losses. Sanitary losses of the people are also divided according to severity: light, heavy, very heavy.

Table 1. Calculation results of the effects of the asteroid impact.

Medical situation	
Name	**Number of people**
Total losses	1,289,010
Irrecoverable losses	408,070–427,070
Sanitary losses *including:* extremely heavy heavy light	825,090–861,940 195,250–207,650 209,630–222,030 420,210–432,260

The obtained data provide basis for preparation and acceptance of administrative decisions on planning and carrying out emergency measures and rescue operations and mitigating consequences of emergencies caused by DCBs falls. It allows determining, in particular, required numbers of lifeguards, medical staff, emergency personnel and technical teams, and also engineering equipment and life support indicators. For the examined case, more than 500 thousand people need to be involved in rescue works (Table. 2).

Table 2. The required resources for carrying out rescue operations.

Name of units	**Number of people**
Mechanised rescue groups	121,940–128,510
Manual blockages dismantling groups	406,470–428,380
First medical aid *by medical personnel*	26,020–26,980 8,670–8,990
Special medical aid *by medical personnel*	67,480–71,620 26,990–28,650

The mitigation of consequences of the accident will require about 16,000 pieces of heavy equipment, 14,000 fire units and 11,000 vehicles, as well as many other engineering equipment. Since the number of people who lost housing could reach more than over 3 million, about 400,000 tents in summer or 800,000 in winter, about 800,000 heating appliances and 25,000 field kitchens will be needed. In addition, the need for basic food and water will amount to several thousand tons per day.

The given quantitative estimates of the disaster scale and resources needed for its elimination are, of course, imposing. It is obvious that such large-scale rescue operation is beyond capabilities of a single state. This is despite the fact that we examined the fall of not the largest hypothetical asteroid. However, even the fall of a smaller object can lead to very serious consequences.

Another example is an asteroid of the Tunguska meteorite type falling on Sicily. We assume the following parameters of this hypothetical asteroid:

- Diameter—50 m.

- Density—2,600 kg/m3
- Velocity—18 km/s
- Angle of incidence—45°.

Unlike the previous version, this asteroid does not reach the Earth surface and explodes at about 9 miles altitude. Possible consequences evaluation for this asteroid fall is presented in Figure 5. The blue circle marks the area of possible irretrievable losses, and the yellow—area of possible total losses.

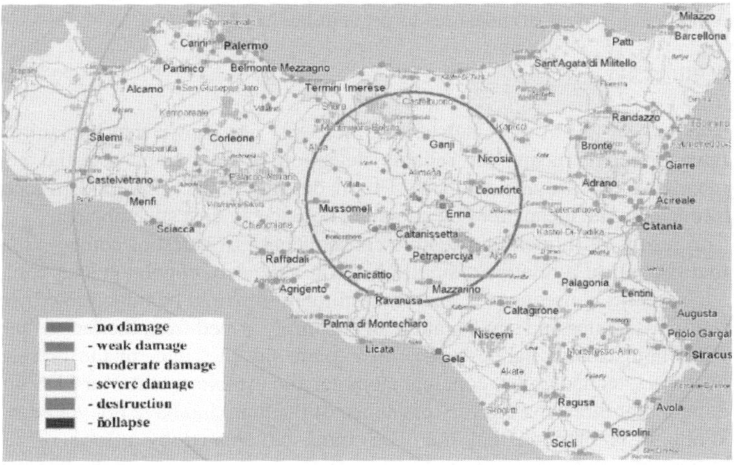

Fig. 5: Consequences of asteroid fall on the territory of Sicily.

Table 3 presents the asteroid impact consequences evaluation, which includes main indicators, characterising the scope of casualties and damage. Number of injured in this accident may exceed 30 thousand people.

Table 3. Results of the asteroid impact consequences evaluation.

Medical situation	
Name	**Number of people**
Total losses	33,302
Irrecoverable losses	180
Sanitary losses	33,122

More than 20 thousand people will be needed for rescue works (Table. 4).

Table. 4. The required resources for carrying out rescue operations.

Name of units	**Number of people**
Mechanised rescue groups	5,000
Manual blockages dismantling groups	16,650
First medical aid	520
Special medical aid	5,520

The mitigation of the accident consequences will require about 1000 units of heavy engineering and automotive equipment and many other equipment—compressor stations, mobile power stations, etc. The number of people who lost housing could exceed 30 thousand. To accommodate them, about 1,000 tents in summer and about 2,000 in winter, as well as approximately 2,000 heaters and 250 field kitchens will be needed. In addition, the need for basic food and water will amount to several tens of tons per day.

Results of modelling possible consequences of DCBs falls and evaluation of required means and forces for carrying out rescue operations and mitigating emergency consequences caused by these falls show the gravity of the asteroid-comet danger and instigate the development of measures to prevent it.

CONCLUSIONS

1. The sudden fall of the Chelyabinsk meteorite has shown that humanity does not have effective means for reliable detection of dangerous celestial bodies and means for defence against them. This is despite the fact that modern technologies allow to create means for detection and defence against such objects.
2. The results of modelling possible consequences of celestial bodies' falls showed that the fall of a relatively small object in a densely populated area could lead to such serious consequences, that their mitigation would be impossible by efforts of individual states and small states can be simply wiped from the face of the Earth.
3. Only unjustifiable carelessness can explain the fact that, in the presence of all necessary technologies, concrete steps for the establishment of a PDS against the danger that threatens the very existence of the mankind have not yet been taken.
4. While it is obvious that the PDS must be created by efforts of the whole humanity, decision of leaders of, at least, the leading states on the implementation of practical steps for the PDS establishment is necessary. If the decision is taken, an echelon for short-term reaction, similar to the "Citadel-1" can be created within 5-7 years. This will provide guaranteed protection of our planet from asteroid and partly from comet impact hazard.
5. Expected financial costs will be negligible compared to possible damage from catastrophic falls of celestial bodies. Estimates show that annual expenses on creation of the system will make about a thousandth part of the mankind expenses on arms that is, in fact—on self-destruction. Having cut down these expenses by a thousandth share, we could get means for defence against one of the most threatening natural hazards.

ACKNOWLEDGEMENTS

We thank Dr. Walter F. Huebner for his support and helpful comments of this work and our colleague Yuri A. Yurishev for assistance in redaction and translation this manuscript.

REFERENCES

1. Medvedev, Yu. D., Sveshnikov, M.L., Sokolsky, A.G., Timoshkova, E.I., Chernetenko, Yu. A., Chernykh, N.S., Shor, V.A. (1996) Asteroid-Comet Hazard. (Ed. A. G. Sokolsky), St.Petersburg, ITA RAS. 1996. 244 pp.
2. Zaitsev, A.V. (1988) "Some Principles of Construction of the System of Prevention of the Earth Collision with Asteroids and Comets." *Proceedings of 23-d readings of K. E. Tsiolkovsky. (Kaluga, Sept. 13-16,1988) Section: Problems of the rocket/space technology. Moscow. IHST of AS of the USSR,* pp. 141 - 147, 1989.
3. Bashilov, A.S., Volk, I.P., Gofin, M.Ja., Zaitsev, A.V., Konyukhov, S.N., Pobedonostsev, K.A., Slyunyaev, N.N. (2010) Possible Approaches to Implementation of "Citadel-1" International Planetary Defense System Project. In "Space for Security and Prosperity on the Peoples. Editors: J.-M. Contant and V. A. Menshikov. M.: A.A. Maksimov Space Systems Research Institute. 2010. pp. 154-163.
4. A.V. Zaitsev (2010) "Defense Against Asteroid-Comet Nucleus Dangers—a Real Task of Today." *International Seminar on Nuclear War and Planetary Emergencies—43rd Session: the Role of Science in the Third Millennium.* "E. Majorana" Centre for Scientific Culture. Erice, Italy, 19-24 August 2010. Series Editor and Chairman: A. Zichichi. Edited by R. Ragaini. World Scientific. pp. 429–441.
5. Koblov, P.I., Simonenko, V.A., Ivanov. Yu.A. et al. (1997) "Main requirements to nuclear explosive device (NED) to be used in system of Earth protection against asteroids and comets." Chelyabinsk Scientific Center News, Special Issue, *Space Protection of the Earth,* 1997, P. 107-110.
6. Grigal. P.V., Zamyshlyaev. B.V., Komarov. A.S., Lyubimov. A.G., Rodionov. S.N., Taranov. A.A., Chistov. V.G. (1996) "On the Use of Kinetic Impact for Destroying an Asteroid." *International Conference "Space Protection of the Earth-96" ("SPE-96"),* 23-27 Sept. 1996, Snezhinsk, Abstracts, pp. 80,81.
7. Huebner, W.F., Boice, D.C., Bradley, P., Chocron, S., Clement, R., Ghosh, A., Giguere, P.T., Goldstein, R., Guzik, J.A., Johnson, L.N., Keady, J.J., Mukherjee, J., Patrick, W., Plesko, C., Tapley, M., Walker, J.D., Weaver, R. P., Wohletz, K. (2010) "The Engagement Space for Countermeasures Against Potentially Hazardous Objects (PHOs)." In "Protecting the Earth against Collisions with Asteroids and Comet Nuclei". Proceedings of the International Conference "Asteroid-Comet Hazard-2009", A. Finkelstein, W. Huebner, V. Shor (Eds). Saint Petersburg, "Nauka", 2010, pp. 337-346.
8. M.A. Wise, J.M. Lafleur, J.H. Saleh *Regression Analysis of Launch Vehicle Payload Capability for Interplanetary Missions,* IAC-10.D2.P06.
9. N.A. Makhutov, V.A. Puchkov, D. Reznikov, A.A, Taranov, A.V. Zaitsev (2009) About Measures on Minimization of Damage from Collisions with Asteroids and Nuclei of Comets. "Protecting the Earth against collisions with asteroids and comet nuclei". *Proceeding of the International Conference Asteroid-Comet Hazard-2009.* A. Finkelstein, W. Huebner, V. Shor (Eds). St. Petersburg, Nauka, Russia. 2010. p.376-380.

10. A.V. Zaitsev, Yu. I. Zetser, O.M. Kochetova, V.I. Larionov, N.A. Makhutov, O.P. Popova, V.A. Puchkov, V.V. Svetsov, V.A. Simonenko, N.A. Skorkin, A.A. Taranov, Yu. A. Chernetenko, V.A. Shor, V.V. Shuvalov (2014) "Forecasting Areas and Consequences of Celestial Bodies' Falls." *Zababakhin Scientific Talks*. Abstracts, P. 26.

THE NEOSHIELD PROJECT FOR NEAR-EARTH OBJECT IMPACT THREAT MITIGATION

LINE DRUBE, ALAN HARRIS AND THE NEOSHIELD CONSORTIUM
Germany Aerospace Center, DLR Institute of Planetary Research
Berlin-Adlershof, Germany

INTRODUCTION

Every day the Earth collects some 100 tonnes of cosmic material, most of it in the form of dust or small rocks, which burn up as meteors in the atmosphere. Sometimes, however, larger objects, asteroids or comets, enter the Earth's atmosphere, which can cause considerable damage. The object that exploded over the Russian city of Chelyabinsk in February 2013 had a diameter of only 17–20 m, yet it produced a blast wave that damaged more than 7000 buildings and some 1600 people sought medical help for injuries (see Figure 1). The Chelyabinsk asteroid entered the Earth's atmosphere with a velocity of 65 000 km/h and, due to the heating and stresses caused by compression of the air, exploded at an altitude of some 25 km releasing energy 30 times that of the Hiroshima bomb. The potentially devastating effects on Earth of a collision with a large asteroid or comet are now well recognized by scientists. So the question is now, can we protect our civilization from the next major impact?

Fig. 1: Collage of images from the Chelyabinsk impact. Top row: a collapsed roof of a building and the broken windows of another are both examples of damage caused by pressure waves from the airburst. Bottom row: the spectacular explosive disintegration of the incoming asteroid. Images from Wikipedia.

An indication that awareness of the impact hazard has reached political circles in Europe was the call for proposals for research projects to address scientific and technical aspects of the impact hazard issued by the European Commission's Seventh Framework Programme in 2010. The NEOShield project, a consortium of 13 partners from 6 countries, won a total of 5.8 million euros funding from January 2012 until May 2015 to investigate mitigation-relevant NEO physical properties and potential mitigation techniques (Harris et al. 2013). The consortium consisted of researchers and engineers from the German Aerospace Centre DLR (the project coordinator), Paris Observatory, French National Centre for Scientific Research, The Open University (United Kingdom), Fraunhofer Ernst-Mach Institute (Germany), Queen's University Belfast (United Kingdom), Astrium (now Airbus Defence and Space) in Germany, UK, and France, Deimos Space (Spain), SETI Institute (USA), Russian Federal Space Agency's TsNIIMash Institute, and The University of Surrey (United Kingdom).

We present a brief overview of the NEOShield project in terms of the four main themes addressed.

MITIGATION DEMONSTRATION MISSION DESIGNS

NEOShield paved the way for future demonstration missions by providing detailed designs of technically and financially realistic space missions to demonstrate the effectiveness of mitigation techniques. The NEOShield technical partners, with the support of the science team, designed three feasible mission concepts that will facilitate the rapid development of actual test missions at a later stage, whenever the requisite funding becomes available. The three deflection techniques considered are:

- The kinetic impactor, in which a spacecraft transfers momentum to an asteroid by impacting it at a very high velocity.
- Blast deflection, in which an explosive, such as a nuclear device, is detonated near, on, or just beneath the surface of the object.
- The gravity tractor, in which a spacecraft hovering under power in close proximity to an asteroid uses the gravitational force between the asteroid and itself to tow the asteroid onto a safe trajectory relative to the Earth ("slow-pull" method).

The above three methods are currently considered the most realistic techniques, depending on the asteroid size and the length of time available for the mitigation attempt (see Figure 2).

The NEOShield work also included the compilation of a prioritized list of accessible and realistic asteroid targets for test missions (Figure 3) and the corresponding mission trajectory calculations.

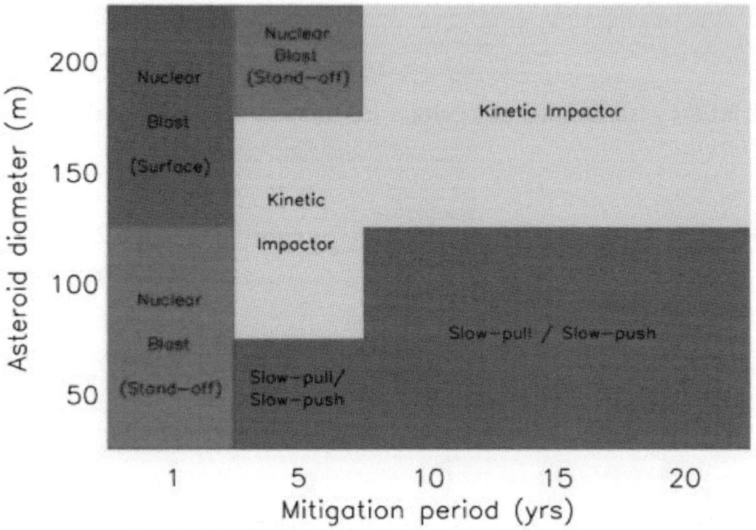

Fig. 2: The results of a NEOShield trade-off study by Airbus Defence and Space (UK), showing the best choice of mitigation methods for different asteroid diameters and the time available for the mitigation attempt. For asteroids smaller than 50 meters, an evacuation of the impact area is judged more appropriate than a deflection attempt. Figure credit: S. Eckersley (NEOShield, Airbus Defence and Space Limited, UK) and D. Perna (NEOShield, Observatoire de Paris, LESIA). (Drube et al. 2014)

NEOShield mission target selection lists

SEVENTH FRAMEWORK PROGRAMME

Rendezvous targets			Fly-by targets		
Selected	All (Delta-v)	All (Name)	Selected	All (Delta-v)	All (Name)

All NEOs with Rendezvous delta-v < 5 km/sec, ordered by delta-v

Colour	MPC Orbital Uncertainty Code
White	0 - 1
Green	2 - 3
Orange	4 - 6

Warnings: CEU is an RMS uncertainty, the error ellipse may be significantly larger along the major axis.
Numbers in brackets () are inferred from other parameters and have not been directly measured.

	Number	Name	H	delta-v (km/s)	U (MPC)	CEU (")	D (km)	a/b	Albedo	Taxonomy	Period (hr)	Tumbling?	Binary?	PHA?
1		2006RH120	29.5	3.94	1	8.7		3.0			0.0458;0.03			
2		2012TF79	27.4	3.94	3	228								
3		2009BD	28.1	3.95	0	1.89	0.004		.45					
4		2010UE51	28.3	4.00	2	135								
5		2007UN12	28.7	4.00	3	294								
6		2008HU4	28.2	4.02	3	480								
7		2010VQ98	28.2	4.09	3	132								
8		2014UV210	26.9	4.09	3	11.1								
9		2015HP116	31.2	4.10	4	0								
10		2014WX202	29.6	4.12	2	12.3								

Fig. 3: A screen shot of the start of the continuously updated list of potential target asteroids for deflection test missions, listed in order of increasing delta-v (a measure of the accessibility of the asteroid). The list, maintained *by Alan Fitzsimmons of Queen's University Belfast, can be found on: http://star.pst.qub.ac.uk/%7Eaf/lowdv_neos/.*

TECHNOLOGY DEVELOPMENT

Further development of some key technologies is a prerequisite to carrying out a deflection test on a small asteroid, such as improvements to spacecraft guidance, navigation and control systems.

In the case of the blast deflection, NEOShield work concentrated on theoretical studies of the effects of a nuclear explosion on a typical small asteroid (Meshcheryakov and Lipnitskii, 2015).

Other NEOShield results include solutions to issues concerning the operation of gravity tractor and kinetic impactor NEO deflection missions (Ummen and Lappas, 2014; Chapuy et al. 2014, respectively), and a discussion of the importance of high precision astrometry of NEOs for asteroid threat assessment and mitigation (Eggl et al. 2013).

MITIGATION-RELEVANT PHYSICAL PROPERTIES OF ASTEROIDS

Project partners also carried out research into the mitigation-relevant physical properties of NEOs, including the analysis of available observational data, laboratory experiments on asteroid analog materials, and computer modelling and impact simulations. The aim of the scientific work was to facilitate predictions of the outcome of deflection attempts using different techniques on a variety of NEO types.

The main objective of the analysis of the observational data was to investigate the most likely properties of future potential impactors that could trigger a space-borne mitigation action. By narrowing the range of expected properties (e.g. albedo, shape, spin rate, mineralogy, internal structure), a rational basis for the choice of objects to serve as targets in deflection test missions can be provided.

In the course of the work a new method was discovered for helping to identify metal-rich asteroids. Observations with infrared telescopes have revealed that the surfaces of some asteroids appear cooler than others at similar distances from the Sun. The reason appears to be related to the thermal conductivity of the asteroid material: Asteroids orbit the Sun absorbing incident sunlight which they re-emit as heat radiation in the infrared part of the spectrum. A rotating asteroid with a high thermal conductivity, e.g. due to a metal-rich surface, will absorb solar energy deeper into its surface and re-emit some of it as thermal-infrared radiation with a time delay on the night side of the object. Due to the conservation of energy, the energy re-emitted on the day side is reduced and the temperature is consequently lower. By comparing the emitted heat radiation with radar reflectance measurements, we have established that the lower temperatures of many asteroids are due to their high metal content (see Figure 4). The relevance of this result to mitigation derives from the fact that metallic asteroids are denser and presumably much more robust than other asteroid types and can therefore penetrate deeper into the atmosphere and cause more damage. The Tunguska airburst, which flattened some 2000 km^2 of forest in Siberia in 1908, is attributed to a stony object with a diameter in the range 30 m – 50 m, whereas a similarly sized metallic asteroid is thought to be responsible for hitting the ground and forming the 1.2 km – diameter Barringer crater in Arizona. In a heavily populated region a crater-forming event would presumably cause much greater devastation than an airburst. Metal-rich asteroids may differ significantly from their metal-poor stony or carbonaceous counterparts in terms of strength, density, and internal structure, characteristics that could seriously affect the outcome of a deflection attempt. Knowledge of the mineralogy of a threatening object would therefore be of vital importance to mitigation planners.

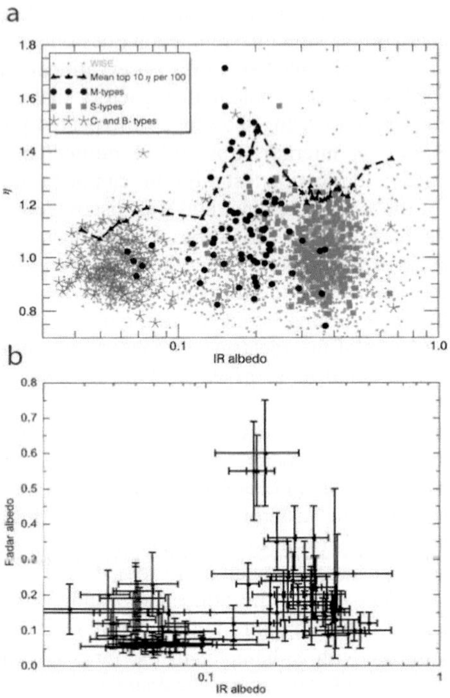

Fig. 4:(a) η values versus infrared albedo (η is a fitting parameter derived by fitting the Near-Earth Asteroid Thermal Model developed by Harris, 1998, which was used in the analysis of WISE thermal-infrared flux data). Basic taxonomic types are shown as bullets, with M-types being asteroids with possible high metallicity. Higher values of η imply lower temperatures on the sunward side of the observed asteroids. The peak in η-values coincides with the location in the plot of M-type asteroids and a peak in the distribution of radar albedos (b). High radar reflectivity is characteristic of surfaces with a high metal content. Therefore the model fitting parameter, η, can be used as an indicator of metal-rich asteroids (Harris and Drube, 2014).

Modelling work and computer simulations (see Figure 5 for an example of the results) accompanied the data analysis, with the aim of improving our understanding of the internal structure of small NEOs and their response to deflection attempts.

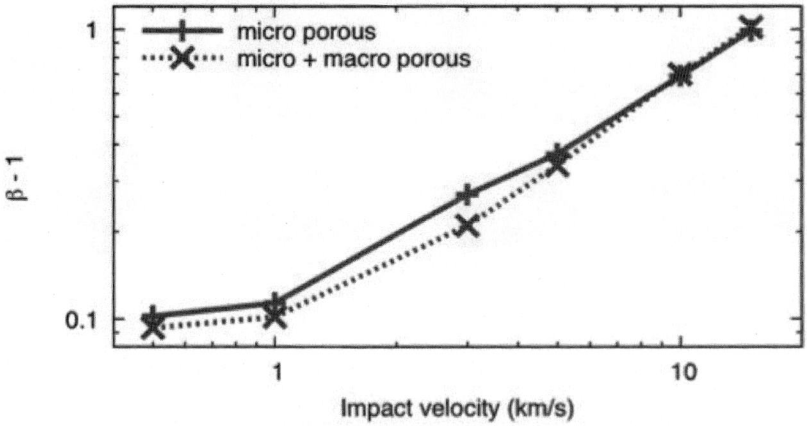

Fig. 5: The momentum multiplication factor, β, versus impact velocity of a projectile. *β is defined as the ratio between the change in the asteroids momentum after the impact and the momentum of the projectile. Material ejected from the asteroid during the impact will enhance the asteroid's momentum. This graph shows that there is a difference in β depending on the porosity of the material. It also shows that at velocities of more than 5 km/s, the effect of porosity becomes negligible. (Jutzi and Michel, 2014)*

In order to validate the models used in the numerical simulations at small scales, impact experiments were carried out in the laboratory using high-velocity light gas guns and targets consisting of asteroid analog materials (Figure 6). Preliminary work carried out within the NEOShield project indicates that the momentum transfer in the case of a porous body is significantly less efficient than in the case of a non-porous body.

Much work remains to be done, however, to enable reliable predictions to be made as to how an asteroid with a particular set of physical parameters would respond to an impacting spacecraft.

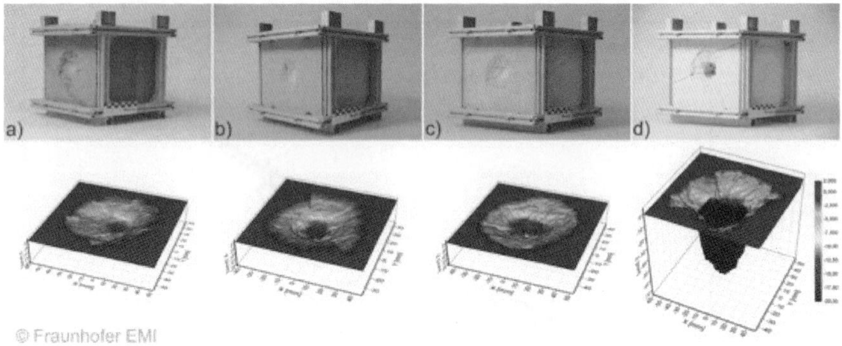

Fig. 6: Impact craters formed by hypervelocity impacts into targets with different porosities using a light-gas gun. Porosity rises towards the right. The results show that large craters are created in material with high porosity, but the ejected mass is small, which would reduce the momentum enhancement of the asteroid due to ejecta in a mitigation attempt. Images credit: Fraunhofer Institute for High Speed Dynamics, Ernst-Mach-Institut, EMI. (Drube et al. 2014)

GLOBAL RESPONSE CAMPAIGN ROADMAP

Software tools have been developed for NEO impact-risk mitigation within the NEOShield project (Cano et al. 2013). The software tools address the issues of impact risk assessment, and the evaluation of the required deflection and the most efficient deflection strategy. These tools serve as an aid in the selection of the most suitable deflection mission(s) given the circumstances of the potential impact scenario. In the event of a serious impact threat, deciding how to handle the situation would be a major challenge for current political structures, with the danger that considerable time would be wasted before an international who-does-what agreement could be reached. The United Nations sanctioned two groups in 2014 as part of this endeavour: the International Asteroid Warning Network (IAWN) and the Space Missions Planning Advisory Group (SMPAG). NEOShield partners take part in the activities of the UN-sanctioned groups with a view to setting up a global roadmap to assist decision makers in responding effectively to a confirmed serious impact threat.

SUMMARY

The NEOShield project was a broad undertaking and, at the time of writing, results are becoming available in many scientific and technical areas addressed by the project. Results to date include:

- Detailed mission designs based on several types of deflection concept.
- A trade-off analysis of mitigation methods for different asteroid sizes and periods of warning time available for the mitigation attempt.
- A continuously updated prioritized list of asteroids suitable for use as targets in mitigation test missions.

- The discovery of a novel means to identify asteroids with high metal content. Metal content, and associated higher density and robustness would increase the likelihood of an incoming object reaching the ground and causing serious damage. Moreover a threatening object with high metal content and density would require greater effort to deflect.
- Analyses of laboratory experiments in which hypervelocity projectiles are fired into blocks of asteroid analogue materials, showing how the momentum transfer depends on the porosity, impact velocity, material strength, and other parameters
- Computer modelling of high-velocity impacts into asteroids, which show that the momentum enhancement effect due to ejecta is reduced with increased porosity, higher tensile strength, and lower strength against crushing. Results also revealed that the shape of the impacting spacecraft can play an important role.

The article started out with the question "Can we protect our civilization from the next major impact?" While the work of NEOShield has improved our chances, there is still much research needed before we can feel confident in our ability to deflect an asteroid. There are many unknown factors involved in a mitigation attempt. Laboratory work, numerical modelling, and computer simulations are essential but not sufficient to enable robust predictions to be made of the outcome of a deflection attempt: It is vitally important to carry out test missions to deflect suitable representative NEO targets before we are forced to apply deflection technology in a hurry in a real impact-prevention scenario.

A successor project to NEOShield, called NEOShield-2, has been granted funding from the European Commission's Horizon 2020 Programme to further develop many aspects of NEOShield research and to increase the readiness level of vital technologies. The work began in March 2015 and will run for 2.5 years. In contrast to NEOShield, NEOShield-2 includes a large work package for observational work on mitigation-relevant NEO targets and associated data reduction and analysis.

ACKNOWLEDGMENTS

The research leading to these results has received funding from the European Union's Seventh Framework Programme (FP7/2007-2013) under grant agreement n° 282703 (NEOShield).

REFERENCES

1. Cano, J. L., Bellei, G., and Martin, J. (2013) Integrated end-to-end NEO threat mitigation software suite. Proc. 64th International Astronautical Congress, Beijing, China, 23-27 Sept, 2013 (IAC-13/B5/2/8).
2. Chapuy, M., Vernis, P., Despré, N., and Capolupo, F. (2014) NEOShield project: GNC design and performance of a liquid-fuelled asteroid kinetic impactor. GNC 2014: 9th International ESA Conf. on Guidance, Navigation, and Control Systems, Porto, Portugal, 2-6 June, 2014.
3. Drube, L., A.W. Harris, T. Hoerth, P. Michel, D. Perna, F. Schäfer, (2014) NEOShield—A global approach to near-Earth object impact threat mitigation,

Handbook of Cosmic Hazards and Planetary Defense, ed. J. Pelton, F. Allahdadi, Springer, in production. DOI: 10.1007/978-3-319-03952-7_61.

4. Eggl, S., Ivantsov, A., Hestroffer, D., Perna, D., Bancelin, D, and Thuillot, W. (2013) "High precision astrometry in asteroid mitigation—the NEOShield perspective." *Proc. Société Française d'Astronomie et a'Astrophysique (SF2A) 2013*. Montpellier, France, 4-7 June, 2013.
5. Harris, A.W. and Drube, L. (2014) "How to find metal-rich asteroids." *Astrophysical Journal Letters*. 785, L4, doi:10.1088/2041-8205/785/1/L4
6. Harris, A.W., Barucci, M.A., Cano, J.L., Fitzsimmons, A., Fulchignoni, M., Green, S.F., Hestroffer, D., Lappas, V., Lork, W., Michel, P., Morrison, D., Payson, D. and Schäfer, F. (2013) "The European Union funded NEOShield project: a global approach to near-Earth object impact threat mitigation." *Acta Astronautica*. 90(1) pp. 80–84. doi:10.1016/j.actaastro.2012.08.026.
7. Harris, A.W. (1998) "A thermal model for near-Earth asteroids." *Icarus*, 131, 291-301. doi:10.1006/icar.1997.5865.
8. Jutzi, M., P. Michel (2014) "Hypervelocity impacts on asteroids and momentum transfer I. Numerical simulations using porous targets." *Icarus* 229, 247–253. DOI:10.1016/j.icarus.2013.11.020.
9. Meshcheryakov, S.A. and Lipnitskii, Y.M. (2015) "Estimated efficiency of the deflection of a dangerous space object using an explosion or impact." *Technical Physics*, 60:26-30. DOI: 10.1134/S1063784215010181.
10. Ummen, N. and Lappas, V. (2014) "Polyhedron tracking and gravity tractor asteroid deflection." *Acta Astronautica*, 104:106-124. DOI: 10.1016/j.actaastro.2014.07.024.

PHYSICAL AND DYNAMICAL CHARACTERIZATION OF NEAR-EARTH OBJECTS: THE ROLE OF THE GAIA MISSION

ALBERTO CELLINO
Osservatorio Astronomico di Pino Torinese
Pino Torinese, Italy

P. TANGA
Observatoire de la Côte d'Azur
Nice, France

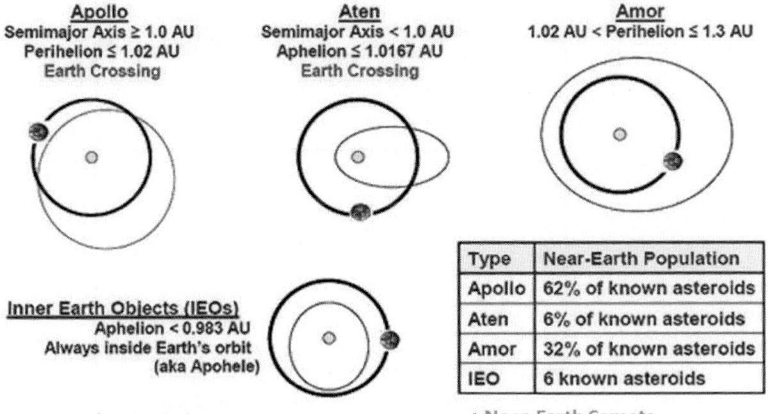

Type	Near-Earth Population
Apollo	62% of known asteroids
Aten	6% of known asteroids
Amor	32% of known asteroids
IEO	6 known asteroids

+ Near-Earth Comets

(Courtesy of NASA JPL)

The orbits evolve under gravitational and non-gravitational perturbations.

Potentially Hazardous Objects (PHA): Minimum Orbit Intersection Distance (MOID) ≤ 0.05 AU, and Absolute Magnitude $H \leq 22 \rightarrow$ (size > 150 m).

There are currently (Aug 17, 2015) 1605 known PHAs (out of >12,000 NEAs).

What are the first problems to be solved to set up a credible defense against possible impacts by near-Earth objects (asteroids and comets)?

Even before the development of any possible technique to avoid a possible catastrophic collision, as a first step we have to:

1. Discover possible impactors. This means not only to detect them, but to compute their orbits and orbital evolution.
2. Physically characterize them ("How big is it? What it is made of?", etc.)

First Problem: in spite of big improvements in recent years (availability of new dedicated telescope time), many dangerous objects may be difficult to detect from the Earth.

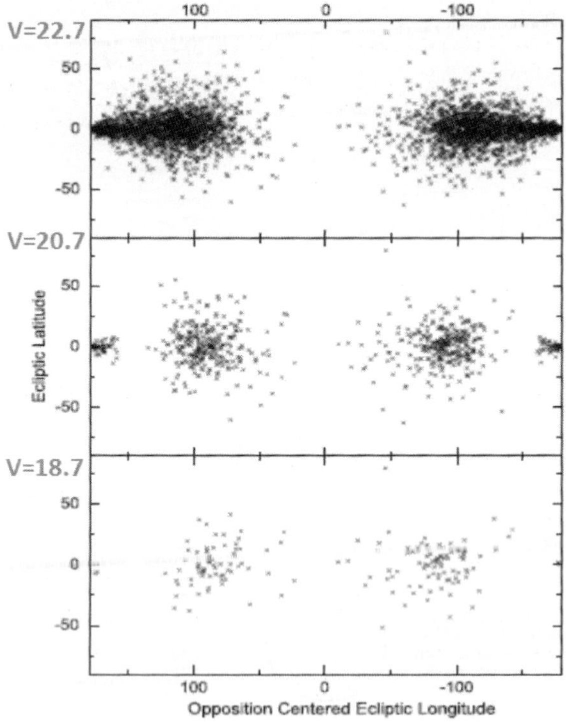

130,000 simulated impactors, their locations 20 years before impact. Veres et al. Icarus *203:472, 2009.*

In addition to the problem of discovering the dangerous objects, there is also the problem of obtaining for them an adequate physical characterization.

This requires the use of observing techniques (spectroscopy, polarimetry) that require large ground-based telescopes.

The problem is that one single measurement of the brightness of a newly discovered object is not sufficient to determine its orbit nor to assess its size and composition. The albedo of the objects varies much, as well as their surface properties.

The Gaia mission of the ESA has been launched on December 19, 2013, and has reached its final orbits in the L2 Lagrangian point some months later.

- Duration of the mission: 5 years. Complete survey of the sky, results for all branches of Astrophysics.
- Gaia will observe ~350,000 asteroids down to apparent magnitude G = 20.7.
- 60-70 transits per object (during 5 years).
- "*Short Term*" reduction chain to process objects newly discovered by Gaia (mostly asteroids with orbits interior to the Earth's). Data made public to allow prompt *follow-up* from the ground.
- "*Long Term*" reduction chain for all known objects.
- Astrometry, Photometry, Spectroscopy.

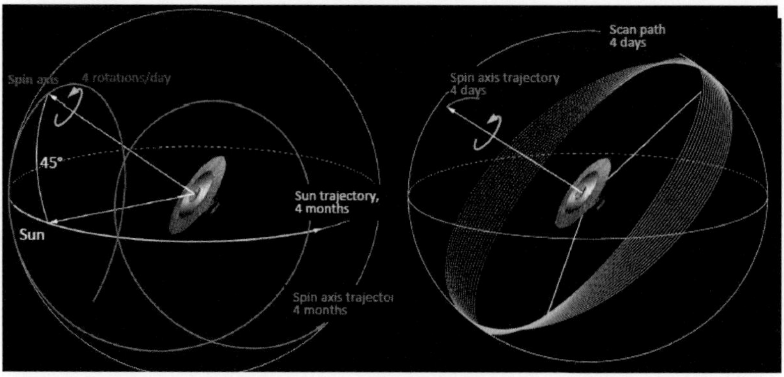

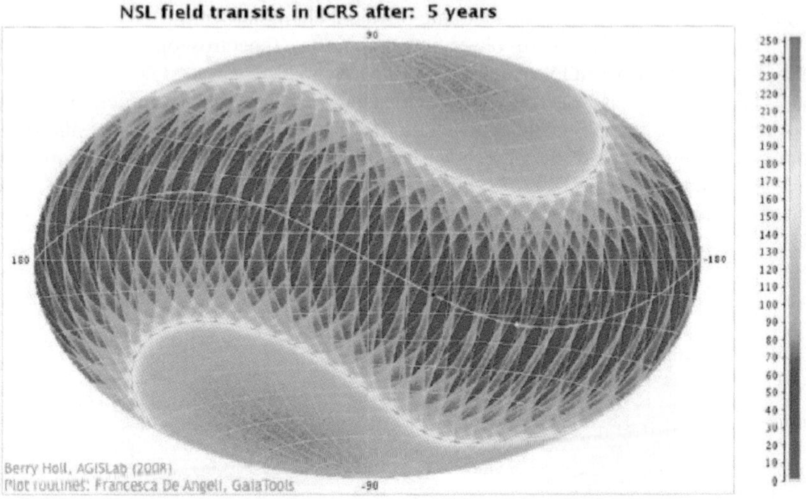

EXPLOITING THE UNPRECEDENTED ASTROMETRIC ACCURACY OF GAIA

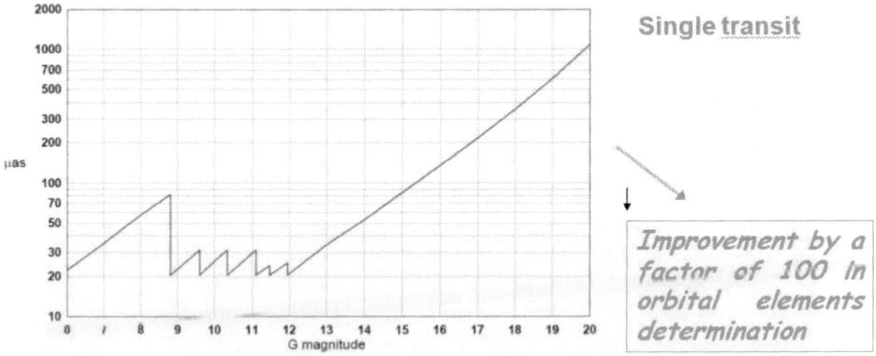

Very weak dynamical effects will be measurable, making it possible to obtain mass measurements based on deflections produced by asteroid mutual close encounters

JUST AN EXAMPLE: (99942) APOPHIS. THE ROLE OF GAIA

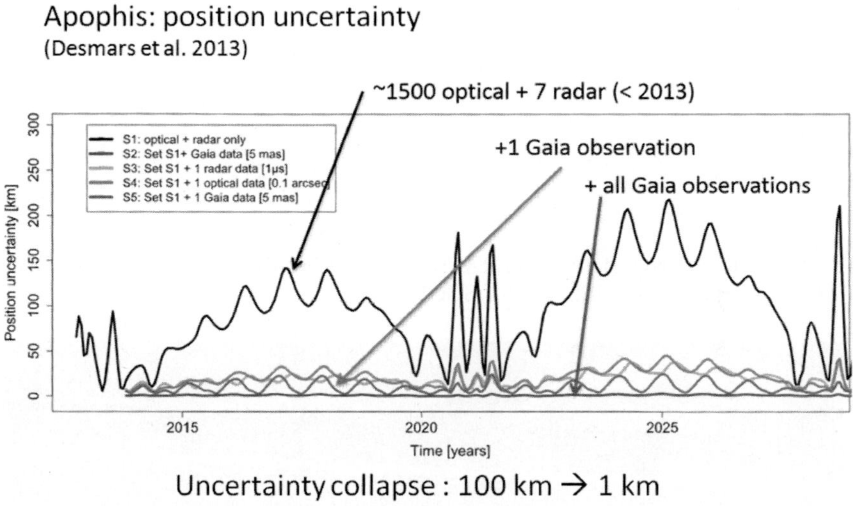

GAIA SPECTRO-PHOTOMETRIC CAPABILITY

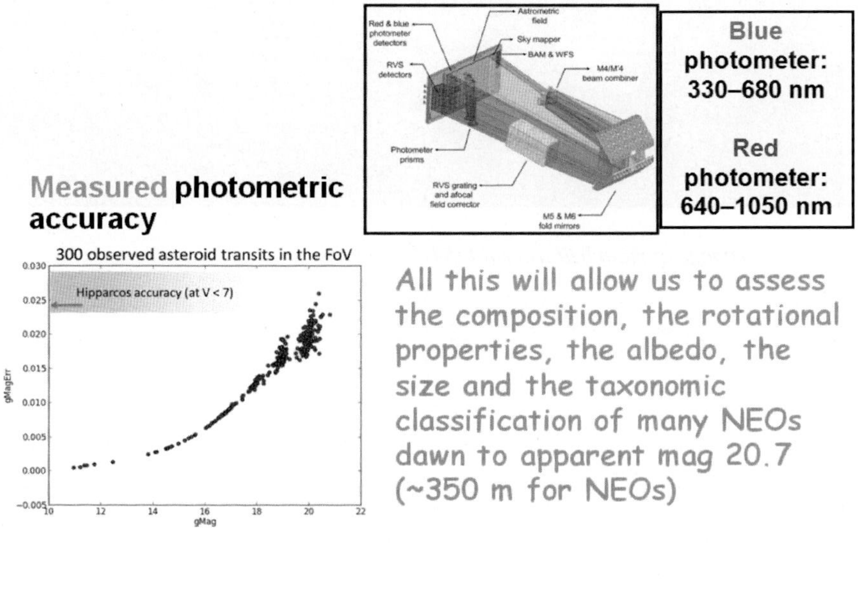

GAIA STARTED NOMINAL OPERATIONS AND WORKS! SHORT-TERM DATA REDUCTION CHAIN WILL START SOON.

(Image of main belt asteroid (54) Alexandra)

SESSION 14

WATER POLLUTION AND DEVELOPMENT ISSUES—EU ENERGY TRANSITION PLANS AND SHIP-BASED NUCLEAR ENERGY

THE STRUGGLE FOR WATER IN AN URBANIZING WORLD: CONSTRUCTING HOLISTIC POLICIES

IGNACIO ARMILLAS

Retired Director of the United Nations Centre for Human Settlements (Habitat), Fukuoka Office, Mexico, DF, Mexico

ABSTRACT

The world is in the midst of a period of hyper-urbanization that started last century and will continue well into the present one. This process is resulting in the formation of an increasing number of mega-cities that concentrate tens of millions of inhabitants into vast urbanized regions. One of the main challenges of this phenomenon, if not the main one, is to meet the water need of all thc inhabitants of these conurbations. A challenge that is both complex and costly, and yet inescapable. Exacerbating this challenge is the fact that practically all of the unprecedented urban growth is taking place in less developed regions, those least able to cope with the challenges of urbanization. This paper contends that the old paradigm of simply transporting water from distant sources to the emerging mega-cities is not only prohibitive in terms of financial, environmental and political costs, but not feasible in terms of the enormous need for water these urban areas concentrate. It is suggested that what is required are holistic water policies based on a new paradigm that takes into account the multiplicity of ways water is used within the urban metabolism.

THE URBANIZATION OF THE PLANET[1]

In mid-twentieth the world's population stood at roughly 2.5 billion.[2] Of this total 70.6 percent was considered rural, while the remaining 29.4 percent was considered urban.[3] Although the economy and the cultural and political life of most nations centered in large, dense cities, most people lived in small, low-density, settlements dispersed in the countryside.

By the end of the century the world's population had increased by some 3.6 billion people, surpassing the 6 billion mark. Such significant increment in population would present major challenges to the environment and development efforts under any circumstances. However, two major tendencies came into play which exacerbated the already problematic situation. The first tendency was that practically 90 percent of the

---

[1] Except when noted otherwise all data on population shown in this paper is from *World Population Prospects: The 2010 Revision* and *World Urbanization Prospects: The 2011 Revision*, United Nations Department of Economic and Social Affairs/Population Division, United Nations, New York, 2012. Accessed Friday, February 15, 2013; 8:36:33 PM.

[2] The term billion will be used in this paper to denote one thousand million as is customary in North American English.

[3] Countries differ in how they classify population as "urban" or "rural." Settlements with a population of a few thousand are considered urban. Sometimes it also depends on whether they are incorporated as a municipality or other form of local governing structure. A listing of country definitions is published annually in the United Nations Demographic Yearbook.

population increase occurred in the less developed regions[4] of the world. The population in these regions grew from 1.7 billion in 1950 to over 4.9 billion by the end of the century. Consequently, these regions had to absorb a total of over 3.2 billion people; this amounted to almost a three-fold increase in population in the fifty-year span. The other tendency that exacerbated the situation was that over half of the population increase concentrated in the urban areas of developing regions. While the urban population of the more developed regions doubled in size from 441,845 to 881,344, the cities in the less developed regions multiplied their combined population by five and one-half times from 303,650 to 1,977,289 inhabitants. (Annex #1 contains relevant population tables for the period 1950-2050).

The unprecedented surge of urban populations can best be described as hyper-urbanization, a process that is characterized by the massive and uncontrolled growth of cities. Urban areas all over the less developed world were transformed. Cities had to absorb vast amounts of population, some of it being the product of natural internal growth, but the major part being poor migrants that rural areas could not accommodate. And this was happening under the most unfavorable of conditions. The growth of cities in the less developed regions was not the result of a process of industrialization and economic expansion creating employment in urban areas, as had been the case in the more developed regions during their transformation into urban societies. In fact, the process of industrialization and economic growth was nonexistent or lagged considerably behind the rate of urbanization in all the countries of the less developed regions.

THE EMERGENCE OF MEGA-CITIES

The main product of hyper-urbanization has been the formation of larger more complex urbanized regions holding tens of millions of inhabitants. These vast continuous urban areas are in essence a new more complex form of human settlement, a new tier at the top of the hierarchy of settlements. The term mega-city[5] has come into common use to denote such large settlements. All-told, the world's population living in mega-cities expanded from 23.6 million to 252.8 million over the fifty-year period from 1950 to 2000, a more than ten-fold increase. All told, mega-cities comprised 10% of the world's urban population by the turn of the century.[6] (Annex #2 lists the twenty largest urban agglomerations in the years 1950, 1975 and 2000 and highlights the mega-cities.)

---

[4] The term less developed regions (LDR's) is used to include all countries in Africa, Asia (excluding Japan), and Latin America and the Caribbean, and the regions of Melanesia, Micronesia, and Polynesia. The term more developed regions (MDR's) include all countries in Europe, North America, Japan, Australia, and New Zealand.

[5] The United Nations Human Settlements Programme (Habitat) defines a single large urban agglomeration that may or may not be a conurbation of several cities and towns that exceeds 10 million population as a mega-city or megalopolis and those exceeding 20 million as meta-cities. UN-Habitat, *The State of the World's Cities —2006/2007: 30 Years of Shaping the Habitat Agenda*, Nairobi, 2006.

[6] WHO and UN-Habitat, *Hidden Cities: Unmasking and Overcoming Health Inequities in Urban Settings,* Geneva, 2010.

THE URBAN CONDITION IN 2000

This precipitous demographic growth of cities in developing regions has far exceeded the limited capacities and resources of national and local governments to manage urban development in an efficient, equitable and sustainable manner. As a result living conditions are particularly harsh for large segments of the urban population who live in slums and squatter settlements and other low-income areas where clean water and other public services are either deficient or lacking altogether.

Slum and squatter areas overwhelm major cities everywhere in the less developed regions, in some instances accounting for 40, 50 and even 60 percent of the population of the city. Of the nearly two billion inhabitants of urban areas in the less developed regions in the year 2000 over 800 million people or 39% of the total urban population lived in slums and squatter areas.[7] Although the proportion of slum dwellers declined during the first decade of this century, the absolute numbers of people living in slum and squatter areas has actually grown considerably, and will continue to rise in the foreseeable future.[8]

In very broad terms this was the situation at the dawn of the twenty-first century after five decades of demographic growth unparalleled in human history.

PROJECTIONS FOR THE COMING DECADES

The process of hyper-urbanization is expected to continue well into the middle of the present century. The projections are that in between 2000 and 2050 the total world population will increase by about 3.2 billion. Moreover, the urban population will actually increase by an even greater 3.4 billion. This is because the rural population worldwide is expected to decrease by some 210 million during this period. This means that, over the first half of this century, urban areas will absorb all the total population increase as well as the decrease from rural areas. While the urban population in developed regions will remain essentially unchanged, practically all of the projected urban growth will continue to occur in human settlements located in the less developed regions.

As has been pointed out earlier, the urban areas in the less developed regions were severely punished by the demographic growth experienced in the latter half of the twentieth century. These urban areas are now expected to have to absorb twice as much population in the first half of the present century as they did in the previous fifty years. In actual numbers the urban areas of the less developed regions absorbed close to 1.7 billion people in the period 1950-2000; it is projected that in the first half of this century these areas will have to absorb twice that amount, which is 3.4 billion people. The projected increase will bring the total urban population in the less developed regions to more than 5.1 billion. This represents a total urban population two and a half times greater than what it was in the year 2000; and seventeen times greater than what it was in 1950.

As a result of hyper-urbanization, by 2050 about 70 percent[9] of humanity will be living in urban settlements of all sizes. However, it is expected that urban growth will

---

[7] UN-Habitat, *The State of the World's Cities—2010/2011: Bridging the Urban Divide*, Nairobi, 2010.

[8] Ibid

[9] United Nations Human Settlements Programme (UN-Habitat). *State of the World's Cities 2012/2013: Prosperity of Cities*, UN-Habitat, Nairobi, 2012.

continue to favor the larger metropolitan areas, thus the larger metropolitan areas will grow at faster rates. For this reason although the majority of the urban population will be living in cities with less than ten million inhabitants, the most significant development in the first half of this century, will be the increasing number of mega-cities and the further expansion of already existing ones.[10] These emerging vast amorphous conurbations are, in effect, uncharted worlds.

THE STRUGGLE FOR WATER

There are many issues concerning the impact of hyper-urbanization on the social, political and environmental landscape, but perhaps the most critical issue is that of water. Water is a particularly complex element that is essential not only for human consumption and hygiene, but also a vital element for manufacturing and other economic activities in urban areas, as well as for producing the food and energy required by cities. In essence cities concentrate demand for water and the footprints of mega-cities are trans-regional, international and in some aspects global.[11] This is perhaps the main reason that providing these vast conurbations with water defies practically all of the prevailing paradigms of water sourcing and management.

Moreover, because of its relative scarcity and uneven geographical distribution over the surface of the planet there are major social and political dimensions to the use and sharing of available supplies.[12]

Beyond the challenges of bringing sufficient water to urban areas there is also an internal struggle for water within cities. A struggle that is completely hidden in per capita consumption statistics. Marked differences exist in large cities across much of the less developed regions with regards to access by different groups of the population to potable water. These differences pretty much reflect the distribution of wealth and in some instances social/ethnic differences in the society. Not only do high-income areas generally have served by piped water while low-income areas have limited or no access to such services; and invariably, high-income households use far more water than poor households. In cities such as Mexico City, Mumbai and Dar-es-Salaam per capita water consumption can be 15 times or more in high-income areas than in the slum and squatter settlements in the urban fringe. In addition, not only do the poorest sectors have much more restricted access to water, but that water may not be as clean since inadequate distribution methods are prone to contamination. Not having access to piped water also means higher costs per unit of water. Slum dwellers in Jakarta, Manila or Nairobi pay, on average, five to ten times more for water than people living in areas serviced by the city's water system.[13]

---

http://www.unhabitat.org/pmss/listItemDetails.aspx?publicationID=3387

[10] China presents a mixed picture. Government policies are directing urban growth towards hundreds of middle-sized cities which are growing at a faster rate than the mega-cities.

[11] The footprints of urban areas are certainly global if we take virtual water into account.

[12] Although agriculture is by far the leading user of water, accounting for an estimated 72 percent of water withdrawals worldwide and over 90 percent in low-income developing countries urban and industrial demands are growing at a faster rate. See Rosengrant, M.W. and C. Ringler, "Impact on food security and rural development of transferring water out of agriculture", *Water Policy* 1(6), pages 567-586. 1998.

[13] United Nations Development Programme, *Human Development Report 2006 —Beyond scarcity: Power, poverty and the global water crisis*, New York, 2006.

Understandably different countries with different social, environmental, and economic situations have different capabilities in tackling urban problems. But most large conurbations are emerging in regions that are capital poor and not institutionally and politically robust enough to cope with the challenges of hyper-urbanization. We need only to look at the bursting mega-cities in South Asia and in Africa, and to a similar degree in Latin America, to appreciate the enormity and complexity of the challenges ahead as the less developed regions struggle to accommodate ever increasing numbers of urban populations. This accelerating growth of urban regions will have significant negative repercussions for a considerable portion of humankind in the coming decades.

As population increases and urban areas expand water scarcity and shortages will become more and more widespread; this being the combined product of increased demand within urban areas and of competition with other uses beyond the urban fringe.

In a recent article a group of researchers of The Nature Conservancy,[14] estimated that early in this century somewhere in the neighborhood of 150 million people lived in urban areas with perennial water shortage. They define water shortage as the situation where there is on average less than one hundred liters of water available per person per day. They further estimated that an additional 886 million lived in urban areas subject to seasonal water shortages; which they describe as having an insufficient supply of water during at least one month of the year. Taking into account the accelerating rate of hyper-urbanization, the same study forecasts that by the middle of the century 993 million urban residents—that is 1 in 6 persons—will be living in cities with perennial water shortages and another 3.1 billion urban residents—that is 1 in 2 persons—will be confronting seasonal water shortages of varying proportions.

HOW ARE MEGA-CITIES COPING

The prevailing approach to meeting the water demands of growing urban regions is one that can best be labeled as "supply oriented"; that is if there is shortage of water in the metropolis simply increase the supply. Consequently, as the demand for water resulting from hyper-urbanization outstrips sources within the proximate hydrological basin the supply oriented mind-set will lead to the construction of major public works projects to increase the flow of water to the urban region. These include local undertakings such as building desalination plants in coastal areas and dams where feasible. But in the case of mega-cities water supply projects almost always involve the building of long distance inter-basin water transfer systems. At the beginning of this century over twelve percent of large cities worldwide obtained their water through such inter-basin transfers and the percentage is expected to grow rapidly in the coming decades.[15]

A prime example of the supply-oriented approach leading to massive inter-basin transfers is the metropolitan area of Mexico City. This metropolitan area lies within a

---

[14] McDonald, Robert I., Pamela Green, Deborah Balk, Balazs M. Fekete, Carmen Revenga, Megan Todd, and Mark Montgomery, "Urban growth, climate change, and freshwater availability", PNAS - Proceedings of the National Academy of Sciences of the United States of America, vol. 108 no. 15, pages 6312–6317. (Published online 28 March 2011).

[15] McDonald, Robert I., et al. "Water on an urban planet: Urbanization and the reach of urban water infrastructure", in Global Environmental Change, Elsevier B.V., Volume 27, July 2014, Pages 96-105, accessed 9 June 2014, *www.sciencedirect.com/science/journal/09593780.*

large endorheic basin located some 2,250 meters above sea level and surrounded by high mountains. Two thirds of the water supply for the urban area is obtained from sources within the basin. The main water sourcs are aquifers within the basin. Although the aquifers are extensive they are being over exploited as the present extraction volume is well over twice their natural recharge rates.[16] The other internal sources are surface water and springs, but these sources are somewhat limited. Thus the Mexico City metropolitan area must obtain one third of its water from sources outside the basin through long distance water transfers. These transfers come from two neighboring basins, the Lerma-Balsas and the Cutzamala river systems. The Lerma-Balsas inter-basin transfer system was built 70 years ago and brings water over the mountains from a distance of some 60 kilometers. The first of three phases of the Cutzamala system was constructed in the 1970's. Consecutive phases were built in the 80s and 90s. This second system brings water from as far away as 150 kilometers and an altitude of 1100 meters below that of Mexico City. It is one of the largest water supply systems in the world, in terms of both the total quantity of water supplied - some 485 million cubic meters per year—and in terms of the difference in elevation that has to be overcome. The pumping required to lift the water consumes a significant amount of energy, variously estimated at between 1.3 and 1.8 terawatt hours per year, equivalent to about 0.6% of the country´s total annual energy consumption.[17]

As the Mexico City conurbation continues to grow the government is preparing proposals for further water transfer from other neighboring basins. One scheme that is being detailed would bring water from the Amacuzac river basin. This water would have to be transported along a 160 kilometer long aqueduct and pumped to a height some 1825 meters above the source. It is estimated that the energy consumption of this system would be 5 percent of national annual electricity production.[18]

Beijing is another example of how mega-cities are turning to mega-projects to meet the increasing demand for water as their populations grow. At present the total annual water demand of the Beijing metropolitan area exceeds available renewable resources by about some twenty percent. The difference is made up by the overexploitation of local aquifers that provide two thirds of the water supply consumed in the urban region. The remaining one third comes from surface water sources within the region. However, these sources are waning due partly to deforestation and partly to pollution.

In an attempt to supply more water to the cities as well as to industry and agriculture in the northern plains in which Beijing is located, the government is building the ambitious South-North Water Transfer Project. When completed this project is expected to divert 44.8 billion cubic meters of water per year from the Yangtze River Basin in southern China to the Yellow River Basin in the northern part of the country. The project consists of three water routes, each over a thousand kilometers long. The Eastern and Central Routes are expected to be on-line shortly while it is estimated that the Western Route will take until 2050 to be complete.

---

[16] Burns, Elena, et al. *Repensar la Cuenca: la gestión de ciclos del agua en el Valle de México*. Universidad Autónoma Metropolitana: Centro para la Sustentabilidad, 160p. Mexico, DF, 2009.
[17] Ibid
[18] Ibid

A third example of supply oriented approaches to water scarcity is the Sao Paulo Metropolitan Region. This urban area has relied, up to now, on water resources within the Alto Tietê Basin in which it is located. The state water utility and the various municipal utilities obtain all their water from surface sources, which account for about 80 percent of water consumed in the Metropolitan area. The remaining 20 percent is obtained from individual groundwater sources. As the population of the conurbation increases these sources are no longer sufficient to meet present needs. As a result the government has taken a number of measures to reduce water demand with some success. Per capita consumption has decreased some 20 percent in recent years. But local sources are being exploited to the limit, thus long term plans to meet the future water needs of the expanding Sao Paulo conurbation include inter-basin transfers. The overall plan, known as the Macro-Metropolis Plan, includes a water diversion project which will bring water from the Ribeira de Iguape River Valley some 80 kilometers south to the Sao Paulo area. The system is expected to be operational by 2017.

THE LIMITS OF SUPPLY ORIENTED APPROACHES TO MEETING URBAN WATER DEMAND.

The three cases discussed illustrate the persistence of the supply-oriented mindset in governments. But the construction of massive infrastructure projects to bring greater amounts of water from more and more distant sources to expanding urbanized regions is neither sustainable nor environmentally, economically or socially desirable. The engineering projects required for such transfers are nearly always daunting in scale and their financial costs are similarly large. Additionally, water transfer projects tend to ignite political and social conflicts between the people in the areas from which the water is being taken and the people in the areas benefiting from the water transfer. And, of course, water cannot be subtracted from a region without affecting the local environment; there are invariably negative impacts on the "donor" environment. Often the environmental degradation resulting from the transfer of water from "donor" areas curtails agriculture and other economic activities provoking migration from the now withered country-side to the cities that have taken the water.

Moreover, while inter-basin transfers and other supply-oriented measures may alleviate some of the need for urban water, supply-oriented approaches alone will not be adequate to meet the scale of demand of the mega-cities that are now emerging. As was pointed out in the first part of this paper the scale and pace of hyper-urbanization is projected to result in conurbations with populations reaching into the several tens of millions. Even if we consider satisfying only minimum requirements at 100 or so liters per capita per day and we multiply this number by say 30 or 40 million people, which will be the population of many mega-cities, we can appreciate the vast amount of water these conurbations will require. It is unrealistic to think that needs of this scale can be met through supply-oriented means. Moreover, the problems of obtaining water are compounded if we take into account that many of these mega-cities are emerging in developing regions that are already water stressed or are suffering from water scarcity.

If supply oriented approaches are insufficient to address urban water needs it has been reasoned that new thinking on meeting urban water needs must emerge. Thinking that moves from the conventional ''bring more water at whatever cost'' mindset to holistic multidimensional approaches that balance environmental, social and economic considerations with technology and new perceptions of water based on the many ways it flows through the urban metabolism.

IMPROVED WATER MANAGEMENT AND CONSERVATION MEASURES

Without any doubt a fundamental component of any water management strategy must be conservation; reducing consumption in every possible way, including finding alternatives to the use of water for sanitation and other high consumption uses. Some efforts in this regard are being undertaken in urban areas around the world. These efforts include the use of technological approaches ranging from mandating the installation of simple hardware such as water-saving valves, aerators, and low flush toilets to reduce water consumption in the homes, the utilization of gray water to irrigate parks and peri-urban agricultural fields, the recycling and reusing of water in industry, to the construction of complex high technology systems to regulate pressure in water distribution systems to reduce leakage, and so on.

On the "soft" side education and awareness programs aimed at reducing water consumption and wastage are also becoming more widely accepted. Educational campaigns can result in some reductions in the demand for water and should be part of any set of policies to promote conservation.

Economic and fiscal incentives and instruments are another set of tools being applied in cities in many parts of the world to lower water demand. These incentives can be either of a carrot or a stick nature. In some instances penalties are placed on excessive consumption and hefty taxes are imposed on extravagant uses of water such as swimming pools and lawns. On the other hand proper maintenance of water systems within a property and installation of water saving devices are rewarded. Research findings from both developed and developing regions indicates that a mix of vigorous regulatory and economic-fiscal incentives does lead to more efficient water use. In some instances reductions in the order of 20 to 30 percent in water use have been noted.[19]

In addition to including measures for reducing consumption water management strategies sometimes include the tapping of sources beyond the traditional ones. One greatly underutilized unconventional source of water is the collection of rainfall through small and large scale harvesting techniques. Small scale low technology methods are already being used in some parts of the world, particularly in low-income settlements that do not have adequate access to water and settlements in areas under severe water stress. In wealthier regions rain water collecting systems are increasingly being incorporated into the design of large surface buildings such as covered stadiums. Capturing storm water from streets and public spaces, before it is mixed with waste water in the sewers and becomes a liability instead of an asset, is very rare and limited to a few cities in developed countries.

Another potential source of water for urban areas is water that is already there. Ironically most cities are spending large sums of money to discard their sewerage and runoff water while they spend additional large sums to obtain new raw water from distant locations. Instead of discarding sewerage it should be considered a source of water for the

[19] Engel, Katalina, D. Jokiel, A. Kraljevic, M. Geiger, K. Smith, *Big Cities. Big Water. Big Challenges — Water in an Urbanizing World*, WWF Germany, Berlin, 2011.

city. Of course treating sewerage for urban needs is costly, but so is the trans-basin transportation of water. The funds and energy spent on constructing and operating pumping stations, canals and tunnels required to move water over long distances could well be used for water treatment plants close to the urban area to treat and reuse the water that is already there.

The various approaches towards water conservation mentioned in the previous paragraphs can and do lead to better utilization of water in urban areas. Unfortunately the application of these water conservation tools is at best spotty and disjointed at present. There are very few examples of countries with water management policies that integrate a broad variety of the economic and fiscal incentives, technological hardware, education and awareness programs, and tapping of unconventional sources, into a comprehensive set of coordinated programs. Japan and Israel are among the few examples of countries developing and implementing holistic approaches to maximize the use of what is a very limited resource in their respective countries.

THE NEED FOR A NEW PARADIGM

In order to meet urban water requirements of the large conurbations that are emerging there will be a need to go beyond current water management and conservation measures. These approaches do help but have already proven inadequate in addressing the overall needs of mega-cities in developing regions. In order to do this what is needed is a new way to think about water, a new paradigm.

At the present time the tendency is to perceive water as a single element which flows through the urban system. Water is collected from somewhere beyond the city, transported to the urban area, all of it is treated to produce a homogeneous high quality product. The potable water is then distributed through a single system of pipes to be used for a variety of purposes, ranging from human consumption to fighting fires. *Once* used it is again all mixed altogether and expelled from the city through another unitary system of pipes, most of this sewerage going on to pollute rivers, lakes and seas.

This thinking is not adequate to meet the needs of present day mega-cities and it certainly will not be adequate in the future. The dimensions and pace of hyper-urbanization in the coming decades will necessitate a shift from the current view of water to a new paradigm based on a broader perception of water. A perception that takes into account the multiplicity of ways water is used within the urban metabolism. Under such a paradigm water becomes plural, there are different needs for water in an urban area, and thus there different types of waters required. Each type can be supplied through different networks, and certainly priced differently.

As part of this new paradigm water regarded as an element that once introduced into the urban system it can be used multiple times for different purposes. It can flow in cycles from one use to another cascading from higher to lower quality levels. Conversely, if it is treated, it can be recycled "upward" and reused in levels requiring higher quality of water. Broadening or diversifying the concept of water offers the opportunity to advance solutions better targeted to the various needs for water in urban areas.

A final thought. Beyond the challenge of providing an adequate and sustainable supply of water to urban areas holistic water policies and programs must also include ways and means to ensure equality of access to water to all the population. At least equality of access to meet basic human needs.

ANNEX # 1

URBAN—RURAL POPULATION 1950-2050 (THOUSANDS)

World

Year	Total Population	Rural Population	Total %	Urban Population	Total %
1950	2 532 229	1 786 734	70.6	745 495	29.4
1960	3 038 413	2 018 775	66.4	1 019 638	33.6
1970	3 696 186	2 343 767	63.4	1 352 419	36.6
1980	4 453 007	2 699 779	60.6	1 753 229	39.4
1990	5 306 425	3 025 020	57.0	2 281 405	43.0
2000	6 122 770	3 264 138	53.3	2 858 632	46.7
2010	6 895 889	3 337 311	48.4	3 558 578	51.6
2020	7 656 528	3 366 710	44.0	4 289 818	56.0
2030	8 321 380	3 337 472	40.1	4 983 908	59.9
2040	8 874 041	3 237 815	36.5	5 636 226	63.5
2050	9 306 128	3 053 953	32.8	6 252 175	67.2

Source of data for all the annexes is: Population Division of the Department of Economic and Social Affairs of the United Nations Secretariat, *World Population Prospects: The 2010 Revision* and *World Urbanization Prospects: The 2011 Revision.* Accessed Friday, February 15, 2013.

More Developed Regions (MDR's)

Year	Total Population	Rural Population	Total %	Urban Population	Total %
1950	811 187	369 341	45.5	441 845	54.5
1960	913 330	357 360	39.1	555 970	60.9
1970	1 006 421	335 848	33.4	670 573	66.6
1980	1 081 094	323 119	29.9	757 975	70.1
1990	1 144 404	317 306	27.7	827 098	72.3
2000	1 188 809	307 465	25.9	881 344	74.1
2010	1 235 900	278 649	22.5	957 251	77.5
2020	1 273 439	255 074	20.0	1 018 365	80.0
2030	1 296 089	231 799	17.9	1 064 290	82.1
2040	1 306 885	207 619	15.9	1 099 266	84.1
2050	1 311731	184 510	14.1	1 127 222	85.9

Less Developed Regions (LDR's)

Year	Total Population	Rural Population Total	%	Urban Population Total	%
1950	1 721 042	1 417 393	82.4	303 650	17.6
1960	2 125 083	1 661 415	78.2	463 668	21.8
1970	2 689 765	2 007 919	74.7	681 846	25.3
1980	3 371 913	2 376 660	70.5	995 253	29.5
1990	4 162 021	2 707 714	65.1	1 454 307	34.9
2000	4 933 961	2 956 673	59.9	1 977 289	40.1
2010	5 659 989	3 058 663	54.0	2 601 326	46.0
2020	6 383 089	3 111 636	48.7	3 271 453	51.3
2030	7 025 290	3 105 672	44.2	3 919 618	55.8
2040	7 567 156	3 030 196	40.0	4 536 960	60.0
2050	7 994 397	2 869 443	35.9	5 124 953	64.1

ANNEX # 2

20 LARGEST URBAN AGGLOMERATIONS RANKED BY POPULATION SIZE
1950, 1975 & 2000

MEGA-CITIES: CITIES WITH 10 MILLION INHABITANTS OR MORE
1950, 1975 & 2000

1950			**1975**			**2000**		
1	**New York**[20]	**12.3**	**1**	**Tokyo**	**26.6**	**1**	**Tokyo**	**34.4**
2	**Tokyo**	**11.3**	**2**	**New York**	**15.9**	**2**	**Mexico City**	**18.0**
3	London	8.4	**3**	**Mexico City**	**10.7**	**3**	**New York**	**17.8**
4	Paris	6.5	4	Osaka-Kobe	9.8	**4**	**Sao Paulo**	**17.1**
5	Shanghai	6.1	5	Sao Paolo	9.6	**5**	**Mumbai**	**16.4**
6	Moscow	5.4	6	Los Angeles	8.9	**6**	**Delhi**	**15.7**
7	Buenos Aires	5.1	7	Buenos Aires	8.7	**7**	**Shanghai**	**14.0**
8	Chicago	5.0	8	Paris	8.6	**8**	**Kolkata**	**13.1**
9	Kolkata	4.5	9	Kolkata	7.9	**9**	**Buenos Aires**	**11.8**
10	Beijing	4.3	10	Moscow	7.6	**10**	**Los Angeles**	**11.8**
11	Osaka-Kobe	4.1	11	Rio de Janeiro	7.6	**11**	**Osaka-Kobe**	**11.2**
12	Los Angeles[21]	4.0	12	London	7.5	**12**	**Rio de Janeiro**	**10.8**
13	Berlin	3.4	13	Chicago	7.2	**13**	**Dhaka**	**10.3**
14	Philadelphia	3.1	14	Mumbai	7.1	**14**	**Cairo**	**10.2**
15	Rio de Janeiro	3.0	15	Seoul	6.8	**15**	**Beijing**	**10.2**
16	St. Petersburg	2.9	16	Cairo	6.4	**16**	**Karachi**	**10.0**
17	Mexico City	2.9	17	Shanghai	5.6	**17**	**Moscow**	**10.0**
18	Mumbai	2.9	18	Manila	5.0	**18**	**Manila**[22]	**10.0**
19	Detroit	2.8	19	Beijing	4.8	**19**	Seoul	9.9
20	Boston	2.6	20	Jakarta	4.8	20	Paris	9.7

[20] Includes the New York—Newark urban area.
[21] Includes the Los Angeles—Long Beach—Santa Ana urban area.
[22] Manila's population in 2000 was actually 9,958,000, short of the ten million mark to qualify as a mega-city, but in the rounding process it does appear as ten million.

THE WORLD ECONOMIES IN A CUBE[1]

SIMONE BASTIANONI
DEEPS Department of Earth, Environmental and Physical Sciences, University of Siena, Siena, Italy

F.M. PULSELLI, L. COSCIEME, L. NERI, G. BETTI AND A. LEMMI
University of Siena
Siena, Italy

If we want to measure progress towards a more sustainable state of our societies we need to move from the typical habit of just using GDP as the only important indicator for policy. The typical representations of sustainability focus on three aspects: environment, society and economy. These are not three equivalent components (as it could be understood from Figure 1), since they have very different characteristics: while the economic system changes rapidly and something economically unfavorable can become feasible changing taxes or incentives, the social system is much slower in changes and for the environment there is nothing we can do to make favorable something that is harmful.

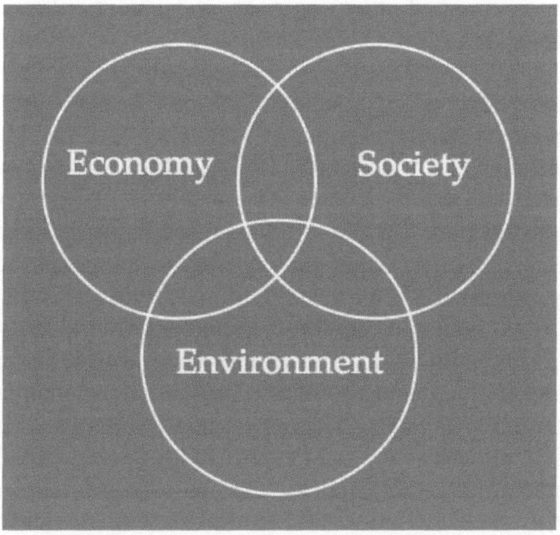

Fig. 1: The "usual" representation of sustainability (Barbier, 1987).

We prefer a representation that gives also the idea of the relational order that links the three dimensions: the environment provides the goods and services that are used by societies for their well-being; an organized society has a useful economic production that contributes to the well-being as well.

---

[1] This presentation is based on the paper F.M. Pulselli et al. 2015. *The World Economy in a cube: a more rational structural representation of Sustainability*. Global Environmental Change, 2015 (submitted).

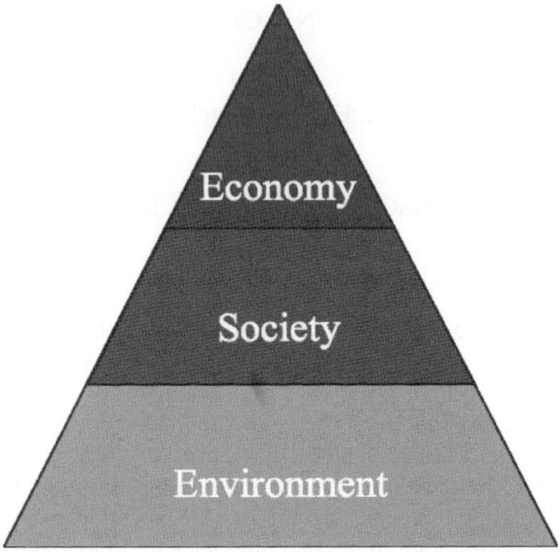

Fig. 2: A pyramid represents better sustainability showing that a society is based on resources and produces economic activity. There are feedbacks from Economy to Society and Environment.

To understand the level of sustainability of a system the simple use of GDP cannot be enough; nonetheless the number of indicators used has to be low enough to be easily understandable: this is a "minimal" representation of the sustainability status of a national system and it can be represented by a 3-axis diagram where three different indicators are used to account for resource use, societal organization, and goods and services produced, respectively (Pulselli et al. 2015). This approach is a trade-off that aims at maximizing information with the minimum number of indicators to illustrate the sustainability level of a system. The three indicators should be representative of the system as a whole. The aim is to depict the three different dimensions of system sustainability independently, ensuring that every indicator maintains its identity and complementary informative capacity. This representation also avoids the use of hundreds of indicators very difficult to interpret and to be used for actual policies.

The three indicators we selected for the analysis are: the energy flow per capita (Odum, 1988; 1996); the Gini index of income distribution and the Gross Domestic Product per capita (GDP). Other choices could have been made for representative indicators, for example: Ecological Footprint, indicators derived from Material Flow Analysis on the environmental axis; indicators of Social Capital or of labor or education for the social axis; indicators like the Index of Sustainable Economic Welfare (ISEW) or Genuine Progress Indicator (GPI) on the economic one.

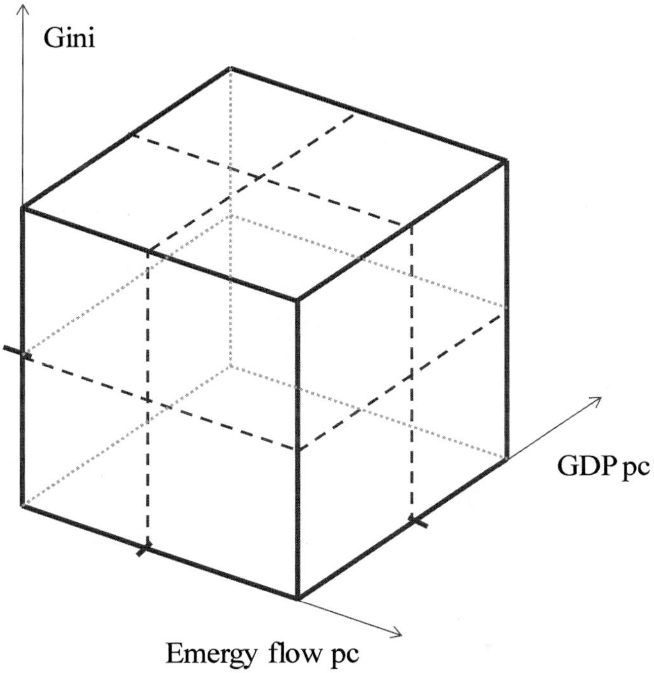

Fig. 3: The cube and sub-cubes within which we represent the national economies.

Energy accounts for the quantity of solar energy that has been directly or indirectly used to produce a flow or a product. Energy (and the energy flow, i.e., energy per unit time), is defined as the equivalent amount of solar energy "memorized" in different types of energy and matter forms.

The Gini coefficient is a measure of inequality ranging from 0 (in the situation of perfect equality, where every unit has the same income) to 1 (in the situation of greatest inequality, where only one unit receives the whole income). Even if, strictly speaking, the Gini Index is an economic indicator it has been shown that inequality is strongly correlated with health and social problems (Wilkinson and Pickett, 2009).

GDP is the sum of the market value of the overall set of goods and services produced by an economic system in a given period of time (generally 1 year). It can thus be intended as an indicator of the economic output.

We have examined 99 national economies, grouping them into 8 sub-cubes separated by the median value of the three axes (Pulselli et al. 2015). Data are referred to the year 2008, since it is the last one available for energy.

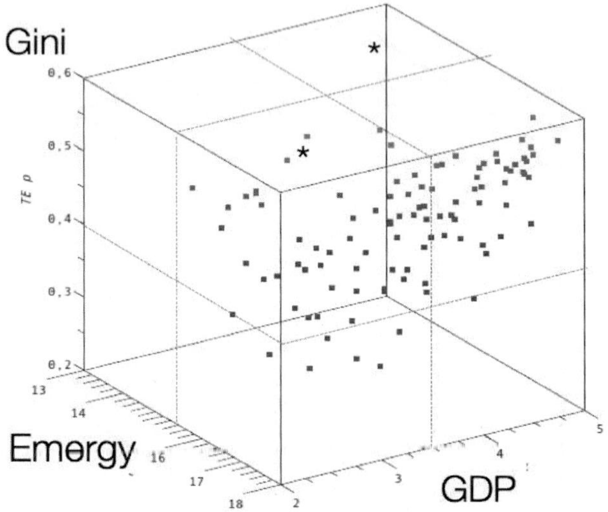

Fig. 4: The 99 Economies in the cube. Each segment is divided using the median value.

Most of the economies (85) fall in four of the eight possible sub-cubes, while the other 14 points fall in less populated sub-cubes, but in zones that are very close to median values of at least one of the indicators.

From the results a strong relationship between resource use per capita and GDP per capita emerges, pointing out how economic growth drives, and depends on, an increasing requirement of energy and matter to be transformed by the economic system, while the level of inequality on societies is quite independent of the other two metrics (Pulselli et al. 2015).

The area of the cube where there should be nations with low level of energy and materials use and high GDP is practically empty indicating that dematerialization is a goal hard to be reached.

A cluster analysis has been carried out confirming the classification in sub-cubes, with the exception of a cluster that collects the countries close to the median values.

The temporal dimension is a key factor of this representation: more than the ranking of the nations in one year, the evolution of a single point in the diagram is meaningful in order to follow the behavior of a nation in time (see Figure 5). This can be useful to evaluate the effects of national policies in economic but also in social and environmental terms (Pulselli et al. 2015).

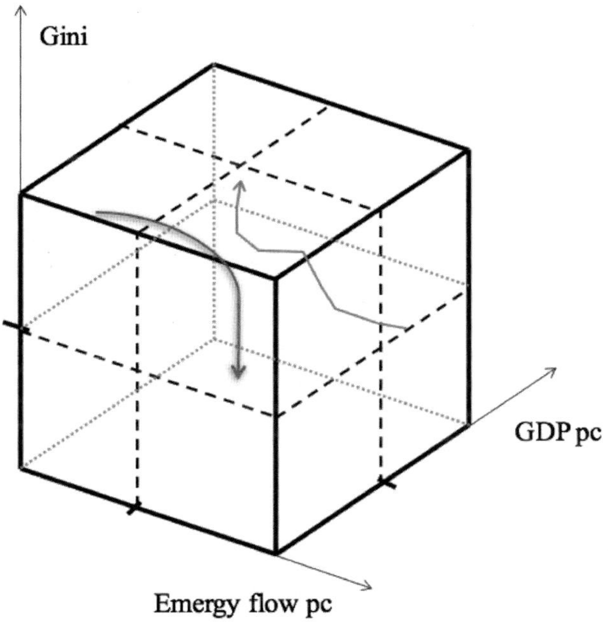

Fig. 5: Evolution of a nation within the cube framework.

REFERENCES

1. Barbier, E. (1987) "The Concept of Sustainable Economic Development." *Environmental Conservation*. 14:101-110.
2. Bastianoni, S., Coscieme, L., Pulselli, F.M. (2014) "The input-state-output model and related indicators to investigate the relationships among environment, society and economy." *Ecological Modelling. in press,* DOI: 10.1016/j.ecolmodel. 2014.10.015.
3. MA (2005) *Millennium Ecosystem Assessment*. Washington DC, Island Press.
4. Odum, H.T. (1988) "Self-Organization, Transformity, and Information." *Science*. 242:1132-1139.
5. Odum, H.T. (1996) *Environmental Accounting. Energy and Environmental Decision Making*, John Wiley and Sons, New York.
6. Sweeney, S., Cohen, M.J., King, D., Brown, M.T. (2007) Creation of a global energy database for standardized national energy synthesis, in: Brown, M. (Ed.), Energy Synthesis 4: Theory and application of energy methodology. Gainesville, FL, 23.1-23.15.
7. The World Bank (2014) The World Bank Group. http://worldbank.org.
8. Wilkinson, R., and Pickett, K. (2009) *The Spirit Level: Why More Equal Societies Almost Always Do Better*. London: Allen Lane.

NEAR TERM APPLICATIONS AND BENEFITS OF SHIP-BASED NUCLEAR ENERGY SYSTEMS

ROBERT PETROSKI
TerraPower LLC
Bellevue, WA, USA

NICHOLAS TOURAN
TerraPower LLC
Seattle, WA, USA

LOWELL WOOD
Hoover Institution, Stanford University, University of California and Lawrence Livermore National Laboratory, Stanford, CA, USA

ABSTRACT

Modern day shipyard construction techniques have enabled the deployment of extremely large floating facilities, including large scale (100s of MW-electric to >1 GW) nuclear energy systems. These ship-based energy systems appear to offer a variety of unique benefits relative to present modes of energy production, centered on minimizations of environmental footprint, effects on human health, total system costs, and attainable rates-&-latencies of deployment. Not only can they offer the exceptionally low externalities of modern nuclear power generation, but they can do so with far lower and more predictable costs, shorter lead times, and reduced financial risks to both vendors and customers. Well-designed versions of such systems can achieve an unparalleled level of manifest safety while strengthening resistance against proliferation and diversion.

Quantitative estimates of health and economic benefits are provided for the following application areas:

- Providing fresh water and reducing water consumption in water-stressed areas
- Improving public health by drastic curtailment of fossil fuel combustion
- Conservation of fluid hydrocarbon resources
- Novel marine applications
- Accelerating socioeconomic advanced in developing countries

Sizes and locations for each application area are illustrated, forming a picture of the potential global range of ship-based nuclear energy systems. Total quantities of avoided greenhouse gas and pollutant emissions are also estimated. Special consideration is given to prevention and mitigation of accidents and malicious behaviors, emphasizing avoidance of all harms to the public and minimization of economic losses.

It is concluded that ship-based nuclear energy systems have potential applicability that extends far beyond conventional land-based nuclear and fossil power plants in meeting planetary-scale energy needs, moreover in the next few decades, when societal needs are greatest. By acting as a source of reliable and readily-dispatchable baseload energy, they can complement other renewable energy systems to comprise a sustainable energy infrastructure that minimizes negative health and environmental consequences while reducing costs and conserving natural resources.

INTRODUCTION AND BACKGROUND

At the 2013 Session of these Seminars, the authors introduced a proposal for large-scale floating nuclear energy systems,[1] distinguished by high power ratings (from hundreds of megawatts electric to gigawatt scale), a readily deployable and re-locatable ship-based platform, and offshore siting far away from land. The key benefits of this approach relative to traditional land-based nuclear power generation are markedly-improved economics—via a combination of economies of scale and economies of mass production—and greatly-enhanced safety—in terms of drastic reductions in both accident frequencies and accident consequences.

Following this initial publication, additional researchers have also begun detailing the benefits of large-scale floating nuclear energy systems, and have begun developing innovative concepts at a range of power levels.[2] The idea of offshore floating nuclear energy systems has also attracted positive attention and interest from both the media and the nuclear engineering community.

This paper aims to build on this previous work by better defining the benefits that can be gained via deployment of ship-based nuclear energy systems. Five application areas are evaluated:

1. Providing fresh water and reducing water consumption in water-stressed areas
2. Improving public health by reducing combustion of fossil fuels
3. Conserving oil and gas resources
4. Developing new marine applications
5. Accelerating socioeconomic development

For each of these applications, geographic regions in which benefits can be realized are identified, and the degrees of benefit are quantified, in terms of total energy services provided, improvements in human health (measured in DALYs—disability-adjusted life years), and economic gains.

The individual application areas are then combined to give an estimate of the total applicability of ship-based nuclear energy systems. Positive and negative externalities of such systems are quantified, including avoidance of greenhouse gas emissions and diminishing the costs in life-&-health of supplying requisite energy to power human civilization, e.g., via reducing the frequencies and consequences of potential mishaps.

WATER

Water is an absolutely basic need for human life and civilization, and water crises have been identified by the World Economic Forum as the greatest global risk we face in the next 10 years.[3] Areas of water scarcity are highlighted in Figure 1, and include the major population centers of Northern China and India, as well as portions of North and South America, Northern Africa and the Middle East, and Europe.

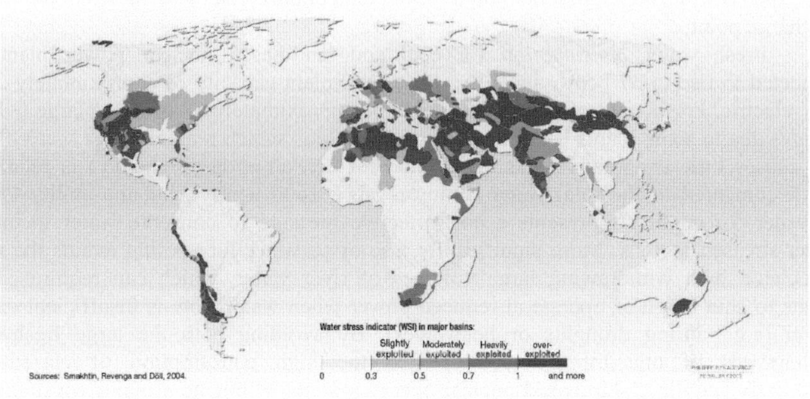

Fig. 1: Water stress indicator worldwide (ratio of withdrawals to renewable supplies).[4]

The 2030 Water Resources Group report *Charting Our Water Future* projects that without efficiency gains or increases in supply, water demand would become 64% higher than water supply by 2030: 6900 billion cubic meters per year (bcm/y) vs. 4200 bcm/y, a shortfall of 2700 bcm/y.[5] Even factoring in historical gains in efficiency and water supply, 60% of this shortfall, ~1600 bcm/y, would still need to be made up to close the gap between supply and demand.

While a perhaps feasible approach to closing this gap would be to increase supply through seawater desalination, this would be a very high cost remedy, requiring on the order of $1T per year, even with optimistic cost assumptions for producing and distributing desalinated water. Energy requirements for this approach would also be very large, of the order of 1000 GW electric, similar in magnitude to the capacity that China is projected to add over the next two decades.

Instead, the 2030 Water Resources Group report provides dozens of disparate measures to reduce water use and enhance supply, and proposes that the lowest cost measures be implemented first. The lowest cost measures (e.g., improvements in industrial water-use efficiency) actually bear negative costs. These measures, which vary from region to region, focus largely on agricultural and industrial efficiency improvements, since these represent the largest and the fastest growing sectors of water use respectively.

Ship-based nuclear power plants can help address water challenges in several ways. First, by virtue of their offshore siting, they do not require withdrawals or consumption of fresh water, since they can instead use seawater for cooling. This is directly analogous to current cooling practices in floating oil and gas facilities, which also have desalination equipment on board to provide fresh water used by the facility. Like these oil and gas facilities, a ship-based nuclear energy system can draw from cooler seawater below the surface to avoid creating thermal pollution and to very readily comply with the World Bank guideline of a <3°C temperature rise 100 m from the water discharge point.

Fresh water *consumption* for coal and land-based nuclear power plants is projected to rise to 37 bcm/y in 2035,[6] for a total plant capacity of approximately 2500 GW electric. Fresh water *withdrawals* would be approximately 50 times as large (~1700 bcm/y) due to some plants using once-through cooling, which requires large water flows but causes little evaporative loss. The value for consumption (37 bcm/y) is relatively small compared to the total water resource gap (~1600 bcm/y), but the ability to use seawater for cooling represents a cost-effective measure to conserve water in highly water-stressed regions. More significantly, use of seawater for cooling avoids the risks associated with withdrawing large volumes of river water, which can require power plants to shut down or operate at reduced power when water flow is insufficient or too warm (e.g., during droughts or heatwaves). By avoiding both the large freshwater withdrawals of once-through cooling and the high consumption of closed-loop evaporative cooling, ship-based nuclear energy systems eliminate a major constraint on electricity supply growth.

Energy from ship-based nuclear energy systems can be used to increase fresh water supplies in highly water-stressed regions, by providing energy to recycle wastewater and to desalinate seawater. Wastewater treatment and reuse are expected to increase markedly; the 2030 Water Resources Group anticipates the market for low-pressure membrane treatment to increase by over 5000% by 2030 to over 85 bcm/y in China alone. Energy requirements would be of the order of 1 kWh-electric per cubic meter, translating to approximately 10 GW electric of continuously operating capacity. Desalination is similarly a quickly-growing application, with capacity projected to more than triple to 280 million cubic meters per day (~100 bcm/y) by 2030. This desalination capacity corresponds to a generating capacity of approximately 60 GW electric, representing a significant new demand for electricity. Ship-based nuclear energy systems appear ideal for meeting this additional demand, by allowing desalination plants to operate at high capacity factors while requiring no fresh water themselves, in addition to being congenial with modest brine concentration factors which both diminish power requirements per unit of fresh water generated and also minimize concentrated brine-release impact on the local environment. Furthermore, if combined with floating desalination facilities, the two facilities can be co-located to provide desalination services to different areas as water shortfalls develop.

HEALTH

The large-scale negative health consequences of fossil fuel combustion, particularly coal, are quite substantial and are extensively documented. Health burdens due to fossil fuel combustion are caused in large part by fine particulate matter pollution, the inhalation of which engenders pulmonary and cardiovascular disease; the toxic gas fractions (e.g., sulfur oxides) of 'raw' (unscrubbed) stack gases often have major, independent health impacts. These health effects have been quantified in both bottom-up mechanistic modeling[7,8,9] and in statistical studies of air pollution, coal consumption and mortality.[10,11,12]

Worldwide, coal is used for 40% of electricity production, and in low- and middle-income countries coal is also commonly used for indoor heating and cooking, which results in indoor air pollution with similarly harmful effects. Figure 2 shows the fraction of coal electricity generation in each country. The map bears a remarkable, and non-coincidental resemblance to Figure 3, which shows the change in mortality rate due to outdoor fine particulate pollution from 1850 to 2000. While there are other sources of outdoor particulate pollution (e.g., transportation and industrial use), fossil fuel-burning power plants are the dominant contributor to outdoor air pollution.

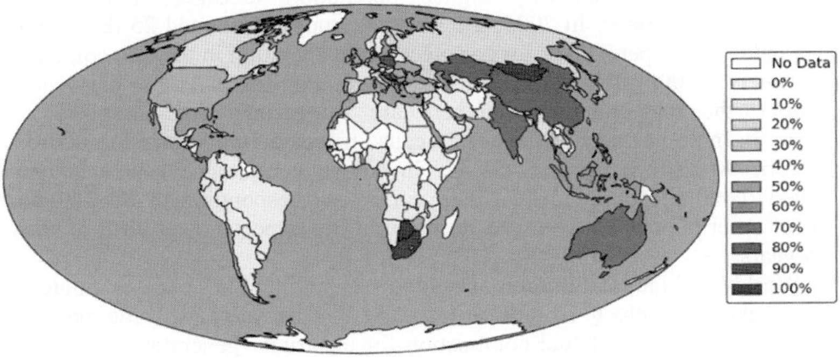

Fig. 2: Electricity production from coal in 2012, data from IEA, 2015.[13]

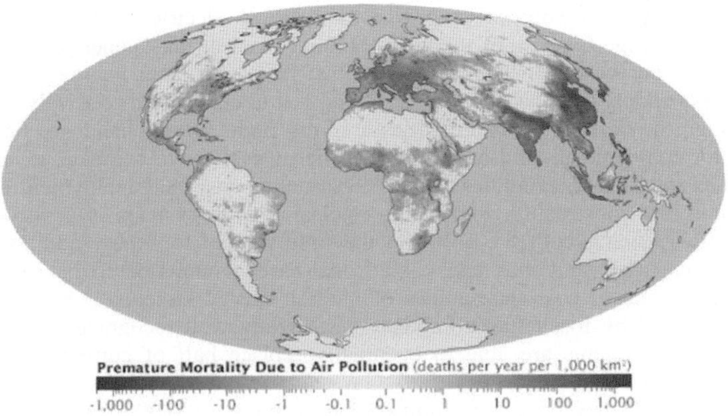

Fig. 3: Change in premature mortality between 1850 and 2000 due to outdoor air pollution.[14]

The resulting human health costs and resulting economic costs are immense. In the U.S., where coal generation is subject to stringent emissions standards, mortality from coal power generation has decreased from 13,200 deaths per year in 2010 to 7,500 deaths per year in 2014 (approximately 60,000 DALYs lost per year), or 5 deaths per TW-hr of generation.[9] The monetary costs of coal's health impacts alone are massive, ~$100 billion per year or 6 cents/kWh, comparable to the cost of producing electricity. Most extensive accounting of the externalities of coal electricity production in the U.S. in 2005 has yielded cost-estimates upward of $300 billion.[15]

Costs to human life and economic damage in other countries are higher still, due to lower emissions standards, higher population densities, and less-favorable meteorological conditions. In 2005, the ExternE study[16] estimated 25 deaths per TW-hr for coal and 4 deaths per TW-hr for gas electricity generation in the EU, corresponding to approximately 26,000 premature deaths (and ~200,000 lost DALYs) per year. In India and China, the corresponding numbers are staggering. Coal-energized electricity generation in India has been estimated to cause between 70,000 and 115,000 deaths per year,[17,18] or 90–140 deaths per TW-hr. In China, comparable estimates range from 300,000–670,000 premature deaths per year,[19,20] corresponding to 80–180 deaths per TW-hr, or a net population-averaged decrease in life expectancy at birth of between 2.5 and 5.5 years.

Worldwide, the total burden of outdoor air pollution has been calculated to be ~2 million premature deaths (~16 million lost DALYs) per year,[7] with the majority of this figure attributable to fossil fuel combustion for electricity generation. Even if dramatic measures are taken to reduce coal emissions and bring the average death per TW-hr down to levels equal that of the present-day EU, the number of premature deaths due to coal pollution would still number approximately 250,000 per year.

In order to reduce these health impacts, coal power generation would need to be replaced with cleaner sources of power to energize electricity supply, which is an enormous challenge because of the sheer amount of electricity generated by coal: over 9,000 TW-hr per year, considerably more than from any other source on Earth. Coal generation is widely used because it is easy and fast to deploy and relatively inexpensive, even though the costs of its externalities are very high. These challenges have made it difficult for conventional land-based nuclear energy systems to compete, given the large magnitudes and high uncertainties of their construction costs and deployment times.

Shipyard construction of floating nuclear energy systems is a very promising approach to reducing both the cost and the time needed to deploy additional nuclear generating capacity, and thus can enable coal-based generation to be replaced more quickly and at lower cost. The ability of nuclear energy systems to reduce premature mortality caused by fossil fuel combustion is already well recognized.[21] If ship-based nuclear energy systems are able to eventually replace half of current coal electricity generation worldwide (requiring ~600 GW electric, or roughly double present day nuclear capacity), they have the promise of preventing ~500,000 premature deaths per year and associated economic damages in the trillions of dollars.

RESOURCES

Ship-based nuclear plants have the potential to ease resource consumption in areas other than just electricity production from coal. Besides coal, which made up 40% of total electricity production in 2012, the other fossil fuel sources of electricity are natural gas

(22% of total) and oil (5%). While coal is primarily used for electricity production (72% of all coal consumption), gas and oil have numerous additional applications, with only 38% of gas and 8% of oil used to generate electricity.[13]

Outside of electricity production, coal is primarily used in industry (e.g., as coke for smelting iron), while natural gas is more versatile and is used as a clean domestic cooking and heating fuel, a transportation fuel, and a chemical feedstock—most famously to fix nitrogen into ammonia for fertilizer via the Haber–Bosch process, which consumes ~3% of all presently-produced gas (in contrast to marine propulsion, which consumes ~10% of all petroleum produced). Even within the area of electricity production, natural gas is especially useful as a fuel for peaking power plants, the Brayton-cycles generators for which feature low capital costs and the ability to rapidly change power outputs, i.e., very similarly to the performance characteristics of jet engines from which such generators are directly descended. Petroleum is a very high value resource and provides the majority of energy used in modern civilization, primarily in the form of transportation, aviation, and industrial fuels, and with a secondary application as a versatile chemical feedstock.

Because of these additional and high-flexibility uses, natural gas and oil have correspondingly higher market value per unit of energy than coal. Coal prices ranged from 2.5-3.6 dollars per million BTU (mmBtu) in different markets.[22] Natural gas, which is less widely distributed and much more expensive to transport, sees prices ranging from 3 dollars per mmBtu in North America, where it is currently notably abundant, to over 16 dollars per mmBtu in Japan, which imports it as liquefied natural gas. Oil prices are currently in a trough, at ~\$50 per barrel or 8 dollars per mmBtu, although barely a year ago it exceeded \$110 per barrel or ~20 dollars per mmBtu. Resource availability, expressed as the ratio of more-or-less proven, economically recoverable global reserves to production, is also lower for oil and gas, at 52.5 years and 54.1 years respectively, compared to 110 years for coal.[22] These timeframes would grow shorter as world energy demand increases, unless there is an associated increase in resource price to increase the amounts deemed to be economically recoverable. The 2014 World Energy Outlook[23] projects a 50% increase in energy demand by 2040 under current policies, or a 37% increase with policies designed to curb energy use and emissions, and meanwhile increasing population and rising standards of living will cause energy demand to eventually double or quite possibly more than triple. Reducing the amount of oil and gas used for electricity is therefore beneficial not only from a resource conservation standpoint, but also for quite fundamental strategic, energy security, and economic reasons.

Figures 4 and 5 depict the fraction of electricity produced with oil and natural gas, respectively. Areas that use a high fraction of oil include oil-producing regions as well as countries with relatively small or comparatively-inaccessible electricity grids. For these areas, electricity from petroleum, while expensive (10–20 cents per kW-hr electric for the crude oil alone, before it is refined into fuel) is nevertheless attractive because the resulting diesel fuel is easily transported and stored.

For countries with relatively low energy demand, conventional land based nuclear plants are less attractive, because the fixed costs of establishing a nuclear infrastructure—a regulatory body, a trained nuclear workforce, and nuclear construction and materials handling capabilities—are too costly to merit the construction of a small number of nuclear power plants. Furthermore, the nuclear fuel cycle presents additional challenges, including the need to source enriched fuel, guard against materials diversion, and dispose of nuclear wastes.

Ship-based nuclear energy systems appear uniquely suited to at least partly meet the energy needs of countries currently dependent on petroleum for their electricity, because they can be readily and swiftly transported and deployed following comparatively inexpensive serial manufacturing in shipyards accustomed to such endeavors, and furthermore allow the fixed costs of nuclear energy to be more effectively and equitably shared among many different countries. With identical units being operated in the same manner in the same environments, international regulatory cooperation can be more effectively implemented, similar to the current Western European Nuclear Regulators Association (WENRA). Furthermore, far less construction infrastructure would be needed—just enough for an onshore switchyard. Plant operation (and associated liability) can also be made the responsibility of the power plant owner. The customer country would never need to avail itself of any nuclear fuel cycle capabilities or services nor to stand up and operate a nuclear power infrastructure, since all of these could be bundled and made-portable by a forever-offshore supplier and included as part of a power purchase agreement (PPA). Perhaps most important, employing a PPA would mean that the customer country would not need to bear the upfront costs and risks of a nuclear construction project. Instead, this risk would be borne by the plant owner, which drastically minimizes per-plant risk because the owner-operator would develop extensive experience constructing multiple copies of the plant in a shipyard and then in operating them at various customer "power-drop points." The only requirement for the customer country would be to allow support for the security of the plant should the need arise, similar to that routinely provided for any other critical national infrastructure component.

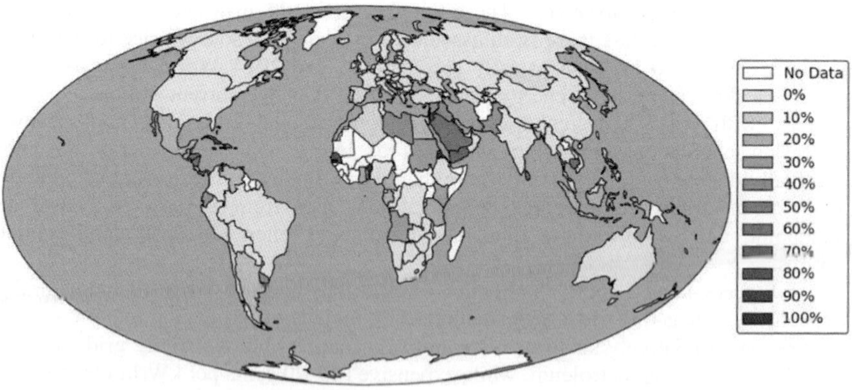

Fig. 4: Electricity production from oil in 2012.

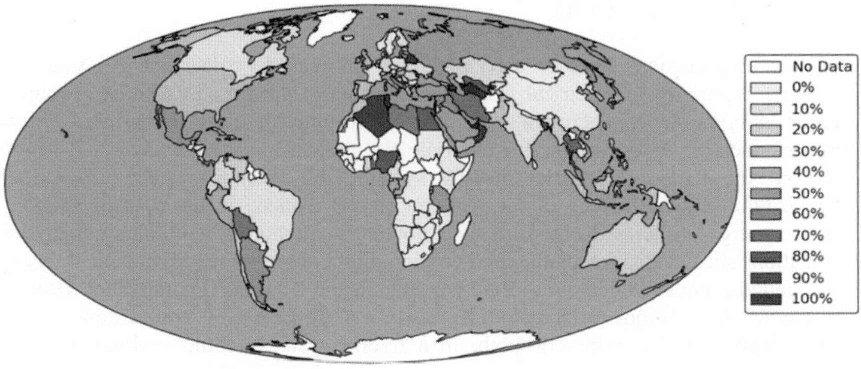

Fig. 5: Electricity production from natural gas in 2012.

Ship-based nuclear energy systems can similarly help replace natural gas as a fuel for electricity, which in some regions is an expensive option, e.g., fuel alone may cost more than 12 cents per kW-hr electric in Asia. Natural gas differs from oil in that is more expensive to transport, allowing greater global price variations to develop, so the applicability of nuclear energy would not be as uniform: for example, it would be more difficult to compete on basis of cost with natural gas in North America, which current incurs fuel costs of just 4 cents per kW-hr electric for natural gas-fueled combined cycle plants. However, this difficulty of transport can also carry energy security implications, since a country purchasing natural gas becomes more dependent on the countries near-or-far from which it purchases the (pipelined or liquefied) gas. The options associated with ship-based nuclear energy systems would intrinsically give such countries a more diversified generating portfolio and improved energy security.

If comparatively low-cost ship-based nuclear energy systems—with cost-structures associated with serial production in the planet's super-scale shipyards—were to become available, they could be expected to replace a significant fraction of current petroleum-based electricity generation for economic reasons alone. Electricity from oil totaled approximately 1100 TW-hr in 2012, nearly all of it in countries with substantial to large coastlines. Replacing half of this energy production with baseload nuclear generation would result in approximately 70 GW-electric of new capacity.

Because of natural gas's lower cost and its use in peaking plants, ship-based nuclear energy systems would likely not replace it quite as readily, at least in the most gas-favoring locales. Electricity generation from natural gas amounted to over 5000 TW-hr in 2012, approximately 5 times more than was realized from oil. Excluding North America (~30% of the total), where natural gas is currently very inexpensive, and taking 1/3 of the remainder implies ~1200 TW-hr per year, or approximately 150 GW-electric of new nuclear capacity.

NEW APPLICATION AREAS

The preceding sections have focused on using ship-based nuclear energy systems for supplying electricity to electrical grids because this is the primary use of present-day nuclear energy systems. There are some examples of nuclear energy being used for other purposes such as cogeneration; for example the Beznau plant in Switzerland supplies steam to the REFUNA district heating system, and weapons plutonium-producing reactors famously heat much of the Russian city of Tomsk. In the case of ship-based nuclear energy systems located many kilometers offshore, traditional cogeneration applications such as district heating and cooling might not be as readily practical, because of the added expense of deploying fluid conduits to and from the floating plant. However, because of their unique offshore siting attributes, several new application areas for nuclear heat and electricity—or both, in a hybrid system—could become potentially available.

One example of a new application is using produced heat or electricity to power offshore facilities, most notably offshore oil-and-gas systems that extract and process hydrocarbon fuels, modern versions of which have very extensive power, heating, and cooling requirements. One example of such a proposed system is the OPera concept by DNV,[24] a 700 MW, natural gas fueled combined cycle plant. A ship-based nuclear energy system could provide a similar service while potentially reducing cost and emissions.

Another proposed application for nuclear heat is to pyrolyze underground oil shale *in situ* to enable petroleum production.[25] Offshore nuclear energy systems can be used for this purpose to allow production from offshore oil shale resources that might otherwise be effectively inaccessible. In a similar but more advanced application, nuclear electricity and heat could also be used to produce fuels from future marine biomass farms that grow and harvest algae to supply carbon-neutral energy. A simpler hybrid energy application would be to use waste heat generated by a ship-based nuclear power plant to provide thermal regulation for aquaculture applications, the surprisingly-great utility of which is extensively documented.

While not a new application area *per se*, the ability to relocate ship-based nuclear power plants allows for new operating modes to be employed. One example of such an application is to better meet seasonal changes in energy demand. In more temperate climates, energy and electricity usage peaks in winter months (to provide heating), while in warmer tropical climate, usage peaks in summer (to provide cooling). Therefore, improved system efficiency and cost reductions can be realized by shifting generation from temperate to tropical climes in summer, and back in winter. Such a system would likely not entail power plants shuffling back and forth every season, which would negatively impact their capacity factors, but more likely involve a subset of power plants changing location as part of their normal maintenance outage schedule.

Because these new application areas are each more speculative than is the conventional application of supplying electricity, it is difficult to estimate the potential scale of each area. The degree to which ship-based nuclear energy systems will be employed will depend on both their cost and the demand for each application. Therefore, in totaling the potential applicability of ship-based nuclear energy systems, these new application areas are conservatively not included.

SOCIO-ECONOMIC DEVELOPMENT

A former Chairman of the Atomic Energy Commission, David Lilienthal, once proclaimed: "Energy is part of a historic process, a substitute for the labor of human beings. As human aspirations develop, so does the demand for and use of energy grow and develop".[26] With one billion people still lacking access to any electricity at all,[27] and many more with drastically-limited access—i.e., a few dozen Watts per capita in much of the Developing World compared to 0.5-1.5 kWe in the Developed World—this process continues today. In the Developed World, modern fuels and electrical appliances have replaced the strenuous activities of gathering wood for cooking and heating, hand-washing clothes and diapers, and fetching-&-conveniently supplying water of secured quality for human ingestion. Improved access to energy would enable families in the Developing World to reclaim these substantial opportunity costs and to better spend their performing easier, healthier, and higher-value tasks—as well as in presently far-too-scarce educational, innovative and recreational activities.

The causality associated with the positive correlations between GDP and energy consumption has been long debated, often from the perspective of understanding implications of energy conservation policies in the Developing World. Conclusions range from forward to reverse causality, including bidirectional feedback and neutral causality.[28] Additionally, results vary by country, and often by methodology. Thus, it is difficult to unequivocally and incontrovertibly predict and quantitate the direct effect of enhanced energy supply on GDP.

Consistency is found, however, when studying the relative change of energy demand due to changes in regional energy prices, as measured by the price elasticity of energy demand. Energy demand is generally inelastic worldwide in the short term, largely because energy use is set by fixed capital.[29] However, long term price elasticity is consistently found to be negative, meaning that as electricity prices fall, electricity demand increases. Studies have indicated that Namibia, Kuwait, South Africa, Thailand, Kazakhstan, Brazil, and Mexico have price elasticities more negative than -0.8, while many more, including India, are below -0.2.[30] Notably, China shows very near-zero price elasticity. Worldwide, the price elasticity of crude oil demand from 1918-2004 is -1.59, and -0.45 for natural gas, so as prices change, consumption patterns change in response, with both short term and longer term effects.[29] These figures suggest that were low cost ship-based nuclear electricity to arrive in Brazil or India, local energy use would certainly increase. With it, more productive tasks will be accomplished with less human effort.

The Human Development Index (HDI)—a composite of long life, good education, high standard of living, environmental sustainability, human rights, and gender equality—is strongly and positively correlated with energy use, in essentially every circumstance-studied. Thus, while it's not proven that quality of life will improve *because* of increased *per capita* energy demand, more energy is rather necessary to accommodate such improvements. As a basic example, increased electrification allows for lower reliance on indoor biomass- and coal-burning for space-heating and cooking, therefore leading to major improvements in indoor air quality and respiratory health, in addition to the benefits sketched in Section 3.[31]

The coexistence of negative price elasticity of electricity demand in many parts of the world and the global positive correlation between the HDI and energy consumption suggests that the introduction of low-price, low-emissions ship-based nuclear energy systems in developing areas will, at the least, accommodate an improvement in quality of life, if not directly impel it. If improved energy resources can in fact improve GDP and/or local values of HDI, then such an introduction will be directly responsible for substantially improving the lives of billions.

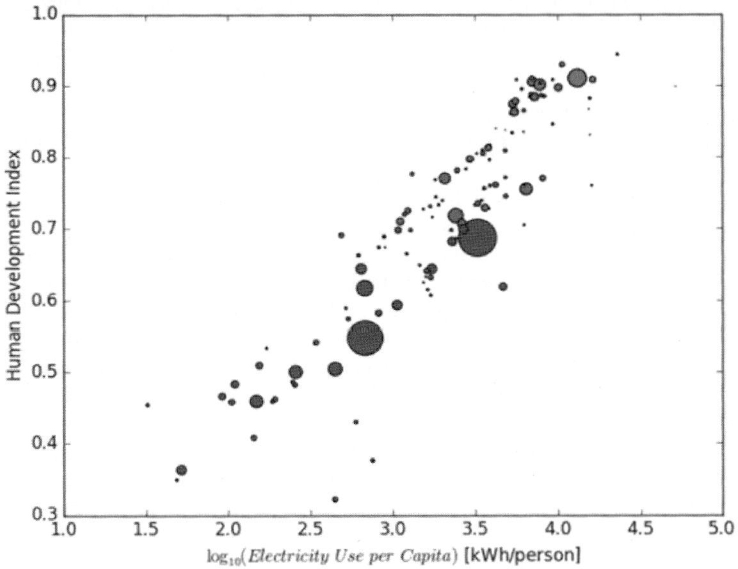

Fig. 6: Human development index as a function of electricity use, with marker sizes proportional to country populations. Data from Human Development Report, 2014.[32]

The trend around the world naturally shows this development taking place. Figure 7 shows the growth rate in electricity generation between the years 2000 and 2012, which has been extremely high (up to 15% per year) in developing countries, and averaged 3.4% per year worldwide. While energy use is projected to increase by 37% from 2012 to 2040 in the IEA "new policies" scenario,[23] electricity use is expected to grow by 78% over the same period. In addition to new capacity additions, existing generating capacity will be retired and require replacement. There will therefore be an enormous demand for new electrical generating capacity over the next several decades, one that ship-based nuclear energy systems can help meet with very low costs, low environmental impacts, and strictly-minimized health-&-safety consequences.

A key concern of increased energy intensity in the developing world is the perceived increase in environmental damage, which is an important component of development. However, environmental impact is related to energy density, and it has been observed that transitioning from utilization of biomass to more energy-dense fuels actually reduces threats to forests, land, and groundwater. A ship-based package with a PPA-like arrangement uniquely provides access to nuclear electricity to any energy-demanding population regardless of the local availability of a high-tech nuclear labor force, infrastructure and regulatory expertise. While nuclear energy is often ignored as a possible contributing solution for many developing parts of the world, ship-based nuclear power supplies make it an intriguing, indeed eminently-realistic possibility.

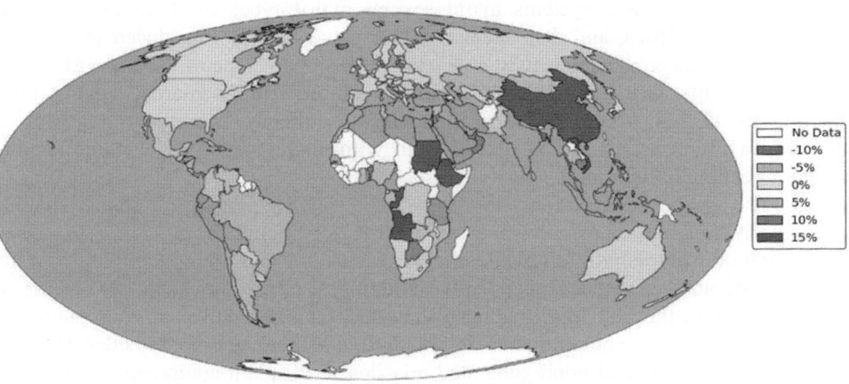

Fig. 7: Annual growth in electricity generation from 2000 to 2012.

TOTAL SYSTEM CHARACTERISTICS

In the preceding sections, ship-based nuclear power plants have been postulated as a replacement for significant fractions of present day fossil fuel-based electricity generating capacity, as well as an attractive path for adding major new capacity. In total, it has been noted that of the order of 820 GW-electric of capacity can be employed while yielding major human health and economic benefits—and that without taking any credit for water desalination, HVAC and potential new applications. This figure could exceed 1,500 GW after factoring in growth of electrical power generation in the coming decades, which would enable ship-based nuclear energy systems to supply ~12,000 TWh of electricity out of a total 40,000 TWh generated worldwide in 2040.

Meeting this level of demand will require on the order of one thousand to several thousand ship-based nuclear energy systems. This significant fleet would clearly be sufficient to allow economies of mass production to be reliably accessed, and also would support the coexistence of multiple system vendors competing to best serve a global market in terms of cost, delivery time, and operational flexibility.

Safety

A key question about such a fleet is how safe it would be. The U.S. Navy has demonstrated that marine nuclear reactors can be operated extremely safely, with 6200 reactor years of accident-free operation extending across five decades and 526 reactors cores,[33] in operational circumstances far more demanding than would be experienced by stationary floating nuclear power plants. Modern commercial reactor designs are extremely safe, with probabilistic risk assessments estimating core damage frequencies due to plant operation on the order of 5 x 10^{-7}/y.[34] At these low frequencies, risk becomes dominated by rare but extreme external events (that can occur with higher frequency) that can simultaneously challenge many systems at once, such as earthquakes, fires, and floods.

Floating nuclear systems avoid several major types of these events, such as earthquakes, wildfires, and dam breaks. While new risks are introduced by the marine environment—large storms, waves, and vessel collisions—none of these are as violent or as unpredictable as major earthquakes, and are more readily guarded against through appropriate design, siting, and operation. This is already commonly done for large floating facilities, which can be designed to withstand 10,000-year cyclone conditions without needing to unmoor.[35]

The proposed locations for ship-based nuclear energy systems should be particularly suitable for avoiding marine hazards: far enough from shore (~20–100 km, depending on water depth) to obviate effects of tsunamis and running aground, but not so far as to be subject to deep ocean wave conditions. Since they will not require large storage tanks in the hull like floating oil-and-gas facilities do, plant equipment can be located in protected compartments in the hull, so there are no topsides that would vulnerable to storms and wind-generated missiles. With plant equipment within the hull and below the waterline, it is possible to use surrounding ocean water as a limitless source of containment cooling, conferring additional safety and allowing for an indefinite outage-coping time with no power or operators present.

With these safety advantages, achieving a core-damage frequency of 1 x10^{-6}/y or lower appears eminently achievable, meaning a fleet of ship-based nuclear energy systems supplying 1/3rd of planetary electricity might suffer one accident every several centuries or millennium of operation. The majority of such accidents would be contained completely within the reactor containment, meaning that the only damage sustained in such cases would be the economic loss of the plant.

Even if radioactivity were to be released from containment, the consequences of such an accident would be remarkably-benign relative to a similar accident at a land-based plant: there would be no evacuations, no contaminated land, and because water naturally traps radionuclides, no air plume. Any radioactivity released into water would naturally disperse and decay, for example, radioactivity released by the Fukushima accident into the ocean fell to far below background levels within the first year: <100 Bq/m^3 of Cs-137 compared to 12,000 Bq/m^3 of naturally occurring radioactivity in the same locale.[36]

Security

Ship-based nuclear energy systems would employ measures similar to those used by the oil and gas industry to protect its multi-billion dollar, environmentally-sensitive, and strategically-important assets. These measures include establishing an exclusion zone around the vessel, equipment to detect and deter anyone approaching the vessel (including from below the surface), support boats, and procedures to lock down important areas, such as the reactor containment and control room. Large floating facilities are challenging to assault due to their distance from shore, high freeboards, and their abilities to automatically detect assailants approaching from far away. The large size, mass, and extensive compartmentalization of the vessel would prevent direct attacks (e.g., via explosive rockets, explosives-laden boats, or ship/aircraft collisions) from threatening the integrity of the vessel or its reactor systems. Even if they could be successfully assaulted, offshore energy systems would not make appealing targets for attack simply because any consequences would be primarily economic, with no harm (other than loss-of-services) imposed on even the most proximately-located portions of the public.

Proliferation

Another hypothetical scenario is one in which assailants attempt to steal nuclear materials from a ship-based nuclear power plant. Such a scenario would be at least as unlikely as for a land-based plant. The assailants would have to defeat plant security and deploy specialized heavy equipment to move highly radioactive fuel assemblies off of the vessel, while somehow avoiding military counteractions. With a once-through fuel cycle, the low-enrichment fresh fuel transported to the ship-based plant would be completely unsuitable for use in a weapon.

Ship-based nuclear energy systems would also help guard against surreptitious diversion of nuclear materials, because a floating platform natural provides an easily monitored spatial boundary that is only accessed at discrete intervals: i.e., when a service ship or helicopter meets with the vessel. Spent fuel would be stored in a spent fuel pool on board the vessel, and would not be removable without specialized external equipment located at a maintenance shipyard.

Perhaps most important, ship-based nuclear energy systems would be able to help discourage the proliferation of nuclear technology, particularly fuel cycle technology, since they provide a pathway for nations to take full advantage of nuclear energy without needing to set up domestic nuclear construction and nuclear fuel infrastructures. With conventional deployment of land based reactors, a nation may be incentivized to develop fuel cycle capabilities to protect their sunken investments in nuclear power plants, but with ship-based reactors financed through PPAs, both nuclear capital risks and fuel cycle risks are centralized to a far-smaller number of much more readily-monitored suppliers.

Climate change

Because ship-based nuclear energy systems would primarily displace existing and new coal power generation as a source of affordable baseload and dispatchable electricity, they will naturally induce major reductions in greenhouse gas emissions. Producing 12,000 TWh/y of electricity with nuclear energy instead of with fossil fuels (coal, oil, and gas are displaced in a 3/4–1/12–1/6 ratio) would avoid 10 gigatons of CO_2 emissions per year, about 32% of total emissions (31.6 Gt) in 2012.

This degree of reduction in fossil-fired generation is similar to the difference between the IEA "New Policies" and "450" scenarios, where the 450 scenario refers to the concentration in ppmv of atmospheric CO_2-equivalents that corresponds to the international target of 2°C warming relative to pre-industrial levels. In the 450 scenario, CO_2 emissions peak before 2020 and decrease to 19 Gt/y, instead of rising to 38 Gt/y as in the New Policies scenario. Ship-based nuclear energy systems can comprise a major fraction of this needed reduction, and thus represent an important, low-to-negative cost option for addressing climate change alongside increased energy efficiency and use of other renewable energy sources having non-baseload and non-dispatchable characteristics.

Renewability and sustainability

In the context of climate change and the philosophical desire for 'sustainable' energy sources, it's appropriate to invite consideration of the undisputed facts that the ocean contains several hundred thousand years of uranium at full planetary use-rates in reactor-types of unquestioned feasibility, and that extraction of this gigantic store at unit costs not much greater than contemporary marketplace ones has already been repeatedly demonstrated, and that this inventory is continually renewed by the action of rainfall on (especially, granitic) mountain ranges via riverine drainage, a renewal process which will persist until the Sun grows cold. It's also interesting to note that already-mined-&-refined uranium sufficient to power the entire planet at present energy-use rates for more than a thousand years already exists in the storage-yards of the major nuclear powers as 'depleted' but still entirely energy-potent uranium. This constitutes indefinite sustainability in all practical terms.

CONCLUSIONS

The challenges presently faced in practically/realistically replacing both baseload-quality and dispatchable electricity generation currently powered with pollution-intensive, large carbon-footprint means on time-scales of years vs. decades and at affordable costs are very well-known and extensively documented, if not discussed extensively in debates too often rich in high-intensity wishful and non-quantitative thinking.

The swift pace of technological advance has eliminated all but the traditional relatively high-cost disadvantage of nuclear fission power, although the huge gains in safety and minimized release of radioactivity of modern nuclear power plants are remarkably little-recognized.

Ship-based nuclear power offers the realistic prospect of obviating the last significant impediment of nuclear power, its comparatively high cost and associated long time-from-commitment to power on-line. Sustainability and renewability issues simply don't exist in the *physical* world of nuclear power (but only in some portions of the as-perceived one).

Thus, truly-modern nuclear power systems—including ones providing societal essentials other than electricity, such as potable water and HVAC services—combined with marine basing can provide-at-full-scale the full set civilization-energizing desirements, with essentially none of the traditional burdens and hazards.

The primary remaining challenge is a successful demonstration of offshore ship-based nuclear energy systems being deployed in a timely, economic, safe, and secure manner. Such a demonstration would be instrumental in overcoming outmoded perceptions of nuclear energy systems as being expensive, slow, or high risk.

ACKNOWLEDGEMENTS

The authors would like to acknowledge Abdulla Alhajri for his aid in collecting data and preparing the world map figures in this work.

REFERENCES

1. Petroski, R., Wood, L. (2013) "Ship-based Nuclear Energy Systems for Accelerating Developing World Socioeconomic Advance." *The 46th Session of the International Seminars on Planetary Emergencies*. Erice, Italy, August 20-23 2013.
2. Buongiorno, J., Jurewicz, J., Golay, M., Todreas, N. (2015) "Light-Water reactors on Offshore Floating Platforms: Scalable and Economic Nuclear Energy." *The 2015 International Congress on Advances in Nuclear Power Plants (ICAPP 2015)*. Nice, France. May 3-6 2015.
3. World Economic Forum (2015) Global Risks 2015, 10th Edition. Geneva.
4. United Nations Environmental Program, GRID-Arendal. (2015) Vital Water Graphics: An Overview of the state of the World's Fresh and Marine Waters, 2nd Edition. 2008. Available online: http://www.unep.org/dewa/vitalwater/ article77.html, accessed July 1, 2015.
5. 2030 Water Resources Group (2009) *Charting Our Water Future: Economic frameworks to inform our decision making.*
6. World Energy Council (2010) *Water for Energy*. London, United Kingdom. ISBN: [978-0-946121-10-6]
7. Silva, R.A., West, J.J., et al. (2013) "Global premature mortality due to anthropogenic outdoor air pollution and the contribution of past climate change." *Environmental Research Letters*. 8 034005. July 2013. Available online: stacks.iop.org/ERL/8/034005.
8. Markandya, A., Wilkinson, P. (2007) "Electricity Generation and Health." *The Lancet*. vol, 370 pp 979-990. September 15, 2007.
9. Clean Air Task Force. (2010) *The Toll from Coal: An Updated Assessment of Death and Disease from America's Dirtiest Energy Source*. Boston, MA, September 2010.
10. Chen, Y., Ebenstein, A., Greenstone, M., Li, H. (2013) "Evidence on the impact of sustained exposure to air pollution on life expectancy from China's Huai River policy." *Proceedings of the National Academy of Sciences*. vol. 110 no. 32, pp 12936–12941. August 6, 2013. Available online: www.pnas.org/cgi/doi/10.1073/ pnas.1300018110.
11. Gohlke, J.M., Thomas, R. et al. (2011) "Estimating Global Public Health Implications of Electricity and Coal Consumption." *Environmental Health Perspectives*. vol. 119, no. 6. June 2011.
12. World Health Organization (2009) *Global Health Risks: Mortality and burden of disease attributable to selected major risks*. Geneva, Switzerland. ISBN: 978 92 4 156387 1.
13. International Energy Agency (2015) Electricity Information database (2015 edition). July 2015.

14. NASA Earth Observatory (2015) "The Global Toll of Fine Particulate Matter." September 19, 2013. Available online: http://earthobservatory.nasa.gov/IOTD/view.php?id=82087, accessed July 1, 2015. Image by Robert Simmon.
15. Epstein, P.R., Buonocore, J.J. el al. (2011) Full cost accounting for the life cycle of coal in "Ecological Economics Reviews." Annals of the New York Academy of Sciences 1219: 73-89.
16. European Commission (2005) ExternE: Externalities of Energy: Methodology 2005 Update. EUR 21951. Brussels.
17. Lagarde, C (2012) International Monetary Fund. Back to Rio–the Road to a Sustainable Economic Future. Washington DC, June 12 2012. Text available online: https://www.imf.org/external/np/speeches/2012/061212.htm.
18. Conservation Action Trust, Urban Emissions (India), Greenpeace India (2013) *Coal Kills: An Assessment of Death and Disease caused by India's Dirtiest Energy Source*. March 12, 2013.
19. Eliason, B., Lee, Y.Y., eds. (2003) Integrated Assessment of Sustainable Energy Systems in China, The China Energy Technology Program: A Framework for Decision Support in the Electric Sector of Shandong Province. Springer Netherlands. ISBN: 978-1-4020-1199-3
20. Jing, L. (2014) South China Morning Post. 670,000 smog-related deaths a year: the cost of China's reliance on coal. November 5, 2014. Available online: http://www.scmp.com/news/china/article/1632163/670000-deaths-year-cost-chinas-reliance-coal?page=all, accessed July 1 2015.
21. Kharecha, P.A., Hansen, J.E. (2013) "Prevented mortality and greenhouse gas emissions from historical and projected nuclear power." *Environmental Science & Technology* 47.9:4889-4895.
22. BP. (2015) Statistical Review of World Energy 2015. June, 2015. Available online: www.bp.com/statisticalreview, accessed July 1 2015.
23. International Energy Agency (2014) World Energy Outlook 2014. IEA.
24. DNV GL. OPera: Offshore Power system for the new era. DNV Innovation Project 2012.
25. Curtis, D., Forsberg, C., Garcia, H. (2015) Development of Nuclear Renewable Oil Shale Systems for Flexible Electricity and Reduced Fossil Fuel Emissions. The 2015 International Congress on Advances in Nuclear Power Plants (ICAPP 2015). Nice, France, May 3-6 2015.
26. Lilienthal, D.E. (1980) Atomic energy: A new start.
27. World Bank. (2015). Data retrieved July, 2015, from World Development, Indicators Online (WDI) database.
28. Lee, C.C. (2005) "Energy consumption and GDP in developing countries: a cointegrated panel analysis." *Energy Economics*. 27.3:415-427.
29. Krichene. N. (2005) A Simultaneous Equations Model for World Crude Oil and Natural Gas Markets, International Monetary Fund, WP/05/32.
30. Khanna, Madhu, and Narasimha, D. Rao (2009) "Supply and demand of electricity in the developing world." *Annu. Rev. Resour. Econ.* 1.1:567-596.
31. Anderson, D. (2000) Energy and economic prosperity. UNDP, World Energy Assessment. Energy and the Challenge of Sustainability. New York: UNDP/UNDESA/WEC, pp394-413.

32. Human Development Report (2014) "Sustaining Human Progress: Reducing Vulnerabilities and Building Resilience". HDRO (Human Development Report Office) United Nations Development Programme.
33. World Nuclear Association. Nuclear-Powered Ships. May 2015. Available online http://www.world-nuclear.org/info/Non-Power-Nuclear-Applications/Transport/Nuclear-Powered-Ships/, accessed July 1, 2015.
34. U.S. Nuclear Regulatory Commission. Final Safety Evaluation Report Related to Certification of the AP1000 Standard Design (NUREG-1793, Initial Report), Chapter 19, "Severe Accidents." September 2004. Available online: http://www.nrc.gov/reading-rm/doc-collections/nuregs/staff/sr1793/initial/
35. Shell Development (Australia) Proprietary Limited. Prelude Floating LNG Project Draft Environmental Impact Statement. EPBC 2008/4046, October 2009. Available online: http://s05.static-shell.com/content/dam/shell-new/local/country/aus/downloads/about-shell/prelude/execsummstandalonelowres.pdf
36. Behrens, E., Schwarzkopf, F.U., Lübbecke, J.F., Böning, C.W. (2012) "Model simulations on the long-term dispersal of ^{137}Cs released into the Pacific Ocean off Fukushima." *Environ. Res. Lett.* vol. 7 no. 3.

EUROPEAN ENERGY POLICY AND GLOBAL REDUCTION OF CO_2 EMISSIONS: TOWARDS AN EFFECTIVE SUSTAINABLE ELECTRICITY PRODUCTION IN EUROPE

JEF ONGENA
Plasmaphysics Laboratory, Ecole Royale Militaire
Brussels, Belgium

THE POSITION OF THE ENERGY GROUP OF THE EPS

The forthcoming United Nations Climate Change Conference (Paris, December 2015) will be held with the objective of achieving a binding and global agreement on climate-related policy from all nations of the world.

This conference, seeking to protect the climate, will be a great opportunity to find solutions in the human quest for sustainable energy as a global endeavour. The Energy Group of the European Physical Society (EPS) welcomes the energy policy of the European Union (EU) to promote renewable energies for electricity generation, together with energy efficiency measures. This policy needs to be implemented by taking into account the necessary investments and the impact on the economical position of the EU in the world. Since the direct impact of any EU energy policy on world CO_2 emissions is rather limited, the best strategy is to take the lead in mitigating climate change and in developing an energy policy that offers an attractive and economically viable model with reduced CO_2 emissions and lower energy dependence.

The Energy Group of the EPS has the following observations on the **presently** planned energy transformation in Europe:

i. **Europe alone cannot curb the increase in global CO_2 emissions. Global action is required.**
The specific emissions due to power generation in Europe (352 gCO_2/kWh in 2011[1]) are already only half as large as those in e.g., China and the EU contribution to the global CO_2 emissions is currently only about 11%. So, the recently proposed implementation[2] of the 2050 Energy Roadmap,[3] aiming, *inter alia*, at a reduction by 40 to 50% in the total CO_2 emissions of the EU by 2030, would correspond to an estimated saving of a mere 4 to 5% in global CO_2 emissions,[4] even after decades of enormous investments. The impact of reductions in CO_2 emissions in the EU will therefore remain marginal and will be massively overcompensated by rising emissions elsewhere, unless other nations contribute their share in reducing CO_2 emissions.

ii. **An efficient expansion of renewable energy sources requires solutions for intermittency and storage.**
The substantial CO_2 reductions recommended in the 2050 Energy Roadmap for the electrical power sector (as compared to 1990: a reduction of 57-63% by 2030 and 93-99% by 2050) implies a vast transformation of the existing power supply system in the EU (in which about 50% of the electricity production currently comes from fossil fuels) into either renewable or nuclear power or some combination. Already at the present level of penetration of

renewables (~23% of the electricity production), solutions are urgently needed to tackle adequately the problem of intermittency. This will demand a combination of high capacity non-intermittent (e.g., nuclear, oil and coal fired) and flexible backup plants (e.g., gas, hydropower), large electrical energy storage capacities and substantial expansion of electricity distribution grids, including smart grids. Furthermore, the integration of this variable electricity supply is expected to become even more difficult as its percentage rises to above 30-40%. The expected change in the energy system is a colossal challenge (Wagner, 2014) and is expected to raise major technological and political issues requiring long-term perspectives and sustained investment in research and development.

iii. **The EU energy policy ought to be framed to strengthen Europe's economic position in the world.**
The level of subsidies for carbon-free energy sources ought to be set to assist in promoting the necessary research and to provide a competitive environment for developments in new energy technologies. Excessive subsidies are currently causing high electricity prices that are far beyond those in other regions of the world, contributing to the on-going relocation of industry from Europe into other regions of the world with less expensive energy. There is growing concern over the accelerating relocation from electricity intensive industries in the EU into countries with lower environmental standards,[6] leading not only to increased European unemployment but also to increases in global greenhouse gas emissions.

iv. **Europe can lead the global effort in CO_2 reductions by proposing and demonstrating an attractive and economically viable model.**
In Europe we are facing a global challenge that requires a multi-decade approach, going beyond the present "European Energy Transition Model". This will require a coherent and sustained energy policy that strengthens the mutually beneficial relationship between education, research and innovation.

THE ENERGY GROUP OF THE EPS THEREFORE RECOMMENDS THE FOLLOWING

1. **A review of current electrical energy subsidies in the EU.**
A review of current electrical energy regulations in Europe (subsidies, feed-in tariffs, etc.) is urgently needed and should focus on supporting the internal energy market in providing clean energy at competitive prices. This implies in particular enhanced efforts in research and innovation for sustainable energy technologies.

2. **The inclusion of external costs when pricing electricity from all supply options.**
The cost of various electricity production options should be correctly quantified by taking into account external costs, including an adequate carbon price, costs linked to societal and health risks and the additional costs inherent in the use of large amounts of intermittent power generation. This last point will become very important when implementing the 2050 Energy Roadmap,

as it will be essential to have parallel development programmes for energy storage technologies (batteries, power-to-gas, etc.), back up power solutions, upgrades of distribution grids including smart grid technologies and energy efficiency measures (heating, cooling, light,…).

3. **An increase in R&D funding for sustainable technologies with a focus on developing an effective and economically viable low carbon electricity system.**
   R&D in improved technologies for the electricity sector together with Carbon Capture and Storage is a key to achieving a sustainable electricity future. The return on this investment will be a competitive advantage in the rapidly growing world market for sustainable technologies.
4. **Consideration of all non-fossil electricity options when discussing the EU energy future.**
   The present focus on the deployment of intermittent renewables should be complemented by developing and implementing other low carbon options, which are able to provide base load and dispatchable power to the grid (such as second generation biomass, geothermal, nuclear fission at present and fusion in the longer term). The objective should be to provide a balanced energy mix, which is economically optimized and ensures security and diversity of supply. Although the present policies on nuclear fission differ in individual EU countries, this option and the related know-how, remains globally important. An adequate level of research and technological competence must be maintained to keep open the option of building next generation fission reactors, in addition to fusion technology at a later stage.
5. **Negotiation of a global agreement for a reduction in worldwide CO_2 emissions.**
   The implementation of a globally relevant and economically viable future energy policy, agreed upon by all European nations, will strengthen the position of Europe as a leading party in on-going international negotiations towards a worldwide agreement on reducing global CO_2 emissions, together with other major greenhouse gases. Global emission reductions can only be achieved if all nations in the world assume their responsibility for mitigating greenhouse gas emissions.
6. **A revisit of the 2050 Energy Roadmap.**
   What is needed is a Roadmap that not only sets targets with defined dates in terms of amounts of installed electrical power, energy production and demand and emission reductions, but also takes into account technology development times, safety and socio-economic aspects of various energy technologies. As low carbon options become more competitive, the choice of the technology mix should increasingly be left to the market, taking into account national and regional differences.
7. **Fostering scientifically and factually based educational programmes for students and the general public on energy use and energy technologies.**
   The EPS calls for discussions on the EU Energy Policy and on these recommendations in all relevant contexts. The EPS also encourages strengthening the voice of scientific and technical knowledge in the field of

energy, with the aim of assisting in the definition of a farsighted and effective energy policy.

REFERENCES

1. CO_2 Emissions from fuel combustion, Highlights, IEA Statistics, 2013 Edition.
2. "A policy framework for climate and energy in the period 2020 to 2030", COM (2014) 015, EC Brussels, 22 Jan 2014.
3. "Energy Roadmap 2050", COM (2011) 885, EC Brussels, 15 Dec 2011.
4. EDGAR, Emission Database for Global Atmospheric Research (http://edgar.jrc.ec.europa.eu).
5. F. Wagner (2014) "Considerations for an EU-wide use of renewable energies for electricity generation". *Eur. Phys. J. Plus.* 129:219.
6. "Resolution and Report on renewable energies and economic competitiveness", EEA Consultative Committee, Ref. 1130862, Oslo, 8 May 2014; "Impact of energy and feedstock costs on the competitiveness of the chemical industry in the ARA-cluster", Essenscia, Brussels, 8 May 2014; "Development and integration of renewable energy: lessons learned from Germany", Finadvice, Aldiswil, June 2014.

PMP REPORTS AND CONCLUSIONS

CLIMATE, SCIENCE, AND PEOPLE: CLIMATE PMP REPORT

CHRISTOPHER ESSEX
Department of Applied Mathematics, University of Western Ontario
London, ON, Canada

Leading up to yet another highly-promoted United Nations event on climate in December, climate as a topic remains filled with daunting scientific and human problems.

THE SCIENTIFIC PROBLEM

At the heart of the scientific problem is "complexity," in the technical sense, which signals fundamental theoretical, computational, and experimental challenges to modern science. It is a ubiquitous part of our lives that can be found nearly anywhere, if one simply looks.

Figure 1 shows a green laser sheet made by rapidly moving the beam of a green laser pointer in a plane, then intersecting it with the smoke plume of a simple campfire. The resulting image shows well-known features and shapes of complex turbulent motion, which are repeated on all scales of space and time in the atmosphere and oceans. The existence of these phenomena ensures that the climate problem is not only unsolved, but also remains ill defined in any rigorous sense. We have only been able to nibble on the margins of this topic. That core truth has not changed since the beginning of the computer age.

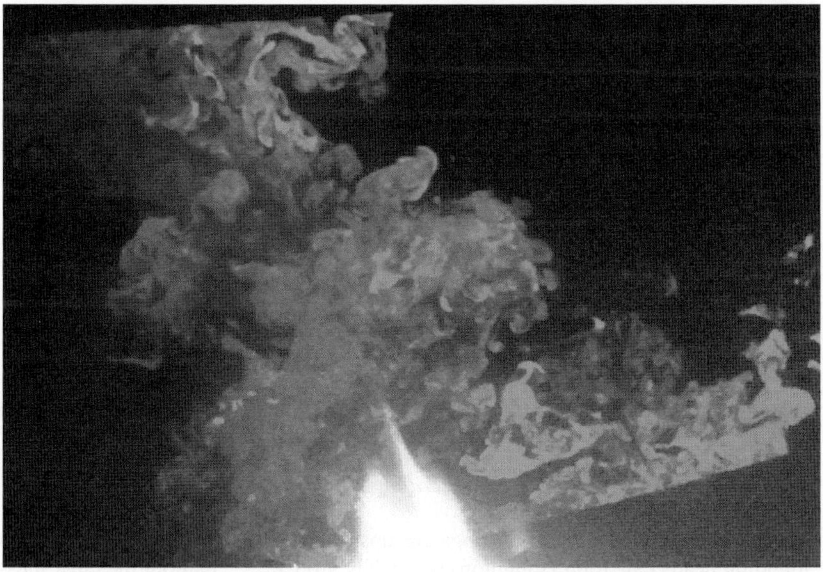

Fig. 1: Turbulence revealed in a campfire using a laser pointer.

THE HUMAN PROBLEM

I Can Calculate the Motion of the Heavenly Bodies, but Not the Madness of People—Isaac Newton

However for nearly 30 years governments, politicians, lobbyists, and many academics have operated as if this core problem does not exist, or that it is of no significance. Many in fact have no idea that there is any sort of scientific difficulty at all, let alone one that challenges modern science to its core. Others sense there is something there but proceed carefree without further exploration.

In this intellectual vacuum, they pretend to formulate and propose policy to engineer a physical system that is not understood well enough to even be an engineering problem. Social engineering becomes the substitute for actual engineering and the science that they have never learned. The deep physical, mathematical, and computational issues of the scientific problem are simply lost from discourse in favor of environmental morality and social engineering. These usurpers fancy that by pushing science into the background, like bullies, they can claim the credibility of science for their other agendas.

In contrast, the premise of these meetings here in Erice is that rigorous rational thought about science has to be the sound basis of solving human problems concerning the natural world. All policies with this aim must be set upon a sound foundation of science (as in Figure 2) in the first instance to avoid directives that contradict the laws of nature, and to be aware of what scientifically the sound options actually are.

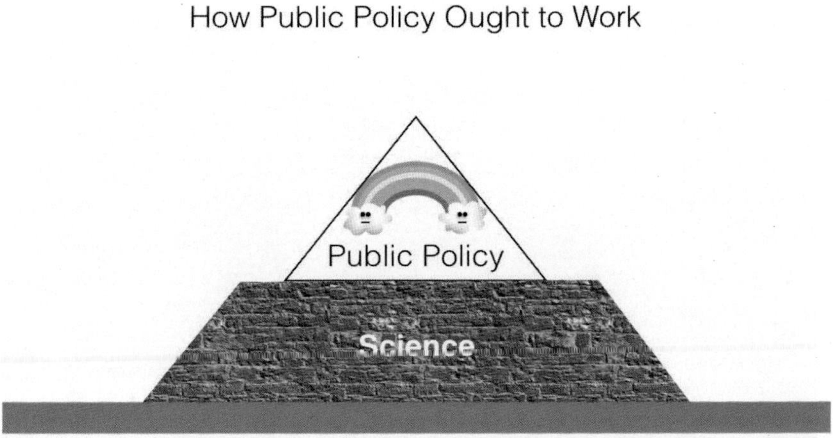

Fig. 2: Policy with a Strong Foundation in Science.

But in the field of climate things are upside down, as was made clear by one of the founding heads (Sir John Houghton) of the UN's IPCC: "...without the clear message which came from the world's scientists, orchestrated by the IPCC, the world's leaders would never have agreed to sign…"

Figure 3 depicts the meaning of this statement, which thoroughly characterizes all aspects of the relationship between science and policy in connection with climate.

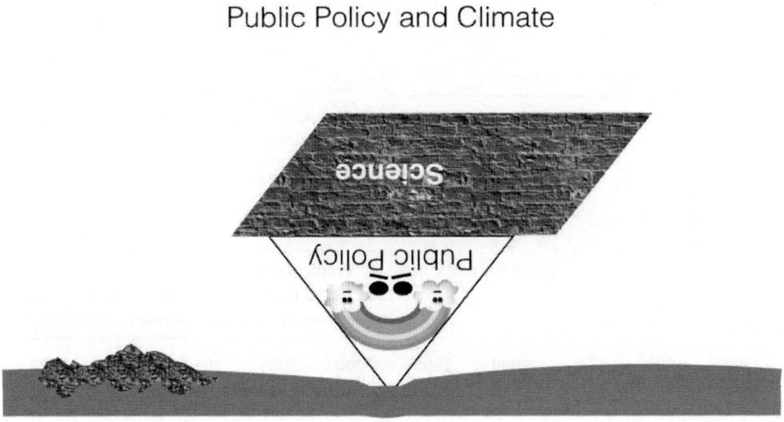

Fig. 3: Silly Science based on Poor Politics.

Climate is a modern example of an ancient problem for humans. They have trouble distinguishing their innate political nature from their often hard won scientific mode of thought. Science and politics are different, but confusion between them produces poor politics, bad policy, and silly science.

PERSISTENT WRONG IDEAS AND BAD INTUITION

There are many seriously defective ideas in the subject of climate that are all but uncorrectable because of the inverted paradigm of climate depicted in Figure 3. Here are only a few, and their absurd consequences. Figure 4 shows the growing divergence between a standard global temperature index computed from models and the same computed from balloon and satellite temperature data as plotted by Dr. Roy Spencer. This divergence is emphasized by the sequence of red arrows. These arrows were added to the figure by Dr. Howard Hayden. Each arrow also represents when the release of one of the UN climate reports occurred. To complete the picture, the UN's expression of confidence in each report was presented in a box with each arrow. Not only are there scientifically meaningless probability numbers used, but the trend in confidence is absurdly counterfactual.

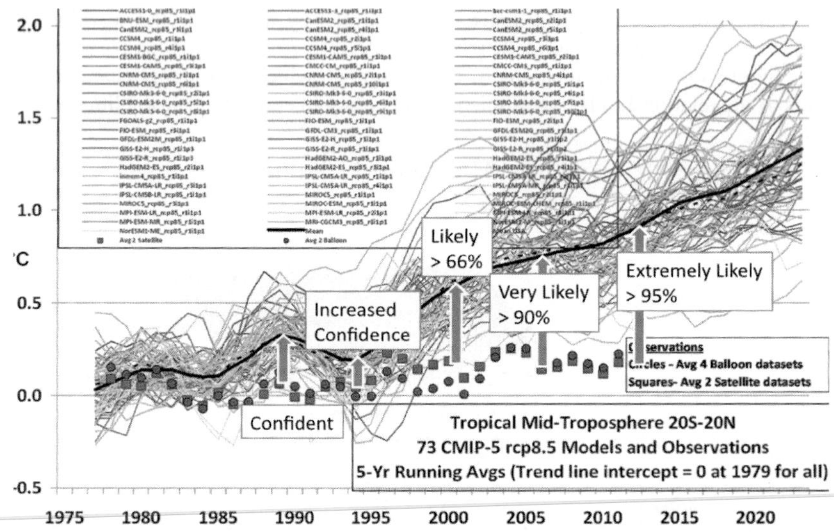

Fig. 4: Growing Confidence with Weakening Performance.

But counterfactual thinking is the norm for this subject. In part this because of bad intuition. When thinking of climate, many people do not comprehend how much cold there is in the world. Figure 5 shows two recent scenes from Canada. The left shows common daily life in many locations in winter, where it is so very cold that if you do not plug your car into a electrical socket routinely provided in parking lots (in posts in the picture) to power your installed engine block heater, your car will not go anywhere without a tow truck till spring if you don't. The right hand picture is in the Canadian Arctic this July where an icebreaker was needed to help deliver supplies to remote Arctic towns.

Fig. 5: Canada.

BAD IDEAS AND FALSE INTUITION FROM TEMPERATURE

The surface temperature version of Figure 4 is Figure 6 from the UN IPCC.

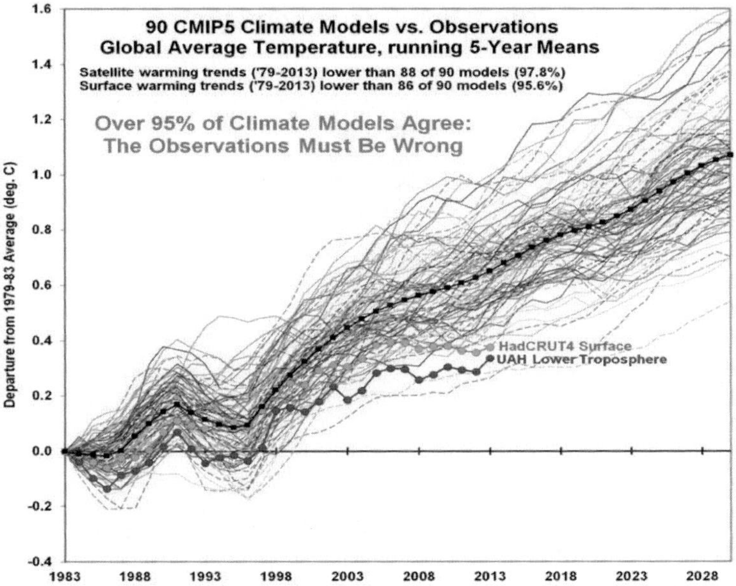

Fig. 6: The UN Recognizes the "Pause".

It shows the same divergence between computation and observations as in Figure 4. Look at the units of the ordinate. It seems like a temperature in degrees Celsius. But, while the tick marks have Celsius measure, these are statistical anomaly coordinates and as such are not actual temperatures. The distinction here is that there is no constant conversion of units between these numbers and actual temperatures. Anomaly coordinates are measured relative to the current time mean, which can and does vary in time, so a conversion to proper temperature is not strictly legitimate as the conversion rule is a function of time.

This is no big loss if one realizes that the Earth does not actually have a temperature at any instant or over time. It is an out-of-equilibrium system, so "temperature," as computed for climate, is strictly a creation of statistics and not thermodynamics. There is no global thermometer. Of course there is a price to pay for the epiphany that this is a statistical index: when it goes up, one cannot claim warming. Thermodynamical intuition about temperature is very misleading here.

But let's set these cautions aside and suppose that there is a global thermometer. What value should we set it at? The preceding paragraphs establish that Figure 6 does not tell us the answer. But it is customary to employ a value of about 15°C in characterizations of climate, corresponding to several (outdoor) international conventions of STP (Standard Temperature and Pressure). This in no way makes an actual temperature exist; it is just a design standard. So if there were an actual global thermometer it might look something like Figure 7.

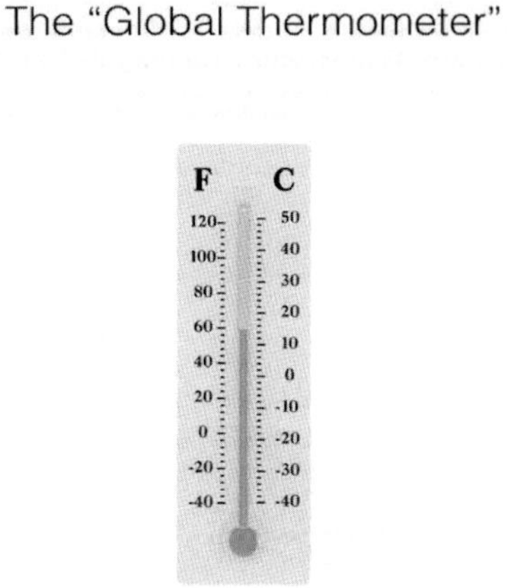

Fig. 7: Global Thermometer.

Let's use this to explore the meaning of some claims swirling around climate. A few years back a scientist exclaimed at a global warming talk, "0.4 degrees per decade is the highest rate increase of global temperature in history. Do you have any idea what that means?!" He was referring to the rise in the surface temperature index from about 1988 to 1998. Figure 8 shows what a change of 0.4°C looks like on the global thermometer of Figure 7.

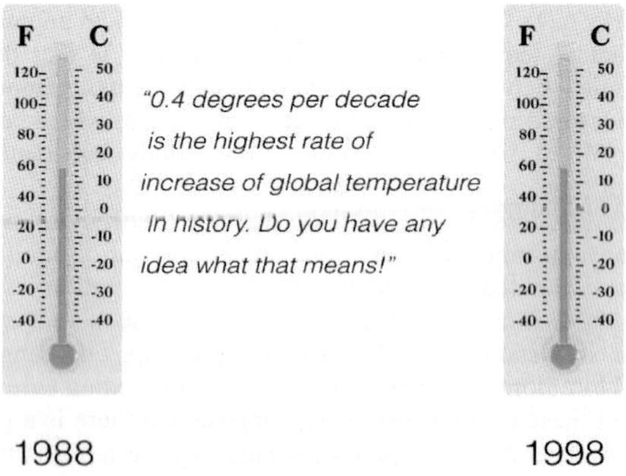

Fig. 8.

Clearly this dramatic change is not dramatic enough to be visible on our imaginary thermometer. There is nothing here in terms of the normal human intuitions about temperature. What about the global thermometer since 1998? Since then it has been virtually unchanged; that is what the "pause" means. This is depicted in Figure 9.

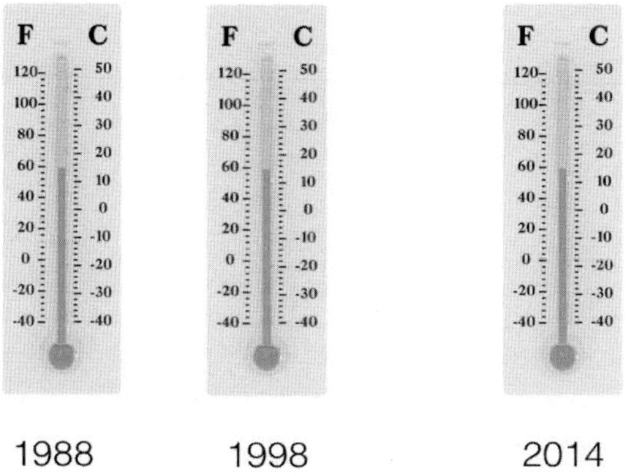

Fig. 9.

2015 would be more up to date, but 2014 was used because it was cited in various news stories as the "warmest year" in history! It illustrates the dramatic claims suggesting a significance that does not stand up to simple investigation.

DIRECT EXPERIENCE VERSUS CLIMATE MYTHS

The symbolic thermometer captures little of the structure exhibited in the complex graphs in Figure 6. So let us look at this issue graphically. The abscissa of Figure 6 is measured in steps of 0.1°C. This is not the scale that human beings experience temperature on. Using the same strategy of mapping anomaly coordinates to temperature coordinates again to set a force a pseudo temperature scale onto Figure 6, let us re-plot everything from Figure 6 on a scale humans actually encounter. Figure 10 uses a human experience scale chosen for temperatures actually experienced on a day in February 2014 when the daytime high temperature in London Canada was -25°C, while it was 40 °C in Melbourne Australia.

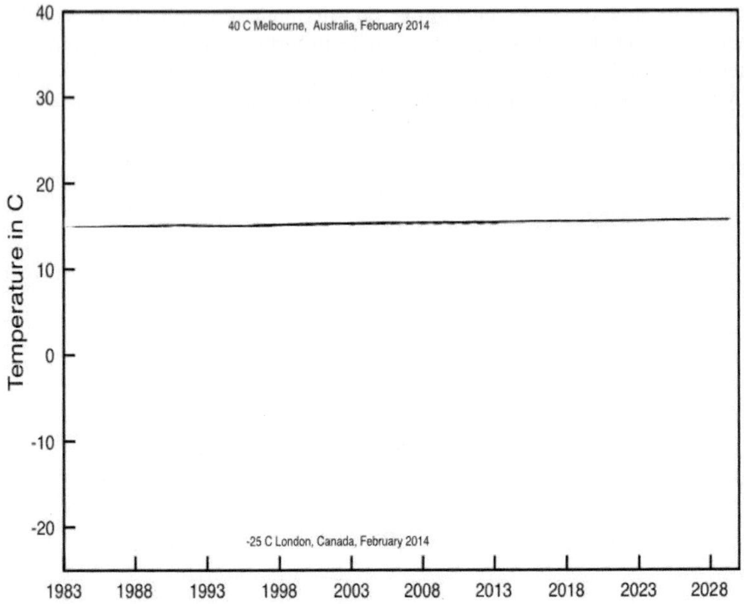

Fig. 10: Figure 6 in Terms of Human Experience.

The structure of Figure 6 simply vanishes on a scale range that people actually experience. The loss of this structure for actual experience means there is little alarming to see in this graph.

It is worth noting that -25°C is 248 K, while gives 288K, which is the canonical value used for many standards in climate discourse. 288K is 15°C which is much higher than the estimated values from blackbody calculations of about 255K, which is thus widely employed to argue that the atmosphere irrevocably warms by the greenhouse effect. But this is naïve and misleading, because what is going on other than the greenhouse effect is the deep scientific challenge from the beginning of this article. That there is more going on by this reasoning can be seen in that London Canada was colder on that day in February than it would be without an atmosphere! Thus by the usual arguments the atmosphere cooled the surface—some greenhouse.

ABSOLUTE CLIMATE "TEMPERATURE" AND "IMPACTS"

It may seem prejudicial to choose a scale reflecting such extreme ranges of temperature, as an individual would rarely experience such a range, even if humans do so collectively nearly every day. But the question of experience is only food for intuition; it is not meant to be decisive. What is decisive is what the laws of physics call for. And what they call for is even less sympathetic than Figure 10. Temperatures in physics are absolute. Figure 10 is re-plotted accordingly in Figure 11 using Kelvin units.

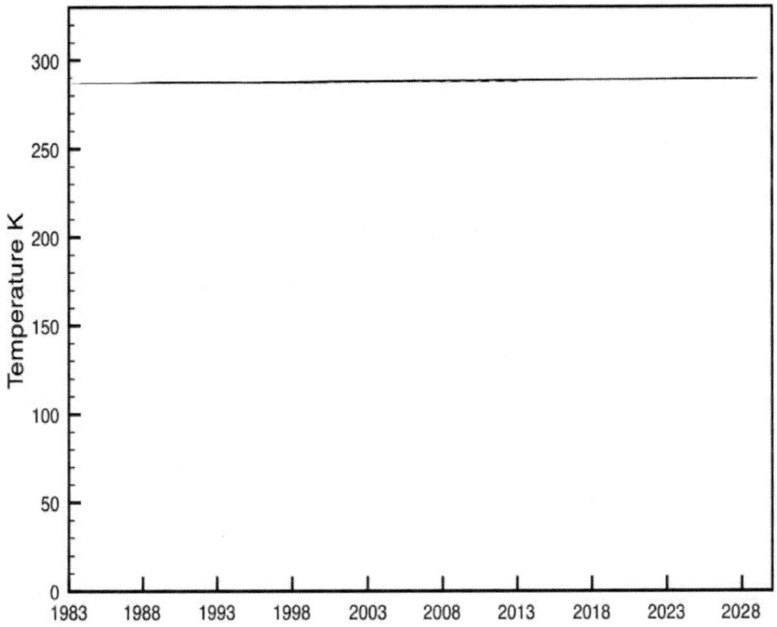

Fig. 11: Figure 6 in its Most Physically Meaningful Form.

This is as physically coherent as the pseudo-temperature anomaly scale from Figure 6 can be made to be. If the index plotted in Figure 6 can possibly have any physical consequences on its own merit, it must be seen in this light because these are the values that go into physical formulas and nothing else. Two facts are dramatically clear. First, models are stunningly accurate, if their job is to produce this statistic alone. Second, if this pseudo-temperature represents climate (and there is no compelling theoretical reason why it should!), climate is nearly homeostatic.

Homeostasis is a dynamical condition where nothing or very little happens. It is hardly a regime where "impacts" are envisioned with dread. Yet impacts are the only reason for any non-scientific interest in this subject at all. Indeed many extraordinary effects are endlessly proffered as a result of what happens in Figure 11. They fall like rain in the press and in official press releases. In Figure 12 only some of the anticipated impacts are depicted ranging from imaginable to silly, to absurd. There is no physical theory of any kind connecting trends in Figure 11 to any of these! All of it is speculation at best and pure hand waving or just madness at worst.

Fig. 12: Impact Madness.

CONCLUSION

No respectable scientist would support such things. Most laugh when openly presented with them. Scientists, comfortable with the status quo, can afford to shrug them off as silly. But these absurdities, when ignored, act as obstacles to any scientific questions raised, forcing questioners to expend time and effort to dispatch them one by one in order to get to genuine scientific issues.

But the people who make policy, said to be based on science, who should be following the model depicted by Figure 2, are flowing Figure 3 instead. For them the detritus of Figure 12 is not a joke, but the scientific state of the art. It is what the delegates in Paris COP 21 have in the back of their minds when making deals that aim to affect us all. It is their intuition and their passion. The madness of Figure 12 is also the foundation of the imbecilic moral case against scientists who question received wisdom. It is the heart of the defective but overweening survey research done on climate scientists aimed at shutting down dissent.

Until people do a more disciplined job at separating politics from science, and start thinking more deeply for themselves, we will be stuck with this.

FINAL OUT BRIEF PMP MITIGATION OF TERRORIST ACTS

ANNETTE SOBEL
Critical Infrastructure Protection and Health Security Initiatives, Texas Tech University
Lubbock, TX, USA

SALLY LEIVESLEY
Newrisk Limited London, Mitigation of Terrorists Acts PMP, World Federation of Scientists, London, UK

- Professor Annette Sobel (Co-Chair)
- Dr. Sally Leivesley (Co-Chair)
- Professor Robert Duncan
- Dr. Richard Garwin
- Professor Pervez Hoodboy
- Dr. Vasiliy Krivokhiza
- Professor Richard Lanza
- Dr. Alan (Leigh) Moore
- Professor Friedrich Steinhausler
- Professor Richard Wilson (Absent)
- Exercise scenario and Advisors: Dr. Grayman, Dr. Kroll, Dr. Budnitz, Ambassador Wegener

HIGHLIGHTS OF PMP ACTIVITIES

- Pre-Erice Smart Cities exercise planning and development.
- Table-Top Smart Cities Exercise to identify cross-disciplinary cyber-physical-cultural vulnerabilities.
- Joint meetings w/Motivation, Pollution, Energy PMPs.
- Technology Mitigation and Sharing.
- Fragility of Energy Network Analysis.
- Smart Cities Response Applied Technology.
- Overviews of Terrorism Concerns: USSR and Pakistani perspectives.

<u>Section I. Technology Mitigation and Sharing</u>

- Technology Landscape
  - Innovation occurring faster than governments can control.
  - Rapid unrestricted distribution using the Internet.
  - Pressure for academics—publish, innovate, commercialize
- Obstacles to Sharing
  - National security concerns.
    - Mitigate security risk by controlled sharing and open-source pubs.
    - Example: Counter-Improvised Explosive Devices.
    - Enforcement/awareness of Dual-use technologies of concern.
  - Cost of technology development and national expenditures.

    - Protect/leverage national investment to achieve ROI
  - Economic support and reinvestment of national defense industrial base.

Section II. Critical Infrastructure Protection: Energy

- Recognition of cyber-physical system vulnerabilities, especially components and resupply times for custom parts.
- Consideration of environmental ecosystem of cyber-physical infrastructure (e.g., Fukushima).
- Gas links to electricity as well as industrial heat and processing.

Section III. Smart Cities, Response, and Applied Technology

- Response operations are complex and goal is resilience, sustainability, and capacity building.
- Rapid recovery/prepositioning of resources/personnel minimizes impact to human life, property, and services.
- Need redundancies (physical or mechanical) within critical infrastructure (back up to cyber).
- Programs for youth education and engagement in cyber protection are effective adjunct.
- Social media integration into response plans.

Large scale incidents require special countermeasures in Smart City

- Physical attacks (server farm, transformer stations, power pylons).
- Software attack on Smart Grid.
- Self-sustaining fire storm in urban district.
- Tornado (F 4), earthquake (Richter scale > 6).
- Terror attack by several teams (Mumbay-style scenario).
- Terror attack deploying RDD or CB device.
- Drinking water contamination.
- LNG fuel explosion (e.g., tanker in harbor).
- Toxic chemical plume due to uncontrolled release (e.g., chlorine).

Smart Responses: Layered Response

- *First layer of response:*

  City Operational Centre (COC)

- *Second layer of response:*

  Civil Defense Central Command & Control (CD3C)

- *Third layer of response:*

  Smart Technologies

Examples of Applied Technologies

1. Advanced Situation Awareness/augmented VR
2. Dynamic Victim Tagging System (eTRIAGE)
3. Recovered family member bar-coding (IRC)
4. Information fusion for decision-making
5. Multilayer Master Display System
6. Smartphone-based Helpbeacon
7. UAS/Situation Awareness Resource Management
8. 3-D modeling/ map generation for responders

Section V. Summary

- Energy components and protracted resupply times (e.g., transformers) are key vulnerabilities in cyber-physical systems.
- Means to improve resilience of the electric power grid against natural phenomena (i.e., solar storms)/attacks exist.
- Hubs of cities may serve as distribution centers to remote sites when loss of critical service support occurs.
- Prioritized safeguards and security includes controls and human safety with deployed technologies.
- UAS's clearly enable terrorist entities and countermeasures are lagging.
- Actively engage youth in cyber-protection to mitigate against terrorist recruitment.

- Social media, UAS, automated messaging garnering an increasing role in maintaining confidence in government services/continuity of operations and information dissemination/validation.

NEXT STEPS

- Continue interdisciplinary focus of Smart Cities' vulnerabilities to terrorist attacks and standards, strategy, policy recommendations.
- Potential joint research/interim meeting with Center for the Resolution of Intractable Conflict (Lord Alderdice, Director).
- Assess and review rapid emergence of UAS capabilities.

The PMP is very grateful to Dr. Sally Leivesley for her years of co-leadership.

LIMITS OF DEVELOPMENT PMP: WATER POLLUTION AND LIMITS OF DEVELOPMENT ISSUES

MBARECK DIOP
Science and Technology Advisor to the President of Senegal
Dakar, Senegal

SESSION DEVELOPMENT

- This year we had three active Members and one visitor.
- Our Chairman for the last three years, Professor Alberto González-Pozo from Mexico, could not attend because of health problem. We all wish to him a quick recovery and express our recognition for his careful guidance and his help.
- The group accepted the participation of Dr. Abdel Nasser TAWFIK, Director of Egyptian Center For Theoretical Physics (Cairo) as Observer.

CONTRIBUTIONS

- "Fresh Water Preservation in Korea." Professor Sun Kun Oh, Physics, Konkuk University, Korea.
- "The Struggle for Water in an Urbanizing World: Constructing Holistic Policies." Professor Ignacio Armillas, Urban Planning and Development, Mexico.
- "Critical Water Issues in Africa." Colonel Mbareck Diop, Former Science and Technology Advisor to the President of the Republic of Senegal.

“FRESH WATER PRESERVATION IN KOREA”—Prof. Sun Kun OH, Konkuk University, Seoul, Korea

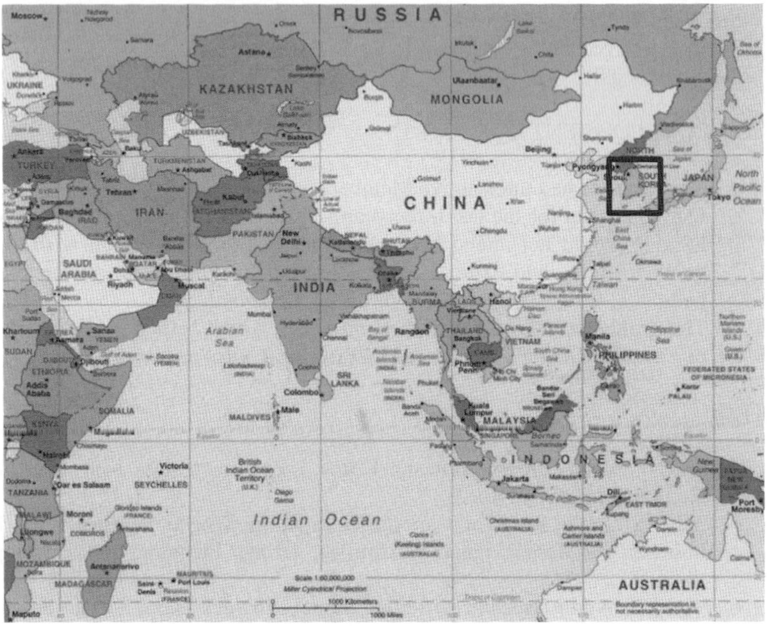

Rainfall in Korea

- The 5-year average is gradually increasing, probably due to climate change.

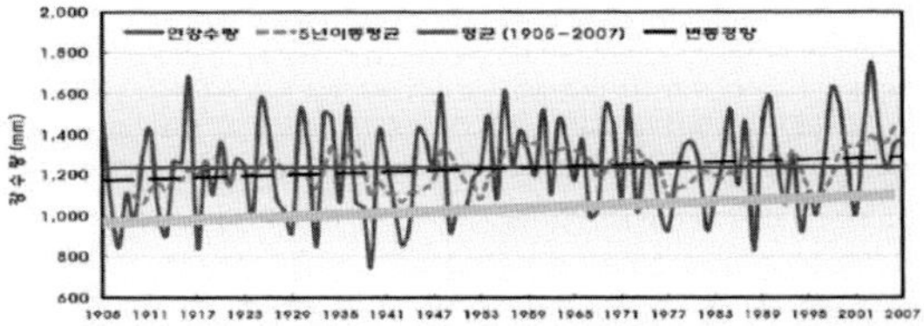

1905-2007

- Seasonal variation and topographical variation are large.

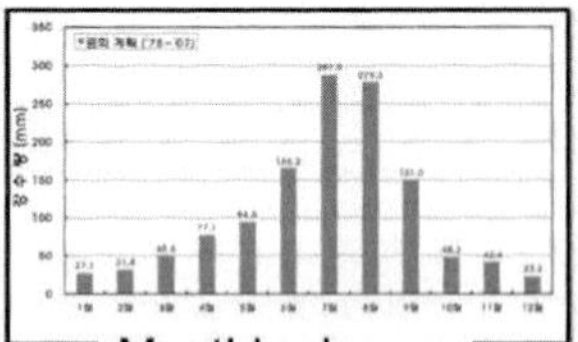

Monthly change

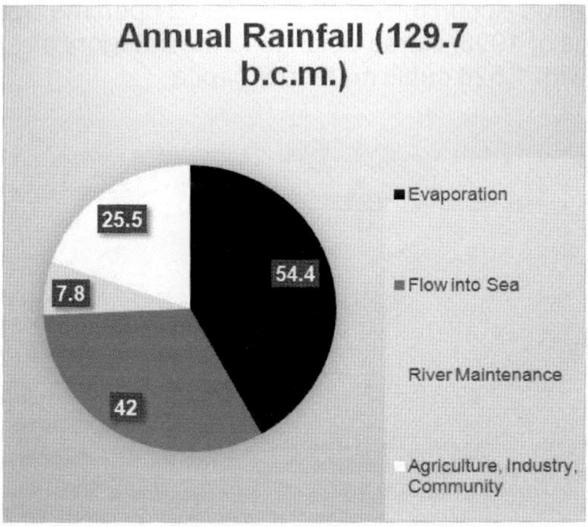

- Evaporation = 42 % of income
- Renewable water = 75.3 billion cubic meter = 58 % of income
- Flow into sea = 42 billion cubic meter = 32 % of income
- At hand = 33.3 billion cubic meter = 26 % of income
- Among them, 7.8 billion cubic meter for river maintenance; and only 25.5 billion cubic meter for human use (= agriculture, industry, community)
- 25.5/75.3 = 34 % = water exploitation rate

Korea is water-stressed

- The ratio of water usage (agriculture, industry, community) to available water per capita is: 517 cubic meter to 1,526 cubic meter = 34 %.

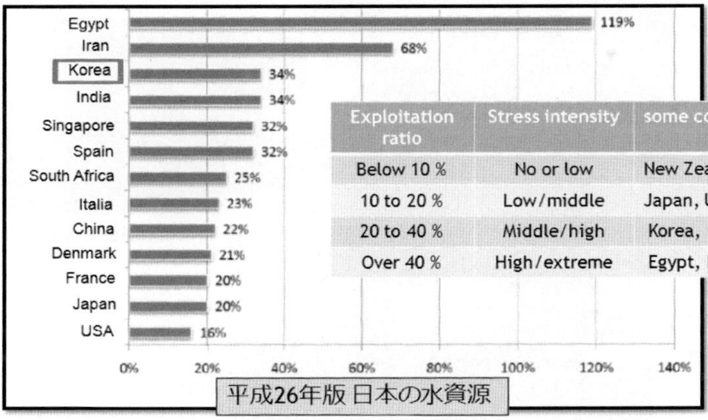

Exploitation ratio	Stress intensity	some countries
Below 10 %	No or low	New Zealand, Canada, Russia
10 to 20 %	Low/middle	Japan, USA, UK, France
20 to 40 %	Middle/high	Korea, China, Italy, India
Over 40 %	High/extreme	Egypt, Iran

- Also, according to PAI (Population Action International), based upon the UN Medium Population Projections of 1998, Korea belongs to the category of the water-stressed countries with 1,526 cubic meter per capita.

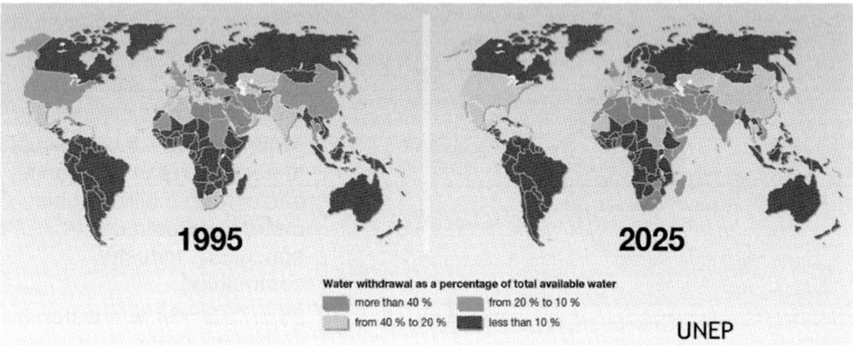

4 Major River Project

- Construction of 14 weirs along the 4 major rivers of Korea mainly in order to restore the economy and vitality of rivers.
- As the flow slows down, water quality is degraded and environmental side effects take place.

Pectinatella magnifica

algal bloom

“THE STRUGGLE FOR WATER IN AN URBANIZING WORLD: CONSTRUCTING HOLISTIC POLICIES”—Professor Ignacio Armillas, Urban Planning and Development, Mexico.

Dr. Ignacio Armillas presented a paper entitled: He started by outlining the process of rapid urbanization the world is experiencing which he referred to as hyper-urbanization.

Population increase 1950-2000

Urban-rural distribution (in millions)

Total 3.591
Rural 1.477
Urban 2.113

WORLD TOTAL POPULATION 1950-2000 Rural – Urban Breakdown (Thousands)					
Year	World Total	Rural Total	Rural %	Urban Total	Urban %
1950	2,532,229	1,786,734	70.6	745,495	29.4
1960	3,038,413	2,018,775	66.4	1,019,638	33.6
1970	3,696,186	2,343,767	63.4	1,352,419	36.6
1980	4,453,007	2,699,779	60.6	1,753,229	39.4
1990	5,306,425	3,025,020	57.0	2,281,405	43.0
2000	6,122,770	3,264,138	53.3	2,858,632	46.7

Projected Population Increase 2000-2050

Urban-rural distribution (in millions)

Total 3.183
Rural -210
Urban 3.394

WORLD TOTAL POPULATION 2000-2050 Rural – Urban Breakdown (Thousands)					
Year	World Total	Rural Total	Rural %	Urban Total	Urban %
2000	6,122,770	3,264,138	53.3	2,858,632	46.7
2010	6,895,889	3,337,311	48.4	3,558,578	51.6
2020	7,656,528	3,366,710	44.0	4,289,818	56.0
2030	8,321,380	3,337,472	40.1	4,983,908	59.9
2040	8,874,041	3,237,815	36.5	5,636,226	63.5
2050	9,306,128	3,053,953	32.8	6,252,175	67.2

Outcomes of Hyper-Urbanization

- By 2050 some 6.252.175.000 people will be living in urban areas.
- Most significant development will be the growing number of mega-cities and expansion of existing ones.
- A critical issue for mega-cities is the provision of *water*.

How are Mega-Cities Coping?

- Policy makers follow a "supply-oriented" approach, simply get more water.
- Which leads to over exploitation of surface sources and aquifers.
- Results in increasing reliance on inter-basin transfers.
- This approach is simply not sustainable, nor environmentally, economically, socially or politically desirable.

What is needed is a holistic multidimensional approach adapted to local circumstances.

- Better water management
  - Reduce consumption
    - Hard—tech means
    - Soft—education/awareness/social incentives
    - Economic and fiscal instruments and incentives
    - Draconian measures such as rationing consumption
  - Improve storage and distribution methods
    - Reduce water loss
    - Transportation and storage
- Alternative sources to increase supply
  - Tapping unconventional sources
    - Harvesting rainfall within urban area
    - Capturing peri-urban runoff
    - Condensate and cooling tower blowdown
  - Reuse
    - Separation of used water—reuse of gray water
    - Treating and reuse of water already circulating within the urban area
  - Substitute
    - Wider use of dry toilets and urinals
    - Use spray instead of flowing water
    - Landscaping without water

"CRITICAL WATER ISSUES IN AFRICA" —Colonel (Ret.) Mbareck Diop, Former Science and Technology Advisor to the President of the Republic of Senegal

- Around 748 million people today still do not have access to an improved source of drinking water, and water demand for manufacturing is expected to increase by 400 per cent between 2000 and 2050 globally.

- Water is essential for promoting inclusive sustainable development, as it supports human communities, maintains the functions of ecosystems and ensures economic development.
- Translating this vision into reality requires efforts by all, through concrete and interrelated actions that go from establishing the legal and institutional framework to ensure sustainable water management and increasing investments and financial support for water development to enhancing and improving access to water supply, sanitation and hygiene services.
- The sustainable use and management of water is vital for welfare of all humanity today.
- Demand for freshwater is growing.
- Unless the balance between demand and finite supplies is restored, the world will face an increasingly severe global water deficit.
- Global water demand is largely influenced by population growth, urbanization, food, energy security policies and macro-economic processes such as trade globalization, changing diets and increasing consumption.
- By 2050, global water demand is projected to increase by 55%, mainly due to growing demands from manufacturing, thermal electricity generation and domestic use.
- Wastewater also has impacts on the global environment as wastewater-related emissions of methane, a powerful global warming gas, and nitrous oxide could rise by 50% and 25%, respectively, between 1990 and 2020 (Corcoran et al. 2010).
- There is a need for more innovative options for such as decentralized wastewater treatment solutions and biogas production for reusing and recycling wastewater and reducing the cost of wastewater management.

<u>Inadequate Access to Safe Water and Sanitation</u>

- The world's slum population is expected to reach 889 million by 2020 (UN-Habitat, 2010).
- As slum dwellers are generally more likely to suffer inadequate access to safe water and sanitation and are also more vulnerable to the impacts of extreme weather events, water management in cities, particularly slum settlements, will be a major challenge in the future.
- In some informal settlements, however, local communities and the private sector have come up with innovative solutions.
- In Mombasa, for example, where only about 15% of the people have access to piped water supply, more than 80% have access to an improved water source because they receive water from kiosks.

<u>Climate Change and Water-Related Disasters</u>

- Because the impacts of climate change are complex and unpredictable the availability of and demand for water are highly likely to be affected.

- Water and sanitation infrastructure may be at risk because of extreme events and sea level rise.
- With increased urbanization, there is also an urgent need for more sustainable urban drainage systems to address the issues of inundation and water contamination.

AFRICA CHALLENGES

- Africa has a problem of rapid population growth, which can be seen as an opportunity to transform African economies to sustainable and inclusive growth for the next half century.
- This population must be fed, educated, and kept healthy and productive.
- It is in this context that water for food, water for health and water for energy is critical to the continent's sustained development.
- Important progress has been made towards attaining the MDG target for access to improved drinking water (except for the most fragile states), especially for urban populations, but much less progress has been made with respect to sanitation.

Key Water Challenges Related to Sustainable Development in Africa (The Sharm el-Sheikh Commitments by the AU, 2014)

1. Water infrastructure for economic growth;
2. Managing and protecting water resources;
3. Achieving water supply and sanitation MDGs;
4. Global changes and risk management in Africa;
5. Water governance and management;
6. Financing water and sanitation sector;
7. Education, knowledge, capacity development and water information.

African Union (AU) Role

- At the highest policymaking level, the African Union (AU) has recognized the critical role of water in various Summit Declarations (AU, 2004 and 2008; African Ministerial Council in charge of Water—AMCOW, 2008) and has adopted the African Water Vision 2025 as the fundamental policy instrument for the management of Africa's water resources for sustainable development.
- The shared vision is for: "An Africa where there is an equitable and sustainable use and management of water resources for poverty alleviation, socio-economic development, regional cooperation, and the environment" (UNECA, 2000).

African Water Vision 2025

- As a policy instrument, the African Water Vision 2025 identifies the key challenges for water resources development in Africa.

- The water-food-energy nexus in Africa Water security: The availability of an acceptable quantity and quality of water for health, livelihoods, ecosystems and production, coupled with an acceptable level of water-related risks to people, environment and economies (Grey and Sadoff, 2007).
- Food Security: When all people at all times have access to sufficient, safe, nutritious food to maintain a healthy and active life (WHO, n.d.).
- Energy Security: The uninterrupted physical availability (of energy) at a price which is affordable, while respecting environment concerns (IEA, 2011).
- Water Security: When all people at all times have access to sufficient, safe water and sanitation facilities.

Water, Food and Energy

- Of particular importance for Africa is the critical nexus between water, food and energy.
- Water availability, access and optimal use are essential for transforming the vicious cycle of insecurities to a virtuous upward spiral towards inclusive green (and blue) growth.
- Currently, only 5% of the Africa's potential water resources are developed and average per capita storage is 200 m3 compared to 6,000 m3 in North America.
- Only 5% of Africa's cultivated land is irrigated and less than 10% of hydropower potential is utilized for electricity generation (Sperling and Bahri, 2014).

Managing Africa Water Resources—Regional Cooperation

- In 2012, three years before the end of MDGs, on average about 36% of the population did not have access to improved water resources and 70% still did not have access to improved sanitation. (WHO and UNICEF, 2014).
- In addition to these statistics, climate variability and change is likely to restrain progress in managing Africa water resources.
- Regional cooperation is especially needed due to the multiplicity of transboundary water resources (more than 80 international river basins and aquifers) which must be managed coherently and equitably to meet regional and national goals and evolving sectoral needs.

Need of Regional Integration and Cooperation

- The important African potential is dependent on developing the infrastructure and expertise for harnessing the abundant water resources for irrigation, power generation, domestic use and value adding industries to traditional export commodities.
- Such water infrastructure development is dependent on a deeper level of regional integration and cooperation due to the transboundary nature of most of Africa's water resources.

<u>The Grand Inga Project</u>

- The Grand Inga project offering 40,000 MW of potential hydropower, is a project on Africa's largest river, the Congo River, which may provide a potential answer to meeting Africa's regional energy needs.
- This makes the Grand Inga hydropower project a truly regional project and could help improve regional integration using a renewable energy source, water, as the prime instrument.

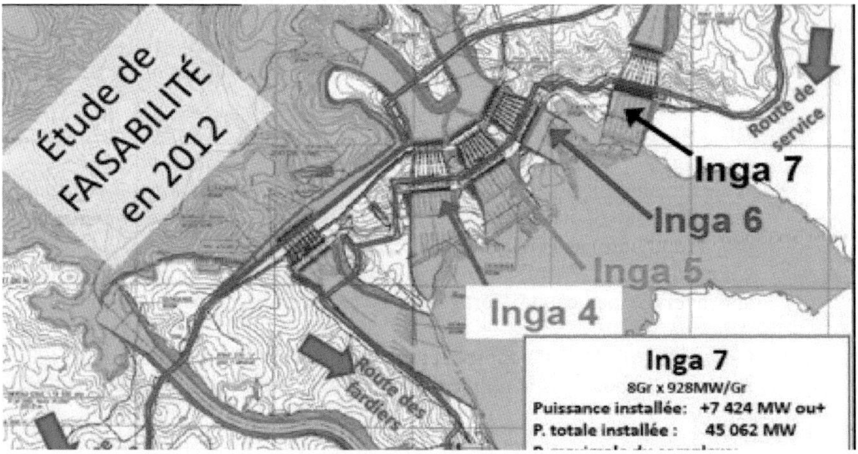

Water Storage and Management

- In terms of agriculture, Africa's response, led by the African Development Bank (AfDB), is to promote interventions which focus on water storage and management to overcome the low, unreliable rainfall during the cropping season, in situ rainwater management, water harvesting or runoff harvesting, and water management for crop growth in wetlands.
- Additional interventions will also focus on infrastructure for livestock, fisheries and rural energy (AfDB, 2010).
- To achieve this, regional—and especially transboundary—cooperation in the development of Africa's water resources will be critical in the coming decades.

CONCLUSION

Among the main challenges for water issue in the world it is important to take into account:

- Hyper urbanization,
- Pollution,
- Sustainable management to secure the resource,
- More cooperation for transboundary rivers and aquifers,
- More R&D for innovative solutions.

What is needed is a holistic multidimensional approach adapted to local circumstances. The Manhattan Project could be an opportunity to this purpose.

THANKS! GRAZIE! DIEUREUDIEUF!

POLLUTION PMP REPORT

LORNE EVERETT
Chancellor Emeritus, Lakehead University, Thunder Bay, Canada and L. Everett and Associates, LLC, Santa Barbara, CA, USA

POLLUTION PANEL ACTIVITIES 2015

Award Winning New Book: *Continuous Soil Gas Measurements: Worst Case Parameters* by L. Everett and M. Kram, ASTM International, 2015.

Pollution PMP Paper: "Concerns Over use of Glyphosate-based Herbicides and Hazards Associated with Exposures: a Consensus Statement."

- Pollution Plenary Session: Our Dying Oceans
- Pollution Plenary Session: The Endocrine Disrupting Herbicide Glyphosate
- Pollution Joint Plenary Session: Pollution and Limits of Development
- Pollution Workshop: Widespread Contamination—Herbicide Glyphosate
- Pollution Joint Workshop: Pollution and Terrorism
- Pollution Joint PMP meeting: Pollution and Energy
- 6 Pollution PMP Papers and Reports, 19 August
- 14 Invited Guests for Plenary, PMP meetings and Workshops

Spread of Hypoxia

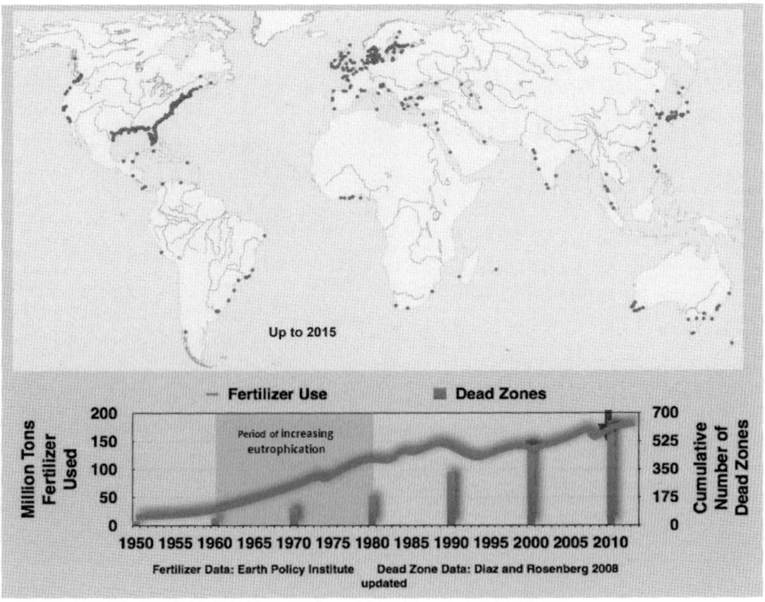

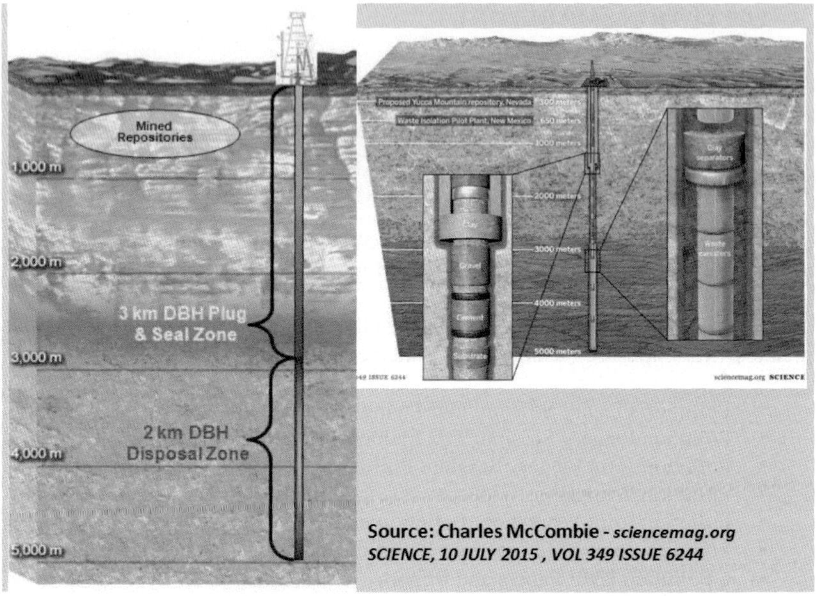

Source: Charles McCombie - *sciencemag.org*
SCIENCE, 10 JULY 2015, VOL 349 ISSUE 6244

Kosovo's Capitol City's Water Supply Reservoir Targeted by Terrorists

THE NEW YORK TIMES **INTERNATIONAL** MONDAY, JULY 13, 2015

Kosovo Charges 5 People In Plot to Poison Water

By RICK LYMAN

WARSAW — Five people were in custody in Kosovo on Sunday, suspected of a terrorist plot to contaminate the water supply in the capital, Pristina, after exhortations on recent videos released by the Islamic State to poison the country's food and water supplies.

Local news reports said that three people were arrested early Saturday morning while acting suspiciously in the vicinity of the Badovac reservoir, which supplies Pristina with nearly half its water. Weapons, ammunition and Islamic State propaganda were found in the suspects' two Volkswagens, the reports said.

Two more people were arrested later in Pristina and charged with terrorism in connection with the suspected plot, Kosovo officials said.

Two of the suspects had also been arrested last year on suspicion of traveling to Syria, Reuters reported, citing an unidentified police source.

Kosovo police officers escorting a suspect to court in Pristina, the capital, on Sunday after his arrest the day before.

Pristina water officials decided to cut off water to tens of thousands of the capital's residents on Saturday morning until tests could be conducted to determine whether the water had been contaminated. At first, Kosovo officials attributed the cutoff only to "security issues," but they later said it was linked to the suspected poisoning plot.

By Saturday evening, government health officials said tests had concluded that the water was not tainted and urged the city to restore service.

In the recent Islamic State videos, a man can be heard urging just such a plot, Reuters reported.

"If you can, take poison and put it in their meal or in their drink," the man is heard saying. "Make them die, make them die of poisoning, kill them wherever you are."

Mexico City, Population 20 million, sits on Sole Source Aquifer

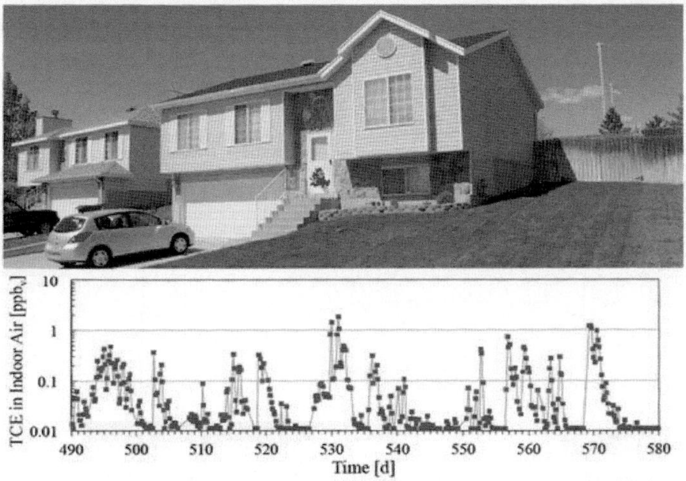

Current vapor intrusion (VI) pathway assessment heavily weights concentrations from infrequent (monthly–seasonal) 24 h indoor air samples. This study collected a long-term and high-frequency data set that can be used to assess indoor air sampling strategies for answering key pathway assessment questions like: "Is VI occurring?", and "Will VI impacts exceed thresholds of concern?". Indoor air sampling was conducted for 2.5 years at 2–4 h intervals in a house overlying a dilute chlorinated solvent plume (10–50 μg/L TCE). Indoor air concentrations varied by 3 orders of magnitude (<0.01–10 ppb_v TCE) with two recurring behaviors. The VI-active

Benzene-a human carcinogen-soil vapor screening level-36.2 μg/m³

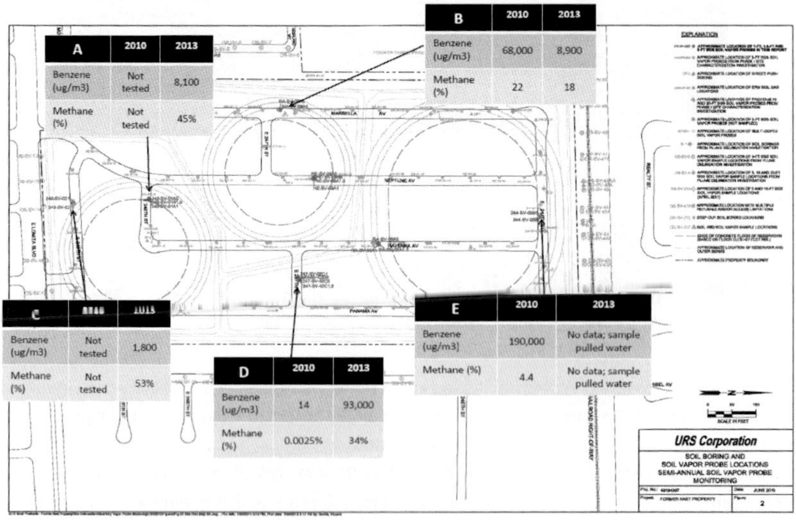

Everett Expert Report Exhibit 7. Base Map from URS, August 1, 2013, First Semi-Annual 2013 Sampling of Existing and Replacement Soil Vapor Probes, Carousel and Monterey Pines Neighborhoods, Former Kast Property, Carson, California. Concentration data added by Dr. Everett. 2013 data is from URS report cited above. 2010 data is from September 30, 2010 URS report titled, Soil Vapor Extraction Pilot Test Report Former Kast Property, Carson, California

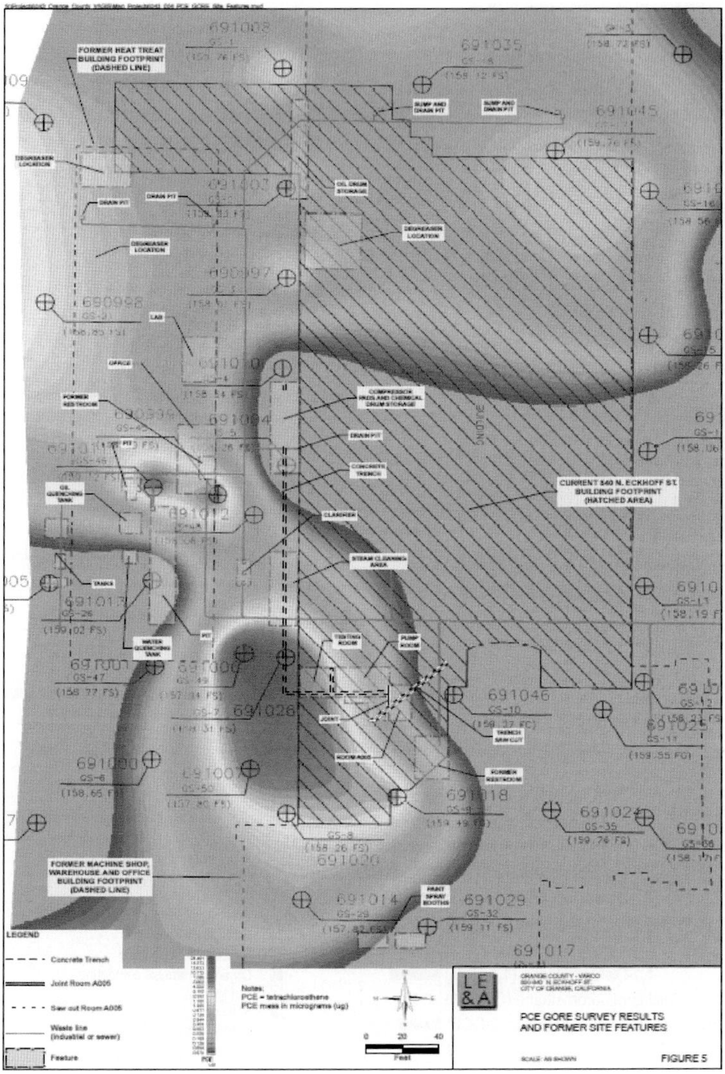
FORMER HEAT TREAT BUILDING FOOTPRINT (DASHED LINE)
CURRENT 540 N. ECKHOFF ST. BUILDING FOOTPRINT (HATCHED AREA)
FORMER MACHINE SHOP, WAREHOUSE AND OFFICE BUILDING FOOTPRINT (DASHED LINE)
LEGEND
Concrete Trench
Joint Room A005
Saw cut Room A005
Waste line (industrial or sewer)
Feature
Notes:
PCE = tetrachloroethene
PCE mass in micrograms (ug)
Feet
PCE GORE SURVEY RESULTS AND FORMER SITE FEATURES
FIGURE 5

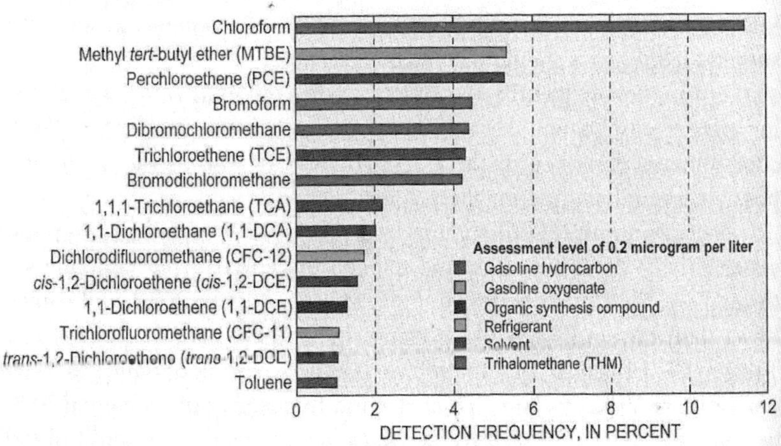

FIGURE 2-3 The 15 most frequently detected VOCs in public supply wells. SOURCE: Zogorski et al. (2006) with illustration provided by USGS National Water Quality Assessment program.

"Human Health Effects of Trichloroethylene: Key Findings and Scientific Issues"
March 1, 2013 Reviews

Conclusions: TCE is carcinogenic to humans by all routes of exposure and poses a potential human health hazard for non-cancer toxicity to the central nervous system, kidney, liver, immune system, male reproductive system, and the developing embryo/fetus.

Environ Health Perspect. 121:303–311 (2013). http://dx.doi.org/10.1289/ehp.1205879 [Online 18 December 2012]

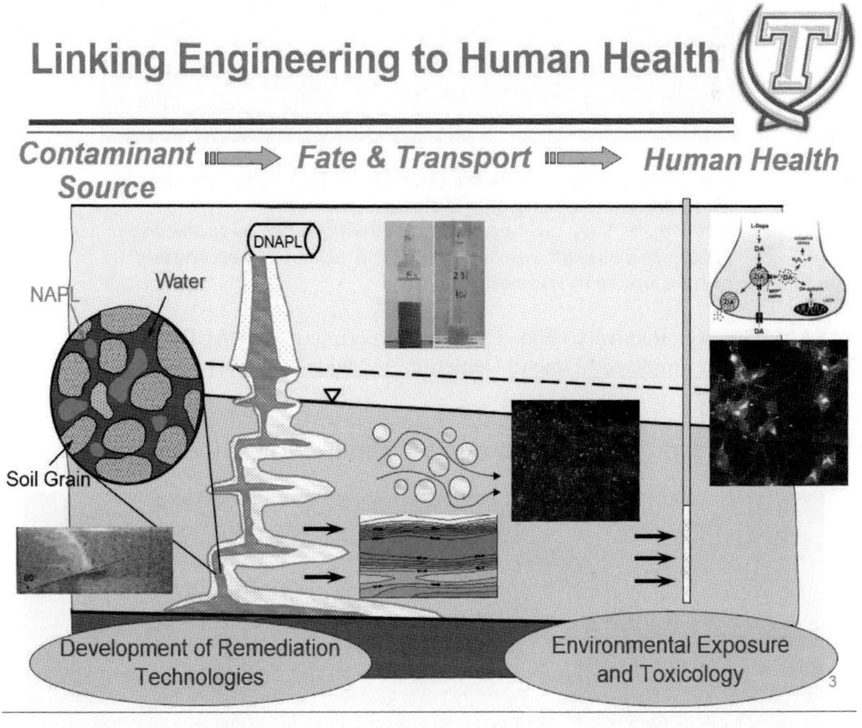

Satellite Tracking of Groundwater Resources

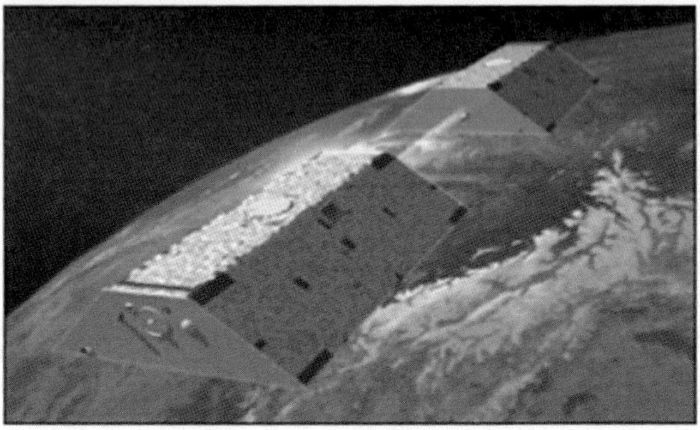

The GRACE mission detects changes in Earth's gravity field by monitoring the changes in distance between the two satellites as they orbit Earth. The drawing is not to scale; the trailing spacecraft would actually be about 220 kilometers behind the lead spacecraft.

NASA's Gravity Recovery and Climate Experiment (GRACE) Satellites (Major Discrepancies in Projected "Time to Depletion"—Major Aquifers)

NEWS | JUNE 16, 2015

Study: Third of Big Groundwater Basins in Distress

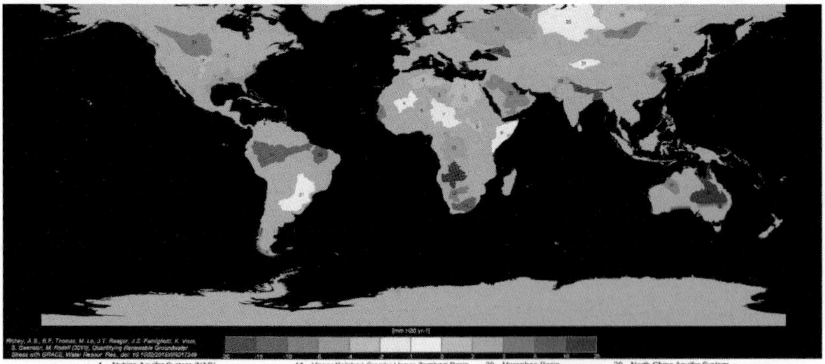

1 Nubian Aquifer System (NAS)
2 Northwestern Sahara Aquifer System (NWSAS)
3 Murzuk-Djado Basin
4 Taoudeni-Tanezrouft Basin
5 Senegalo-Mauritanian Basin
6 Iullemeden-Irhazer Aquifer System
7 Lake Chad Basin
8 Sudd Basin (Umm Ruwaba Aquifer)
9 Ogaden-Juba Basin
10 Congo Basin
11 Upper Kalahari-Cuvelai-Upper Zambezi Basin
12 Lower Kalahari-Stampriet Basin
13 Karoo Basin
14 Northern Great Plains Aquifer
15 Cambro-Ordovician Aquifer System
16 Californian Central Valley Aquifer System
17 Ogallala Aquifer (High Plains)
18 Atlantic and Gulf Coastal Plains Aquifer
19 Amazon Basin
20 Maranhao Basin
21 Guarani Aquifer System
22 Arabian Aquifer System
23 Indus Basin
24 Ganges-Brahmaputra Basin
25 West Siberian Basin
26 Tunguss Basin
27 Angara-Lena Basin
28 Yakut Basin
29 North China Aquifer System
30 Song-Liao Basin
31 Tarim Basin
32 Paris Basin
33 Russian Platform Basins
34 North Caucasus Basin
35 Pechora Basin
36 Great Artesian Basin
37 Canning Basin

Groundwater storage trends for Earth's 37 largest aquifers from UCI-led study using NASA GRACE data (2003 - 2013). Of these, 21 have exceeded sustainability tipping points and are being depleted, with 13 considered significantly distressed, threatening regional water security and resilience. Image Credit: UC Irvine/NASA/JPL-Caltech

California's Drought
July 20th, 2011
August 19th, 2014
Lake Oroville
January 18th, 2013
January 18th, 2014
Sierra Nevada Mountains
3
Source: Justin Sullivan/Getty Images/California Department of Water Resources/NASA Earth Observatory

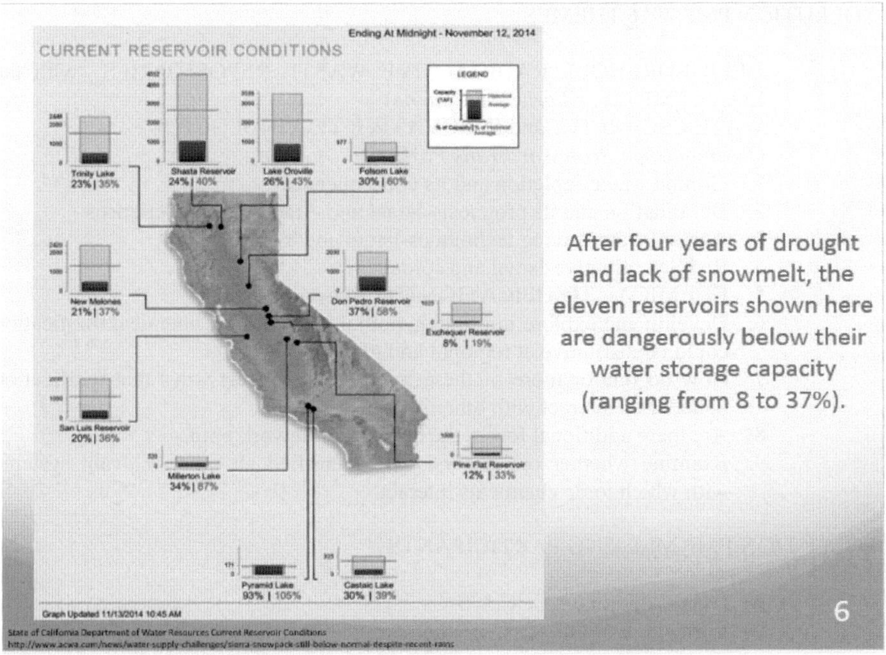
Ending At Midnight - November 12, 2014
CURRENT RESERVOIR CONDITIONS
LEGEND
Trinity Lake
23% | 35%
Shasta Reservoir
24% | 40%
Lake Oroville
26% | 43%
Folsom Lake
30% | 60%
New Melones
21% | 37%
Don Pedro Reservoir
37% | 58%
Exchequer Reservoir
8% | 19%
San Luis Reservoir
20% | 36%
Millerton Lake
34% | 87%
Pine Flat Reservoir
12% | 33%
Pyramid Lake
93% | 105%
Castaic Lake
30% | 39%
Graph Updated 11/13/2014 10:45 AM
After four years of drought and lack of snowmelt, the eleven reservoirs shown here are dangerously below their water storage capacity (ranging from 8 to 37%).
6
State of California Department of Water Resources Current Reservoir Conditions
http://www.acwa.com/news/water-supply-challenges/sierra-snowpack-still-below-normal-despite-recent-rains

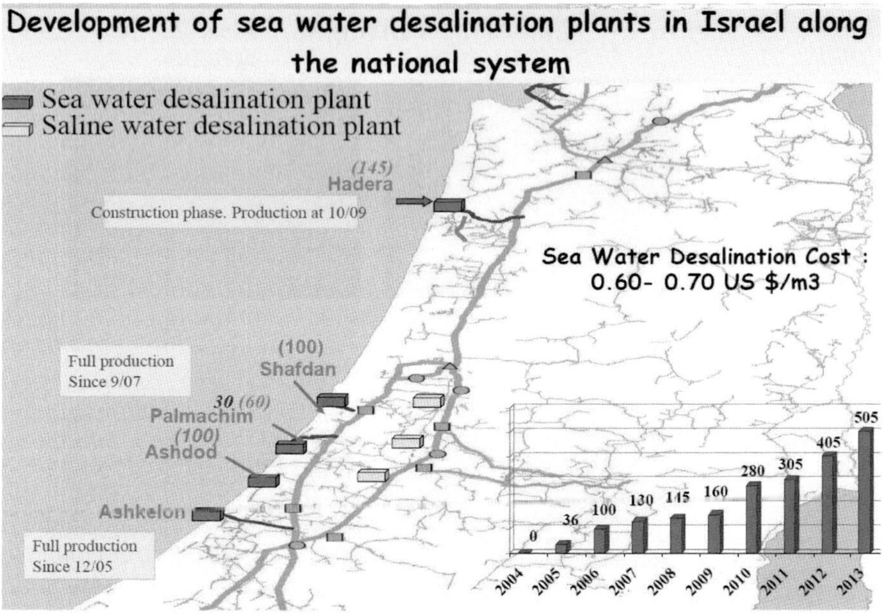

POLLUTION PMP 2016 THEMES

- DEEP BOREHOLE RADIOACTIVE WASTE REPOSITORIES—with the Energy PMP (*Dr. Charles McCombie*)
- WATER SCARCITY and INNOVATIVE TECHNOLOGIES—(*Distinguished Professor Frank Parker)*
  1. Ground water depletion and its consequences
  2. Desalination and its prognosis-Israel and Arab States experiences
  3. Ancient water saving techniques-Israel and Yemen
  4. Red Sea initiative-Israel and Jordan
  5. CHEMICAL TOXIFICATION (*Dr. Peterson Myers)*
  6. Develop and explore multiple hypotheses about how chemical toxification could be disruptive at regional and global scales.
  7. How do one or more of these hypotheses suggest ways that toxification boundaries interact with other systems?
  8. Are there additional forces the current framework omit.
  9. Examine whether our analysis has identified all other relevant systems with which toxic chemicals interact.

POLLUTION PMP MEETING PARTICIPANTS

- Dr. Gina Calderone—ECC, USA
- Dr. Lorne G. Everett—L. Everett & Associates, USA

- Distinguished Professor Frank L. Parker—Vanderbilt Univ. USA
- Professor Fred vom Saal—University of Missouri, USA
- Professor Stephano Parmiagiano—University of Parma, Italy
- Dr. Richard Ragaini—Lawrence Livermore National Lab-retired, USA
- Dr. Dan J. Kroll—Hach Homeland Security Technologies, USA
- Dr. Walter M. Grayman—Consulting Engineer, USA
- Professor Emeritus Robert J. Diaz—College of William and Mary, USA
- Professor Rutger Rosenberg—Goteborg University, Sweden
- Dr. Daniel Pauly—UBC Fisheries Centre, Vancouver, Canada
- Dr. Samuel Dover, DVM—Channel Islands Marine & Wildlife Institute, USA
- Professor Simone Bastianoni—Universita Di Siena, Italy
- Dr. Michael Antoniou—King's College London School of Medicine, London, UK
- Dr. John Peterson Myers—Carnegie Mellon University, USA
- Dr. Andres Carrasco—University of Buenos Aires, Argentina
- Dr. Hans Herren—Millenium Institute, Washington, DC
- Dr. Charles Benbrook—Washington State University, USA
- Dr. Laura Vandenberg—University of Massachusetts, USA
- Dr. Patrick Holden—Sustainable Food Trust, UK
- Dr. Andre' Contin—University of Bologna, Italy
- Dr. Charles McCombie—McCombie Consulting, Switzerland
- Dr. Robin Mesnage—Kings College London School of Medicine, UK
- Dr. Michael Hansen—Consumers Union, USA
- Dr. Bruce Blumberg—UC Irvine, USA
- Dr. Robert Kremer—U Missouri, USA

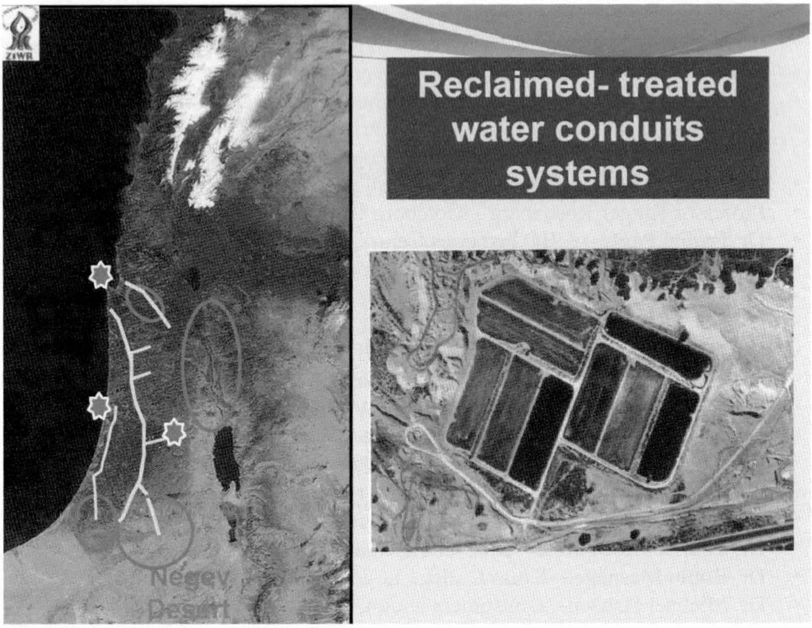

Spatial Variability of Radon in Indoor Air

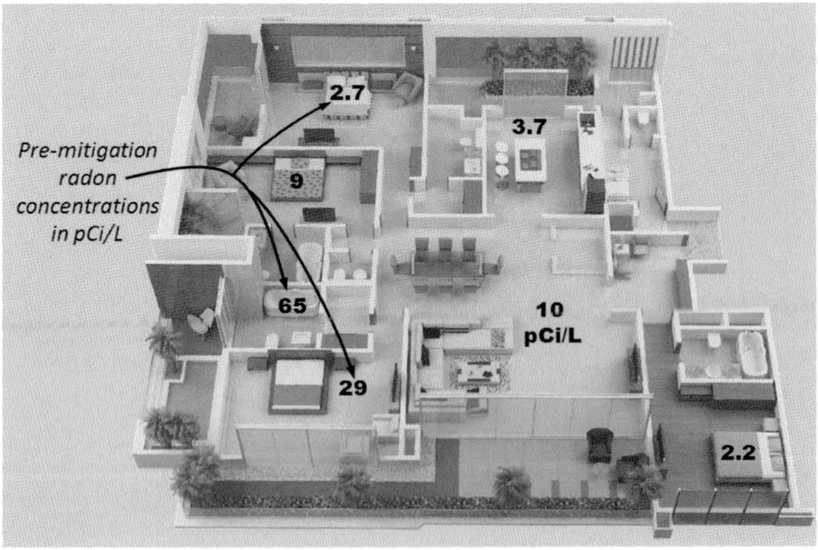

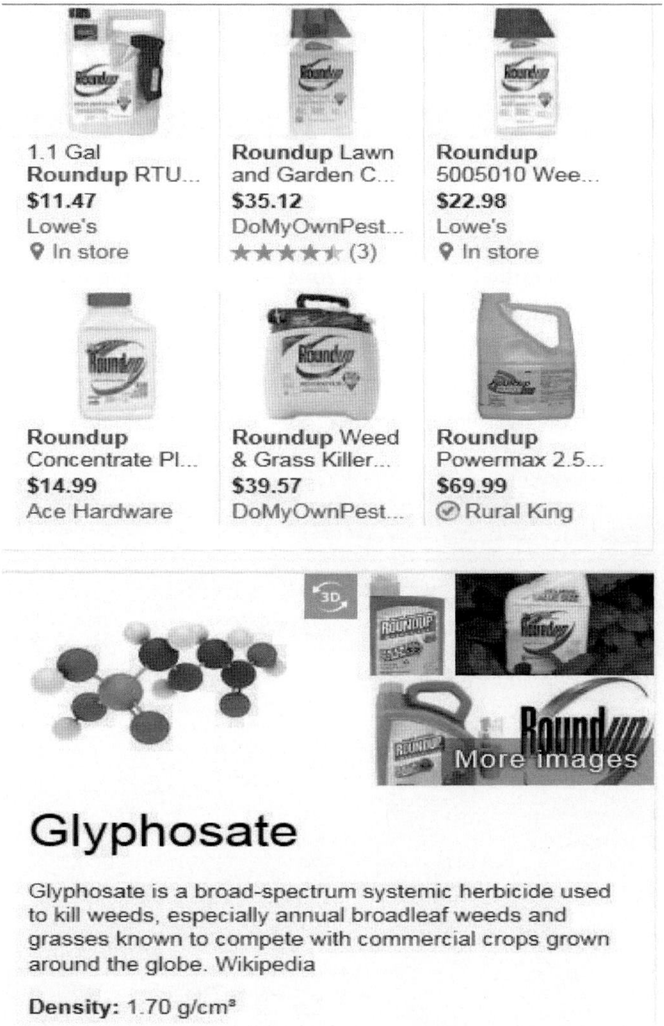

1.1 Gal **Roundup** RTU...
$11.47
Lowe's
In store

Roundup Lawn and Garden C...
$35.12
DoMyOwnPest...
(3)

Roundup 5005010 Wee...
$22.98
Lowe's
In store

Roundup Concentrate Pl...
$14.99
Ace Hardware

Roundup Weed & Grass Killer...
$39.57
DoMyOwnPest...

Roundup Powermax 2.5...
$69.99
Rural King

More images

Glyphosate

Glyphosate is a broad-spectrum systemic herbicide used to kill weeds, especially annual broadleaf weeds and grasses known to compete with commercial crops grown around the globe. Wikipedia

Density: 1.70 g/cm³

Ocean Dead Zones Double Down: The Seas Are the Limit

The world's first identified ocean "dead zone"—a watery region where a combo of nitrogen and phosphorous fertilizer runoff creates monstrous algae blooms that kill off *everything* in the water—was discovered more than 40 years ago at the mouth of the Mississippi River in the Gulf of Mexico. Today, more than 400 dead zones are growing around the globe; the number has doubled every decade.

No fish can live in the dead zones where fertilizer-polluted rivers dump into the sea. (Photo: Ho New/Reuters)

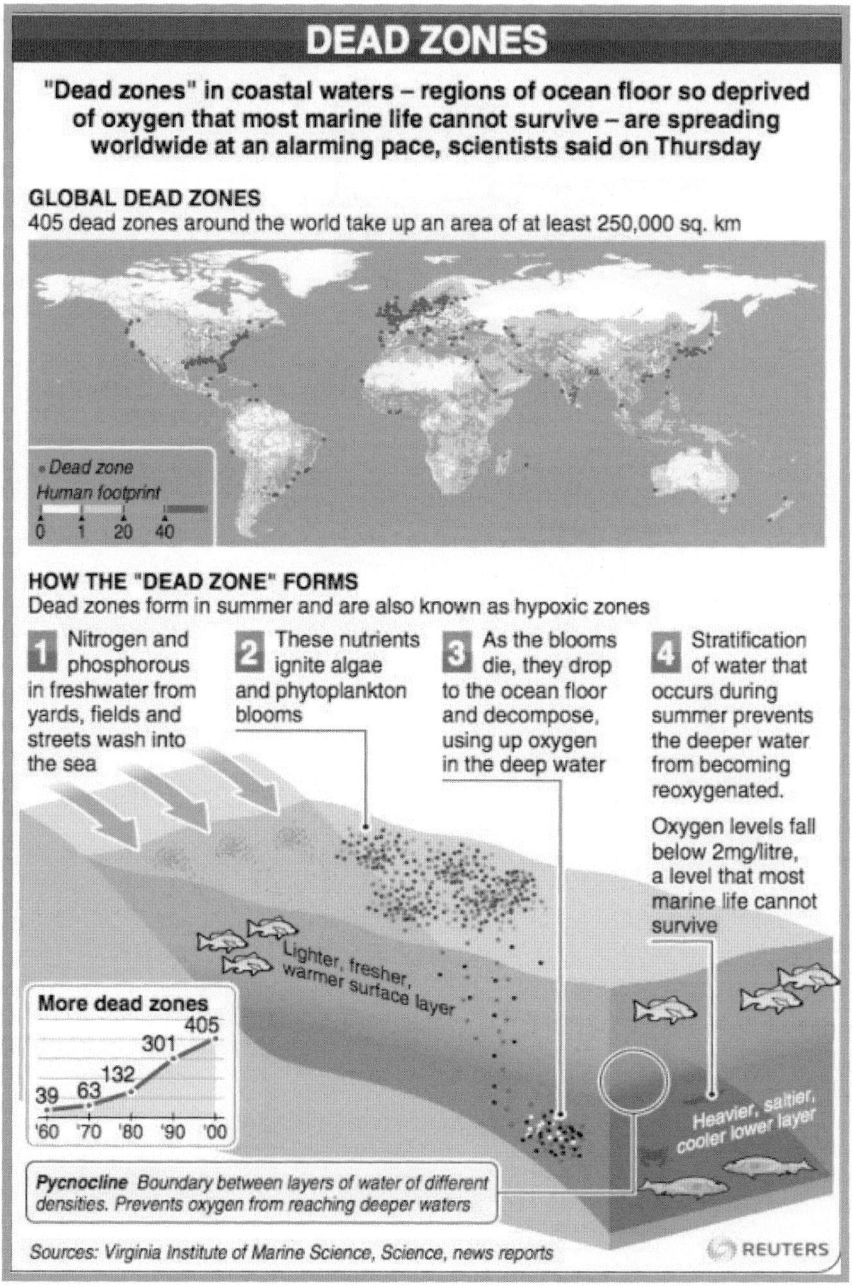
DEAD ZONES
"Dead zones" in coastal waters – regions of ocean floor so deprived of oxygen that most marine life cannot survive – are spreading worldwide at an alarming pace, scientists said on Thursday
GLOBAL DEAD ZONES
405 dead zones around the world take up an area of at least 250,000 sq. km
Dead zone
Human footprint
0 1 20 40
HOW THE "DEAD ZONE" FORMS
Dead zones form in summer and are also known as hypoxic zones
1 Nitrogen and phosphorous in freshwater from yards, fields and streets wash into the sea
2 These nutrients ignite algae and phytoplankton blooms
3 As the blooms die, they drop to the ocean floor and decompose, using up oxygen in the deep water
4 Stratification of water that occurs during summer prevents the deeper water from becoming reoxygenated.
Oxygen levels fall below 2mg/litre, a level that most marine life cannot survive
Lighter, fresher, warmer surface layer
Heavier, saltier, cooler lower layer
More dead zones
39 63 132 301 405
'60 '70 '80 '90 '00
Pycnocline Boundary between layers of water of different densities. Prevents oxygen from reaching deeper waters
Sources: Virginia Institute of Marine Science, Science, news reports
REUTERS

<u>Schematic of Vapor Intrusion</u>

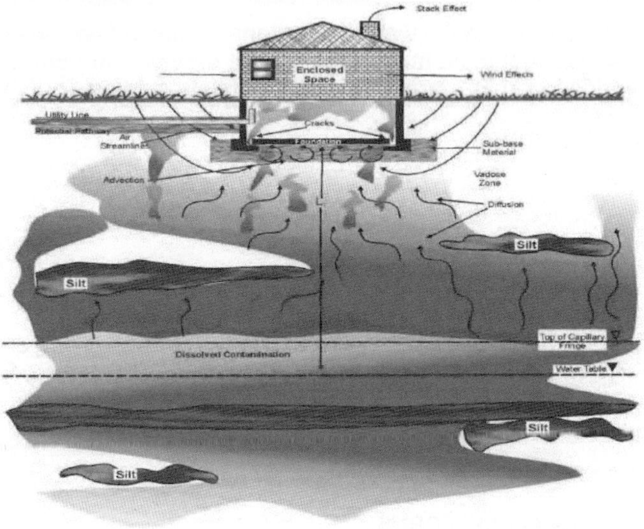

Source: EPA, 2012, Fluctuation of Indoor Radon and VOC Concentrations Due to Seasonal Variations, EPA/600/R-12/673

Scientists fear mass extinction as oceans choke

By Amy Simmons / ABC News

Australian scientists fear the planet is on the brink of another mass extinction as ocean dead zones continue to grow in size and number.

More than 400 ocean dead zones – areas so low in oxygen that sea life cannot survive – have been reported by oceanographers around the world between 2000 and 2008.

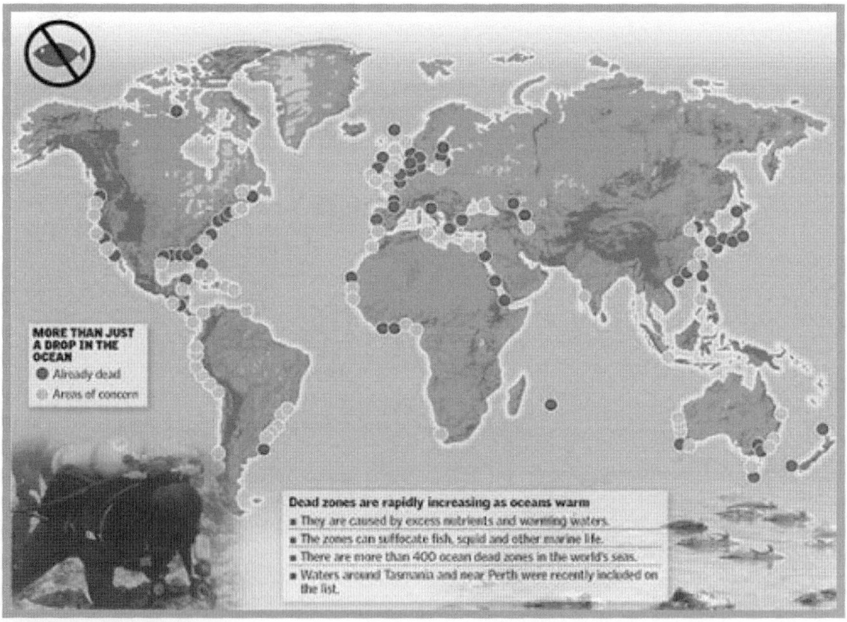

That is compared with 300 in the 1990s and 120 in the 1980s.

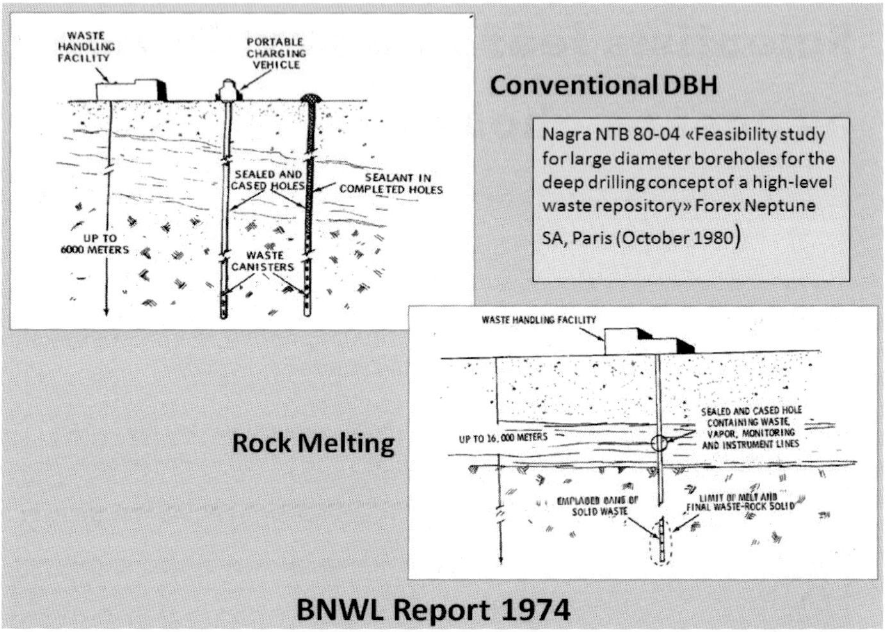
Conventional DBH
WASTE HANDLING FACILITY
PORTABLE CHARGING VEHICLE
SEALED AND CASED HOLES
SEALANT IN COMPLETED HOLES
UP TO 6000 METERS
WASTE CANISTERS
Nagra NTB 80-04 «Feasibility study for large diameter boreholes for the deep drilling concept of a high-level waste repository» Forex Neptune SA, Paris (October 1980)
Rock Melting
WASTE HANDLING FACILITY
UP TO 16,000 METERS
SEALED AND CASED HOLE CONTAINING WASTE, VAPOR, MONITORING AND INSTRUMENT LINES
LIMIT OF MELT AND FINAL WASTE-ROCK SOLID
BNWL Report 1974

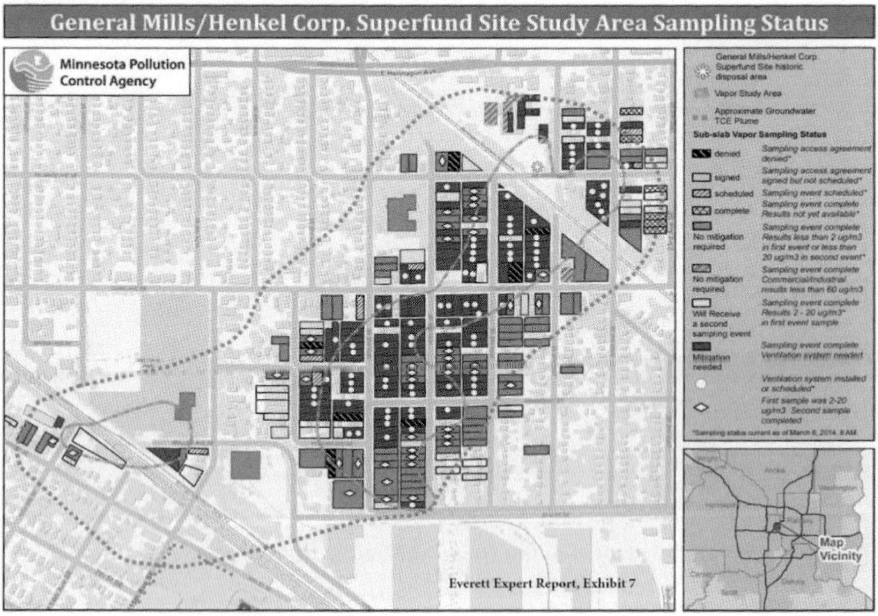
General Mills/Henkel Corp. Superfund Site Study Area Sampling Status
Minnesota Pollution Control Agency
Sub-slab Vapor Sampling Status
Map Vicinity
Everett Expert Report, Exhibit 7

Temporal Variability of Methane

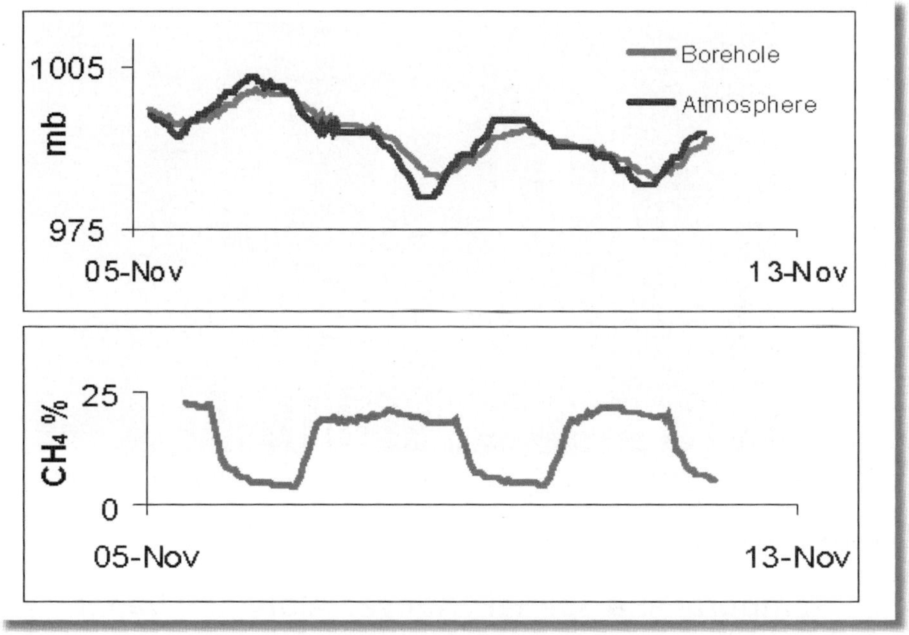

Item #1 – Interim TCE Indoor Air Short-term Response Action Levels and Guidelines

In September 2011, EPA published its *Toxicological Review of Trichloroethylene in Support of the Integrated Risk Information System (IRIS).* Recent findings on TCE conclude that women in the first trimester of pregnancy are one of the most sensitive populations to TCE short-term inhalation exposure due to the potential for heart malformation for the developing fetus.

Because this is a developmental effect, the critical period for exposure is considered to be within an approximate 3-week period in the first trimester of pregnancy during which the heart develops. Scientific information on the exact critical period of exposure for this health impact is not currently available; however, general risk assessment guidelines for developmental effects indicate that exposures over a period as limited as 24 hours[1] may be of concern for some developmental toxicants.

These guidelines identify women of reproductive age as the sensitive population of concern, rather than only pregnant women, because some women may not be aware of their pregnancy during the first trimester.

Temporal Variability of Subslab Benzene at Former Gas Station

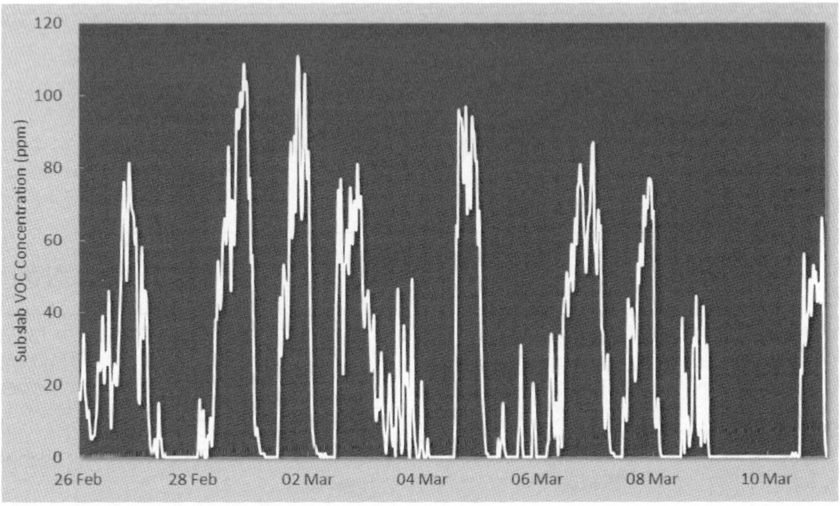

Utah Study finds Temporal Variability for VI

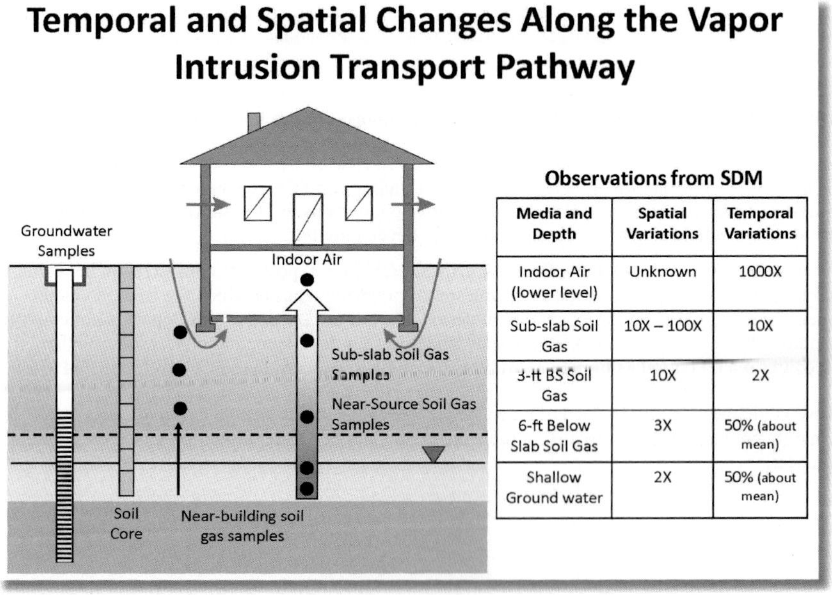

Media and Depth	Spatial Variations	Temporal Variations
Indoor Air (lower level)	Unknown	1000X
Sub-slab Soil Gas	10X – 100X	10X
3-ft BS Soil Gas	10X	2X
6-ft Below Slab Soil Gas	3X	50% (about mean)
Shallow Ground water	2X	50% (about mean)

Source: Johnson, 2014 AEHS Symposium

Antonino Zichichi
Ettore Majorana
His Genius and Long-Lasting Legacy
Grazie a Tutti!

ENERGY PMP REPORT

CARMINE DIFIGLIO
Office of Energy Policy and Systems Analysis, Department of Energy
Washington, DC, USA

ROBERT SCHOCK
World Energy Council, Senior Advisor
London, UK

WILLIAM BARLETTA
U.S. Particle Accelerator School, Department of Physics, Massachusetts
Institute of Technology, Cambridge, MA, USA

RENEWABLE AND NUCLEAR ELECTRICITY: OPPORTUNITIES, CHALLENGES AND POLICY RECOMMENDATIONS

Paper available at: www.energypmp.org

Robert Budnitz
Carmine Difiglio
Michael Jefferson
Hisham Khatib
Tatsuo Masuda
Charles McCombie
Jef Ongena
Lara Pierpoint
Adnan Shihab-Eldin
Bruce Stram

PROJECT FOR MAN

The Energy PMP has provided two Project-for-Man proposals:

- **Centre on Nuclear Power Safety**
  Richard Garwin, Robert Budnitz, Carmine Difiglio, William Barletta, Adnan Shihab-Eldin and Holger Rogner
- **Control of Planetary Emergencies Using the Science of Complex Networks**
  Robert Schock

2015 ENERGY PMP MEETING

- **Global Risks and Energy System Resilience**
  Dr. Tatsuo Masuda

- **U.S. Nuclear Power Plant Closures, Recent and Soon**
  Dr. Robert Budnitz
- **Nuclear Power Operations from the Commercial Side**
  Dr. James Rogers
- **Multinational Developments at the Back-End of the Nuclear Fuel Cycle**
  Dr. Charles McCombie
- **Fusion Update**
  Dr. Jef Ongena and Dr. Yuichi Ogawa
- **Nuclear Waste Disposal in Boreholes (Joint Meeting with the Pollution PMP)**
  Dr. Charles McCombie and Dr. Frank Parker

2015 PLENARY SESSIONS

- **The Energy-PMP Renewable-Nuclear Energy Project**
  Dr. Carmine Difiglio
- **Economics of Renewable and Nuclear Power**
  Dr. Hisham Khatib
- **Energy Storage for Electric Power Generation**
  Dr. Lara Pierpoint
- **Dispatching Renewable and Nuclear Power**
  Dr. Bruce Stram
- **Environmental Impact of Nuclear and Renewables**
  Dr. Charles McCombie and co-author Professor Michael Jefferson
- **Nuclear Safety and Security**
  Dr. Robert Budnitz
- **The Control of Complex Systems: Energy in Cities**
  Professor Kimberly Gray
- **Policies for an Energy System Transformation**
  Dr. Robert Shock
- **Energy is a Solvable Human Planetary Emergency**
  Professor William Barletta
- **Intelligent Grids for the Poor**
  Dr. James Rogers

PARTICIPATION IN 2015 ERICE EXERCISE ON SMART CITY RESILIENCE

Especially, Robert Schock and Tatsuo Masuda.

ENERGY PMP PRIORITIES FOR 2015

- Follow-up on current Project-for-Man proposals, respond to feedback on current proposals and develop new proposals as necessary.
- Assess the factors that will affect the availability of fossil fuels that, notwithstanding policies to reduce GHG emissions, are expected to remain the primary source of world energy supplies through 2050.

- Provide an updated assessment of carbon capture, storage and utilization.
- Organize, with the Pollution PMP, a plenary session on the most effective storage strategies for nuclear waste.
- Assess the prospects for energy efficiency and its contribution to smart cities.
- Evaluate the prospects for commercialization of small modular reactors. What level of government support is required to establish an industry? Is that model appropriate?
- Provide a review of esoteric nuclear energy concepts including any prospects for Low Energy Nuclear Reactions (LENR) fusion research (cold fusion).
- Update aspects of the Renewable-Nuclear paper including:
  - Effectiveness of government expenditures on "new" renewable energy, and their contributions to electricity generation.
  - Shed light on realistic energy supply prospects across the board in the coming decades.

GUIDING PRINCIPLES

The Energy PMP will give priority to activities that will:

- Materially add to the substantial existing body of energy research and analysis;
- Whether the activity could be expected to improve human welfare; and
- Using an interdisciplinary approach that exploits the unique expertise, talents and experience of WFS Members.

MOTHER AND CHILD HEALTH PMP: 2014-2015 ACTIVITIES

NATHALIE CHARPAK
Kangaroo Foundation
Bogotá, Colombia

SOCCORRO DE LEON-MENDOZA
Jose Fabella Memorial Hospital, Neonatology Unit
Manila, Philippines

MATERNAL AND CHILD MORTALITY IS A PLANETARY EMERGENCY

Background: Mission of the PMP

- To recognize new and monitor existing tools that decrease maternal and infant mortality and morbidity. For example, the Kangaroo Mother Care Method (KMC).
- To highlight the impact of the other planetary emergencies on Maternal and Infant mortality and morbidity and, therefore, contribute to the enhancement of the quality of the present life and future generations.

The Kangaroo Mother Care (KMC) Method

How does KMC work? Once stabilized after birth, the premature or Low Birth Weight child is placed in a kangaroo position on the mother's chest in direct skin to skin contact. The position is initiated as soon as possible and maintained as long as possible even after the baby is discharged. Only when the child is capable of breastfeeding, the mother-child dyad can go home and return to the outpatient KMC for a strict follow-up until the child reaches its full-term (40 weeks). The kangaroo position is kept until the child rejects it. Children are daily monitored during the period of physiological weight loss, then until they gain 15 g per kg per day, which is the normal intrauterine weight gain during the last period of pregnancy. If, despite the efforts, the weight gain is inadequate, breast milk will be complemented with artificial formula for prematurity (up to 30% of the daily ration).

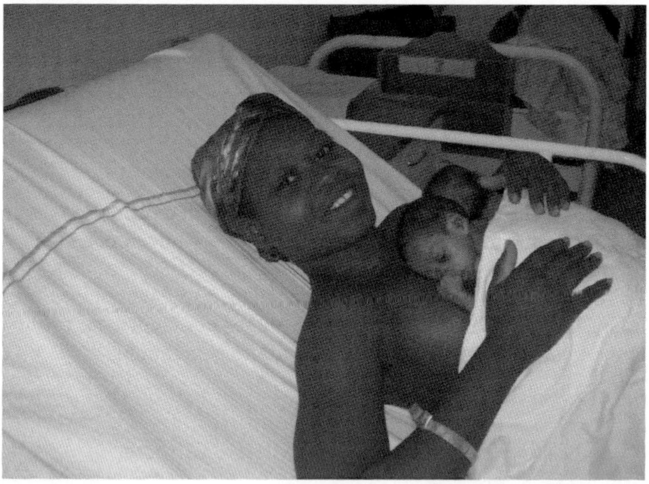

Two questions are waiting for an answer in relation with KMC

- How to accelerate the diffusion of KMC? The main question for all the big NGO and official international organizations and international cooperation.

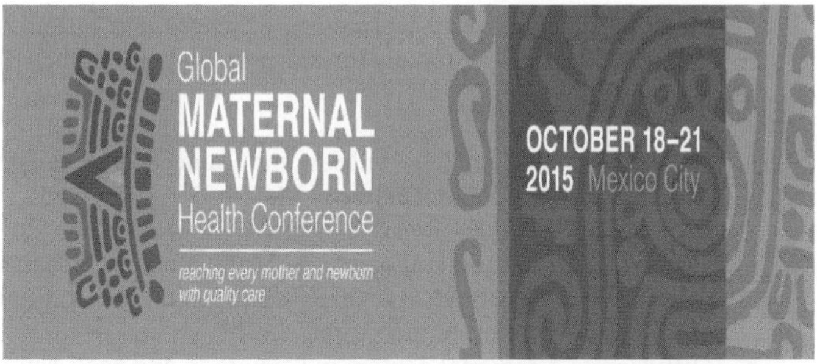

KMC panel Oct 2015

Despite robust evidence for the impact of KMC on mortality and morbidity reduction in low birthweight and preterm newborns, scale-up of quality KMC services has lagged in many parts of the world. This panel examines enablers and barriers to scale, and global initiatives aimed at increasing uptake of KMC.

- What indicators to measure?

Answering these 2 questions is a challenge for the members of our PMP

- How to make examples of how to accelerate the KMC diffusion in the true life of the developing world?
  - The Colombian Kangaroo Foundation won an international call for proposal of GCC (Grand Challenge Canada): to build a KMC E-learning platform and pilot it in Mali and Cameroon through a south-south cooperation. It started in May 2015 with a partnership of CrossKnowledge

Foundation (specialist in e-learning in Sciences and business/Wiley and sons) and Engineers without Borders (Colombian branch specialist in access to clean water).

- We coach 2 KMC centers in Mali and Cameroon to transform them in excellence KMC centers between May to December 2015.
- We support the creation of a "Fondation Kangourou du Cameroon" and an "Association Kangourou du Mali" able to manage the funds of their project. We have the responsibility and we will support them during the execution of the project.
- Diffusion to 5 regional centers in Cameroon and Mali between January 2016 and December 2016.
- Diffusion to 10 primary centers in Cameroon and Mali between January 2017 and June 2018.

– Diffusion of KMC in Colombia at a country scale: 22 centers trained in 2013-2014 with the Health Ministry of Colombia support.
– International Training of trainers as a politic of KMC diffusion of the Kangaroo Foundation: south-south transfer of KMC knowledge (15 days of training in a KMC excellence center) and local support for the translation of this knowledge into practical action (Return mission). 2015: Venezuela, Ghana, Bolivia, Kenya.

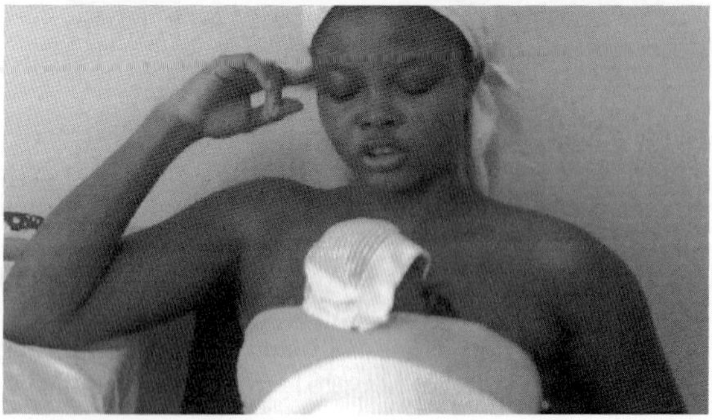

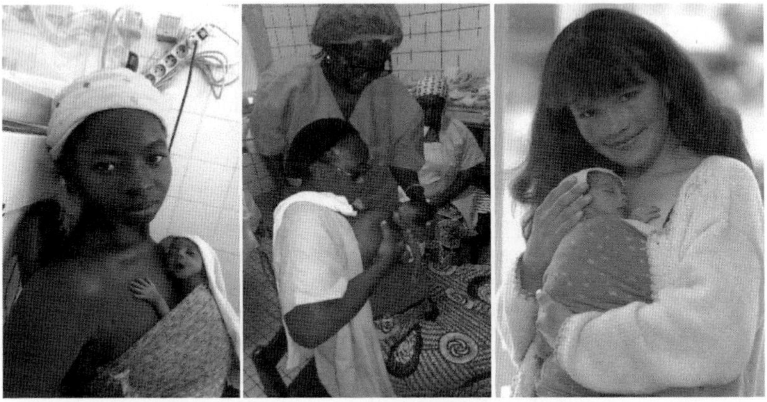

– The Philippines Kangaroo Foundation: acceleration of implementation of KMC excellence centers through a training of trainers program supported by WHO and the health Ministry: at the speed of light!

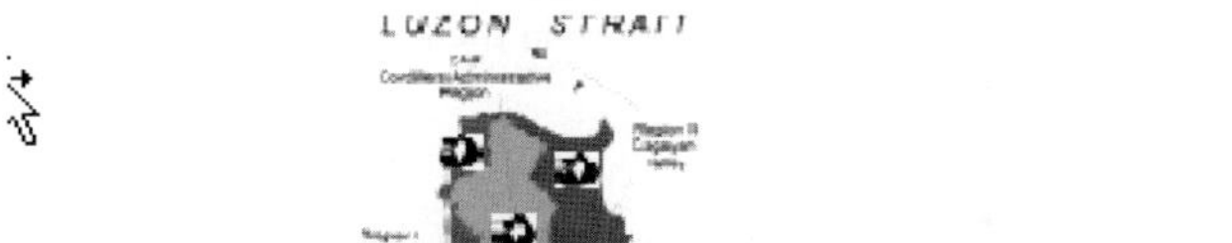

- 16/17 Regions with 52 KMC facilities in various stages of implementation.
- 12 KMC Facilities certified as Centers for Training and Excellence.

- Collaboration with government (Dept. of Health), WHO, UNICEF, USAID, Professional organizations, NGO-KMC Foundation Philippines.
- National Health Insurance system adopting and paying.
- KMC services launching November 2015.

- The KMC research: long term follow up and indicators

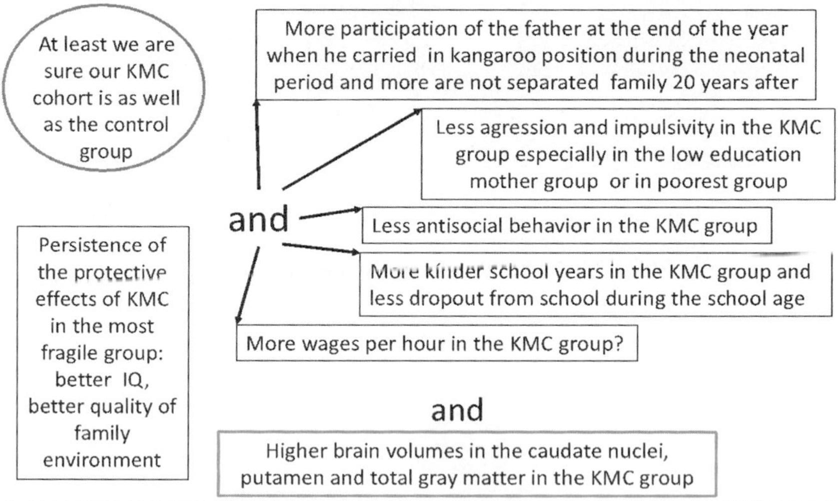

One product of the analytical approach was the development of a new visual analysis tool (Braviz) allowing the integration of interactive visualization of brain imaging, clinical and socioeconomic data with real-time statistical analysis to facilitate exploration and discovery of insights in the study.

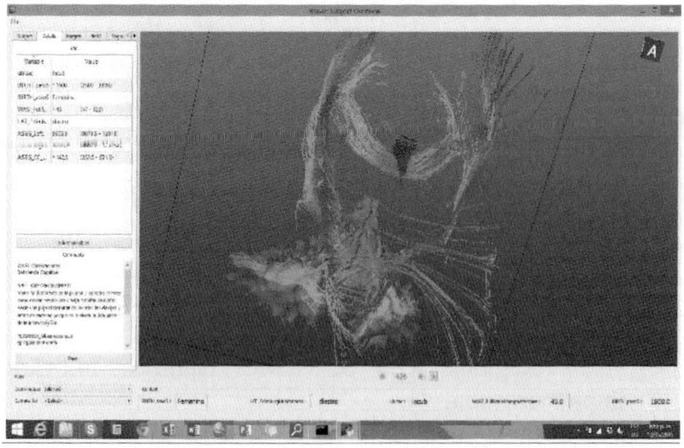

<u>Conclusion: The health of the mother and the newborn are forever a planetary emergency because they are our future.</u>

- KMC a very powerful and efficient, scientifically-based health care intervention which is easily implemented in settings ranging from very restricted to unrestricted access to health care resources.
- Moreover, this intervention is carried out during the neonatal and post-neonatal periods which are very sensitive to brain maturation and parents' involvement.
- Beyond that, KMC decreases neonatal and infantile mortality; KMC has significant, long lasting brain protection effects for preterm infants, not only in the most physically fragile infant but for the most socially at risk also.
- This new knowledge could be used to further support the need and importance to accelerate the KMC coverage to all preterm and low birth weight infants in the world.

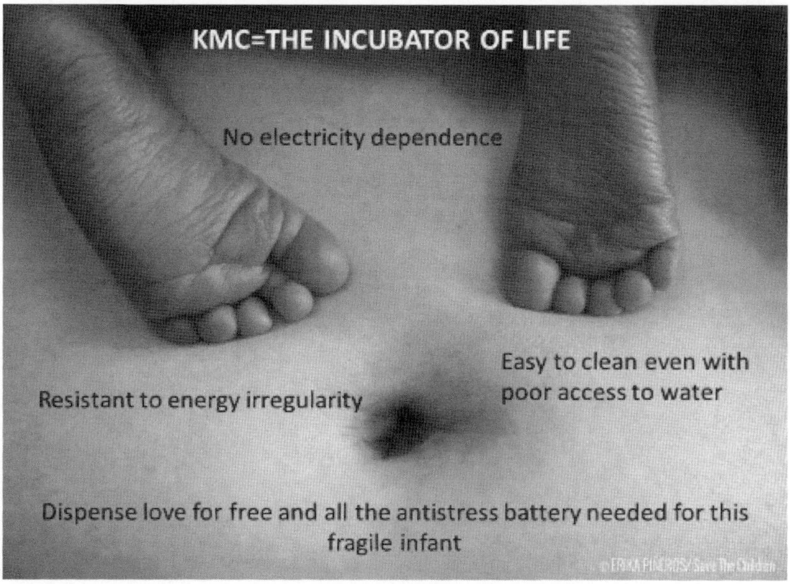

AIDS AND INFECTIOUS DISEASES PMP

FRANCO BUONAGURO
Istituto Nazionale dei Tumori, "Fondazione G. Pascale"
Napoli, Italy

AIDS AND INFECTIOUS DISEASES PMP: 2015 ACTIVITIES

An informal round-table has been held on August 19th

- To define the final details of our session at the plenary component of the Meeting;
- To plan the PMP activities for next year;
- To select the topic for next year plenary session;
- To finalize the yearly report of the activities for the WFS General Meeting of August 23rd.

SESSION 12: MECHANISMS OF SUSCEPTIBILITY/RESISTANCE TO HUMAN PATHOLOGIES
Chairman A. Zichichi—Co-chair F. Buonaguro

16.00 – 17.00
Mechanisms of Susceptibility to Infections: the Human Immunodeficiency Virus (HIV) Paradigm
Dr. Ishwar Gilada
AIDS Society of India, President, Mumbai, India

Etiopathogenic Mechanisms of Immune-related Cancers
Dr. Sam Mbulaiteye
National Institute of Health, Infections and Immunoepidemiology Branch, Bethesda, USA

Deciphering Neurodegeneration: What do Phenotypes, Molecular Biology and Genetics tell us about Pathogenesis?
Dr. Emanuele Buratti
Molecular Pathology Group, International Centre for Genetic Engineering and Biotechnology, Trieste, Italy

Saturday, 22 August 2015

The session was planned to demonstrate the relevance of the proposed Biomedical Project for a Center to study Planetary Emergencies.

WFS Human genome diversity/susceptibility to:

- Infectious diseases;
- Cancer;

- Neurodegenerative diseases.

The project would not target individual diseases, but population-based aspects, critical for health planning and scientific discoveries.

AIDS AND INFECTIOUS DISEASES PMP 2015 PROPOSAL

Immuno-driven diseases:

1. Immunoepidemiology and Cancers (Sam Mbulaiteye, NIH);
2. Role of inflammation in Cancer (Giorgio Trinchieri, NIH);
3. Genetics in immuno-driven Cancers (Riccardo Dalla Favera, Columbia Univ);
4. Pathogens in immuno-driven Cancers (Franco Buonaguro, INT-Napoli);
5. Clinics and therapeutic approaches in immuno-driven Cancers (Ishwar Gilada, Mumbai);
6. Clinics and therapeutic approaches in immuno-driven Cancers in the elderly (Ignazio Carreca, Univ Palermo); eventually along with two Neuroscience presentations:
7. Immuno-driven neuropathologies (Emanuele Buratti, ICGEB); and
8. Genetics in immuno-related neuropathologies (Dino Ghetti, Indiana Univ).

AIDS AND INFECTIOUS DISEASES PMP 2014-15 ACTIVITIES

Report of AIDS and Infectious Diseases related activities:

- AIDS and infectious agents within Global Virus Network GVN—Franco M Buonaguro;
- EMBLEM project and AORTIC activities—Sam M Mbulaiteye; and
- Indian AIDS Society and ASICON annual meeting—Ishwar Gilada.

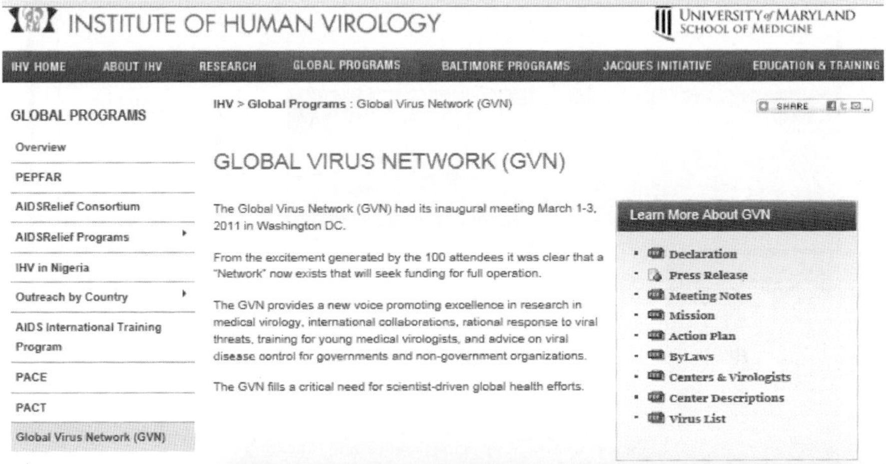

INSTITUTE OF HUMAN VIROLOGY

UNIVERSITY of MARYLAND SCHOOL OF MEDICINE

IHV HOME | ABOUT IHV | RESEARCH | GLOBAL PROGRAMS | BALTIMORE PROGRAMS | JACQUES INITIATIVE | EDUCATION & TRAINING

GLOBAL PROGRAMS

- Overview
- PEPFAR
- AIDSRelief Consortium
- AIDSRelief Programs
- IHV in Nigeria
- Outreach by Country
- AIDS International Training Program
- PACE
- PACT
- Global Virus Network (GVN)

IHV > Global Programs : Global Virus Network (GVN)

SHARE

GLOBAL VIRUS NETWORK (GVN)

The Global Virus Network (GVN) had its inaugural meeting March 1-3, 2011 in Washington DC.

From the excitement generated by the 100 attendees it was clear that a "Network" now exists that will seek funding for full operation.

The GVN provides a new voice promoting excellence in research in medical virology, international collaborations, rational response to viral threats, training for young medical virologists, and advice on viral disease control for governments and non-government organizations.

The GVN fills a critical need for scientist-driven global health efforts.

Learn More About GVN

- Declaration
- Press Release
- Meeting Notes
- Mission
- Action Plan
- ByLaws
- Centers & Virologists
- Center Descriptions
- Virus List

More than 35 of the world's top virologists gathered Thursday, March 3 to ratify the Global Virus Network (GVN).

THE 7TH GVN MEETING HAS BEEN ORGANIZED IN BEIJING (16-19 MAY)

GVN Annual Meeting in Beijing, China 北京 2015!

The Global Virus Network's 7th Annual Meeting is May 16th-19th, 2015 in Beijing, China. This year's meeting is hosted by the Beijing University of Technology of the China GVN under the leadership of Dr. Yi Zeng. The meeting brings together the Directors and leading researchers from GVN's 34 Centers of Excellence around the world plus local guests to discuss the latest research on HIV, Ebola, Influenza, Dengue, and other key viral threats. The conference program is available at: http://gvn.org/beijing-meeting-2015.

Featured Speakers

Dr. Yi Zeng

Dr. Ab Osterhaus

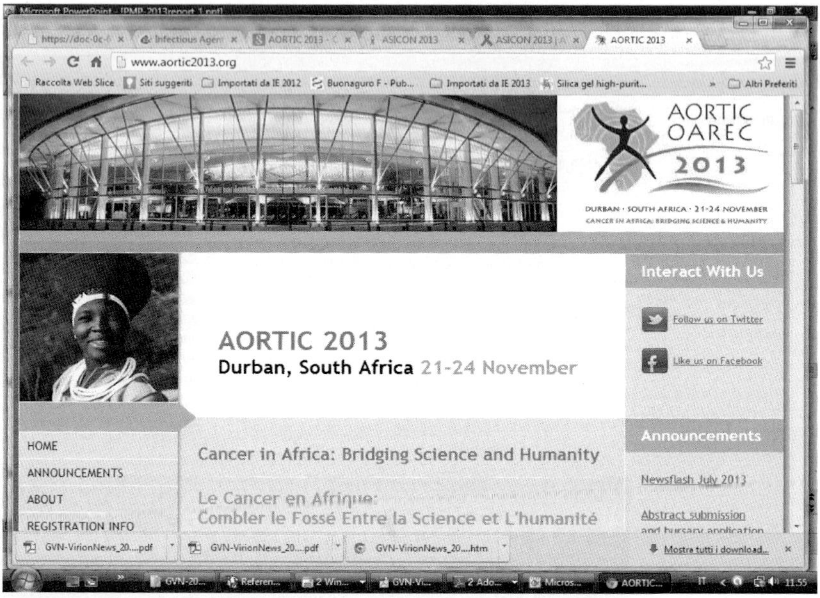
www.aortic2013.org
AORTIC OAREC 2013
DURBAN · SOUTH AFRICA · 21-24 NOVEMBER
CANCER IN AFRICA: BRIDGING SCIENCE & HUMANITY
Interact With Us
Follow us on Twitter
Like us on Facebook
AORTIC 2013
Durban, South Africa 21-24 November
Announcements
HOME
ANNOUNCEMENTS
ABOUT
REGISTRATION INFO
Cancer in Africa: Bridging Science and Humanity
Le Cancer en Afrique:
Combler le Fossé Entre la Science et L'humanité
Newsflash July 2013
Abstract submission

HOME ABOUT CONFERENCES RESEARCH TRAINING PARTNERS CANCER FACTS NEWS MEMBERSHIPS RESOURCES SHOP CONNECT
SCIENCE OF GLOBAL PROSTATE CANCER DISPARITIES IN BLACK MEN CONFERENCE
AORTIC 2015 - MARRAKECH, MOROCCO
AORTIC 2013 - DURBAN, SOUTH AFRICA
OTHER CONFERENCES
CONFERENCES
AORTIC 2015 - MARRAKECH, MOROCCO -
AORTIC OAREC
AORTIC 2015, entitled "AORTIC Roadmap to Cancer Control in Africa", will be held in Marrakech, Morocco from 18 - 22 November 2015. Visit our conference website here: http://www.aorticconference.org. Have any queries? Contact us via e-mail at: info@aorticconference.org

HOME ABOUT CONFERENCES RESEARCH TRAINING PARTNERS CANCER FACTS NEWS MEMBERSHIPS RESOURCES SHOP CONNECT
"DEDICATED TO THE PROMOTION OF CANCER CONTROL IN AFRICA"
YOU CAN MAKE A REAL DIFFERENCE
WELCOME
AORTIC was formed in 1983 by expatriate African cancer care workers, scientists and their friends, and is dedicated to the promotion of cancer control in Africa.
AORTIC's key objectives are to further the research relating to cancers prevalent in Africa, support the management of training programs in oncology for health care workers, deal with the challenges of creating cancer control and prevention programs and raise public awareness of cancer in Africa.
The Executive Council of AORTIC consists of scientists and clinicians with a professional interest in prevention, treatment and study of cancer - volunteering their expertise for the plight of the cancer patient in Africa.
Voir ci-dessous pour le Francais
AORTIC OAREC
WHAT's NEW
2015 AORTIC CONFERENCE WILL BE HELD IN MARRAKECH MOROCCO
click here for more...
DONATE JOIN AORTIC
THANK YOU FOR YOUR SUPPORT

8th NATIONAL CONFERENCE OF
AIDS SOCIETY OF INDIA
30th Oct. - 1st Nov. 2015 | JW Marriott Hotel Mumbai Sahar, Mumbai
Sharing a Vision, Shaping the Future
Home | About us | Contact us
Home
About ASI
Committee
Faculty
Registration
Sponsorship
Tours
Program
Contact Us
Chhatrapati Shivaji Terminus Mumbai
News Update
Downloads
Important Dates
Popular Post
Welcome Message
Related Links
Conference City
Abstract Submission
About Venue
Dear Colleagues,

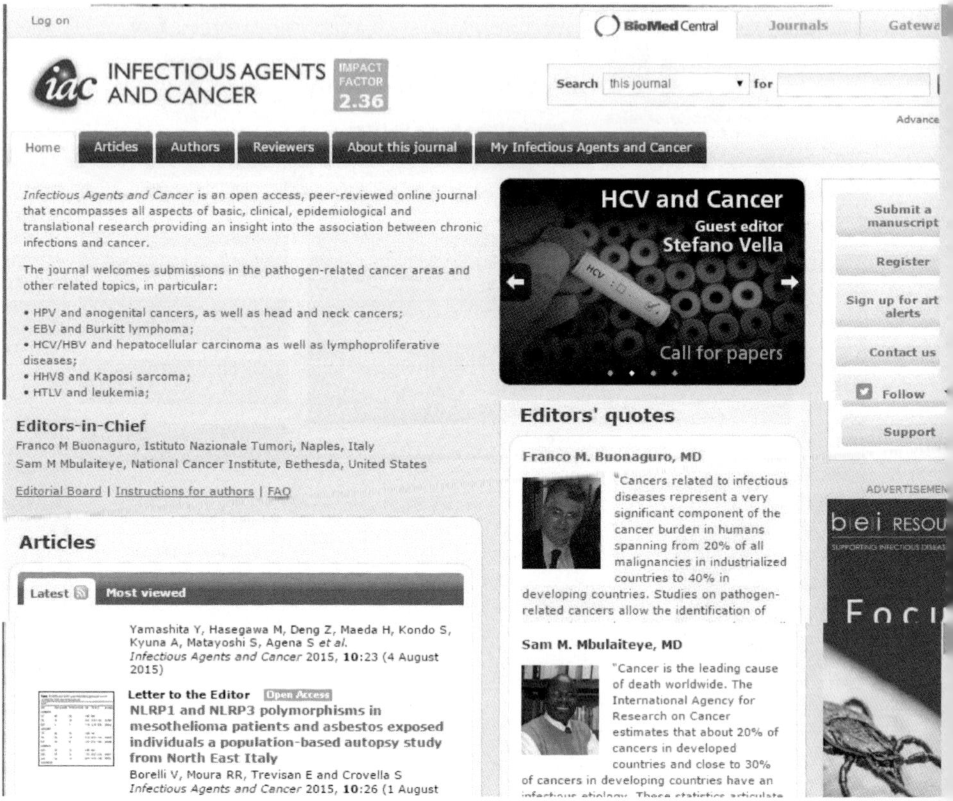
Log on
BioMed Central
Journals
Gatewa
INFECTIOUS AGENTS AND CANCER
IMPACT FACTOR 2.36
Search this journal for
Advance
Home
Articles
Authors
Reviewers
About this journal
My Infectious Agents and Cancer
Infectious Agents and Cancer is an open access, peer-reviewed online journal that encompasses all aspects of basic, clinical, epidemiological and translational research providing an insight into the association between chronic infections and cancer.
The journal welcomes submissions in the pathogen-related cancer areas and other related topics, in particular:
• HPV and anogenital cancers, as well as head and neck cancers;
• EBV and Burkitt lymphoma;
• HCV/HBV and hepatocellular carcinoma as well as lymphoproliferative diseases;
• HHV8 and Kaposi sarcoma;
• HTLV and leukemia;
HCV and Cancer
Guest editor
Stefano Vella
Call for papers
Submit a manuscript
Register
Sign up for art alerts
Contact us
Follow
Support
Editors-in-Chief
Franco M Buonaguro, Istituto Nazionale Tumori, Naples, Italy
Sam M Mbulaiteye, National Cancer Institute, Bethesda, United States
Editorial Board | Instructions for authors | FAQ
Editors' quotes
Franco M. Buonaguro, MD
"Cancers related to infectious diseases represent a very significant component of the cancer burden in humans spanning from 20% of all malignancies in industrialized countries to 40% in developing countries. Studies on pathogen-related cancers allow the identification of
ADVERTISEMEN
bei RESOU
Articles
Latest
Most viewed
Yamashita Y, Hasegawa M, Deng Z, Maeda H, Kondo S, Kyuna A, Matayoshi S, Agena S et al.
Infectious Agents and Cancer 2015, 10:23 (4 August 2015)
Letter to the Editor
Open Access
NLRP1 and NLRP3 polymorphisms in mesothelioma patients and asbestos exposed individuals a population-based autopsy study from North East Italy
Borelli V, Moura RR, Trevisan E and Crovella S
Infectious Agents and Cancer 2015, 10:26 (1 August
Sam M. Mbulaiteye, MD
"Cancer is the leading cause of death worldwide. The International Agency for Research on Cancer estimates that about 20% of cancers in developed countries and close to 30% of cancers in developing countries have an

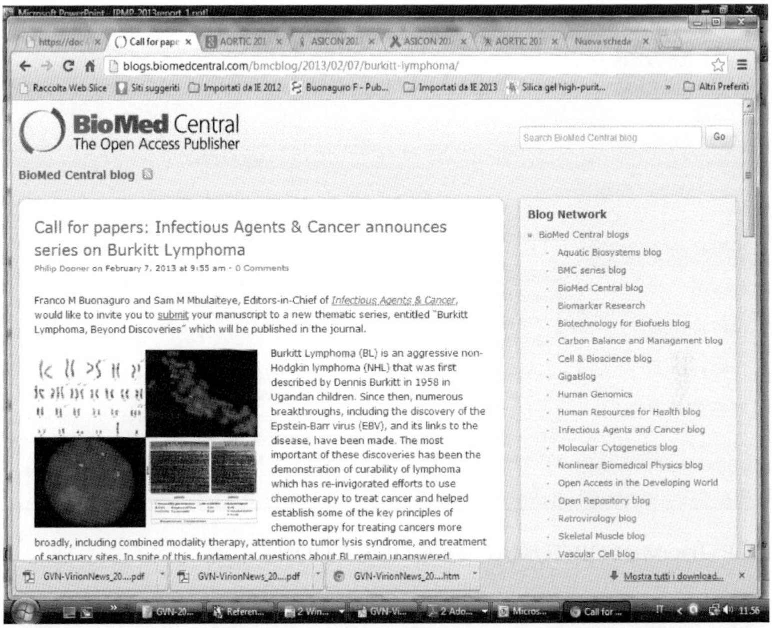
blogs.biomedcentral.com/bmcblog/2013/02/07/burkitt-lymphoma/
BioMed Central
The Open Access Publisher
BioMed Central blog
Call for papers: Infectious Agents & Cancer announces series on Burkitt Lymphoma
Philip Dooner on February 7, 2013 at 9:55 am - 0 Comments
Franco M Buonaguro and Sam M Mbulaiteye, Editors-in-Chief of Infectious Agents & Cancer, would like to invite you to submit your manuscript to a new thematic series, entitled "Burkitt Lymphoma, Beyond Discoveries" which will be published in the journal.
Burkitt Lymphoma (BL) is an aggressive non-Hodgkin lymphoma (NHL) that was first described by Dennis Burkitt in 1958 in Ugandan children. Since then, numerous breakthroughs, including the discovery of the Epstein-Barr virus (EBV), and its links to the disease, have been made. The most important of these discoveries has been the demonstration of curability of lymphoma which has re-invigorated efforts to use chemotherapy to treat cancer and helped establish some of the key principles of chemotherapy for treating cancers more broadly, including combined modality therapy, attention to tumor lysis syndrome, and treatment
Blog Network
BioMed Central blogs
Aquatic Biosystems blog
BMC series blog
BioMed Central blog
Biomarker Research
Biotechnology for Biofuels blog
Carbon Balance and Management blog
Cell & Bioscience blog
GigaBlog
Human Genomics
Human Resources for Health blog
Infectious Agents and Cancer blog
Molecular Cytogenetics blog
Nonlinear Biomedical Physics blog
Open Access in the Developing World
Open Repository blog
Retrovirology blog
Skeletal Muscle blog
Vascular Cell blog

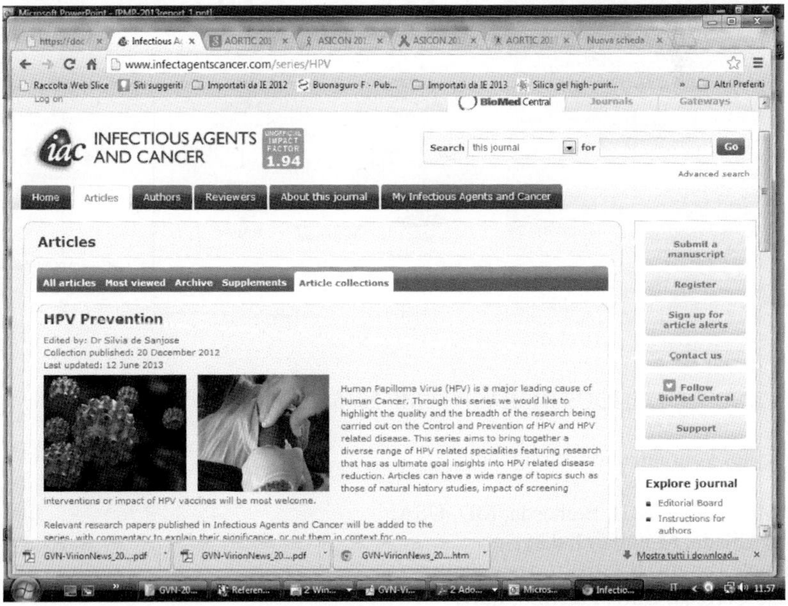
www.infectagentscancer.com/series/HPV
INFECTIOUS AGENTS AND CANCER
1.94
Home
Articles
Authors
Reviewers
About this journal
My Infectious Agents and Cancer
Articles
All articles
Most viewed
Archive
Supplements
Article collections
HPV Prevention
Edited by: Dr Silvia de Sanjose
Collection published: 20 December 2012
Last updated: 12 June 2013
Human Papilloma Virus (HPV) is a major leading cause of Human Cancer. Through this series we would like to highlight the quality and the breadth of the research being carried out on the Control and Prevention of HPV and HPV related disease. This series aims to bring together a diverse range of HPV related specialities featuring research that has as ultimate goal insights into HPV related disease reduction. Articles can have a wide range of topics such as those of natural history studies, impact of screening interventions or impact of HPV vaccines will be most welcome.
Relevant research papers published in Infectious Agents and Cancer will be added to the series
Submit a manuscript
Register
Sign up for article alerts
Contact us
Follow BioMed Central
Support
Explore journal
Editorial Board
Instructions for authors

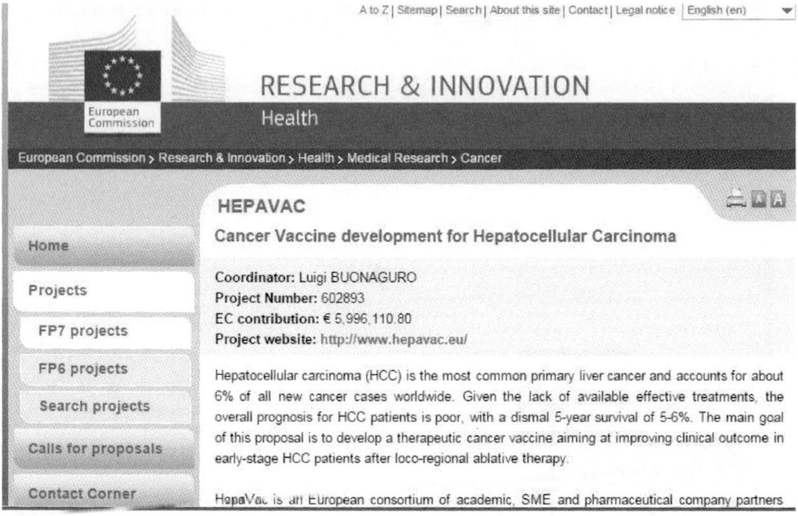

HLA PEPTIDOME FOR TUMAP IDENTIFICATION

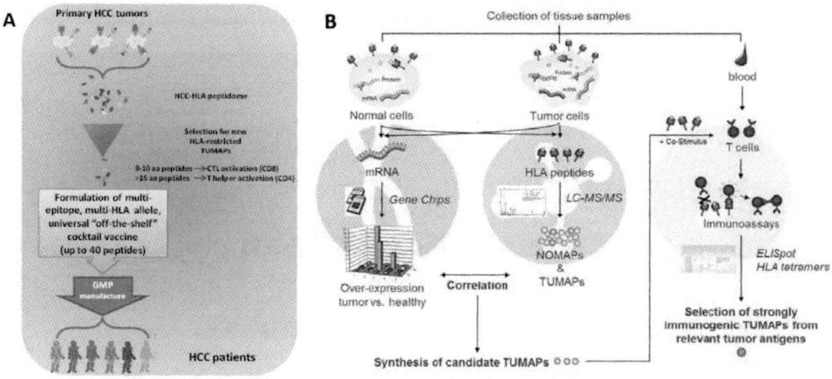

AIDS AND INFECTIOUS DISEASES PMP 2015 PROPOSAL

Thanks for your attention, along with:

Sam Mbulaiteye, NIH, Bethesda, MD, USA;
Ishwar Gilada, Mumbai, India;
Emanuele Buratti, ICGEB, Trieste, Italy;
Ignazio Carreca, Univ. Palermo, Italy.

INFORMATION SECURITY PMP REPORT

HENNING WEGENER
Ambassador of Germany (ret.), Information Security Permanent Monitoring Panel World Federation of Scientists, Madrid, Spain

Our plenary presentation this year, a Round Table entitled "From Bad to Worse: the New Dimension of Cyber Threat" has attempted to present a concise portrait of the alarming current state of global cyber insecurity in its various aspects, emphasizing the extremely rapid growth of all elements that constitute the overall threat landscape. As this debate, conducted less than two days ago, is presumably still fresh on the mind of the audience, I will not summarize it here, the more so as a transcript of the Round Table will appear in the seminar yearbook.

The most visible written output of the Group's work over the last year has been the publication of members of the PMP, jointly with the Secretary General of the ITU, Dr. Hamadoun Toure—a member of the PMP—of the book "The Quest for Cyber Confidence". The book has appeared as a UN/ITU publication and can be downloaded from www.itu.int under *publications*. We took pleasure in presenting a copy of the book to Professor Zichichi during the present meeting. It is a sequel to our earlier book, "The Quest for Cyber Peace", also available at the ITU site.

Among other noteworthy activities of the group is a number of international conferences in which PMP members took an active part. The ITU, under Dr. Touré's chairmanship, organized in October of 2014—a Symposium on cyber space and critical energy infrastructure protection which saw several PMP members as speakers or panel participants. In February of 2014, the Lebanese Ministry of Defense organized in Beirut an International Conference on "National Cybersecurity Strategy: A Must for National Security" in which the Lebanese member of our group was a co-organizer. The WFS co-sponsored the conference, and several members of the group appeared as speakers. Our Indian colleague organized two prestigious conferences on Cyberlaw and Cybercrime in March and October 2014 in New Delhi (www.cyberlawcybercrime.com), also co-sponsored by the WFS, and attended by the PMP Chairman. The presentations given there can be followed on the PMP website www.unibw.de/infosecur.

Otherwise, PMP members oriented their individual work along the lines of a draft annual work program proposed by the Chairman in early 2015, and revised during the PMP meeting a few days ago here in Erice. It looks at current and future cyber developments, as discussed during the group's Plenary Round Table, the problems of the on-going militarization of cyber space, and cyber law developments internationally and in important nations. After the change in the ITU leadership, the group has also established contact with the incoming Secretary General to explore the prospects for continuing the fruitful collaboration with that important organization, a key entity in international cyber policies.

PMP REPORT FOR COSMIC OBJECTS

WALTER HUEBNER
Southwest Research Institute
San Antonio, TX, USA

ONGOING ACTIVITIES OF THE PMP

Discussions centered on methods to increase detectability of small asteroids, in particular asteroids with orbits inside the Earth's orbit. These asteroids are usually not accessible with Earth-based telescopes. A. Cellino described the use of the Gaia Mission, which is ideally suited for this purpose. It can also determine several important properties of these small asteroids with its instruments, such as their size and shape. Furthermore, the mass of the objects can be determined from perturbations of their orbits by the gravitational attraction of nearby planets. Long-term observations are needed for a precise determination of these orbital perturbations. The Gaia Mission is ideally suited for these observations.

The discussion shifted to the consequences of an asteroid or comet nucleus impact on Earth. Very few data exist to assess the potential of fatalities, injuries, population displacement, material, and economic losses caused by such an impact. Material losses include destruction of infrastructure and losses of cultural treasures. P. Bradley and W. Huebner compared the total kinetic energy of an impacting object with the energy releases of underground nuclear tests and earthquakes. Earthquake databases from the insurance industry contain information about fatalities, injuries, and economic losses. It was concluded that a 100 meter asteroid entering the Earth's atmosphere with a speed of only 10 km/s released an energy equivalent to that of an earthquake of magnitude 8 on the Richter scale. Such earthquakes typically cause a few hundred thousand fatalities, up to a million injuries, and economic and other infrastructure losses in the $100 billion range. In comparison, the cost for avoiding such a collision, based on NASA's Deep Impact Mission ($330 million), is of the order of $1 billion, even if three countermeasure missions are needed. The comparison of the impact of a cosmic object with an earthquake depends on the partitioning of the available energy into its components. In an earthquake most of the energy is in seismic waves that could result in a tsunami if the quake is in a coastal region. In case of an asteroid or comet nucleus impact the energy is partitioned to cause also damage resulting from airblast (shock waves), cratering, flying debris, etc. Thus, fatalities, injuries, and economic losses suffered in an impact could be much worse or less severe than in an earthquake. **As an advanced civilization, do we have to wait to suffer and assess the damages of an asteroid or comet nucleus impact on Earth before we seriously prepare for countermeasures to avert such collisions?**

As the next item, L. Drude described the activities of the NEO-Shield consortium. This activity is a multi-year project carried out mostly by European institutes and laboratories under the leadership of the DLR in Berlin. On the scientific and technical side this project considers mainly the deflection of potentially hazardous objects to avert an Earth collision. It also considers political issues such as the involvement of the UN and procedures for evacuation of threatened areas.

Finally, A. Zaitsev presented concepts for the creation of an international planetary defense system "Citadel" as envisioned by the Russian Planetary Defence Centre. The concept centers on global, regional, and local catastrophes caused by an Earth impact of an asteroid or nucleus of a comet. Citadel would include discovery, tracking, determination of properties of an object, and methods for collision avoidance. It also covers organizational, financial, international, and legal implications. It would be an internationally sanctioned and controlled organization with at least two operation centers: East and West.

NEW ACTIVITIES OF THE PMP

Discussions of the PMP included six new issues:

1. Increase and improve collaboration and exchange of information between various institutes and laboratories. This could also lead to joint publications and taking advantage of various forms of expertise.
2. Improve communications with relevant laboratories, institutes, and government agencies.
3. Develop and implement a plan for outreach. This would be directed to news media, the general public, schools, and funding agencies. The purpose is to increase awareness of the asteroid/comet nucleus impact risks. A website would be an important tool for this purpose. It should make references to other relevant NASA, ESA, etc., websites.
4. Define terminology. This is an important aspect for clear communications, particularly when several countries are involved. It would include definitions for:
   - 4.1 Warning time: the difference in time from realization that an object will impact Earth to the predicted time of the impact.
   - 4.2 Structure of an object: monolithic, rubble pile, and binary system.
   - 4.3 Comet nucleus: Ice and dust body. Short period (S-P) comet: One that had its origin in the Kuyper belt, the ecliptic outside of Pluto's orbit. Long period (L-P) comet: One that had its origin in the spherical Oort cloud at distances out to the limits of gravitational attraction of the solar system. Asteroid: Stoney, metallic, or carbonaceous body.
   - 4.4 Properties: Size, mass, dimensions, shape, density, composition, orbital parameters.
   - 4.5 Projected impact area and associated risk.
   - 4.6 NEO: Near-Earth object (NEA or NEC) that comes within 1.3 AU from the Sun.
   - 4.7 NEA: Near-Earth asteroid.
   - 4.8 NEC: Near Earth comet.
   - 4.9 PHO: Asteroid or comet nucleus that passes the Earth's orbit at a minimum orbit intersection distance (MOID) of 0.05 AU or less.
5. International committee of scientists to determine appropriate countermeasures including direction of deflection and who is in charge.
6. Type of countermeasures:
   - 6.1 Hypervelocity impactor.
   - 6.2 Nuclear explosives.
   - 6.3 Slow push/pull techniques in cases of very long warning times (e.g., gravity tractor).

FAST COMPUTING AND INFORMATICS FRONTIERS AND ECOLOGICAL SCIENCE PMP

HORST WENNINGER
CERN, DPH-UAI Division
Geneva, Switzerland

A proposal to set up a permanent Monitoring Panel (PMP) on "Fast Computing and Informatics Frontiers and Ecological Science" has been discussed at the International Seminar on Nuclear War and Planetary Emergencies in August 2012 at ERICE.

It was expected that CERN might provide a valid basis for such a PMP since CERN research combined with technology has great impact on data collection, computing and informatics.

CERN THE INITIATOR OF WEB AND GRID

Indeed, CERN **has specific competences** in topics related to fast computing and informatics. As initiator of Web and promoter of Grids (industrial name CLOUD), CERN is using and developing technics for large scale distributed information via fast networks. CERN - via its OPEN LAB project - is focusing on training and education of students in informatics and collaborating with industrial partners in joint ventures preparing the next generation infrastructures. http://openlab.web.cern.ch/

CERN AND ITS ECOLOGICAL SCIENCE NETWORK

CERN **has no specific competence** in ecological science, but CERN is member of the EIROFORUM* and has access to and close collaboration with all the network partners.

One of the EIROFORUM members, the European Molecular Biology Laboratory, EMBL, has activities in structural biology, in mouse biology, and in bioinformatics. The European Bioinformatics Institute, EBI, is a hub for biological data services and research. Bioinformatics, is an example of interdisciplinary science involving not only biologists and biochemists but also computer scientists and mathematicians. They participate in the project ELIXIR: sustainable infrastructure for biological information in Europe supporting life science research and its translation to: medicine, agriculture, the environment, bio-industries, and society.

Another aspect of ecological science is the observation of climate change: CERN and the International Geneva has of course a frequent information exchange, organizing talks from experts working on global warming and climate extremes and the impact on the weather. A recent presentation at CERN of the secretary General of the World Meteorological Organisation (WMO) outlines its broad programmes and information system.

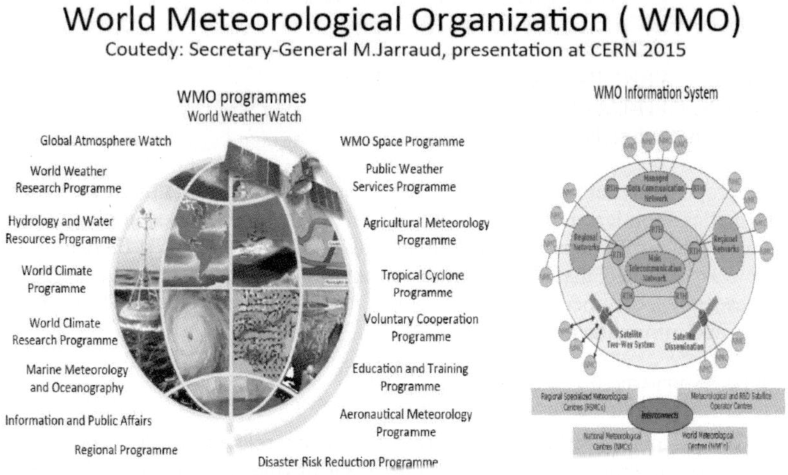

Considering the complexity of processes related to the prediction based on simulations of the world climate development, it can be understood, that for any hypothesis there is often an anti-hypothesis. We can illustrate this by the question of the influence of clouds on the changing climate.

THE CLAW[2] AND THE ANTI-CLAW HYPOTHESIS

From a review of the major developments published by Greg Ayers and Jill Cainey[3] we learn, that understanding the role of clouds in the warming and the cooling of the planet and how that role of clouds would alter in a warming world, is one of the biggest uncertainties climate change researchers are facing.

Important in this regard is the influence of cloud condensation nuclei, aerosols, the tiny atmospheric particles necessary for the nucleation of every single cloud droplet, on cloud properties. The anthropogenic contribution to cloud condensation nuclei is known to be large in some regions through knowledge of pollutant emissions; however, the natural processes that regulate cloud condensation nuclei over large parts of the globe are less well understood.

The CLAW hypothesis provides a mechanism by which plankton may modify climate through the atmospheric sulfur cycle via the provision of sulfate cloud condensation nuclei. The CLAW hypothesis was published over 20 years ago and has stimulated a great deal of research.[3] A feedback loop is proposed that operates between ocean ecosystems and the Earth's climate. The hypothesis specifically proposes that particular phytoplankton that produce dimethyl sulfide are responsive to variations in climate forcing, and that these responses lead to a **negative** feedback loop that acts to stabilise the temperature of the Earth's atmosphere. The anti-CLAW hypothesis (discussed by Lovelock, 2006[3]) proposed that instead of providing negative feedback in the climate system, the components of the CLAW hypothesis may act to create a **positive** feedback loop. Under future global warming, increasing temperature may stratify the world ocean, decreasing the supply of nutrients from the deep ocean to its productive euphotic zone. The consequence of this will be further climate warming which may lead to even less DMS production (and further climate warming).

All these hypotheses rely on simulations and climate modeling. Ecological modelling aims to understand these basic ecosystem functions using mathematical and conceptual modelling, systems analysis, thermodynamics, computer simulations, ecological theory.

The Limits to models in ecology seem to be multiple—improvement will require actions at the capacity-building stage, and the removal of technological bottlenecks, and limitations in the data available to build, drive, and validate the models.[4]

MONITORING USING MEASUREMENTS BASED ON ADEQUATE SENSORS

CERN has no specific competence in ecological science but CERN has great competences in sensor technologies needed to get data to verify ecological modelling. Data mining, data analytics and big data needs valid data input based on measurements. CERN launches a new initiative with focus on industrial and societal value of detection and imaging technology. It is indeed difficult to think of a technology developed for particle physics which is not in connection with detection and imaging. This is schematically shown in the pictures below.

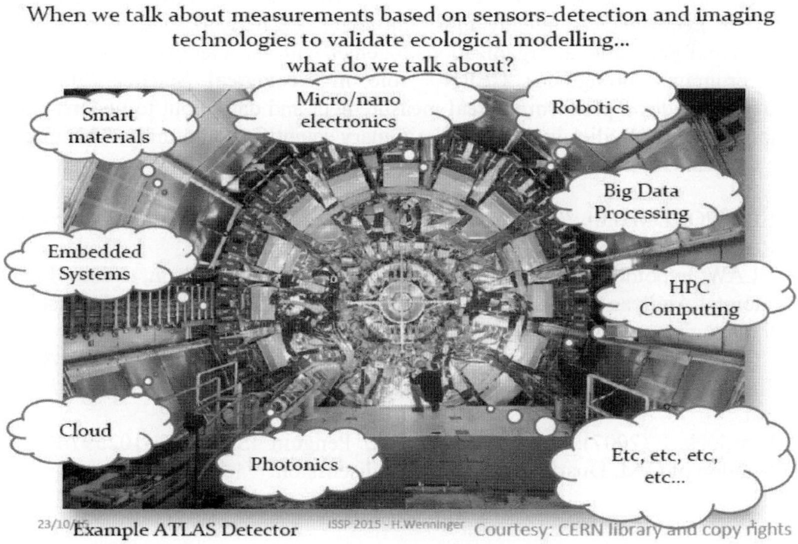

Courtesy: Marcus Nordberg/ CERN

CLOUD FORMATION BY COSMIC RAYS—CERN EXPERIMENT

CERN also contributes directly to the ecological science and climate research with an interdisciplinary experiment, which studies cloud formation initiated by cosmic radiation. The goal is to know more about earth's climate sensitivity and the effect of clouds for the warming of the earth, which is still highly uncertain.

The experimental result adds significant measurement in understanding the climate. As of now, actual measurements and results neither rule out a role for cosmic radiation in cloud formation, nor are the results offering a quick fix for global warming. The CLOUD experiment at CERN is continuing at present.

CERN is housing the CLOUD experiment, using PS beams to simulate cosmic rays: http://cloud.web.cern.ch/cloud.

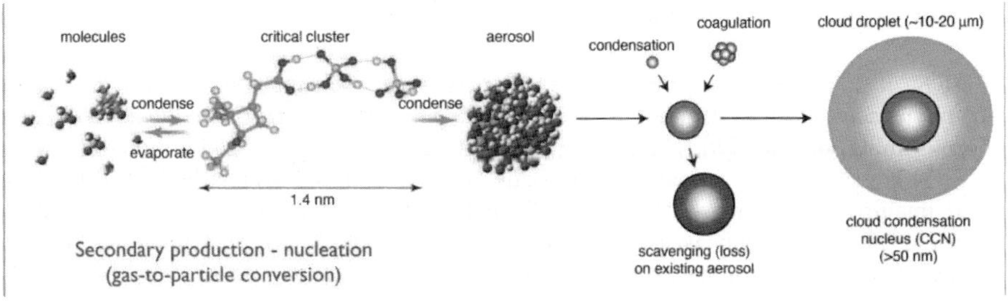

We have discussed that CERN's contribution to "**Fast Computing and Informatics Frontiers**" is already most important. Its network with partner research institutions and international organizations together with its scientific users and students training programs strengthens CERN's role in **Ecological Science** "Its CLOUD experiment provides a platform for real measurement and data input to understand cloud formation processes studies by an interdisciplinary scientific community. This might also provide the basis for a permanent Monitoring Panel (PMP).

1. EIROFORUM. http://eiroforum.org/about/index.html. Forum of eight International Research Organizations.
2. CLAW hypothesis: originally proposed by Robert Charlson, James Lovelock, Meinrat Andreae and Stephen Warren, and takes its acronym from the first letter of their surnames. https://en.wikipedia.org/wiki/CLAW_hypothesis.
3. "The CLAW hypothesis: a review of the major developments." Greg P. Ayers and Jill M. Cainey http://www.publish.csiro.au/?act=view_file&file_id= EN07080. pdf.

   Lovelock, J. (2007). *The Revenge of Gaia.* Penguin. ISBN 0-14-102597-2.
4. Citation of C.M. Duarte et al. in a panel discussion.

CONCLUSIONS

CLOSING LECTURE

ANTONINO ZICHICHI
CERN, Geneva, Switzerland, INFN and University of Bologna, Italy
Centro Enrico Fermi, Rome, Italy

Dear Colleagues and Friends,

Let me close this 48th Session of our Seminars recalling the basic points of our activities, whose target is the New Manhattan Project.

Section 1: The role of our PMPs.

Section 2: The 72 Planetary Emergencies.

Section 3: The Logic of the New Manhattan Project where the ELN collider plays the most important role. A brief summary of the physics motivations is given in Figures 3–12.

Section 4: The status of the EEE Project is illustrated in Figure 13. The purpose of the project is to involve the young generation in one of the frontiers problems of our physics: the origin of cosmic rays.

SECTION 1: THE ROLE OF OUR PMPS

The PMPs are on the following Problems:

- Biotechnology
- Brain and Behaviour
- Climatology
- Defence against Cosmic Objects
- Desertification
- Energy
- Floods and Extreme Weather Events
- Information Security
- Limits of Development
- Missile Proliferation
- Mother & Child
- Pollution
- Motivations for Terrorism
- Mitigation of Terrorist Acts
- Transmissible Spongiform Encephalopathy
- Water

The scope of each PMP is:

- Monitoring the scientific results of research conducted in the relevant fields.

- Organising scientific workshops to reconcile conflicting views on scientific issues and channel research efforts.
- Helping disseminate relevant scientific data and information.
- Organising yearly Group Meetings and reporting on group activities to the WFS Annual General Meeting in Erice.
- Writing recommendations for use by governments and international agencies.

At present each PMP is working on a component of the "New Manhattan Project".

Once the New Manhattan Project receives the financial support from all Governments in the world, each PMP will have the responsibility for the detailed elaboration and the implementation of its PMP-Project.

SECTION 2: THE 72 PLANETARY EMERGENCIES

Overview of the 72 Planetary Emergencies Grouped into 15 Classes

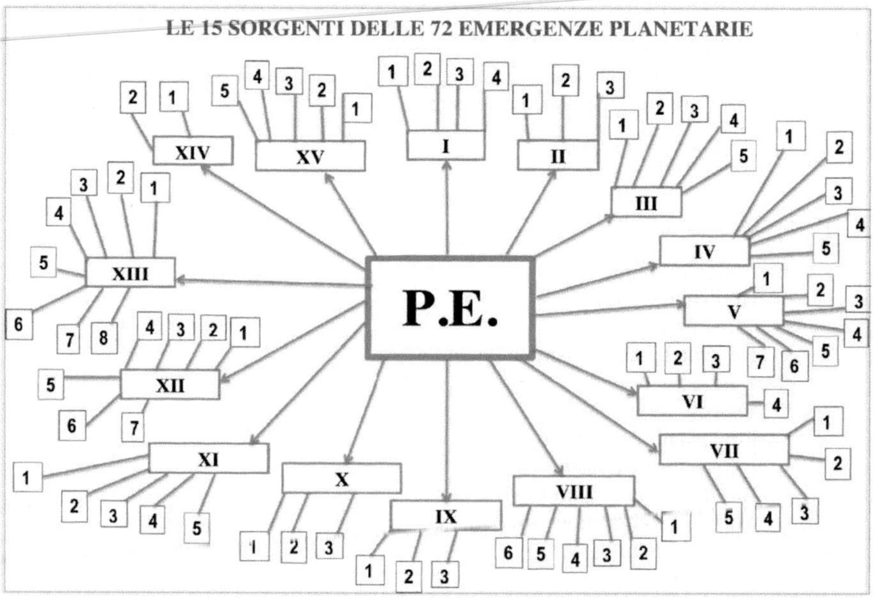

Fig. 1: 72 Planetary Emergencies Grouped into 15 Classes.

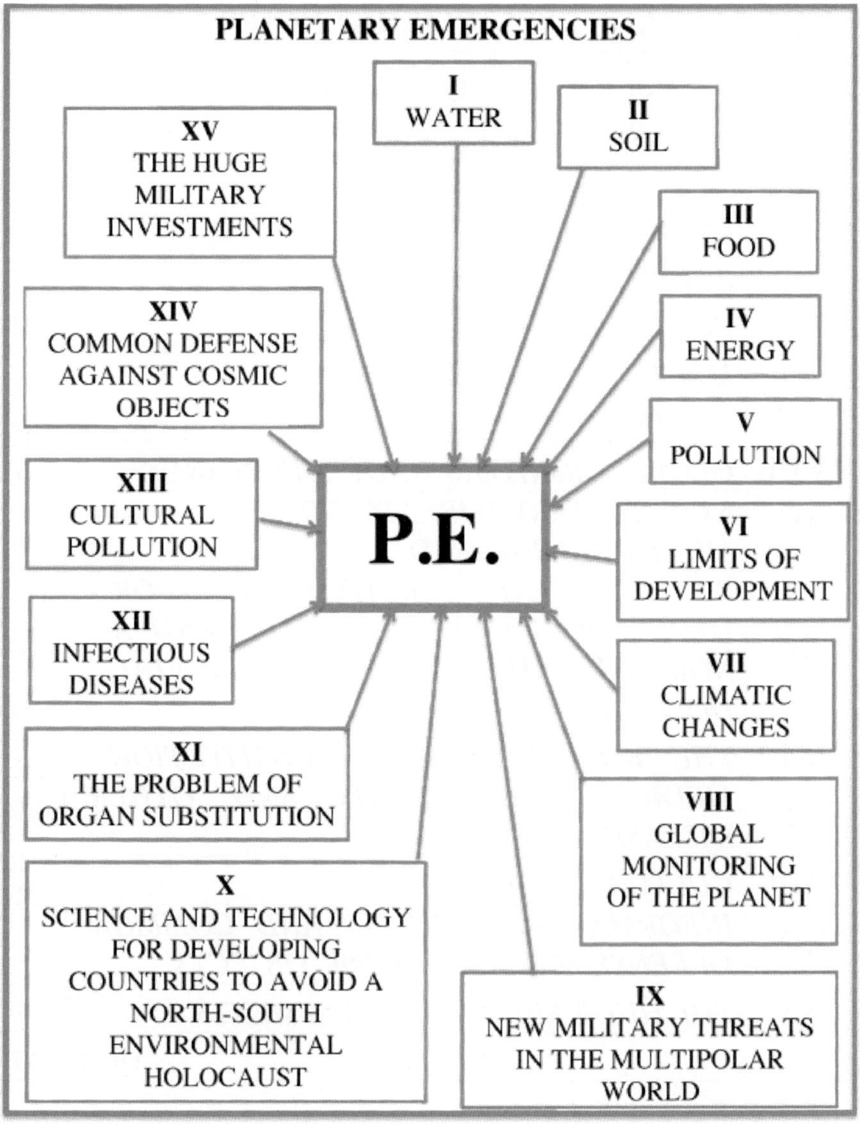

Fig. 2: Planetary Emergencies.

THE 72 PLANETARY EMERGENCIES IDENTIFIED BY THE ERICE SCIENTISTS AND CLASSIFIED INTO 15 BASIC SOURCES		NUMBER OF SOURCES
I	*WATER*	4
II	*SOIL*	3
III	*FOOD*	5
IV	*ENERGY*	5
V	*POLLUTION*	7
VI	*LIMITS OF DEVELOPMENT*	4
VII	*CLIMATIC CHANGES*	5
VIII	*GLOBAL MONITORING OF THE PLANET*	6
IX	*NEW MILITARY THREATS IN THE MULTIPOLAR WORLD*	3
X	*SCIENCE AND TECHNOLOGY FOR DEVELOPING COUNTRIES TO AVOID A NORTH-SOUTH ENVIRONMENTAL HOLOCAUST*	3
XI	*THE PROBLEM OF ORGAN SUBSTITUTION*	5
XII	*MEDICINE, INFECTIOUS AND OTHER DISEASES*	7
XIII	*CULTURAL POLLUTION*	8
XIV	*INFORMATION SECURITY AND COMMON DEFENSE AGAINST COSMIC OBJECTS*	2
XV	*THE HUGE MILITARY INVESTMENTS*	5
	Total	**72**

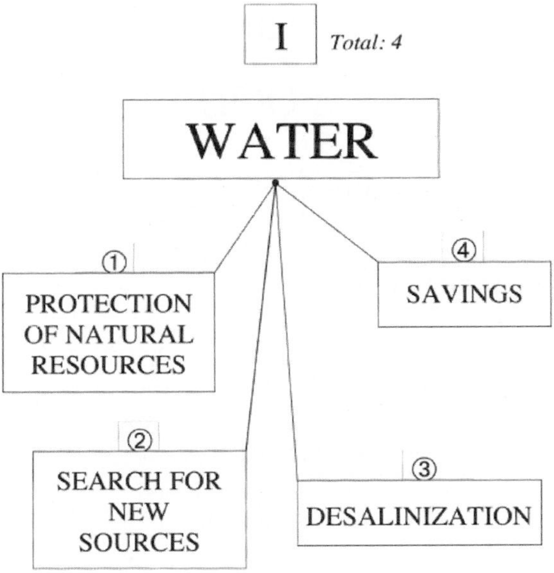
I
Total: 4
WATER
①
PROTECTION OF NATURAL RESOURCES
②
SEARCH FOR NEW SOURCES
③
DESALINIZATION
④
SAVINGS

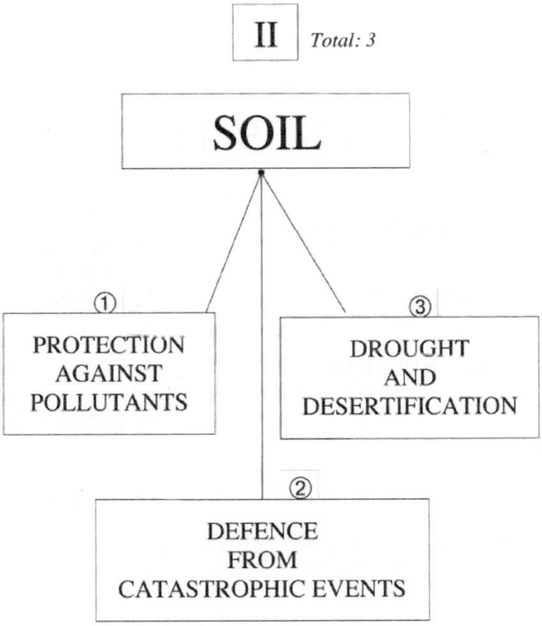
II
Total: 3
SOIL
①
PROTECTION AGAINST POLLUTANTS
②
DEFENCE FROM CATASTROPHIC EVENTS
③
DROUGHT AND DESERTIFICATION

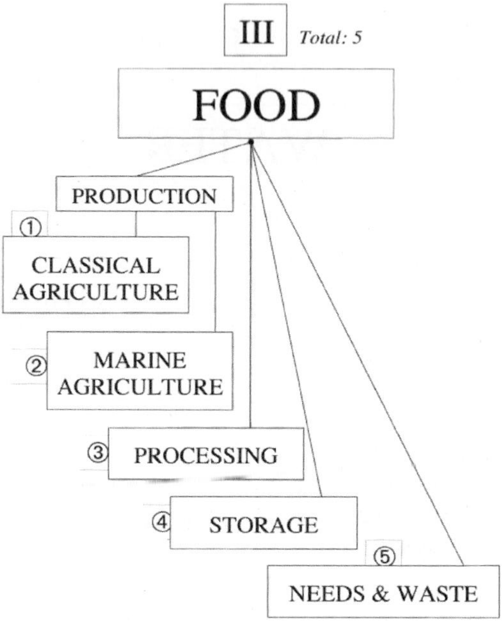
III
Total: 5
FOOD
PRODUCTION
① CLASSICAL AGRICULTURE
② MARINE AGRICULTURE
③ PROCESSING
④ STORAGE
⑤ NEEDS & WASTE

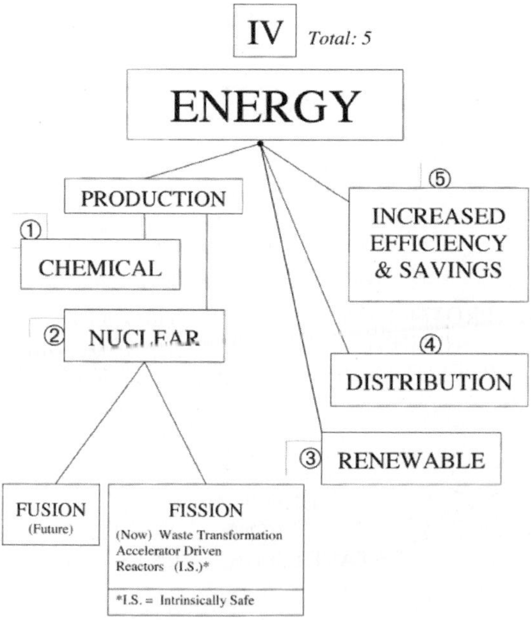
IV
Total: 5
ENERGY
PRODUCTION
① CHEMICAL
② NUCLEAR
FUSION
(Future)
FISSION
(Now) Waste Transformation
Accelerator Driven
Reactors (I.S.)*
*I.S. = Intrinsically Safe
③ RENEWABLE
④ DISTRIBUTION
⑤ INCREASED EFFICIENCY & SAVINGS

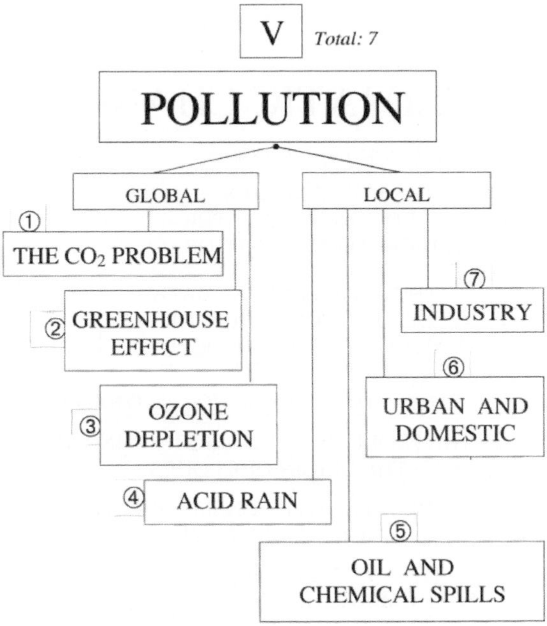
V
Total: 7
POLLUTION
GLOBAL
LOCAL
① THE CO2 PROBLEM
② GREENHOUSE EFFECT
③ OZONE DEPLETION
④ ACID RAIN
⑤ OIL AND CHEMICAL SPILLS
⑥ URBAN AND DOMESTIC
⑦ INDUSTRY

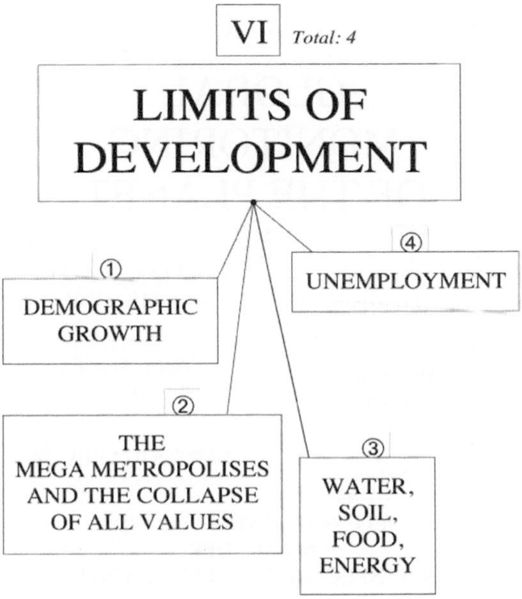
VI
Total: 4
LIMITS OF DEVELOPMENT
① DEMOGRAPHIC GROWTH
② THE MEGA METROPOLISES AND THE COLLAPSE OF ALL VALUES
③ WATER, SOIL, FOOD, ENERGY
④ UNEMPLOYMENT

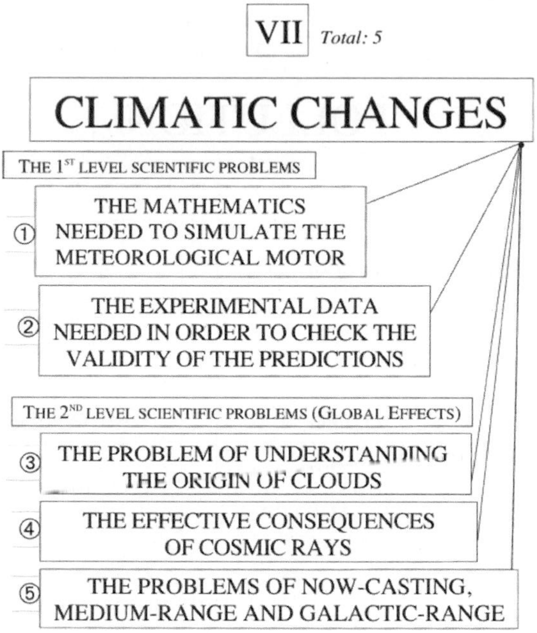
VII Total: 5
CLIMATIC CHANGES
THE 1ST LEVEL SCIENTIFIC PROBLEMS
① THE MATHEMATICS NEEDED TO SIMULATE THE METEOROLOGICAL MOTOR
② THE EXPERIMENTAL DATA NEEDED IN ORDER TO CHECK THE VALIDITY OF THE PREDICTIONS
THE 2ND LEVEL SCIENTIFIC PROBLEMS (GLOBAL EFFECTS)
③ THE PROBLEM OF UNDERSTANDING THE ORIGIN OF CLOUDS
④ THE EFFECTIVE CONSEQUENCES OF COSMIC RAYS
⑤ THE PROBLEMS OF NOW-CASTING, MEDIUM-RANGE AND GALACTIC-RANGE

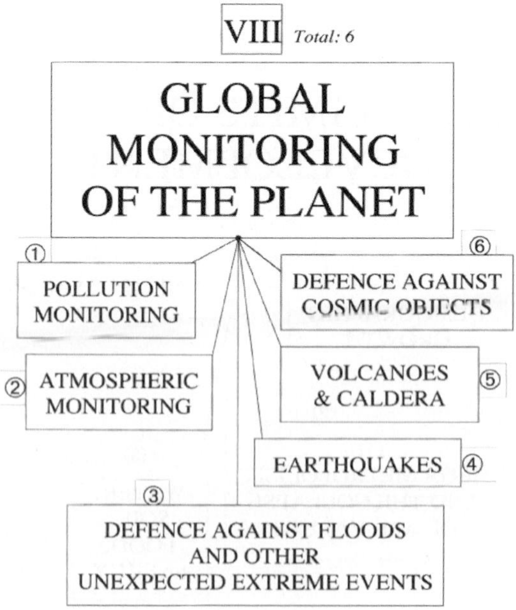
VIII Total: 6
GLOBAL MONITORING OF THE PLANET
① POLLUTION MONITORING
② ATMOSPHERIC MONITORING
③ DEFENCE AGAINST FLOODS AND OTHER UNEXPECTED EXTREME EVENTS
④ EARTHQUAKES
⑤ VOLCANOES & CALDERA
⑥ DEFENCE AGAINST COSMIC OBJECTS

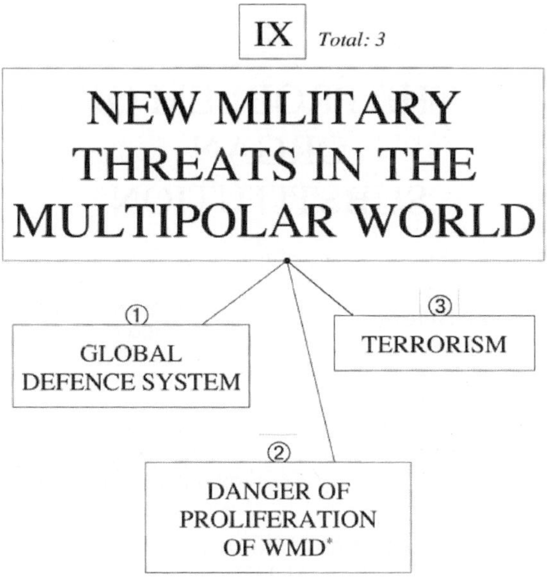

*WMD = Weapons of Mass Destruction

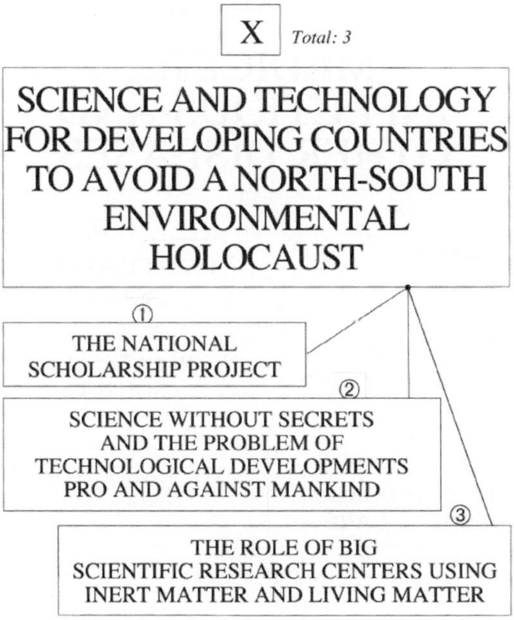

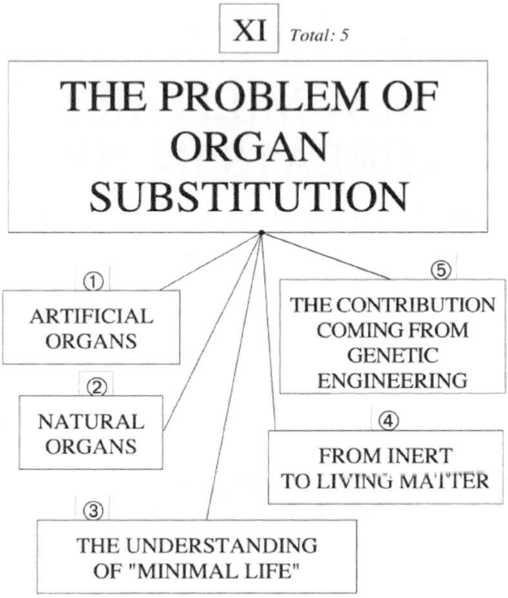
XI
Total: 5
THE PROBLEM OF ORGAN SUBSTITUTION
① ARTIFICIAL ORGANS
② NATURAL ORGANS
③ THE UNDERSTANDING OF "MINIMAL LIFE"
④ FROM INERT TO LIVING MATTER
⑤ THE CONTRIBUTION COMING FROM GENETIC ENGINEERING

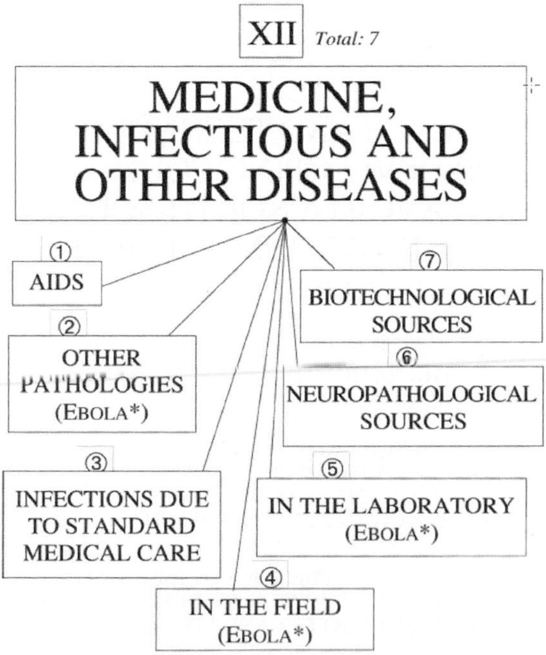
XII
Total: 7
MEDICINE, INFECTIOUS AND OTHER DISEASES
① AIDS
② OTHER PATHOLOGIES (EBOLA*)
③ INFECTIONS DUE TO STANDARD MEDICAL CARE
④ IN THE FIELD (EBOLA*)
⑤ IN THE LABORATORY (EBOLA*)
⑥ NEUROPATHOLOGICAL SOURCES
⑦ BIOTECHNOLOGICAL SOURCES

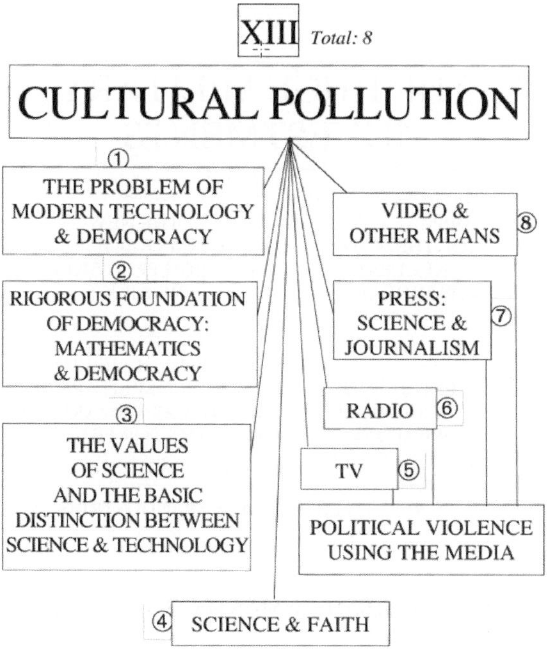
XIII
Total: 8
CULTURAL POLLUTION
1 THE PROBLEM OF MODERN TECHNOLOGY & DEMOCRACY
2 RIGOROUS FOUNDATION OF DEMOCRACY: MATHEMATICS & DEMOCRACY
3 THE VALUES OF SCIENCE AND THE BASIC DISTINCTION BETWEEN SCIENCE & TECHNOLOGY
4 SCIENCE & FAITH
5 TV
6 RADIO
7 PRESS: SCIENCE & JOURNALISM
8 VIDEO & OTHER MEANS
POLITICAL VIOLENCE USING THE MEDIA

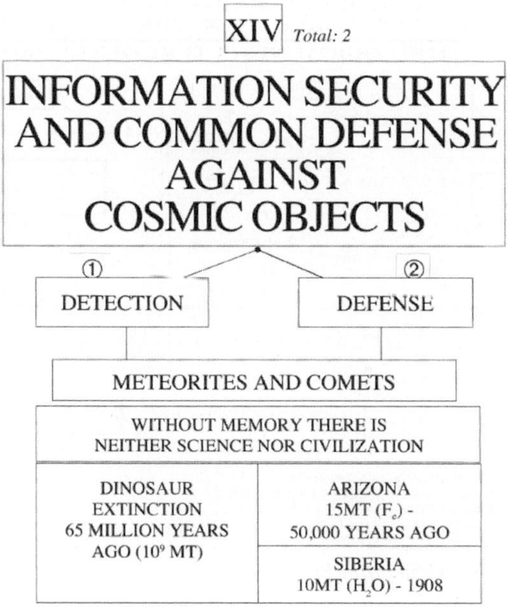
XIV
Total: 2
INFORMATION SECURITY AND COMMON DEFENSE AGAINST COSMIC OBJECTS
1 DETECTION
2 DEFENSE
METEORITES AND COMETS
WITHOUT MEMORY THERE IS NEITHER SCIENCE NOR CIVILIZATION
DINOSAUR EXTINCTION 65 MILLION YEARS AGO (10^9 MT)
ARIZONA 15MT (F_e) - 50,000 YEARS AGO
SIBERIA 10MT (H_2O) - 1908

SECTION 3: THE LOGIC OF THE NEW MANHATTAN PROJECT

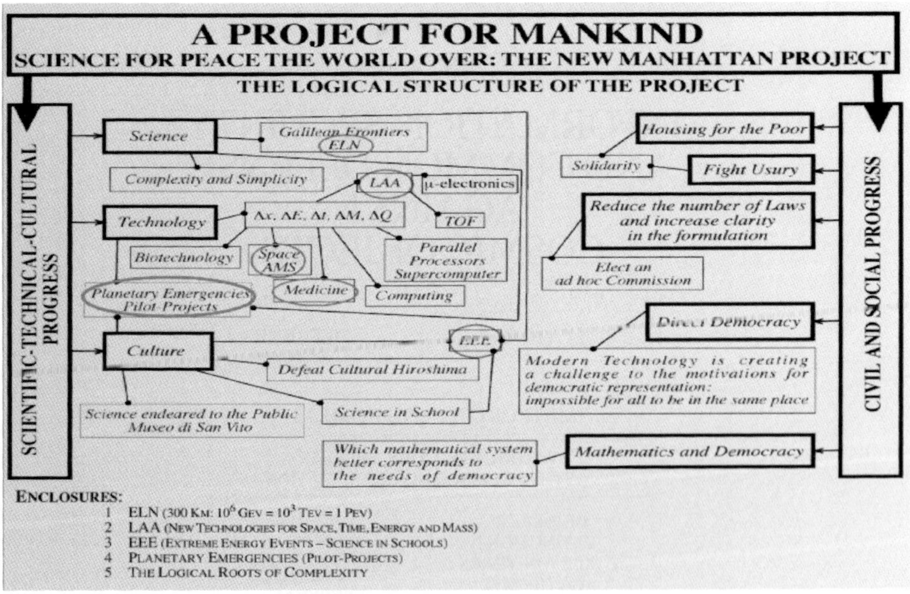

Fig. 3: The New Manhattan Project.

Following the cultural strategy mentioned in the Opening Lecture I have recently written my Testimony on Lord Patrick Blackett.

We cannot forget the roots of our achievements. Our Future is in these roots. The greatest goal for our future in Science is the ELN Project.

Let me show the picture of the 300 km rings of ELN and the Physics reasons why this collider should be implemented.

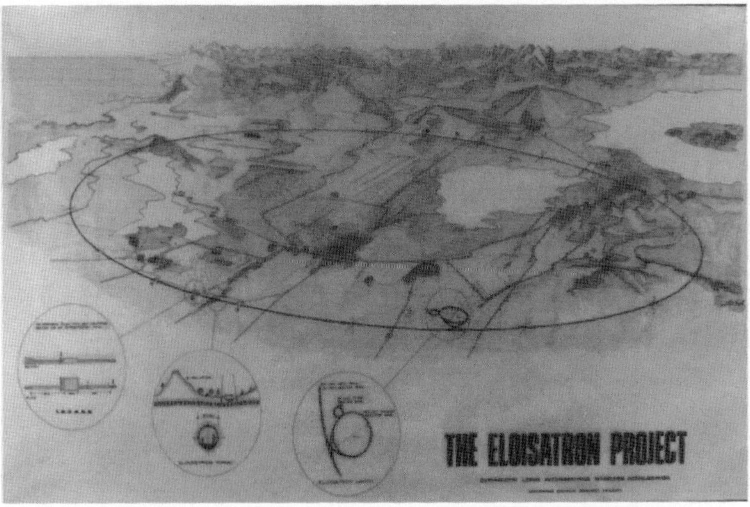

Fig. 4: A view of the ELN 300 km rings.

The reasons why ELN should be implemented.

From 1947 to the present, a large amount of knowledge has been gained. This is expressed in terms of three fundamental forces (Figure 5) and three columns (Figure 6). They are needed in order to build the Standard Model (Figure 7) and Beyond, as reported in (Figure 8).

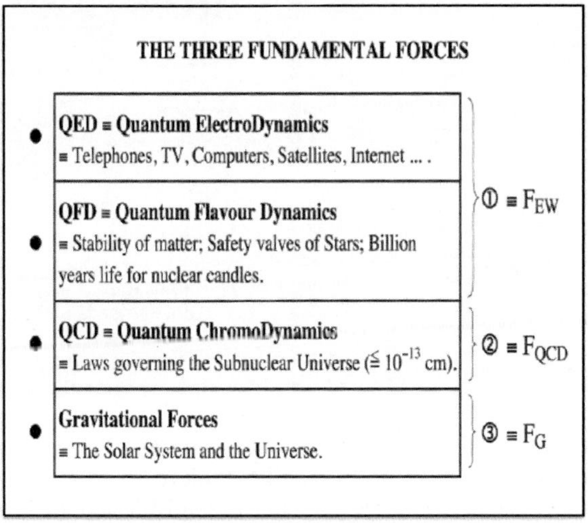

Fig. 5: The Three Fundamental Forces.

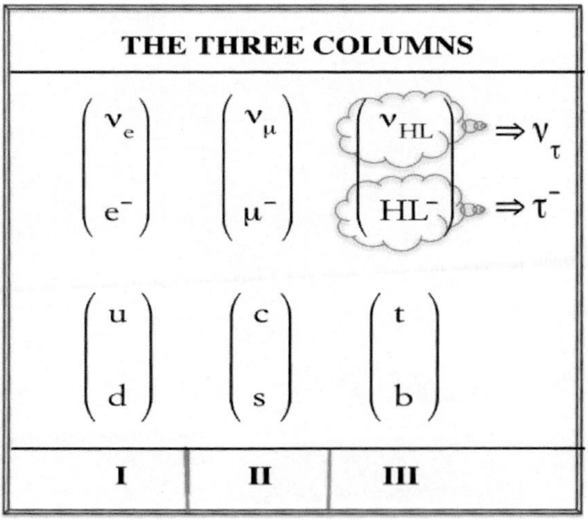

Fig. 6: The Three Columns.

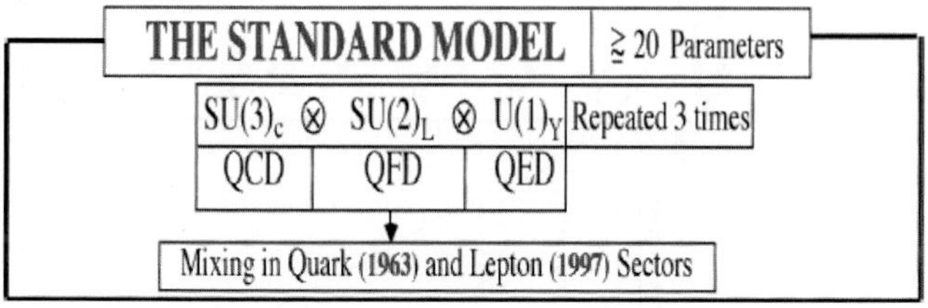

Fig. 7: The Standard Model.

We are sure that it is necessary to go beyond the Standard Model whose five basic steps are reported in Figure 8.

SM&B

THE STANDARD MODEL AND BEYOND

THE FIVE BASIC STEPS NEEDED FOR SM&B

① The renormalization group equations (RGEs) imply that the gauge couplings (α_i) and the masses (m_j) all run with k^2. It is this running which allows GUT, suggests SUSY and produces the need for a non point-like description (RQST) of Physics processes, thus opening the way to quantize gravity.
RGEs (α_i (i = 1, 2, 3); m_j (j = q, l, G, H)) : $f(k^2)$.
- GUT ($\alpha_{GUT} \simeq 1/24$) & GAP ($10^{16} - 10^{18}$) GeV.
- SUSY (to stabilize $m_F/m_P \simeq 10^{-17}$).
- RQST (to quantize Gravity).

② All forces originate in the same way: the gauge principle.
Gauge Principle (hidden and expanded dimensions).
— How a Fundamental Force is generated: SU(3); SU(2); U(1) and Gravity.

③ Imaginary masses play a central role in describing nature: SSB & Confinement.
The Physics of Imaginary Masses: SSB.
— The Imaginary Mass in SU(2)×U(1) produces masses thanks to SSB: ($m_{W^\pm}$; m_{Z^0}; m_q; m_l), including $m_\gamma = 0$.
— The Imaginary Mass in SU(5)⇒SU(3)×SU(2)×U(1) or in any higher (not containing U(1)) Symmetry Group ⇒ SU(3)× SU(2)×U(1) produces Monopoles.
— The Imaginary Mass in $SU(3)_c$ generates Confinement.

④ The mass-eigenstates are mixed when the Fermi forces come in: the matrix describing the mixing is the product of two fundamental matrices.
Flavour Mixings & CP ≠ , T ≠ (direct ≠ , not via SSB).
— Why is the mixing there? No need for it but it is there.

⑤ The Abelian force QED has lost its role of being the guide for all fundamental forces. The non-Abelian gauge forces dominate and have features which are not present in QED.
Anomalies & Instantons.
— Basic Features of all Non-Abelian Forces.

Figure 8

Fig 8: The Standard Model and Beyond.

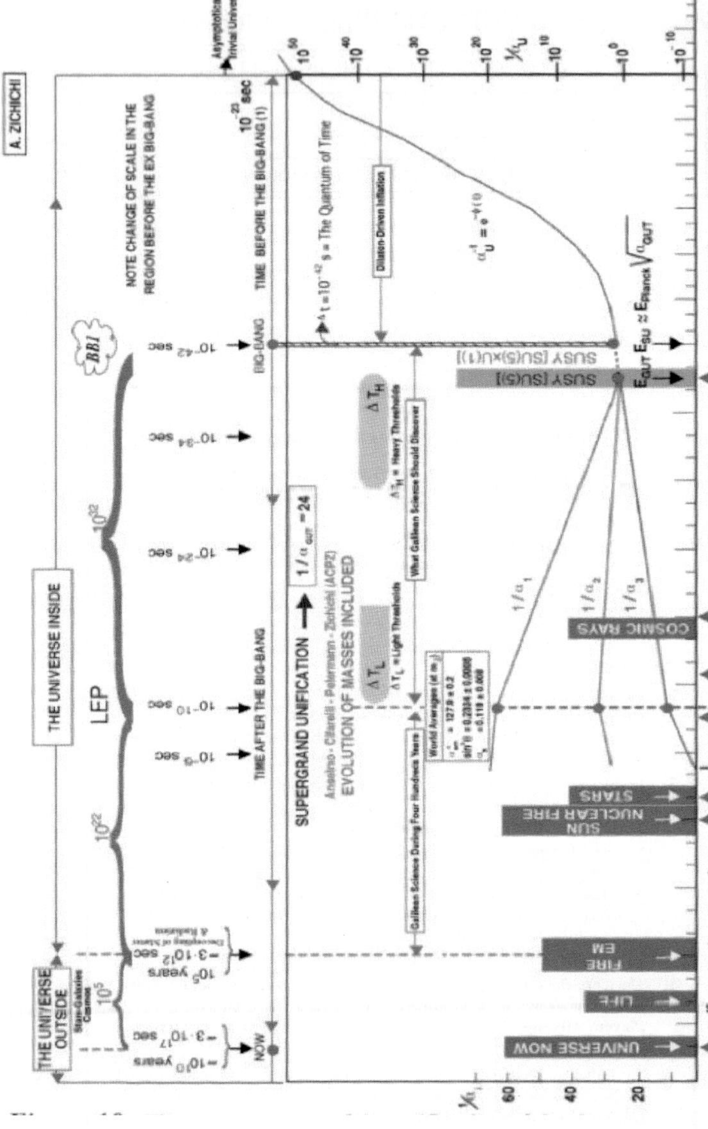

Fig. 10: The present status of the unification of the three gauge couplings α_1 α_2 α_3 is reported.

THE UNIFICATION OF ALL FUNDAMENTAL FORCES

The lines in Figure 10 result from calculations executed with a supercomputer using the following system of equations:

$$\mu \frac{d\alpha_i}{d\mu} = \frac{b_i}{2\pi}\alpha_i^2 + \sum_j \frac{b_{ij}}{8\pi^2}\alpha_i\alpha_j$$

This is a system of coupled non-linear differential equations where the existence of the Superworld is included. This system describes how the gauge couplings (α_1, α_2, α_3) vary with "μ", the basic parameter which depends on the energy of the elementary process, from the maximum level of Energy (Planck Scale) to the energy level of our world. **During more than ten years (from 1979 to 1991), no one had realized that the energy threshold for the existence of the Superworld was strongly dependent on the "running" of the masses.**

This is now called: **the EGM effect** (from the initials of Evolution of Gaugino Masses).

Fig. 11: The Unification of all Fundamental Forces.

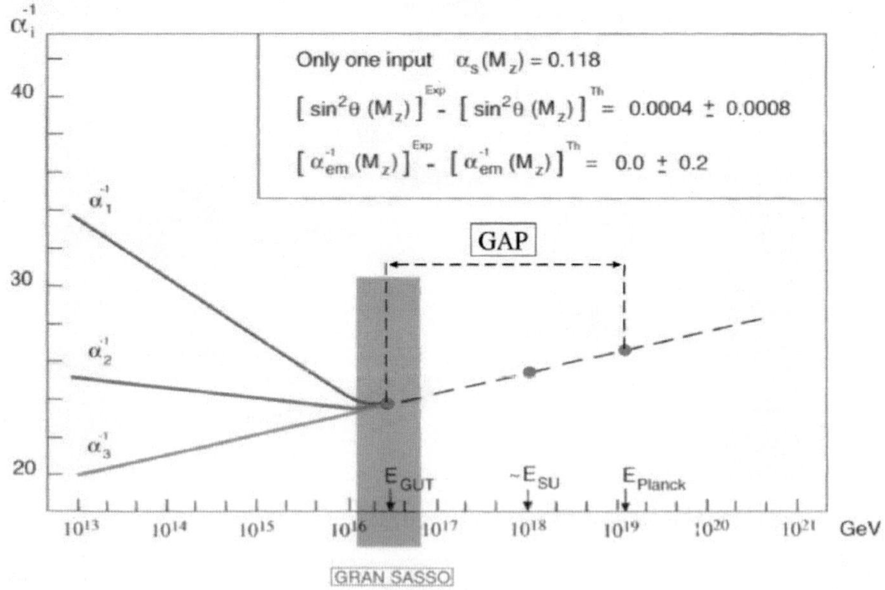

Fig. 12.

SECTION 4: THE STATUS OF THE EEE PROJECT

A large project called EEE (Extreme Energy Events) was launched for high schools in Italy to study the origin of the cosmic rays using the most advanced detectors built at CERN by the students.

The total number of detectors will be one hundred. At the present 50% are operative all over Italy, as shown in Figure 13. Each star indicates the City where one detector is installed. If more than one, the number in a star corresponds to the number of detectors installed in different Institutes of the same City. The two detectors at CERN act as basic reference for the time structure of the cosmic rays showers.

The high school students and their teachers are thus exposed to a frontier physics experiment, to which they can effectively contribute. In fact, the cosmic rays have been discovered by Victor Franz Hess more than 100 years ago. But no one has been able so far to establish where they are coming from. A problem which only the Extremely Energetic Cosmic Rays can hopefully tell us sooner or later.

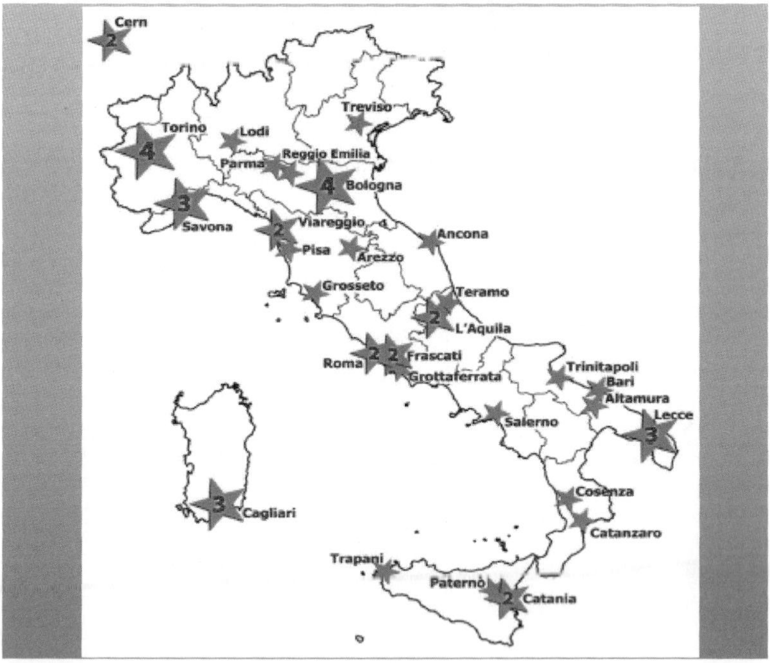

Fig. 13: Map of the EEE telescopes (2015).

PROJECT DIRECTORS AND SCIENTIFIC COMMITTEE MEETING

2014-15 PROGRESS REPORT OF THE MCD-2/7 PROJECT AND 2016 RESEARCH PROJECT

FRANCO BUONAGURO
Istituto Nazionale dei Tumori, "Fondazione G. Pascale"
Napoli, Italy

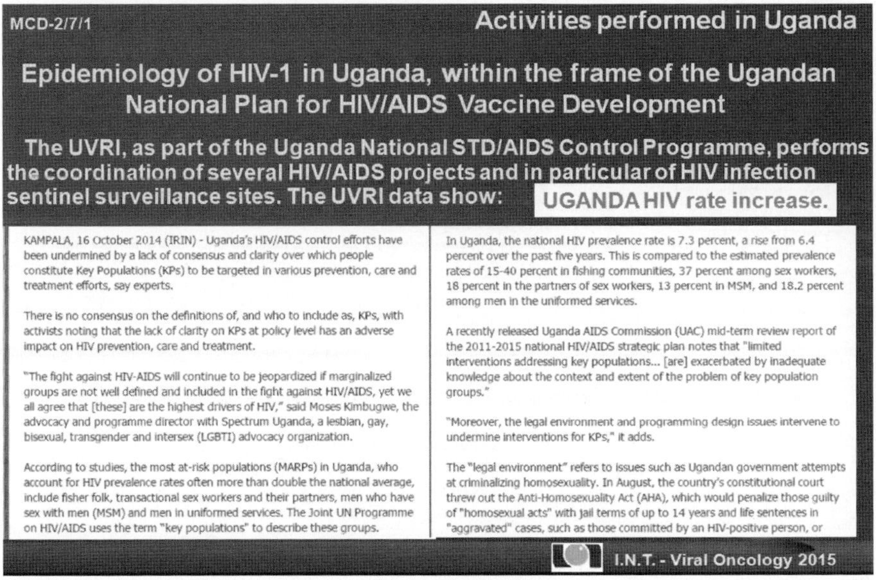

ETIOPATHOGENESIS OF AIDS ASSOCIATED NEOPLASIAS

- Kaposi's sarcoma: Sample collection and virus characterization is still ongoing, with inclusion of low KS incidence countries (i.e., Iran);
- Oropharingeal and Genital Cancer: Virus genotyping and vaccine studies are in progress;
- Conjunctival carcinoma: Genetic susceptibility and Virus analysis, in particular role of p53 mutations in cancers and detection of HHV8;
- Hepatocellular carcinoma: Biomolecular characterization.

In particular studies have focused on:

- HHV8/KSHV and conjunctival carcinoma;
- HCV-role on HCC at molecular level.

BioMed Research International

Impact Factor 1.579

About this Journal | Submit a Manuscript | Table of Contents

Journal Menu

- About this Journal
- Abstracting and Indexing
- Advance Access
- Aims and Scope
- Annual Issues
- Article Processing Charges
- Articles in Press
- Author Guidelines
- Bibliographic Information
- Citations to this Journal
- Contact Information
- Editorial Board

BioMed Research International
Article ID 801353

Review Article

Identification of Human Herpesvirus 8 Sequences in Conjunctiva Intraepithelial Neoplasia and Squamous Cell Carcinoma of Ugandan Patients

Noemy Starita,[1] Clorinda Annunziata,[1] Keith M. Waddell,[2] Luigi Buonaguro,[1] Franco M. Buonaguro,[1] and Maria Lina Tornesello[1]

[1]Molecular Biology and Viral Oncology, Istituto Nazionale Tumori "Fondazione G Pascale", IRCCS, 80131 Naples, Italy
[2]Departments of Ophthalmology and Paediatrics, Mbarara University of Science and

- Abstract
- Full-Text PDF
- Full-Text HTML
- Full-Text ePUB
- Linked References
- How to Cite this Article

Table 3: Distribution of HHV8 in HIV-positive and HIV-negative conjunctival neoplasia samples and controls.

HIV status	Conjunctiva neoplastic tissues N (%)	Conjunctiva control tissue* N† (%)	P value
HIV positive			0.347‡
HHV8-Pos	12 (25.0)	2 (13.3)	
HHV8-Neg	36 (75.0)	13 (86.7)	
HIV negative			1.000‡
HHV8-Pos	2 (8.3)	1 (4.2)	
HHV8-Neg	22 (91.7)	23 (95.8)	

*Two pingueculae, 1 pterygium, and 1 papilloma of the conjunctiva were included in the control group.
†21 control subjects with undetermined HIV serology were not included.
‡Fisher's exact test, two-tailed.

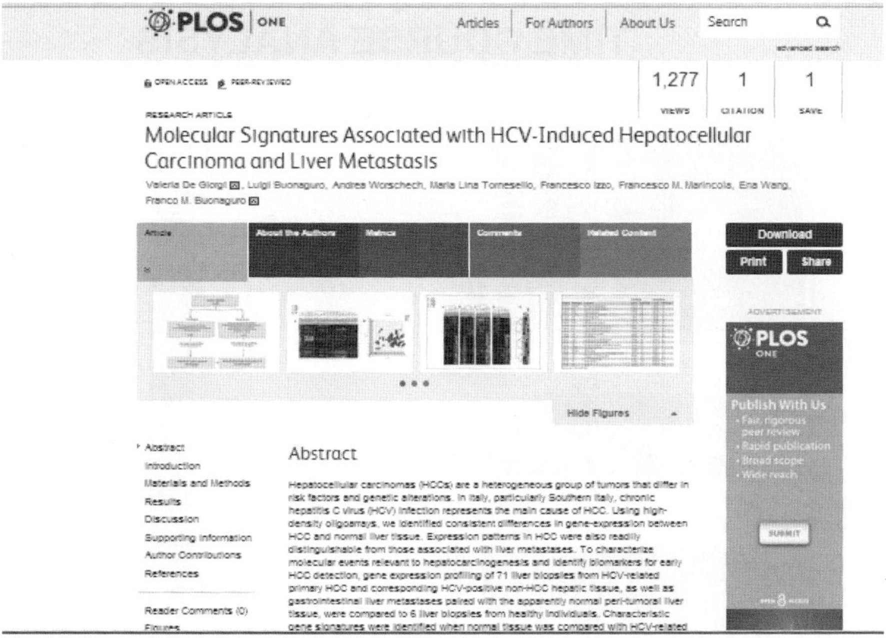

PLOS | ONE

Articles | For Authors | About Us | Search

advanced search

OPEN ACCESS | PEER-REVIEWED

1,277 VIEWS | 1 CITATION | 1 SAVE

RESEARCH ARTICLE

Molecular Signatures Associated with HCV-Induced Hepatocellular Carcinoma and Liver Metastasis

Valeria De Giorgi, Luigi Buonaguro, Andrea Worschech, Maria Lina Tornesello, Francesco Izzo, Francesco M. Marincola, Ena Wang, Franco M. Buonaguro

Article | About the Authors | Metrics | Comments | Related Content

Download | Print | Share

Hide Figures

Abstract
Introduction
Materials and Methods
Results
Discussion
Supporting Information
Author Contributions
References

Reader Comments (0)
Figures

Abstract

Hepatocellular carcinomas (HCCs) are a heterogeneous group of tumors that differ in risk factors and genetic alterations. In Italy, particularly Southern Italy, chronic hepatitis C virus (HCV) infection represents the main cause of HCC. Using high-density oligoarrays, we identified consistent differences in gene-expression between HCC and normal liver tissue. Expression patterns in HCC were also readily distinguishable from those associated with liver metastases. To characterize molecular events relevant to hepatocarcinogenesis and identify biomarkers for early HCC detection, gene expression profiling of 71 liver biopsies from HCV-related primary HCC and corresponding HCV-positive non-HCC hepatic tissue, as well as gastrointestinal liver metastases paired with the apparently normal peri-tumoral liver tissue, were compared to 6 liver biopsies from healthy individuals. Characteristic gene signatures were identified when normal tissue was compared with HCV-related

ADVERTISEMENT

PLOS ONE

Publish With Us
• Fair, rigorous peer review
• Rapid publication
• Broad scope
• Wide reach

SUBMIT

CLASS COMPARISON ANALYSIS

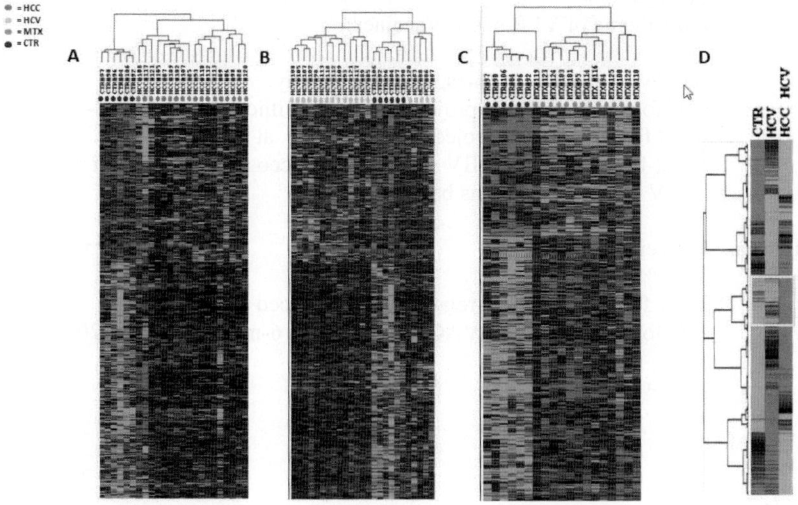

TIME COURSE ANALYSIS

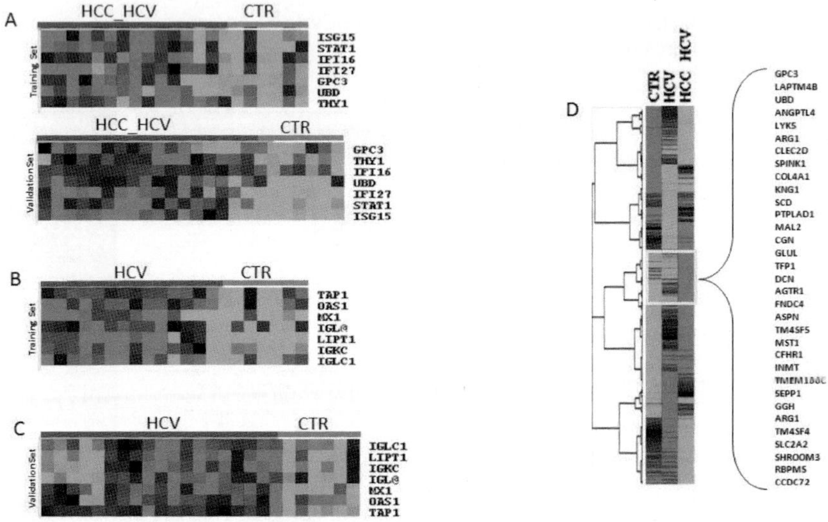

DEVELOPMENT OF ONCOGENIC VIRUS VACCINES

VLP based Vaccines

Within the theme of VLP-based Vaccines:

- Baculovirus-based vaccines, financed also by:
  - EU for the FP7 NGIN project with a 12-million € grant [2008-2013];
  - NIH for a combined project with Dr. Hioe at NYU;
- A plant-based model of HIV vaccine in tobacco along with plant geneticist;
- An HIV vaccine patent has been obtained.

Peptide based Vaccines

- Mainly focused on HCV-related HCC, financed also by:
  - EU for the FP7 HEPAVAC project with a 6-million € grant [2014-2019].

SPECIAL FOCUS I Virus-Like Particles — Foreword

www.expert-reviews.com/toc/erv/12/2

For reprint orders, please contact reprints@expert-reviews.com

The application of virus-like particles to human diseases

Expert Rev. Vaccines 12(2), 99 (2013)

Franco Buonaguro

Author for correspondence: Unit of Molecular Biology and Viral Oncology, Department of Experimental Oncology, Istituto Nazionale Tumori "Fond Pascale", Via Mariano Semmola 142, 80131 Napoli, Italy

Tel.: +39 081 5903 830

Fax: +39 081 5451 276

fm.buonaguro@istitutotumori.na.it

Luigi Buonaguro

Unit of Molecular Biology and Viral Oncology, Department of Experimental Oncology, Istituto Nazionale Tumori "Fond Pascale", Via Mariano Semmola 142, 80131 Napoli, Italy

"...virus-like particles offer several valuable features and represent a very appealing model ... currently considered to be the best alternative vaccine approach for specific viral infections, which may require a large and timely manufacturing readiness."

Most of the successful vaccines against infectious diseases are based on live attenuated or inactivated pathogens that, however, have shown some adverse effects, raising concerns about their safety. In order to address such concerns, several alternative vaccine strategies have been and are constantly developed aiming at increasing the safety without significant loss of immunogenicity.

Among such a broad array of strategies, virus-like particles (VLPs) offer several valuable features and represent a very appealing model. After the first demonstration of Engerix® (HBV) and Cervarix® (HPV) and Merck and Co., Inc.'s (NJ, USA) Recombivax HB® (HBV) and Gardasil® (HPV). This has dramatically boosted the overall VLP-based vaccine field, which is currently considered to be the best alternative vaccine approach for specific viral infections, which may require a large and timely manufacturing readiness (i.e., influenza virus).

In this framework, the present issue contains a series of timely in-depth reviews written by leaders in the field covering a range of current topics in VLP-based vac-

Viral Oncology - Virus e Tumori

sabato 27 aprile 2013

Un Progetto Europeo per lo sviluppo di un vaccino terapeutico per l'epatocarcinoma, coordinato dalla Biologia Molecolare ed Oncogenesi Virale del Pascale

Un finanziamento di 5'800'000 € per lo sviluppo di un vaccino terapeutico per l'epatocarcinoma è stato assegnato dalla Comunità Europea alla Virologia del Pascale.

Un risultato davvero eccezionale per il proponente e coordinatore, il Dr Luigi Buonaguro, medico ricercatore dell'Istituto Nazionale Tumori "Fondazione Pascale", Professore Associato Aggiunto dell'Univerità del Maryland, docente del PhD program in Molecular Medicine del San Raffaele-MI, specializzato in Microbiologia e Virologia, che negli ultimi anni si è focalizzato su studi immunologici vaccinali. Il Dr Buonaguro è uno degli esperti F1000 per i vaccini anti HIV, section editor in Immunologia del J. of Translational Medicine, Membro del Board di Clinical and Vaccine Immunology -Soc. Americana di Microbiologia, Managing Editor di Expert Review of Vaccines. Rilevante è stato l'aver studiato e sviluppato modelli vaccinali anti-HIV, che essendo un retrovirus si integra nel genoma umano e, nelle cellule infettate

PEPFAR Report at IHV - Baltimora, MD

Visualizza il mio profilo completo

Editor in Chief/Guest Editor

- Cancer Diagnostic & Predictive Biomarkers
- VLPs for human diseases
- Plant-derived Vaccines (eBook)
- Pharma Plant in Expert Review of Vaccines
- Infectious Agents and Cancer
- Frontiers in Oncology: diagnostic approaches and therapeutic strategies

HepaVac Consortium: Who We Are
6 Countries;
9 Partners:
7 Academia
1. INT-NA – L. Buonaguro (Coordinator)
2. Univ. Insubria – R. Accolla
3. Univ. Tubingen – H-G. Rammensee
4. Univ. Navarra – B. Sangro
5. INSERM – M. Ayyoub
6. Univ. Birmingham – D. Adams
7. Univ. Antwerp – F. Sven
2 Industrial Partners
1. Immatics – H. Singh/S. Kutscher (Vice – Coordinator)
2. Curevac – Heidenreich
UZA

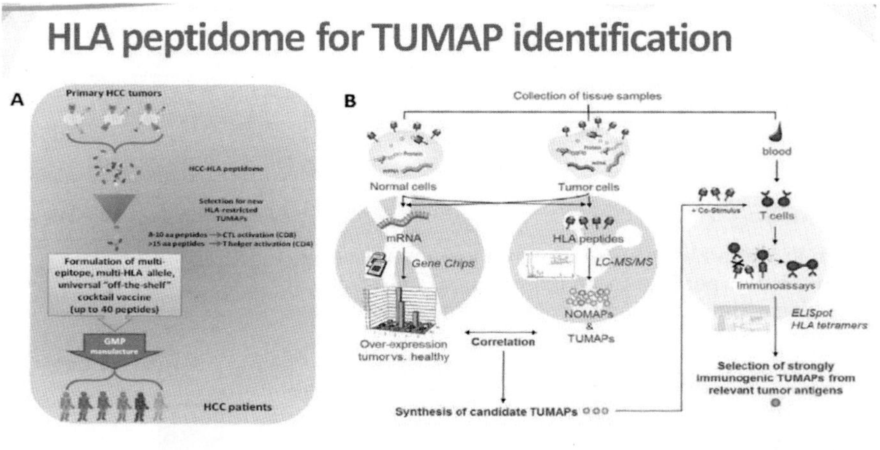
HLA peptidome for TUMAP identification
A
Primary HCC tumors
HCC-HLA peptidome
Selection for new HLA-restricted TUMAPs
8-10 aa peptides → CTL activation (CD8)
>15 aa peptides → T helper activation (CD4)
Formulation of multi-epitope, multi-HLA allele, universal "off-the-shelf" cocktail vaccine (up to 40 peptides)
GMP manufacture
HCC patients
B
Collection of tissue samples
Normal cells
Tumor cells
blood
mRNA
Gene Chips
HLA peptides
LC-MS/MS
NOMAPs & TUMAPs
Over-expression tumor vs. healthy
Correlation
+ Co-Stimulus
T cells
Immunoassays
ELISpot
HLA tetramers
Selection of strongly immunogenic TUMAPs from relevant tumor antigens
Synthesis of candidate TUMAPs

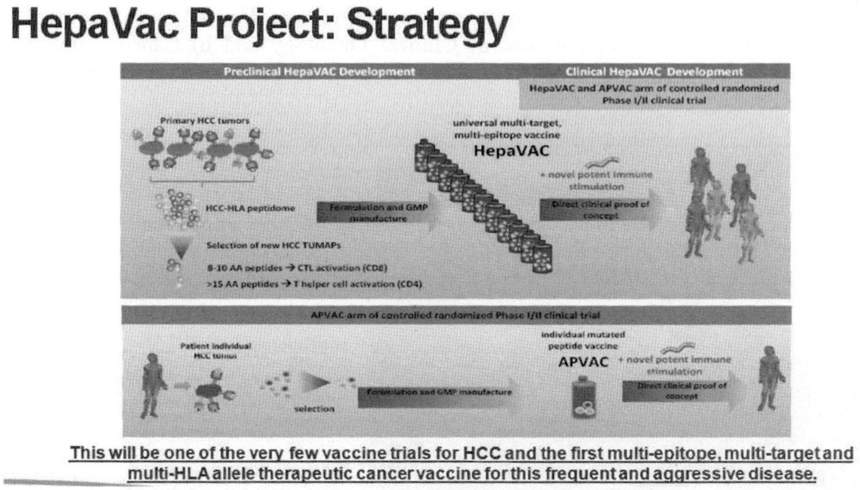

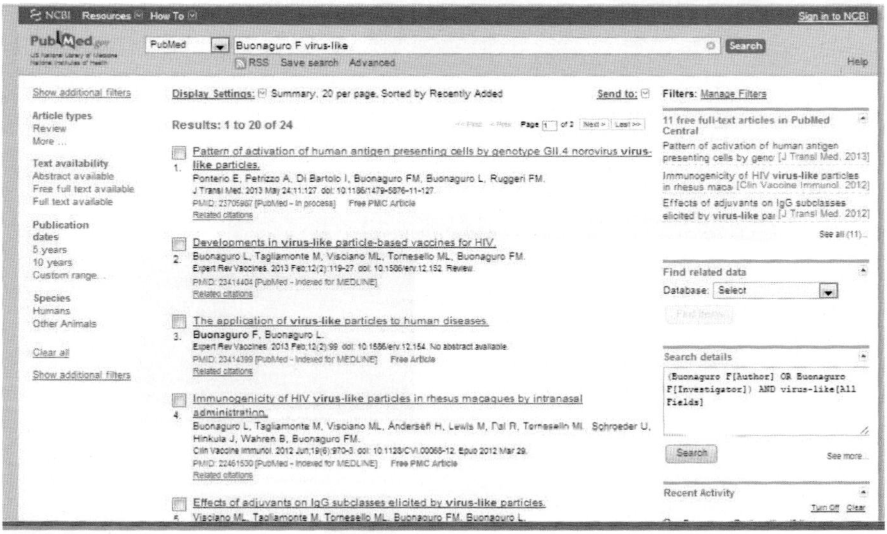

SCIENCE-RELATED SOCIAL ACTIVITIES

Within the theme of social activities:

- Direct support to the GuluNap project the newly established School of Medicine in Northern Uganda with teaching Missions and exchange programs;

- Editorial activity on the *Infectious Agents and Cancer Journal*, with the opening two sections on: a) Clinical Oncology and b) Cancer Centers of LMICs.

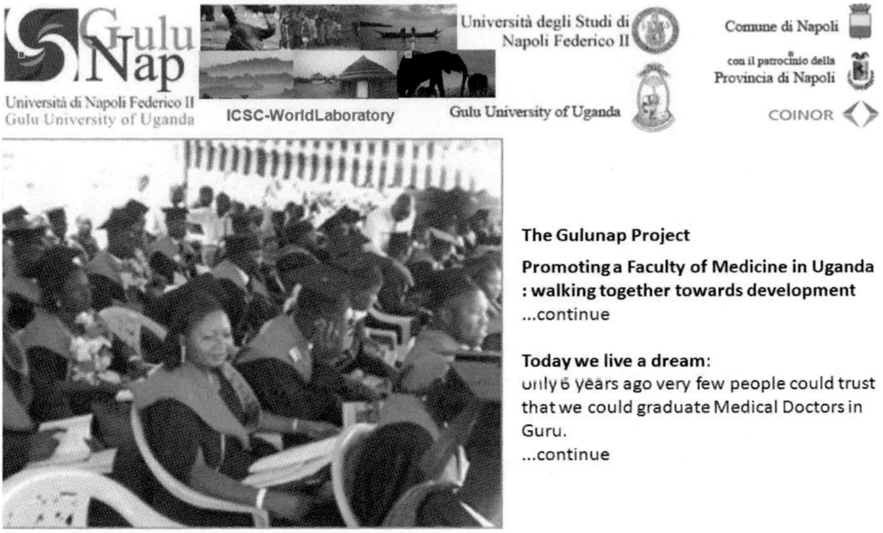

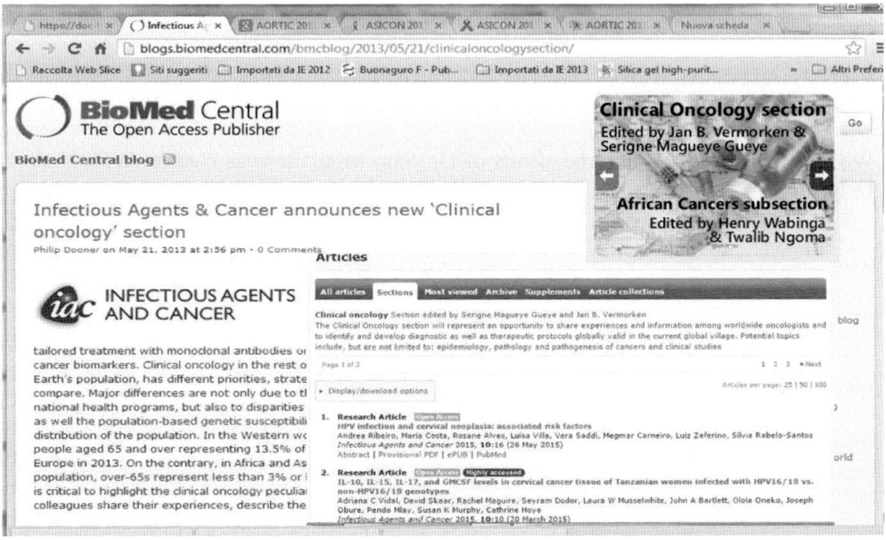

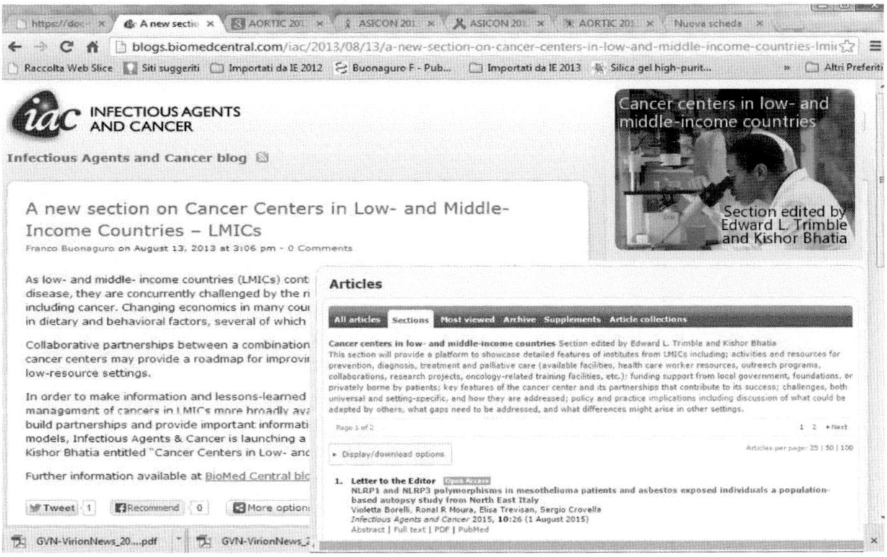

We would like to take this opportunity to express our gratitude to Professor Zichichi and Dr. Manoli for their continuous and relevant support to the EAARC since 1986, which has been very critical for planning and implementation of our current activities.

Finally I would like to confirm my interest in the WFS Center, I proposed last year, whose formal documents will be sent by next week.

WFS – NATIONAL SCHOLARSHIP PROGRAM
UGANDA

We are strongly supporting the efforts of Dr. Sam Mbulaiteye as Uganda WFS National Delegate.

WORLD LABORATORY PROJECT FOR THE R&D TRAINING IN ADVANCED TECHNOLOGY FOR YOUNG SCIENTISTS FROM DEVELOPING COUNTRIES

CRISPIN WILLIAMS
Advanced Technology for Young Scientists from Developing Countries, CERN
Geneva, Switzerland

CORE PROJECTS

- Time-of-Flight detector for the ALICE experiment
- Muon detectors for EEE project
- General R&D gaseous detectors
- Fast timing

PROCEDURE

- Offer World lab fellowships for 1 year (some fellows extended for a maximum of 2 years) to gifted physicists from less developed countries.
- Work in our detector development group (cutting edge of detector development) [e.g. timing electronics].
- Take their new-found knowledge back to home institutes—transfer of technology.

DURING THE LAST YEAR, WE SUPPORTED 3 FELLOWS

Three more will join in the next months.

- Orsolya Visnyei—Hungary
- Woosung Park—South Korea
- Osnan Maragoto Rodriguez—Cuba

BRIEF DESCRIPTION OF SOME OF THE WORK WE HAVE DONE OVER THE LAST YEAR

Some highlights of the life of World Lab fellows.

Brief recap from previous years where I have reported on the development of the Multigap Resistive Plate Chamber for the ALICE TOF detector.

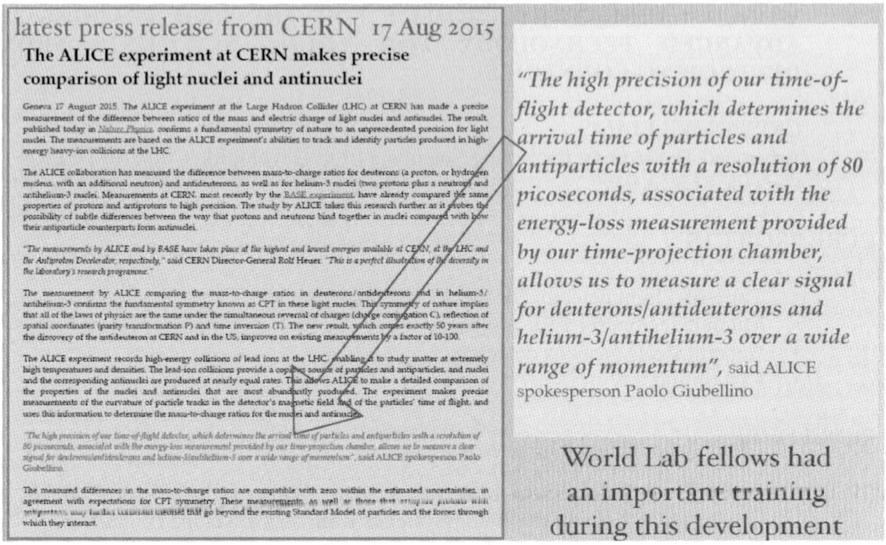

latest press release from CERN 17 Aug 2015

The ALICE experiment at CERN makes precise comparison of light nuclei and antinuclei

Geneva 17 August 2015. The ALICE experiment at the Large Hadron Collider (LHC) at CERN has made a precise measurement of the difference between ratios of the mass and electric charge of light nuclei and antinuclei. The result, published today in Nature Physics, confirms a fundamental symmetry of nature to an unprecedented precision for light nuclei. The measurements are based on the ALICE experiment's abilities to track and identify particles produced in high-energy heavy-ion collisions at the LHC.

The ALICE collaboration has measured the difference between mass-to-charge ratios for deuterons (a proton, or hydrogen nucleus, with an additional neutron) and antideuterons, as well as for helium-3 nuclei (two protons plus a neutron) and antihelium-3 nuclei. Measurements at CERN, most recently by the BASE experiment, have already compared the same properties of protons and antiprotons to high precision. The study by ALICE takes this research further as it probes the possibility of subtle differences between the way that protons and neutrons bind together in nuclei compared with how their antiparticle counterparts form antinuclei.

"The measurements by ALICE and by BASE have taken place at the highest and lowest energies available at CERN, at the LHC and the Antiproton Decelerator, respectively," said CERN Director-General Rolf Heuer. *"This is a perfect illustration of the diversity in the laboratory's research programme."*

The measurement by ALICE comparing the mass-to-charge ratios in deuterons/antideuterons and in helium-3/antihelium-3 confirms the fundamental symmetry known as CPT in these light nuclei. This symmetry of nature implies that all of the laws of physics are the same under the simultaneous reversal of charges (charge conjugation C), reflection of spatial coordinates (parity transformation P) and time inversion (T). The new result, which comes exactly 50 years after the discovery of the antideuteron at CERN and in the US, improves on existing measurements by a factor of 10-100.

The ALICE experiment records high-energy collisions of lead ions at the LHC, enabling it to study matter at extremely high temperatures and densities. The lead-ion collisions provide a copious source of particles and antiparticles, and nuclei and the corresponding antinuclei are produced at nearly equal rates. This allows ALICE to make a detailed comparison of the properties of the nuclei and antinuclei that are most abundantly produced. The experiment makes precise measurements of the curvature of particle tracks in the detector's magnetic field and of the particles' time of flight, and uses this information to determine the mass-to-charge ratios for the nuclei and antinuclei.

"The high precision of our time-of-flight detector, which determines the arrival time of particles and antiparticles with a resolution of 80 picoseconds, associated with the energy-loss measurement provided by our time-projection chamber, allows us to measure a clear signal for deuterons/antideuterons and helium-3/antihelium-3 over a wide range of momentum", said ALICE spokesperson Paolo Giubellino.

The measured differences in the mass-to-charge ratios are compatible with zero within the estimated uncertainties, in agreement with expectations for CPT symmetry. These measurements, as well as those that [illegible] theories that go beyond the existing Standard Model of particles and the forces through which they interact.

The Extreme Energy Events project

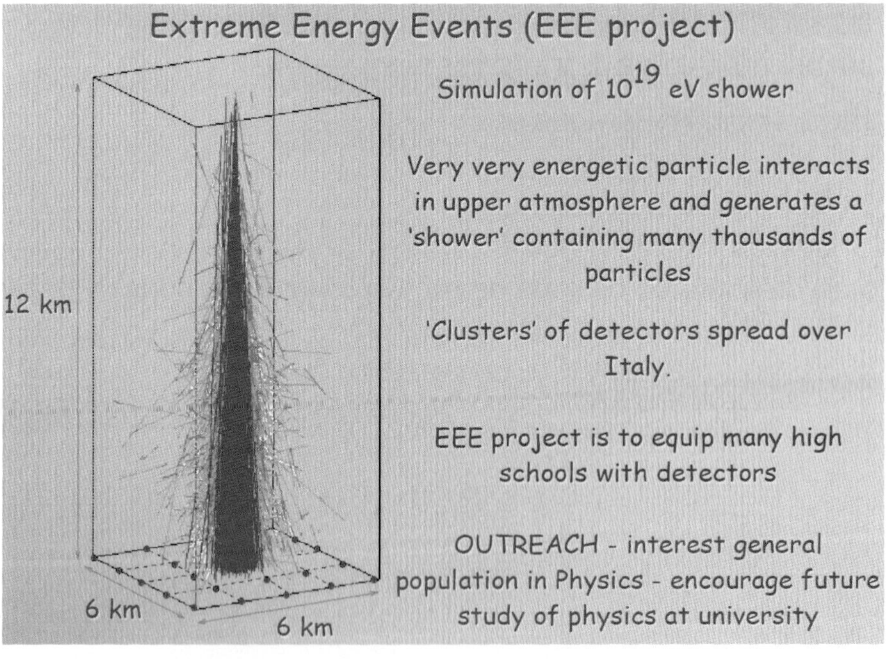

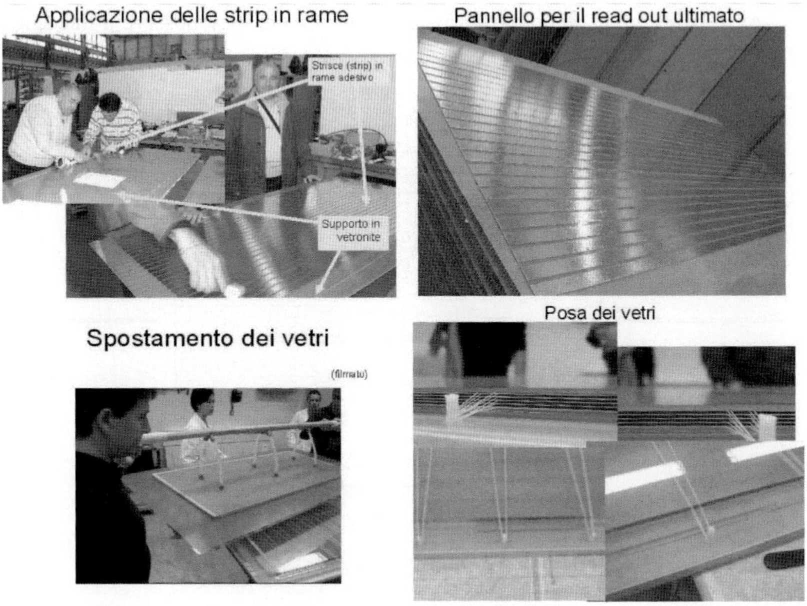

A photograph is taken of each group at the end of the week of chamber construction.

At a given muon station at an Italian high school: they observe a through-going muon:

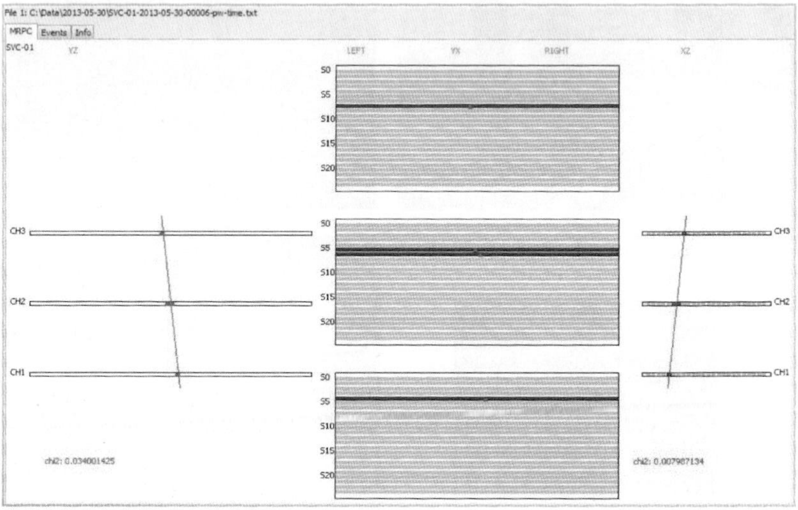

The EEE Project: an extended network of muon telescopes for the study of cosmic rays

EEE Collaboration
Museo Storico della Fisica, Centro Studi e Ricerche E. FERMI, Roma, Italy
Dipartimento di Matematica e Fisica, Università del Salento, Italy

The EEE (Extreme Energy Event) Project [1] is an experiment to study high energy extensive air showers (EAS) over a very large area, using an array of muon telescopes, based on position-sensitive Multigap Resistive Plate Chambers (MRPCs) and distributed over a total area of 3×10^5 km^2. The stations are installed inside Italian high school buildings, INFN sections and at CERN, and most of them have been independently taking data since several years. Young students are directly involved in assembling and monitoring the telescopes, with the aim to introduce them to advanced physic research [2]. A new combined run (RUN 1) has started in February 2015, with more than 30 telescopes taking data simultaneously and it has collected about 4×10^9 reconstructed events.

The EEE Telescope

Three MRPC for particle traking

Single MRPC:

- 6 gas gaps: 2 vetronite panels with 5 floating glass plates, 300μm spaced
- ~10^{13}Ωcm³ volume resistivity - Licron resistive paint on vetronite panels
- $C_2H_2F_4$(98%) and SF_6 (2%) continuously fluxed by (3l/h)
- 24 readout copper strips as electrodes, pitch of 3.2 cm
- HV up to 20 kV (avalanche mode) supplied by 2 DC/DC converters

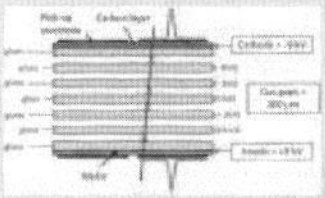

When a charged particle traverses the gas it creates, along its path, ion-electron pairs which are amplified in the Townsend avalanche process. The glass plates terminates the avalanches development in each gaps. The induced signals, sum of the electric charges in all of the gaps, are picked up by the copper strips on both vetronite panels [3] [4].

The EEE Station

- 6 Front-End boards (FEAs) with 24 channels to pick up readout signals on the strips
- 2 multi-hits Time to Digital Converters (TDCs 128 + 64 channels) to reconstruct the particle impact point.
- Trigger Card. A six-fold coincidence of both front-end cards of the three MRPCs generates the data acquisition Trigger.
- GPS unit gets the event time stamp (UTC time) to record and synchronize informations
- VME bridge. DAQ connected to a PC via USB, controlled by LabView program
- Voltage control System (VCS) in the MRPCs DC/DC Converters and FEBs
- Weather Station measures the temperature and the pressure inside and outside the telescopes building

Performance

The particle impact point is reconstructed by the hit strip (x) in one direction and by the difference of signal arrival times at the strip ends (y) measured by TDCs in the other direction [5].

- ✓ 100ps time resolution of the TDCs
- ✓ ~7 mm spatial resolution along both coordinate
- ✓ > 95% MRPC efficiency at the operating voltage of 18 kV
- ✓ 50 ns GPS time resolution

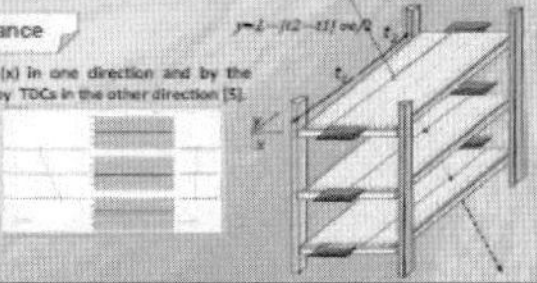

RUN 1

45 days Run

23 February 2015 – 30 April 2015

For the first time a large number of telescopes have been contemporaneously taking data

- 35 telescopes
- 20 –40 Hz muon rate in each one.
- Data are transferred and stored to CNAF where events and tracks are analyzed
- Tracking procedure: "Good" events are selected → Cuts $\chi^2 < 10$, length, ToF
- Particle polar and azimuthal angles of the event are reconstructed

4 x 10⁹ GOOD TRACKS have been collected

Since the beginning of their operation the single EEE telescopes and their clusters in a same metropolitan area have been able to produce significant scientific outcomes. **At the present, data transfer to CNAF**, allowing a direct way to store and access all data, **makes it easier to analyse contemporaneously all the EEE network results**

Study of cosmic rays flux

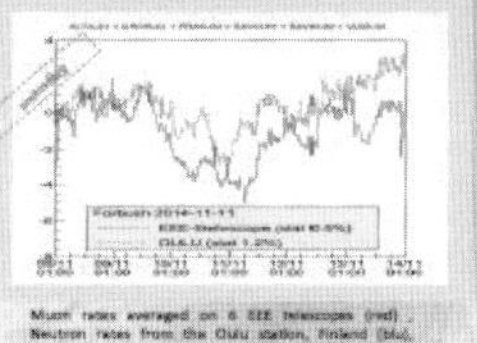

Muon rates averaged on 6 EEE telescopes (red) , Neutron rates from the Oulu station, Finland (blu), during the FD associated to solar flare on the date 7-11-2014

Local value of temperature and pressure influence both cosmic rays flux at the sea level and gas density in the MRPC. Therefore muon rates $R_{measured}$ in each telescope need to be normalized to the effctive rate R_{eff} by means of a *barometric coefficient* β evaluated in each different environment.

$$R_{eff} = (1 - \beta \Delta p)\, R_{measured}$$

Δp is the variation of atmospheric pressure with respect to the average value

Forbush decreases (FDs) are rapid variations of the cosmic rays flux that take place over the course of a few hours, and are associated to solar phenomena as CME and solar flares. FDs have been already observed by single EEE telescopes [8] [9], and also by different telescopes contemporaneously. Adding up data set from different station allows to reduce the *Signal/noise* value.

Search for Coincidences

Extensive Atmospheric Showers (EAS) Detections

The telescopes relative distances are ranging from a few hundred meters for clusters of 2, 3 and 4 telescopes in the same city, to more than 1000 km for the farthest stations. Muons from the same EAS are detected by different stations [6].

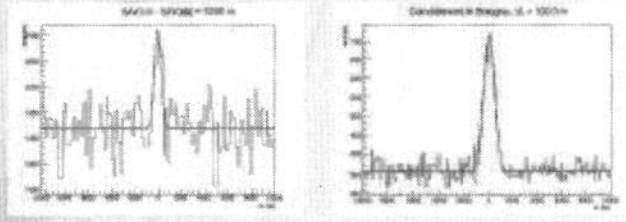

Distance corrections

Arrival time difference of particles in the same EAS depend on their own angular position from the axis shower and on the axis shower direction. Therefore time intervals for checking out muon coincidences are correlated to the distance between the different stations [7].

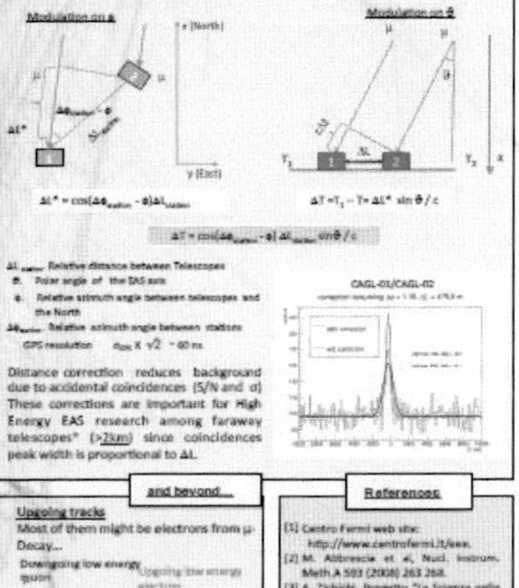

- $\Delta L_{station}$ Relative distance between Telescopes
- θ Polar angle of the EAS axis
- φ Relative azimuth angle between telescopes and the North
- $\Delta\phi_{station}$ Relative azimuth angle between stations

GPS resolution $\sigma_{GPS} \times \sqrt{2} \sim 60$ ns

Distance correction reduces background due to accidental coincidences (S/N and σ) These corrections are important for High Energy EAS research among faraway telescopes* (>2km) since coincidences peak width is proportional to ΔL.

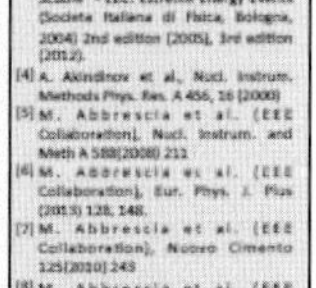

and beyond...

Upgoing tracks

Most of them might be electrons from μ-Decay...

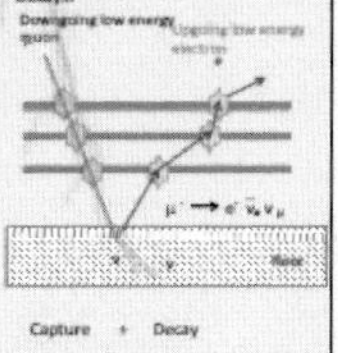

Some open questions

How to evaluate the telescopes time exposure and acceptance? Could be possible to search local anisotropy in primary cosmic ray through the EEE network?

References

[1] Centro Fermi web site: http://www.centrofermi.it/eee.
[2] M. Abbrescia et al, Nucl. Instrum. Meth.A 593 (2008) 263 268.
[3] A. Zichichi, Progetto "La Scienza nelle Scuole" - EEE: Extreme Energy Events (Societa Italiana di Fisica, Bologna, 2004) 2nd edition (2005), 3rd edition (2012).
[4] A. Akindinov et al., Nucl. Instrum. Methods Phys. Res. A 456, 16 (2000)
[5] M. Abbrescia et al. (EEE Collaboration), Nucl. Instrum. and Meth A 588(2008) 211
[6] M. Abbrescia et al. (EEE Collaboration), Eur. Phys. J. Plus (2013) 128, 148.
[7] M. Abbrescia et al. (EEE Collaboration), Nuovo Cimento 125(2010) 243
[8] M. Abbrescia et al. (EEE Collaboration), Eur. Phys. J. Plus (2011) 126,61.
[9] M. Abbrescia et al. (EEECollaboration), Eur. Phys. J. Plus (2013) 128, 62.

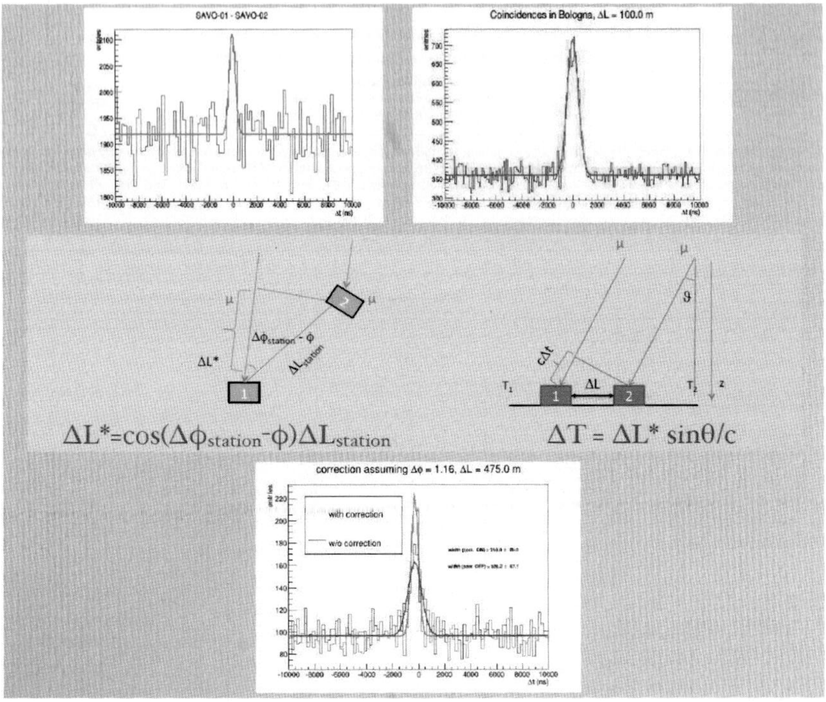
SAVO-01 - SAVO-02
Δt (ns)
Coincidences in Bologna, ΔL = 100.0 m
Δt (ns)
μ
μ
Δφstation - φ
ΔL*
ΔLstation
1
2
μ
μ
ϑ
cΔt
T1
ΔL
1
2
T2
z
ΔL*=cos(Δφstation-φ)ΔLstation
ΔT = ΔL* sinθ/c
correction assuming Δφ = 1.16, ΔL = 475.0 m
with correction
w/o correction
Δt (ns)

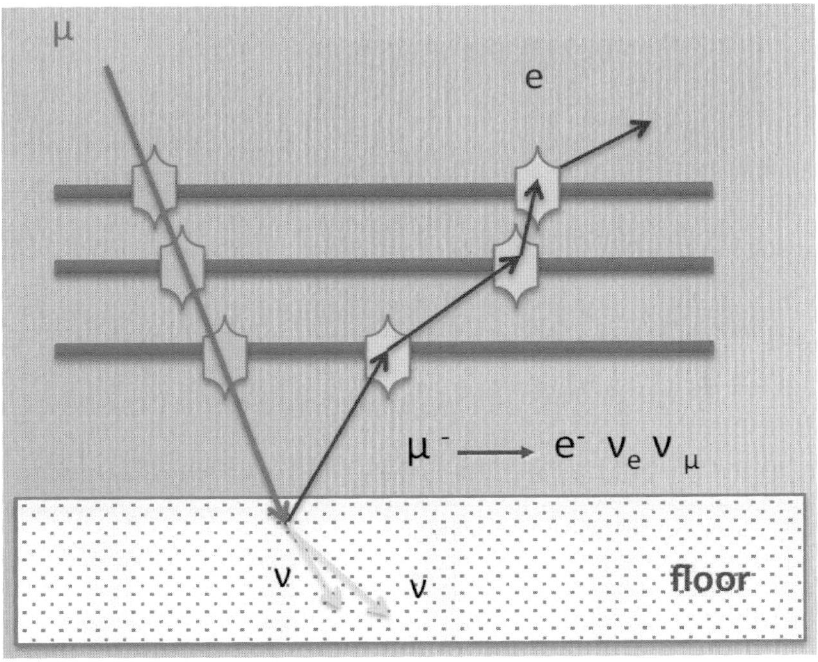

The Multigap Resistive Plate Chamber has good timing characteristics: so direction of particle can be determined. Want to investigate whether upward going particles from muon capture and decay.

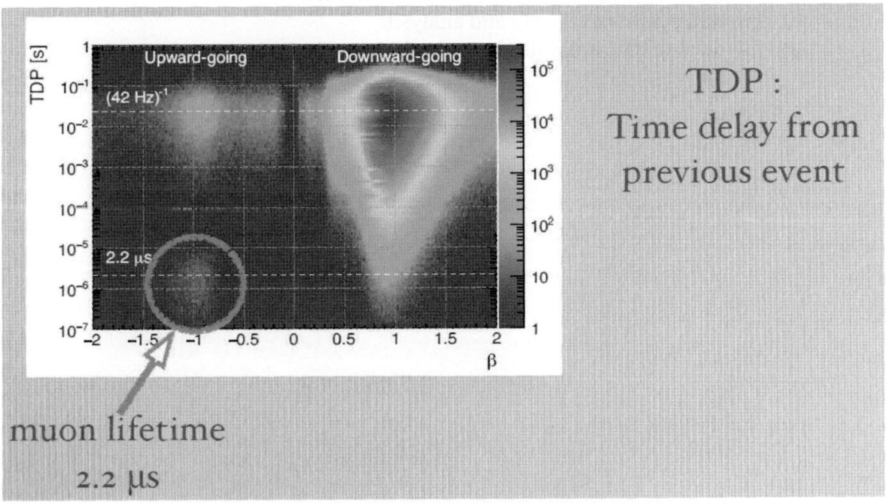

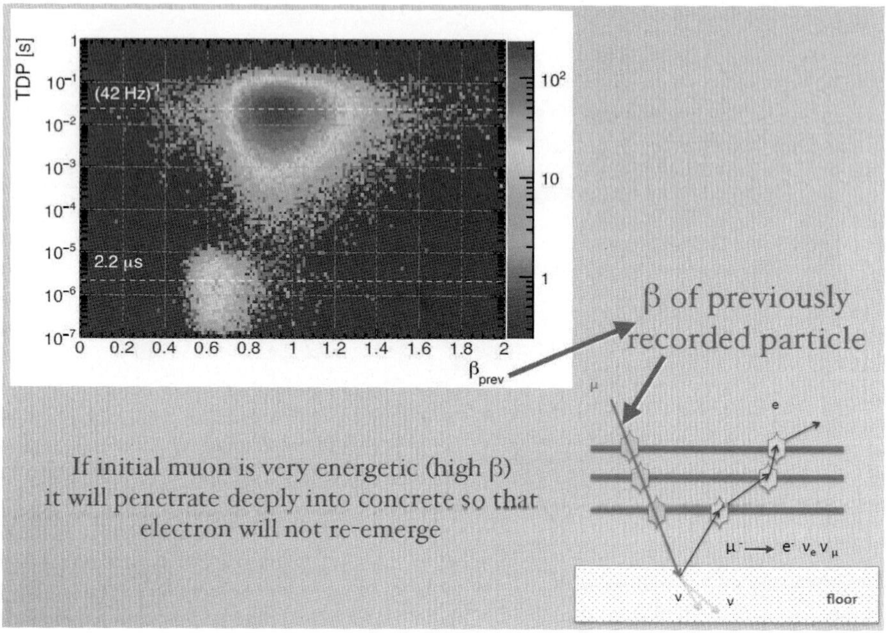

Involvement of World lab fellows in the EE project:

a) Supervision of high school students to build the MRPC chambers for the EEE telescope.
b) Running two EEE telescopes located at CERN.
c) Providing software: DAQ and analysis.
d) Involved in data analysis (large database management).

CHANGE OF TOPIC: HADRON CALORIMETERS

Can the MRPC make a good detection element?

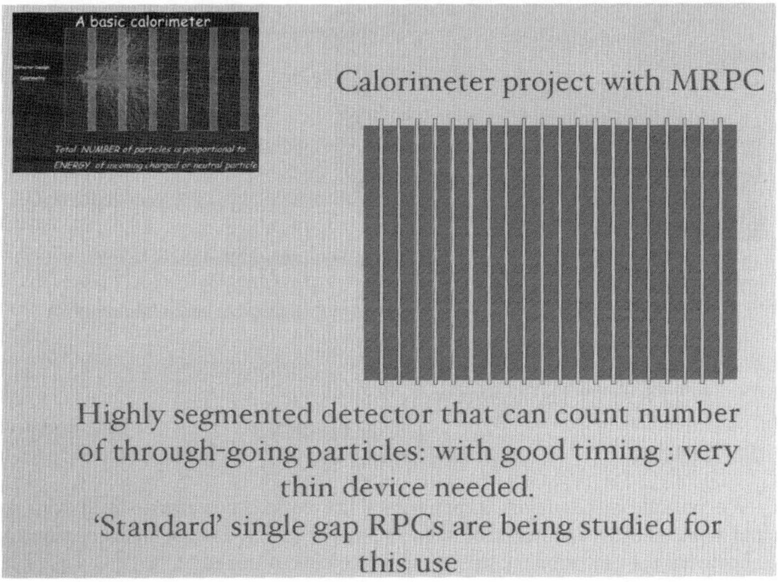

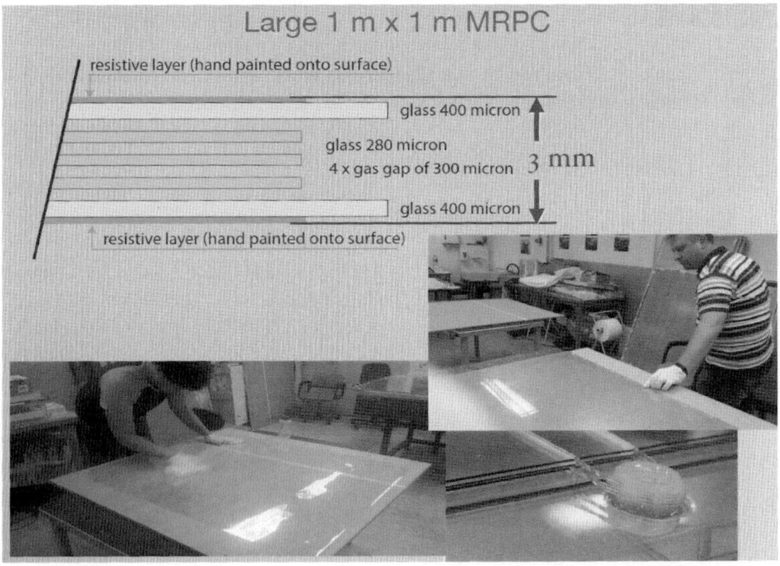

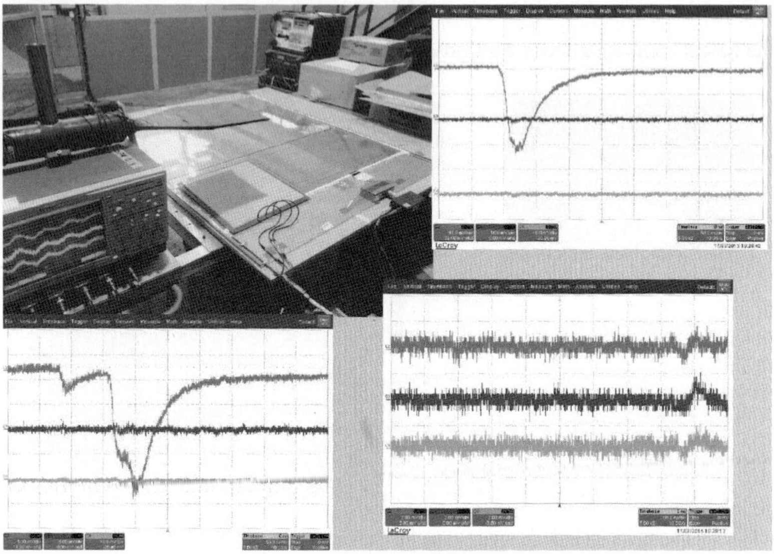

We provide a 'hands-on' experience with particle detectors—built with window glass and fishing line—and operated with Freon gas (same as used in air conditioners for cars)...however the performance of such detectors is cutting edge.

THANK YOU

- Thank you from me (it is a pleasure to work with these fellows).
- Thank you from the fellows for the financial support they receive.

IMPLEMENTATION OF THE IDEAS OF THE WORLD FEDERATION OF SCIENTISTS (WFS) VIA ITS PROGRAMME OF SCHOLARSHIPS FOR LITHUANIAN RESEARCHERS IN 2014-2015

DOMAS KAUNAS
The Lithuanian Academy of Sciences
Vilnius, Lithuania

The WFS programme of one-year scholarships for performing research in the areas that have been defined as Planetary Emergencies by the World Federation of Sciences has been continued in Lithuania in 2014-2015 by the Lithuanian Academy of Sciences. Ten successful applicants have completed their projects in May, 2015.

MS. ŽIVILĖ ČIULADAITĖ
PhD student at Faculty of Medicine, Vilnius University.

Her research, "Clinical significance of *de novo* and inherited copy-number variation in patients with intellectual disability". The research was connected to Planetary Emergency "Medicine and Biotechnology" and supervised by Professor Vaidutis Kučinskas.

Deletions and duplications, termed copy number variants (CNVs), have been implicated in a variety of neurodevelopmental disorders including intellectual disability (ID), autism, and schizophrenia. Analysis of molecular karyotyping results from 76 idiopathic ID trios revealed 17 rare CNVs.

Based on literature data and web-based database (ECARUCA, DECIPHER, ISCA, ClinVar) 7/8 of *de novo* CNVs were considered clearly pathogenic and one *de novo* deletion at 16q22.3, 726 kb in size, was considered as probably pathogenic. Two publications of unique *de novo* cases of this study have been accepted in the web sites of scientific journals.

Interpretation of rare inherited CNVs still remains a challenge. The research expresses the hope that at larger collaborative cohort studies, functional studies will change interpretation of rare unknown clinical significance CNVs.

DR. INGA GRISKOVA-BULANOVA
Researcher at the Department of Neurobiology and Biophysics, Vilnius University.

Her research, "Modulation of electrophysiological correlates of response inhibition by female sex hormones: implications for brain disorders". The research was connected to Planetary Emergency "Medicine and Biotechnology" and supervised by Professor Osvaldas Ruksenas.

The young scholar's research aimed to evaluate the effect the menstrual cycle has on response inhibition in Go/NoGo task in a sample of healthy females and contrast it to males prior to translation of Go/NoGo paradigm into clinical testing. Females were recruited and randomly assigned to participate in the experiment during 1) their early follicular (n-14), 2) late follicular (n=12) and 3) mid-luteal phases of the menstrual cycle (n=13). Concentrations of 1713-estradiol and progesterone in the saliva were measured to retrospectively confirm the menstrual cycle phase. 15 male subjects were also recruited.

An unwarned equiprobable auditory Go/NoGo task was used while EEG was recorded. ERPs from Fz, Cz, Pz electrodes were analysed.

Reaction times did not differ between menstrual cycle phases and genders; percentage of correct responses was lowest in late follicular phase. On the ERP level, two types of effects were observed. Task-related effect was observed as corresponding to previously published results: 1) Go and NoGo r RPs differed in the period of N2 and P3 peaks fronto-centrally; 2) Go and NoGo ERPs differed at slow-wave timing parietally electrode during early follicular and mid-luteal phases. Phase-related effect was observed during both Go and NoGo trials at P2, N2 and P3 amplitudes were most positive during late follicular phase.

The results suggest that inhibitory/excitatory balance is changed during the late follicular phase, causing higher amplitudes of ERP signal.

MS. MARIJA MATULIONYTĖ
Researcher at the National Cancer Institute (former Institute of Oncology, Vilnius University).

Her research "Spectroscopic and structure investigation of biocompatible nanoparticles in various media". The research was connected to Planetary Emergency "Medicine and Biotechnology" and supervised by Professor Ričardas Rotomskis.

Ultra small photoluminescent gold nanoclusters (Au NCs), composed of several atoms with sizes up to a few nanometers, have recently stimulated extensive interest. Unique molecule-like behaviors, low toxicity, and facile synthesis make photoluminescent Au NCs very promising in broad ranges of biomedical applications. However, using gold nanoparticles (Au NPs) for bioimaging might cause their degradation under continuous excitation with UV light, which might result in toxicity. Internalization of Au NCs in cells and organisms and their possible toxic effects have not been fully investigated as well.

In the final report results on photostability of synthesized photoluminescent 2-(N-morpholino) ethanesulfonic acid (MES)-coated gold (Au-IVIES) NCs and cellular uptake and cytotoxicity of two types of photoluminescent Au NCs that differ in coating: Au-MES NCs and bovine serum albumin (BSA)-encapsulated gold (BSA-Au) NCs were presented. The results have shown that synthesized photoluminescent Au-MES NCs are not photostable under irradiation with UV/blue light. Upon irradiation at 402 nm, the size of the photoluminescent Au-MES NCs decreases from ~9 to ~7 gold atoms, while irradiation with 330 nm wavelength light, induced formation of photoluminescent Au-MES NCs, likely from the disrupted nonluminescent Au NPs. The BSA-Au NCs were accumulated in vesicles at the perinuclear region in MCF-7 and MDA-MB-231 cancer cells indicating endocytic uptake. In contrast, Au-MES NCs were distributed homogeneously inside the cells including nuclei. The ultra small size of the Au-MES NCs resulted in significant toxic effects already after 3h of incubation. At the same time BSA-Au NCs revealed no toxic effects even after 24h of incubation. This could be due to increased stability and biocompatibility provided by encapsulation of Au NCs inside the BSA molecule.

MR. ROKAS NAVICKAS
PhD student at Vilnius University Hospital.

His research "Arterial and molecular biomarker study for cardiovascular and diabetes risk prediction". The research was connected to Planetary Emergency "Medicine and Biotechnology" and supervised by Professor Aleksandras Laucevicius.

The study included two parts, one analysing the cardiovascular risk or the investigated cohort using the common validated and widely available risk stratification models, the second part concentrated on novel potential biomarkers, possibly including the microRNA while investigating on any additional predictive value for cardiovascular outcomes within the group.

The correlations between miRNAs and specific components of MetS were investigated. With the defined list of miRNAs and trained staff, miRNA testing started in April 2015, aiming to test ±150 randomly selected patients out of the 400 biobanked by the end of July 2015.

By the end of August 2015 the young scholar aims to finish epigenetic part of the study and concentrate on risk assessment part.

MRS. EGLĖ PREIKSAITIENĖ
PhD student at Faculty of Medicine, Vilnius University.

Her research "Evaluation of the impact of pathogenic copy number variants on the phenome in patients with developmental delay or intellectual disability". The research was connected to Planetary Emergency "Medicine and Biotechnology" and supervised by Professor Vaidutis Kučinskas.

The aim of the research was to determine the phenotypic impact of pathogenic copy number variants and their role in disturbing the development and function of the central nervous system in patients with intellectual disability. Totally 50 patients have been recruited for the research. A blood sample was collected from each patient with sporadic ID and his parents for DNA extraction. From all selected families, array CGH analysis was performed to 32 patients. In 12 patients pathogenic copy number variants have been detected. Detailed analysis of the relationship between genotype and phenotype in patients with unique chromosomal alterations was performed. Identification of causative genomic variants using molecular genetic technologies: in few patients monogenic syndromes were suspected.

DR. EGLĖ SENDŽIKIENĖ
Associate Professor at Aleksandras Stulginskis University.

Her research was "Biodiesel production applying simultaneous process *(in situ)* of oil extraction from ground oilseeds and transesterification". The research was connected to the Planetary Emergency "Energy" and supervised by Professor Violeta Makarevičienė.

The aim of research was to investigate the simultaneous vegetable oil extraction and transesterification with butanol process, to evaluate the properties of the product.

Lipozyme RM IM is the enzyme preparation that was found to have major catalytic activity during oil transesterification. The optimum temperature of the simultaneous oil extraction from rape seeds and transesterification using butanol is 40°C. By applying a response surface methodology and using an *"in-situ"* process of rape seed oil extraction and transesterification, the following optimum conditions were established for the generation of a mineral diesel and 10% biodiesel mixture: a butanol and oil molar ratio of 30:1, a Lipozyme RM IM amount of 5.2%, a duration lasting 19.6 h, and a temperature of 40°C. The application of these conditions enables to obtain the maximum amount of rape seed butyl esters and oil and the maximum degree of transesterification meeting the standard requirements. All of physical and chemical properties agree with standards LST EN 14214 and EN 590 requirements. According to the OECD classification, the mixture with D-RBE could be classified as partially biodegradable material.

MRS. INGA SONGAILIENE
PhD student from the Institute of Biotechnology, Vilnius University.

Her report was "Structural and functional studies of CRISPR systems." The research was connected to Planetary Emergency "Medicine and Biotechnology" and supervised by Professor Virginijus Šikšnys.

Bacteria or archaea can tackle the invader nucleic acids by an adaptive immunity mechanism called CRISPR (clustered regularly interspaced short palindromic repeats)-Cas. The distinct feature of CRISPR-Cas systems is CRISPR loci that consists of short repeats interspaced by short (21-72 nt) spacers that originates from invader genetic material.

The young scholar investigated type I-E CRISPR system from *S. thermophilus*. In St-Cascade, the Cas proteins plus a 61-nt crRNA assemble into an ~400 kDa riboncleorprotein complex in her research. The complex recognizes putative DNA targets by establishing base-specific pairing between the crRNA guide and the protospacer, forming an R-loop. The presence of PAM motif in the proximity to the protospacer is crucial for R-loop formation and serves as an initiation point. The target readout may be determined by a directional R-loop formation or base pairing thermodynamics. The analysis of phages/plasmid escape mutants identified PAM-proximal sequence as crucial for target recognition, but the directional R-loop formation mechanism has not been directly proven so far.

In her study the scholar aimed to investigate the R-loop formation mechanism by I-E type Cascade complex and target interference by Cas3 nuclease/helicase. By single molecule and biochemical methods directional R-loop formation mechanism close to base pair resolution were proved. In this model, R-loop propagation serves as the only verification of a potential target and is key to reject of off-target sites.

DR. PATRIK ŠČAJEV
Researcher at the Institute of Applied Research, Vilnius University.

His research is "Carrier Dynamics in Highly Excited Diamond". The research is connected to the Planetary Emergency "Energy" and was supervised by Professor Kęstutis Jarašiūnas.

Experimental measurements have been performed on three CVD diamond samples. The obtained experimental results can be summarized by the following:

Precise control of photogenerated carrier density was reached by using accurate absorption coefficient values, measured using diffraction-based technique in 80-800 K range. Surface, bulk and nonlinear recombination rates were separated by combining time-resolved differential transmitivity and light induced transient grating measurements at different temperatures.

At high (>10^{17} cm^{-3}) nonequilibrium carrier density many body effects, such as exciton and biexciton formation, electron-hole scattering, band gap renormalization and degeneracy contributed to the ambipolar diffusion coefficient value.

In the bulk CVD sample the low excitation free carrier lifetime (approaching 800 ns) was point defect limited and temperature independent.

At high excitations strong lifetime reduction in T <300 K and T >300 K wings was observed. Temperature and excitation dependences of diffusion coefficient and lifetime provided carrier diffusion length, which varied in range from 1 to 50 i.tm exhibiting decrease with excitation.

MS. EVELINA ZAGORSKAITE
PhD student from the Institute of Biotechnology, Vilnius University.

Her report was "Structure and mechanism of methyl-directed restriction endonucleases". The research was connected to Planetary Emergency "Medicine and Biotechnology" and supervised by Dr. Giedrius Sasnauskas.

The epigenetic DNA modifications 5-methylcytosine (5mC) and 5-hydroxymethylcytosine (5hmC) in eukaryotes are recognized either in the context of double-stranded DNA, or in the flipped-out tate. The SRA-like domains and the base-flipping mechanism for 5(h)mC recognition are also shared by the recently discovered prokaryotic modification-dependent endonucleases of the MspJI and PvuRtsll families. Since the mechanism of modified cytosine recognition by many potential cukaryotic and prokaryotic 5(h)mC "readers" is still unknown, a fast solution based method for the detection of ex trahelical 5(h)mC would be very useful. In her study the young scholar tested base-flipping by MspJI- and PvuRts1I-like restriction enzymes using several solution-based methods, including fluorescence measurements of the cytosine analog pyrrolocytosine and chemical modification of extrahelical pyrimidines with chloroacetaldehyde and KMnO4. It was found that only KMnO4 proved an efficient probe for the positive display of flipped out pyrimidines, albeit the method required either non-physiological pH (4.3) or a substitution of the target cytosine with thymine. The results imply that DNA recognition mechanism of 5(h)mC binding proteins should be tested using a combination of all available methods, as the lack of a positive signal in some assays does not exclude the base flipping mechanism.

DR. JONAS ŽIAUKA
Researcher at the Lithuanian Centre for Agriculture and Forestry, Institute of Forestry.

His research topic was "Management of hybrid poplar organogenetic processes *in vitro.*" The research was connected to the Planetary Emergency "Energy" and supervised by Dr. Sigutė Kuusienė.

This scholarship of WFS was intended for the investigation of new opportunities in the area of vegetative propagation of hybrid poplars that are widely recognized as perspective trees for the establishment of short rotation energetic forest stands.

Increased number of hybrid poplar shoots produced by a single apical explant can be considered among the most important practical outcomes of the culture tube sealing with Parafilm. It was demonstrated that decreasing culture vessel volume shared by a single explant further increased shoot production, and covering of explants from direct light also contributed to the same end. All these observations suggest that space and energy requirements for micropropagation of selected hybrid poplar clones can be considerably reduced if hybrid poplar explants were cultured in tightly packed small-volume culture vessels, which could be put onto illuminated shelves in several layers, thus saving space and electricity. Furthermore, it was proved that culturing of hybrid poplar explants in tightly packed and sealed vessels contributes to significantly increased poplar shoot growth during the subsequent culture when there is a need for a fast establishment of strong shoots which would be suitable for the adaptation to *ex vitro* conditions and subsequent tree production.

The full final reports of the Lithuanian National Scholarships of the World Federation of Scientists recipients were sent to Professor Antonio Zichichi by mail in May, 2015.

We want to take this opportunity and express our gratitude to the World Federation of Scientists for the possibility for talented young Lithuanian scientists to experience the benefits of the scholarships.

VIETNAM AUGER TRAINING LABORATORY (VATLY), 2015 PROGRESS REPORT

PIERRE DARRIULAT
Vietnam Auger Training Laboratory, Institute for Nuclear Science and Technology
Hanoi, Vietnam

SIX WORLDLAB FELLOWS

Pham Ngoc Diep
Pham Tuyet Nhung
Nguyen Thi Thao
Pham Tuan Anh
Do Thi Hoai
Nguyen Thi Phuong

Tuan Anh obtained his PhD at the end of last year.
Hoai will defend her PhD thesis in December.
Phuong is a master student.

The year has seen an important event in the life of our team: we have moved from the Institute of Nuclear Science and Technology (part of the Vietnam Atomic Energy Commission) to the Vietnam National Satellite Centre (VNSC, part of the Vietnam Academy of Science and Technology).

VNSC, of which we are the Department of Astrophysics, is a better environment for us, offering opportunities to meet foreign astrophysicists, in particular in the context of existing bilateral framework agreements, such as with the French CNRS, the Belgian FNRS or the Japanese NAOJ.

We have made important contributions to the study of evolved stars, partly in collaboration with Dr. T. Le Bertre of Paris Observatory. In addition to the work done by Nhung and Hoai during their stay in Paris at the end of last year, we produced two publications in Ha Noi (on the Red Rectangle and EP Aqr) and we are now preparing a third one.

Tuan Anh has been active in the study of ALMA observations of remote galaxies and is presently laying the groundwork for future collaboration with a Japanese team.

Finally, we have completed the transfer of our small radio telescope to the roof of a nearby building where it will be used for the training of students.

A SPECTACULAR POST-AGB STAR: THE RED RECTANGLE

Stars like our Sun spend a few million years on the Asymptotic Giant Branch (AGB) burning alternately hydrogen into helium and helium into carbon, oxygen and other light elements in a gigantic gaseous envelope surrounding a small core (future white dwarf). At the end of their life, the envelope dissolves into the interstellar medium (ISM).

The Red Rectangle is between these two phases. The star in the centre has a companion accreting its wind in a disk perpendicular to the star axis. Direct light from the star is prevented to reach the Earth by the presence of a dense dust torus.

We have analysed publicly accessible ALMA observations in CO(6-5) and CO(3-2) with an order of magnitude better resolution than previously available and have been first (and alone) to publish.

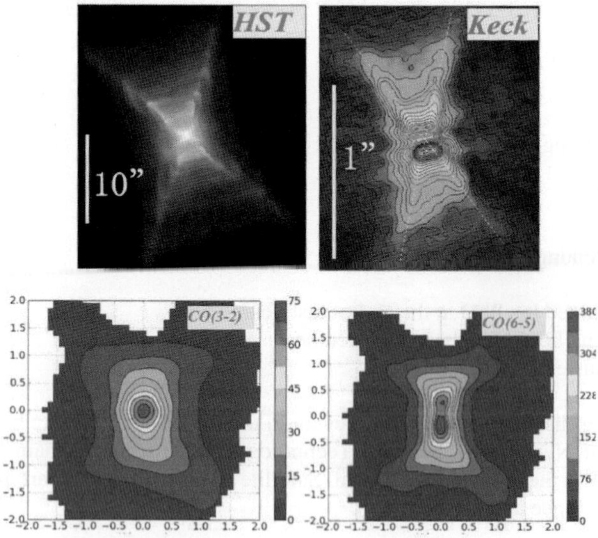

The effective density, combining actual density and temperature, was reconstructed in space under the hypothesis of rotation symmetry about the star axis and assuming that absorption is negligible.

Comparison between the CO(3-2) and CO(6-5) effective densities provides an evaluation of the gas temperature, observed to decrease slowly with distance in the equatorial region, from ~60 K at 0.3" to ~50 K at 2". In the polar region, the temperature takes much higher values at short distance to the star, typically 200 K at 0.3", and decreases steeply with distance to reach some 80 K at 1.5". A crude model of the calculated effective densities has been presented.

The gas kinematics displays a sharp separation between the equatorial rotating region (~1 km/s at 1") and a bipolar outflow (~7 km/s). A very simple model has been proposed, giving quite a good description of the measured Doppler velocities.

Important deviations from a fully symmetric model have been revealed: an excess of the effective density has been observed in both CO(3-2) and CO(6-5) data on the eastern limb of the bicone, particularly strong in the south-east direction.

The continuum observations are consistent with a compact dust source elongated along the equator and with the spectral energy distributions (SED) values available in the literature.

ALMA's open access policy, one year after data collection, is an invaluable asset for developing country teams such as ours.

AN AGB STAR SEEN POLE ON: EP Aqr

EP Aqr is one of the nearest mass-losing AGB stars, and as such one of the best characterized objects of its class. Combined observations have been obtained in 2003 and 2004 with the IRAM 30-m telescope and the Plateau-de-Bure Interferometer in ^{12}CO(1-0) and ^{12}CO(2-1). The spectra, displaying a narrow and broad component, were originally analyzed in terms of multiple isotropic winds.

We have revisited this early analysis, assuming instead a bipolar outflow approximately directed along the line of sight, taking advantage of the experience gained in similar studies of other stars.

The peculiar geometry allows for reconstructing the effective density in space, displaying important fluctuations.

The angular aperture of the bipolar outflow is ~45°.

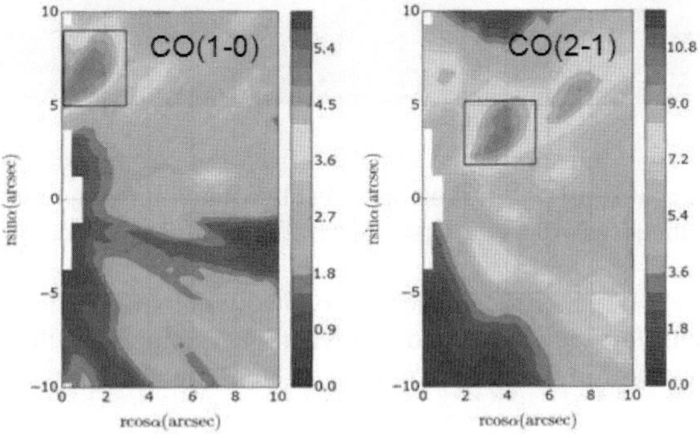

The CO(1-0)/CO(2-1) flux ratio reveals a significant dependence of the temperature on star latitude, smaller and steeper at the poles.

The expansion velocity increases from 2 km/s at the equator to 10 km/s near the poles. The mass loss rate is $\sim 1.7\ 10^{-7}$ solar masses per year.

The new analysis gives a better description of the star than the original one. However, better quality observations would greatly help: we have submitted a proposal to ALMA.

NEUTRAL ATOMIC HYDROGEN (HI) OBSERVATIONS

Far away from the star, carbon monoxide and other molecules are photo-dissociated by the interstellar radiation field. Observations at large distance, in particular of the wind-ISM interaction region, require the use of other tracers, such as dust or atomic species. However, being associated with continuous frequency distributions, they cannot provide information on kinematics.

For this, one needs a tracer sensitive to the gas Doppler velocity, namely featuring a line in its spectrum. Such is the case of HI emission at 21 cm wavelength, essentially unaffected by the interstellar radiation field and offering a measurement of the gas Doppler velocity with excellent spectral resolution.

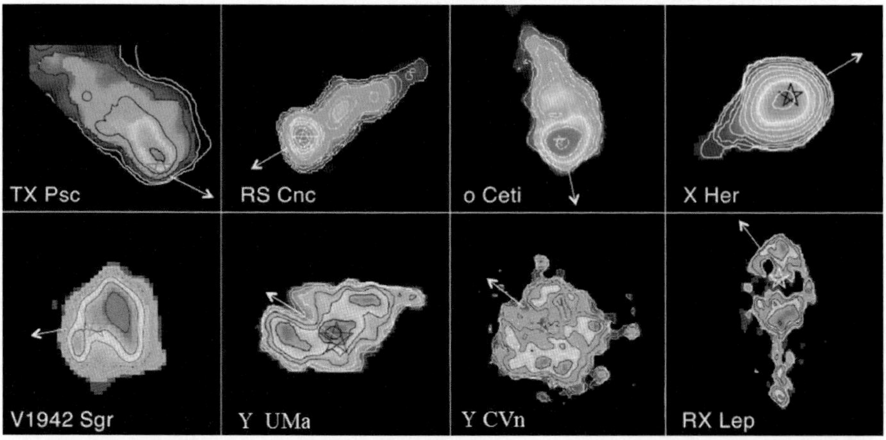

Observing HI emission from the circumstellar envelopes of AGB stars is difficult because of the strong ISM background. Indeed, initial attempts on evolved stars were unsuccessful. The first successful observations came from Nançay some twelve years ago.

The line profiles are typically narrower than close to the star, suggesting that the outflows are slowing down when interacting with the ISM. VLA images often show tails away from the proper motion of the star and hosting a significant velocity gradient.

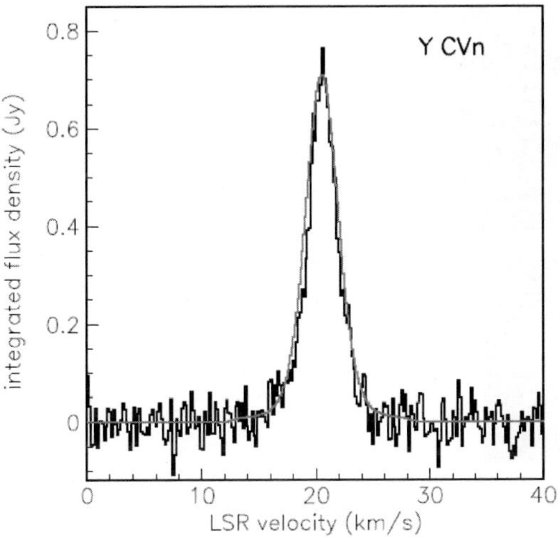

Hoai's contribution to these studies has been to simulate the HI line profiles expected for three different scenarii (a freely expanding isotropic wind, a single detached shell and a dynamical description of circumstellar shells taking wind-wind interactions into account) and to examine how they compare with presently available observations.

Sources with large mass-loss rates have rarely been detected, even at large galactic latitude. But the simulations cannot account for such a result: line profiles should not be seriously affected by background and flux densities are large enough for HI to be detected up to a few kpc.

It seems therefore that, in sources with large mass-loss rates, hydrogen is generally not in atomic, but rather in molecular form.

RX J0911 AND OTHER REMOTE GALAXIES

Tuan Anh defended his PhD thesis at the end of last year. The subject was the study of CO emission from RX J0911, a remote quasar host galaxy (redshift z=2.8) gravitationally lensed by a galaxy in the foreground.

His main contributions have been a rigorous treatment of measurement uncertainties and a detailed analysis of gravitational lensing of an extended source overlapping the lens caustic.

The source was found to be spatially resolved, on the line with good confidence but in the continuum to within two standard deviations only. Its size was evaluated both directly and from a comparison of the relative brightness of the four lensed images. The line image was observed to be elliptic with an eccentricity deviating from zero by four standard deviations.

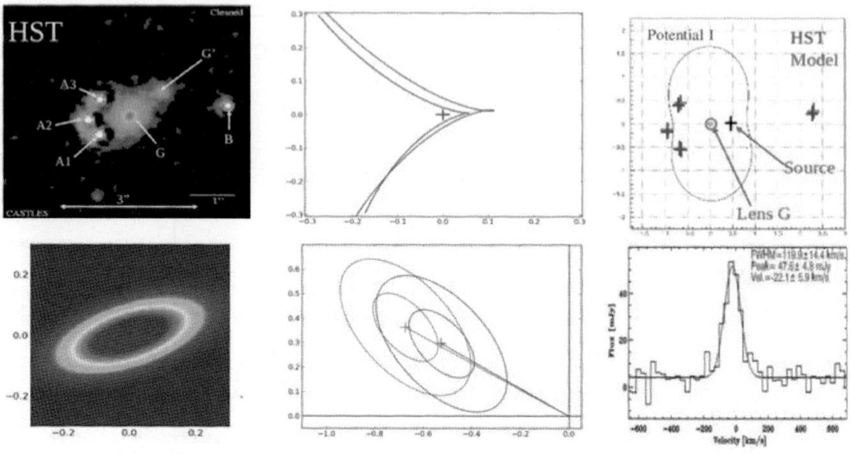

A particularity of this quasar host galaxy is the abnormally narrow width of the line, observed both in CO(1-0) and CO(7-6). Our observations make it unlikely that this be the result of the galaxy being seen face on but rather suggest that it is intrinsic evidence for a very small dynamical mass. The large star formation efficiency suggests that the galaxy has exhausted a large part of its gas reserve following a period of intense star formation and lies now at the boundary between high z and low z quasar hosts.

The study of remote galaxies tells us about the process of star formation at the beginning of the Universe, providing essential information on its early evolution. The brighter of these, which are easier to observe, have often their emission amplified by gravitational lensing, the lens being usually a galaxy in the foreground. Such welcome amplification is however accompanied by a significant distortion, which complicates the analysis of the data.

The field has made a big step forward with the operation in Chile of ALMA (Atacama Large Millimetre/sub-millimetre Array), with unprecedented sensitivity and spatial and spectral resolutions. We are eager to have access to observations of remote galaxies using this instrument. Tuan Anh is currently looking into ALMA observations of gravitationally lensed galaxies that are available on open access, in particular of SDP 81. He has established close relations with the Japanese team of Professor Daisuke Iono in Tokyo and is working toward collaborating with them on such studies.

TRAINING AND OUTREACH

In addition to giving astrophysics lectures at various Ha Noi universities, team members maintain particularly close relations with one of these, USTH, tutoring bachelor and master students on internship in the laboratory.

Together with a geographer, Thao is active in preparing for the outreach mission of the future Space Centre in Hoa Lac, which will include a planetarium and a small optical telescope.

We provide assistance to local astronomy amateur clubs and take part in exchanging information and experience with young colleagues in other South East Asian countries.

CONCLUSION

The team has now reached a critical size and a good visibility in the Vietnamese scientific landscape, being the largest, and indeed the only research team in the country working in astronomy and astrophysics. The association with the Academy of Science and Technology is a clear step forward in the recognition of our achievements. Yet, there is still a long way before approaching the status of a research team in developed countries, mostly because of the lack of understanding of our authorities in matters of fundamental research and its implications.

The support offered to us by the World laboratory is an invaluable and vital asset, allowing the team members to devote full time to their research rather than having to spend time and effort on a second better paid job, as is too often the case in the country. It is wholeheartedly that we express our deepest gratitude to the World Laboratory and, in particular, to Professor Antonino Zichichi.

NATIONAL SCHOLARSHIP PROGRAM (NSP)—KENYA

JANDOUWE VILLINGER
International Centre of Insect Physiology and Ecology (ICIPE)
Nairobi, Kenya

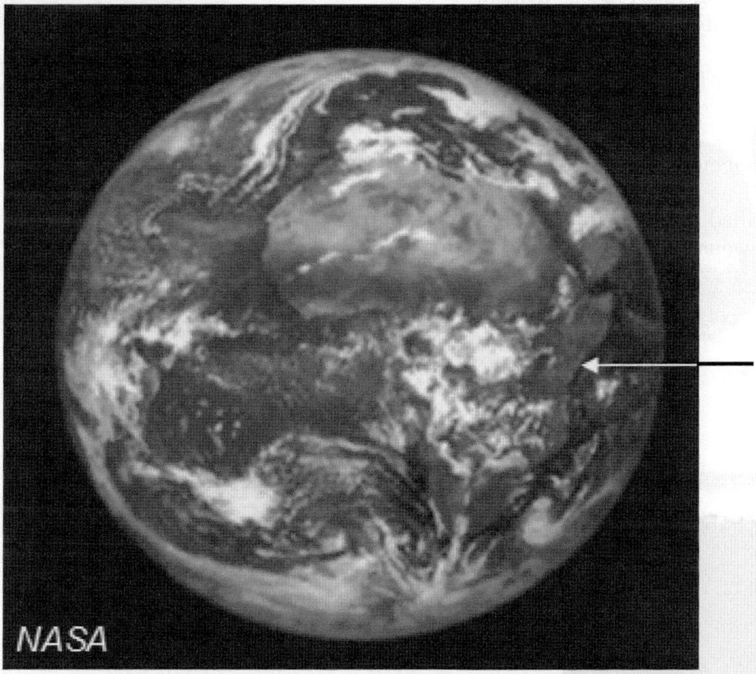

WFS NSP Coordinator: Daniel Masiga

ICIPE representative: Jandouwe Villinger

10 Scholarships awarded in June 2014

PROJECT I

Title: Phylogeny and antibiotic activity of *Xenorhabdus* bacteria isolated from nematode symbionts in Kenya.

Scholar: **Ryan Awori**
(University of Nairobi)
Supervisor: Prof Francis Mwaura
(University of Nairobi)

Planetary Emergency: Human Disease

Impact:
Development of novel antibiotics for treatment of animal and human diseases.

Identified a bacterium, *Xenorhabdus griffiniae*, that produced effective antibiotics for open wound and urinary tract infections in man and mastitis in cattle.

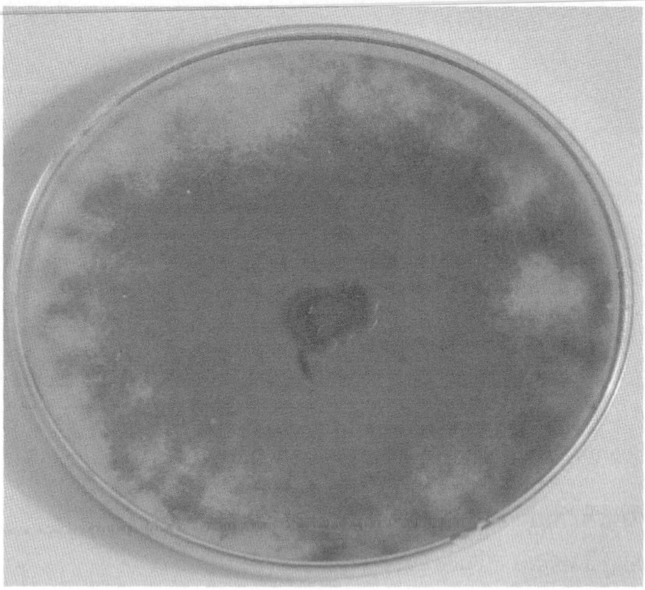

PROJECT II

Title: Molecular survey of tick species and tick borne parasites in cattle and wildebeest in the Maasai-Mara wildlife-livestock interface, Narok County, Kenya.

Scholar: **Sarah Akwabi Mukolwe**
(University of Nairobi)
Supervisor: Dr. Lillian Wambua
(University of Nairobi)

Planetary Emergency: Infectious Disease & Food Security

Impact:
This study ascertained the role played by ticks and wildebeests in the cattle-wildebeest interactions in Kenya specifically the Maasai-Mara ecosystem as a model to provide insights on the importance of the livestock-wildlife interface as a hotspot for emerging and re-emerging diseases of livestock and zoonoses.

Diverse *Theileria* and *Babesia* pathogens were identified in diverse ticks.

PROJECT III

Title: Development of simple and rapid diagnostic assays for detection of wildebeest-associated malignant catarrhal fever in cattle in Kenya.

Scholar: **Peninah Nduku Wambua**
(University of Nairobi)
Supervisor: Dr. Maina Magacha
(University of Nairobi)

Planetary Emergency: Infectious Diseases & Food Security

Transmission is dependent on the presence of a wildlife reservoir (wildebeest).

Impact:
Rapid and sensitive diagnostic assays based on Loop-Mediated Isothermal Amplification (LAMP) and Recombinase Polymerase Amplification (RPA) isothermal technologies were developed for diagnosis of malignant catarrhal fever. These are meant to help in early detection of malignant catarrhal fever so as to diminish its impact by adequate outbreak management.

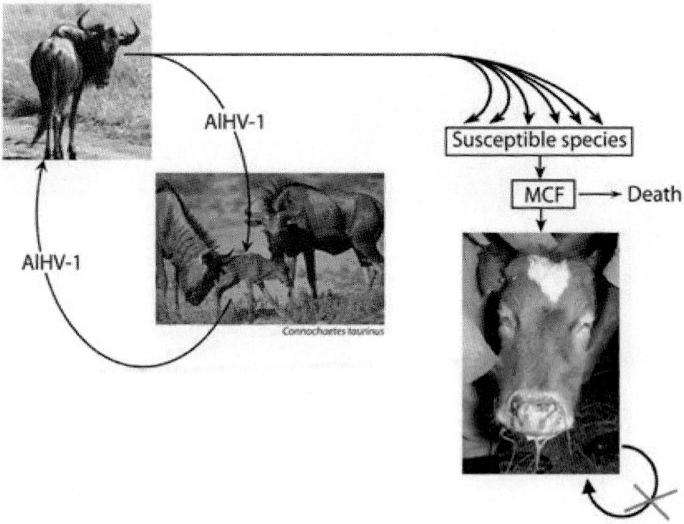

PROJECT IV

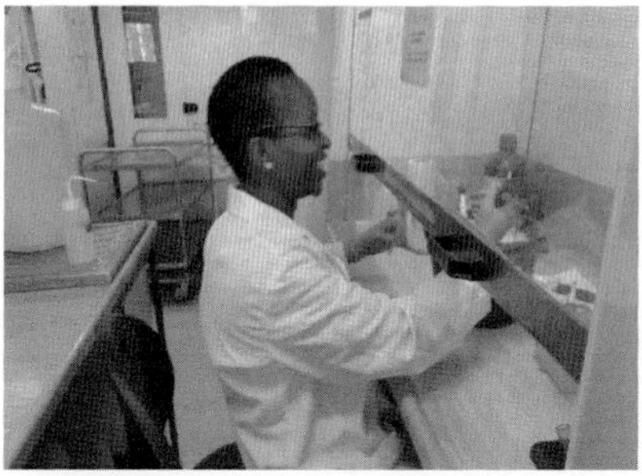

Title: Gene expression profiling of odorant binding proteins (OBPs) in *Glossina brevipalpis*

Scholar: **Mary Murithi**
Jomo Kenyatta University of Agriculture and Technology (JKUAT)
Supervisors: Dr. Steven Ger, Prof. Esther Magiri (JKUAT)

Planetary Emergency: Infectious Disease & Food Security

Impact:
The findings from this study could be used to design efficient and environmentally friendly control techniques that target the tsetse fly vector and reduce African Trypanosomiasis cases.

Some OBPs were expressed in larval and pupal developmental stages while others were expressed in the differentially starved adults. This supports the hypothesis that OBPs may be involved in non-olfactory processes.

PROJECT V

Title: *In silico* analysis of XptA genes of *Xenorhabdus* sp. and efficacy of the bacterium on *Sitophilus zeamais* and *Prostephanus truncatus.*

Scholar: **Ng'ang'a Peter Njenga**
(University of Nairobi)
Supervisor: Dr. Amugune Nelson
(University of Nairobi)

Planetary Emergency: Food Security

Impact:
The *Xenorhabdus* bacterium used in this study showed insecticidal activity against the target pests. It also significantly reduced the rate of consumption of the artificial diet by the larger grain borer. Therefore, the bacterium has potential for use as a biological control agent. With an appropriate application method, this bacterium could be used to control insects in the health and agriculture sectors. This study is the first to report of the use of *Xenorhabdus* to control the storage pests of maize.

Maize weevil

Grain borer

PROJECT VI

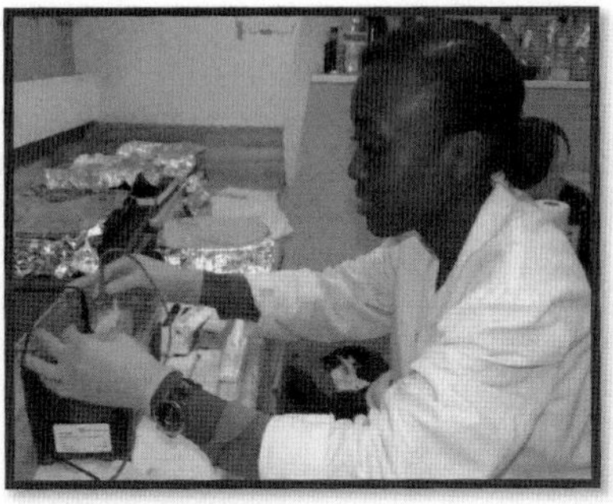

Title: Characterization of *Xenorhabdus* spp. insecticidal toxin genes *XptB1* and *XptC1* and determination of the bacteria's efficacy on the field cereal pests *Chilo partellus* and *Busseola fusca.*

Scholar: **Lorine Nanjala Nyongesa**
(University of Nairobi)
Supervisor: Dr. Nelson Amugune
(University of Nairobi)

Planetary Emergency: Food Security

Impact:
This study investigated novel bacterial toxins that provide potential potent biological control agents against Lepidopteran maize stem borers to augment existing control measures.

Xenorhabdus sp. isolate had potent bioactive agents (>70% mortality when injected into L3-stage larvae). However, only partial characterization of XptB1 and XptC1 toxins was achieved.

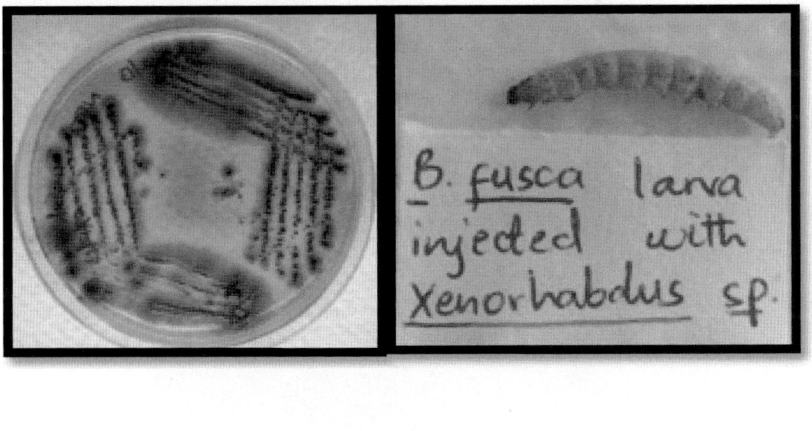

PROJECT VII

Title: Endophyte fungi as first line of defense against the bean stem maggot (BSM), *Ophiomyia spp,* in semi-arid areas of eastern Kenya.

Scholar: **Beritah Mutune**
(JKUAT)
Supervisor: Dr. Nguya K. Maniania
(ICIPE)

Planetary Emergency: Food Security

Impact:
We found that fungal endophytes can be considered as the first line of defense strategy against BSM. Fungus-inoculation of bean plants significantly reduced BSM feeding and oviposition, which in turn affected pupation and adult emergence. Future research should focus on the traceability of the endophytes within plant tissue and their performance in field conditions.

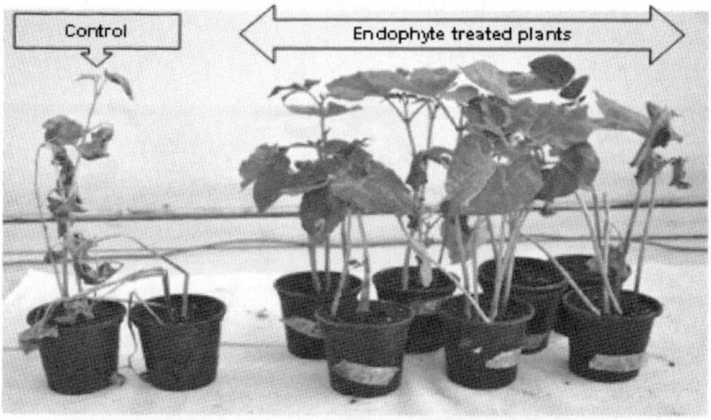

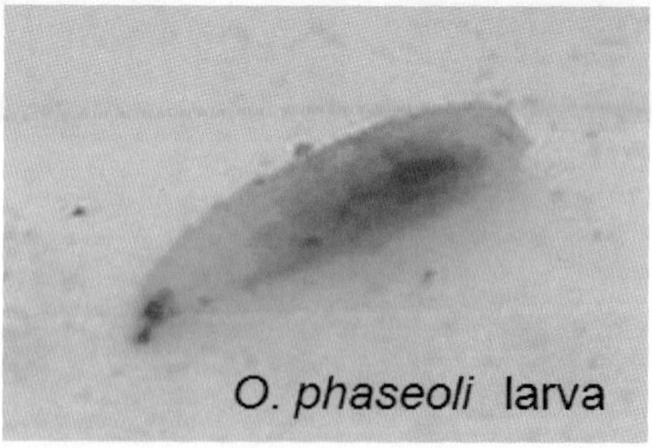

PROJECT VIII

Title: Isolation and characterization of cellulolytic bacteria from soils excavated from geothermal wells along the Kenyan Rift Valley.

Scholar: **Ms. Irene Kiio**
(University of Nairobi)
Supervisor: Dr. Edward Muge
(University of Nairobi)

Planetary Emergency: Environmental Pollution

Impact:
Presence of life in the geothermal wells along the Kenyan Rift Valley was determined by isolation and identification of four bacteria isolates.

Optimization of enzyme activity conditions for the selected isolate gave an optimum temperature of 60°C with no significant difference in activity at 70°C, an optimum pH of 5 and retained 92.2% of its activity after 8 hours of incubation. This isolate was therefore considered to be thermostable and has potential to be used in industrial processes for breaking down celluloses that require high temperatures such as in biorefinery.

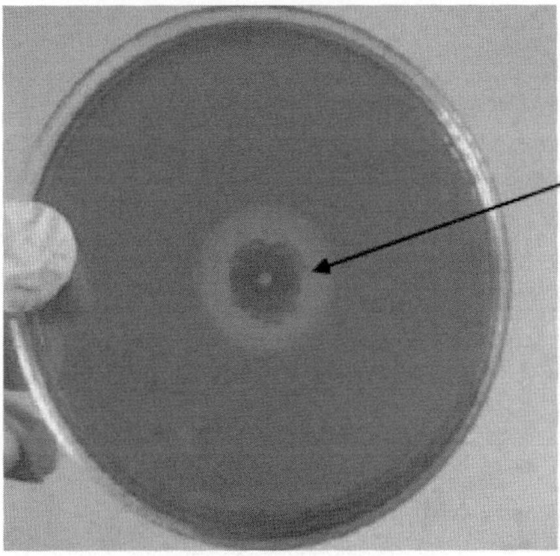

Halo on CMC agar plate stained with Congo red showing cellulolytic activity of an isolate.

PROJECT IX

Title: Adsorption of Methylene Blue and Congo Red dyes from aqueous solutions using banana peelings powder and ash.

Scholar: **Ronald Kemboi**
(University of Nairobi)
Supervisor: John M. Onyari
(University of Nairobi)

Planetary Emergency: Environmental Pollution

Impact:
Low-cost dye removal agents for cleaner water systems.

- Banana peelings are an agricultural waste that is abundantly available and requires minimal processing.
- Banana peelings were effective in the removal of Congo Red and Methylene Blue dyes from water and wastewater systems.
- Its use will result in low cost dye removal from waste water systems and recycling of banana peels leading to a clean environment.

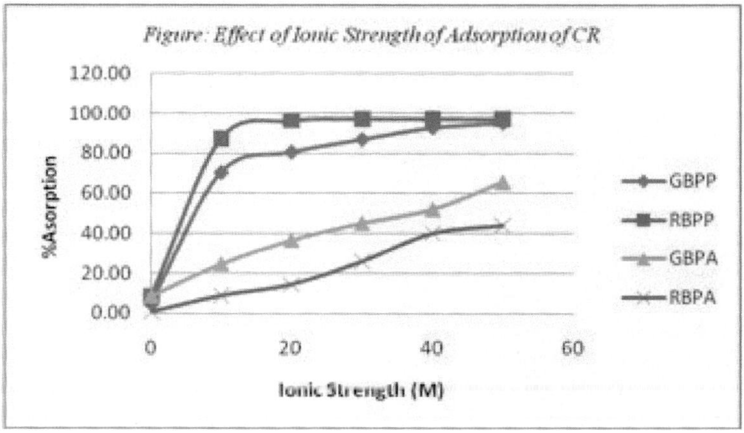

PROJECT X

Title: Heavy metals accumulation in sugarcane: Case study of Nairobi and Nyahururu farms.

Scholar: **Solomon N. Kamau**
(University of Nairobi)
Supervisor: Prof. Lydia W. Njenga
(University of Nairobi)

Planetary Emergency: Environmental Pollution

Impact:
Since some sugarcane samples from the three rivers in Nairobi were found to have high amounts of heavy metals, consumption of sugarcane juice from these areas may pose increased health risks to the local population in the long term.

PROJECT XI

Title: Validation of surface glycoproteins for potential application in diagnosis of *Trypanosoma brucei rhodesiense.*

Scholar: **Simon Ngao Mule**
(JKUAT)
Supervisor: Vincent A. Owino, PhD
(Egerton University)

Planetary Emergency: Infectious Disease

Impact:
Drugs used for treatment suffer from resistance and those used in the treatment of late stage HAT are toxic, causing post treatment encephalitis (PTE) and death. This necessitates early diagnosis, stage determination and treatment, which are presently limited by lack of reliable diagnostic tools and, effective and safe drugs.

This project aimed to validate the application of variable surface glycoproteins (VSGs) in diagnosis. These VSGs had been previously shown to be expressed early on infection hence could used in development of a simple, cheap and sensitive point-of-care diagnostic tool for early detection in resource poor areas in sub-Sahara Africa where the disease is prevalent.

This will enable timely treatment to prevent risks associated with late stage disease.

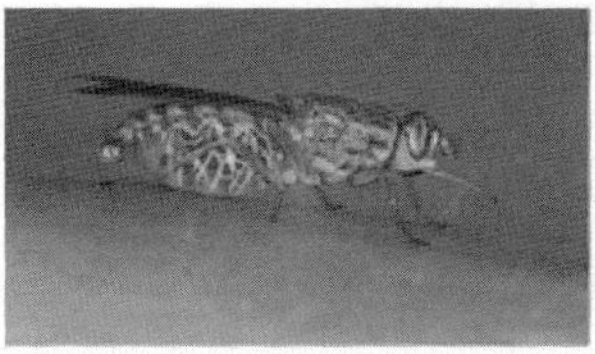

Tsetse fly (Genus: *Glossina*), the sole vector for Trypanosmes in sub-Saharan Africa

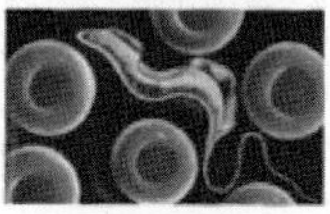

Ttrypanosome (Genus: *Trypanosoma*), unicellular parasites that cause Human African trypanosomiasis (HAT or sleeping sickness) and African Animal Trypanosomiasis (AAT or nagana)

PROJECT XII

Title: Characterization of the selective White-rot fungus Kraft pulp-modifying enzymes to improve selected paper properties and degradation of recalcitrant organic dyes.

Scholar: **Wycliffe Chisutia Wanyonyi**
(University of Nairobi)
Supervisor: Prof. Paul M .Shiundu
(Technical University of Kenya)

Planetary Emergency: Environmental Pollution

Impact:
The natural capabilities of microorganisms and their enzymes in bio-bleaching of Kraft pulp in paper manufacturing and leather processing can reduce product costs, raise product quality and eliminate toxic chemicals used in paper making, thereby lessening its environmental impact (an industry which has traditionally been deemed highly polluting). The project also seeks to apply enzymes in bioremediation of organic dyes-polluted wastewaters. The study will contribute to Kenya's vision 2030 and UN millennium Development Goals (MDGs) on sustainable development and pollution control.

BEYOND 2012-13

"…if we truly seek a peaceful and just world, we must systematically enhance women's access to—and participation in—decision-making processes."

Liberian President Ellen Johnson Sirleaf, October 18, 2006.
http://www.time.com/time/magazine/article/0,9171,1187159,00.html

- Support for women in science
- Expand geographical scope beyond Nairobi
- More planetary emergencies
- More scholarships $\geq$10

ACKNOWLEDGEMENTS

Professor Zichichi and the World Federation of Scientists.

WORLD FEDERATION OF SCIENTISTS NATIONAL SCHOLARSHIP PROGRAMME IN THE REPUBLIC OF SLOVENIA: AUGUST 2014-AUGUST 2015

EDVARD KOBAL
Slovenian Scientific Foundation
Ljubljana, Slovenia

CONTENTS

- Review of achievements and findings by World Federation of Scientists scholarship recipients from Slovenia
- Support activities for better understanding of content related to science that is the focus of the World Federation of Scientists

REVIEW OF ACHIEVEMENTS AND FINDINGS BY WORLD FEDERATION OF SCIENTIST SCHOLARSHIP RECIPIENTS FROM SLOVENIA

Name and Surname	**Status, Employer**	**WFS Planetary Emergency**	**Title of the Research Project**
Maja Šikić-Pogačar (born 1982)	Young researcher, Medical Faculty, University of Maribor	Medicine and Biotechnology and Health	Campylobacter jejuni: resistance and virulence
Maja Šikić-Pogačar (born 1982)	Young researcher, Medical Faculty, University of Maribor	Medicine and Biotechnology and Health	Campylobacter jejuni: resistance and virulence
Maša Čater (born 1986)	Young Researcher, Biotechnical Faculty, University of Ljubljana	Medicine and Biotechnology	Optimisation of biogas production from lignocellulosic substrate
Lea Lukšič (born 1988)	PhD Student	Food	Potential of Tartary buckwheat and barley for food security in climate change conditions
Maja Grundner (born 1983)	PhD Student	Science and Technology	The effect of lysophospolipids on the ostrenolysin A on CHO cells
Maja Gutman (born 1984)	Researcher, Institute for the Protection of Cultural Heritage of Slovenia	Science and Technology	Deterioration of Roman Wall paintings

MAJA ŠIKIĆ-POGAČAR, UNIVERSITY OF MARIBOR
***Campylobacter jejuni:* resistance and virulence.**

Campylobacter spp. Is the most common gastrointestinal bacterial pathogen reported in EU (EFSA, 2012). It is extremely susceptible to a wide variety of food-processing methods and environmental stresses.

- In her study Maja Šikić-Pogačar had presented an alternative strategy for the safe control of *Campylobacter* contamination by targeting their adhesion properties.
- Bioactive plant extracts and their waste material and agricultural by-products can be used as promising novel therapeutic agents, with possible medical and industrial applications.

MAŠA ČATER, UNIVERSITY OF LJUBLJANA
Optimisation of biogas production from lignocellulosic substrate with granular sludge from the UASB reactor.

Maša Čater analysed the ability of four anaerobic hydrolytic bacteria to enhance hydrolysis of brewery spent grain as a model lignocellulosic substrate for a more efficient biogas production:

- The highest net methane production was developed.
- Final methane production and production rates with standard errors for experimental mixture were calculated.

The results of achieved biogas production elevation by adding hydrolytic bacteria or enzyme mixtures were presented on scientific meetings in Sweden and Slovenia. Findings and research were submitted to Springer journal *Applied Biochemistry and Biotechnology* as an original research article.

LEA LUKŠIČ, PHD STUDENT
Potential of Tartary buckwheat and barley for food security in climate change conditions.

Tartary buckwheat (TB) and common buckwheat (CB) are used in different parts of the world to make various food products. The grain of both cultivated buckwheat species contains up to 27% fiber (Bonafaccia et al. 2003). Buckwheat groats were found to be a prebiotic food because it can, for example, affect the increase of lactic acid bacteria in the intestine due to the presence of resistant starch. Buckwheat is an important source of antioxidant activity in functional foods.

Sour bread based on TB and CB sour dough has been prepared. Volume, color, antioxidant activity and concentration of flavonoids, rutin, and quercetin for each bread have been analysed.

MAJA GRUNDNER, PHD STUDENT
The effect of lysophospolipids on the ostrenolysin A on CHO cells

- The binding of OlyA-mCherry to the LUVs doesn't depend on the cholesterol availability but more probably on the distribution of raft domains.
- Lysophosphatidylcholine affects the availability of cholesterol only in LUVs and not also in lipid droplets.
- The binding of OlyA-mCherry to the plasma membrane of CHO cells is affected by the disruption and rearrangement of cytoskeleton elements.
- Lysophosphatidylcholine changes the amount of cytoskeleton elements and caveolae and therefore affects to protein binding.

Hopefully, all these results will give us a better understanding of the binding properties of this new marker for raft domains.

MAJA GUTMAN, INSTITUTE FOR THE PROTECTION OF CULTURAL HERITAGE OF SLOVENIA
Deterioration of Roman wall paintings

- During the period of the financial support of the WFS (World Laboratory) samples of Roman wall paintings were collected from five different archaeological sites (Roman town Emona (present Ljubljana), Roman town Celeia (present Celje), Mošnje, San Simone in Izola, Školarice).
- Paint layers and pigment of the Roman wall paintings, and mineralogical-petrographic, mechanical and micro structural characteristics of mortar layers were analysed using different methods and techniques.
- Results were presented at scientific conferences in Italy and Azerbaijan, in professional articles and in scientific journals.

SUPPORT ACTIVITIES FOR BETTER UNDERSTANDING OF CONTENT RELATED TO SCIENCE THAT IS THE FOCUS OF THE WORLD FEDERATION OF SCIENTISTS

19 September 2014: **Discussion on current scientific developments in Slovenia.**

24-25 September, and 8 October 2014: **Workshops on sustainable development.**

7 October 2014: **Establishment of a living memorial to the victims of the atomic bomb—Hiroshima 1945.**

8 October 2014: **Round table "Appeal of modern science".**

During the 2014 business year, especially in September and October, the Slovenian Science Foundation integrated individual planetary emergencies, as defined by the World Federation of Scientists under the charismatic leadership of Professor Antonino Zichichi, into multiple public events. The Foundation highlighted as the most important: the sustainable development, personal responsibility of researchers, contribution to preserving peace and promoting development, attitudes and relationships between men and women in science, and appeal of modern science for the youth.

19 September 2014: **Discussion on current scientific developments in Slovenia.**

The Slovenian Science Foundation co-organised the discussion "Perspectives of responsibility, ethics, and genders: New relationships in science". It was organised at the initiative of Slovenian women researchers, whereas the support for organising the public event was given by the Foundation, as well as the Slovenian Academy of Sciences and Arts.

Co-organisers: Women in Science Commission at the Ministry of Higher Education, Science and Sport, and the Slovenian Academy of Sciences and Arts.

24- 25 September 2014 and 8 October 2014: **Workshops on sustainable development.**

The Slovenian Science Foundation prepared three workshops for the curious and creative youth. The workshops included a social and natural science parts.

The project was financed by Zavarovalnica Triglav from Ljubljana.

7 October 2014: **Establishment of a living memorial to the victims of the atomic bomb—Hiroshima 1945.**

On the initiative of a Japanese scientist Dr. Shin-ichi Uye from the Hiroshima University and a Slovenian scientist Dr. Alenka Malej from the Marine Biology Station of the National Institute of Biology, the Slovenian Science Foundation coordinated the Slovenian-Japanese project of establishing a live memorial for the victims of Hiroshima. Specifically, as a memorial in the form of trees growing from the seeds of trees that survived the nuclear explosion over Hiroshima on 6 August 1945. This is a memorial to Japan, in Slovenia.

The project was financed by the Slovene National Commission for UNESCO.

8 October 2014: **Round table: "Appeal of modern science"**

Speakers from Slovenian universities and private research institute exchange views, evaluations, and opinions on the differences between public and private understanding of Science.

The project was financed by the L'Oreal Slovenija and the Slovene National Commission for UNESCO.

ACKNOWLEDGMENT

The Slovenian Science Foundation would like to sincerely thank the National Scholarship Programme in Slovenia and the World Federation of Scientists, as well as its President, Professor Antonino Zichichi personally, for the financial support provided.

NATIONAL SCHOLARSHIP PROGRAMME IN SENEGAL, YEAR 2015

MBARECK DIOP
National Coordinator of WFS-WL Activities in Senegal
Dakar, Senegal

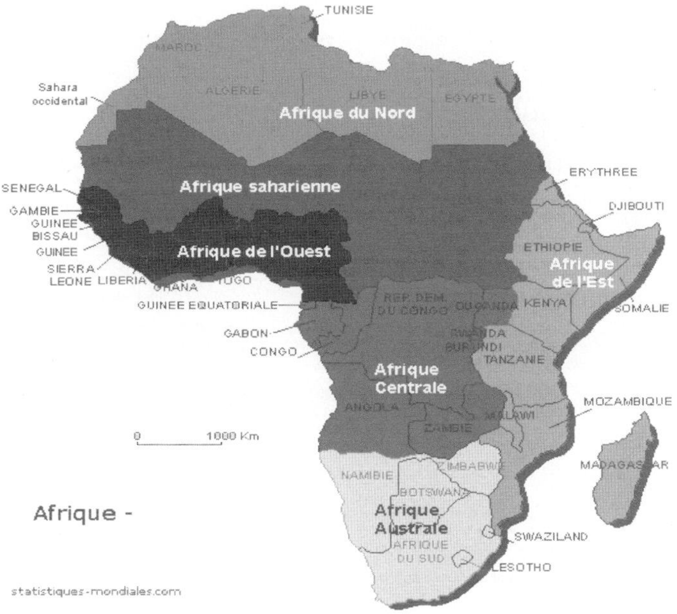

THE CONTEXT

- As a developing country, Senegal faces many challenges: Energy, Environment, Food security, Health, Water, Pollution, Desertification, Waste Management, etc.
- Scientific research is needed to overcome these challenges. Many of them are related to Planetary Emergencies as defined by the WFS.
- The NSP is a good opportunity for young Senegalese Scientists to help them to conduct their research in connection with the priorities of the country under the supervision of Senegalese Professors in their own country, avoiding Brain Drain.
- The NSP started in 2005 in Senegal.
- A total of 101 students awarded from 9 Laboratories in the fields of ENERGY, CLIMATE and FOOD SECURITY as Planetary Emergencies.
- Ministers of Scientific Research and Foreign Affairs from new Senegalese Government sent letters of support.

LETTER OF SUPPORT FROM THE MINISTER OF SCIENTIFIC RESEARCH

Le Ministre
Porte-parole du Gouvernement

Mon Colonel,

Vous avez bien voulu, en votre nom et au nom du Professeur Antonino ZICHICHI, Président de la Fédération mondiale des Scientifiques (FMS), me présenter vos félicitations suite à ma nomination au poste de Ministre de l'Enseignement supérieur et de la Recherche ; je voudrais vous en remercier tous les deux.

Je vous remercie également pour le rapport d'activités 2011 du Programme de bourses nationales de la FMS que vous m'avez transmis et qui a retenu toute mon attention ; il reste entendu que j'aurai sans doute l'occasion de mieux appréhender ce programme qui me semble ambitieux et mérite d'être soutenu.

Dans cette attente, je vous encourage à poursuivre vos efforts pour un bon déroulement de ces activités de recherche et vous invite à me faire tenir toute information qui pourrait sembler utile.

Je vous prie de croire, Mon Colonel, à l'assurance de ma considération distinguée.

République du Sénégal — Le Ministre — Ministère de l'Enseignement Supérieur et de la Recherche

Serigne Mbaye THIAM

Au
Colonel (er) Mbareck DIOP
Ingénieur des Ponts et Chaussées
Coordonnateur des activités de la FMS au Sénégal
Sotrac Mermoz, ancienne piste, N° 25
Dakar

Translation

June, 6th 2012

Ministry of High Education and Scientific Research
The Minister Spokesman of the Government

Dear Colonel,

Thank you for the letter of congratulation sent on behalf of WFS and his President, Pr Antonino ZICHICHI.

Thank you also for sending me the 2011 Annual Report on NSP of WFS, an ambitious program I intend to support.

I encourage you to continue your efforts and please give me all useful information.

Sincerely Yours,
Serigne Mbaye THIAM

To Col. (Ret.)
Mbareck Diop
Coordinator of WFS
Activities in Senegal

STUDENT: AISSATOU FAYE
PhD student in Atmospheric & Oceanic Physics, LPAO-SF/ESP
aichafaye8@gmail.com

Topic:

Cyclogenesis in the Senegalese coast: Analysis of the born conditions of the Tropical Storm Josephine (TSJ) and simulation in WRF regional model (CLIMATE)

Global objective:

Evaluate the capacity of the WRF model to reproduce the sequences of the evolution of a disturbance in the wide of African coast.

Data and Methodology:

- The National Hurricane Center's (NHC): Tropical Cyclone Reports contain comprehensive information on each storm, including synoptic scale, meteorological statistics, casualties and damages, and the post-analysis best track (six-hourly positions and intensities).
- Global Brightness Temperature imagery from the Cloud Archive User Service (CLAUS) project. This project produced a long time-series of global thermal infra-red imagery of the Earth using data from operational meteorological satellites, which was used in validating WRF model.
- ERA-Interim is the latest global atmospheric reanalysis produced by the ECMWF from 1979 to present. Gridded data products include a large variety of 3-hourly surface parameters, describing weather as well as ocean-wave and land-surface conditions, and 6-hourly upper-air parameters covering the troposphere and stratosphere.

Results:

- The satellite observations CLAUS which allowed us to follow the convective system associated with the disturbance was rather low at first but fast became intensified in agreement with the report of NHC.
- ERA-Interim reanalysis data were allowed to see the environmental conditions that are very favorable to the evolution of the cyclonic system and in agreement with the report of NHC.

Perspective:

After these results it would be interesting to testing the robustness of results in the wide of African coasts by producing a sensitivity study with ensemble simulations of the WRF model.

Acknowledgment:

World Federation of Scientists (WFS)

STUDENT: NDÉYE MATY NDIAYE
PhD student in Condensed Matter Physics, UCAD/FST-GPSSM
nmaty.ndiaye@gmail.com

Topic:

VO_2 base smart hybrid thermochromic nano-coatings for solar heat management for energy-efficient windows. (ENERGY)

Global objective:

The aim of this project is to study the thermochromic properties of VO_2 doped for application in smart windows for energy efficiency

Specific Objectives and Methodology:

(a) Objective One:
Theoretical predictions of entirely novel doped VO_2 based materials with tunable transition temperature via computational combinatorial approach.

(b) Objective Two:
Engineering and design of novel doped-VO_2 based multilayered nano-coatings with a "room temperature" TC and large infrared modulation in the NIR and IR spectral range with a visible transmission above 65%

(c) Objective Three:
Target a joint IP and seek furthering this project within a bilateral partnership involving the World Federation of Scientists and the University Cheikh Anta Diop of Dakar. As well as writing publications.

To achieve these scientific objectives of the project, the synthesis of novel doped-VO_2 based multilayered nano-coatings with a "room temperature" TC and large infrared modulation in the NIR and IR spectral range with a visible transmission above 65%, on glass substrate will be done via PLD and the study of their nonlinear optical properties (NLO).

Results:

- The XRD patterns of the films obtained show peaks due to the thin layer at angles of 18.48°, 27.97°, 37.18°, 39.98°, 55.72°, 57.048° and 79.89° and a broad peak due to the glass substrate at an angle around 20 °. Following the calculated pattern description of the monoclinic structure of VO_2 (JCPDS No: 43-1051), these peaks can be indexed as the reflections on (1 0 0), (0 1 1), (2 0 0), (0 2 0), (2 1 1), (0 2 2) and (4 0 0) planes respectively, thus showing the highly oriented growth of these films.
- The VO_2 nanostructures exhibit sharp a-axis diffraction peaks, characteristic of the VO_2 monoclinic phase for the films prepared from 5 a 12 mTorr, which implies that pure highly a-axis textured VO_2 was formed. For the sample prepared under 3 mTorr the films are amorphous while the sample prepared at

an oxygen pressure of 15 mTorr is crystalline. It is interesting to notice the disappearance of the (1 0 0) plan and a change in the texturation of the film, which become highly oriented along (0 1 1).

- The SEM results of the sample prepared at 15 mTorr show net change on the morphology. This could be a direct correlation with the disappearance of the (1 0 0) at the very same pressure. A detailed description of the growth mechanisms and the substrate–film interaction is given.

Perspectives:

The combination of the a-axis preferential orientation with the doping of the films shall induced news properties of the VO2 the future work will concentration on the doping and the study of the effect of deposition substrate temperature on the optical properties of the films as well as the on its transition temperature.

Acknowledgement:

Ndeye Maty Ndiaye is very grateful for financial support from WSF.

STUDENT: YOUNOUSS DIEYE
PhD student in Energy System and Environment, UCAD/ESP
youdi85@yahoo.fr

Topic:

Thermophysical characterization of construction materials based typha (ENERGY)

Global objective:

Reduction of Greenhouse Gases (GHG) emissions in the building sector by the diminution of energy consumption.

Methodology of semester activities:

- Preparation of sample: Earth-typha sample are manufactured from fibers of cattail leaves and clay binder. The binder, which is a mixture of clay and water, is kneaded until a viscous liquid adhesive.
- Characterization method: For the mechanical tests, compressive and tensile test are made on dimensions 4x4x16cm3. The thermal properties such as thermal conductivity and thermal effusivity are measured by the hot plane method.

Results:

- The tensile and compressive strength of the samples increases with increasing binder content. The tensile strength ranges from 0,34MPa at 0,976MPa and the compressive strength of 0,265MPa at 0,969MPa.
- For the thermal properties, the thermal conductivity varies of 0,123W/m.K at 0,165W/m.K and the thermal effusivity of 231J.m-2.°C-1.s-1/2 at 315J.m-2.°C-1.s-1/2.

Perspective:

- Study the influence of others binders such as cement or lime on cattail fibers.
- Study the thermomechanical performances of materials 100% vegetal in typha-vegetal binder.
- Make studies humidity samples typha.

Acknowledgment:

World Federation of Scientists (WFS)

STUDENT: MOUSTAPHA DIOP
PhD student in Energetics Systems and Environment, UCAD / ESP/LER
moustaphadiop1987@gmail.com

Topic:

Design and command of static converters for mills with three-phase asynchronous machines (ENERGY)

Global objective:

Design and implement a static interface between a photovoltaic single-phase mini-network and a mill with three-phase asynchronous motor

Objective of semester activities:

- Study and characterization of the mill
- Study of the device for the interfacing of the three-phase mill and the mini-network

Methodology of semester activities:

- The characterization consists in determining characteristic mechanics of the mill. So the experimental determination of the curves of variation of the speed according to the flow and to the torque according to the speed.
- For the interfacing, a study of the various electronic devices which can serve as interface is made with their advantages and inconveniences. The choice of the device is validated by simulation with the characteristics of the mill

Results:

- Determination of the mechanical characteristics to assure an optimal functioning of the system of grinding and endowed these mills local of data sheets indicating their mechanical and electric characteristics.
- Results satisfying on the implementation and the simulation of the device. She allowed adapting the single-phase source and the three-phase mill, to manage the shape of wave of the absorbed current, the current to the starting up, to pilot the speed and to improve the quality according to the product of grinding.

Perspective:

- Study completes of the static interface with its command.
- Simulation of the mill with other strategies of command more sophisticated.

Acknowledgment:

World Federation of Scientists (WFS)

STUDENT: FATOUMA FALL
Supervisors: Pr Amadou Ba and Pr Diégane Diouf, Laboratoire Commun de Microbiologie (LCM) IRD/ISRA/UCAD (Dakar)

Topic:

Role of *Sporobolus robustus* Kunth in the strategy of regeneration of *Acacia seyal* Del. et *Prosopis juliflora* (Swartz) DC) on salt affected soils, Senegal (FOOD SECURITY).

Main objective:

The aim of the present study is to determine the effect of *Sporobolus* on germination and growth of two moderately salt tolerant legumes trees (*A. seyal* and *P. juliflora*) in saline soils

Specific objectives and methodology:

1. Diversity of arbuscular mycorrhizal fungi (AMF) associated with *Sporobolus* and two legumes trees: Evaluate spores density (Gerdemann and Nicolson, 1963), AMF roots colonisation (Phillips and Hayman 1970; Trouvelot et al. 1986), mycorrhizal potential of saline soils (Sieverding 1991) and pyrosequencing.
2. Diversity of rhizobia associated with *P. juliflora* and *A. seyal:* Rhizobia potential of saline soils, genetic identification of Rhizobia (genotyping and sequencing).
3. Impact of *S. robustus* on the germination and growth of *A. seyal* and *P. juliflora* in cotrolled conditions: Germination and growth test of *A. seyal* and *P. juliflora* seeds on saline soils with *Sporobolus* and without *Sporobolus.*
4.

Results:

- Salinity showed significant inhibitor effect on mycorrhizal and rhizobia potential (MPN), and the spore numbers of *Sporobolus* and legumes trees rhizospheres soils along the salinity gradient.
- Germination rates of *A. seyal* and *P. juliflora,* were significantly greater on soil with *S. robustus.*
- The presence of *S. robustus* improves growth of *A. seyal* and *P. juliflora* on salt affected soils.
-

Conclusions and Perspectives:

Sporobolus could promote regeneration of *P. juliflora* and *A. seyal* seeds by reducing salt content of soils in saline conditions and roots colonization by arbuscular mycorrhizal (AM) fungi that improve plant nutrition.

- Genetic identification of AMF associated to *Sporobolus* by genotyping and sequencing.
- Genetic identification of Rhizobia associated to *P. juliflora* and *A. seyal* by genotyping and sequencing.
-

Acknowledgement:

World Federation of Scientists (WFS).

STUDENT: MARIAMA NGOM
PhD student in Plant and Germ Biotechnologies, UCAD/FST-BV

Topic:

Symbiotic characterization of *Frankia* strains for greater efficiency in reforestation programs in Sahelian zones (FOOD SECURITY)

Global objective:

Investigate mechanisms involved in salt stress tolerance of *Frankia* strains and to identify suitable symbiotic partners (*Casuarina* – *Frankia* strain) that will be proposed in reforestation programs.

Methodology of semester activities:

- Impact of salt stress on *C. glauca* plants nodulation: Several concentrations [0, 50, 100, 200, 300, 400 & 500 mM] of NaCl were first applied gradually then plants were inoculated separately with *Frankia* strains *CcI3* (salt sensitive) and *CeD* (salt tolerant).
- Effect of inoculation with selected *Frankia* strains on *Casuarina* plants salt tolerance: *Cauarina glauca* plants were inoculated separately with two *Frankia* strains (*CcI3/CeD*) then the salt stress was gradually applied. Physiological and morphological parameters like root and shoot dry weight were assessed.

Results:

- Salt stress inhibited nodulation of *C. glauca* plants by these two contrasting *Frankia* strains. No nodule was observed beyond 100 mM and 200 mM for *Frankia* strains *CcI3* and *CeD*, respectively.
- However, prior inoculation with both *Frankia* strains improved significantly *C. glauca* plants salt tolerance. Physiological parameters like proline and chlorophyll contents were significantly higher in plants inoculated by these strains compared to un-inoculated plants. They showed also better height growth, shoot and root dry weight.

Perspective:

- Effect of inoculation with *CcI3/CeD* strains on *C. equisetifolia* plants salt tolerance will be completed.
- Impact of salt stress on *C. glauca* plants root hairs deformation is ongoing in hydroponic conditions to see if this inhibition of nodulation occurs during the early stages of the establishment of the symbiosis.

Acknowledgment:

World Federation of Scientists (WFS)

STUDENT: MOHAMED IZZEDINE SERGE ADJIBADE
PhD student at Applied Energy Laboratory /ESP-UCAD
amihad007@yahoo.fr

Topic:

Energetic and exergetic analysis of cogeneration system to product power and cold with 10 kW internal combustion engine run by biogas in sub Saharan area for the conservation of dairy product (ENERGY)

Global objective:

This work aims to make energy and exergy of internal combustion engine, absorption chiller and global modeling of the coupled system to find out the coupling capabilities of both systems for cold production for the conservation of dairy products in sub Saharan area.

Methodology of semester activities:

- Development of mathematical models of the various components of the absorption machine: it is to develop the different operating state equation of the system, mass balance, energetic balance, exergetic balance and exergetic efficiency
- Simulation of the absorption refrigeration machine with Matlab and exergetic optimization: this is to determine the different temperatures of the system for optimal exergetic efficiency

Results

As shown in the figure below, at temperature of heat source, the increase of the condensation temperature decreases the exergetic efficiency. Indeed, the increase of the condensing temperature leads to the increase of the temperature difference between the coolant and the refrigerant in the condenser. The temperature of the heat source and the exergetic efficiency increased together until a maximum temperature. Indeed, this temperature elevation will cause a water condensation problem condenser which will reduce the cooling output.

Definitively, too high condensation temperature (>37°C) and heat source temperature (>140°C) causes a reduction of the exergetic efficiency.

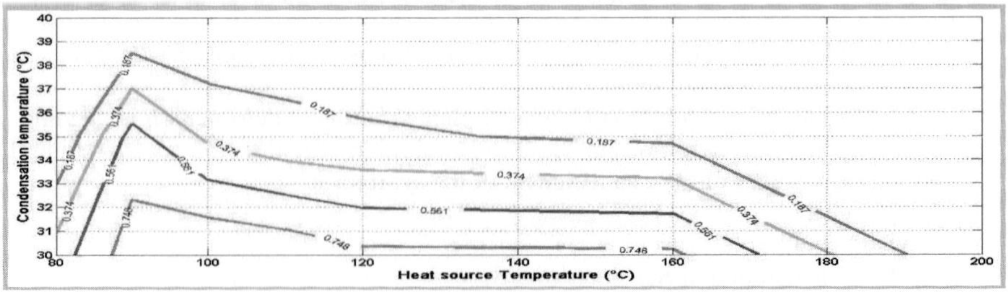

Perspectives:

- Internal combustion engine modeling
- Simulation with Matlab of internal combustion engine
- Coupling the internal combustion engine and absorption chiller for exergetic optimization

Acknowledgment:
World Federation of Scientists (WFS)

STUDENT: AHMADOU TIDIANE DIABY
PhD student in Plant and Applied Energy Laboratory, Polytechnic Higher School, University Cheikh Anta Diop. tigediaby@gmail.com

Topic:

Experimental and simulated study of a heat pump for the production of cold and desalination. (ENERGY)

Global objective:

The objective of this thesis is to use a heat pump for simultaneous production of cold and desalination.

Methodology of semester activities:

To attain objective of this thesis a number of work must be conducted:

- Bibliographic study on heat pump for simultaneous heating and cooling, desalination systems, transfers in porous media, membranes.
- Desalination unit simulation coupled with a heat pump for simultaneous heating and cooling and solar panels, parametric study, performance evaluation

Results:

- The choice of refrigerant is based on technical and economic considerations.
- The plate heat exchangers seem to be well indicated for heat pumps using hydrocarbons.
- Desalination systems using heat pumps produce much more fresh water than without heat pump systems.

Perspectives:

- Studies of the performance of natural refrigerants
- A comparative study of the performance of natural refrigerants with R1234yf
- The realization of a prototype of a heat pump connected to a refrigerating cupboard and a desalination unit

Acknowledgment:

World Federation of Scientists (WFS)

STUDENT: DANIEL GOUBA
Research Master in Renewable Energy and Electric Systems, CIFRES / ESP / UCAD gouba_daniel@yahoo.fr

Topic:

Contribution to the development of an experimental platform to study the performance of photovoltaic solar modules in Sahelian Africa. (ENERGY)

Global Aim:

The current study concerns the performance study of photovoltaic modules in Sahelian Africa. It consists of purposing an experimental platform which can enable the PV modules performance study.

Methodology of semester activities:

- Detailed study of photovoltaics and photovoltaic modules: we identified all technologies of PV modules and their performance parameters.
- Balance sheet of degradation suffered by the operating modules under real conditions: determination of modules degradation factors, forms of degradation and study of their impacts on the performance of the modules.
- Bibliographic review of PV module performance studies: here, we have done the state of art on the module performance analysis techniques.

Results:

- Electrical performance parameters of a PV module: Short-circuit current (Isc), the open circuit voltage (Voc), the form factor (FF), the maximum output power (Pmax) and efficiency (η).
- PV modules degradation factors: temperature, humidity, irradiation, dust, mechanical collisions.
- Degradation forms: discolouration, delamination, corrosion, cells cracking and breaking.
- Technics of modules' performance study: Pan models, Exponential, Zimmerman, accelerated tests, measurements under real conditions.

Perspectives:

- Study of the platforms and different existing configurations.
- Proposition of one platform configuration for module performance study.

Acknowledgment:

World Federation of Scientists (WFS).

STUDENT: OUSMANE WANE
PhD student in thermal, Energy System and Environment
CIFRES/ESP/UCAD

Topic:

Evaluation and Mapping Photovoltaic Solar Potential in Senegal: Application to the production of electricity network connection or remote sites. (ENERGY)

Objective:

The project aims to analyze and evaluate the potential of energetic production of solar origin (photovoltaic principally) generatable in Senegal. One proposes the estimation of the solar radiation incident to regional level (map of solar resource) and the subsequent application to the calculation of the electric potential of production (map of potential).

Methodology of semester activities:

To achieve this overall objective it is proposed to structure the project according to the following steps for the first half year of our doctoral thesis:

- State of the art. Review status associated with previous works that have been among the targets of renewable resource studies at the regional level knowledge. This allows us to better situate the problematic of our research topic.
- Localize the existent radiometric stations in the study area that serve for the control of the uncertainty of radiation map.
- Determine the daily global solar radiation on monthly average estimated from geostationary satellite with a maximal spatial resolution.
- Regionalize the study area using the k-means algorithm (climatic zoning).

Results of semester activities:

The archived results till now within the framework of our research works are:

- Establishment of the bibliographical References and a data base of monthly and annually average daily global solar irradiation on horizontal surface maps for Senegal.
- Regionalization of Senegal using the meteorological parameters (Temperature, precipitation…): climatic zoning.

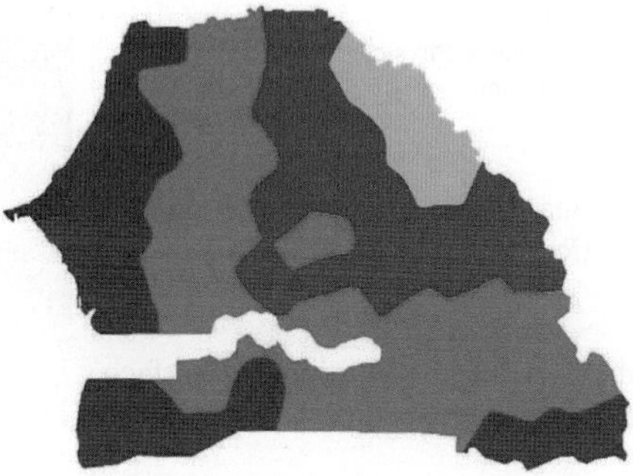

Regionalization of Senegal with 5 clustering

<u>Perspective for the second half-year:</u>

The particular way of thinking for the second half-year is:

- Validate/Criticize the time series data of radiometric stations (ground measurements),
- Map of Solar Radiation for Senegal describing the yearly and monthly average solar radiation that can be exploited for the production of electricity, and for research and educational purposes.

STUDENT: DAHIROU WANE
PhD student in Atmospheric and Oceanic Physics, LPAO-SF/ESP/UCAD mahasimw@gmail.com

Topic:

Influence of the sea surface temperature on the pluviometry in Sahel, particularly in Dakar (CLIMATE)

Global objective:

Estimate the impact of the oceanic conditions of surface on the variability of the Sahelian pluviometry, particularly in Dakar.

Methodology of semester activities:

- Method of the composite analysis: this method is used for verifying the existing relation between the Sea Surface Temperature Anomalies (SSTA) and those precipitations (PRECIPA) in the Sahel West. This method involves extracting a complex environment signals or interactions can be observed if a predefined signal in a reference variable is included in other variables.
- Method of the linear correlations: this statistical method, correlation, allows us to study the intensity of the relationship that can exist between two or more random variables. In this work, the relationship is studies between the SSTA and PRECIPA. This intensity is estimate by calculating the linear coefficient of correlation

Results:

- The Sahelian rainfall depends to the SST anomalies in the Guinea Gulf (GG).
- The warm water, positive SSTA, (cold water, negative SSTA) in GG cause a decrease (increase) of Sahelian rainfall and particularly in Dakar.
- The significance of the tile evolution of the linear correlation shows that the influence of the SSTA in GG on the rainfall is not stationary. The strongest and significant correlations were obtained during the wet period.

Perspectives:

- Extend our field of study throughout Senegal;
- Study the influence of the other oceanic basins on the variability of the rainfall in this Sahelian region; the mechanisms of interaction associated with it.
- Use observation data of rainfall stations, to study the sensitivity of our results, and climate models to more advanced studies in the field of forecasting rainfall.

Acknowledgment:

World Federation of Scientists (WFS)

DISTRIBUTION OF 11 AWARDED STUDENTS IN 2015 AND THE TOTAL OF 90 STUDENTS FROM 2005 TO 2014

Laboratory	Year 2015	Years 2005-2014
Plant Biotech. Lab (Food)	2	15
Animal Biotech. Lab (Food)	0	1
LPA (Climate Change)	1	16
Dept. of Physics (Energy)	1	15
CIFRES (Solar)	2	13
LEA (Applied Energy)	4	23
Materiaux & Hydr. Lab		2
Renewable Energy Lab (Energy)	1	4
Nuclear Tech. Institute (Energy)		1

LETTER OF THANKING FROM THE MINISTER OF SCIENTIFIC RESEARCH OF SENEGAL

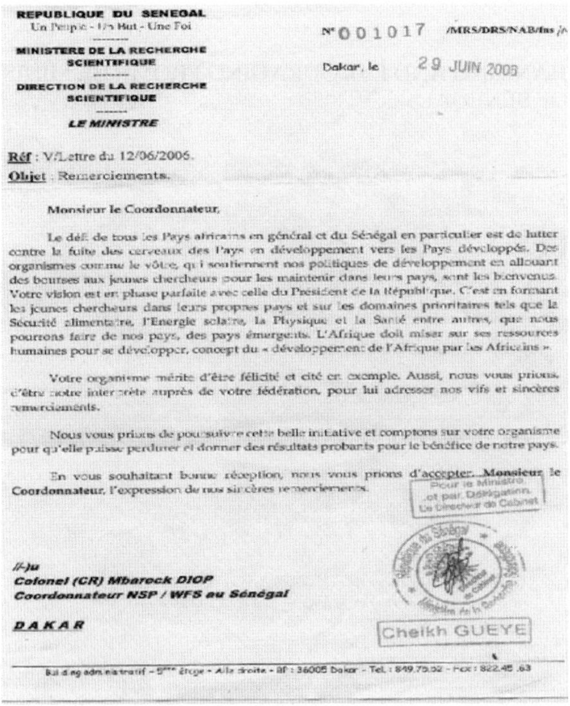

REPUBLIQUE DU SENEGAL
Un Peuple - Un But - Une Foi

MINISTERE DE LA RECHERCHE SCIENTIFIQUE

DIRECTION DE LA RECHERCHE SCIENTIFIQUE

LE MINISTRE

N° 001017 /MRS/DRS/NAB/fns

Dakar, le 29 JUIN 2006

Réf : V/Lettre du 12/06/2006.
Objet : Remerciements.

Monsieur le Coordonnateur,

Le défi de tous les Pays africains en général et du Sénégal en particulier est de lutter contre la fuite des cerveaux des Pays en développement vers les Pays développés. Des organismes comme le vôtre, qui soutiennent nos politiques de développement en allouant des bourses aux jeunes chercheurs pour les maintenir dans leurs pays, sont les bienvenus. Votre vision est en phase parfaite avec celle du Président de la République. C'est en formant les jeunes chercheurs dans leurs propres pays et sur les domaines prioritaires tels que la Sécurité alimentaire, l'Energie solaire, la Physique et la Santé entre autres, que nous pourrons faire de nos pays, des pays émergents. L'Afrique doit miser sur ses ressources humaines pour se développer, concept du « développement de l'Afrique par les Africains ».

Votre organisme mérite d'être félicité et cité en exemple. Aussi, nous vous prions d'être notre interprète auprès de votre fédération, pour lui adresser nos vifs et sincères remerciements.

Nous vous prions de poursuivre cette belle initative et comptons sur votre organisme pour qu'elle puisse perdurer et donner des résultats probants pour le bénéfice de notre pays.

En vous souhaitant bonne réception, nous vous prions d'accepter, **Monsieur le Coordonnateur**, l'expression de nos sincères remerciements.

Pour le Ministre,
et par Délégation,
Le Directeur de Cabinet

Cheikh GUEYE

//-)u
Colonel (CR) Mbarock DIOP
Coordonnateur NSP / WFS au Sénégal

DAKAR

Building administratif - 5ème étage - Aile droite - BP : 36005 Dakar - Tel. : 849.75.02 - Fax : 822.45 .63

Translation

Dear Coordinator,

The challenge of all African countries in general and for Senegal in particular is to struggle against brain drain from developing countries to developed countries. Such Institutions like yours, who support our development policy by giving scholarships to young researchers in order to keep them in their countries, are welcome.

Training young scientists in their own countries in priority sectors like Food Security, Solar Energy, Physics and Health within others, is the best way to get our countries into emerging countries.

Africa must stake on its own human resources to develop itself, the concept of "Development of Africa by African".

Your organization deserves to be congratulated and cited as an example.

Thus, we beg you to speak to your Federation on our behalf, to address to them our warmest and sincere thanks.

We beg you to continue this wonderful initiative and we do rely on your organization to endure and give convincing results for the benefit of our country.

Sincerely Yours.

LETTER OF THANKING AND ENCOURAGING FROM THE MINISTER OF EDUCATION OF SENEGAL

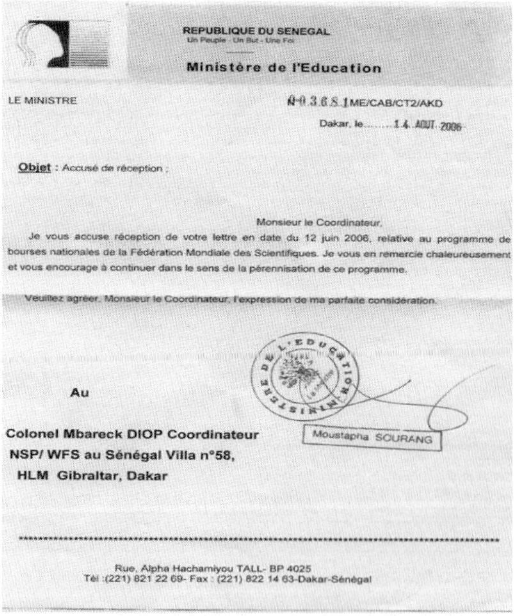
REPUBLIQUE DU SENEGAL
Un Peuple - Un But - Une Foi

Ministère de l'Education

LE MINISTRE

0 0 3 6 8 1 ME/CAB/CT2/AKD

Dakar, le .... 14 AOUT 2006

Objet : Accusé de réception ;

Monsieur le Coordinateur,

Je vous accuse réception de votre lettre en date du 12 juin 2006, relative au programme de bourses nationales de la Fédération Mondiale des Scientifiques. Je vous en remercie chaleureusement et vous encourage à continuer dans le sens de la pérennisation de ce programme.

Veuillez agréer, Monsieur le Coordinateur, l'expression de ma parfaite considération.

MINISTERE DE L'EDUCATION

Moustapha SOURANG

Au

Colonel Mbareck DIOP Coordinateur
NSP/ WFS au Sénégal Villa n°58,
HLM Gibraltar, Dakar

Rue, Alpha Hachamiyou TALL- BP 4025
Tél :(221) 821 22 69- Fax : (221) 822 14 63-Dakar-Sénégal

Translation

Dear Coordinator,

I acknowledge receipt of your letter dated June 12th 2006, concerning the National Scholarship Program of the World Federation of Scientists.

I warmly thank you and I encourage you to continue in the way of perpetuation of this programme.

Sincerely yours.

CONCLUSION

- The NSP is at its first step in Senegal, and it is getting more and more successful: 101 Students have been awarded (20 Females and 81 Males) for 70 Masters and 31 PhD.
- In the Planetary Emergencies related to Food (21 Students), Climate Change (24 Students) and Energy (56 Students).
- It is a remarkable way to fight against brain drain and a significant support of the WL to the scientific Senegalese community in connection with the priorities of development of the country.
- According to the success of this program and the real interest of the highest Senegalese authorities it would be useful to maintain or increase the number of annual Scholarships if possible.
- The driving force of the World Laboratory in Senegal is CIFRES (Center for Training and Research on Solar Energy) created and launched on October 2nd 2002 by WFS Representatives (Dr. Manoli and Pr. Vivargent) and the former President of the Republic of Senegal, Hon. Abdoulaye Wade.
- I take the opportunity to pay a tribute to Pr. Marcel Vivargent and greet his memory by testifying of our recognition for its efforts aiming at the creation of the CIFRES.
- More support is needed for CIFRES which is a Centre of excellence for Solar Energy for Western Africa.
- Thanks to Dr. Manoli and the Staff for their support.
- I would extend the gratitude of all awarded Senegalese students, their Supervisors and the Senegalese Government (represented by both Ministers in charge of Education and Scientific Research) to Professor Antonino Zichichi.

THANKS !
GRAZIE !
DIEUREUDIEUF !

SCHOLARSHIP PROGRAMME OF THE WORLD FEDERATION OF SCIENTISTS IN THE REPUBLIC OF MOLDOVA

GHEORGHE DUCA
The Academy of Sciences of Moldova
Chisinau, Republic of Moldova

It is a privilege for the research community of Moldova to be part of an organization with more than 10,000 (ten thousand) members from 110 (a hundred and ten) member countries. Thank you for the opportunity to present the latest developments of the cooperation between Moldova and the World Federation of Scientist.

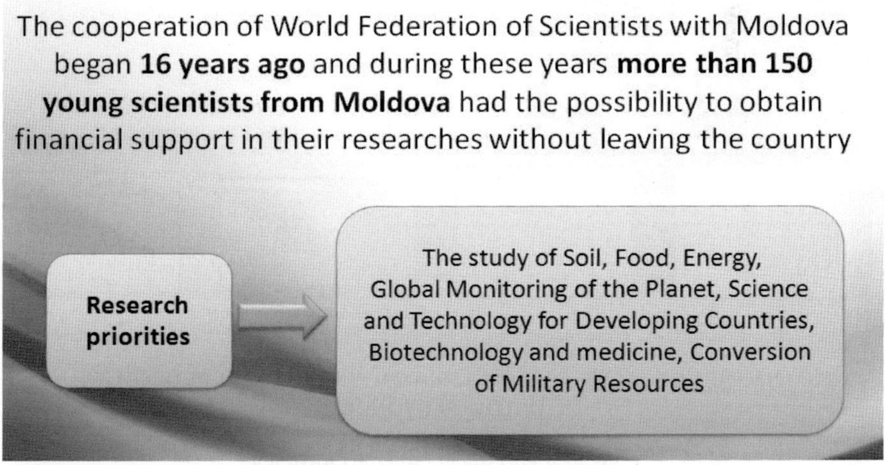

GRATITUDE

I would like to express my gratitude towards Professor Antonino Zichichi. Your efforts have helped consolidate the scientific community of Moldova. Thank you Professor Zichichi!

Next, I would like to present briefly the results obtained under the Federation scholarship programme.

What are the scientific fields where young researchers from Moldova have won awards?

1. *Planetary direction* - ***Soil***

"STUDYING THE PARTICULARITIES OF THE SOIL FOR THE ELABORATION OF SOIL PROTECTION MEASURES"

A scholarship was awarded:
3rd year PhD Student **Olesea COJOCARU**
Institute of Pedology, Agrochemistry and Soil Protection "N. Dimo"
Scientific advisor, Dr. Hab. Prof. **Valerian CERBARI**

Objectives:

- Prevention of soil erosion
- The determination and establishment in the laboratory of soil analysis

Results:

Results of the project were presented and published in more then 3 proceedings

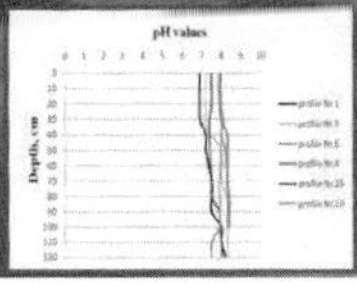

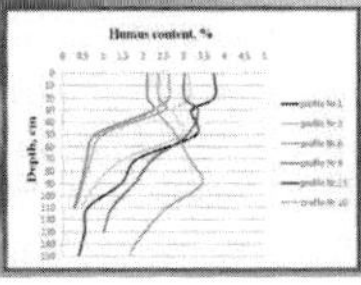

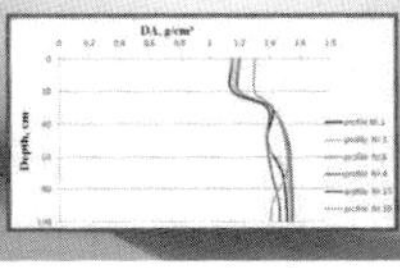
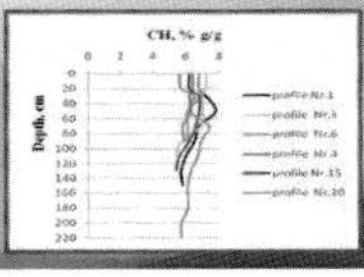

2. *Planetary direction* - ***Food***

"IMPACT OF LACTIC ACID BACTERIA ON THE TEXTURAL PROPERTIES OF LOW FAT YOGURT"

A scholarship was awarded:
3rd year PhD student **Anatoli CARTAȘEV**
Institute of Microbiology and Biotechnology
Scientific advisor, Academician **Valeriu RUDIC**

The samples of yoghurt

Objective:

Production of samples of lyophilized yogurt.

Results:

- Results were presented at 2 international conferences
- **This invention is patented - the method for producing a powder fermented dairy product. MD 865 Y 2015.01.31.**

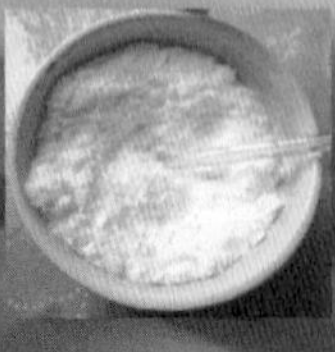

Freeze-dried yogurt

Freeze-dried bacteria concentrate yogurt starter Sy1 and Sy2

EPS-producing strain of *Streptococcus thermophilus*

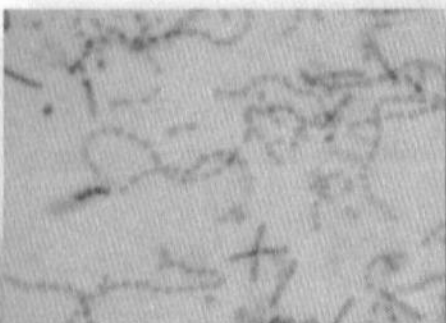

The symbiotic EPS-producing cultures for yoghurt

3. *Planetary direction – **Energy***

"SYNTHESIS AND STUDY OF 8TH GROUP COMPLEX COMPOUNDS WITH C, N, O DONOR CONTAINING LIGANDS"

A scholarship was awarded:

3rd year PhD Student **Olesea Cuzan**

Institute of Chemistry

Scientific supervisor, Dr. Hab., Prof., Academician **Constantin Turta**

X-ray structure of Bis[2-(2H-benzotriazol-2-yl)-4,6-ditert-pentylphenolato] palladium(II) moiety.

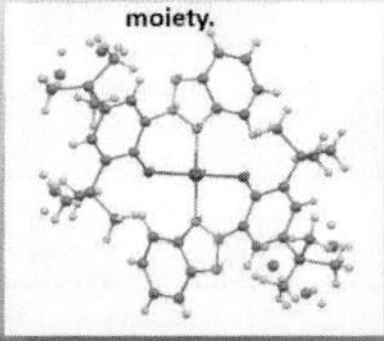

Objectives:

- ways to obtain species that efficiently convert sunlight into the necessary electrochemical potential, and
- producing catalysts in order to facilitate H—H and O—O bond formation.

Results:

- Two new coordinative Pd containing compounds were synthesized as a potential catalyst for H_2 production/oxidation.
- Published in 3 journals

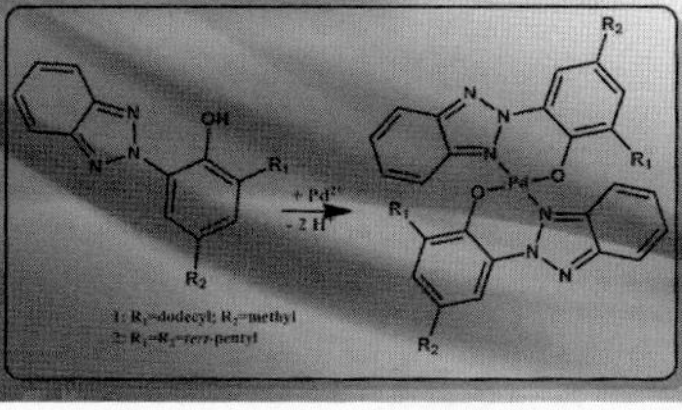

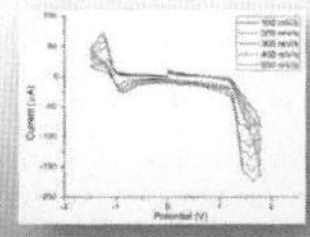

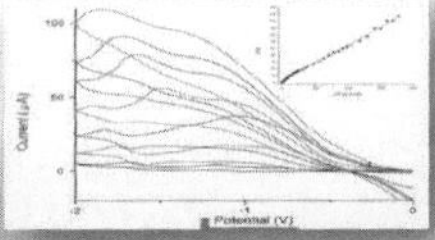

Reaction pathway of Bis[2-(2H-benzotriazol-2-yl)-6-dodecyl-4-methylphenolato]palladium(II) and Bis[2-(2H-benzotriazol-2-yl)-4,6-ditert-pentylphenolato] palladium(II) synthesis.

4. *Planetary direction – **Global Monitoring of the Planet***

"DETERMINATION OF THERMO-TOLERANCE Of Tree OAK (*QUERCUS ROBUR* L.) DESCENDENTS FROM DIFFERENT ECOLOGICAL ZONES"

A scholarship was awarded:

3rd year PhD student **Ion AGAPI**

Institute of Ecology and Geography

Scientific advisor, Assoc. Prof. PhD. Hab. **Petru CUZA**

Pedunculate oak (Quercus robur L) comparative cultures, Scientific Reservation "Plaiul Fagului"

Objective:

Study the comparative plant state depending on the genetic characteristics of the pedunculate oak (Quercus robur L) descendents at the same age, different ecological origin, and growing in the Scientific Reservation "Plaiul Fagului"

Results:

Several publications and presented at 3 conferences

5. Planetary direction – ***Science and Technology for Developing Countries***

"OBTAINING AND CHARACTERIZATION OF ULTRATHIN GAN MEMBRANES FOR SENSOR APPLICATIONS"

A scholarship was awarded:

2nd year PhD student **Fiodor BRANIȘTE**

Technical University of Moldova

Scientific advisor, Academician **Ion TIGHINEANU**

Objectives:

- Obtaining of ultrathin GaN membranes;
- Membranes characterization;
- Integrate ultrathin membranes in a device for UV and gas sensing applications.

Results:

- Continuous and nanoperforated GaN ultrathin membranes were obtained and visualized
- Published in one publication

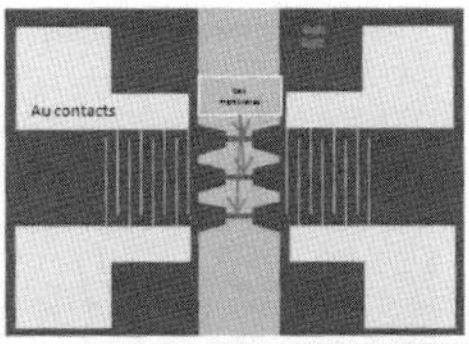

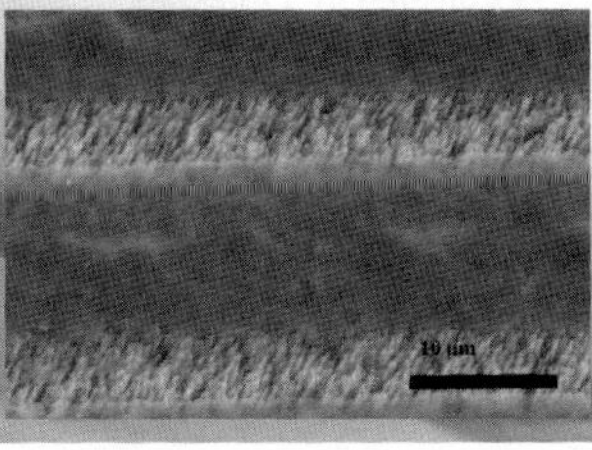

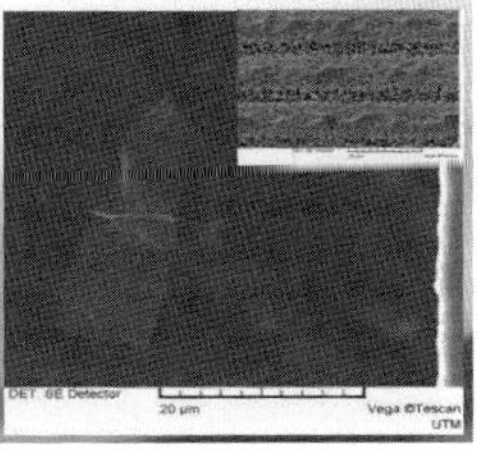

"SCIENTIFIC RESEARCH OF LANGUAGE PARADIGMS IN PHILOSOPHICAL WORK"

A scholarship was awarded:

3rd year PhD student **Liliana RUSU**

Institute of History

Scientific advisor dr. hab. **Gheorghe BOBÎNĂ**

Objective:

Conceptual analysis, based on the idea of rational reconstruction of philosophical work.

Results:

Published 2 research articles and presented at 3 conferences

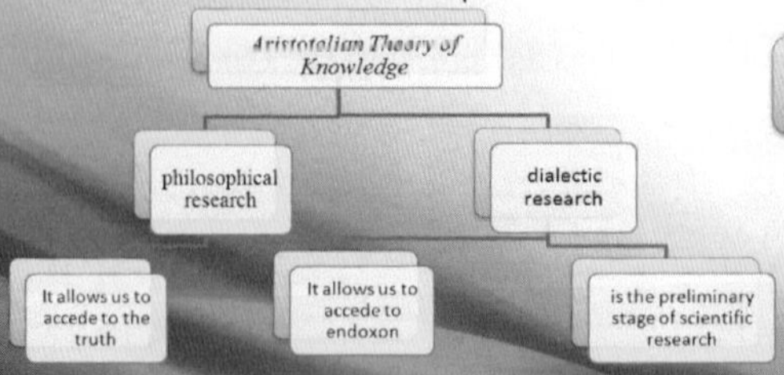

Schematic representation of Aristotelian Theory of Knowledge

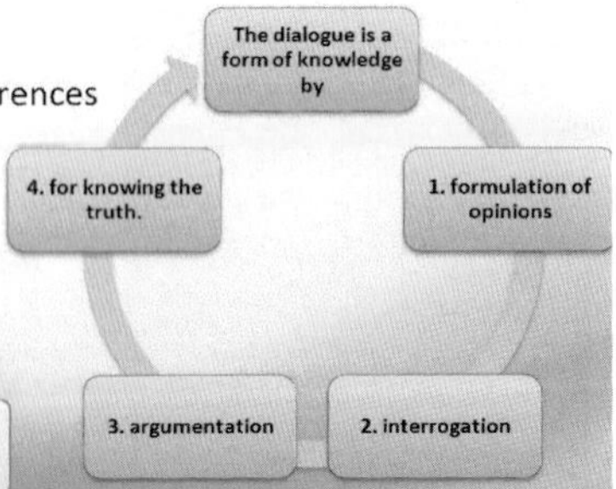

Schematic representation of the Dialectics of Reason

"THE MANAGERIAL-ECONOMIC INSTRUMENTS OF EFFICIENT ENERGY USE IN THE MOLDOVAN ENTERPRISES OF FOOD INDUSTRY"

A scholarship was awarded:
3rd year PhD student **Corina GRIBINCEA**
National Institute for Economic Research
Scientific advisor, Prof. PhD **Alexandru STRATAN**

Objective:

Study the managerial-economic instruments of efficient energy use in the enterprises of Moldovan food industry

Results:

- A new scheme was proposed **new management system of energy cost** in the production process that could be considered as an effective benchmark.
- Published in several scientific journals

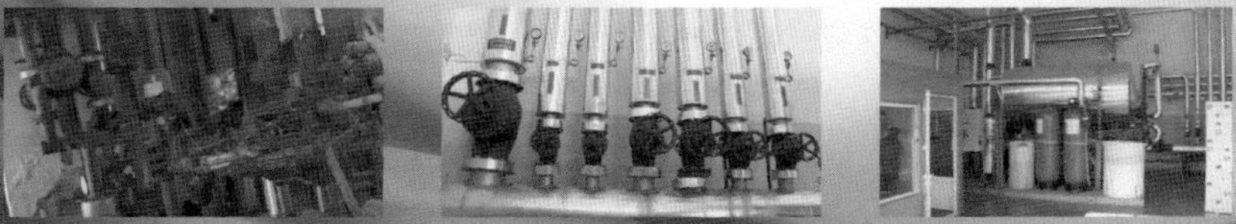

Figure 1,2,3. Before and after the implementation of efficient management of costs, by changing equipment for sustainable use of energy in Moldovan food industry

6. Planetary direction – __Biotechnology and medicine__

"THE IMPACT OF NEOADJUVANT CHEMOTHERAPY ON THE BREAST CANCER"

A scholarship was awarded:
3rd year PhD student **Ludmila CHISTRUGA**
Institute of Oncology
Scientific advisor Assoc. Prof., Dr. Hab. **Vasile JOVMIR**

Objectives:

- The use of haemostasis indices for the foundation of thromboembolic complications risk group of patients with breast cancer
- Decrease of the rate of thromboembolic complications at patients who are expected to have surgical treatment after neo-adjuvant chemotherapy.

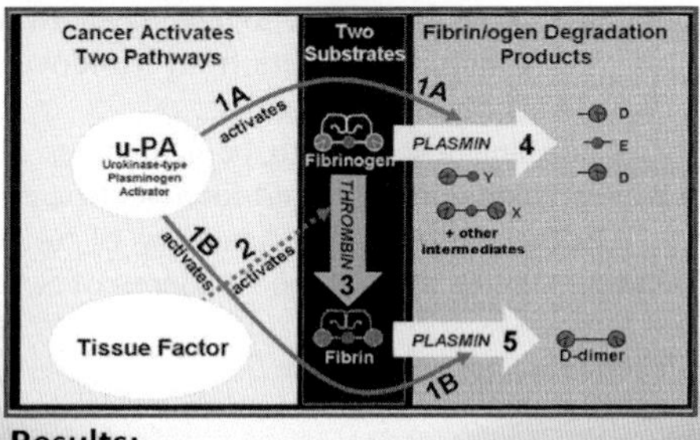

Results:

One published, one submitted and two articles are under preparation; two abstracts presented at international conferences.

7. Planetary direction – ***Conversion of Military Resources***

"THE RESEARCH OF EXHAUSTIVE POLLING MODELS WITH SEMI-MARKOV VACATIONS AND PRIORITIES"

A scholarship was awarded:
3rd year PhD Student **Lilia MITEV**
Institute of Mathematics and Computer Science
Scientific advisor, Acad., Dr. hab., Prof. **Gheorghe MISHKOY**

Objectives:

- Research of a new class of mathematical models from Queueing Theory: models with semi-Markov switching of states, called generalized models, especially a particular case of these models – the exhaustive polling models with semi-Markov "vacations"

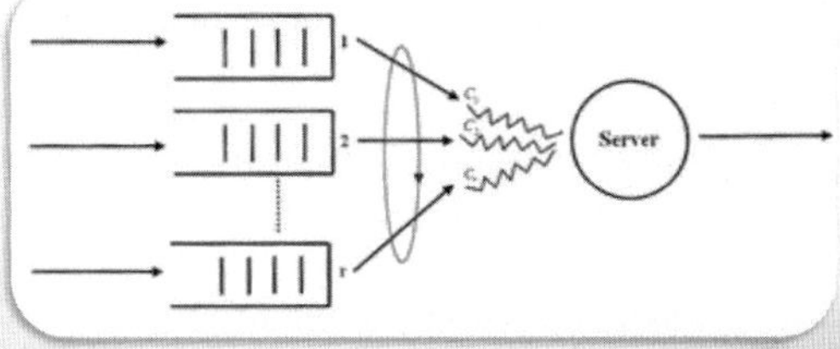

Queueing model of polling type with semi-Markov vacations

Results:
Published in 3 journals and presented at an international conferences.

DECLARATION

Dear Antonino,

You contributed substantially for the development of science in the world, you are the author of several scientific discoveries, you have popularized science, formed many organizations which have consolidated researchers, you organized 126 schools and delivered 1738 courses in various domains of science, which have been attended by 120,680 researchers from 932 universities and research institutes from 140 countries, including 134 participants who have been awarded the Nobel Prize.

150 young researchers from Moldova have been awarded 150 research grants by the Word Federation of Scientists.

You are deserving to be a member of all academies from across the world and receive all distinctions offered for scientific excellence, including Nobel Prize.

In the name of the Academy of Sciences of Moldova, I ask you to accept the Honorary Membership of the Academy of Sciences of Moldova.

LETTER OF INVITATION

The Academy of Sciences of Moldova would like to cordially extend an invitation for you to become an honorary member of the Academy for the outstanding scientific reputation, excellence in research in the field of physics and nuclear physics and for the organization of international cooperation and management of science.

For the aforementioned, as well as many more achievements and for the consolidation of international cooperation with Moldova, we would like to award you with the highest title of the Academy of Sciences of Moldova—Honorary Member of the Academy.

ACADEMIA DE ŞTIINŢE A MOLDOVEI

Bd. Ştefan cel Mare şi Sfânt,1
MD – 2001, Chişinău,
Republica Moldova
tel: (+373 22) 27-14-78
Fax: (+373 22) 54-28-23
E-mail: consiliu@asm.md

ACADEMY OF SCIENCES OF MOLDOVA

1, Ştefan cel Mare şi Sfânt Ave.
MD – 2001, Chişinău,
Republic of Moldova
tel: (+373 22) 27-14-78
Fax: (+373 22) 54-28-23
E-mail: consiliu@asm.md

23rd of August 2015

Professor Antonino ZICHICHI,
President of the World Federation of Scientists

Your Excellency prof. Antonino ZICHICHI,

The Academy of Sciences of Moldova would like to cordially extend an invitation for **you to become an honorary member of the Academy** for the outstanding scientific reputation, excellence in research in the field of physics and nuclear physics and for the organization of international cooperation and management of science.

The Academy of Sciences of Moldova, representing the Republic of Moldova, as a member of the World Federation of Scientists praises the activity of the organization, which, under the coordination of yourself, has grown to include more than 10,000 scientists drawn from 110 countries. Furthermore, the organization has successfully harmonized research priorities in the manner that all members, including Moldova, share the same aims and ideals and contributes to the development of science.

The cooperation of World Federation of Scientists with Moldova began 16 years ago and during these 16 years, more than 150 young scientists from Moldova had the possibility to obtain financial support in their research without leaving the country. The results of the research has enabled youth to engage in science for the benefit of society, cooperate with peers from across the world, publish their work in international journals and patent their ideas.

As a renowned scientist, you have collaborated on several important discoveries in the field of subnuclear physics and worked in some of the most important research laboratories in the world. You have popularized science, established several organizations which have consolidated young researchers potential, organized 126 schools and delivered 1738 courses in various domains of science, which were attended by 120, 680 researchers from 932 universities and research institutes from 140 countries, including 134 participants who were awarded the Nobel Prize.

For the aforementioned, as well as many more achievements and for the consolidation of international cooperation with Moldova, we would like to **award you with the highest title of the Academy of Sciences of Moldova – Honorary Member of the Academy.**

With high regard,

President, academician **Gheorghe DUCA**

Igor Serotila
+373 (22) 27 03 39

Thank you for your attention!

NATIONAL SCHOLARSHIPS PROGRAMME IN GEORGIA

MERAB ELIASHVILI
Facility of Exact and Natural Sciences, Tbilisi State University
Tbilisi, Georgia

APPLICATIONS OF THE OCTONIONIC QUASI-GROUPS IN PARTICLE PHYSICS

Otar Sakhelashvili
(Supervisor Merab Gogberashvili)

Iv. Javakhishvili Tbilisi state University
Department of Fundamental Physics
1, Chavchavdze Ave., 0179 Tbilisi, Georgia

Physical signals is described with a split octonion

$$s = \omega + \lambda^n J_n + x^n j_n + ctI$$

They have norm:

$$ss^+ = \omega^2 - \lambda^n\lambda^n + x^n x^n - (ct)^2.$$

Limit: Minkowsky space:

$$\omega \to 0,\ \lambda^n \to 0$$

Automorphisms of the octonions: G2 Exceptional Group

Classical result for the light aberration is modified

$$\Delta\vartheta = \frac{V}{c}\sin\vartheta + \frac{V}{c^2}\dot{\lambda}.$$

For 1 particle state, suggesting:

$$\lambda_i = \frac{\hbar p_i}{p^2}.$$

For free relativistic particle Lagrangian one gets Quantum Correction:

$$L = -mc^2\sqrt{1 - \frac{v^2}{c^2} + \frac{\hbar^2 \dot{p}^2}{p^4 c^2}}.$$

BAND-INSULATOR-METAL TRANSITION IN ONE-DIMENSIONAL HUBBARD MODEL WITH SPIN-DEPENDED HOPPING

Luka Jibuti
(Supervisor George I. Japaridze)

Ilia State University
Faculty of Natural Sciences and Engineering
Cholokashvili ave. 3-5, 0162, Tbilisi, Georgia

The system is described by the T-B Hamiltonian

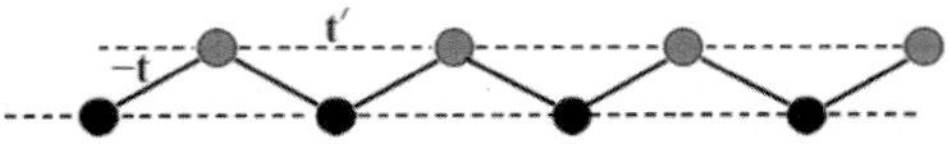

$$H = -\sum_{n,\sigma}\left[-t_\sigma\left(c^+_{n,\sigma}c_{n+1,\sigma}+H.c.\right)+t'_\sigma\left(c^+_{n,\sigma}c_{n+2,\sigma}+H.c.\right)\right]+\frac{\Delta}{2}\sum_{n,\sigma}c^+_{n,\sigma}c_{n,\sigma} + \sum_{n,\sigma,\sigma'}\left(i\gamma_R^{(1)}c^+_{n,\sigma}\,i\sigma^y_{\sigma\sigma'}c_{n+1,\sigma'}-i\gamma_R^{(2)}c^+_{n,\sigma}\,i\sigma^y_{\sigma\sigma'}c_{n+2,\sigma'}+H.c.\right)-\mu\sum_{n,\sigma}c^+_{n,\sigma}c_{n,\sigma}$$

$c^+_{n,\sigma}(c_{n,\sigma})$ - the creation (annihilation) operators, t and t' - the hopping parameters, Δ is the amplitude of the single-electron on-site energy modulation, μ is he chemical potential, $\gamma_R^{(1)}$ and $\gamma_R^{(2)}$ are amplitudes of the uniform Rashba spin-orbit interaction and $i\sigma^y_{\sigma\sigma'}$ is the Pauli matrix for the y-direction.

The energy spectrum of the system is as follows:

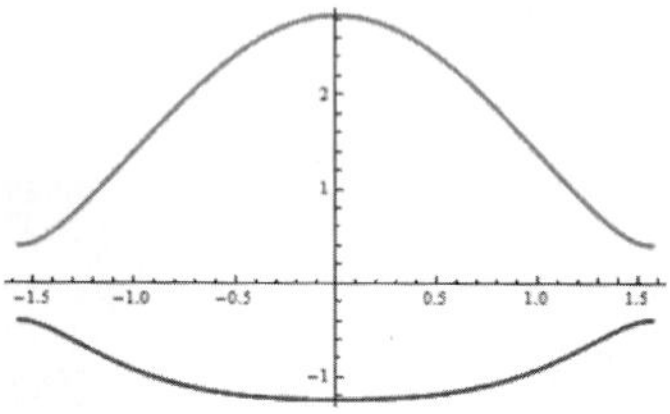

The system is an insulator

$$t'_{\sigma} = t'_{\sigma,c} = \frac{\Delta}{8} + \sqrt{4t^2 + \frac{\Delta^2}{4}}$$

At critical value of t'_{σ} system will undergo phase transition

The system is a metal

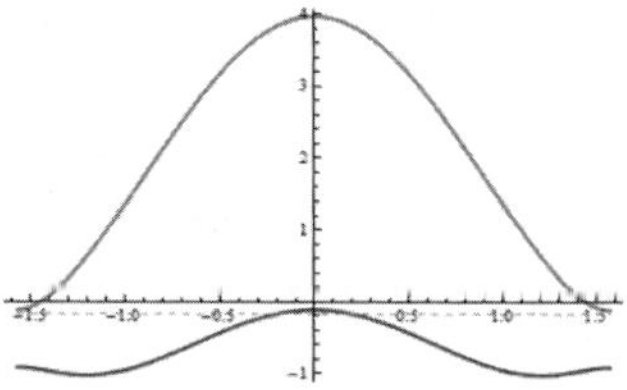

When hopping parameter t depends on the spin orientation

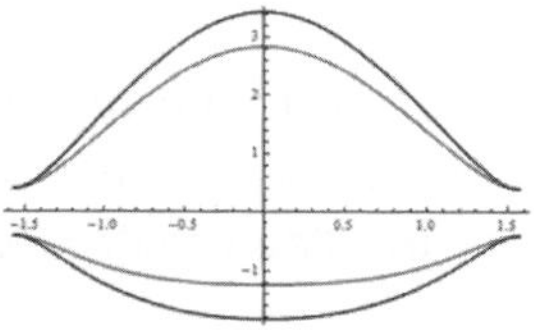

The system is an insulator

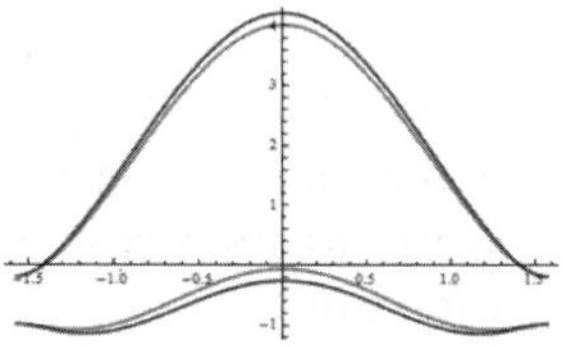

The system is a metal

$$t' = t'_c = \sqrt{(t_{\uparrow} + t_{\downarrow})^2 + \frac{\Delta^2}{4}} + \frac{\Delta}{8}$$

critical value of t'_{σ} when system undergoes phase transition

When hopping parameter t' depends on spin orientation

$$t'_{-} = t'_{+} + \frac{\Delta}{2}$$

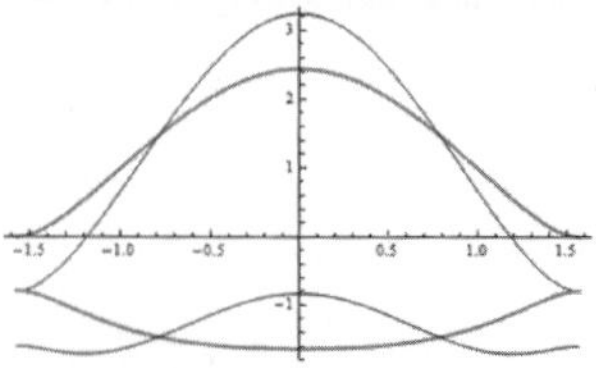

The system is a metal

STUDY OF CANDIDATE COMPOUNDS FOR MAGNETOCALORIC MATERIALS

Mamuka Chikovani
(Supervisor Alexander Shengelaya)

Iv. Javakhishvili Tbilisi state University
Department of Fundamental Physics
1, Chavchavdze Ave., 0179 Tbilisi, Georgia

Magnetocaloric effect (MCE)

Magnetocaloric effect has three prominent advantages:

- Environmentally-friendly with no harmful gases involved
- More energy efficient
- Generates much less noise compared to its counterparts

[1]

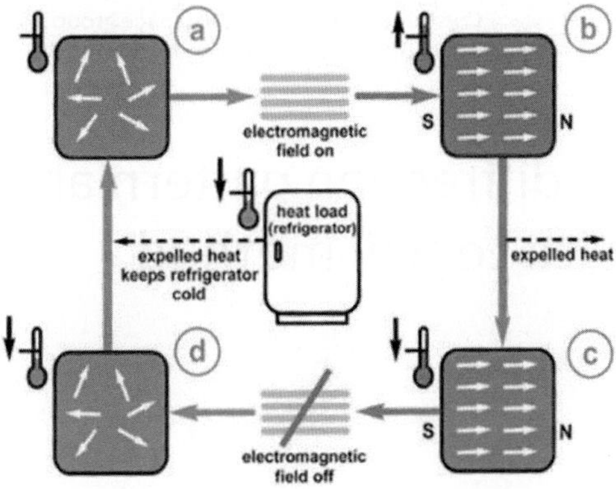

Scheme of magnetocaloric refrigeration cycle

$Mn_{2-x}Fe_xSb$ series – possible compounds for MCE

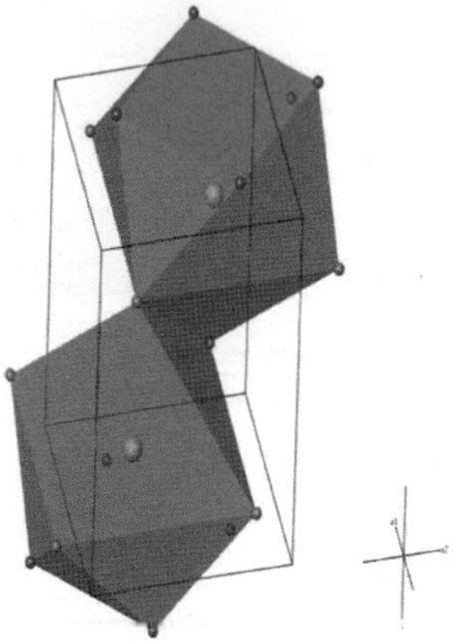

- Affordability
- Easy to handle

oThe parent compound of Mn_2Sb with Cu_2Sb structure type exhibits the ferrimagnetic ordering at 550K. [3]
oIron atoms replace Mn [2]

Crystal structure of Mn_2Sb (space group P4/nmm, a=b=4.08 (A), c=6.56 (A).) The polyhedra around Sb are indicated

Powder x-ray diffraction pattern analyzed with Le Bail method

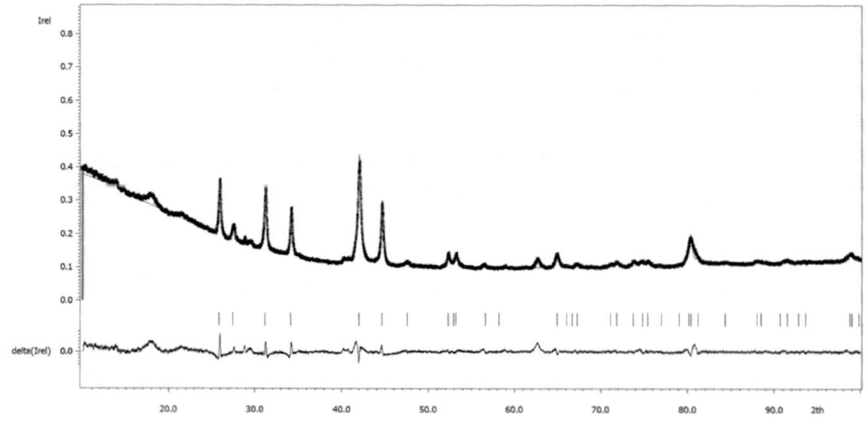

Magnetization measurements and MCE calculations

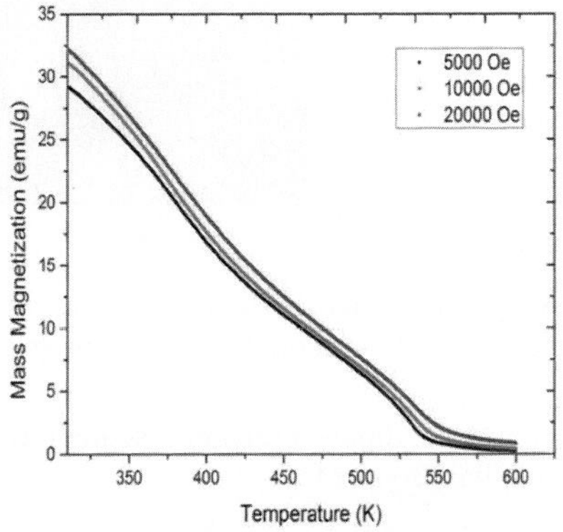

High temperature magnetization measurements of $Mn_{1.8}Fe_{0.2}Sb$ with the field cooling option

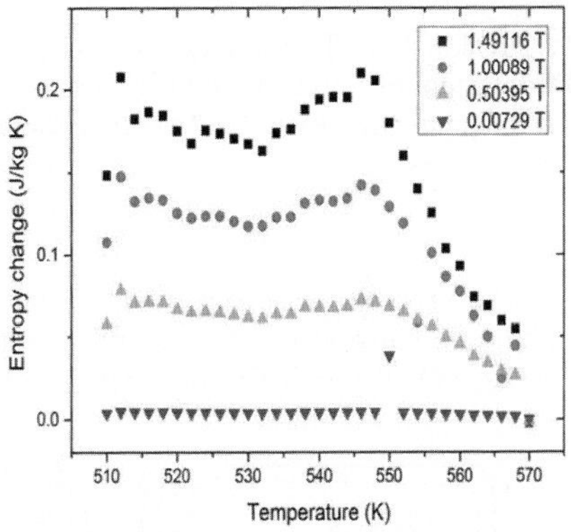

Calculated MCE in $Mn_{1.8}Fe_{0.2}Sb$

Conclusions and Outlook

- Good starting point for investigating the MCE in Cu_2Sb type structures.
- Small MCE effect observed in a sample with small amount of Fe (x=0.2) content
- Specific heat measurement to be performed to clarify possible phase transitions.
- Further study of compounds with different iron concentrations is being carried out.

References:

[1] http://green.blogs.nytimes.com/tag/magnetocaloric effect/

[2] Ryzhkovskii, V.M.; Dymont, V.P.; Erofeenko, E.L.; Nekrashevich, E.M. Inorganic Materials (1997) 33, (8) p781-p783

[3] M. K. Wilkinson, N. S. Gingrich and C. G. Shull: J. Phys. Chem. Solids 2 (1957) 289-300.

HUMAN EM EXPOSURE STUDY FOR SOME BIG SCENARIOS (ROOM WITH WINDOW) USING MAS

Mariam Tsverava
(Supervisor Revaz Zaridze)

Iv. Javakhishvili Tbilisi state University
Department of Fundamental Physics
1, Chavchavadze Ave., 0179 Tbilisi, Georgia

The problem is solved by the Method of Auxiliary Sources (MAS) which gives ability to solve efficiently complicated and big cases, such as room with window and walls, with different ***T*** transparency coefficient.

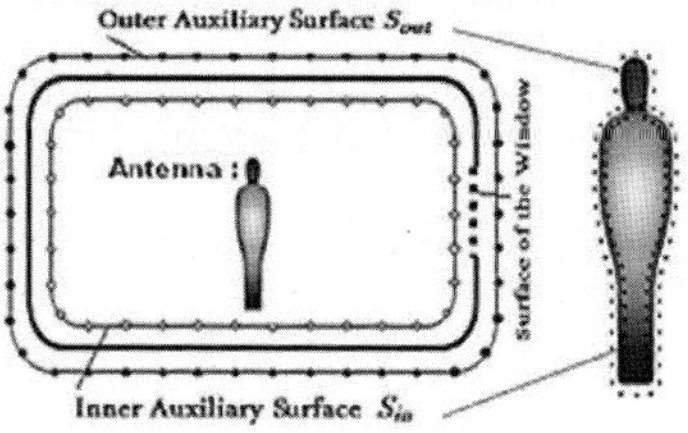

The combined dipole (Huygens source) is used as auxiliary source for solution.

$$\vec{E}_{comb} = \vec{E}_{el} + \sqrt{\mu/\varepsilon}\,\vec{E}_{mag}$$

$$\vec{H}_{comb} = \vec{H}_{el} + \sqrt{\mu/\varepsilon}\,\vec{H}_{mag}$$

The dependency of the ***R*** reflection and ***T*** transparency coefficients on the collocation point's density with N/λ^2:

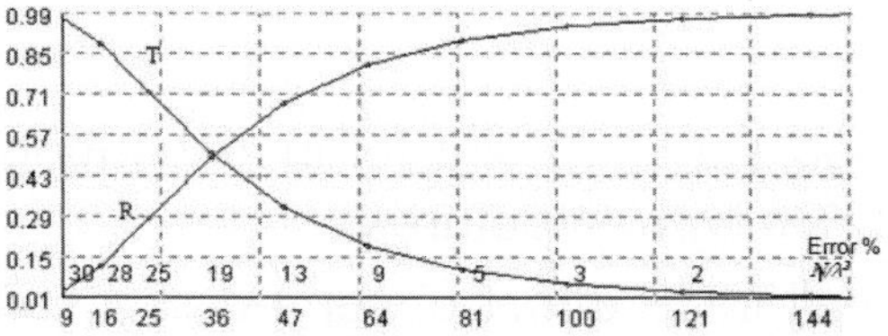

Transmission and reflection coefficients dependence on the collocation points and the calculation error, when d=0.1

EM Field Distribution in the Room at 300 MHz with different transparency coefficients: a) T=0.75; b) T=0.22

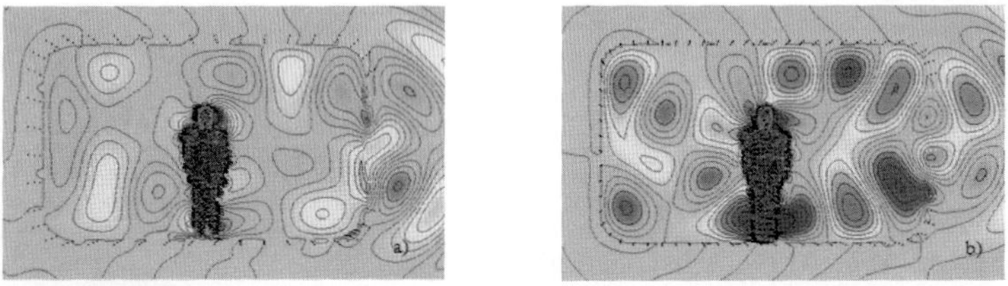

Near field distribution for two different cases at 900 MHz: a) user is located far from the window, b) user is located near the window, T=0.22.

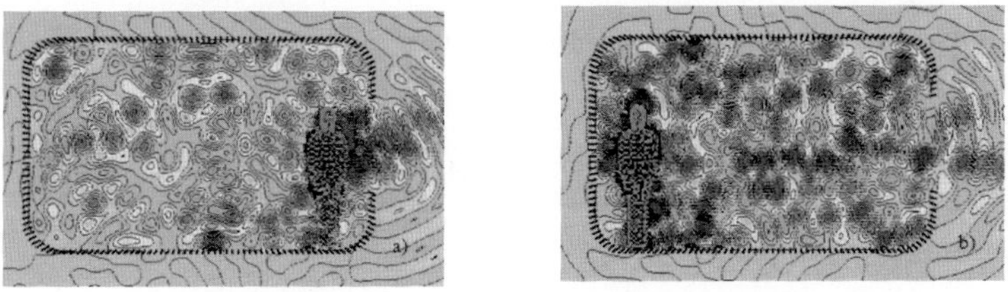

It is preferred to use telephone near the window.

OPTIMAL ANTENNA SYNTHESIS PROBLEM BY THE PREDETERMINED FAR FIELD PATTERN

Beka Poniava
(Supervisor: Revaz Zaridze)

Iv. Javakhishvili Tbilisi state University
Department of applied Physics
1, Chavchavdze Ave., 0179 Tbilisi, Georgia

Using the theorem of summation, the value of the Hankel function in the point A

can be expressed as a sum of high order Hankel function of integer-value, with the poles in the

$$H_0^{(2)}(k\rho_1)=\sum_{m=-\infty}^{+\infty} J_m(k\rho_0)H_m^{(2)}(k\rho)e^{im\varphi}$$

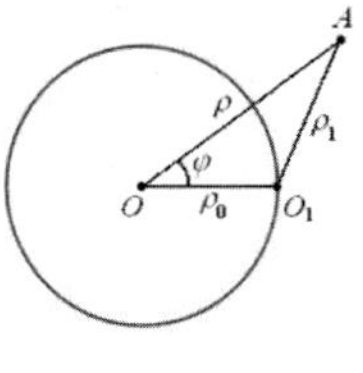

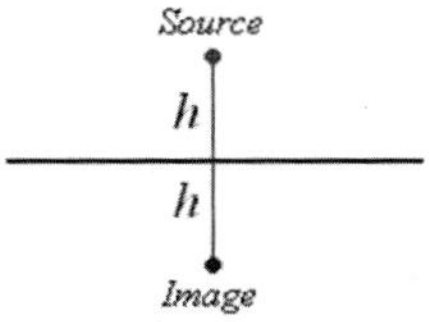

If the point of observation is inside of the circle than the value of the field is determined with the Bessel functions:

$$H_0^{(2)}(k\rho_1)=\sum_{m=-\infty}^{+\infty} H_m^{(2)}(k\rho_0)J_m(k\rho)e^{im\varphi}$$

Suppose the observed field outside of the circle is created by of the multiples in the point O. For this field we have the series:

$$\sum_{m=-\infty}^{+\infty} J_m(k\rho_0)H_m^{(2)}(k\rho)e^{im\varphi}$$

GaN HETEROSTRUCTURE BARRIER VARACTOR DIODE MODELLING

Giorgi Aroshvili
(Supervisor Giorgi Ghvedashvili)

Iv. Javakhishvili Tbilisi state University
Department of Electrical and Electronics Engineering
1, Chavchavdze Ave., 0179 Tbilisi, Georgia

OUTLINE

- Device operations principle
- Results
- Next Steps

GaN/AlGaN/GaN Structure

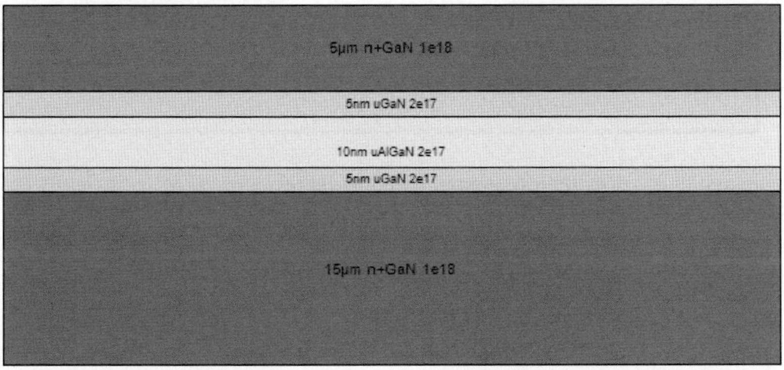

Model Structure

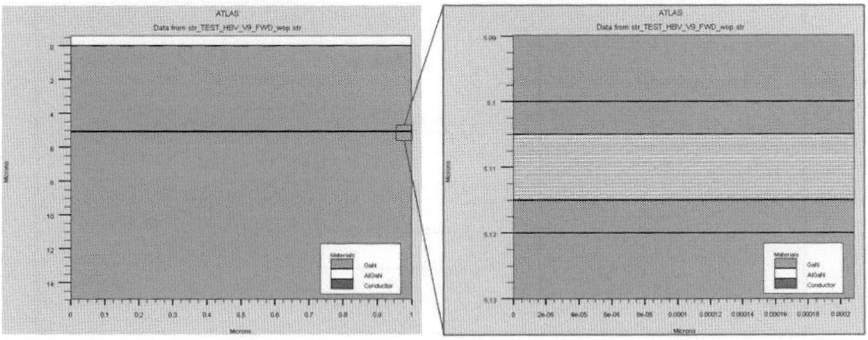

Donor Concentration and Net Polarization Charge at the interface

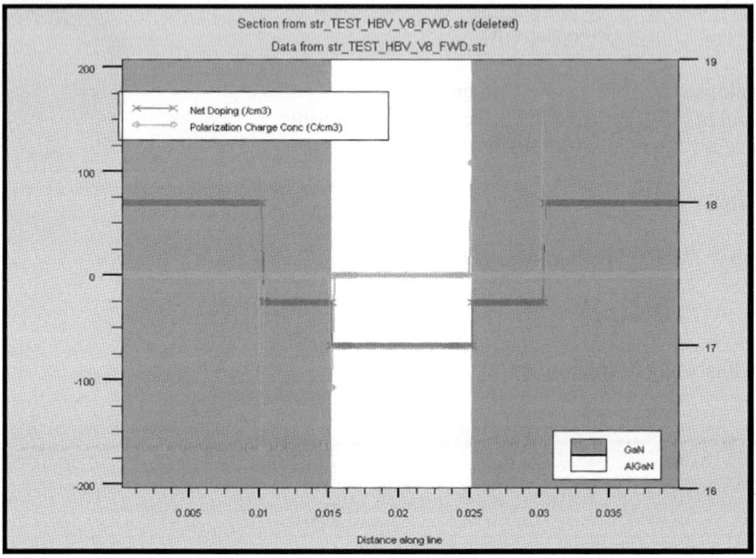

Bandgap and donor distribution w/o polarization charge

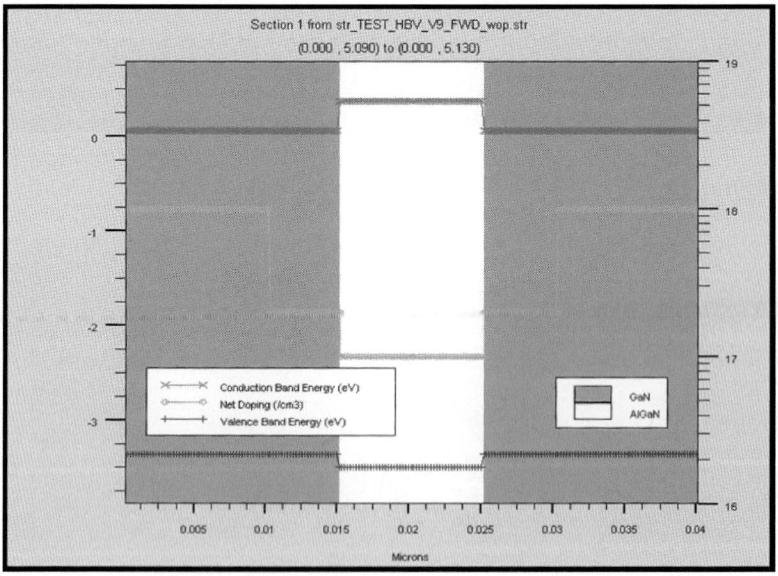

Bandgap with polarization charge

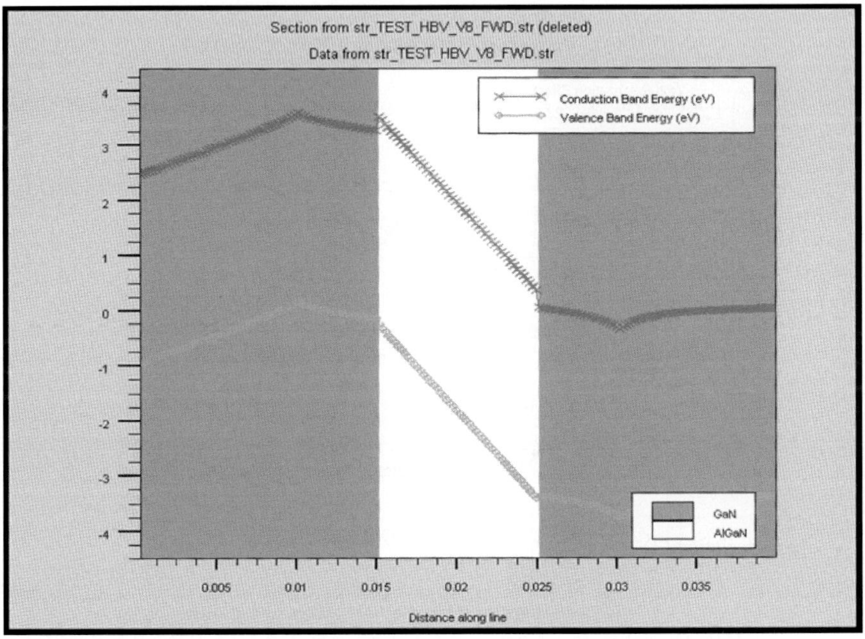

CV Characteristics (1)

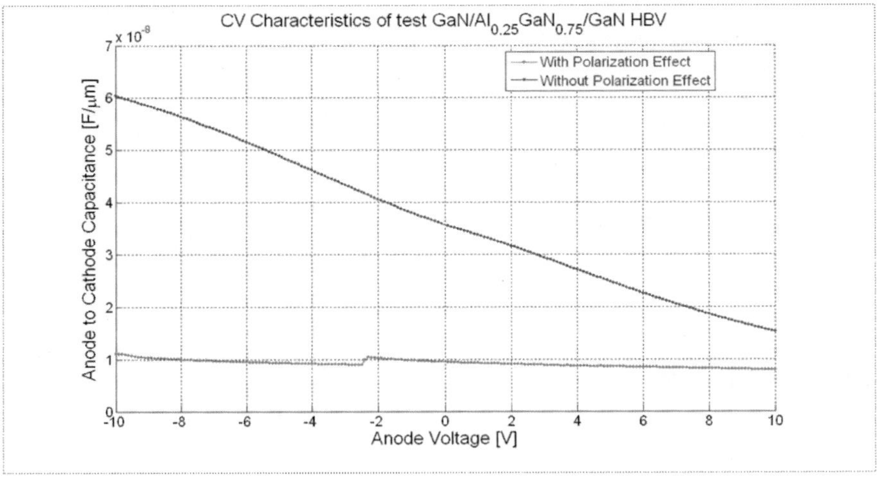

CV Characteristics (2)

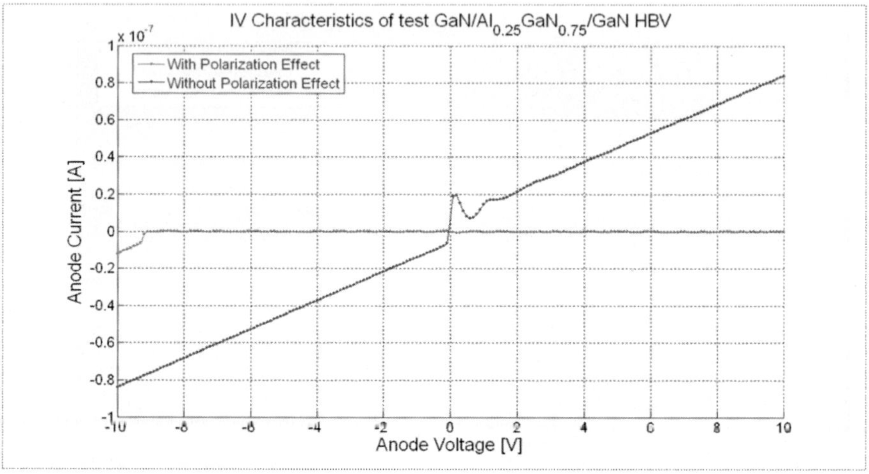

CV Characteristics (3)

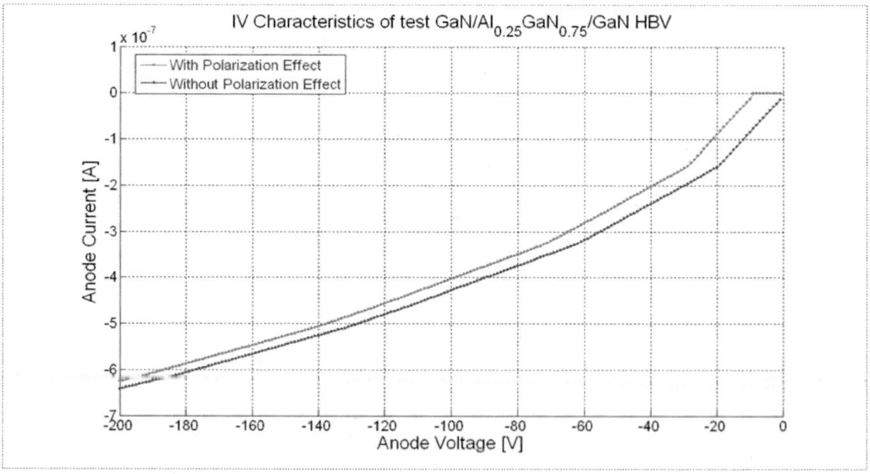

CV Characteristics (4)

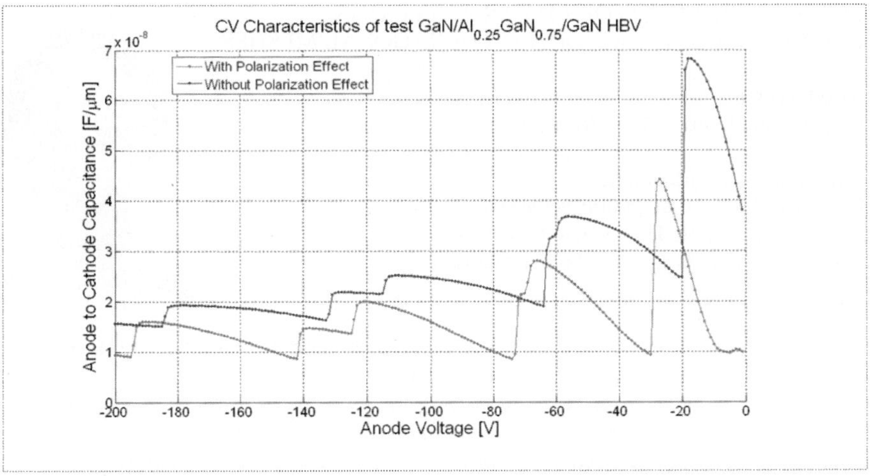

NEXT STEPS

- InGaN/GaN, InAlN/GaN or InGaN/ other III-Nitrides.
- Research into extremely high conductance.
- Confirmation of polarization effects.
- CV Asymmetry.

MESH ADAPTATION FOR LINEAR ADVECTION EQUATION AND ITS ADJOINT ON HEXAGONAL MESHES

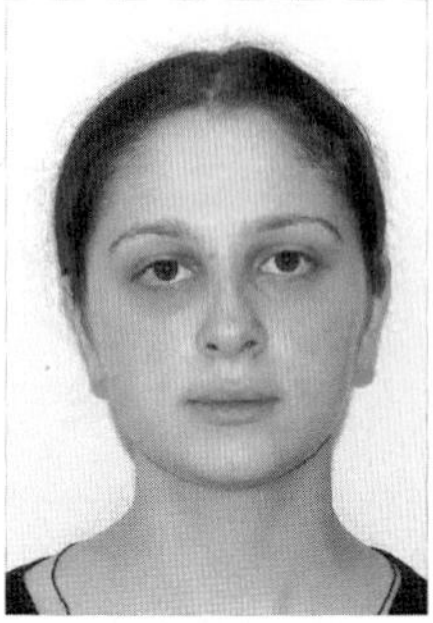

Ana Dolidze
(Supervisor Botchorishvili Ramaz)

Department of Mathematics,
Faculty of Exact and Natural Sciences,
Ivane Javakhishvili Tbilisi state University.
1, Chavchavdze Ave., 0179 Tbilisi, Georgia

Mesh adaptation for linear advection equation and its adjoint on hexagonal meshes

Finite Volume Multischeme approach allows to couple two hexagonal meshes with different numerical schemes or different step sizes. Here multischeme is considered for linear advection equation ($\frac{\partial u}{\partial t} + \frac{\partial(uv_1)}{\partial x} + \frac{\partial(uv_2)}{\partial y} = 0$) *In two dimensional space. Hexagonal meshes are more complicated then Cartesian meshes. In order to understand multischeme on hexagonal meshes scheme behavior on Cartesian meshes is studied by means of calculation of test problem with refined moving domain.*

Multischemes with two different meshes In two-dimensional space

Main idea of multischeme algorithm is adaptive coupling of suitable finite volume schemes and patches for a given region of computational domain depending of properties of solution. consider two different meshes-refined and rough with the same numerical scheme on both meshes(see Figure 1). If the space step on the rough mesh is $\Delta x = \Delta y = h$, *then on the refined mesh we get space step refined by factor k(because refined mesh is received by means of refining the rough mesh by factor k) and* $\frac{\Delta x}{k} = \frac{\Delta y}{k} = \frac{h}{k}$ *, the same is true for the time steps, since we consider explicit schemes and for stability reasons time step size is calculated from CFL condition on account of space step size.*

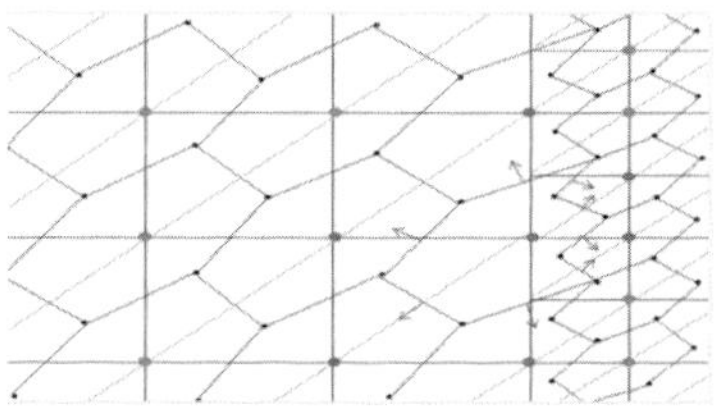

Numerical test:

Test problem:
Refined rectangular mesh embedded in the rectangular domain changes the direction dynamically as the cone moves.

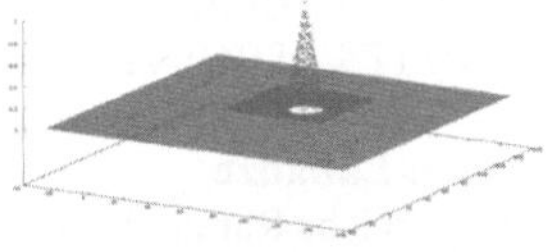

Initial cone

Goal:
Refined moving domain, space-time adaptation. Study accuracy of dynamically moving refined rectangular mesh algorithm.

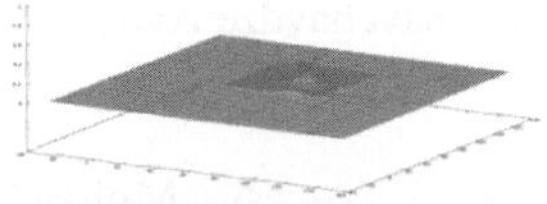

Cone after one rotation

Initial data:
cone $(4-4/15\sqrt{(x-70^2)+(y-70)^2})$
specific velocity field ensuring rotation of cone around prescribed point $\vec{V}(-0,1(y-50);0,1(x-50))$

- Test results

Rough Mesh			Small Mesh			Combination			Refined Moving Domain			
									Initial Point		Domain Size	
Step X	Step Y	Time Step	Step X	Step Y	Time Step	Second Norm	Infinitive Norm	Program Execution Time (In Seconds)	Initial Point X	Initial Point Y	Domain length	Domain Width
2	2	0.08	1	1	0.04	4.89497	0.760392	12.523	30	40	30	30

MULTISCHEME FOR LINEAR ADVECTION EQUATION AND ITS ADJOINT ON CARTESIAN MESHES

Maka Labadze
(Supervisor Ramaz Botchorishvili)

Department of Mathematics,
Faculty of exact and natural sciences,
Ivane Javakhishvili Tbilisi state University
1, Chavchavdze Ave., 0179 Tbilisi, Georgia

Multischeme algorithm with two different meshes

- Linear advection equation takes the following form in two space dimensions: $\frac{\partial u}{\partial t}+\frac{\partial(uv_1)}{\partial x}+\frac{\partial(uv_2)}{\partial y}=0$

  where (t,x,y) are independent variables, u(t,x,y) is unknown function and $v_1(t,x,y)$, $v_2(t,x,y)$ are components of velocity vector $\vec{v}=(v_1,v_2)^T$. It is known as conservative form of linear advection equation.

- Multischeme approach allows to couple mesh adaptation with different numerical schemes(Finite Volume schemes) on Cartesian meshes of different step sizes.
- Two different meshes-refined and rough with the same numerical scheme on both meshes

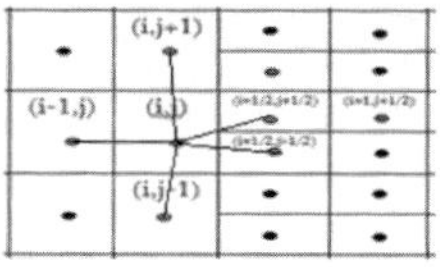

Figure 1

- If the space step on the rough mesh is Δx=Δy=h, then on the refined mesh we get space step refined by factor k(because refined mesh is received by means of refining the rough mesh by factor k) and $\frac{\Delta x}{k}=\frac{\Delta y}{k}=\frac{h}{k}$, the same is true for the time steps, since we consider explicit schemes and for stability reasons time step size is calculated from CFL condition on account of space step size. For k=2 meshes are given on Figure 1 as well as stencil of Multischeme near boundary between two meshes.

Numerical Test

➢ Test problem:

- Linear advection equation, rotating cone data
- First order numerical scheme
- Two different meshes, fine mesh step size is smaller by factor 2, 4 and 8 compared to rough mesh step size. Two different meshes - rough and fine are combined with one numerical scheme

➢ Goals:
 1)study accuracy of mesh refinement algorithm

Rough Mesh			Small Mesh			Combination		
Step X	Step Y	Time Step	Step X	Step Y	Time Step	Second Norm	Infinitive Norm	Program Execution Time(In Seconds)
2	2	0.08	1	1	0.04	31.9557	3.23008	29.562
2	2	0.08	0.5	0.5	0.02	30.709	3.13476	223.195
2	2	0.08	0.25	0.25	0.01	28.915	2.99413	1631.88

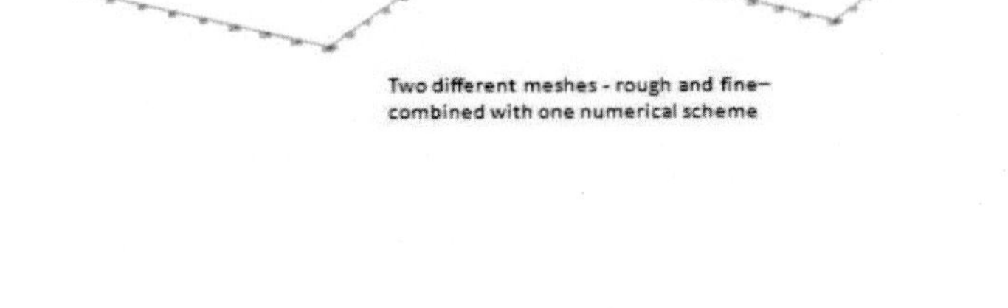

Two different meshes - rough and fine– combined with one numerical scheme

COOPERATIVE EFFORT BETWEEN THE AZERBAIJAN NATIONAL ACADEMY OF SCIENCES (ANAS) AND WORLD FEDERATION OF SCIENTISTS (WFS)

ADALAT HASANOV
National Academy of Sciences
Baku, Azerbaijan

WFS NATIONAL SCHOLARSHIP PROGRAMME REPORT—2014-2015

Greetings from Baku—the oldest cultural and economic centers of Azerbaijan.

https://youtu.be/hFkpqblp1ls

In Baku in 2015 were the first in the history the European Games 2015. By the way, the decision was taken in December 2012 in Rome.

№№	Teams	Gold	Silver	Bronze	Total
1	Russia	79	40	45	164
2	Azerbaijan	21	15	20	56
3	United Kingdom	18	10	19	47
4	Germany	15	16	33	64
5	Italy	10	25	10	45
6	Belorussia	10	11	22	43
7	France	9	11	18	38
8	Ukraine	8	14	23	45
9	Netherlands	8	11	9	28
10	Hungary	8	4	8	20

Baku 2015 European Games were an international multi-sport event for athletes representing the National Olympic Committees of Europe and including almost 6,000 athletes from 50 countries competing in 20 sports.

In Baku is located the main scientific organization in the Azerbaijan the Azerbaijan National Academy of Sciences (ANAS)

Currently ANAS has 4 branches in main regions of Azerbaijan and include about 40 scientific organizations with 10,000 employees.

To date, a total of 57 Azerbaijani post-graduate students have been awarded by the WFS National Scholarships. All students successfully completed their researches in time.

In 2015 to the WFS were submitted next 6 applications of students from different organizations of ANAS.

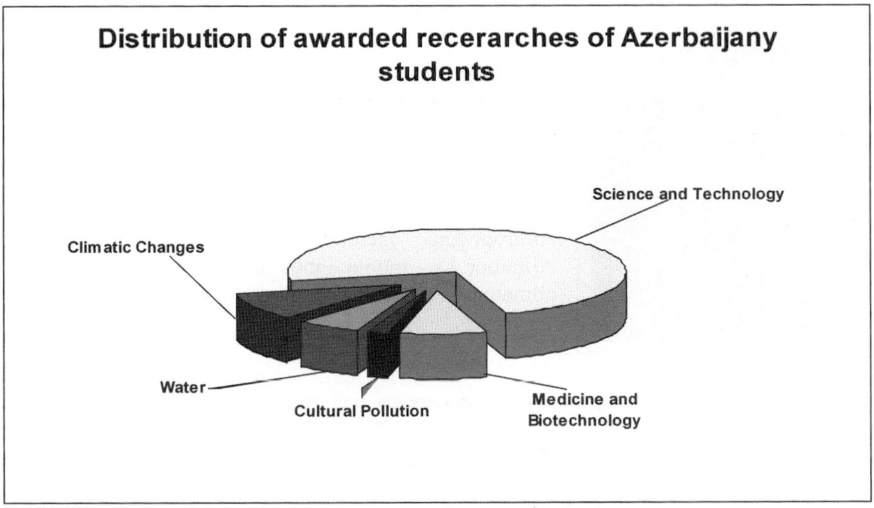

The general statistics of distribution awarded researches according of Planetary Emergencies identified by the WFS.

№№	Students	Institutes	Title of researches
1	Afsana Huseynova	Institute of Microbiology	Research dedicated to biotechnology basis of bioconversion of same plant waste.
2	Ferid Seyfullayev	Central Botanical Garden	Research dedicated to applied dendrochronology for investigation of the past environmental condition in Azerbaijan by using annual tree-rings.
3	Azad Dadashov	Institute of geology&geophysics	Research dedicated to fluid dynamic and petrophysical parameters of reservoir and influence their on the distribution of hydrocarbon reserves.
4	Khayala Gasanova	Institute of Physics	Research dedicated to grow up mono crystals of semimetals and she received uniform in length and made on structure of mono crystals of $Bi_{1-x}Sb_x$ solid solutions and their alloying.
5	Vusala Eminova	Institute of Physics	Research dedicated to determination of effective masses of electrons and holes in $TlIn_{1-x}Yb_xTe_2$ solid solutions from the measurement of the kinetic coefficients in weak magnetic fields
6	Shahla Khalilova	Institute of Physics	Research dedicated to searching for the mixed phase of the strongly interacting matter in relativistic and ultra relativistic nucleus-nucleus collisions.
7	Tural Naghiyev	Institute of Physics	Research dedicated to Luminescent properties of $Ca_xBa_{1-x}Ga_2S_4$ solid solutions activated with rare-earth elements.
8	Lala Ahmadova	Institute of Geography	Research dedicated to evaluation and optimization of anthropogenic degradation processes in the lower mountainous Ajinohur area in Azerbaijan.
9	Gulnara Hasanova	Institute of Microbiology	Research dedicated to growing of fungi from genus Aspergillus and Aspergillus niger.

List of post-graduate students completed WFS National Scholarships in 2014-2015.

LALA AHMADOVA

The ways of evaluation and optimization of anthropogenic degradation processes in the lower mountainous Ajinohur area.

Soil degradation map has been prepared in the 1:100000 scale for the first time taking into account natural and anthropogenic factors in 188 thousand hectares of Ajinohur low mountainous and adjoining areas in arid climate and complicated relief.

DEGRADATION MAP OF THE AJINOHUR LOW MOUNTAINOUS REGION
Scale 1:100 000

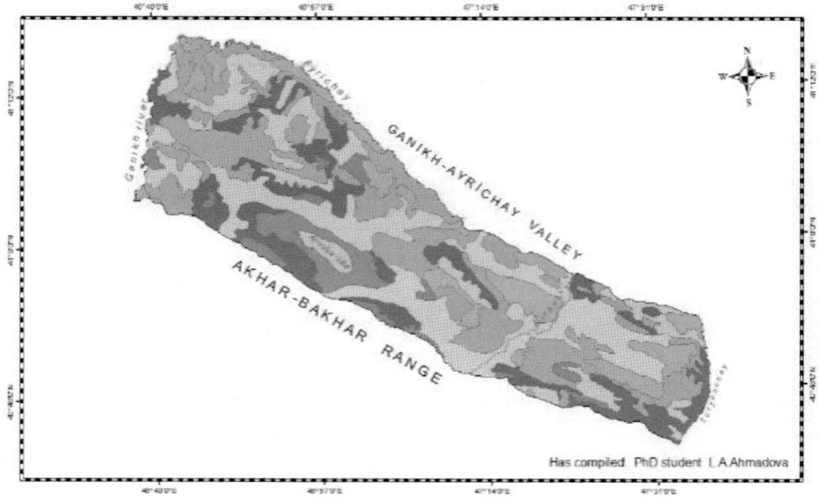

TURAL NAGHIYEV

Luminescent properties of $Ca_xBa_{1-x}Ga_2S_4$ solid solutions activated with rare-earth elements

The dependence of photoluminescence efficiency was determined as the ratio of input power to pump power density. The results are shown in figure. All dependencies show the comparable units in photoluminescence intensity, and accordingly, the efficiency for photoluminescence.

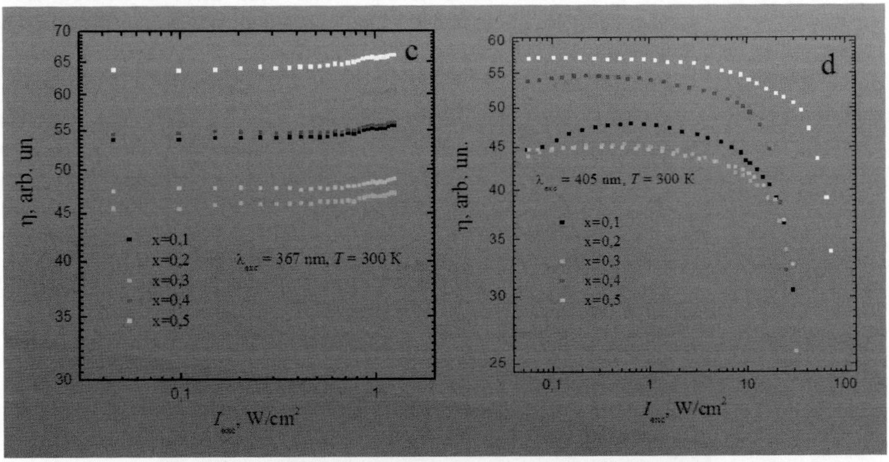

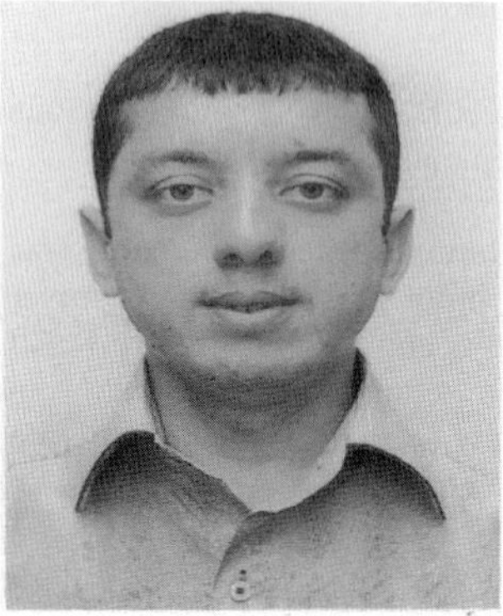

FERID SEYFULLAYEV

Applied dendrochronology for investigation of the past environmental condition in Azerbaijan by using annual tree-rings.

Dendrochronology is used as for exact dating of tree-ring width series of wood. The principle is straightforward: variation in tree-ring width is to a great extent influenced by climate. This makes that tree-ring patterns of trees of the same species growing in a certain climate region can be cross-dated and averaged into so-called master chronologies. By overlapping and cross-dating tree-ring series from living trees with those measured from historical, archaeological and geological context long chronologies were constructed that reach back for thousands of years. Meanwhile a network of hundreds of master chronologies of different tree species has been built up for Europe.

SHAHLA KHALILOVA

Searching for the mixed phase of the strongly interacting matter in relativistic and ultra relativistic nucleus-nucleus collisions.

In results of conducted analyses are obtained the average characteristics (multiplicity, momentum and transverse momentum) of light nuclei as a function of centrality in ^{12}CC-interaction and their comparison with models. The regions ($7 < N_p < 12$) and ($N_p \geq 12$) are identified where all these characteristics show a change in behavior. These changes indicate a mechanism of light nuclei formation other than the fragmentation of the colliding nuclei, which could possibly be nuclear coalescence effect.

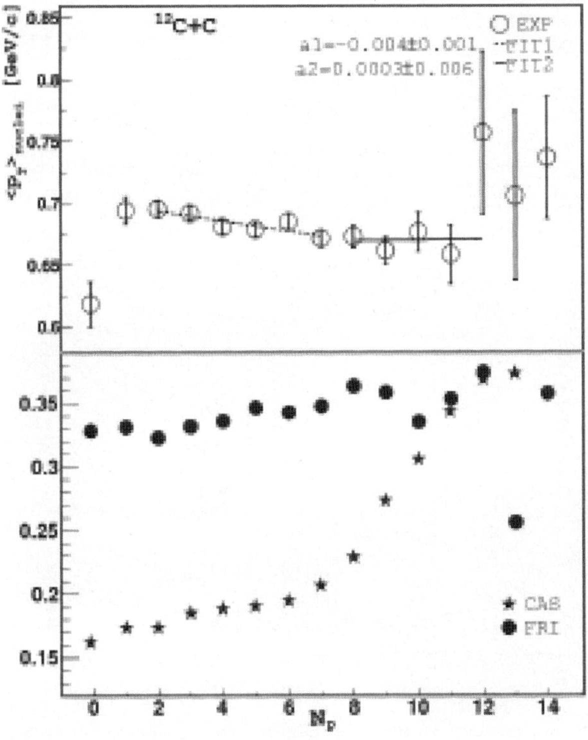

KHAYALA GASANOVA

Grow up mono crystals of semimetals and receiving uniform in length and on structure of mono crystals of $Bi_{1-x}Sb_x$ solid solutions and their alloying.

We're studying and identifying microorganisms with PCR method. Real-time quantitative PCR is an analytical method for the rapid and potentially sensitive enumeration of broad and specific microbial populations in environmental samples. For bioaerosol analysis, this method allows for detection and enumeration independent of culturing, thereby circumventing the significant concerns surrounding the unculturability of environmental microorganisms and loss of culturability due to aerosol sampling.

EGYPT NATIONAL SCHOLARSHIP PROGRAM

ABDEL NASSER TAWFIK
ECTP Founding Director
MTI University
Cairo, Egypt

Agenda

NSP since 2011

Membership of STAR Collaboration (RHIC at BNL)

Lists of Publications/Projects/Conference, etc.

Currently Running NSP (2014-2015)

Future Plans

First NSP (2011-2012)

Three scholarships: one Lady and two Gentlemen

Miss Reham Abdel Rahman

Relativistic Non-Equilibrium Thermodynamics at LHC Energy
Now: PhD-Program

Mohamed Abdallah

Hadronic States in Deconfinement Phase: Hadronization Mechanism
Now: M.Sc. Program

Ahmed Ismael Soudi

Task Scheduling Algorithms for Grid Computing Systems
Now: Saudi Arabia

Second NSP (2013-2014)

Hend Magdy

Universal Freeze-out Conditions and Chiral Symmetry Breaking

Hadeer Nasr El Din

Signals for Chiral Phase Transitions at Large Net-Baryon Density

Ehab Gamal Abbas

Particle Ratios and dynamical Fluctuations in the ALICE Experiment

Abdel Magied Diab

Acceleration in the Vicinity of Black Holes

Mohamed Sayed Marzouk *Now: PhD Program*

Task Scheduling Algorithms for LCG Grid Site in Egypt

Mahmoud Hanafy Nassar

On the correlations and critical phenomenon at LHC energies

Ibrahim El Mashad *Now: PhD Program*

Properties of Colored Quarks and Gluons in ALICE experiment

Third NSP (2014-Now)

Hend Magdy

Universal Freeze-out Conditions and Chiral Symmetry Breaking

Hadeer Nasr El Din

Signals for Chiral Phase Transitions at Large Net-Baryon Density

Ehab Gamal Abbas

Particle Ratios and dynamical Fluctuations in the ALICE Experiment

Abdel Magied Diab

Acceleration in the Vicinity of Black Holes

Mahmoud Hanafy Nassar

On the correlations and critical phenomenon at LHC energies

Miss Hend Magdy Baza

MSc.-Thesis: Relativistic Viscous Fluid under Extreme Conditions

- *Enrolling for M.Sc. Degree at Cairo University,*
- *Participation in various seminars/workshops/conferences,*
- *Organization of various seminars/workshops/conferences,*

- *Publications: five papers,*
- ***Summer-School, CERN 2013***
- ***African School for HEP, Senegal 2014***
- ***Egyptian School, Dubna 2014***

Mrs Hadeer Nasr El Din

Topic: *MSc.-Thesis: Characteristics of Particle Production in different interacting systems at various energies*

- *Absolved the pre-master courses,*
- *Enrolling M.Sc. Degree at Ain Shams University,*
- *Participated in various seminars/workshops/conferences,*
- *Organization of various seminars/workshops,*

- *Work in-progress: Two papers,*

Ehab Gamal Abbas

Topic: PhD-Thesis: Multi-particle production at High Energy collisions

- Enrolling Ph.D. Degree at Aim Shams Univ.,
- Participated in various workshops/conferences,

- Publications: 10 papers (5 ALICE Collaboration),
- Schools and Trainings at CERN, JINR, GSI
- Expertise in AliROOT, mpdROOT and cbmROOT

Mahmoud Hanafy Nassar

Topic: PhD-Thesis: On the Correlations and Critical Phenomenon at LHC energies

- Enrolling PhD Degree at Benha University,
- Participated in various seminars/workshops/conferences,

- Paper in-progress: two papers,
- Training: JINR at Dubna

Abdel Magied Diab

Topic: MSc.-Thesis: Electromagnetic Properties of Quark-Gluon Plasma

- ***Enrolling MSc-Degree at Cairo Univ.***
- ***Participated in various workshops/conferences,***

- ***Publications: 5 published (5 in presse)***
- ***Schools and Trainings: CERN and JINR***

Publications since June 2013

Inspireshep.net

Total number of papers analyzed:	32	20
Total number of citations:	***87***	***78***
Average citations per paper:	***2.7***	***3.9***
Known papers (10-49)	*1*	*1*
Less known papers (1-9)	*22*	*15*
Unknown papers (0)	*9*	*4*
h_{HEP} *index*	*5*	*5*

Publications Since June 2013

1) Centrality and transverse momentum dependence of elliptic flow of multi-strange hadrons and ϕ meson in Au+Au collisions at $\sqrt{s_{NN}}$ = 200 GeV.
By** STAR Collaboration **(L. Adamczyk et al.). [arXiv:1507.05247 [nucl-ex]].

2) Beam-Energy Dependence of Charge Balance Functions from Au+Au Collisions at RHIC.
By** STAR Collaboration **(L. Adamczyk et al.). arXiv:1507.03539 [nucl-ex]].

3) Probing Parton Dynamics of QCD Matter with Ω and ϕ Production.
By** STAR Collaboration **(L. Adamczyk et al.). [arXiv:1506.07605 [nucl-ex]].

4) Azimuthal anisotropy in U+U and Au+Au collisions at RHIC.
By** STAR Collaboration **(L. Adamczyk et al.). arXiv:1505.07812 [nucl-ex]].

5) Observation of charge asymmetry dependence of pion elliptic flow and the possible chiral magnetic wave in heavy-ion collisions.
By** STAR Collaboration **(L. Adamczyk et al.). [arXiv:1504.02175 [nucl-ex]].
*10.1103/PhysRevLett.114.252302. **Phys.Rev.Lett. 114 (2015) 25, 252302.***

Publications Since June 2013

6) Measurements of Dielectron Production in Au$+$Au Collisions at $\sqrt{s_{\rm NN}}$ = 200 GeV from the STAR Experiment.
By** STAR Collaboration **(L. Adamczyk et al.).
[arXiv:1504.01317 [hep-ex]].

7) Observation of Transverse Spin-Dependent Azimuthal Correlations of Charged Pion Pairs in $p^\uparrow+p$ at $\sqrt{s}=200$ GeV.
By** STAR Collaboration **(L. Adamczyk et al.).
[arXiv:1504.00415 [hep-ex]].

8) Leptonic decay of the phi(1020) meson in Au+Au collisions at $\sqrt{s_{\rm NN}}$ = 200 GeV measured with the STAR experiment.
By** STAR Collaboration **(L. Adamczyk et al.).
[arXiv:1503.04217 [hep-ex]].

9) Long-range pseudorapidity dihadron correlations in d+Au collisions at $\sqrt{s_{\rm NN}}=200$ GeV.
By** STAR Collaboration **(L. Adamczyk et al.).
[arXiv:1502.07652 [nucl-ex]].
*10.1016/j.physletb.2015.05.075. **Phys.Lett. B747 (2015) 265-271.***

Publications Since June 2013

10) Black Hole Corrections due to Minimal Length and Modified Dispersion Relation.
By Abdel Nasser Tawfik, Abdel Magied Diab.
[arXiv:1502.04562 [gr-qc]].
10.1142/S0217751X15500591. Int.J.Mod.Phys. A30 (2015) 12, 1550059.

11) Energy dependence of acceptance-corrected dielectron excess mass spectrum at mid-rapidity in Au+Au collisions at $\sqrt{s_{NN}} = 19.6$ and 200 GeV.
By STAR Collaboration (L. Adamczyk et al.).
[arXiv:1501.05341 [hep-ex]].

12) SU(3) Polyakov linear-σ model in magnetic fields: Thermodynamics, higher-order moments, chiral phase structure, and meson masses.
By Abdel Nasser Tawfik, Niseem Magdy.
[arXiv:1501.01124 [hep-ph]].
10.1103/PhysRevC.91.015206. Phys.Rev. C91 (2015) 015206.

13) An Estimate of the Thermodynamic Pressure in High-Energy Collisions.
By Abdel Nasser Tawfik.
[arXiv:1501.00812 [hep-ph]].
10.1142/S0217751X1550027X. Int.J.Mod.Phys. A30 (2015) 07, 1550027.

Publications Since June 2013

14) Effect of event selection on jetlike correlation measurement in d+Au collisions at $\sqrt{s_{\rm{NN}}}=200$ GeV.
By STAR Collaboration (L. Adamczyk et al.).
[arXiv:1412.8437 [nucl-ex]].
10.1016/j.physletb.2015.02.068.
Phys.Lett. B743 (2015) 333-339.

15) Polyakov SU(3) extended linear- σ model: Sixteen mesonic states in chiral phase structure.
By Abdel Nasser Tawfik, Abdel Magied Diab.
[arXiv:1412.2395 [hep-ph]].
10.1103/PhysRevC.91.015204.
Phys.Rev. C91 (2015) 1, 015204.

16) Thermodynamics and higher order moments in SU(3) linear σ-model with gluonic quasi-particles.
By Abdel Nasser Tawfik, Niseem Magdy.
[arXiv:1411.1871 [hep-ph]].
10.1088/0954-3899/42/1/015004.
J.Phys. G42 (2015) 1, 015004.

Publications Since June 2013

17) Generalized Uncertainty Principle and Recent Cosmic Inflation Observations.
By Abdel Nasser Tawfik, Abdel Magied Diab.
[arXiv:1410.7966 [gr-qc]].
Electron.J.Theor.Phys. 12 (2015) 9-30.

18) Energy Dependence of K/π, p/π, and K/p Fluctuations in Au+Au Collisions from $\rm \sqrt{s_{NN}}$ = 7.7 to 200 GeV.
By STAR Collaboration (N.M. Abdelwahab et al.).
[arXiv:1410.5375 [nucl-ex]].
10.1103/PhysRevC.92.021901. Phys.Rev. C92 (2015) 2, 021901.

19) Di-Hadron Correlations with Identified Leading Hadrons in 200 GeV Au+Au and d+Au Collisions at STAR.
By STAR Collaboration (N.M. Abdelwahab et al.).
[arXiv:1410.3524 [nucl-ex]].

20) Generalized Uncertainty Principle: Approaches and Applications.
By Abdel Nasser Tawfik, Abdel Magied Diab.
[arXiv:1410.0206 [gr-qc]].
10.1142/S0218271814300250. Int.J.Mod.Phys. D23 (2014) 12, 1430025.

Publications Since June 2013

21) Isolation of Flow and Nonflow Correlations by Two- and Four-Particle Cumulant Measurements of Azimuthal Harmonics in $\sqrt{s_{_{\rm NN}}} =$ 200 GeV Au+Au Collisions.
By STAR Collaboration (N.M. Abdelwahab et al.). [arXiv:1409.2043 [nucl-ex]].
10.1016/j.physletb.2015.04.033. Phys.Lett. B745 (2015) 40-47.

22) Equilibrium statistical-thermal models in high-energy physics.
By Abdel Nasser Tawfik. [arXiv:1410.0372 [hep-ph]].
10.1142/S0217751X1430021X. Int.J.Mod.Phys. A29 (2014) 17, 1430021.

23) SU(3) Polyakov linear-σ model in an external magnetic field.
By Abdel Nasser Tawfik, Niseem Magdy.
[arXiv:1406.7488 [hep-ph]].
10.1103/PhysRevC.90.015204. Phys.Rev. C90 (2014) 1, 015204.

24) Polyakov linear SU(3) σ model: Features of higher-order moments in a dense and thermal hadronic medium.
By A. Tawfik, N. Magdy, A. Diab.
[arXiv:1405.0577 [hep-ph]].
10.1103/PhysRevC.89.055210. Phys.Rev. C89 (2014) 5, 055210.

WLCAPP World Laboratory Cosmology And Particle Physics

Publications Since June 2013

25) Degree of Chemical Non-equilibrium in Central Au-Au Collisions at RHIC energies.
By Abdel Nasser Tawfik, M.Y. El-Bakry, D.M. Habashy, M.T. Mohamed, Ehab Abbas.
[arXiv:1401.5715 [nucl-th]].
10.1142/S0218301315500676. Int.J.Mod.Phys. E24 (2015) 1550067.

26) Koppe's Work of 1948: A fundamental for non-equilibrium rate of particle production.
By Abdel Nasser Tawfik. [arXiv:1312.1968 [nucl-th]].
10.5560/ZNA.2013-0077. Z.Naturforsch. 69 (2014) 106-107.

27) Thermal description of particle production in Au-Au collisions at RHIC energies (STAR).
By Abdel Nasser Tawfik, Ehab Abbas. arXiv:1311.7508 [nucl-th]].
10.1134/S154747711504024X. Phys.Part.Nucl.Lett. 12 (2015) 4, 521-531.

28) Some Intensive and Extensive Quantities in High-Energy Collisions.
By A. Tawfik.
[arXiv:1310.0718 [hep-ph]].
10.5506/APhysPolBSupp.7.17. Acta Phys.Polon.Supp. 7 (2014) 1, 17-26.

WLCAPP World Laboratory Cosmology And Particle Physics

Publications Since June 2013

29) Constant Trace Anomaly as a Universal Condition for the Chemical Freeze-Out.
By A. Tawfik.
[arXiv:1308.1712 [hep-ph]].
10.1103/PhysRevC.88.035203. Phys.Rev. C88 (2013) 035203.

30) Impacts of Generalized Uncertainty Principle on Black Hole Thermodynamics and Salecker-Wigner Inequalities.
By A. Tawfik.
[arXiv:1307.1894 [gr-qc]].
10.1088/1475-7516/2013/07/040. JCAP 1307 (2013) 040.

31) Chemical Freeze-Out and Higher Order Multiplicity Moments.
By A. Tawfik.
[arXiv:1306.1025 [hep-ph]].
10.1016/j.nuclphysa.2013.12.008. Nucl.Phys. A922 (2014) 225-236.

32) Hadronic Equation of State and Speed of Sound in Thermal and Dense Medium.
By Abdel Nasser Tawfik, Hend Magdy.
[arXiv:1206.0901 [hep-ph]].
10.1142/S0217751X14501528. Int.J.Mod.Phys. A29 (2014) 27, 1450152

Workshops and Conference

- ICTP External Scientific Meeting (SMR2619)
  Advanced Workshop on LHC Physics and Cosmology,
  February 3rd-14th 2014, Cairo-Egypt.

- 10th International MasterClasses in ALICE-Physics
  Nuclear Modification Factor and Strangeness Enhancement
  April 1st and 5th 2014, Cairo-Egypt.

- Third WLCAPP International Conference on Primordial QCD Matter in LHC Era: Implications of QCD Results on the Early Universe, August 24th-28th, 2014, Cairo-Egypt.

- 11th International MasterClasses in ALICE-Physics
  Nuclear Modification Factor and Strangeness Enhancement
  March 13th and 28th 2015, Cairo-Egypt.

Training/Orientation: Undergraduates

First Summer Bachelor-Training in Theoretical Physics

Aug. 2nd – 13th 2015

Target group:	3rd-level Physics-Students at Egyptian Public Universities
Nomination:	Head of the Physics Department
Duration:	70 hours
Expected Output:	Successful Working-Out a Concrete Physical Topic with Complete Documentation
Tactic Goals:	Attracting Young Generation to Currently-Running Theoretical Physics Research, Supplying Fresh-Bloods to Currently-Active Research Teams, and Maintaining Currently-Operating Research Activities,
Strategic Goals:	Fair Orientation for Young Generation, Interdisciplinary-Balance in Favor of Theoretical Physics, Uncovering Horizon of Theoretical Physics and Opportunities, Spreading Scientific Spirit, Indorsed Outreach to Elites, etc.

First Bachelor Training in Theoretical Physics: Target group: 3rd-level-physics-students at public Universities, Nomination: only head of physics department

1st week:

Day / Time	Sunday August 02, 2015	Monday August 03, 2015	Tuesday August 04, 2015	Wednesday August 05, 2015	Thursday August 6, 2015
10:00 - 11:00	Welcome	Dirk Rischke	Dirk Rischke	Sonia Kabana	Newly declared National Vacation Launching an extension to Sues Chanel
11:00 – 12:00	Introducing ECTP: Abdel Nasser Tawfik		Jun'ichi Yokoyama	Helmut Satz	
12:00 –1:00	Ernst-Michael Ilgenfritz				
1:00 – 2:00	Coffee Break				
2:00 –3:00	C-Language, Hayam Yassin	ALICE Measurement	ALICE Measurement	Virtual Visit ALICE Control Room	
3:00 –4:00	Gnuplot: Abdel Magied Diab	R_AA	Strangeness		
4:00 –5:00	Latex: Abdel Nasser Tawfik	Michael Tannenbaum	Michael Tannenbaum		

First Bachelor Training in Theoretical Physics: Target group: 3rd-level-physics-students at public Universities, Nomination: only head of physics department

2nd week:

Day / Time	Sunday August 09, 2015	Monday August 10, 2015	Tuesday August 11, 2015	Wednesday August 12, 2015	Thursday August 13, 2015
10:00 - 11:00	Ernst-Michael Ilgenfritz		Ludwik Dobrzynski	Ahmandjon Abdeljabbarov	Ludwik Dobrzynski
11:00 – 12:00	Amine Ahriche	David Blaschke	شمسي	Volker Friese	Hendrik van Hees
12:00 –1:00		David Blaschke	Kate Shaw	Ludwik Dobrzynski	
1:00 – 2:00	Coffee Break				
2:00 –3:00					
3:00 –4:00	Zhangbu Xu		Kate Shaw	Jean Cleymans	
4:00 –5:00	Ahmed M. Hamed	Paul Sorensen	Ahmed M. Hamed	Ahmed M. Hamed	

Foreigner Lecturers	Introductory Talks	Experimental measurement	Research Projects	Student presentations	coffee Break

WLCAPP World Laboratory Cosmology And Particle Physics

Future Plans

- Fourth WLCAPP International Conference on Primordial QCD Matter in LHC Era: Implications of QCD Results on the Early Universe, March 3rd - 7th, 2016, Cairo-Egypt.
- 12th International MasterClasses in ALICE-Physics *Nuclear Modification Factor and Strangeness Enhancement* Spring 2016, Cairo-Egypt.
- Light-Flavor Analysis for **ALICE** Collaboration
- Direct-Photon Analysis for **STAR** Collaboration
- Strangeness Analysis for **CBM** Collaboration
- Reconstruction and Models for mpdROOT (**MPD** Collaboration)

WLCAPP World Laboratory Cosmology And Particle Physics

Grazie!

Thanks!

شكرا!

UGANDA WFS SCHOLARSHIP PROGRAM REPORT 2013/2014

SAM MBULAITEYE
National Institutes of Health/NCI/DCEG, Infections and Immunoepidemiology Branch, Bethesda, MD, USA

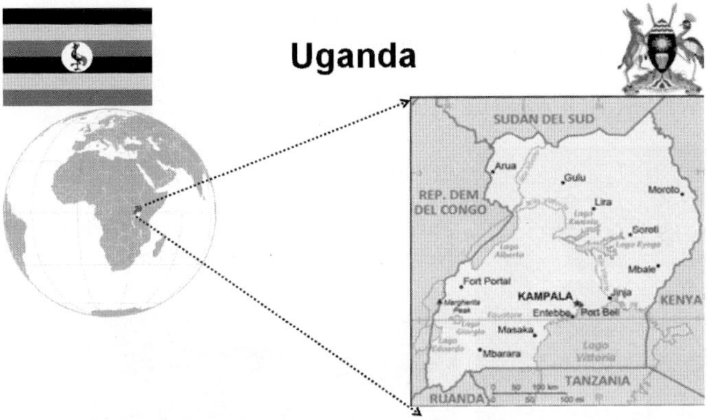

- Population: 35.0 million (2015)
- Annual growth rate 3.03% (projected to reach 47.4 million in 2025)
- Expanding pyramid structure
  - 49.9% <15 years
  - 48.1% 15-64 years
  - 2.1% 64+ years

UGANDA WFS SCHOLARSHIP PROGRAM

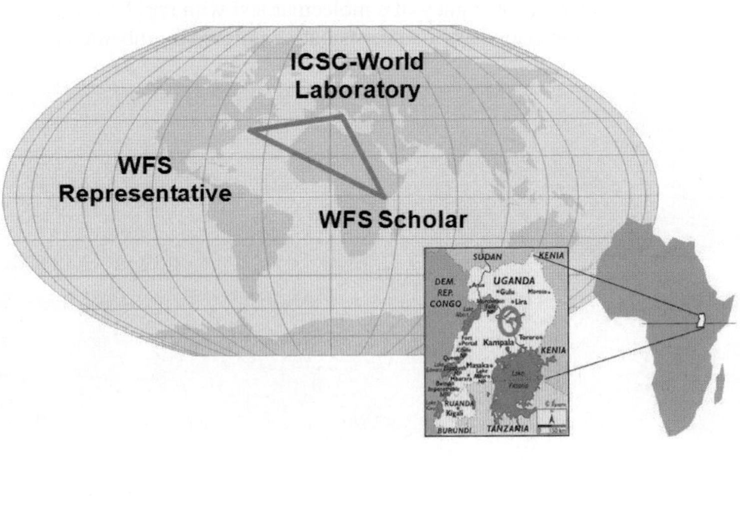

1. Science and technologies in developing countries
2. Medicine and biotechnology

October 2010	—	Nominated WFS Representative for Uganda
January 2011	—	Procedure to identify scholars
May 2011	—	First scholar nominated to WFS
May 2012	—	End of Year 1 for WFS Scholarship
2013-2014	—	Announce new call for 3 scholars

UGANDA WFS SCHOLARSHIP PROGRAM—YEAR 1

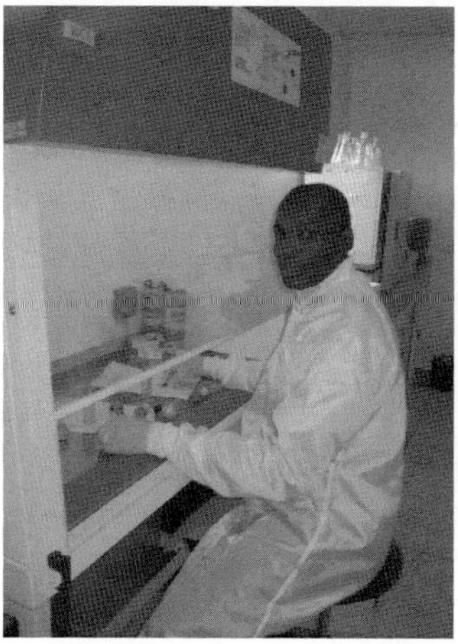

Simon Walusimbi

PhD student Makerere University/Karolinska Institute
New diagnostics technologies for Tuberculosis

Publication:
Meta-analysis to compare the accuracy of a molecular test with rapid culture and the WHO 2007 algorithm for diagnosis of smear-negative pulmonary tuberculosis.

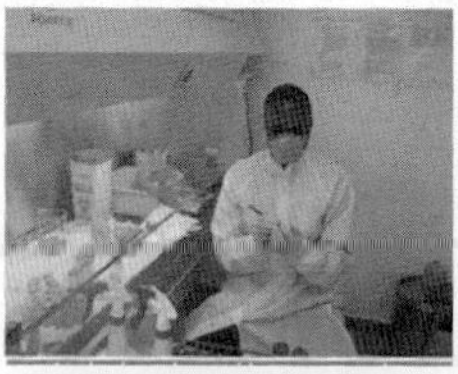

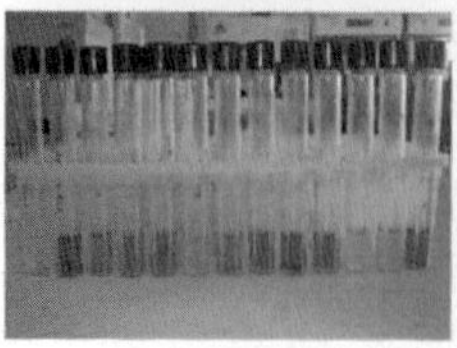

UGANDA WFS SCHOLARSHIP PROGRAM—YEAR 2

- September 2012—Call for Applicants
- 11 applicants were vetted locally
  - 6 females; 5 males
  - 3 biology; 1 veterinary medicine; 1 medical laboratory technology; 6 medicine
  - PhD in immunology, neonatology, parasitology, and medicinal plants
  - 3 were <35 years!
  - 3 Masters level applicants

UGANDA WFS SCHOLARSHIP PROGRAM—YEAR 1

Samuel Kirimunda

Medical Lab Technology
MPH student on PhD track Makerere University
HLA control in Tuberculosis
Application under review

Claire Aguttu

Veterinary Medicine PhD track Makerere University
Identification of hypothetical proteins of plasmodium as a vaccine
Waiting for Letter of support

LESSONS AND SOLUTIONS

1. Improve vetting process
   - Age
   - Document completion
2. Increase outreach to universities
   - Gulu, Mbarara
3. Consider Masters level projects

TAJIKISTAN NATIONAL SCHOLARSHIP PROGRAM: 2014-2015

NUSRAT RAJABOV
Department of Mechanics and Mathematics
Tajik State University
Dushanbe, Tajikistan

Tajikistan World Federation of Scientists National Scholarship Program at the lines began in October 2001.

Other former national scholarship fellows have become, in their time, well known leaders.

An example is Guljehra Kachorova, from Astrophysics Research Institute of the Academy of Science of the Republic of Tajikistan, who became director the Astrophysics Research Institute. Lutfaya Rajabova from Tajik Technical University began work in Majlisi Namoyandagon Majlisi Oli (Parliament) the Republic of Tajikistan, Director of Committee by Science Educations, Culture and Youth political Majlisi Namoyandagon Majlisi Oli (Parliament) the Republic of Tajikistan.

The following Tajikistan National Scholarship receivers were represented by the World Federation of Scientists and successfully finished their represented project.

SHUKUFA DAVLATOVA
PhD student
Research Institute of the Tajik National University.

She presented the following project at the problem of “Specific the features generation of the optoacoustic signals of the first and second sounds in super fluid solution$He^3 - He^4$ by different forms of the beam of laser impulse”.

(from 1 August, 2014, to 30 July 2015)

She’s supervised by Professor Salikhof Tafoymurat.

In this problem she obtained the following results:

1. The feature of the generation of optoacoustic (OA) of the pulses of the first and second sounds in super fluid solution $He^3 - He^4$ by the pulse square-shaped laser beam has been investigated.
2. The future of the generation of the acoustic waves of the first and second sound in super fluid $He^3 - He^4$ solution by Gaussian laser pulse through a thermal mechanism a theoretically has been investigated.
3. The specific features generation of the OA signals of the first and second sound in super fluid $He^3 - He^4$ solution by the non Gaussian form of the laser pulse and thermal mechanism has been investigated.

Thus under the draft of the present project nonlinear theory generation (OA) Optoacoustic signal (fundamental and second harmonic) by bilayer samples with the first transparent and the second opaque layers have been developed. For the important cases simple formulas which describing the dependence of the characteristics of the signal from the thermal parameters, their thermal coefficients, modulation frequency and intensity of the incident beam has been obtained.

In this problem she has published series of works. In particular the following scientific works:

1. T. Kh Salikhov, Sh. Davlatjonova, O. Sh. Odilov. Excitation of the optoacoustic pulses of the first and second sound in the super fluid solution $He^3 - He^4$ by rectangular impulse of laser. Vestnik TNU, № 1/1(№ 102.), Series of Natural Science, pp. 106-110, 2013(www.tnu.tj).
2. T. Kh .Salikhov, Sh. Davlatjonova, O. The feature of the generation of the acoustic waves of the first and second sound in the super fluid $He^3 - He^4$ solution by Gaussian leaser pulse. Thermal mechanism./Proceedings of the international scientific- practical conference dedicated to the 1150th anniversary of the Persian-Tajik scholar, scientist, doctor, philosopher and alchemist Abu Bakribn Muhammed Zakir Razi, Dushanbe, 2015, 27-28 May, Nikitin name the chemical institute of the Tajik Academy of Science, pp.76-77.
3. T. Kh .Salikhov, Sh. Davlatjonova, O. Excitation of optooacoustic waves of the first and second sound in super fluid solution $He^3 - He^4$ by the Gaussian laser pulse and the thermal for mechanism.// Report of the Tajik Academy of science (Docladi AN of Republic of Tajikistan, accepted publication, March, 2015).
4. T. Kh .Salikhov, Sh. Davlatjonova. The generation of the first and second sounds in the superfluid solution $He^3 - He^4$ by thenon – Gaussian pulse the laser beam. Thermal mechanism. Vestnic TNU, Series of Natural science, www.thu.tj, accepted for publication, May, 2015.

SUHROB ZARIPOV
Post graduate student
Research Institute of the Tajik National University

He presented the following project at the problem of: “Two-dimensional symmetric Volterra type Integral Equation with interior singular and super-singular point” (1.08.2014-31.07.2015).

His supervisor is Professor Nusrat Rajabov.

March 25-29, 2013

Attended the International Conference “Function Spaces. Differential Operators. General Topology. Problems of Mathematical Education” dedicated to Corresponding Member of RAS, Member of European Academy of Sciences Professor L.D. Kudryavtsev on the occasion of the 90th anniversary (Moscow RFUR, March 25-29).

This conference had a lecture by Professor A.S. Aparsyn and I.V Sidler on "Application of no classic Volterra equations of the first kind for modeling of development strategy of electric power systems". This problem reduce to investigation the first kind Volterra type Integral Equation, when upper limit and lower limit is function. In this connection are important investigations of a second kind of Volterra type Integral Equation, when upper limit and lower limits are variable.

The project Suhrob Zaripov dedicate to problem investigate two-dimensional Volterra type integral equation, when one the integrals have lower and upper integration limits.

Let D_0 denote the rectangle $D_0 = \{(x,y): -a < x < a, 0 < y < b\}$, correspondingly denote $D_0^- = \{(x,y): -a < x < 0, 0 < y < b\}$,
$D_0^+ = \{(x,y): 0 < x < a, 0 < y < b\}, \Gamma_0 = \{-a < x < a, y = o\}$,
$\Gamma_1 = \{x = 0, 0 < y < b\}$.

In the domain $D = D_0 \backslash \Gamma_1$ we shall consider the following two-dimensional Volterra type integral equation

$$\varphi(x,y) + \int_{-x}^{x} \frac{A(t)\varphi(t,y)}{|t|^\alpha} dt + \int_{0}^{y} \frac{B(s)\varphi(x,s)}{s^\beta} ds +$$
$$+\int_{-x}^{x} \frac{dt}{|t|^\alpha} \int_0^y \frac{E(t,s)\varphi(t,s)}{s^\beta} ds = f(x,y), \tag{1}$$

where $A(x)$, $B(y)$, $E(x,y)$ – are given function in Γ_0, Γ_1, D, $\varphi(x,y)$ – is the desired function $\alpha =$ const.>0, $\beta =$ const.>0 . Moreover the function $A(x)$ in point $x = 0$ may be have first king singularity.

The solution of the integral equation (1) is sough in the class of functions $(x,y) \in C(\overline{D})$, $\varphi(0,0) = 0$ wiyh the asymptotic behavior
$\varphi(x,y) = o[|x|^\varepsilon y^\varepsilon] \varepsilon > 0$at $\alpha = 1, \beta = 1$ and $(x,y) \to (0,0)$,

And

$\varphi(x,y) = o\left[|x|^{\delta_1} y^{\delta_2}\right], \delta_1 > \alpha, \delta_2 > \beta$

at $\alpha > 1, \beta > 1$ and $(x,y) \to (0,0)$.

Investigation one-dimensional, two-dimensional and other cases many dimensional Volterra type integral equation with one and two singular lines and domains, dedicate:

1. Rajabov N. *Volterra-type Integral Equation with Boundary and Interior Fixed Singularity and Super-singularity Kernels and Their Application.* LAP LAMBERT Academic Publishing, Germany, 2011, 282p.
2. Rajabov N. "About New Class of Volterra type Integral Equation with Boundary Singularity in Kernels, Advances in Applied Mathematics and

Approximation Theory." *Springer Proceedings in Mathematics and Statistics* 41, 2013, USA, 41-60 p.
3. Rajabov, N. "To theory one class of Volyerra type Integral Equation with fixed boundary singularity in Kernel." *Doklady Academy of Sciences*, 2014, v. 459, № 1, pp. 1-5.
4. Rajabov, N., Rajabova, L. *Introduction to theory many-dimensional Volterra type integral equation with fixed singularity and super-singularity kernels and its applications.* LAP LAMBERT. Leipzig: Acad. Publ. 2012. 502p.
5. Rajabov, N. "Many-dimensional Volterra type integral equation with boundary singular domains in kernels." *Russian Academy of Science.* Doklady, 2011, v. 473, №2бззю 1-3ю.

In this project investigate integral equation (1), when function present in among themselves connection that is, when, $E(x,y) = A(x)B(y)$.

1). In the cases when $B(0) < 0$, $A(+0) - A(-0) < 0$, we have the following confirmation.

Theorem 1.

Let in integral equation (1), $E(x,y) = A(x)B(y)$, $A(x) \in C(\Gamma_0 \backslash \{0\})$ *and in point* $x = 0$, $A(x)$ *satisfied Holder condition and* $A(+0) - A(-0) < 0$. Function $B(y)$ in neighborhood point $= 0$, *satisfy Holder condition and* $B(0) < 0$. *Function* $(x,y) \in C(\overline{D})$, $f(0,0) = 0$ with

asymptomatic behavior

$f(x,y) = o\left[|y|^{\delta_3} y^{\delta_4}\right]$, $\delta_3 > |A(+0) - A(-0)|$, $\delta_4 > |B(0)|$ *at* $(x,y) \to (0,0)$.

Then general solution integral equation (1) in the class $C(\overline{D})$ *given by formula*

$$\varphi(x,y) = \begin{cases} K_1[c(x), c_1(y)] + K_2[f(x,y)] \\ K_3[c(x), c_1(y)] + K_4[f(x,y)] \end{cases},$$

wherec$(x), c_1(y) -$ *arbitrary function point* Γ_0, Γ_1, $K_1 (1 \le j \le 4)$ it is well knoun integral operators.*Moreover* $c(x) \in C(\overline{\Gamma_0})$, $c(0) = 0$ *with asymptotic behavior*

$c(x) = 0\left[|x|^{\delta_5}\right]$, $\delta_5 > |A(+0) - A(-0)|$, *at* $x \to 0$,
$c_1(y) \in C(\overline{\Gamma_1})$, $c_1(0) = 0$ *with asymptotic behavior*
$c_1(y) = 0[y^{\varepsilon}]$, $\varepsilon > 0$, *at* $y \to 0$.

Remark 1.

Analogous results similar to theorem 1 obtained in the following cases:
1.) $A(+0) - A(-0) > 0$, $B(0) > 0$; 2) $A(+0) - A(-0) < 0$, $B(0) > 0$; 3) $A(+0) - A(-0) > 0$, $B(0) < 0$.

In the case, when $\alpha = 1$ and general solution integral equation (1) consider two or one arbitrary function one variable, for integral equation (1) stand and investigate different boundary value problems.

Example in the case, when $A(+0) - A(-0) < 0$, $B(0) < 0$ stand and investigate the following problem:

Problem R_1.

Is required found the solution integral equation (1) at $(+0) - A(-0) < 0$, $B(0) < 0$ by following boundary conditions:

$$\begin{cases} \left(y^{B(0)}\varphi(x,y)\right)_{y=+0} = D_1^+(x), 0 < x < a \\ \left(y^{B(0)}\varphi(-x,y)\right)_{y=-0} = D_2^-(x), -a < x < 0 \\ \left(|x|^{A(+0)-A(-0)}\varphi(x,y)\right)_{x-0} = D_3^+(y), 0 < y < b \end{cases},$$

where $D_1^+(x)$, $D_2^-(x)$, $D_3^+(y)$ - are given function points Γ_0, Γ_1 .

2.) In the cases, when$B(0) < 0$, $A(+0) - A(-0) < 0$, $\alpha =$ const.>1, $\beta =$ const.>1 we have the following confirmation.

Theorem 2.

Let in integral equation (1), $E(x,y) = A(x)B(y)$, $\alpha =$ const.>1, $\beta =$ const.>1,$A(x) \in C(\Gamma_0\backslash\{0\})$ *and in point* $x = 0$ *have first kind singularity,* $A(x)$ *satisfied the following condition*

$$|A(x) - A(\pm 0)| \le H_1|x|^\gamma, \gamma > \alpha - 1$$

and

$A(+0) - A(-0) < 0$. Function $B(y)$ in neighborhood point$y = 0$, *satisfy the following condition*

$$|B(y) - B(0)| \le H_2|y|^\delta, \delta > \beta - 1$$

and

$B(0) < 0$. *Function* $(x,y) \in C(\overline{D})$, $f(0,0) = 0$*with asymptotic behavior* $f(x,y) = o\left[|x|^{\delta_6}y^{\delta_7}\right]$, $\delta_6 > \alpha - 1, \delta_7 > \beta - 1$ at $(x,y) \to (0,0)$.

Then general solution equation (1) from class $C(\overline{D})$ *given by formula*

$$\varphi(x,y) = \begin{cases} K^1_{\alpha,\beta}[c(x), c_1(y)] + K^2_{\alpha,\beta}[f(x,y)] \\ K^3_{\alpha,\beta}[c(x), c_1(y)] + K^4_{\alpha,\beta}[f(x,y)] \end{cases},$$

where $c(x), c_1(y) -$ *arbitrary function point* Γ_0, Γ_1, $K^j_{\alpha,\beta}(1 \leq j \leq 4)$ it is well known integral operators. *Moreover* $c(x) \in C(\overline{\Gamma_0})$, $c(0) = 0$*with asymptotic behavior*

$c(x) = 0\left[|x|^{\delta_8}\right]$, $\delta_8 > \alpha - 1$, *at* $x \to 0$,
$c_1(y) \in C(\overline{\Gamma_1})$, $c_1(0) = 0$ *with asymptotic behavior*
$c_1(y) = 0\left[y^{\delta_9}ul\right]$, $\delta_9 > \beta - 1$, *at* $y \to 0$.

Remark 1.

Analogous results similar to theorem 2 obtained in the following cases: 1) $A(+0) - A(-0) > 0, B(0) > 0$; 2) $A(+0) - A(-0) < 0$, $B(0) > 0$; 3) $A(+0) - A(-0) > 0, B(0) < 0$

at

$\alpha > 1, \beta > 1.$

In this case, when general solution integral equation (1) at $\alpha > 1$, $\beta > 1$ contains arbitrary functions one variable, for equation (1) it is possible stand and investigate different boundary value problems.

Example, when $A(+0) - A(-0) > 0$, $B(0) < 0$ for equation (1) at $\alpha > 1$, $\beta > 1$stand the following problem

Problem R_2.

Is required found the solution integral equation (1) at$\alpha > 1$, $\beta > 1$, $A(+0) - A(-0) < 0$, $B(0) < 0$ by following boundary conditions

$$\left(exp\left[-(A(+0) - A(-0)\omega_\alpha(x))\right]\varphi(x,y)\right)_{x-0} = D_1^{+,\alpha}(y),$$

where$D_1^{+,\alpha}(y)$ —are given function point Γ_1.

At this problem Mr. Suhrob Zaripov has published series of works. In particular, the following works:

1. Rajabov, N., Zaripov, S. "About one two-dimensional Volterra-type Integral Equation with one boundary and one interior singular lines// Materials 11-young school scientific." Russian Federation, Russia, Trescol, 4-8 December, 2013, pp. 52-55.
2. Rajabov, N., Zaripov, S. To theory one class of symmetrical two-dimensional Volterra-type integral equation with one boundary and one interior singular lines//Report of the Academy of sciences of the Republic of Tajikistan, v. 56, № 12, 2013, pp. 962-970.
3. Rajabov, N., Zaripov, S. "Boundary value problems for one class of two-dimensional Volterra-type integral equation with two singular lines, symmetrical with one of the variables// Modern problems of mathematics and

teaching." Proceedings of the International scientific conference dedicated to the 20th anniversary of the constitution Republic of Tajikistan and 60th anniversary mathematical Learners A. Muhsinov, A.B. Nazimov, S. Bayzoev, D. Ocimi, K. Tuchliev. Learned a report Khujand State University, №2(29)б28Огту 2014, p. 5-13.

4. Rajabov, N., Zaripov, S. To theory one class of symmetrical two-dimensional Volterra type Integral Equation with one boundary and one interior super-singular lines//Report of the Academy of sciences of the Republic of Tajikistan, v. 58, № 1, 2015, pp. 5-13.
5. Rajabov, N., Zaripov S. About one two-dimensional Volterra type Integral Equation with one boundary and one interior singular lines//Modern problem differential equation and its Application. Materials the Republic Conference. Dushanbe, 23 May, 2015. p. 111-113.

2014-2015 REPORT ON WFS ROMANIAN NATIONAL SCHOLARSHIP PROGRAMME

GHEORGHE STRATAN
National Institute of Physics and Nuclear Engineering
Bucharest-Magurele, Romania

SUMMARY RESULTS

WFS Scholarship recipients for 2014-2015: 10 (eight for one year, two for six months). The present report deals with their final results.

WFS Scholarship recipients for 2015-2016: 10 (all of them for one year). They are in the first stage of the scholarship and their final results will be reported next year.

BUJDEI ANDREEA-ANCUŢA

(MS, 6 months WFS support) Scientific Supervisor, Assoc. Prof. Radu Silaghi-Dumitrescu

Interaction of globins with sulphide (S^{2-}) and with sulphur oxides and oxyanions ($AxOy^{Z-}$).

- The objective of research was the study of the interaction of globins (oxiHb, metHb, deoxiHb) from bovine blood Hb with sulphide, sulphur oxides and oxyanions. The sulphides bind irreversible to the metals like ferrous (Fe^{2+}) and ferric (Fe^{3+}) iron from haemoglobin in the red blood cells. This can stop the haemoglobin from transporting oxygen to the cells.
- The modifications of globines at different concentrations of sulphide and sulphur oxides and interaction times were studied using the UV and visible spectroscopy

DEAC AGLAIA RALUCA

(PhD, 1year support) Scientific Supervisor, Prof. Dr. Liana Mureşan

Nanomaterials with biomedical and environmental protection applications.

Bi impregnated carbon xerogel (BiCXe), is presented as a promising alternative to electrode materials used to replace mercury electrode in detection of heavy metals ions from aqueous solutions. The result of this work is the preparation and characterization of BiCXe from both morpho-structural and electrochemical points of view in order to obtain a sensor for Cd^{2+}, Pb^{2+} detection. (1g BiCXe = 900sqm).

DEAK NOEMI
(PhD student, 1 year WFS) Scientific Supervisor, Prof. Dr. Luminiţa Silaghi-Dumitrescu

Towards new metallylenes of group 14 elements (Ge, Sn): Synthesis, characterization and evaluation of possible applications in catalysis.

The metallylenes, the heavier analogs of carbenes were studied. Their stabilization was realized with different organic ligands with steric and electronic effect. Pincer-type ligands were used to stabilize metallylenes. These ligands are versatile; their properties can be easily modified by changing the donor group, the length of the arm, the substituents on the aromatic ring. A new sulfur based pincer-type ligand was synthetized. The ligand was then used for obtaining new metallylenes and their transition-metal complexes.

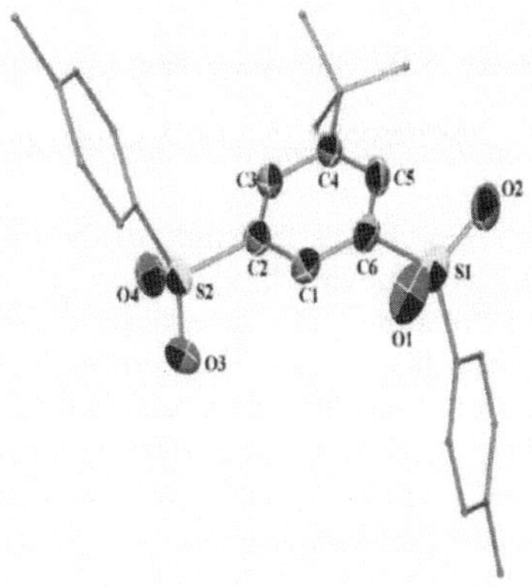

Solid state structure of the new bis-sulfonyl ligand.

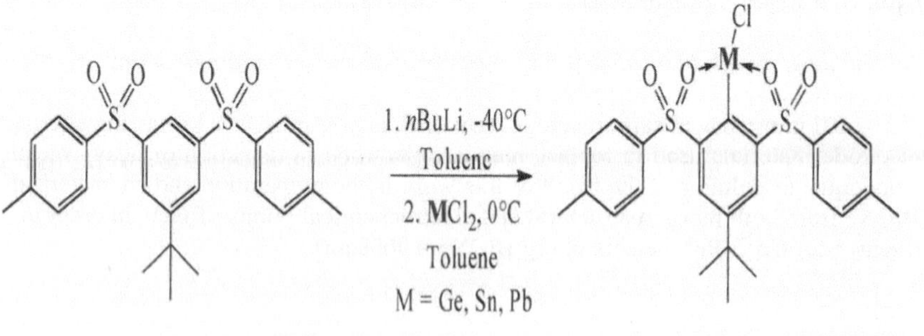

Synthesis of the new metallylenes.

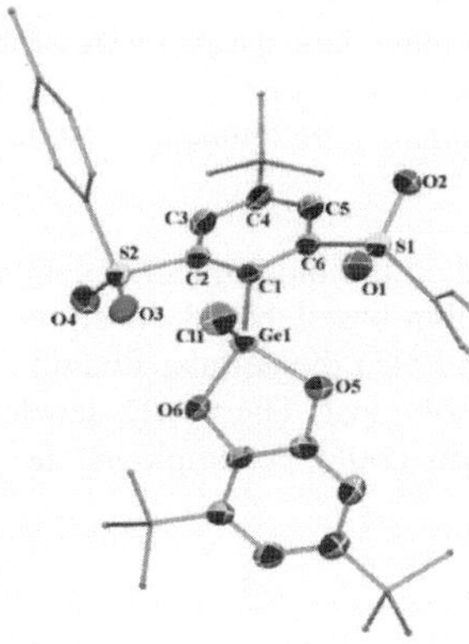

Solid state structure of the germylene-ortho-benzoquinone cycloadduct.

MIHÁLY RÉKA
(PhD, 1 year) Scientific Supervisor, Dr. Tamás Lovász

Synthesis of New Chiral Aminonaphtols and their Applications in Catalysis

Direct aminoalkylation has great interest in synthetic organic chemistry and considerable importance for the synthesis of drugs, pesticides and natural products.

Investigation of phosphination reactions of the alkylated aminonaphthols and measurement of the catalytic activity in the case of aminonaphthols was achieved.

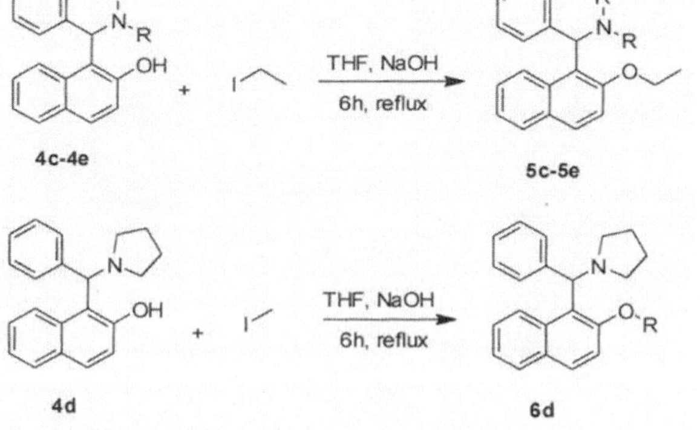

Synthesis of alkylated aminonaphthols 5c-5e, and 6d.

ANCA-DANIELA FARCAŞ
(MS, 1 year) Scientific Supervisor, Assoc. Prof. Radu Silaghi-Dumitrescu

Biochemical and acid-base parameters in animals exposed to newly-developed blood substitutes.

The WFS fellow investigated the blood biochemical and acid-base parameters in Wistar rats exposed to hemoglobin-based blood substitutes in order to examine the biochemical status of the body and also the cellular functional integrity. There are some modifications, but not physiopathological. The newly-developed blood substitutes, both with hemoglobin and hemerythrin could be employed in order to demonstrate their clinical potential.

ALEXANDRINA ZUZA
(PhD, 1 year)
Scientific Supervisor, Prof. Dr. Paul Şerban Agachi

Modeling and simulation of anaerobic digesters in the production of biogas.

Anaerobic digestion is a biological process that converts the organic matter present in various types of wastes, activated sludge from the wastewater treatment facilities respectively, into biogas (methane, carbon dioxide, nitrogen and oxygen). The mathematical modeling, simulation and control practices have helped in gaining a better insight of the process. In this research, modeling and simulation, energy efficiency and techno-economical investigation of anaerobic digestion technology for the CHP (Combined Heat and Power) cogeneration unit, has been done to detect maximum concentrations of methane and to minimize the costs at Cluj-Napoca Municipal Wastewater Treatment Plant.

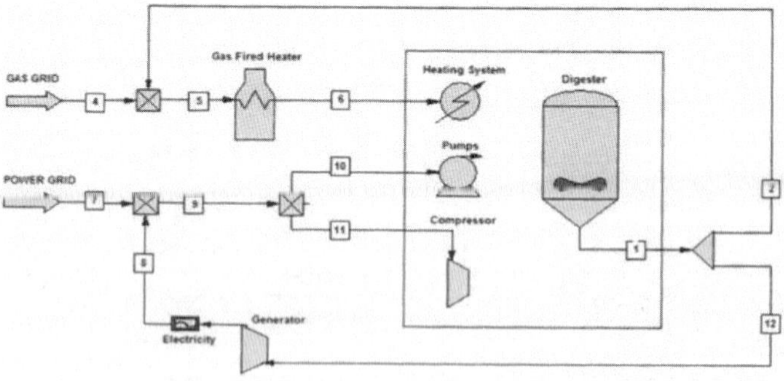

CHP cogeneration unit schematic.

CARMEN MÂNZATU
(PhD, 1 year) Scientific Supervisor: Associate Prof. Majdik Cornelia

*Biosorption study of organic and inorganic pollutants using Romanian pine cones (*Pinus sylvestris*)*

Biosorption study of untreated pine cones (*Pinus sylvestris*) biomass has proven to be a highly effective and promising adsorbent for the removal of cadmium and zinc ions from aqueous solutions. This biosorbent is widely available as a natural material, is mechanically stable, and most important, it is eco-friendly. Investigating the biosorption equilibrium in function of contact time at varying initial Cd(II) concentrations, it can be concluded that maximum adsorption capacities were achieved almost in the first 80 min of contact.

Removal efficiency

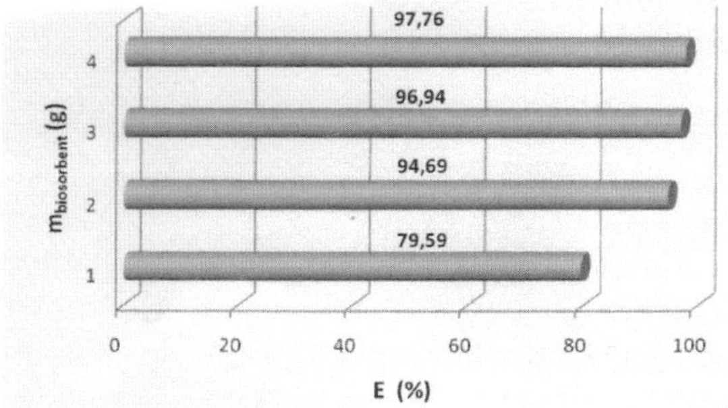

The effect of the pine cone quantity on Cd(II) biosorption over the removal efficiency; C_i = 49 mg Cd(II)/L, d = 200-400 μm, 296 K, 5.41pH and 500 rpm.

Evolution in time

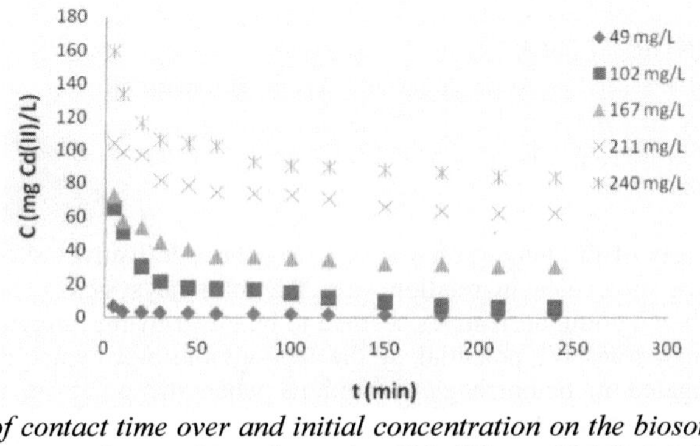

Effect of contact time over and initial concentration on the biosorption of Cd (II) ions on fir cone biosorbent; d = 200-400 μm, 296 K, 5.4pH, 500 rpm.

ÉVA ANDREA MOLNÁR
(MS, 1 year)
Scientific Supervisor, Lect. Dr. Tamás Lovász

Phenothiazinyl—substituted porphyrins.

Porphyrins, the organic macrocyclic compounds with aromatic properties, are potentially useful as chemical sensors, nonlinear optical materials, molecular wires, unidimensional (semi)conductors and organic solar cells and photosensitizers for artificial photosynthesis and photodynamic therapy of skin cancer.

The synthesis of four phenothiazine derivatives, four new phenothiazinyl-porphyrins, and their metal complexes was achieved. The obtained products were characterized by NMR measurements. The optical properties of these new compounds were investigated by UV and visible spectroscopy methods.

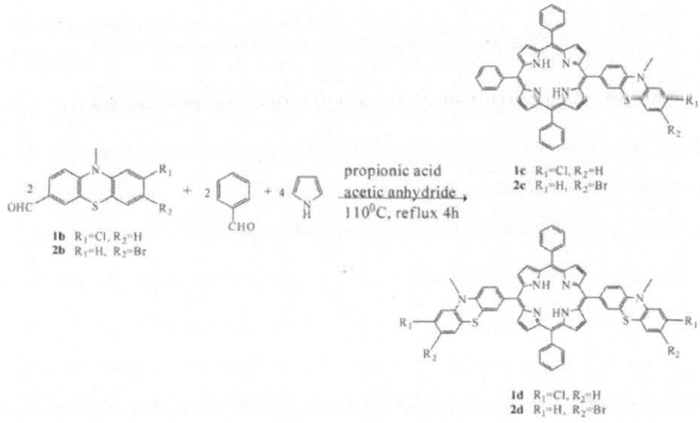

Synthesis of phenothiazinyl-porphyrins.

VLAD-ALEXANDRU TOMA
(MS, 1 year) Scientific Supervisor, Assoc. Professor Radu Silaghi-Dumitrescu

Immunological Study of the Reaction to Newly-Developed Blood Substitutes on Laboratory Animals.

In vivo tests of the hemoglobin and hemerythrin derivatives showed a different behavior of these molecules in relation with the immune system. The immunogenic potential of the hemoglobin derivatives seemed to be lower than hemerythrin derivatives. However, the high transport potential of the hemoglobin and hemerythrin derivatives must be investigated in hemorrhagic conditions when the organism requires among others, respiratory gas transport.

TÖRÖK ANAMARIA

(PhD student, 6 months) Scientific Supervisor, Associate Prof. Majdik Cornelia

The phytoextraction of heavy metals with selected aquatic plants and determination of the plant biochemical response

Heavy metals in the aquatic medium may originate from wastewater of many industries such as tanneries, electrical, batteries, fertilizers, pesticides or mining etc. plant cultures with the hyperaccumulator aspect such as rockcress *(Arabis paniculata, Arabidopsis halleri),* stonecrops (*Sedum alfredii),*[3] yellowtuft (*Alyssum murale), or wonder berry (Solanum nigrum).*

CONCLUDING REMARKS

- The 2014-2015 group of ten receivers of WFS Scholarship Programme shows a very close connection with the main directions of WFS Planetary Emergencies, often with 2-3 of them.
- The high efficiency demonstrated by the receivers is a valuable argument to continue in the next years the WFS National Scholarship Programme for Romania.
- For all of the receivers, the work accomplished was the beginning of a very fruitful scientific career.
- It is now our duty to acknowledge the WFS for the generous support of Romanian applicants for the WFS Programme, and first of all, the WFS President, Professor Antonino Zichichi. Our gratitude goes also to all collaborators of WFS.

WFS UKRAINIAN H.E.P. NATIONAL SCHOLARSHIP PROGRAMME: ALICE LHC ACTIVITY 2013–2014

GENNADY ZINOVJEV
Bogolyubov Institute for Theoretical Physics, National Academy of Sciences of Ukraine, Kiev, Ukraine

2013: TWO GRANTEES

Anton Alkin—Processing and analysis of the ALICE data.

Violetta Sagun—Analysis of the ALICE data and developing scenario of relativistic heavy ion collisions, equation of state of strongly interacting matter.

2014: TWO GRANTEES

Volodymyr Chelnokov—Lattice field theory simulations, equation of state for strongly interacting matter.

Oleksandr Kovalenko—Analysis of proton-proton (antiproton) and proton-nucleus collisions with the ALICE, ATLAS, CMS and TOTEM data.

NEED

To increase our Grid pledges in WLCG and ALICE.

Studying the phase structure of strongly interacting matter is one of the most complicated tasks of up-to-date theoretical development and any theoretical achievement here is highly non-trivial being very indicative and guiding for planning and designing next round of physics experiments.

It concerns not only the heavy ion experiments which are very instrumental to analyze phase diagram of QCD matter at high T and low baryon number density but also the fundamental astrophysical research as being very informative to trace phase diagram at low T and very high baryon number densities.

Investigation of color superconductivity problem within such a strategy provide us with new look at the QCD ground state (quark condensate) and shed a new light on possible QCD phase structure.

- Theory of liquids and gases relies on number theory and geometry.
- Most important geometric representation is based on the equation of state that is treated as a two-dimensional surface in a three-dimensional contact manifold.
- The latter is constructed from four-dimensional phase space in which intensive variables (pressure P, temperature T and chemical potential μ) are taken for coordinates.

- The momentum corresponding to the coordinates (P, T, μ) are the volume V, negentropy -S and number of particles N.
- Passage to the three-dimensional contact space is related to the fact that actually V and N have never been considered separately but always as a density ratio $\rho = N/V$.
- Equation of state gives rise to the two-dimensional Lagrangian surface that describes dependence of two variables, say, N and -S on other two μ and T.

As a rule, Lagrangian surface has singular points of the projection from four-dimensional symplectic manifold or three dimensional contact manifold in which it is embedded.

These points are referred to as caustics or focal points (optics) and turning points (quantum mechanics).

Then well-known spinodals from the geometric view point are none other than simple caustics while the point of intersection of two caustics (gas and liquid branches of the spinodal) is just the critical point.

OUTPUT

It was shown that the Bose-Einstein distribution describes the van der Waals gas while the Fermi-Dirac distribution describes the van der Waals liquid. Liquid-to-crystalline transition in disperse system occurs as liquid-to-amorphous-solid (glass) transition and only then glass is transformed into crystalline.

Why is it of great importance for ultra relativistic heavy ion collisions?

- Scenario of a collision process
- Initial conditions in hydrodynamical approach
- Phase diagram
- Three-critical point
- Crossover
- Quark condensates
- Structure of the QCD ground state
- Mixed phase of baryonic matter and QCD vacuum
- Whether QCD is really deconfining (infinitely increasing correlation length)

WORLD FEDERATION OF SCIENTISTS SCHOLARSHIP PROGRAM IN BULGARIA: REPORT (2013-2014)

STEFAN HADJITODOROV
Bulgarian Academy of Sciences
Sofia, Bulgaria

- Established in late 1998
- Started in 1999
- So far 103 young Bulgarian Scholars have been granted a scholarship in the field of 13 Planetary Emergencies

2013/2014—9 scholars

- The research conducted have been supervised by eminent scientists/instructors from different Institutes of the BAS; University of Sofia and Technical University
- The final selection—made by the Scientific Secretariat of BAS

SCHOLARS DISTRIBUTION

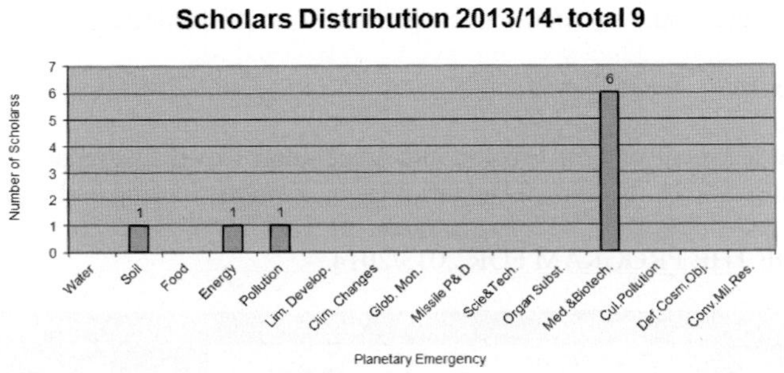

Females – 7 Males – 2

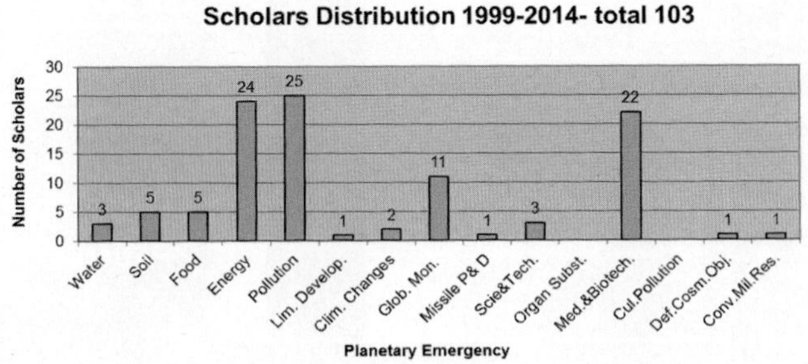

Females – 61 Males – 42

MEDICINE AND BIOTECHNOLOGY—6 scholars

- The effect of the non-histone HMGB1 protein on the antitumor activity of the drug cis-platinum tested in different human tumor cell lines;
- Differential scanning calorimetry of blood plasma—application for cancer diagnostics;
- Diagnosis of colon and rectum cancer. Optical spectroscopy of lower tract gastrointestinal tumors and metastatic lymph nodes;
- Early diagnosis of skin neoplasia;
- Effects of culture conditions on the expression of genes responsible for the synthesis of polysaccharide antigens of E. coli 0157;
- The role of melatonin system on disturbed diurnal behavioral rhythms in kainite model of temporal lobe epilepsy.

POLLUTION—1 scholar

- Trace elements in rocks and minerals from mining affected regions in Bulgaria as source for potential environmental pollution.

ENERGY—1 scholar

- Synthesis and structural investigations of multifunctional materials.

SOIL—1 scholar

- Assessment of Soil Erosion and Vegetation Cover Degradation in Chuprene Reserve (Northwestern Bulgaria) Using Remote Sensing and GIS.

RESULTS OF THE PROGRAM FOR 2013/2014

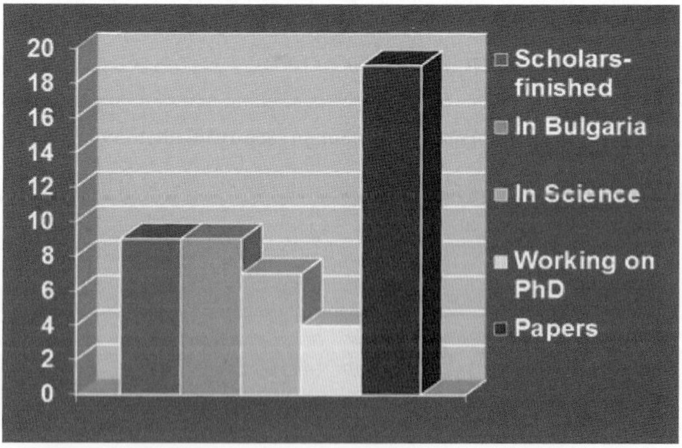

From 9 Scholars: 9 remained in Bulgaria, 7 working in the field of science, 4 currently working on their PhD thesis, 19 published papers (5 in Journals, 14 in Conference Proceedings).

CE(X) = X / Number of scholars

- Coeff. Efficiency (usefulness) (Bulg.) = 1.00
- Coeff. usefulness (Sci.) = 0.77
- Coeff. usefulness (Ph.D.) = 0.44
- Coeff. usefulness (Papers) =2.11

CONCLUSION

- The Program is efficient and successful.
- Simple and flexible rules.
- Effective control—6 months report.
- Modest support, but advantage and merit of being delivered immediately.

THANK YOU FOR YOUR ATTENTION!

STATUS UPDATE ON WFS IN SOUTH AFRICA

ZEBLON VILAKAZI
University of the Witwatersrand
Johannesburg, South Africa

MAJOR ACHIEVEMENT 50 YEARS AGO

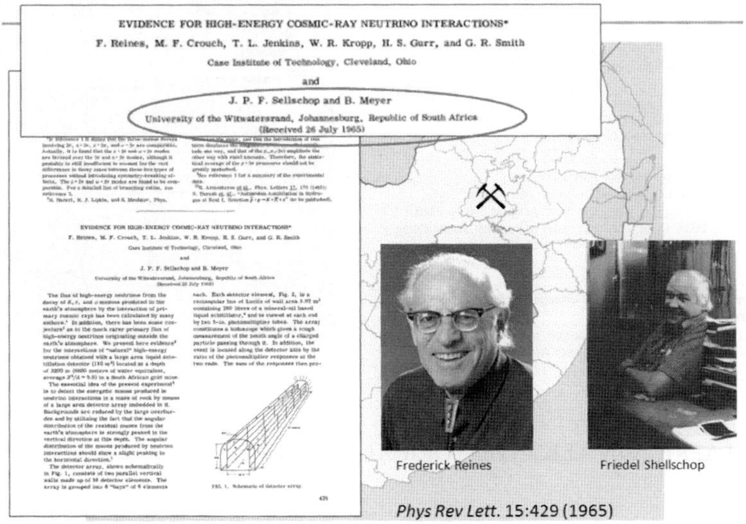

AFRICA AT CERN

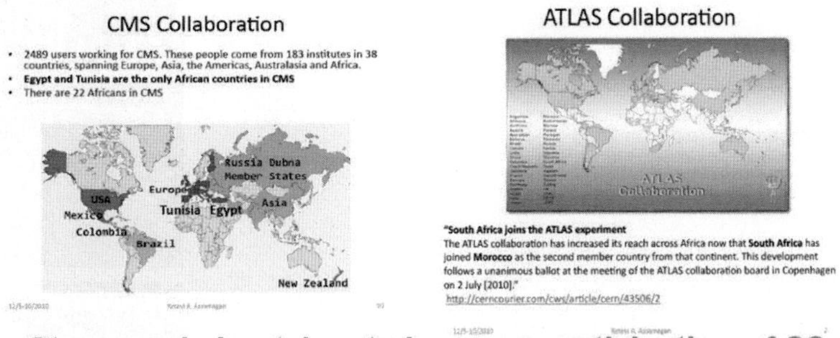

Steps are being taken to increase participation of SS Africa in this global scientific enterprise–with South Africa (and other African partners; e.g., Egypt) as a pivot/gateway

OBJECTIVES

- **Increase capacity in fundamental physics and related applications**
  - Organizing a biennial school rotating in different African Countries
  - Networking among African Students
  - Sharing of information on high education opportunities around the world
  - Part of the global Science for Peace and *Physicien sans frontier* project
- **First edition of the school, ASP2010**
  - August 1-21, 2010 (3 weeks) in Stellenbosch, South Africa. Attended by 65 students from 17 countries
- **Second edition of the school, ASP2012**
  - At the Kwame Nkrumah University of Science and Technology (KNUST), Kumasi Ghana. July 15 – August 8, 2012 (3.5 weeks). Attended by 50 students from 15 countries
- **Third edition of the school ASP2014**
  - Was hosted by the Cheikh Anta Diop, Dakar University, Senegal, from **August 3rd to 23rd, 2014**, to benefit from some of the best scientists in high energy physics and share their knowledge and expertise
- **School participation is free of charge to the selected students**
  - For ASP2012 for example, we received 132 applications from all over Africa, of which we selected 50

Ketevi A. Assamagan

SOUTH AFRICA IN ALICE: HISTORICAL EVENTS

- UCT joined 2001, became UCT-CERN Research Centre in 2003.

- iThemba LABS joined in 2008

- Launch of the National SA-CERN Programme in 2008.

SOUTH AFRICA GROUP

A Large Ion Collider Experiment

Tom Dietel (UCT) Team-Leader, Member of the Collaboration Board and Conference Committee

Jean Cleymans (UCT and iThemba LABS) Deputy-Team Leader, Member of Institutional Board (Muons), Chairman of SA-CERN Program

Zinhle Buthelezi (iThemba LABS) Member

Siegfried Förtsch (iThemba LABS) Deputy-Team Leader

Deon Steyn (iThemba LABS) Member

Francesco Bossù (iThemba LABS) Post-doctoral Member, Convener of PAG-HFM

Z. Vilakazi Wits joined ALICE 28 March 2014 in association with UCT

SOUTH AFRICA GROUP: STUDENT ACTIVITIES

A Large Ion Collider Experiment

Sean Murray (UCT and iThemba LABS) PhD, Interim System Administrator of CHPC
Research topic: *High-Level Trigger for the ALICE Muon Detector*

Kgotlaesele Johnson Senosi (UCT and iThemba LABS) PhD
Research topic: *W production in pp, P-Pb and Pb-Pb collisions at LHC energies*
WRF funding – ended in 2013

Sibaliso Mhlanga (UCT and iThemba LABS) MSc
Research topic: *Heavy Flavour production vs. multiplicity at forward rapidity with ALICE*
Attended CERN Summer School 16 June - 8 August 2014 (not funded)

A Large Ion Collider Experiment

Di-lepton and Quarkonia Physics (Upsilon)

ALICE

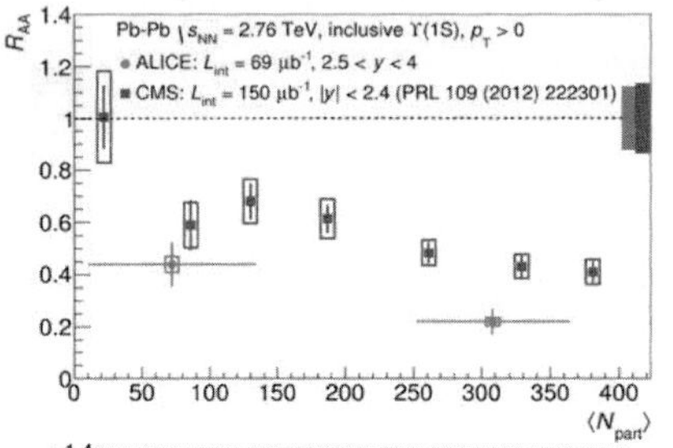

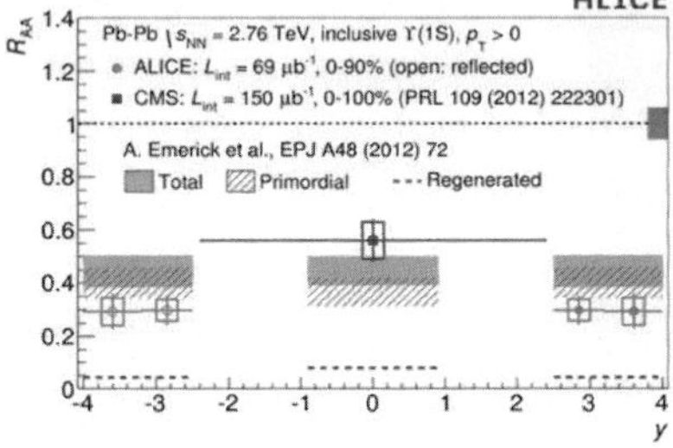

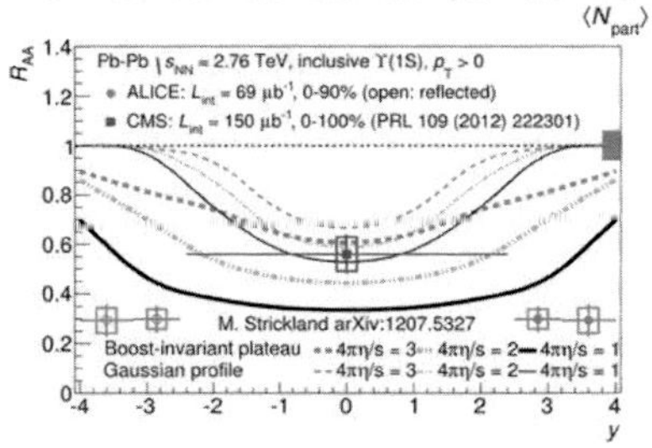

Published: Phys Lett B 738 (2014) 361–372
Upsilon-study: *Suppression of Y(1S) at forward rapidity in Pb-Pb collisions at $\sqrt{s_{NN}}$ = 2.76 TeV*

- Strong suppression in forward direction
- Not reproduced by models

A Large Ion Collider Experiment

Heavy Flavour Muons:

- W boson production measurement in p-Pb collisions with ALICE

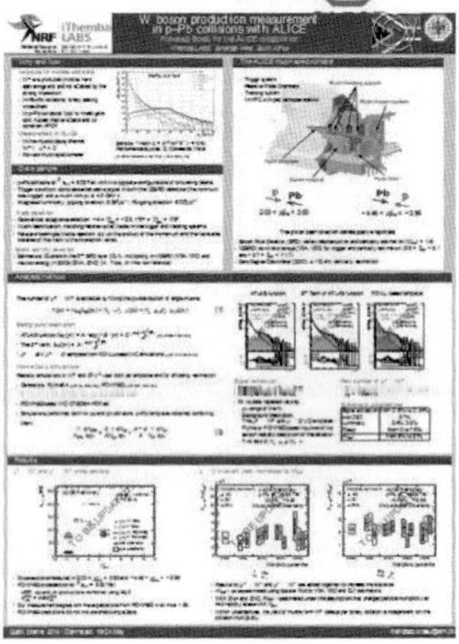

- $W^{\pm}$ are produced in initial hard scatterings and are not affected by the strong interaction.
- In Pb-Pb collisions: test the scaling of hard processes with the number of binary nucleon-nucleon collisions.
- In p-Pb collisions: investigate cold nuclear matter effects and constrain nuclear PDF.

Measurement in ALICE

- In the muonic decay channel $W^{\pm} \rightarrow \mu^{\pm} + \nu$
- Forward muon spectrometer

Preliminary results presented at QM2014, Darmstadt, 19-24 May 2014

- Yield of W versus event activity → test the dependence of this hard process with the event activity.
- Cross-section → is sensitive to cold nuclear matter effects (shadowing).
- Comparison with theoretical predictions based on pQCD NLO with and without shadowing shown in JHEP 1103 (2011) 071 → Shadowing effects are more pronounced at forward rapidity.

Report on student progress WFS Scholarship (I)

- Title: "*Analysis of W bosons with ALICE: Effect of alignment on W Bosons analysis*" : Pieter d[illegible] (University of Pretoria)

- Thesis was submitted and [illegible] successfully in 2013.

Project Closed

- Mr Du Toit [illegible] changed fields (moved to mate[illegible]) and moved on.

Report on student progress WFS Scholarship (II)

Title: "*Production of W bosons in the semi-muonic channel at forward rapidity in ALICE*". (University of Cape Town)

Mr Johnson Senosi attended the "Ettore Majorana International School of Sub-nuclear Physics", held in Erice, Sicily, 23 June - 2 July, 2012.

July 2012 – December 2012
- Final analysis before writing up M Sc dissertation.

February 2013 – June 2013
- Writing Dissertation submitted for examination

March 2014 (MSc awarded)

Mr Senosi continues with this project as part of his PhD

CONCLUSIONS

•p_T distribution from PYTHIA simulations of $u_\pm \leftarrow W_\pm$ obtained at forward rapidity in pp collisions at 8 TeV

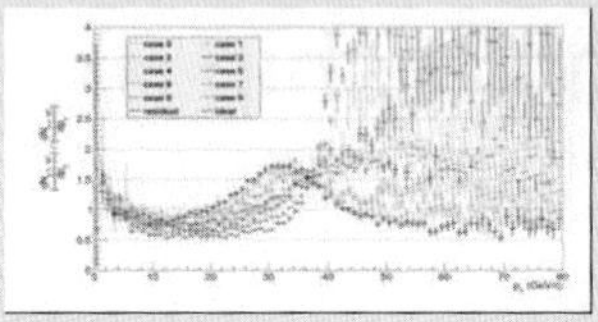

Ratio obtained from W-> u_+ / W- -> u_- is plotted as a function of p_T here shown for the range $0 < p_T < 80$ GeV/c to emphasize the charge asymmetry in this region

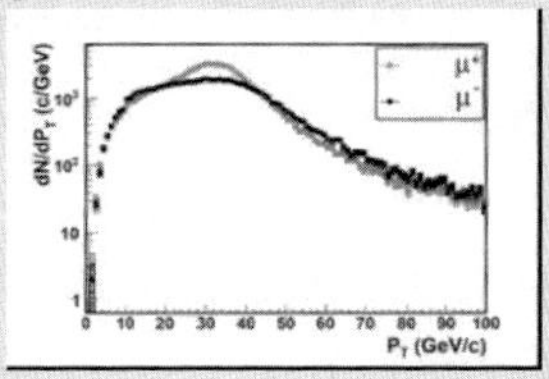

As part of their MSc projects electroweak W± bosons were studied by Messrs KJ Senosi and PWJ Du Toit via the single muon decay channels at forward rapidity in proton-proton (pp) collisions at LHC energies with ALICE.

part of ongoing investigations motivated by previous predictions to produce W± bosons in the forward rapidity ($-4.0 < y < -2.5$) of ALICE in pp, Pb-Pb and pPb / Pbp collisions at centre of mass energies available at the LHC. of the detector.

The spread (variation) in the trends is indicative of a systematic uncertainty and might be the result of "misaligned" tracking chambers or detector elements, which has to be taken into account when conducting a single muon analysis, especially in high *pT studies, (pT > 20 GeV/c).*

These results demonstrate the crucial role played by the analysis of u± ← W± production as it also forms an important test of the performance of the detector

OUTREACH

A Large Ion Collider Experiment

International Master Class on 5 April 2014

ALICE

"Looking for Strange Particles in ALICE"

- Organized by SA-ALICE and iThemba LABS Community, Interaction and Training
- 24 students from 6 schools
- International video discussion with Cairo, Geneva and Warsaw

WITS

Thank you

Support from WFS has been – and/or will be – invaluable in playing a key role of supporting South Africa (and African continent) in developing its HEP human capacity in frontier level research. Continuation of support would be greatly appreciated.

UCT

iThemba LABS

UNIVERSITY OF JOHANNESBURG

UKZN

SEMINAR PARTICIPANTS

Professor Mona Al-Achkar Jabbour	Faculty of Law, Lebanese University Pan Arab Observatory for Cyber Security Beirut, Lebanon
Lord John Alderdice	House of Lords London, UK
Dr. Michael Antoniou	King's College London School of Medicine, Department of Medical and Molecular Genetics London, UK
Professor Ignacio Armillas	Retired Director of the United Nations Centre for Human Settlements (Habitat), Fukuoka Office Mexico, DF, Mexico
Dr. Scott Atran	Centre National de la Recherche Scientifique, Institut Jean Nicod Paris, France
Professor William A. Barletta	U.S. Particle Accelerator School, Department of Physics, Massachusetts Institute of Technology Cambridge, MA, USA
Professor Simone Bastianoni	DEEPS Department of Earth Environmental and Physical Sciences University of Siena Siena, Italy
Professor Charles Benbrook	Center for Sustaining Agriculture and Natural Resources, Washington State University Enterprise, OR, USA
Dr. Bruce Blumberg	Department of Pharmaceutical Sciences University of California Irvine, CA, USA
Dr. Paul A. Bradley	Los Alamos National Laboratory Los Alamos, NM, USA

Dr. Vladimir B. Britkov	Institute for Systems Analysis Russian Academy of Sciences Moscow, Russia
Dr. Robert J. Budnitz	Earth Sciences Division Lawrence Berkeley National Laboratory Berkeley, CA, USA
Dr. Shashi Buluswar	Institute for Globally Transformative Technologies, Lawrence Berkeley National Lab Berkeley, CA, USA
Dr. Franco Buonaguro	Istituto Nazionale dei Tumori "Fondazione G. Pascale" Napoli, Italy
Dr. Emanuele Buratti	International Centre for Genetic Engineering and Biotechnology Trieste, Italy
Dr. Gina M. Calderone	Environmental Chemical Corporation (ECC) Marlborough, MA, USA
Professor Ignazio Carreca	Policlinico "P. Giaccone" University of Palermo Palermo, Italy
Professor Robert Carter	James Cook University Townsville, Australia
Professor Alberto Cellino	Osservatorio Astronomico di Pino Torinese Pino Torinese, Italy
Dr. Nathalie Charpak	Kangaroo Foundation Bogotá, Colombia
Dr. Monica Chiarini-Contin	University of Bologna Bologna, Italy
Professor Luisa Cifarelli	Department of Physics University of Bologna Bologna, Italy

Professor Andrea Contin	Department of Physics University of Bologna Bologna, Italy
Professor Pierre Darriulat	Vietnam Auger Training Laboratory Institute for Nuclear Science and Technology Hanoi, Vietnam
Dr. Richard Davis	Davis Energy Washington DC, USA
Dr. Socorro de Leon-Mendoza	Jose Fabella Memorial Hospital Neonatology Unit Manila, Philippines
Professor Robert J. Diaz	Virginia Institute of Marine Science College of Williams and Mary Gloucester Point, VA, USA
Dr. Carmine Difiglio	Office of Energy Policy and Systems Analysis, Department of Energy Washington, DC, USA
Dr. Mbareck Diop	Former Science and Technology Advisor to the President of Senegal Dakar, Senegal
Mr. Samuel Dover	Channel Islands Marine and Wildlife Institute Santa Barbara, CA, USA
Dr. Line Drube	Germany Aerospace Center DLR Institute of Planetary Research Berlin-Adlershof, Germany
Professor Gheorghe Duca	The Academy of Sciences of Moldova Chisinau, Republic of Moldova
Dr. Pavan Duggal	Associates Cyberlaw India New Delhi, India
Professor Robert Duncan	Vice President Research Texas Tech University Lubbock, TX, USA

Professor Merab Eliashvili	Tbilisi State University Tbilisi, Georgia
Professor Christopher Essex	Department of Applied Mathematics University of Western Ontario London, ON, Canada
Dr. Lorne Everett	Chancellor Emeritus, Lakehead University Thunder Bay, Canada and L. Everett and Associates, LLC Santa Barbara, CA, USA
Dr. Patrick Frank	SLAC, Stanford University Palo Alto, CA, USA
Dr. Bertil Galland	Writer and Historian Buxy, France
Professor Richard L. Garwin	Thomas J. Watson Research Center IBM Research Division Yorktown Heights, NY, USA
Dr. Ishwar Gilada	President of AIDS Society of India Chennai, India
Professor Kimberly A. Gray	Department of Civil and Environmental Engineering, Northwestern University Evanston, IL, USA
Dr. Walter M. Grayman	Consulting Engineer Cincinnati, OH, USA
Professor Stefan Hadjitodorov	Bulgarian Academy of Sciences Sofia, Bulgaria
Dr. Michael Hansen	Consumers Union Yonkers, NY, USA
Professor William Happer	Cyrus Fogg Brackett, Physics Department Princeton University Princeton, NJ, USA
Dr. Adalat Hasanov	National Academy of Sciences Baku, Azerbaijan

Dr. Patrick Holden	Sustainable Food Trust Bristol, UK
Professor Pervez Hoodbhoy	Physics Department Quaid-i-Azam University Islamabad, Pakistan
Dr. Walter F. Huebner	Southwest Research Institute San Antonio, TX, USA
Professor Michael Jefferson	ESCP Europe, Business School London, UK
Professor Domas Kaunas	The Lithuanian Academy of Sciences Vilnius, Lithuania
Dr. Hisham Khatib	World Energy Council Amman, Jordan
Dr. Do-Won Kim	CERN PH-UAI Division Geneva, Switzerland
Dr. Erwin Knupp	TOTH SA Geneva Geneva, Switzerland
Dr. Edvard Kobal	Slovenian Scientific Foundation Ljubljana, Slovenia
Dr. Vasiliy Krivokhizha	International Department, Federal Assembly of the Russian Federation Moscow, Russia
Dr. Dan J. Kroll	Hach Homeland Security Technologies Loveland, CO, USA
Dr. Valery P. Kukhar	Ukrainian Academy of Sciences Laboratory of Fine Organic Synthesis Kiev, Ukraine
Mrs. Donna Laframboise	Journalist and a Senior Research Fellow at the Frontier Centre for Public Policy Toronto, Canada

Dr. Richard C. Lanza	Dept. of Nuclear Science and Engineering Massachusetts Institute of Technology Cambridge, MA, USA
Professor Axel Lehmann	Institute for Technical Computer Sciences German Armed Forces University Munich Neubiberg, Germany
Dr. Sally Leivesley	Newrisk Limited London, Mitigation of Terrorists Acts PMP World Federation of Scientists London, UK
Dr. Giovanni Levi	National Scientific Research Centre Evolution of Endocrine Regulation Paris, France
Professor Richard S. Lindzen	Earth, Atmospheric and Planetary Sciences Massachusetts Institute of Technology Cambridge, USA
Professor Tatsuo Masuda	The Graduate School, Nagoya University Commerce and Business Tokyo, Japan
Professor Sam Mbulaiteye	National Institutes of Health/NCI/DCEG Infections and Immunoepidemiology Branch Bethesda, MD, USA
Dr. Charles McCombie	McCombie Consulting Arius Association Gipf-Oberfrick, Switzerland
Professor Stephen McIntyre	University of Guelph Toronto, Canada
Dr. Robin Mesnage	King's College London School of Medicine. Department of Medical and Molecular Genetics London, UK

Lord Christopher Monckton	The Viscount Monckton of Brenchley Monckton Enterprises Ltd. Carie, Rannoch, UK
Dr. Alan Leigh Moore	Research Institute, Von Braun Research Hall, The University of Alabama in Huntsville Huntsville, AL, USA
Professor Mirzayusuf Musakhanov	Uzbekistan National University Tashkent, Uzbekistan
Dr. John Peterson Myers	Environmental Health Sciences Charlottesville, VA, USA
Dr. André Naudi	Director of Finance and Human Resources (Former) CERN, Geneva, Switzerland
Professor Yuichi Ogawa	University of Tokyo Graduate School of Frontier Sciences Chiba Prefecture, Japan
Professor SunKun Oh	Department of Physics Konkuk University Seoul, Korea
Dr. Jef Ongena	Plasmaphysics Laboratory Ecole Royale Militaire Brussels, Belgium
Professor Gennady Palshin	World Laboratory Branch Ukraine Kiev, Ukraine
Professor Frank Leon Parker	Environmental Engineering Vanderbilt University Nashville, TN, USA
Professor Stefano Parmigiani	Evolutive and Functional Biology Università di Parma Parma, Italy
Dr. Daniel Pauly	UBC Fisheries Centre Vancouver, BC, Canada

Dr. Benny Peiser	The Global Warming Policy Foundation London, UK
Dr. Robert Petroski	Nuclear Innovation Engineer TerraPower LLC Bellevue, WA, USA
Dr. Lara Pierpoint	Office of Energy Policy and Systems Analysis, U.S. Department of Energy Washington, DC, USA
Professor Richard Ragaini	Department of Environmental Protection University of California Livermore, CA, USA
Professor Nusrat Rajabov	Department of Mechanics and Mathematics Tajik State University Dushanbe, Tajikistan
Dr. Anil Razdan	Energy Consultant, Principal Scientific Advisor New Delhi, India
Mr. James E. Rogers	Duke Energy Corporation Charlotte, NC, USA
Dr. Rutger Rosenberg	Department of Biological and Environmental Sciences, University of Gothenburg Fiskebäckskil, Sweden
Professor Howard Schmidt	Ridge Schmidt Cyber Washington, DC, USA
Dr. Robert N. Schock	World Energy Council Senior Advisor London, UK
Dr. Giusto Sciarabba	The Academy of Sciences for the Developing World, TWAS Trieste, Italy

Dr. Adnan Shihab-Eldin	Kuwait Foundation for the Advancement of Sciences Safat, Kuwait
Professor Annette Sobel	Critical Infrastructure Protection and Health Security Initiatives, Texas Tech University Lubbock, TX, USA
Dr. Willie Soon	Harvard's Center for Astrophysics Harvard, MA, USA
Professor Friedrich Steinhäusler	Physics and Biophysics Division University of Salzburg Salzburg, Austria
Dr. Bruce Stram	Element Markets Houston, TX, USA
Professor Gheorghe Stratan	National Institute of Physics and Nuclear Engineering Bucharest-Magurele, Romania
Dr. Carlo Strenger	Department of Psychology Tel Aviv University Tel Aviv, Israel
Professor Henrik Svensmark	Center for Sun-Climate Research Danish National Space Center Technical University of Denmark Copenhagen, Denmark
Professor Michael J. Tannenbaum	Physics Department Brookhaven National Laboratory Upton, NY, USA
Professor Abdel Nasser Tawfik	ECTP Founding Director MTI University Cairo, Egypt
Professor Anastasios A. Tsonis	Department of Mathematical Sciences University of Wisconsin-Milwaukee Milwaukee, WI, USA

Dr. Laura N. Vandenberg	University of Massachusetts School of Public Health and Division of Environmental Health Science Amherst, MA, USA
Professor Zeblon Vilakazi	University of the Witwatersrand Johannesburg, South Africa
Dr. Jandouwe Villinger	International Centre of Insect Physiology and Ecology (ICIPE) Nairobi, Kenya
Professor Frederick S. vom Saal	University of Missouri Division of Biological Sciences Columbia, MO, USA
Dr. Henning Wegener	Ambassador of Germany (ret.) Information Security Permanent Monitoring Panel World Federation of Scientists Madrid, Spain
Dr. Horst Wenninger	CERN, DPH-UAI Division Geneva, Switzerland
Dr. Crispin Williams	Advanced Technology for Young Scientists from Developing Countries CERN Geneva, Switzerland
Professor Lydia Wilson	Centre for the Resolution of Intractable Conflict Harris Manchester College Oxford, UK
Dr. Lowell Wood	Hoover Institution, Stanford University, University of California and Lawrence Livermore National Laboratory Stanford, CA, USA
Dr. Anatoly V. Zaitsev	Planetary Defense Center Moscow, Russia

Professor Antonino Zichichi	CERN, Geneva, Switzerland INFN and University of Bologna, Italy Centro Enrico Fermi, Rome, Italy
Professor Gennady Zinovjev	Bogolyubov Institute for Theoretical Physics, National Academy of Sciences of Ukraine Kiev, Ukraine